AKSHAY PATEL

Statistics

Statistics

Fourth Edition

Murray R. Spiegel, Ph.D.
Former Professor and Chairman
Mathematics Department, Rensselaer Polytechnic Institute
Hartford Graduate Center

Larry J. Stephens, Ph.D.
Full Professor
Mathematics Department
University of Nebraska at Omaha

Schaum's Outline Series

New York Chicago San Francisco Lisbon London Madrid
Mexico City Milan New Delhi San Juan Seoul
Singapore Sydney Toronto

The late **MURRAY R. SPIEGEL** received his M.S. degree in Physics and Ph.D. in Mathematics from Cornell University. He had positions at Harvard University, Columbia University, Oak Ridge and Rensselaer Polytechnic Institute, and served as a mathematical consultant at several large companies. His last position was professor and chairman of mathematics at Rensselaer Polytechnic Institute, Hartford Graduate Center. He was interested in most branches of mathematics, especially those involving applications to physics and engineering problems. He was the author of numerous journal articles and 14 books on various topics in mathematics.

LARRY J. STEPHENS is professor of mathematics at the University of Nebraska at Omaha, where he has taught since 1974. He has also taught at the University of Arizona, Gonzaga University, and Oklahoma State University. He has worked for NASA, Lawrence Berkeley National Laboratory, and Los Alamos National Laboratory. He has consulted widely, and spent 10 years as a consultant and conducted seminars for the engineering group at 3M in Valley, Nebraska. Dr. Stephens has over 35 years of experience teaching statistical methodology, engineering statistics, and mathematical statistics. He has over 50 publications in professional journals, and has written books in the Schaum's Outlines Series as well as books in the Utterly Confused and Demystified series published by McGraw-Hill.

3 4 5 6 7 8 9 10 CUS/CUS 1 9 8 7 6 5 4 3 2 1

ISBN 978-0-07-175549-8
MHID 0-07-175549-7

To the memory of my Mother and Father, Rosie and Johnie Stephens

L.J.S.

This fourth edition, completed in 2007, contains new examples, 130 new figures, and output from five computer software packages representative of the hundreds or perhaps thousands of computer software packages that are available for use in statistics. All figures in the third edition are replaced by new and sometimes different figures created by the five software packages: EXCEL, MINITAB, SAS, SPSS, and STATISTIX. My examples were greatly influenced by *USA Today* as I put the book together. This newspaper is a great source of current uses and examples in statistics.

Other changes in the book include the following. Chapter 18 on the analyses of time series was removed and Chapter 19 on statistical process control and process capability is now Chapter 18. The answers to the supplementary problems at the end of each chapter are given in more detail. More discussion and use of p-values is included throughout the book.

ACKNOWLEDGMENTS

As the statistical software plays such an important role in the book, I would like to thank the following people and companies for the right to use their software.

MINITAB: Ms. Laura Brown, Coordinator of the Author Assistance Program, Minitab Inc., 1829 Pine Hall Road, State College, PA 16801. I am a member of the author assistance program that Minitab sponsors. "Portions of the input and output contained in this publication/book are printed with the permission of Minitab Inc. All material remains the exclusive property and copyright of Minitab Inc. All rights reserved." The web address for Minitab is www.minitab.com.

SAS: Ms. Sandy Varner, Marketing Operations Manager, SAS Publishing, Cary, NC. "Created with SAS software. Copyright 2006. SAS Institute Inc., Cary, NC, USA. All rights reserved. Reproduced with permission of SAS Institute Inc., Cary, NC." I quote from the Website: "SAS is the leader in business intelligence software and services. Over its 30 years, SAS has grown – from seven employees to nearly 10,000 worldwide, from a few customer sites to more than 40,000 – and has been profitable every year." The web address for SAS is www.sas.com.

SPSS: Ms. Jill Rietema, Account Manager, Publications, SPSS. I quote from the Website: "SPSS Inc. is a leading worldwide provider of predictive analytics software and solutions. Founded in 1968, today SPSS has more than 250,000 customers worldwide, served by more than 1200 employees in 60 countries." The Web address for SPSS is www.spss.com.

STATISTIX: Dr. Gerard Nimis, President, Analytical Software, P.O. Box 12185, Tallahassee, FL 32317. I quote from the Website: "If you have data to analyze, but you're a researcher, not a statistician, Statistix is designed for you. You'll be up and running in minutes-without programming or using a manual. This easy to learn and simple to use software saves you valuable time and money. Statistix combines all the basic and advanced statistics and powerful data manipulation tools you need in a single, inexpensive package." The Web address for STATISTIX is www.statistix.com.

EXCEL: Microsoft Excel has been around since 1985. It is available to almost all college students. It is widely used in this book.

I wish to thank Stanley Wileman for the computer advice that he so unselfishly gave to me during the writing of the book. I wish to thank my wife, Lana, for her understanding while I was often day dreaming and thinking about the best way to present some concept. Thanks to Chuck Wall, Senior Acquisitions Editor, and his staff at McGraw-Hill. Finally, I would like to thank Jeremy Toynbee, project manager at Keyword Publishing Services Ltd., London, England, and John Omiston, freelance copy editor, for their fine production work. I invite comments and questions about the book at Lstephens@mail.unomaha.edu.

LARRY J. STEPHENS

In preparing this third edition of *Schaum's Outline of Statistics*, I have replaced dated problems with problems that reflect the technological and sociological changes that have occurred since the first edition was published in 1961. One problem in the second edition dealt with the lifetimes of radio tubes, for example. Since most people under 30 probably do not know what radio tubes are, this problem, as well as many other problems, have been replaced by problems involving current topics such as health care issues, AIDS, the Internet, cellular phones, and so forth. The mathematical and statistical aspects have not changed, but the areas of application and the computational aspects of statistics have changed.

Another important improvement is the introduction of statistical software into the text. The development of statistical software packages such as SAS, SPSS, and MINITAB has dramatically changed the application of statistics to real-world problems. One of the most widely used statistical packages in academia as well as in industrial settings is the package called MINITAB (Minitab Inc., 1829 Pine Hall Road, State College, PA 16801). I would like to thank Minitab Inc. for granting me permission to include Minitab output throughout the text. Many modern statistical textbooks include computer software output as part of the text. I have chosen to include Minitab because it is widely used and is very friendly. Once a student learns the various data file structures needed to use MINITAB, and the structure of the commands and subcommands, this knowledge is readily transferable to other statistical software. With the introduction of pull-down menus and dialog boxes, the software has been made even friendlier. I include both commands and pull-down menus in the Minitab discussions in the text.

Many of the new problems discuss the important statistical concept of the p-value for a statistical test. When the first edition was introduced in 1961, the p-value was not as widely used as it is today, because it is often difficult to determine without the aid of computer software. Today p-values are routinely provided by statistical software packages since the computer software computation of p-values is often a trivial matter.

A new chapter entitled "Statistical Process Control and Process Capability" has replaced Chapter 19, "Index Numbers." These topics have many industrial applications, and I felt they needed to be included in the text. The inclusion of the techniques of statistical process control and process capability in modern software packages has facilitated the implementation of these techniques in many industrial settings. The software performs all the computations, which are rather burdensome. I chose to use Minitab because I feel that it is among the best software for SPC applications.

I wish to thank my wife Lana for her understanding during the preparation of the book; my friend Stanley Wileman for all the computer help he has given me; and Alan Hunt and the staff at Keyword Publishing Services Ltd., London, England, for their fine production work. Finally, I wish to thank the staff at McGraw-Hill for their cooperation and helpfulness.

LARRY J. STEPHENS

PREFACE TO THE SECOND EDITION

Statistics, or statistical methods as it is sometimes called, is playing an increasingly important role in nearly all phases of human endeavor. Formerly dealing only with affairs of the state, thus accounting for its name, the influence of statistics has now spread to agriculture, biology, business, chemistry, communications, economics, education, electronics, medicine, physics, political science, psychology, sociology and numerous other fields of science and engineering.

The purpose of this book is to present an introduction to the general statistical principles which will be found useful to all individuals regardless of their fields of specialization. It has been designed for use either as a supplement to all current standard texts or as a textbook for a formal course in statistics. It should also be of considerable value as a book of reference for those presently engaged in applications of statistics to their own special problems of research.

Each chapter begins with clear statements of pertinent definitions, theorems and principles together with illustrative and other descriptive material. This is followed by graded sets of solved and supplementary problems which in many instances use data drawn from actual statistical situations. The solved problems serve to illustrate and amplify the theory, bring into sharp focus those fine points without which the students continually feel themselves to be on unsafe ground, and provide the repetition of basic principles so vital to effective teaching. Numerous derivations of formulas are included among the solved problems. The large number of supplementary problems with answers serve as a complete review of the material of each chapter.

The only mathematical background needed for an understanding of the entire book is arithmetic and the elements of algebra. A review of important mathematical concepts used in the book is presented in the first chapter which may either be read at the beginning of the course or referred to later as the need arises.

The early part of the book deals with the analysis of frequency distributions and associated measures of central tendency, dispersion, skewness and kurtosis. This leads quite naturally to a discussion of elementary probability theory and applications, which paves the way for a study of sampling theory. Techniques of large sampling theory, which involve the normal distribution, and applications to statistical estimation and tests of hypotheses and significance are treated first. Small sampling theory, involving Student's t distribution, the chi-square distribution and the F distribution together with the applications appear in a later chapter. Another chapter on curve fitting and the method of least squares leads logically to the topics of correlation and regression involving two variables. Multiple and partial correlation involving more than two variables are treated in a separate chapter. These are followed by chapters on the analysis of variance and nonparametric methods, new in this second edition. Two final chapters deal with the analysis of time series and index numbers respectively.

Considerably more material has been included here than can be covered in most first courses. This has been done to make the book more flexible, to provide a more useful book of reference and to stimulate further interest in the topics. In using the book it is possible to change the order of many later chapters or even to omit certain chapters without difficulty. For example, Chapters 13–15 and 18–19 can, for the most part, be introduced immediately after Chapter 5, if it is desired to treat correlation, regression, times series, and index numbers before sampling theory. Similarly, most of Chapter 6 may be omitted if one does not wish to devote too much time to probability. In a first course all of Chapter 15 may be omitted. The present order has been used because there is an increasing tendency in modern courses to introduce sampling theory and statistical influence as early as possible.

I wish to thank the various agencies, both governmental and private, for their cooperation in supplying data for tables. Appropriate references to such sources are given throughout the book. In particular,

I am indebted to Professor Sir Ronald A. Fisher, F.R.S., Cambridge; Dr. Frank Yates, F.R.S., Rothamsted; and Messrs. Oliver and Boyd Ltd., Edinburgh, for permission to use data from Table III of their book *Statistical Tables for Biological, Agricultural, and Medical Research*. I also wish to thank Esther and Meyer Scher for their encouragement and the staff of McGraw-Hill for their cooperation.

MURRAY R. SPIEGEL

CHAPTER 1 **Variables and Graphs** 1
Statistics 1
Population and Sample; Inductive and Descriptive Statistics 1
Variables: Discrete and Continuous 1
Rounding of Data 2
Scientific Notation 2
Significant Figures 3
Computations 3
Functions 4
Rectangular Coordinates 4
Graphs 4
Equations 5
Inequalities 5
Logarithms 6
Properties of Logarithms 7
Logarithmic Equations 7

CHAPTER 2 **Frequency Distributions** 37
Raw Data 37
Arrays 37
Frequency Distributions 37
Class Intervals and Class Limits 38
Class Boundaries 38
The Size, or Width, of a Class Interval 38
The Class Mark 38
General Rules for Forming Frequency Distributions 38
Histograms and Frequency Polygons 39
Relative-Frequency Distributions 39
Cumulative-Frequency Distributions and Ogives 40
Relative Cumulative-Frequency Distributions and Percentage Ogives 40
Frequency Curves and Smoothed Ogives 41
Types of Frequency Curves 41

CHAPTER 3 **The Mean, Median, Mode, and Other Measures of Central Tendency** 61
Index, or Subscript, Notation 61
Summation Notation 61
Averages, or Measures of Central Tendency 62

The Arithmetic Mean 62
The Weighted Arithmetic Mean 62
Properties of the Arithmetic Mean 63
The Arithmetic Mean Computed from Grouped Data 63
The Median 64
The Mode 64
The Empirical Relation Between the Mean, Median, and Mode 64
The Geometric Mean G 65
The Harmonic Mean H 65
The Relation Between the Arithmetic, Geometric, and Harmonic Means 66
The Root Mean Square 66
Quartiles, Deciles, and Percentiles 66
Software and Measures of Central Tendency 67

CHAPTER 4 The Standard Deviation and Other Measures of Dispersion 95
Dispersion, or Variation 95
The Range 95
The Mean Deviation 95
The Semi-Interquartile Range 96
The 10–90 Percentile Range 96
The Standard Deviation 96
The Variance 97
Short Methods for Computing the Standard Deviation 97
Properties of the Standard Deviation 98
Charlier's Check 99
Sheppard's Correction for Variance 100
Empirical Relations Between Measures of Dispersion 100
Absolute and Relative Dispersion; Coefficient of Variation 100
Standardized Variable; Standard Scores 101
Software and Measures of Dispersion 101

CHAPTER 5 Moments, Skewness, and Kurtosis 123
Moments 123
Moments for Grouped Data 123
Relations Between Moments 124
Computation of Moments for Grouped Data 124
Charlier's Check and Sheppard's Corrections 124
Moments in Dimensionless Form 124
Skewness 125
Kurtosis 125
Population Moments, Skewness, and Kurtosis 126
Software Computation of Skewness and Kurtosis 126

CHAPTER 6 Elementary Probability Theory 139
Definitions of Probability 139
Conditional Probability; Independent and Dependent Events 140
Mutually Exclusive Events 141
Probability Distributions 142
Mathematical Expectation 144
Relation Between Population, Sample Mean, and Variance 144
Combinatorial Analysis 145
Combinations 146
Stirling's Approximation to *n*! 146
Relation of Probability to Point Set Theory 146
Euler or Venn Diagrams and Probability 146

CHAPTER 7 The Binomial, Normal, and Poisson Distributions 172
The Binomial Distribution 172
The Normal Distribution 173
Relation Between the Binomial and Normal Distributions 174
The Poisson Distribution 175
Relation Between the Binomial and Poisson Distributions 176
The Multinomial Distribution 177
Fitting Theoretical Distributions to Sample Frequency Distributions 177

CHAPTER 8 Elementary Sampling Theory 203
Sampling Theory 203
Random Samples and Random Numbers 203
Sampling With and Without Replacement 204
Sampling Distributions 204
Sampling Distribution of Means 204
Sampling Distribution of Proportions 205
Sampling Distributions of Differences and Sums 205
Standard Errors 207
Software Demonstration of Elementary Sampling Theory 207

CHAPTER 9 Statistical Estimation Theory 227
Estimation of Parameters 227
Unbiased Estimates 227
Efficient Estimates 228
Point Estimates and Interval Estimates; Their Reliability 228
Confidence-Interval Estimates of Population Parameters 228
Probable Error 230

CHAPTER 10 **Statistical Decision Theory** 245
Statistical Decisions 245
Statistical Hypotheses 245
Tests of Hypotheses and Significance, or Decision Rules 246
Type I and Type II Errors 246
Level of Significance 246
Tests Involving Normal Distributions 246
Two-Tailed and One-Tailed Tests 247
Special Tests 248
Operating-Characteristic Curves; the Power of a Test 248
p-Values for Hypotheses Tests 248
Control Charts 249
Tests Involving Sample Differences 249
Tests Involving Binomial Distributions 250

CHAPTER 11 **Small Sampling Theory** 275
Small Samples 275
Student's t Distribution 275
Confidence Intervals 276
Tests of Hypotheses and Significance 277
The Chi-Square Distribution 277
Confidence Intervals for σ 278
Degrees of Freedom 279
The F Distribution 279

CHAPTER 12 **The Chi-Square Test** 294
Observed and Theoretical Frequencies 294
Definition of χ^2 294
Significance Tests 295
The Chi-Square Test for Goodness of Fit 295
Contingency Tables 296
Yates' Correction for Continuity 297
Simple Formulas for Computing χ^2 297
Coefficient of Contingency 298
Correlation of Attributes 298
Additive Property of χ^2 299

CHAPTER 13 **Curve Fitting and the Method of Least Squares** 316
Relationship Between Variables 316
Curve Fitting 316
Equations of Approximating Curves 317
Freehand Method of Curve Fitting 318
The Straight Line 318
The Method of Least Squares 319

The Least-Squares Line 319
Nonlinear Relationships 320
The Least-Squares Parabola 320
Regression 321
Applications to Time Series 321
Problems Involving More Than Two Variables 321

CHAPTER 14 Correlation Theory 345
Correlation and Regression 345
Linear Correlation 345
Measures of Correlation 346
The Least-Squares Regression Lines 346
Standard Error of Estimate 347
Explained and Unexplained Variation 348
Coefficient of Correlation 348
Remarks Concerning the Correlation Coefficient 349
Product-Moment Formula for the Linear Correlation Coefficient 350
Short Computational Formulas 350
Regression Lines and the Linear Correlation Coefficient 351
Correlation of Time Series 351
Correlation of Attributes 351
Sampling Theory of Correlation 351
Sampling Theory of Regression 352

CHAPTER 15 Multiple and Partial Correlation 382
Multiple Correlation 382
Subscript Notation 382
Regression Equations and Regression Planes 382
Normal Equations for the Least-Squares Regression Plane 383
Regression Planes and Correlation Coefficients 383
Standard Error of Estimate 384
Coefficient of Multiple Correlation 384
Change of Dependent Variable 384
Generalizations to More Than Three Variables 385
Partial Correlation 385
Relationships Between Multiple and Partial Correlation Coefficients 386
Nonlinear Multiple Regression 386

CHAPTER 16 Analysis of Variance 403
The Purpose of Analysis of Variance 403
One-Way Classification, or One-Factor Experiments 403

Total Variation, Variation Within Treatments, and Variation Between Treatments 404
Shortcut Methods for Obtaining Variations 404
Mathematical Model for Analysis of Variance 405
Expected Values of the Variations 405
Distributions of the Variations 406
The *F* Test for the Null Hypothesis of Equal Means 406
Analysis-of-Variance Tables 406
Modifications for Unequal Numbers of Observations 407
Two-Way Classification, or Two-Factor Experiments 407
Notation for Two-Factor Experiments 408
Variations for Two-Factor Experiments 408
Analysis of Variance for Two-Factor Experiments 409
Two-Factor Experiments with Replication 410
Experimental Design 412

CHAPTER 17 Nonparametric tests 446
Introduction 446
The Sign Test 446
The Mann–Whitney *U* Test 447
The Kruskal–Wallis *H* Test 448
The *H* Test Corrected for Ties 448
The Runs Test for Randomness 449
Further Applications of the Runs Test 450
Spearman's Rank Correlation 450

CHAPTER 18 Statistical Process Control and Process Capability 480
General Discussion of Control Charts 480
Variables and Attributes Control Charts 481
X-bar and *R* Charts 481
Tests for Special Causes 484
Process Capability 484
P- and *NP*-Charts 487
Other Control Charts 489

Answers to Supplementary Problems 505

Appendixes 559
I Ordinates (*Y*) of the Standard Normal Curve at *z* 561
II Areas Under the Standard Normal Curve from 0 to *z* 562
III Percentile Values (t_p) for Student's *t* Distribution with ν Degrees of Freedom 563
IV Percentile Values (χ_p^2) for the Chi-Square Distribution with ν Degrees of Freedom 564

V	95th Percentile Values for the F Distribution	565
VI	99th Percentile Values for the F Distribution	566
VII	Four-Place Common Logarithms	567
VIII	Values of $e^{-\lambda}$	569
IX	Random Numbers	570

Index **571**

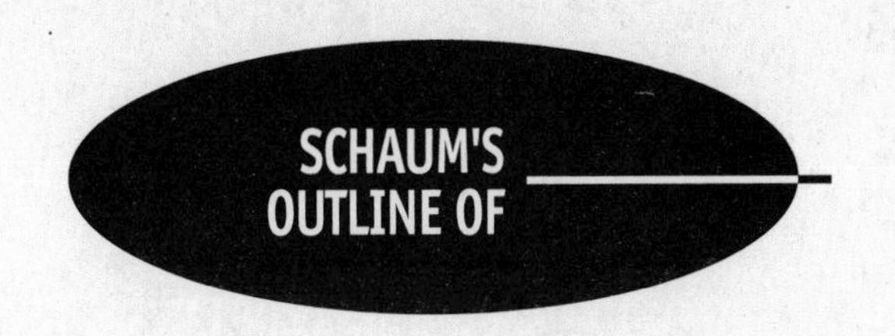

Theory and Problems of

STATISTICS

Variables and Graphs

STATISTICS

Statistics is concerned with scientific methods for collecting, organizing, summarizing, presenting, and analyzing data as well as with drawing valid conclusions and making reasonable decisions on the basis of such analysis.

In a narrower sense, the term *statistics* is used to denote the data themselves or numbers derived from the data, such as averages. Thus we speak of employment statistics, accident statistics, etc.

POPULATION AND SAMPLE; INDUCTIVE AND DESCRIPTIVE STATISTICS

In collecting data concerning the characteristics of a group of individuals or objects, such as the heights and weights of students in a university or the numbers of defective and nondefective bolts produced in a factory on a given day, it is often impossible or impractical to observe the entire group, especially if it is large. Instead of examining the entire group, called the *population*, or *universe*, one examines a small part of the group, called a *sample*.

A population can be *finite* or *infinite*. For example, the population consisting of all bolts produced in a factory on a given day is finite, whereas the population consisting of all possible outcomes (heads, tails) in successive tosses of a coin is infinite.

If a sample is representative of a population, important conclusions about the population can often be inferred from analysis of the sample. The phase of statistics dealing with conditions under which such inference is valid is called *inductive statistics*, or *statistical inference*. Because such inference cannot be absolutely certain, the language of *probability* is often used in stating conclusions.

The phase of statistics that seeks only to describe and analyze a given group without drawing any conclusions or inferences about a larger group is called *descriptive*, or *deductive*, *statistics*.

Before proceeding with the study of statistics, let us review some important mathematical concepts.

VARIABLES: DISCRETE AND CONTINUOUS

A *variable* is a symbol, such as X, Y, H, x, or B, that can assume any of a prescribed set of values, called the *domain* of the variable. If the variable can assume only one value, it is called a *constant*.

A variable that can theoretically assume any value between two given values is called a *continuous variable*; otherwise, it is called a *discrete variable*.

EXAMPLE 1. The number N of children in a family, which can assume any of the values 0, 1, 2, 3, ... but cannot be 2.5 or 3.842, is a discrete variable.

EXAMPLE 2. The height H of an individual, which can be 62 inches (in), 63.8 in, or 65.8341 in, depending on the accuracy of measurement, is a continuous variable.

Data that can be described by a discrete or continuous variable are called *discrete data* or *continuous data*, respectively. The number of children in each of 1000 families is an example of discrete data, while the heights of 100 university students is an example of continuous data. In general, *measurements* give rise to continuous data, while *enumerations*, or *countings*, give rise to discrete data.

It is sometimes convenient to extend the concept of variable to nonnumerical entities; for example, color C in a rainbow is a variable that can take on the "values" red, orange, yellow, green, blue, indigo, and violet. It is generally possible to replace such variables by numerical quantities; for example, denote red by 1, orange by 2, etc.

ROUNDING OF DATA

The result of rounding a number such as 72.8 to the nearest unit is 73, since 72.8 is closer to 73 than to 72. Similarly, 72.8146 rounded to the nearest hundredth (or to two decimal places) is 72.81, since 72.8146 is closer to 72.81 than to 72.82.

In rounding 72.465 to the nearest hundredth, however, we are faced with a dilemma since 72.465 is *just as far* from 72.46 as from 72.47. It has become the practice in such cases to round to the *even integer* preceding the 5. Thus 72.465 is rounded to 72.46, 183.575 is rounded to 183.58, and 116,500,000 rounded to the nearest million is 116,000,000. This practice is especially useful in minimizing *cumulative rounding errors* when a large number of operations is involved (see Problem 1.4).

SCIENTIFIC NOTATION

When writing numbers, especially those involving many zeros before or after the decimal point, it is convenient to employ the scientific notation using powers of 10.

EXAMPLE 3. $10^1 = 10$, $10^2 = 10 \times 10 = 100$, $10^5 = 10 \times 10 \times 10 \times 10 \times 10 = 100{,}000$, and $10^8 = 100{,}000{,}000$.

EXAMPLE 4. $10^0 = 1$; $10^{-1} = .1$, or 0.1; $10^{-2} = .01$, or 0.01; and $10^{-5} = .00001$, or 0.00001.

EXAMPLE 5. $864{,}000{,}000 = 8.64 \times 10^8$, and $0.00003416 = 3.416 \times 10^{-5}$.

Note that multiplying a number by 10^8, for example, has the effect of moving the decimal point of the number eight places *to the right*. Multiplying a number by 10^{-6} has the effect of moving the decimal point of the number six places *to the left*.

We often write 0.1253 rather than .1253 to emphasize the fact that a number other than zero before the decimal point has not accidentally been omitted. However, the zero before the decimal point can be omitted in cases where no confusion can result, such as in tables.

Often we use parentheses or dots to show the multiplication of two or more numbers. Thus $(5)(3) = 5 \cdot 3 = 5 \times 3 = 15$, and $(10)(10)(10) = 10 \cdot 10 \cdot 10 = 10 \times 10 \times 10 = 1000$. When letters are used to represent numbers, the parentheses or dots are often omitted; for example, $ab = (a)(b) = a \cdot b = a \times b$.

The scientific notation is often useful in computation, especially in locating decimal points. Use is then made of the rules

$$(10^p)(10^q) = 10^{p+q} \qquad \frac{10^p}{10^q} = 10^{p-q}$$

where p and q are any numbers.

In 10^p, p is called the *exponent* and 10 is called the *base*.

EXAMPLE 6. $(10^3)(10^2) = 1000 \times 100 = 100{,}000 = 10^5$ i.e., 10^{3+2}

$$\frac{10^6}{10^4} = \frac{1{,}000{,}000}{10{,}000} = 100 = 10^2 \quad \text{i.e., } 10^{6-4}$$

EXAMPLE 7. $(4{,}000{,}000)(0.0000000002) = (4 \times 10^6)(2 \times 10^{-10}) = (4)(2)(10^6)(10^{-10}) = 8 \times 10^{6-10}$

$$= 8 \times 10^{-4} = 0.0008$$

EXAMPLE 8. $\frac{(0.006)(80{,}000)}{0.04} = \frac{(6 \times 10^{-3})(8 \times 10^4)}{4 \times 10^{-2}} = \frac{48 \times 10^1}{4 \times 10^{-2}} = \left(\frac{48}{4}\right) \times 10^{1-(-2)}$

$$= 12 \times 10^3 = 12{,}000$$

SIGNIFICANT FIGURES

If a height is accurately recorded as 65.4 in, it means that the true height lies between 65.35 and 65.45 in. The accurate digits, apart from zeros needed to locate the decimal point, are called the *significant digits*, or *significant figures*, of the number.

EXAMPLE 9. 65.4 has three significant figures.

EXAMPLE 10. 4.5300 has five significant figures.

EXAMPLE 11. $.0018 = 0.0018 = 1.8 \times 10^{-3}$ has two significant figures.

EXAMPLE 12. $.001800 = 0.001800 = 1.800 \times 10^{-3}$ has four significant figures.

Numbers associated with enumerations (or countings), as opposed to measurements, are of course exact and so have an unlimited number of significant figures. In some of these cases, however, it may be difficult to decide which figures are significant without further information. For example, the number 186,000,000 may have 3, 4, ..., 9 significant figures. If it is known to have five significant figures, it would be better to record the number as either 186.00 million or 1.8600×10^8.

COMPUTATIONS

In performing calculations involving multiplication, division, and the extraction of roots of numbers, the final result can have no more significant figures than the numbers with the fewest significant figures (see Problem 1.9).

EXAMPLE 13. $73.24 \times 4.52 = (73.24)(4.52) = 331$

EXAMPLE 14. $1.648/0.023 = 72$

EXAMPLE 15. $\sqrt{38.7} = 6.22$

EXAMPLE 16. $(8.416)(50) = 420.8$ (if 50 is exact)

In performing additions and subtractions of numbers, the final result can have no more significant figures after the decimal point than the numbers with the fewest significant figures after the decimal point (see Problem 1.10).

EXAMPLE 17. $3.16 + 2.7 = 5.9$

EXAMPLE 18. $83.42 - 72 = 11$

EXAMPLE 19. $47.816 - 25 = 22.816$ (if 25 is exact)

The above rule for addition and subtraction can be extended (see Problem 1.11).

FUNCTIONS

If to each value that a variable X can assume there corresponds one or more values of a variable Y, we say that Y is a *function* of X and write $Y = F(X)$ (read "Y equals F of X") to indicate this functional dependence. Other letters (G, ϕ, etc.) can be used instead of F.

The variable X is called the *independent variable* and Y is called the *dependent variable*.

If only one value of Y corresponds to each value of X, we call Y a *single-valued function* of X; otherwise, it is called a *multiple-valued function* of X.

EXAMPLE 20. The total population P of the United States is a function of the time t, and we write $P = F(t)$.

EXAMPLE 21. The stretch S of a vertical spring is a function of the weight W placed on the end of the spring. In symbols, $S = G(W)$.

The functional dependence (or correspondence) between variables is often depicted in a table. However, it can also be indicated by an equation connecting the variables, such as $Y = 2X - 3$, from which Y can be determined corresponding to various values of X.

If $Y = F(X)$, it is customary to let $F(3)$ denote "the value of Y when $X = 3$," to let $F(10)$ denote "the value of Y when $X = 10$," etc. Thus if $Y = F(X) = X^2$, then $F(3) = 3^2 = 9$ is the value of Y when $X = 3$.

The concept of function can be extended to two or more variables (see Problem 1.17).

RECTANGULAR COORDINATES

Fig. 1-1 shows an EXCEL scatter plot for four points. The *scatter plot* is made up of two mutually perpendicular lines called the X and Y *axes*. The X axis is horizontal and the Y axis is vertical. The two axes meet at a point called the *origin*. The two lines divide the *XY plane* into four regions denoted by I, II, III, and IV and called the first, second, third, and fourth *quadrants*. Four points are shown plotted in Fig. 1-1. The point (2, 3) is in the first quadrant and is plotted by going 2 units to the right along the X axis from the origin and 3 units up from there. The point (−2.3, 4.5) is in the second quadrant and is plotted by going 2.3 units to the left along the X axis from the origin and then 4.5 units up from there. The point (−4, − 3) is in the third quadrant and is plotted by going 4 units to the left of the origin along the X axis and then 3 units down from there. The point (3.5, − 4) is in the fourth quadrant and is plotted by going 3.5 units to the right along the X axis and then 4 units down from there. The first number in a pair is called the *abscissa* of the point and the second number is called the *ordinate* of the point. The abscissa and the ordinate taken together are called the *coordinates* of the point.

By constructing a Z axis through the origin and perpendicular to the XY plane, we can easily extend the above ideas. In such cases the coordinates of a point would be denoted by (X, Y, Z).

GRAPHS

A *graph* is a pictorial presentation of the relationship between variables. Many types of graphs are employed in statistics, depending on the nature of the data involved and the purpose for which the graph is intended. Among these are *bar graphs, pie graphs, pictographs*, etc. These graphs are sometimes

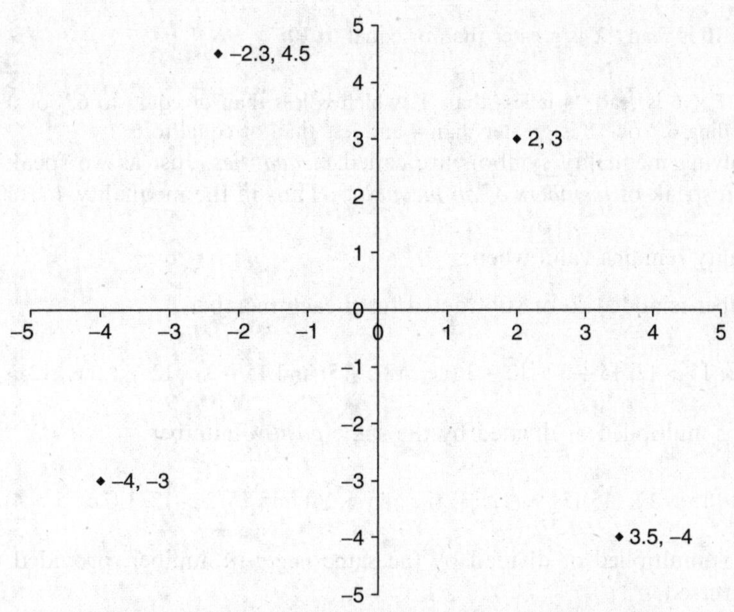

Fig. 1-1 An EXCEL plot of points in the four quadrants.

referred to as *charts* or *diagrams*. Thus we speak of bar charts, pie diagrams, etc. (see Problems 1.23, 1.24, 1.25, 1.26, and 1.27).

EQUATIONS

Equations are statements of the form $A = B$, where A is called the *left-hand member* (or *side*) of the equation, and B the *right-hand member* (or *side*). So long as we apply the *same* operations to both members of an equation, we obtain *equivalent equations*. Thus we can add, subtract, multiply, or divide both members of an equation by the same value and obtain an equivalent equation, the only exception being that *division by zero is not allowed*.

EXAMPLE 22. Given the equation $2X + 3 = 9$, subtract 3 from both members: $2X + 3 - 3 = 9 - 3$, or $2X = 6$. Divide both members by 2: $2X/2 = 6/2$, or $X = 3$. This value of X is a *solution* of the given equation, as seen by replacing X by 3, obtaining $2(3) + 3 = 9$, or $9 = 9$, which is an *identity*. The process of obtaining solutions of an equation is called *solving* the equation.

The above ideas can be extended to finding solutions of two equations in two unknowns, three equations in three unknowns, etc. Such equations are called *simultaneous equations* (see Problem 1.30).

INEQUALITIES

The symbols $<$ and $>$ mean "less than" and "greater than," respectively. The symbols $\leq$ and $\geq$ mean "less than or equal to" and "greater than or equal to," respectively. They are known as *inequality symbols*.

EXAMPLE 23. $3 < 5$ is read "3 is less than 5."

EXAMPLE 24. $5 > 3$ is read "5 is greater than 3."

EXAMPLE 25. $X < 8$ is read "X is less than 8."

EXAMPLE 26. $X \geq 10$ is read "X is greater than or equal to 10."

EXAMPLE 27. $4 < Y \leq 6$ is read "4 is less than Y, which is less than or equal to 6," or Y is between 4 and 6, excluding 4 but including 6," or "Y is greater than 4 and less than or equal to 6."

Relations involving inequality symbols are called *inequalities*. Just as we speak of members of an equation, so we can speak of *members of an inequality*. Thus in the inequality $4 < Y \leq 6$, the members are 4, Y, and 6.

A valid inequality remains valid when:

1. The same number is added to or subtracted from each member.

EXAMPLE 28. Since $15 > 12$, $15 + 3 > 12 + 3$ (i.e., $18 > 15$) and $15 - 3 > 12 - 3$ (i.e., $12 > 9$).

2. Each member is multiplied or divided by the same *positive* number.

EXAMPLE 29. Since $15 > 12$, $(15)(3) > (12)(3)$ (i.e., $45 > 36$) and $15/3 > 12/3$ (i.e., $5 > 4$).

3. Each member is multiplied or divided by the same *negative* number, provided that the inequality symbols are reversed.

EXAMPLE 30. Since $15 > 12$, $(15)(-3) < (12)(-3)$ (i.e., $-45 < -36$) and $15/(-3) < 12/(-3)$ (i.e., $-5 < -4$).

LOGARITHMS

For $x > 0$, $b > 0$, and $b \neq 1$, $y = \log_b x$ if and only if $\log b^y = x$. A logarithm is an exponent. It is the power that the base b must be raised to in order to get the number for which you are taking the logarithm. The two bases that have traditionally been used are 10 and e, which equals 2.71828182.... Logarithms with base 10 are called *common logarithms* and are written as $\log_{10} x$ or simply as $\log(x)$. Logarithms with base e are called *natural logarithms* and are written as $\ln(x)$.

EXAMPLE 31. Find the following logarithms and then use EXCEL to find the same logarithms: $\log_2 8$, $\log_5 25$, and $\log_{10} 1000$. Three is the power of 2 that gives 8, and so $\log_2 8 = 3$. Two is the power of 5 that gives 25, and so $\log_5 25 = 2$. Three is the power of 10 that gives 1000, and so $\log_{10} 1000 = 3$. EXCEL contains three functions that compute logarithms. The function LN computes natural logarithms, the function LOG10 computes common logarithms, and the function LOG(x,b) computes the logarithm of x to base b. =LOG(8,2) gives 3, =LOG(25,5) gives 2, and =LOG10(1000) gives 3.

EXAMPLE 32. Use EXCEL to compute the natural logarithm of the integers 1 through 5. The numbers 1 through 5 are entered into B1:F1 and the expression =LN(B1) is entered into B2 and a click-and-drag is performed from B2 to F2. The following EXCEL output was obtained.

X	1	2	3	4	5
LN(x)	0	0.693147	1.098612	1.386294	1.609438

EXAMPLE 33. Show that the answers in Example 32 are correct by showing that $e^{\ln(x)}$ gives the value x. The logarithms are entered into B1:F1 and the expression $e^{\ln(x)}$, which is represented by =EXP(B1), is entered into B2 and a click-and-drag is performed from B2 to F2. The following EXCEL output was obtained. The numbers in D2 and E2 differ from 3 and 4 because of round off error.

LN(x)	0	0.693147	1.098612	1.386294	1.609438
x = EXP(LN(x))	1	2	2.999999	3.999999	5

Example 33 illustrates that if you have the logarithms of numbers ($\log_b(x)$) the numbers (x) may be recovered by using the relation $b^{\log_b(x)} = x$.

EXAMPLE 34. The number e may be defined as a limiting entity. The quantity $(1+(1/x))^x$ gets closer and closer to e when x grows larger. Consider the EXCEL evaluation of $(1+(1/x))^x$ for $x = 1, 10, 100, 1000, 10000, 100000$, and 1000000.

x	1	10	100	1000	10000	100000	1000000
(1+1/x)^x	2	2.593742	2.704814	2.716924	2.718146	2.718268	2.71828

The numbers 1, 10, 100, 1000, 10000, 100000, and 1000000 are entered into B1:H1 and the expression =(1+1/B1)^B1 is entered into B2 and a click-and-drag is performed from B2 to H2. This is expressed mathematically by the expression $\lim_{x\to\infty}(1+(1/x))^x = e$.

EXAMPLE 35. The balance of an account earning compound interest, n times per year is given by $A(t) = P(1+(r/n))^{nt}$ where P is the principal, r is the interest rate, t is the time in years, and n is the number of compound periods per year. The balance of an account earning interest continuously is given by $A(t) = Pe^{rt}$. EXCEL is used to compare the continuous growth of $1000 and $1000 that is compounded quarterly after 1, 2, 3, 4, and 5 years at interest rate 5%. The results are:

Years	1	2	3	4	5
Quarterly	1050.95	1104.49	1160.75	1219.89	1282.04
Continuously	1051.27	1105.17	1161.83	1221.4	1284.03

The times 1, 2, 3, 4, and 5 are entered into B1:F1. The EXCEL expression =1000*(1.0125)^(4*B1) is entered into B2 and a click-and-drag is performed from B2 to F2. The expression =1000*EXP(0.05*B1) is entered into B3 and a click-and-drag is performed from B3 to F3. The continuous compounding produces slightly better results.

PROPERTIES OF LOGARITHMS

The following are the more important properties of logarithms:

1. $\log_b MN = \log_b M + \log_b N$
2. $\log_b M/N = \log_b M - \log_b N$
3. $\log_b M^P = p \log_b M$

EXAMPLE 36. Write $\log_b(xy^4/z^3)$ as the sum or difference of logarithms of x, y, and z.

$$\log_b \frac{xy^4}{z^3} = \log_b xy^4 - \log_b z^3 \quad \text{property 2}$$

$$\log_b \frac{xy^4}{z^3} = \log_b x + \log_b y^4 - \log_b z^3 \quad \text{property 1}$$

$$\log_b \frac{xy^4}{z^3} = \log_b x + 4\log_b y - 3\log_b z \quad \text{property 3}$$

LOGARITHMIC EQUATIONS

To solve logarithmic equations:

1. Isolate the logarithms on one side of the equation.
2. Express a sum or difference of logarithms as a single logarithm.
3. Re-express the equation in step 2 in exponential form.
4. Solve the equation in step 3.
5. Check all solutions.

EXAMPLE 37. Solve the following logarithmic equation: $\log_4(x+5)=3$. First, re-express in exponential form as $x+5=4^3=64$. Then solve for x as follows. $x=64-5=59$. Then check your solution. $\log_4(59+5)=\log_4(64)=3$ since $4^3=64$.

EXAMPLE 38. Solve the following logarithmic equation. $\log(6y-7)+\log y=\log(5)$. Replace the sum of logs by the log of the products. $\log(6y-7)y=\log(5)$. Now equate $(6y-7)y$ and 5. The result is $6y^2-7y=5$ or $6y^2-7y-5=0$. This quadratic equation factors as $(3y-5)(2y+1)=0$. The solutions are $y=5/3$ and $y=-1/2$. The $-1/2$ is rejected since it gives logs of negative numbers which are not defined. The $y=5/3$ checks out as a solution when tried in the original equation. Therefore our only solution is $y=5/3$.

EXAMPLE 39. Solve the following logarithmic equation:

$$\ln(5x)-\ln(4x+2)=4.$$

Replace the difference of logs by the log of the quotient, $\ln(5x/(4x+2))=4$. Apply the definition of a logarithm: $5x/(4x+2)=e^4=54.59815$. Solving the equation $5x=218.39260x+109.19630$ for x gives $x=-0.5117$. However this answer does not check in the equation $\ln(5x)-\ln(4x+2)=4$, since the log function is not defined for negative numbers. The equation $\ln(5x)-\ln(4x+2)=4$ has no solutions.

Solved Problems

VARIABLES

1.1 State which of the following represent discrete data and which represent continuous data:

(*a*) Numbers of shares sold each day in the stock market
(*b*) Temperatures recorded every half hour at a weather bureau
(*c*) Lifetimes of television tubes produced by a company
(*d*) Yearly incomes of college professors
(*e*) Lengths of 1000 bolts produced in a factory

SOLUTION

(*a*) Discrete; (*b*) continuous; (*c*) continuous; (*d*) discrete; (*e*) continuous.

1.2 Give the domain of each of the following variables, and state whether the variables are continuous or discrete:

(*a*) Number G of gallons (gal) of water in a washing machine
(*b*) Number B of books on a library shelf
(*c*) Sum S of points obtained in tossing a pair of dice
(*d*) Diameter D of a sphere
(*e*) Country C in Europe

SOLUTION

(*a*) *Domain:* Any value from 0 gal to the capacity of the machine. *Variable:* Continuous.

(*b*) *Domain:* 0, 1, 2, 3, ... up to the largest number of books that can fit on a shelf. *Variable:* Discrete.

(*c*) *Domain:* Points obtained on a single die can be 1, 2, 3, 4, 5, or 6. Hence the sum of points on a pair of dice can be 2, 3, 4, 5, 6, 7, 8, 9, 10, 11, or 12, which is the domain of S. *Variable:* Discrete.

(*d*) *Domain:* If we consider a point as a sphere of zero diameter, the domain of D is all values from zero upward. *Variable:* Continuous.

(*e*) *Domain:* England, France, Germany, etc., which can be represented numerically by 1, 2, 3, etc. *Variable:* Discrete.

ROUNDING OF DATA

1.3 Round each of the following numbers to the indicated accuracy:

(*a*)	48.6	nearest unit	(*f*)	143.95	nearest tenth
(*b*)	136.5	nearest unit	(*g*)	368	nearest hundred
(*c*)	2.484	nearest hundredth	(*h*)	24,448	nearest thousand
(*d*)	0.0435	nearest thousandth	(*i*)	5.56500	nearest hundredth
(*e*)	4.50001	nearest unit	(*j*)	5.56501	nearest hundredth

SOLUTION

(*a*) 49; (*b*) 136; (*c*) 2.48; (*d*) 0.044; (*e*) 5; (*f*) 144.0; (*g*) 400; (*h*) 24,000; (*i*) 5.56; (*j*) 5.57.

1.4 Add the numbers 4.35, 8.65, 2.95, 12.45, 6.65, 7.55, and 9.75 (*a*) directly, (*b*) by rounding to the nearest tenth according to the "even integer" convention, and (*c*) by rounding so as to increase the digit before the 5.

SOLUTION

	(*a*)	(*b*)	(*c*)
	4.35	4.4	4.4
	8.65	8.6	8.7
	2.95	3.0	3.0
	12.45	12.4	12.5
	6.65	6.6	6.7
	7.55	7.6	7.6
	9.75	9.8	9.8
Total	52.35	52.4	52.7

Note that procedure (*b*) is superior to procedure (*c*) because *cumulative rounding errors* are minimized in procedure (*b*).

SCIENTIFIC NOTATION AND SIGNIFICANT FIGURES

1.5 Express each of the following numbers without using powers of 10:

(*a*) 4.823×10^7 (*c*) 3.80×10^{-4} (*e*) 300×10^8

(*b*) 8.4×10^{-6} (*d*) 1.86×10^5 (*f*) $70{,}000 \times 10^{-10}$

SOLUTION

(*a*) Move the decimal point seven places to the right and obtain 48,230,000; (*b*) move the decimal point six places to the left and obtain 0.0000084; (*c*) 0.000380; (*d*) 186,000; (*e*) 30,000,000,000; (*f*) 0.0000070000.

1.6 How many significant figures are in each of the following, assuming that the numbers are recorded accurately?

(*a*) 149.8 in (*d*) 0.00280 m (*g*) 9 houses

(*b*) 149.80 in (*e*) 1.00280 m (*h*) 4.0×10^3 pounds (lb)

(*c*) 0.0028 meter (m) (*f*) 9 grams (g) (*i*) 7.58400×10^{-5} dyne

SOLUTION

(*a*) Four; (*b*) five; (*c*) two; (*d*) three; (*e*) six; (*f*) one; (*g*) unlimited; (*h*) two; (*i*) six.

1.7 What is the maximum error in each of the following measurements, assuming that they are recorded accurately?

(*a*) 73.854 in (*b*) 0.09800 cubic feet (ft^3) (*c*) 3.867×10^8 kilometers (km)

SOLUTION

(*a*) The measurement can range anywhere from 73.8535 to 73.8545 in; hence the maximum error is 0.0005 in. Five significant figures are present.

(*b*) The number of cubic feet can range anywhere from 0.097995 to 0.098005 ft^3; hence the maximum error is 0.000005 ft^3. Four significant figures are present.

(*c*) The actual number of kilometers is greater than 3.8665×10^8 but less than 3.8675×10^8; hence the maximum error is 0.0005×10^8, or 50,000 km. Four significant figures are present.

1.8 Write each number using the scientific notation. Unless otherwise indicated, assume that all figures are significant.

(*a*) 24,380,000 (four significant figures)
(*b*) 0.000009851
(*c*) 7,300,000,000 (five significant figures)
(*d*) 0.00018400

SOLUTION

(*a*) 2.438×10^7; (*b*) 9.851×10^{-6}; (*c*) 7.3000×10^9; (*d*) 1.8400×10^{-4}.

COMPUTATIONS

1.9 Show that the product of the numbers 5.74 and 3.8, assumed to have three and two significant figures, respectively, cannot be accurate to more than two significant figures.

SOLUTION

First method

$5.74 \times 3.8 = 21.812$, but not all figures in this product are significant. To determine how many figures are significant, observe that 5.74 stands for any number between 5.735 and 5.745, while 3.8 stands for any number between 3.75 and 3.85. Thus the smallest possible value of the product is $5.735 \times 3.75 = 21.50625$, and the largest possible value is $5.745 \times 3.85 = 21.11825$.

Since the possible range of values is 21.50625 to 22.11825, it is clear that no more than the first two figures in the product can be significant, the result being written as 22. Note that the number 22 stands for any number between 21.5 and 22.5.

Second method

With doubtful figures in italic, the product can be computed as shown here:

$$\begin{array}{r} 5.7\mathit{4} \\ 3\mathit{8} \\ \hline \mathit{4592} \\ 17\mathit{22} \\ \hline 2\mathit{1.812} \end{array}$$

We should keep no more than one doubtful figure in the answer, which is therefore 22 to two significant figures. Note that it is unnecessary to carry more significant figures than are present in the least accurate factor; thus if 5.74 is rounded to 5.7, the product is $5.7 \times 3.8 = 21.66 = 22$ to two significant figures, agreeing with the above results.

In calculating without the use of computers, labor can be saved by not keeping more than one or two figures beyond that of the least accurate factor and rounding to the proper number of significant figures in the final answer. With computers, which can supply many digits, we must be careful not to believe that all the digits are significant.

1.10 Add the numbers 4.19355, 15.28, 5.9561, 12.3, and 8.472, assuming all figures to be significant.

SOLUTION

In calculation (*a*) below, the doubtful figures in the addition are in italic type. The final answer with no more than one doubtful figure is recorded as 46.2.

(a)		(b)
	4.19355	4.19
	15.2*8*	15.28
	5.956*1*	5.96
	12.*3*	12.3
	8.47*2*	8.47
	46.*20165*	46.20

Some labor can be saved by proceeding as in calculation (*b*), where we have kept one more significant decimal place than that in the least accurate number. The final answer, rounded to 46.2, agrees with calculation (*a*).

1.11 Calculate 475,000,000 + 12,684,000 − 1,372,410 if these numbers have three, five, and seven significant figures, respectively.

SOLUTION

In calculation (*a*) below, all figures are kept and the final answer is rounded. In calculation (*b*), a method similar to that of Problem 1.10(*b*) is used. In both cases, doubtful figures are in italic type.

(*a*) 475,*000,000* + 12,68*4,000* = 487,*684,000*; 487,*684,000* − 1,372,410 = 486,*311,590*

(*b*) 475,*000,000* + 12,7*00,000* = 487,*700,000*; 487,*700,000* − 1,*400,000* = 486,*300,000*

The final result is rounded to 486,000,000; or better yet, to show that there are three significant figures, it is written as 486 million or 4.86×10^8.

1.12 Perform each of the indicated operations.

(*a*) 48.0×943

(*b*) $8.35/98$

(*c*) $(28)(4193)(182)$

(*d*) $\dfrac{(526.7)(0.001280)}{0.000034921}$

(*e*) $\dfrac{(1.47562 - 1.47322)(4895.36)}{0.000159180}$

(*f*) If denominators 5 and 6 are exact, $\dfrac{(4.38)^2}{5} + \dfrac{(5.482)^2}{6}$

(*g*) $3.1416\sqrt{71.35}$

(*h*) $\sqrt{128.5 - 89.24}$

SOLUTION

(*a*) $48.0 \times 943 = (48.0)(943) = 45{,}300$

(*b*) $8.35/98 = 0.085$

(*c*) $(28)(4193)(182) = (2.8 \times 10^1)(4.193 \times 10^3)(1.82 \times 10^2)$
$= (2.8)(4.193)(1.82) \times 10^{1+3+2} = 21 \times 10^6 = 2.1 \times 10^7$

This can also be written as 21 million to show the two significant figures.

(*d*) $$\frac{(526.7)(0.001280)}{0.000034921} = \frac{(5.267 \times 10^2)(1.280 \times 10^{-3})}{3.4921 \times 10^{-5}} = \frac{(5.267)(1.280)}{3.4921} \times \frac{(10^2)(10^{-3})}{10^{-5}}$$
$$= 1.931 \times \frac{10^{2-3}}{10^{-5}} = 1.931 \times \frac{10^{-1}}{10^{-5}}$$
$$= 1.931 \times 10^{-1+5} = 1.931 \times 10^4$$

This can also be written as 19.31 thousand to show the four significant figures.

(e) $$\frac{(1.47562 - 1.47322)(4895.36)}{0.000159180} = \frac{(0.00240)(4895.36)}{0.000159180} = \frac{(2.40 \times 10^{-3})(4.89536 \times 10^{3})}{1.59180 \times 10^{-4}}$$

$$= \frac{(2.40)(4.89536)}{1.59180} \times \frac{(10^{-3})(10^{3})}{10^{-4}} = 7.38 \times \frac{10^{0}}{10^{-4}} = 7.38 \times 10^{4}$$

This can also be written as 73.8 thousand to show the three significant figures. Note that although six significant figures were originally present in all numbers, some of these were lost in subtracting 1.47322 from 1.47562.

(f) If denominators 5 and 6 are exact, $\frac{(4.38)^2}{5} + \frac{(5.482)^2}{6} = 3.84 + 5.009 = 8.85$

(g) $3.1416\sqrt{71.35} = (3.1416)(8.447) = 26.54$

(h) $\sqrt{128.5 - 89.24} = \sqrt{39.3} = 6.27$

1.13 Evaluate each of the following, given that $X = 3$, $Y = -5$, $A = 4$, and $B = -7$, where all numbers are assumed to be exact:

(a) $2X - 3Y$

(b) $4Y - 8X + 28$

(c) $\frac{AX + BY}{BX - AY}$

(d) $X^2 - 3XY - 2Y^2$

(e) $2(X + 3Y) - 4(3X - 2Y)$

(f) $\frac{X^2 - Y^2}{A^2 - B^2 + 1}$

(g) $\sqrt{2X^2 - Y^2 - 3A^2 + 4B^2 + 3}$

(h) $\sqrt{\frac{6A^2}{X} + \frac{2B^2}{Y}}$

SOLUTION

(a) $2X - 3Y = 2(3) - 3(-5) = 6 + 15 = 21$

(b) $4Y - 8X + 28 = 4(-5) - 8(3) + 28 = -20 - 24 + 28 = -16$

(c) $$\frac{AX + BY}{BX - AY} = \frac{(4)(3) + (-7)(-5)}{(-7)(3) - (4)(-5)} = \frac{12 + 35}{-21 + 20} = \frac{47}{-1} = -47$$

(d) $X^2 - 3XY - 2Y^2 = (3)^2 - 3(3)(-5) - 2(-5)^2 = 9 + 45 - 50 = 4$

(e) $$2(X + 3Y) - 4(3X - 2Y) = 2[(3) + 3(-5)] - 4[3(3) - 2(-5)]$$
$$= 2(3 - 15) - 4(9 + 10) = 2(-12) - 4(19) = -24 - 76 = -100$$

Another method

$$2(X + 3Y) - 4(3X - 2Y) = 2X + 6Y - 12X + 8Y = -10X + 14Y = -10(3) + 14(-5)$$
$$= -30 - 70 = -100$$

(f) $$\frac{X^2 - Y^2}{A^2 - B^2 + 1} = \frac{(3)^2 - (-5)^2}{(4)^2 - (-7)^2 + 1} = \frac{9 - 25}{16 - 49 + 1} = \frac{-16}{-32} = \frac{1}{2} = 0.5$$

(g) $$\sqrt{2X^2 - Y^2 - 3A^2 + 4B^2 + 3} = \sqrt{2(3)^2 - (-5)^2 - 3(4)^2 + 4(-7)^2 + 3}$$
$$= \sqrt{18 - 25 - 48 + 196 + 3} = \sqrt{144} = 12$$

(h) $$\sqrt{\frac{6A^2}{X} + \frac{2B^2}{Y}} = \sqrt{\frac{6(4)^2}{3} + \frac{2(-7)^2}{-5}} = \sqrt{\frac{96}{3} + \frac{98}{-5}} = \sqrt{12.4} = 3.52 \quad \text{approximately}$$

FUNCTIONS AND GRAPHS

1.14 Table 1.1 shows the number of bushels (bu) of wheat and corn produced on the Tyson farm during the years 2002, 2003, 2004, 2005, and 2006. With reference to this table, determine the year or years during which (a) the least number of bushels of wheat was produced, (b) the greatest number of

Table 1.1 Wheat and Corn Production from 2002 to 2006

Year	Bushels of Wheat	Bushels of Corn
2002	205	80
2003	215	105
2004	190	110
2005	205	115
2006	225	120

bushels of corn was produced, (*c*) the greatest decline in wheat production occurred, (*d*) equal amounts of wheat were produced, and (*e*) the combined production of wheat and corn was a maximum.

SOLUTION

(*a*) 2004; (*b*) 2006; (*c*) 2004; (*d*) 2002 and 2005; (*e*) 2006

1.15 Let W and C denote, respectively, the number of bushels of wheat and corn produced during the year t on the Tyson farm of Problem 1.14. It is clear W and C are both functions of t; this we can indicate by $W = F(t)$ and $C = G(t)$.

(*a*) Find W when $t = 2004$.
(*b*) Find C when $t = 2002$.
(*c*) Find t when $W = 205$.
(*d*) Find $F(2005)$.
(*e*) Find $G(2005)$.
(*f*) Find C when $W = 190$.
(*g*) What is the domain of the variable t?
(*h*) Is W a single-valued function of t?
(*i*) Is t a function of W?
(*j*) Is C a function of W?
(*k*) Which variable is independent, t or W?

SOLUTION

(*a*) 190
(*b*) 80
(*c*) 2002 and 2005
(*d*) 205
(*e*) 115
(*f*) 110
(*g*) The years 2002 through 2006.
(*h*) Yes, since to each value that t can assume, there corresponds one and only one value of W.
(*i*) Yes, the functional dependence of t on W can be written $t = H(W)$.
(*j*) Yes.
(*k*) Physically, it is customary to think of W as determined from t rather than t as determined from W. Thus, physically, t is the dependent variable and W is the independent variable. Mathematically, however, either variable can be considered the independent variable and the other the dependent. The one that is assigned various values is the independent variable; the one that is then determined as a result is the dependent variable.

1.16 A variable Y is determined from a variable X according to the equation $Y = 2X - 3$, where the 2 and 3 are exact.

(*a*) Find Y when $X = 3$, -2, and 1.5.
(*b*) Construct a table showing the values of Y corresponding to $X = -2, -1, 0, 1, 2, 3$, and 4.
(*c*) If the dependence of Y on X is denoted by $Y = F(X)$, determine $F(2.4)$ and $F(0.8)$.
(*d*) What value of X corresponds to $Y = 15$?
(*e*) Can X be expressed as a function of Y?

(*f*) Is Y a single-valued function of X?

(*g*) Is X a single-valued function of Y?

SOLUTION

(*a*) When $X = 3$, $Y = 2X - 3 = 2(3) - 3 = 6 - 3 = 3$. When $X = -2$, $Y = 2X - 3 = 2(-2) - 3 = -4 - 3 = -7$. When $X = 1.5$, $Y = 2X - 3 = 2(1.5) - 3 = 3 - 3 = 0$.

(*b*) The values of Y, computed as in part (*a*), are shown in Table 1.2. Note that by using other values of X, we can construct many tables. The relation $Y = 2X - 3$ is equivalent to the collection of *all* such possible tables.

Table 1.2

X	-2	-1	0	1	2	3	4
Y	-7	-5	-3	-1	1	3	5

(*c*) $F(2.4) = 2(2.4) - 3 = 4.8 - 3 = 1.8$, and $F(0.8) = 2(0.8) - 3 = 1.6 - 3 = -1.4$.

(*d*) Substitute $Y = 15$ in $Y = 2X - 3$. This gives $15 = 2X - 3$, $2X = 18$, and $X = 9$.

(*e*) Yes. Since $Y = 2X - 3$, $Y + 3 = 2X$ and $X = \frac{1}{2}(Y + 3)$. This expresses X *explicitly* as a function of Y.

(*f*) Yes, since for each value that X can assume (and there is an indefinite number of these values) there corresponds one and only one value of Y.

(*g*) Yes, since from part (*e*), $X = \frac{1}{2}(Y + 3)$, so that corresponding to each value assumed by Y there is one and only one value of X.

1.17 If $Z = 16 + 4X - 3Y$, find the value of Z corresponding to (*a*) $X = 2$, $Y = 5$; (*b*) $X = -3$, $Y = -7$; (*c*) $X = -4$, $Y = 2$.

SOLUTION

(*a*) $Z = 16 + 4(2) - 3(5) = 16 + 8 - 15 = 9$

(*b*) $Z = 16 + 4(-3) - 3(-7) = 16 - 12 + 21 = 25$

(*c*) $Z = 16 + 4(-4) - 3(2) = 16 - 16 - 6 = -6$

Given values of X and Y, there corresponds a value of Z. We can denote this dependence of Z on X and Y by writing $Z = F(X, Y)$ (read "Z is a function of X and Y"). $F(2, 5)$ denotes the value of Z when $X = 2$ and $Y = 5$ and is 9, from part (*a*). Similarly, $F(-3, -7) = 25$ and $F(-4, 2) = -6$ from parts (*b*) and (*c*), respectively.

The variables X and Y are called *independent variables*, and Z is called a *dependent variable*.

1.18 The overhead cost for a company is \$1000 per day and the cost of producing each item is \$25.

(*a*) Write a function that gives the total cost of producing x units per day.

(*b*) Use EXCEL to produce a table that evaluates the total cost for producing 5, 10, 15, 20, 25, 30, 35, 40, 45, and 50 units a day.

(*c*) Evaluate and interpret $f(100)$.

SOLUTION

(*a*) $f(x) = 1000 + 25x$.

(*b*) The numbers 5, 10, . . . , 50 are entered into B1:K1 and the expression =1000 + 25*B1 is entered into B2 and a click-and-drag is performed from B2 to K2 to form the following output.

x	5	10	15	20	25	30	35	40	45	50
f(x)	1025	1050	1075	1100	1125	1150	1175	1200	1225	1250

(*c*) $f(100) = 1000 + 25(100) = 1000 + 2500 = 3500$. The cost of manufacturing $x = 100$ units a day is 3500.

1.19 A rectangle has width x and length $x+10$.

(a) Write a function, $A(x)$, that expresses the area as a function of x.

(b) Use EXCEL to produce a table that evaluates $A(x)$ for $x=0, 1, \ldots, 5$.

(c) Write a function, $P(x)$, that expresses the perimeter as a function of x.

(d) Use EXCEL to produce a table that evaluates $P(x)$ for $x=0, 1, \ldots, 5$.

SOLUTION

(a) $A(x)=x(x+10)=x^2+10x$

(b) The numbers 0, 1, 2, 3, 4, and 5 are entered into B1:G1 and the expression =B1^2+10*B1 is entered into B2 and a click-and-drag is performed from B2 to G2 to form the output:

X	0	1	2	3	4	5
A(x)	0	11	24	39	56	75

(c) $P(x)=x+(x+10)+x+(x+10)=4x+20$.

(d) The numbers 0, 1, 2, 3, 4, and 5 are entered into B1:G1 and the expression =4*B1+20 is entered into B2 and a click-and-drag is performed from B2 to G2 to form the output:

X	0	1	2	3	4	5
P(x)	20	24	28	32	36	40

1.20 Locate on a rectangular coordinate system the points having coordinates (a) (5, 2), (b) (2, 5), (c) (−5, 1), (d) ((1, − 3), (e) (3, − 4), (f) (−2.5, − 4.8), (g) (0, − 2.5), and (h) (4, 0). Use MAPLE to plot the points.

SOLUTION

See Fig. 1-2. The MAPLE command to plot the eight points is given. Each point is represented by a circle.

L : = [[5, 2], [2, 5], [−5, 1], [1, −3], [3, −4], [−2.5, −4.8], [0, −2.5], [4, 0]];
pointplot (*L*, *font* = [*TIMES*, *BOLD*, *14*], *symbol* = *cirlce*);

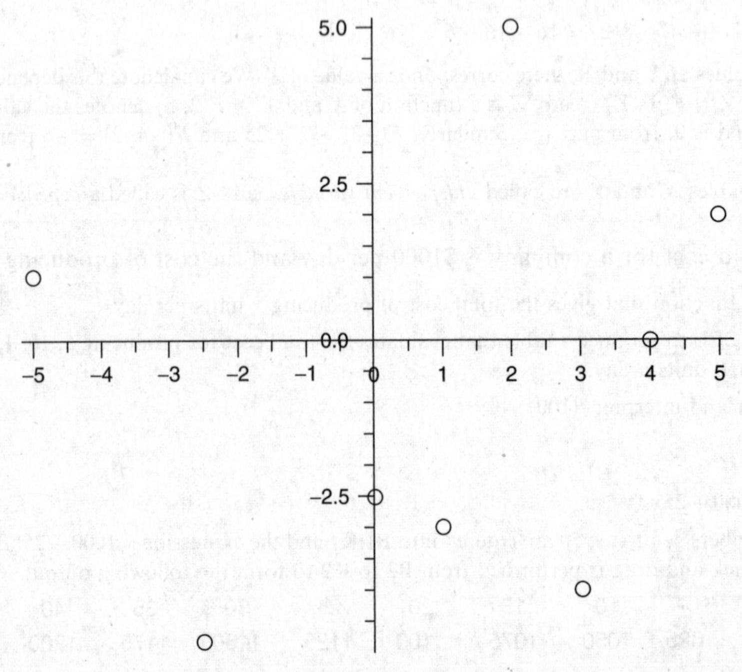

Fig. 1-2 A MAPLE plot of points.

1.21 Graph the equation $Y = 4X - 4$ using MINITAB.

SOLUTION

Note that the graph will extend indefinitely in the positive direction and negative direction of the x axis. We arbitrarily decide to go from -5 to 5. Figure 1-3 shows the MINITAB plot of the line $Y = 4X - 4$. The pull down "**Graph ⇒ Scatterplots**" is used to activate scatterplots. The points on the line are formed by entering the integers from -5 to 5 and using the MINITAB calculator to calculate the corresponding Y values. The X and Y values are as follows.

X	−5	−4	−3	−2	−1	0	1	2	3	4	5
Y	−24	−20	−16	−12	−8	−4	0	4	8	12	16

The points are connected to get an idea of what the graph of the equation $Y = 4X - 4$ looks like.

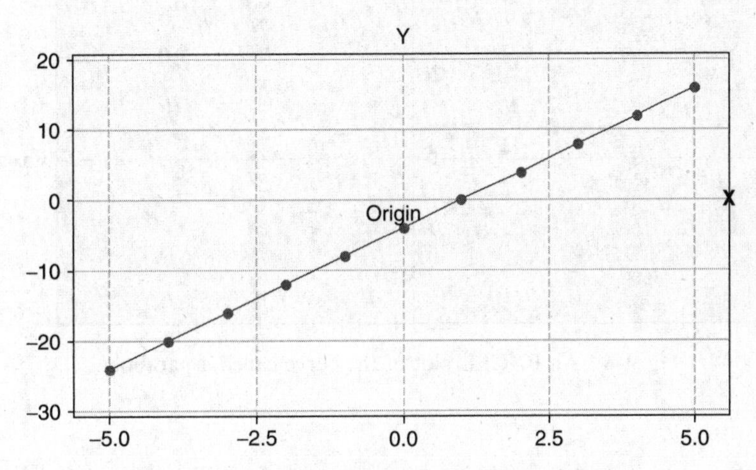

Fig. 1-3 A MINITAB graph of a linear function.

1.22 Graph the equation $Y = 2X^2 - 3X - 9$ using EXCEL.

SOLUTION

Table 1.3 EXCEL-Generated Values for a Quadratic Function

X	−5	−4	−3	−2	−1	0	1	2	3	4	5
Y	56	35	18	5	−4	−9	−10	−7	0	11	26

EXCEL is used to build Table 1.3 that gives the Y values for X values equal to $-5, -4, \ldots, 5$. The expression =2*B1^2-3*B1-9 is entered into cell B2 and a click-and-drag is performed from B2 to L2. The chart wizard, in EXCEL, is used to form the graph shown in Fig. 1-4. The function is called a *quadratic function*. The *roots* (points where the graph cuts the x axis) of this quadratic function are at $X = 3$ and one is between -2 and -1. Clicking the chart wizard in EXCEL shows the many different charts that can be formed using EXCEL. Notice that the graph for this quadratic function goes of to positive infinity as X gets larger positive as well as larger negative. Also note that the graph has its lowest value when X is between 0 and 1.

1.23 Table 1.4 shows the rise in new diabetes diagnoses from 1997 to 2005. Graph these data.

Table 1.4 Number of New Diabetes Diagnoses

Year	1997	1998	1999	2000	2001	2002	2003	2004	2005
Millions	0.88	0.90	1.01	1.10	1.20	1.25	1.28	1.36	1.41

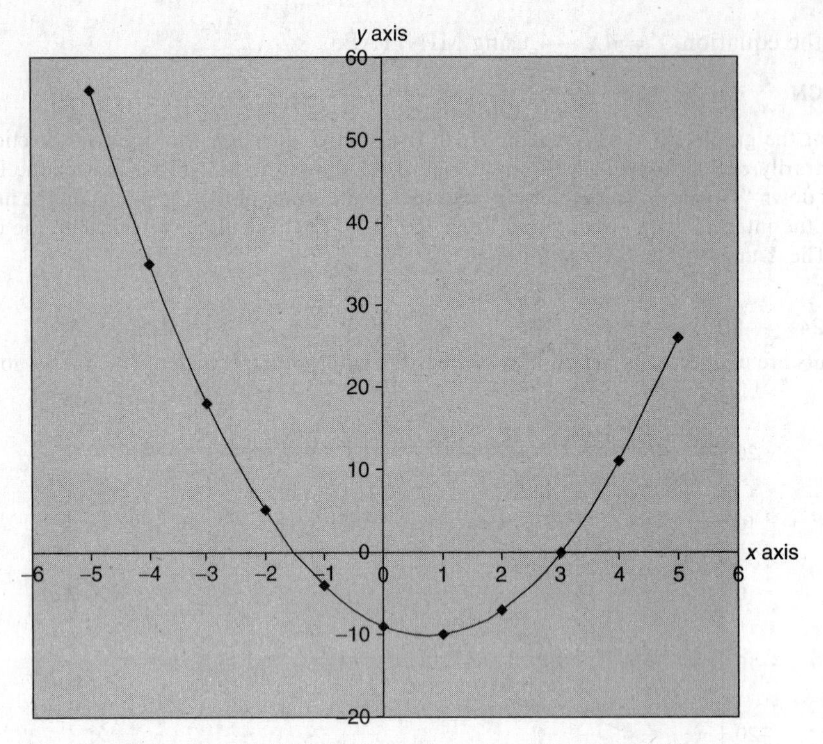

Fig. 1-4 An EXCEL plot of the curve called a parabola.

SOLUTION

First method

The first plot is a *time series plot* and is shown in Fig. 1-5. It plots the number of new cases of diabetes for each year from 1997 to 2005. It shows that the number of new cases has increased each year over the time period covered.

Second method

Figure 1-6 is called a *bar graph, bar chart*, or *bar diagram*. The width of the bars, which are all equal, have no significance in this case and can be made any convenient size as long as the bars do not overlap.

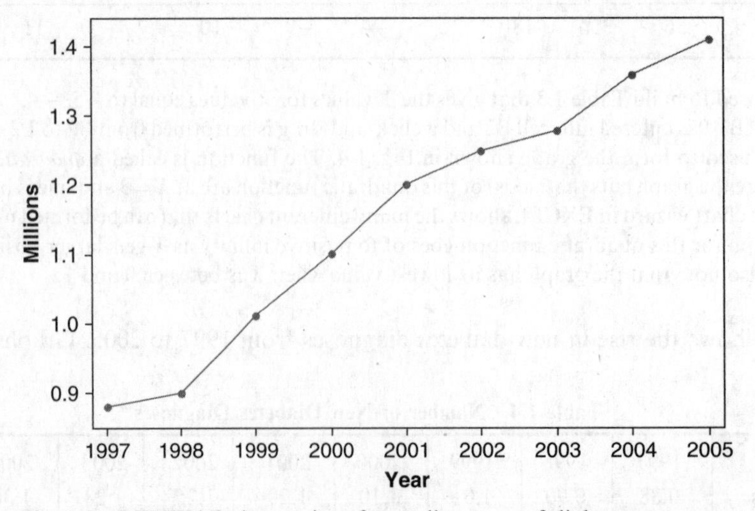

Fig. 1-5 MINITAB time series of new diagnoses of diabetes per year.

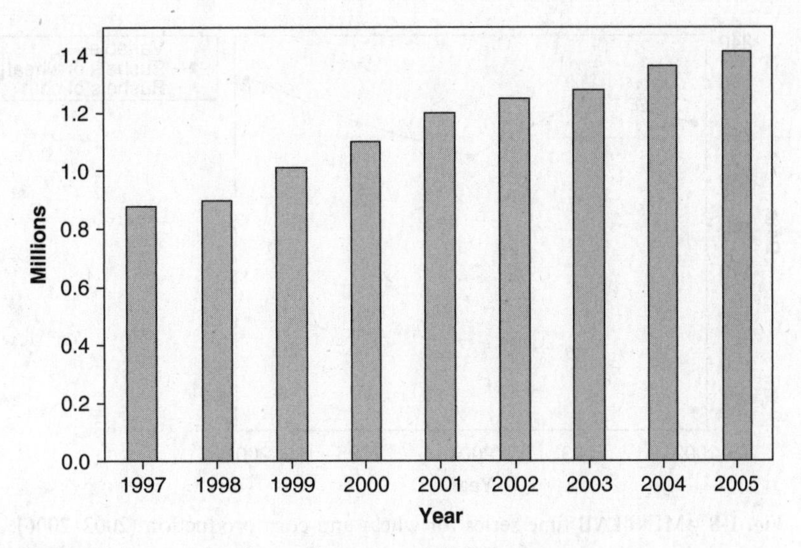

Fig. 1-6 MINITAB bar graph of new diagnoses of diabetes per year.

Third method

A bar chart with bars running horizontal rather than vertical is shown in Fig. 1-7.

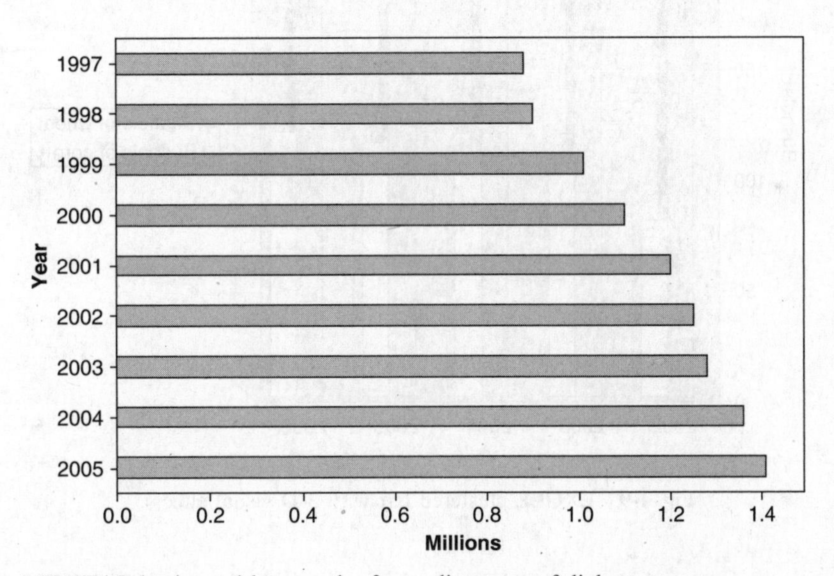

Fig. 1-7 MINITAB horizontal bar graph of new diagnoses of diabetes per year.

1.24 Graph the data of Problem 1.14 by using a time series from MINITAB, a clustered bar with three dimensional (3-D) visual effects from EXCEL, and a stacked bar with 3-D visual effects from EXCEL.

SOLUTION

The solutions are given in Figs. 1-8, 1-9, and 1-10.

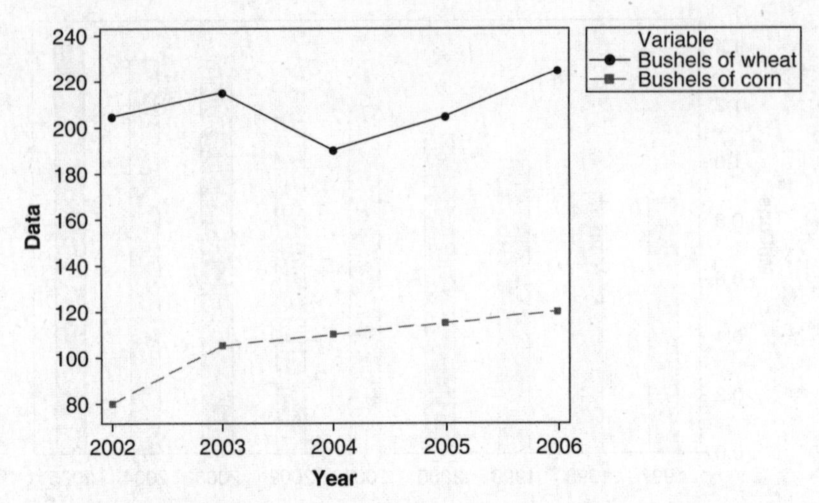

Fig. 1-8 MINITAB time series for wheat and corn production (2002–2006).

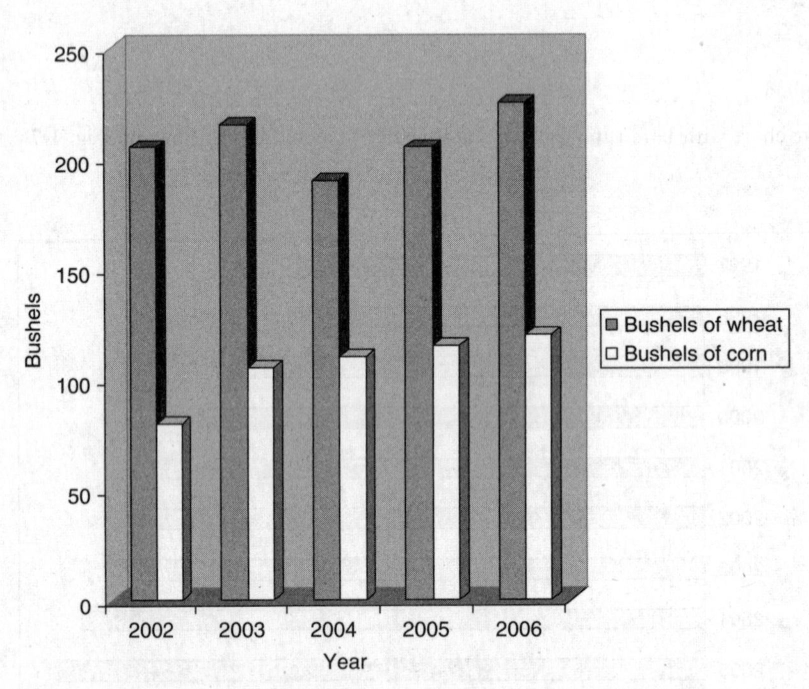

Fig. 1-9 EXCEL clustered bar with 3-D visual effects.

1.25 (*a*) Express the yearly number of bushels of wheat and corn from Table 1.1 of Problem 1.14 as percentages of the total annual production.

(*b*) Graph the percentages obtained in part (*a*).

SOLUTION

(*a*) For 2002, the percentage of wheat $= 205/(205 + 80) = 71.9\%$ and the percentage of corn $= 100\% - 71.9\% = 28.1\%$, etc. The percentages are shown in Table 1.5.

(*b*) The *100% stacked column* compares the percentage each value contributes to a total across categories (Fig. 1-11).

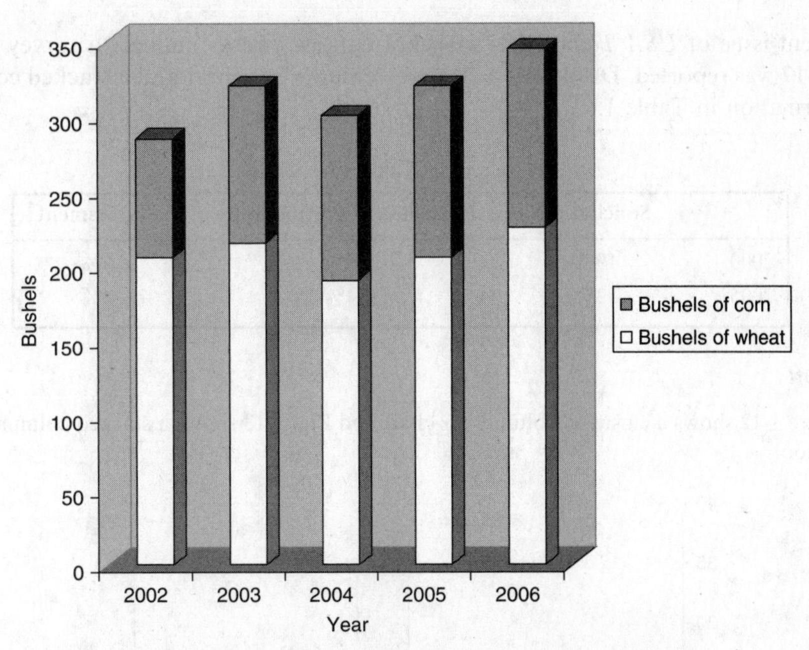

Fig. 1-10 EXCEL stacked bar with 3-D visual effect.

Table 1.5 Wheat and corn production from 2002 till 2006

Year	Wheat (%)	Corn (%)
2002	71.9	28.1
2003	67.2	32.8
2004	63.3	36.7
2005	64.1	35.9
2006	65.2	34.8

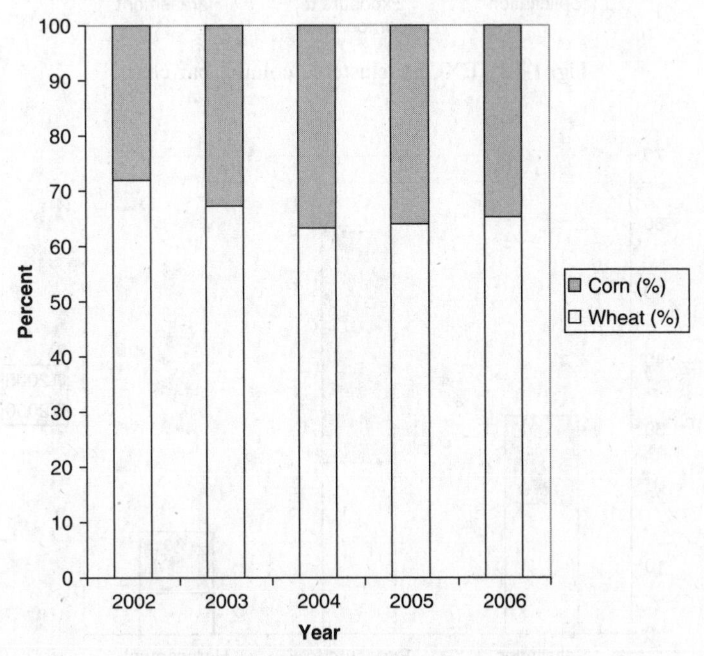

Fig. 1-11 EXCEL 100% stacked column.

1.26 In a recent issue of *USA Today*, in a snapshot entitled "Perils online", a survey of 1500 kids of ages 10–17 was reported. Display as a clustered column bar chart and a stacked column bar chart the information in Table 1.6.

Table 1.6

	Solicitation	Exposure to Pornography	Harassment
2000	19%	25%	6%
2005	13%	34%	9%

SOLUTION

Figure 1-12 shows a clustered column bar chart and Fig. 1-13 shows a stacked column bar chart of the information.

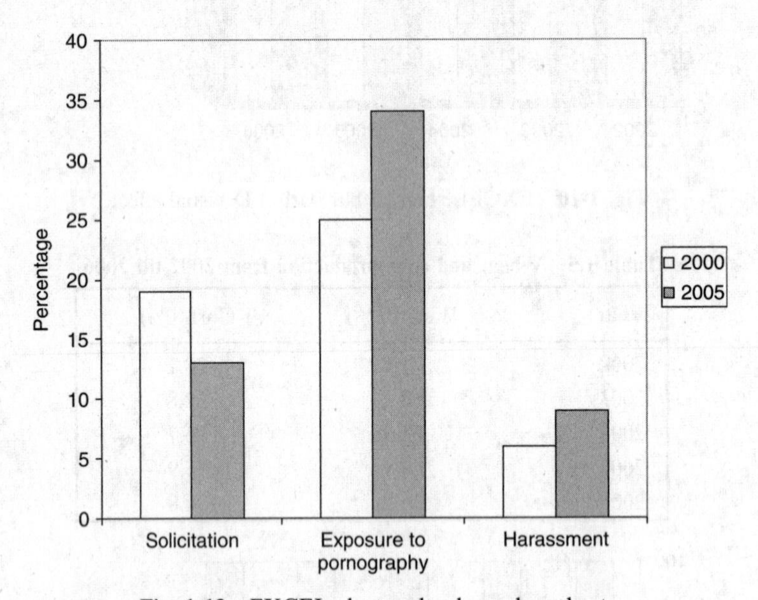

Fig. 1-12 EXCEL clustered column bar chart.

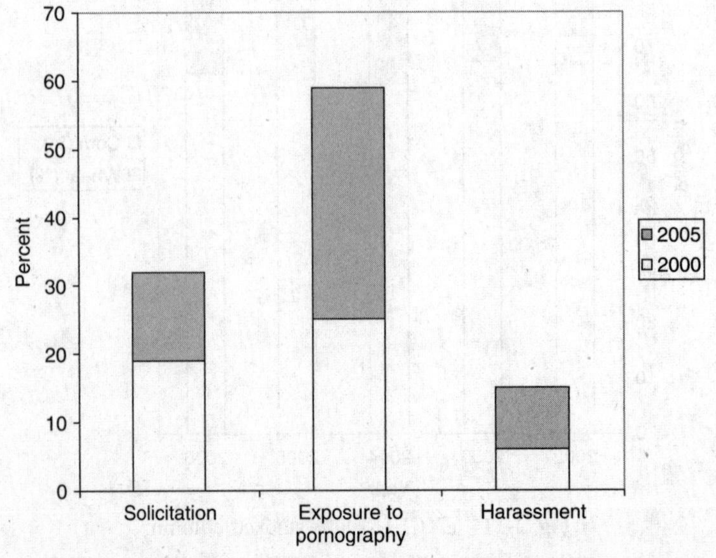

Fig. 1-13 EXCEL stacked column bar chart.

1.27 In a recent *USA Today* snapshot entitled "Where the undergraduates are", it was reported that more than 17.5 million undergraduates attend more than 6400 schools in the USA. Table 1.7 gives the enrollment by type of school.

Table 1.7 Where the undergraduates are

Type of School	Percent
Public 2-year	43
Public 4-year	32
Private non-profit 4-year	15
Private 2- and 4-year	6
Private less than 4-year	3
Other	1

Construct a 3-D bar chart of the information in Table 1.7 using EXCEL and a bar chart using MINITAB.

SOLUTION

Figures 1-14 and 1-15 give 3-D bar charts of the data in Table 1.6.

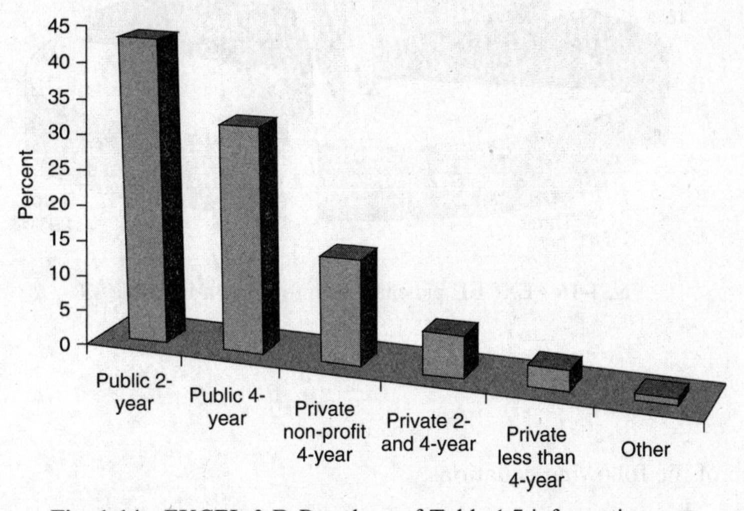

Fig. 1-14 EXCEL 3-D Bar chart of Table 1.7 information.

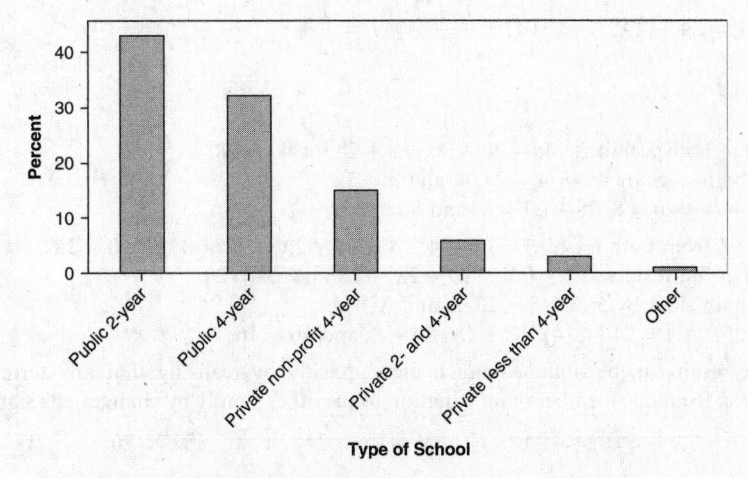

Fig. 1-15 MINITAB bar chart of Table 1.7 information.

1.28 Americans have an average of 2.8 working televisions at home. Table 1.8 gives the breakdown. Use EXCEL to display the information in Table 1.8 with a pie chart.

Table 1.8 Televisions per household

Televisions	Percent
None	2
One	15
Two	29
Three	26
Four	16
Five +	12

SOLUTION

Figure 1-16 gives the EXCEL pie chart for the information in Table 1.8.

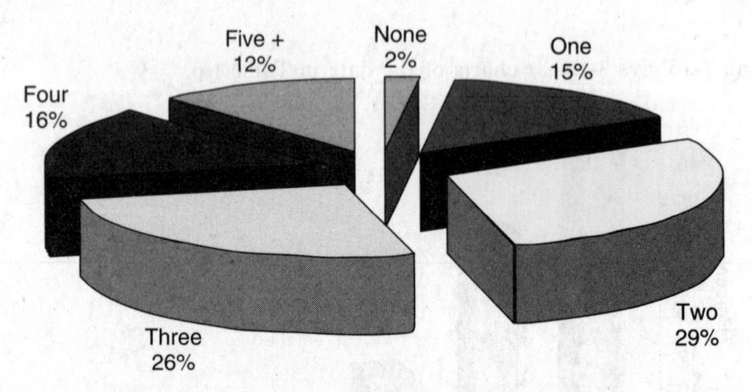

Fig. 1-16 EXCEL pie chart of information in Table 1.8.

EQUATIONS

1.29 Solve each of the following equations:

(*a*) $4a - 20 = 8$ (*c*) $18 - 5b = 3(b + 8) + 10$

(*b*) $3X + 4 = 24 - 2X$ (*d*) $\dfrac{Y+2}{3} + 1 = \dfrac{Y}{2}$

SOLUTION

(*a*) Add 20 to both members: $4a - 20 + 20 = 8 + 20$, or $4a = 28$.
Divide both sides by 4: $4a/4 = 28/4$, and $a = 7$.
Check: $4(7) - 20 = 8$, $28 - 20 = 8$, and $8 = 8$.

(*b*) Subtract 4 from both members: $3X + 4 - 4 = 24 - 2X - 4$, or $3X = 20 - 2X$.
Add $2X$ to both sides: $3X + 2X = 20 - 2X + 2X$, or $5X = 20$.
Divide both sides by 5: $5X/5 = 20/5$, and $X = 4$.
Check: $3(4) + 4 = 24 - 2(4)$, $12 + 4 = 24 - 8$, and $16 = 16$.

The result can be obtained much more quickly by realizing that any term can be moved, or *transposed*, from one member of an equation to the other simply by changing its sign. Thus we can write

$$3X + 4 = 24 - 2X \qquad 3X + 2X = 24 - 4 \qquad 5X = 20 \qquad X = 4$$

(*c*) $18 - 5b = 3b + 24 + 10$, and $18 - 5b = 3b + 34$.
Transposing, $-5b - 3b = 34 - 18$, or $-8b = 16$.
Dividing by -8, $-8b/(-8) = 16/(-8)$, and $b = -2$.
Check: $18 - 5(-2) = 3(-2 + 8) + 10$, $18 + 10 = 3(6) + 10$, and $28 = 28$.

(*d*) First multiply both members by 6, the lowest common denominator.

$$6\left(\frac{Y+2}{3}+1\right) = 6\left(\frac{Y}{2}\right) \qquad 6\left(\frac{Y+2}{3}\right) + 6(1) = \frac{6Y}{2} \qquad 2(Y+2) + 6 = 3Y$$

$$2Y + 4 + 6 = 3Y \qquad 2Y + 10 = 3Y \qquad 10 = 3Y - 2Y \qquad Y = 10$$

Check: $\dfrac{10+2}{3} + 1 = \dfrac{10}{2}$, $\dfrac{12}{3} + 1 = \dfrac{10}{2}$, $4 + 1 = 5$, and $5 = 5$.

1.30 Solve each of the following sets of simultaneous equations:

(*a*) $3a - 2b = 11$
$5a + 7b = 39$

(*b*) $5X + 14Y = 78$
$7X + 3Y = -7$

(*c*) $3a + 2b + 5c = 15$
$7a - 3b + 2c = 52$
$5a + b - 4c = 2$

SOLUTION

(*a*)

Multiply the first equation by 7:	$21a - 14b = 77$	(*1*)
Multiply the second equation by 2:	$10a + 14b = 78$	(*2*)
Add:	$31a = 155$	
Divide by 31:	$a = 5$	

Note that by multiplying each of the given equations by suitable numbers, we are able to write two *equivalent equations*, (*1*) and (*2*), in which the coefficients of the unknown b are numerically equal. Then by addition we are able to *eliminate* the unknown b and thus find a.

Substitute $a = 5$ in the first equation: $3(5) - 2b = 11$, $-2b = -4$, and $b = 2$. Thus $a = 5$, and $b = 2$.
Check: $3(5) - 2(2) = 11$, $15 - 4 = 11$, and $11 = 11$; $5(5) + 7(2) = 39$, $25 + 14 = 39$, and $39 = 39$.

(*b*)

Multiply the first equation by 3:	$15X + 42Y = 234$	(*3*)
Mutiply the second equation by -14:	$-98X - 42Y = 98$	(*4*)
Add:	$-83X = 332$	
Divide by -83:	$X = -4$	

Substitute $X = -4$ in the first equation: $5(-4) + 14Y = 78$, $14Y = 98$, and $Y = 7$.
Thus $X = -4$, and $Y = 7$.
Check: $5(-4) + 14(7) = 78$, $-20 + 98 = 78$, and $78 = 78$; $7(-4) + 3(7) = -7$, $-28 + 21 = -7$, and $-7 = -7$.

(*c*)

Multiply the first equation by 2:	$6a + 4b + 10c = 30$	
Multiply the second equation by -5:	$-35a + 15b - 10c = -260$	
Add:	$-29a + 19b = -230$	(*5*)
Multiply the second equation by 2:	$14a - 6b + 4c = 104$	
Repeat the third equation:	$5a + b - 4c = 2$	
Add:	$19a - 5b = 106$	(*6*)

We have thus eliminated c and are left with two equations, (*5*) and (*6*), to be solved simultaneously for a and b.

Multiply equation (*5*) by 5:	$-145a + 95b = -1150$
Multiply equation (*6*) by 19:	$361a - 95b = 2014$
Add:	$216a = 864$
Divide by 216:	$a = 4$

Substituting $a = 4$ in equation (*5*) or (*6*), we find that $b = -6$.
Substituting $a = 4$ and $b = -6$ in any of the given equations, we obtain $c = 3$.

Thus $a = 4$, $b = -6$, and $c = 3$.
Check: $3(4) + 2(-6) + 5(3) = 15$, and $15 = 15$; $7(4) - 3(-6) + 2(3) = 52$, and $52 = 52$; $5(4) + (-6) -4(3) = 2$, and $2 = 2$.

INEQUALITIES

1.31 Express in words the meaning of each of the following:

(*a*) $N > 30$ (*b*) $X \leq 12$ (*c*) $0 < p \leq 1$ (*d*) $\mu - 2t < X < \mu + 2t$

SOLUTION

(*a*) N is greater than 30.
(*b*) X is less than or equal to 12.
(*c*) p is greater than 0 but less than or equal to 1.
(*d*) X is greater than $\mu - 2t$ but less than $\mu + 2t$.

1.32 Translate the following into symbols:

(*a*) The variable X has values between 2 and 5 inclusive.
(*b*) The arithmetic mean $\bar{X}$ is greater than 28.42 but less than 31.56.
(*c*) m is a positive number less than or equal to 10.
(*d*) P is a nonnegative number.

SOLUTION

(*a*) $2 \leq X \leq 5$; (*b*) $28.42 < \bar{X} < 31.56$; (*c*) $0 < m \leq 10$; (*d*) $P \geq 0$.

1.33 Using inequality symbols, arrange the numbers 3.42, −0.6, −2.1, 1.45, and −3 in (*a*) increasing and (*b*) decreasing order of magnitude.

SOLUTION

(*a*) $-3 < -2.1 < -0.6 < 1.45 < 3.42$
(*b*) $3.42 > 1.45 > -0.6 > -2.1 > -3$

Note that when the numbers are plotted as points on a line (see Problem 1.18), they increase from left to right.

1.34 In each of the following, find a corresponding inequality for X (i.e., solve each inequality for X):

(*a*) $2X < 6$ (*c*) $6 - 4X < -2$ (*e*) $-1 \leq \dfrac{3 - 2X}{5} \leq 7$

(*b*) $3X - 8 \geq 4$ (*d*) $-3 < \dfrac{X - 5}{2} < 3$

SOLUTION

(*a*) Divide both sides by 2 to obtain $X < 3$.
(*b*) Adding 8 to both sides, $3X \geq 12$; dividing both sides by 3, $X \geq 4$.
(*c*) Adding −6 to both sides, $-4X < -8$; dividing both sides by −4, $X > 2$. Note that, as in equations, we can transpose a term from one side of an inequality to the other simply by changing the sign of the term; from part (*b*), for example, $3X \geq 8 + 4$.
(*d*) Multiplying by 2, $-6 < X - 5 < 6$; adding 5, $-1 < X < 11$.
(*e*) Multiplying by 5, $-5 \leq 3 - 2X \leq 35$; adding −3, $-8 \leq -2X \leq 32$; dividing by −2, $4 \geq X \geq -16$, or $-16 \leq X \leq 4$.

LOGARITHMS AND PROPERTIES OF LOGARITHMS

1.35 Use the definition of $y = \log_b x$ to find the following logarithms and then use EXCEL to verify your answers. (Note that $y = \log_b x$ means that $b^y = x$.)

(*a*) Find the log to the base 2 of 32.
(*b*) Find the log to the base 4 of 64.
(*c*) Find the log to the base 6 of 216.
(*d*) Find the log to the base 8 of 4096.
(*e*) Find the log to the base 10 of 10,000.

SOLUTION

(*a*) 5; (*b*) 3; (*c*) 3; (*d*) 4; (*e*) 4.

The EXCEL expression =LOG(32,2) gives 5, =LOG(64,4) gives 3, =LOG(216,6) gives 3, =LOG(4096,8) gives 4, and =LOG(10000,10) gives 4.

1.36 Using the properties of logarithms, re-express the following as sums and differences of logarithms.

(*a*) $\ln\left(\dfrac{x^2y^3z}{ab}\right)$ (*b*) $\log\left(\dfrac{a^2b^3c}{yz}\right)$

Using the properties of logarithms, re-express the following as a single logarithm.

(*c*) $\ln(5) + \ln(10) - 2\ln(5)$ (*d*) $2\log(5) - 3\log(5) + 5\log(5)$

SOLUTION

(*a*) $2\ln(x) + 3\ln(y) + \ln(z) - \ln(a) - \ln(b)$
(*b*) $2\log(a) + 3\log(b) + \log(c) - \log(y) - \log(z)$
(*c*) $\ln(2)$
(*d*) $\log(625)$

1.37 Make a plot of $y = \ln(x)$ using SAS and SPSS.

SOLUTION

The solutions are shown in Figures 1-17 and 1-18.

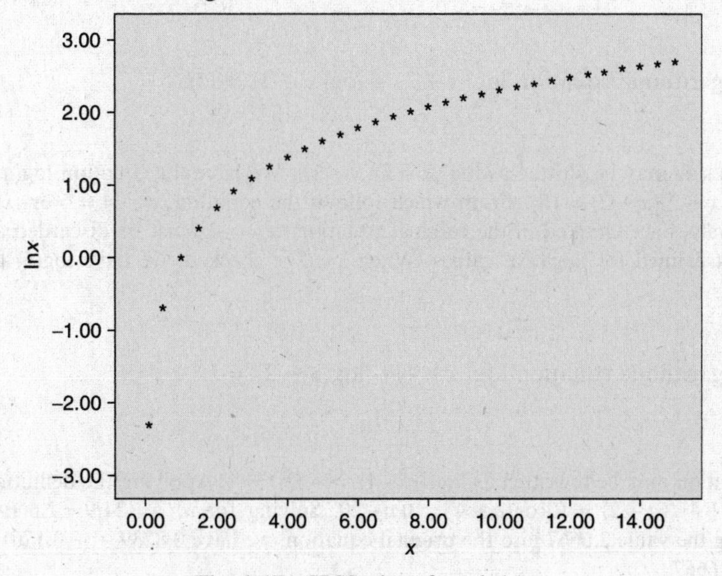

Fig. 1-17 SPSS plot of $y = \ln(x)$.

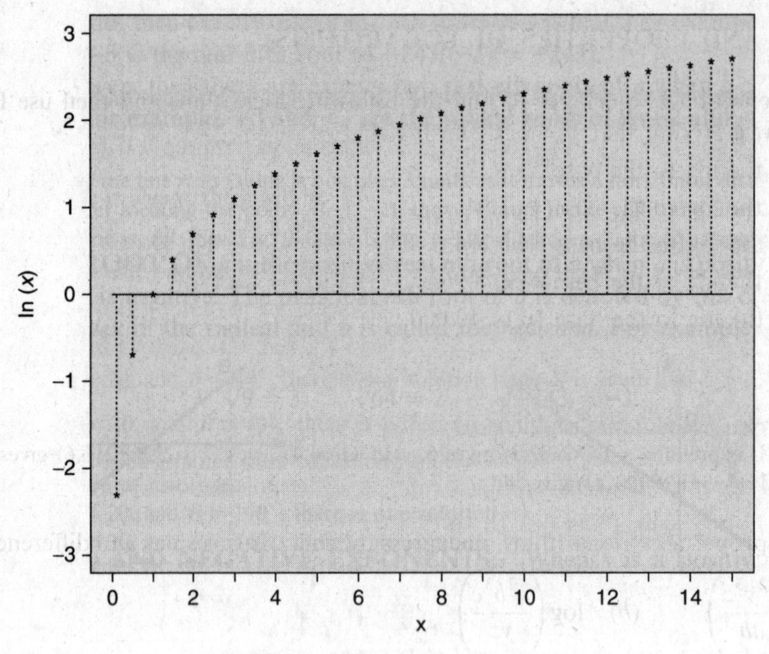

Fig. 1-18 SAS plot of $y = \ln(x)$.

A plot of the curve $y = \ln(x)$ is shown in Figures 1-17 and 1-18. As x gets close to 0 the values for $\ln(x)$ get closer and closer to $-\infty$. As x gets larger and larger, the values for $\ln(x)$ get closer to $+\infty$.

LOGARITHMIC EQUATIONS

1.38 Solve the logarithmic equation $\ln(x) = 10$.

SOLUTION

Using the definition of logarithm, $x = e^{10} = 22026.47$. As a check, we take the natural log of 22026.47 and get 10.00000019.

1.39 Solve the logarithmic equation $\log(x+2) + \log(x-2) = \log(5)$

SOLUTION

The left-hand side may be written as $\log[(x+2)(x-2)]$. We have the equation $\log(x+2)(x-2) = \log(5)$ from which $(x+2)(x-2) = (5)$. From which follows the equation $x^2 - 4 = 5$ or $x^2 = 9$ or $x = -3$ or 3. When these values are checked in the original equation, $x = -3$ must be discarded as a solution because the log is not defined for negative values. When $x = 3$ is checked, we have $\log(5) + \log(1) = \log(5)$ since $\log(1) = 0$.

1.40 Solve the logarithmic equation $\log(a+4) - \log(a-2) = 1$.

SOLUTION

The equation may be rewritten as $\log((a+4)/(a-2)) = 1$. Applying the definition of a logarithm, we have $(a+4)/(a-2) = 10^1$ or $a + 4 = 10a - 20$. Solving for a, $a = 24/9 = 2.6$ (with the 6 repeated). Checking the value 2.6667 into the original equation, we have $0.8239 - (-0.1761) = 1$. The only solution is 2.6667.

1.41 Solve the logarithmic equation $\ln(x)^2 - 1 = 0$.

SOLUTION

The equation may be factored as $[\ln(x)+1][\ln(x)-1]=0$. Setting the factor $\ln(x)+1=0$, we have $\ln(x)=-1$ or $x=e^{-1}=0.3678$. Setting the factor $\ln(x)-1=0$, we have $\ln(x)=1$ or $x=e^{1}=2.7183$. Both values are solutions to the equation.

1.42 Solve the following equation for x: $2\log(x+1) - 3\log(x+1) = 2$.

SOLUTION

The equation may be rewritten as $\log[(x+1)^2/(x+1)^3] = 2$ or $\log[1/(x+1)] = 2$ or $\log(1) - \log(x+1) = 2$ or $0 - \log(x+1) = 2$ or $\log(x+1) = -2$ or $x+1 = 10^{-2}$ or $x = -0.99$. Substituting into the original equation, we find $2\log(0.01) - 3\log(0.01) = 2$. Therefore it checks.

1.43 The software package MAPLE may be used to solve logarithmic equations that are not easy to solve by hand. Solve the following equation using MAPLE:

$\log(x+2) - \ln(x^2) = 4$

SOLUTION

The MAPLE command to solve the equation is "solve(log 10($x+2$) − ln(x^2) = 4);" The solution is given as −0.154594.

Note that MAPLE uses log10 for the common logarithm.

To check and see if the solution is correct, we have $\log(1.845406) - \ln(0.023899)$ which equals 4.00001059.

1.44 EXCEL may be used to solve logarithmic equations also. Solve the following logarithmic equation using EXCEL: $\log(x+4) + \ln(x+5) = 1$.

SOLUTION

Figure 1.19 gives the EXCEL worksheet.

−3	−0.30685	LOG10(A1+4)+LN(A1+5)−1
−2	0.399642	
−1	0.863416	
0	1.211498	
1	1.490729	
2	1.724061	
3	1.92454	
4	2.100315	
5	2.256828	

−3	−0.30685	LOG10(A11+4)+LN(A11+5)−1
−2.9	−0.21667	
−2.8	−0.13236	
−2.7	−0.05315	
−2.6	0.021597	
−2.5	0.092382	
−2.4	0.159631	
−2.3	0.223701	
−2.2	0.284892	
−2.1	0.343464	
−2	0.399642	

Fig. 1-19 EXCEL worksheet for Problem 1.44.

The iterative technique shown above may be used. The upper half locates the root of $\log(x+4)+\ln(x+5)-1$ between -3 and -2. The lower half locates the root between -2.7 and -2.6. This process may be continued to give the root to any desired accuracy. Click-and-drag techniques may be used in doing the technique.

1.45 Solve the equation in Problem 1.44 using the software package MAPLE.

SOLUTION

The MAPLE command "> **solve(log10(*x* + 4) + ln(*x* + 5) = 1);**" gives the solution -2.62947285. Compare this with that obtained in Problem 1.44.

Supplementary Problems

VARIABLES

1.46 State which of the following represent discrete data and which represent continuous data:

(*a*) Number of inches of rainfall in a city during various months of the year
(*b*) Speed of an automobile in miles per hour
(*c*) Number of \$20 bills circulating in the United States at any time
(*d*) Total value of shares sold each day in the stock market
(*e*) Student enrollment in a university over a number of years

1.47 Give the domain of each of the following variables and state whether the variables are continuous or discrete:

(*a*) Number W of bushels of wheat produced per acre on a farm over a number of years
(*b*) Number N of individuals in a family
(*c*) Marital status of an individual
(*d*) Time T of flight of a missile
(*e*) Number P of petals on a flower

ROUNDING OF DATA, SCIENTIFIC NOTATION, AND SIGNIFICANT FIGURES

1.48 Round each of the following numbers to the indicated accuracy:

(*a*) 3256 nearest hundred
(*b*) 5.781 nearest tenth
(*c*) 0.0045 nearest thousandth
(*d*) 46.7385 nearest hundredth
(*e*) 125.9995 two decimal places
(*f*) 3,502,378 nearest million
(*g*) 148.475 nearest unit
(*h*) 0.000098501 nearest millionth
(*i*) 2184.73 nearest ten
(*j*) 43.87500 nearest hundredth

1.49 Express each number without using powers of 10.

(*a*) 132.5×10^4
(*b*) 418.72×10^{-5}
(*c*) 280×10^{-7}
(*d*) 7300×10^6
(*e*) 3.487×10^{-4}
(*f*) 0.0001850×10^5

1.50 How many significant figures are in each of the following, assuming that the numbers are accurately recorded?

(*a*) 2.54 cm
(*b*) 0.004500 yd
(*c*) 3,510,000 bu
(*d*) 3.51 million bu
(*e*) 10.000100 ft
(*f*) 378 people
(*g*) 378 oz
(*h*) 4.50×10^{-3} km
(*i*) 500.8×10^5 kg
(*j*) 100.00 mi

1.51 What is the maximum error in each of the following measurements, assumed to be accurately recorded? Give the number of significant figures in each case.

(*a*) 7.20 million bu
(*b*) 0.00004835 cm
(*c*) 5280 ft
(*d*) 3.0×10^8 m
(*e*) 186,000 mi/sec
(*f*) 186 thousand mi/sec

1.52 Write each of the following numbers using the scientific notation. Assume that all figures are significant unless otherwise indicated.

(*a*) 0.000317
(*b*) 428,000,000 (four significant figures)
(*c*) 21,600.00
(*d*) 0.000009810
(*e*) 732 thousand
(*f*) 18.0 ten-thousandths

COMPUTATIONS

1.53 Show that (*a*) the product and (*b*) the quotient of the numbers 72.48 and 5.16, assumed to have four and three significant figures, respectively, cannot be accurate to more than three significant figures. Write the accurately recorded product and quotient.

1.54 Perform each indicated operation. Unless otherwise specified, assume that the numbers are accurately recorded.

(*a*) 0.36×781.4

(*b*) $\dfrac{873.00}{4.881}$

(*c*) $5.78 \times 2700 \times 16.00$

(*d*) $\dfrac{0.00480 \times 2300}{0.2084}$

(*e*) $\sqrt{120 \times 0.5386 \times 0.4614}$ (120 exact)

(*f*) $\dfrac{(416,000)(0.000187)}{\sqrt{73.84}}$

(*g*) $14.8641 + 4.48 - 8.168 + 0.36125$

(*h*) $4,173,000 - 170,264 + 1,820,470 - 78,320$ (numbers are respectively accurate to four, six, six, and five significant figures)

(*i*) $\sqrt{\dfrac{7(4.386)^2 - 3(6.47)^2}{6}}$ (3, 6, and 7 are exact)

(*j*) $4.120\sqrt{\dfrac{3.1416[(9.483)^2 - (5.075)^2]}{0.0001980}}$

1.55 Evaluate each of the following, given that $U = -2$, $V = \frac{1}{2}$, $W = 3$, $X = -4$, $Y = 9$, and $Z = \frac{1}{6}$, where all numbers are assumed to be exact.

(*a*) $4U + 6V - 2W$

(*b*) $\dfrac{XYZ}{UVW}$

(*c*) $\dfrac{2X - 3Y}{UW + XV}$

(*g*) $\sqrt{\dfrac{(W-2)^2}{V} + \dfrac{(Y-5)^2}{Z}}$

(*h*) $\dfrac{X - 3}{\sqrt{(Y-4)^2 + (U+5)^2}}$

(*i*) $X^3 + 5X^2 - 6X - 8$

(*d*) $3(U-X)^2+Y$

(*e*) $\sqrt{U^2-2UV+W}$

(*f*) $3X(4Y+3Z)-2Y(6X-5Z)-25$

(*j*) $\dfrac{U-V}{\sqrt{U^2+V^2}}[U^2V(W+X)]$

FUNCTIONS, TABLES, AND GRAPHS

1.56 A variable Y is determined from a variable X according to the equation $Y = 10 - 4X$.

(*a*) Find Y when $X = -3, -2, -1, 0, 1, 2, 3, 4$, and 5, and show the results in a table.

(*b*) Find Y when $X = -2.4, -1.6, -0.8, 1.8, 2.7, 3.5$, and 4.6.

(*c*) If the dependence of Y on X is denoted by $Y = F(X)$, find $F(2.8)$, $F(-5)$, $F(\sqrt{2})$, and $F(-\pi)$.

(*d*) What value of X corresponds to $Y = -2, 6, -10, 1.6, 16, 0$, and 10?

(*e*) Express X explicitly as a function of Y.

1.57 If $Z = X^2 - Y^2$, find Z when (*a*) $X = -2$, $Y = 3$, and (*b*) $X = 1$, $Y = 5$. (*c*) If using the functional notation $Z = F(X, Y)$, find $F(-3, -1)$.

1.58 If $W = 3XZ - 4Y^2 + 2XY$, find W when (*a*) $X = 1$, $Y = -2$, $Z = 4$, and (*b*) $X = -5$, $Y = -2$, $Z = 0$. (*c*) If using the functional notation $W = F(X, Y, Z)$, find $F(3, 1, -2)$.

1.59 Locate on a rectangular coordinate system the points having coordinates (*a*) (3, 2), (*b*) (2, 3), (*c*) (−4, 4), (*d*) (4, −4), (*e*) (−3, −2), (*f*) (−2, −3), (*g*) (−4.5, 3), (*h*) (−1.2, −2.4), (*i*) (0, −3), and (*j*) (1.8, 0).

1.60 Graph the equations (*a*) $Y = 10 - 4X$ (see Problem 1.56), (*b*) $Y = 2X + 5$, (*c*) $Y = \frac{1}{3}(X - 6)$, (*d*) $2X + 3Y = 12$, and (*e*) $3X - 2Y = 6$.

1.61 Graph the equations (*a*) $Y = 2X^2 + X - 10$, and (*b*) $Y = 6 - 3X - X^2$.

1.62 Graph $Y = X^3 - 4X^2 + 12X - 6$.

1.63 Table 1.9 gives the number of health clubs and the number of members in millions for the years 2000 through 2005. Use software to construct a time-series plot for clubs and construct a separate time-series plot for members.

Table 1.9

Year	2000	2001	2002	2003	2004	2005
Clubs	13000	13225	15000	20000	25500	28500
Members	32.5	35.0	36.5	39.0	41.0	41.3

1.64 Using the data of Table 1.9 and software, construct bar charts for clubs and for members.

1.65 Using the chart wizard of EXCEL, construct a scatter plot for the clubs as well as the members in Table 1.9.

1.66 Table 1.10 shows the infant deaths per 1000 live births for whites and non-whites in a Midwestern state for the years 2000 through 2005. Use an appropriate graph to portray the data.

Table 1.10

Year	2000	2001	2002	2003	2004	2005
White	6.6	6.3	6.1	6.0	5.9	5.7
Non-white	7.6	7.5	7.3	7.2	7.1	6.8

1.67 Table 1.11 shows the orbital velocities of the planets in our solar system. Graph the data.

Table 1.11

Planet	Mercury	Venus	Earth	Mars	Jupiter	Saturn	Uranus	Neptune	Pluto
Velocity (mi/sec)	29.7	21.8	18.5	15.0	8.1	6.0	4.2	3.4	3.0

1.68 Table 1.12 gives the projected public school enrollments (in thousands) for *K* through grade 8, grades 9 through 12, and college for the years 2000 through 2006. Graph the data, using line graphs, bar graphs, and component bar graphs.

1.69 Graph the data of Table 1.12 by using a percentage component graph.

Table 1.12

Year	2000	2001	2002	2003	2004	2005	2006
K—grade 8	33,852	34,029	34,098	34,065	33,882	33,680	33,507
Grades 9–12	13,804	13,862	14,004	14,169	14,483	14,818	15,021
College	12,091	12,225	12,319	12,420	12,531	12,646	12,768

Source: U.S. National Center for Educational Statistics and Projections of Education Statistics, annual.

1.70 Table 1.13 shows the marital status of males and females (18 years and older) in the United States as of the year 1995. Graph the data, using (*a*) two pie charts having the same diameter and (*b*) a graph of your own choosing.

Table 1.13

Marital Status	Male (percent of total)	Female (percent of total)
Never married	26.8	19.4
Married	62.7	59.2
Widowed	2.5	11.1
Divorced	8.0	10.3

Source: U.S. Bureau of Census—Current Population Reports.

1.71 Table 1.14 gives the number of inmates younger than 18 held in state prisons during the years 2001 through 2005. Graph the data using the appropriate types of graphs.

Table 1.14

Year	2001	2002	2003	2004	2005
number	3,147	3,038	2,741	2,485	2,266

1.72 Table 1.15 gives the total number of visits to the Smithsonian Institution in millions during the years 2001 through 2005. Construct a bar chart for the data.

Table 1.15

Year	2001	2002	2003	2004	2005
number	32	26	24	20	24

1.73 To the nearest million, Table 1.16 shows the seven countries of the world with the largest populations as of 1997. Use a pie chart to illustrate the populations of the seven countries of the world with the largest populations.

Table 1.16

Country	China	India	United States	Indonesia	Brazil	Russia	Pakistan
Population (millions)	1,222	968	268	210	165	148	132

Source: U.S. Bureau of the Census, International database.

1.74 A *Pareto chart* is a bar graph in which the bars are arranged according to the frequency values so that the tallest bar is at the left and the smallest bar is at the right. Construct a Pareto chart for the data in Table 1.16.

1.75 Table 1.17 shows the areas of the oceans of the world in millions of square miles. Graph the data, using (*a*) a bar chart and (*b*) a pie chart.

Table 1.17

Ocean	Pacific	Atlantic	Indian	Antarctic	Arctic
Area (million square miles)	63.8	31.5	28.4	7.6	4.8

Source: United Nations.

EQUATIONS

1.76 Solve each of the following equations:

(*a*) $16 - 5c = 36$ (*c*) $4(X - 3) - 11 = 15 - 2(X + 4)$ (*e*) $3[2(X + 1) - 4] = 10 - 5(4 - 2X)$

(*b*) $2Y - 6 = 4 - 3Y$ (*d*) $3(2U + 1) = 5(3 - U) + 3(U - 2)$ (*f*) $\frac{2}{5}(12 + Y) = 6 - \frac{1}{4}(9 - Y)$

1.77 Solve each of the following simultaneous equations:

(*a*) $2a + b = 10$, $7a - 3b = 9$

(*b*) $3a + 5b = 24$, $2a + 3b = 14$

(*c*) $8X - 3Y = 2$, $3X + 7Y = -9$

(*d*) $5A - 9B = -10$, $3A - 4B = 16$

(*e*) $2a + b - c = 2$, $3a - 4b + 2c = 4$, $4a + 3b - 5c = -8$

(*f*) $5X + 2Y + 3Z = -5$, $2X - 3Y - 6Z = 1$, $X + 5Y - 4Z = 22$

(*g*) $3U - 5V + 6W = 7$, $5U + 3V - 2W = -1$, $4U - 8V + 10W = 11$

1.78 (*a*) Graph the equations $5X + 2Y = 4$ and $7X - 3Y = 23$, using the same set of coordinate axes.

(*b*) From the graphs determine the simultaneous solution of the two equations.

(*c*) Use the method of parts (*a*) and (*b*) to obtain the simultaneous solutions of equations (*a*) to (*d*) of Problem 1.77.

1.79 (*a*) Use the graph of Problem 1.61(*a*) to solve the equation $2X^2 + X - 10 = 0$. (*Hint:* Find the values of X where the parabola intersects the X axis: that is, where $Y = 0$.)

(*b*) Use the method in part (*a*) to solve $3X^2 - 4X - 5 = 0$.

1.80 The solutions of the quadratic equation $aX^2 + bX + c = 0$ are given by the *quadratic formula*:

$$X = \frac{-b \pm \sqrt{b^2 - 4ac}}{2a}$$

Use this formula to find the solutions of (*a*) $3X^2 - 4X - 5 = 0$, (*b*) $2X^2 + X - 10 = 0$, (*c*) $5X^2 + 10X = 7$, and (*d*) $X^2 + 8X + 25 = 0$.

INEQUALITIES

1.81 Using inequality symbols, arrange the numbers -4.3, -6.15, 2.37, 1.52, and -1.5 in (*a*) increasing and (*b*) decreasing order of magnitude.

1.82 Use inequality symbols to express each of the following statements:

(*a*) The number N of children is between 30 and 50 inclusive.

(*b*) The sum S of points on the pair of dice is not less than 7.

(*c*) X is greater than or equal to -4 but less than 3.

(*d*) P is at most 5.

(*e*) X exceeds Y by more than 2.

1.83 Solve each of the following inequalities:

(*a*) $3X \geq 12$

(*b*) $4X < 5X - 3$

(*c*) $2N + 15 > 10 + 3N$

(*d*) $3 + 5(Y - 2) \leq 7 - 3(4 - Y)$

(*e*) $-3 \leq \frac{1}{5}(2X + 1) \leq 3$

(*f*) $0 < \frac{1}{2}(15 - 5N) \leq 12$

(*g*) $-2 \leq 3 + \frac{1}{2}(a - 12) < 8$

LOGARITHMS AND PROPERTIES OF LOGARITHMS

1.84 Find the common logarithms:

(*a*) $\log(10)$ (*b*) $\log(100)$ (*c*) $\log(1000)$ (*d*) $\log(0.1)$ (*d*) $\log(0.01)$

1.85 Find the natural logarithms of the following numbers to four decimal places:

(*a*) $\ln(e)$ (*b*) $\ln(10)$ (*c*) $\ln(100)$ (*d*) $\ln(1000)$ (*e*) $\ln(0.1)$

1.86 Find the logarithms:

(*a*) $\log_4 4$ (*b*) $\log_5 25$ (*c*) $\log_6 216$ (*d*) $\log_7 2401$ (*e*) $\log_8 32768$

1.87 Use EXCEL to find the following logarithms. Give the answers and the commands.

(*a*) $\log_4 5$ (*b*) $\log_5 24$ (*c*) $\log_6 215$ (*d*) $\log_7 8$ (*e*) $\log_8 9$

1.88 Use MAPLE to Problem 1.87. Give the answers and the MAPLE commands.

1.89 Use the properties of logarithms to write the following as sums and differences of logarithms: $\ln((a^3b^4)/c^5)$

1.90 Use the properties of logarithms to write the following as sums and differences of logarithms: $\log((xyz)/w^3)$

1.91 Replace the following by an equivalent logarithm of a single expression: $5\ln(a) - 4\ln(b) + \ln(c) + \ln(d)$

1.92 Replace the following by an equivalent logarithm of a single expression: $\log(u) + \log(v) + \log(w) - 2\log(x) - 3\log(y) - 4\log(z)$

LOGARITHMIC EQUATIONS

1.93 Solve $\log(3x - 4) = 2$

1.94 Solve $\ln(3x^2 - x) = \ln(10)$

1.95 Solve $\log(w - 2) - \log(2w + 7) = \log(w + 2)$

1.96 Solve $\ln(3x + 5) + \ln(2x - 5) = 12$

1.97 Use MAPLE or EXCEL to solve $\ln(2x) + \log(3x - 1) = 10$

1.98 Use MAPLE or EXCEL to solve $\log(2x) + \ln(3x - 1) = 10$

1.99 Use MAPLE or EXCEL to solve $\ln(3x) - \log(x) = \log_2 3$

1.100 Use MAPLE or EXCEL to solve $\log_2(3x) - \log(x) = \ln(3)$

Frequency Distributions

RAW DATA

Raw data are collected data that have not been organized numerically. An example is the set of heights of 100 male students obtained from an alphabetical listing of university records.

ARRAYS

An *array* is an arrangement of raw numerical data in ascending or descending order of magnitude. The difference between the largest and smallest numbers is called the *range* of the data. For example, if the largest height of 100 male students is 74 inches (in) and the smallest height is 60 in, the range is 74 − 60 = 14 in.

FREQUENCY DISTRIBUTIONS

When summarizing large masses of raw data, it is often useful to distribute the data into *classes*, or *categories*, and to determine the number of individuals belonging to each class, called the *class frequency*. A tabular arrangement of data by classes together with the corresponding class frequencies is called a *frequency distribution*, or *frequency table*. Table 2.1 is a frequency distribution of heights (recorded to the nearest inch) of 100 male students at XYZ University.

Table 2.1 Heights of 100 male students at XYZ University

Height (in)	Number of Students
60–62	5
63–65	18
66–68	42
69–71	27
72–74	8
	Total 100

The first class (or category), for example, consists of heights from 60 to 62 in and is indicated by the range symbol 60–62. Since five students have heights belonging to this class, the corresponding class frequency is 5.

Data organized and summarized as in the above frequency distribution are often called *grouped data*. Although the grouping process generally destroys much of the original detail of the data, an important advantage is gained in the clear "overall" picture that is obtained and in the vital relationships that are thereby made evident.

CLASS INTERVALS AND CLASS LIMITS

A symbol defining a class, such as 60–62 in Table 2.1, is called a *class interval*. The end numbers, 60 and 62, are called *class limits*; the smaller number (60) is the *lower class limit*, and the larger number (62) is the *upper class limit*. The terms *class* and *class interval* are often used interchangeably, although the class interval is actually a symbol for the class.

A class interval that, at least theoretically, has either no upper class limit or no lower class limit indicated is called an *open class interval*. For example, referring to age groups of individuals, the class interval "65 years and over" is an open class interval.

CLASS BOUNDARIES

If heights are recorded to the nearest inch, the class interval 60–62 theoretically includes all measurements from 59.5000 to 62.5000 in. These numbers, indicated briefly by the exact numbers 59.5 and 62.5, are called *class boundaries*, or *true class limits*; the smaller number (59.5) is the *lower class boundary*, and the larger number (62.5) is the *upper class boundary*.

In practice, the class boundaries are obtained by adding the upper limit of one class interval to the lower limit of the next-higher class interval and dividing by 2.

Sometimes, class boundaries are used to symbolize classes. For example, the various classes in the first column of Table 2.1 could be indicated by 59.5–62.5, 62.5–65.5, etc. To avoid ambiguity in using such notation, class boundaries should not coincide with actual observations. Thus if an observation were 62.5, it would not be possible to decide whether it belonged to the class interval 59.5–62.5 or 62.5–65.5.

THE SIZE, OR WIDTH, OF A CLASS INTERVAL

The size, or width, of a class interval is the difference between the lower and upper class boundaries and is also referred to as the *class width*, *class size*, or *class length*. If all class intervals of a frequency distribution have equal widths, this common width is denoted by c. In such case c is equal to the difference between two successive lower class limits or two successive upper class limits. For the data of Table 2.1, for example, the class interval is $c = 62.5 - 59.5 = 65.5 - 62.5 = 3$.

THE CLASS MARK

The *class mark* is the midpoint of the class interval and is obtained by adding the lower and upper class limits and dividing by 2. Thus the class mark of the interval 60–62 is $(60 + 62)/2 = 61$. The class mark is also called the *class midpoint*.

For purposes of further mathematical analysis, all observations belonging to a given class interval are assumed to coincide with the class mark. Thus all heights in the class interval 60–62 in are considered to be 61 in.

GENERAL RULES FOR FORMING FREQUENCY DISTRIBUTIONS

1. Determine the largest and smallest numbers in the raw data and thus find the range (the difference between the largest and smallest numbers).

2. Divide the range into a convenient number of class intervals having the same size. If this is not feasible, use class intervals of different sizes or open class intervals (see Problem 2.12). The number of class intervals is usually between 5 and 20, depending on the data. Class intervals are also chosen so that the class marks (or midpoints) coincide with the actually observed data. This tends to lessen the so-called *grouping error* involved in further mathematical analysis. However, the class boundaries should not coincide with the actually observed data.
3. Determine the number of observations falling into each class interval; that is, find the class frequencies. This is best done by using a *tally*, or *score sheet* (see Problem 2.8).

HISTOGRAMS AND FREQUENCY POLYGONS

Histograms and frequency polygons are two graphic representations of frequency distributions.

1. A *histogram* or *frequency histogram*, consists of a set of rectangles having (*a*) bases on a horizontal axis (the X axis), with centers at the class marks and lengths equal to the class interval sizes, and (*b*) areas proportional to the class frequencies.
2. A *frequency polygon* is a line graph of the class frequencies plotted against class marks. It can be obtained by connecting the midpoints of the tops of the rectangles in the histogram.

The histogram and frequency polygon corresponding to the frequency distribution of heights in Table 2.1 are shown in Figs. 2-1 and 2-2.

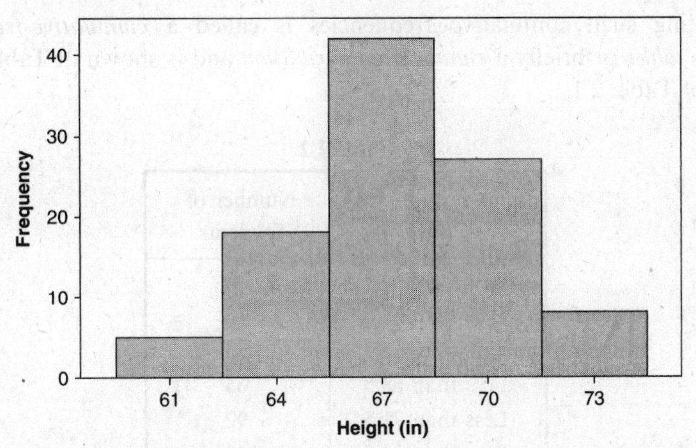

Fig. 2-1 A MINITAB histogram showing the class midpoints and frequencies.

Notice in Fig. 2-2 how the frequency polygon is tied down at the ends, that is, at 58 and 76.

RELATIVE-FREQUENCY DISTRIBUTIONS

The *relative frequency* of a class is the frequency of the class divided by the total frequency of all classes and is generally expressed as a percentage. For example, the relative frequency of the class 66–68 in Table 2.1 is $42/100 = 42\%$. The sum of the relative frequencies of all classes is clearly 1, or 100%.

If the frequencies in Table 2.1 are replaced with the corresponding relative frequencies, the resulting table is called a *relative-frequency distribution*, *percentage distribution*, or *relative-frequency table*.

Graphic representation of relative-frequency distributions can be obtained from the histogram or frequency polygon simply by changing the vertical scale from frequency to relative frequency, keeping exactly the same diagram. The resulting graphs are called *relative-frequency histograms* (or *percentage histograms*) and *relative-frequency polygons* (or *percentage polygons*), respectively.

CUMULATIVE-FREQUENCY DISTRIBUTIONS AND OGIVES

The total frequency of all values less than the upper class boundary of a given class interval is called the *cumulative frequency* up to and including that class interval. For example, the cumulative frequency up to and including the class interval 66–68 in Table 2.1 is $5+18+42=65$, signifying that 65 students have heights less than 68.5 in.

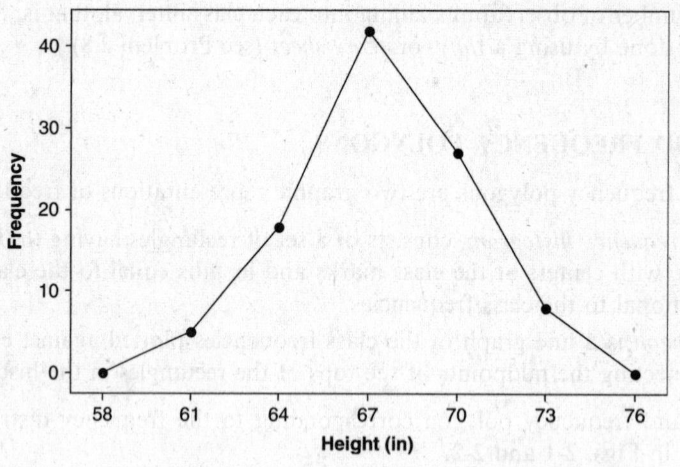

Fig. 2-2 A MINITAB frequency polygon of the student heights.

A table presenting such cumulative frequencies is called a *cumulative-frequency distribution, cumulative-frequency table*, or briefly a *cumulative distribution* and is shown in Table 2.2 for the student height distribution of Table 2.1.

Table 2.2

Height (in)	Number of Students
Less than 59.5	0
Less than 62.5	5
Less than 65.5	23
Less than 68.5	65
Less than 71.5	92
Less than 74.5	100

A graph showing the cumulative frequency less than any upper class boundary plotted against the upper class boundary is called a *cumulative-frequency polygon* or *ogive*. For some purposes, it is desirable to consider a cumulative-frequency distribution of all values greater than or equal to the lower class boundary of each class interval. Because in this case we consider heights of 59.5 in or more, 62.5 in or more, etc., this is sometimes called an *"or more" cumulative distribution*, while the one considered above is a *"less than" cumulative distribution*. One is easily obtained from the other. The corresponding ogives are then called "or more" and "less than" ogives. Whenever we refer to cumulative distributions or ogives without qualification, the "less than" type is implied.

RELATIVE CUMULATIVE-FREQUENCY DISTRIBUTIONS AND PERCENTAGE OGIVES

The *relative cumulative frequency*, or *percentage cumulative frequency*, is the cumulative frequency divided by the total frequency. For example, the relative cumulative frequency of heights less than 68.5 in is $65/100=65\%$, signifying that 65% of the students have heights less than 68.5 in. If the relative cumulative frequencies are used in Table 2.2, in place of cumulative frequencies, the results are called *relative cumulative-frequency distributions* (or *percentage cumulative distributions*) and *relative cumulative-frequency polygons* (or *percentage ogives*), respectively.

FREQUENCY CURVES AND SMOOTHED OGIVES

Collected data can usually be considered as belonging to a sample drawn from a large population. Since so many observations are available in the population, it is theoretically possible (for continuous data) to choose class intervals very small and still have sizable numbers of observations falling within each class. Thus one would expect the frequency polygon or relative-frequency polygon for a large population to have so many small, broken line segments that they closely approximate curves, which we call *frequency curves* or *relative-frequency curves*, respectively.

It is reasonable to expect that such theoretical curves can be approximated by smoothing the frequency polygons or relative-frequency polygons of the sample, the approximation improving as the sample size is increased. For this reason, a frequency curve is sometimes called a *smoothed frequency polygon*.

In a similar manner, *smoothed ogives* are obtained by smoothing the cumulative-frequency polygons, or ogives. It is usually easier to smooth an ogive than a frequency polygon.

TYPES OF FREQUENCY CURVES

Frequency curves arising in practice take on certain characteristic shapes, as shown in Fig. 2-3.

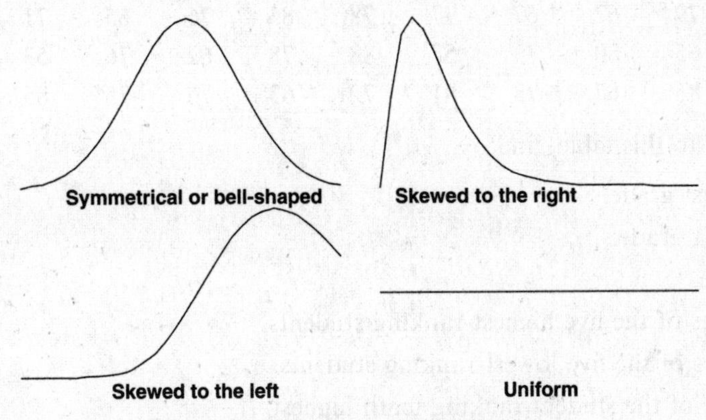

Fig. 2-3 Four common distributions.

1. *Symmetrical* or *bell-shaped* curves are characterized by the fact that observations equidistant from the central maximum have the same frequency. Adult male and adult female heights have bell-shaped distributions.
2. Curves that have tails to the left are said to be *skewed to the left*. The lifetimes of males and females are skewed to the left. A few die early in life but most live between 60 and 80 years. Generally, females live about ten years, on the average, longer than males.
3. Curves that have tails to the right are said to be *skewed to the right*. The ages at the time of marriage of brides and grooms are skewed to the right. Most marry in their twenties and thirties but a few marry in their forties, fifties, sixties and seventies.
4. Curves that have approximately equal frequencies across their values are said to be *uniformly distributed*. Certain machines that dispense liquid colas do so uniformly between 15.9 and 16.1 ounces, for example.
5. In a *J-shaped* or *reverse J-shaped* frequency curve the maximum occurs at one end or the other.
6. A *U-shaped* frequency distribution curve has maxima at both ends and a minimum in between.
7. A *bimodal* frequency curve has two maxima.
8. A *multimodal* frequency curve has more than two maxima.

Solved Problems

ARRAYS

2.1 (*a*) Arrange the numbers 17, 45, 38, 27, 6, 48, 11, 57, 34, and 22 in an array.
(*b*) Determine the range of these numbers.

SOLUTION

(*a*) In ascending order of magnitude, the array is: 6, 11, 17, 22, 27, 34, 38, 45, 48, 57. In descending order of magnitude, the array is: 57, 48, 45, 38, 34, 27, 22, 17, 11, 6.

(*b*) Since the smallest number is 6 and the largest number is 57, the range is $57 - 6 = 51$.

2.2 The final grades in mathematics of 80 students at State University are recorded in the accompanying table.

68	84	75	82	68	90	62	88	76	93
73	79	88	73	60	93	71	59	85	75
61	65	75	87	74	62	95	78	63	72
66	78	82	75	94	77	69	74	68	60
96	78	89	61	75	95	60	79	83	71
79	62	67	97	78	85	76	65	71	75
65	80	73	57	88	78	62	76	53	74
86	67	73	81	72	63	76	75	85	77

With reference to this table, find:

(*a*) The highest grade.
(*b*) The lowest grade.
(*c*) The range.
(*d*) The grades of the five highest-ranking students.
(*e*) The grades of the five lowest-ranking students.
(*f*) The grade of the student ranking tenth highest.
(*g*) The number of students who received grades of 75 or higher.
(*h*) The number of students who received grades below 85.
(*i*) The percentage of students who received grades higher than 65 but not higher than 85.
(*j*) The grades that did not appear at all.

SOLUTION

Some of these questions are so detailed that they are best answered by first constructing an array. This can be done by subdividing the data into convenient classes and placing each number taken from the table into the appropriate class, as in Table 2.3, called an *entry table*. By then arranging the numbers of each class into an array, as in Table 2.4, the required array is obtained. From Table 2.4 it is relatively easy to answer the above questions.

(*a*) The highest grade is 97.

(*b*) The lowest grade is 53.

(*c*) The range is $97 - 53 = 44$.

(*d*) The five highest-ranking students have grades 97, 96, 95, 95, and 94.

Table 2.3

50–54	53
55–59	59, 57
60–64	62, 60, 61, 62, 63, 60, 61, 60, 62, 62, 63
65–69	68, 68, 65, 66, 69, 68, 67, 65, 65, 67
70–74	73, 73, 71, 74, 72, 74, 71, 71, 73, 74, 73, 72
75–79	75, 76, 79, 75, 75, 78, 78, 75, 77, 78, 75, 79, 79, 78, 76, 75, 78, 76, 76, 75, 77
80–84	84, 82, 82, 83, 80, 81
85–89	88, 88, 85, 87, 89, 85, 88, 86, 85
90–94	90, 93, 93, 94
95–99	95, 96, 95, 97

Table 2.4

50–54	53
55–59	57, 59
60–64	60, 60, 60, 61, 61, 62, 62, 62, 62, 63, 63
65–69	65, 65, 65, 66, 67, 67, 68, 68, 68, 69
70–74	71, 71, 71, 72, 72, 73, 73, 73, 73, 74, 74, 74
75–79	75, 75, 75, 75, 75, 75, 75, 76, 76, 76, 76, 77, 77, 78, 78, 78, 78, 78, 79, 79, 79
80–84	80, 81, 82, 82, 83, 84
85–89	85, 85, 85, 86, 87, 88, 88, 88, 89
90–94	90, 93, 93, 94
95–99	95, 95, 96, 97

(*e*) The five lowest-ranking students have grades 53, 57, 59, 60, and 60.

(*f*) The grade of the student ranking tenth highest is 88.

(*g*) The number of students receiving grades of 75 or higher is 44.

(*h*) The number of students receiving grades below 85 is 63.

(*i*) The percentage of students receiving grades higher than 65 but not higher than 85 is $49/80 = 61.2\%$.

(*j*) The grades that did not appear are 0 through 52, 54, 55, 56, 58, 64, 70, 91, 92, 98, 99, and 100.

FREQUENCY DISTRIBUTIONS, HISTOGRAMS, AND FREQUENCY POLYGONS

2.3 Table 2.5 shows a frequency distribution of the weekly wages of 65 employees at the P&R Company. With reference to this table, determine:

(*a*) The lower limit of the sixth class.

(*b*) The upper limit of the fourth class.

(*c*) The class mark (or class midpoint) of the third class.

(*d*) The class boundaries of the fifth class.

(*e*) The size of the fifth-class interval.

(*f*) The frequency of the third class.

(*g*) The relative frequency of the third class.

(*h*) The class interval having the largest frequency. This is sometimes called the *modal class interval*; its frequency is then called the *modal class frequency*.

Table 2.5

Wages	Number of Employees
\$250.00–\$259.99	8
260.00–269.99	10
270.00–279.99	16
280.00–289.99	14
290.00–299.99	10
300.00–309.99	5
310.00–319.99	2
	Total 65

(*i*) The percentage of employees earning less than \$280.00 per week.

(*j*) The percentage of employees earning less than \$300.00 per week but at least \$260.00 per week.

SOLUTION

(*a*) \$300.00.

(*b*) \$289.99.

(*c*) The class mark of the third class $= \frac{1}{2}(\$270.00 + \$279.99) = \$274.995$. For most practical purposes, this is rounded to \$275.00.

(*d*) Lower class boundary of fifth class $= \frac{1}{2}(\$290.00 + \$289.99) = \$289.995$. Upper class boundary of fifth class $= \frac{1}{2}(\$299.99 + \$300.00) = \$299.995$.

(*e*) Size of fifth-class interval = upper boundary of fifth class − lower boundary of fifth class = \$299.995 − \$289.985 = \$10.00. In this case all class intervals have the same size: \$10.00.

(*f*) 16.

(*g*) $16/65 = 0.246 = 24.6\%$.

(*h*) \$270.00–\$279.99.

(*i*) Total number of employees earning less than \$280 per week $= 16 + 10 + 8 = 34$. Percentage of employees earning less than \$280 per week $= 34/65 = 52.3\%$.

(*j*) Number of employees earning less than \$300.00 per week but at least \$260 per week $= 10 + 14 + 16 + 10 = 50$. Percentage of employees earning less than \$300 per week but at least \$260 per week $= 50/65 = 76.9\%$.

2.4 If the class marks in a frequency distribution of the weights of students are 128, 137, 146, 155, 164, 173, and 182 pounds (lb), find (*a*) the class-interval size, (*b*) the class boundaries, and (*c*) the class limits, assuming that the weights were measured to the nearest pound.

SOLUTION

(*a*) Class-interval size = common difference between successive class marks $= 137 - 128 = 146 - 137 =$ etc. $= 9$ lb.

(*b*) Since the class intervals all have equal size, the class boundaries are midway between the class marks and thus have the values

$$\tfrac{1}{2}(128 + 137), \tfrac{1}{2}(137 + 146), \ldots, \tfrac{1}{2}(173 + 182) \quad \text{or} \quad 132.5, 141.5, 150.5, \ldots, 177.5 \text{ lb}$$

The first class boundary is $132.5 - 9 = 123.5$ and the last class boundary is $177.5 + 9 = 186.5$, since the common class-interval size is 9 lb. Thus all the class boundaries are given by

$$123.5, 132, 141.5, 150.5, 159.5, 168.5, 177.5, 186.5 \text{ lb}$$

(*c*) Since the class limits are integers, we choose them as the integers nearest to the class boundaries, namely, 123, 124, 132, 133, 141, 142, ... Thus the first class has limits 124–132, the next 133–141, etc.

2.5 The time in hours spent on cell phones, per week, by 50 college freshmen was sampled and the SPSS pull-down "**Analyze ⇒ Descriptive Statistics ⇒ Frequencies**" resulted in the following output in Fig. 2-4.

Time

		Frequency	Percent	Valid Percent	Cumulative Percent
Valid	3.00	3	6.0	6.0	6.0
	4.00	3	6.0	6.0	12.0
	5.00	5	10.0	10.0	22.0
	6.00	3	6.0	6.0	28.0
	7.00	4	8.0	8.0	36.0
	8.00	4	8.0	8.0	44.0
	9.00	3	6.0	6.0	50.0
	10.00	4	8.0	8.0	58.0
	11.00	2	4.0	4.0	62.0
	12.00	2	4.0	4.0	66.0
	13.00	3	6.0	6.0	72.0
	14.00	1	2.0	2.0	74.0
	15.00	2	4.0	4.0	78.0
	16.00	5	10.0	10.0	88.0
	17.00	2	4.0	4.0	92.0
	18.00	1	2.0	2.0	94.0
	19.00	2	4.0	4.0	98.0
	20.00	1	2.0	2.0	100.0
	Total	50	100.0	100.0	

Fig. 2-4 SPSS output for Problem 2.5.

(*a*) What percent spent 15 or less hours per week on a cell phone?

(*b*) What percent spent more than 10 hours per week on a cell phone?

SOLUTION

(*a*) The cumulative percent for 15 hours is 78%. Thus 78% spent 15 hours or less on a cell phone per week.

(*b*) The cumulative percent for 10 hours is 58%. Thus 58% spent 10 hours or less on a cell phone per week. Therefore, 42% spent more than 10 hours on a cell phone per week.

2.6 The smallest of 150 measurements is 5.18 in, and the largest is 7.44 in. Determine a suitable set of (*a*) class intervals, (*b*) class boundaries, and (*c*) class marks that might be used in forming a frequency distribution of these measurements.

SOLUTION

The range is $7.44 - 5.18 = 2.26$ in. For a minimum of five class intervals, the class-interval size is $2.26/5 = 0.45$ approximately; and for a maximum of 20 class intervals, the class-interval size is

2.26/20 = 0.11 approximately. Convenient choices of class-interval sizes lying between 0.11 and 0.45 would be 0.20, 0.30, or 0.40.

(*a*) Columns I, II, and III of the accompanying table show suitable class intervals, having sizes 0.20, 0.30, and 0.40, respectively.

I	II	III
5.10–5.29	5.10–5.39	5.10–5.49
5.30–5.49	5.40–5.69	5.50–5.89
5.50–5.69	5.70–5.99	5.90–6.29
5.70–5.89	6.00–6.29	6.30–6.69
5.90–6.09	6.30–6.59	6.70–7.09
6.10–6.29	6.60–6.89	7.10–7.49
6.30–6.49	6.90–7.19	
6.50–6.69	7.20–7.49	
6.70–6.89		
6.90–7.09		
7.10–7.29		
7.30–7.49		

Note that the lower limit in each first class could have been different from 5.10; for example, if in column I we had started with 5.15 as the lower limit, the first class interval could have been written 5.15–5.34.

(*b*) The class boundaries corresponding to columns I, II, and III of part (*a*) are given, respectively, by

I 5.095–5.295, 5.295–5.495, 5.495–5.695, ..., 7.295–7.495

II 5.095–5.395, 5.395–5.695, 5.695–5.995, ..., 7.195–7.495

III 5.095–5.495, 5.495–5.895, 5.895–6.295, ..., 7.095–7.495

Note that these class boundaries are suitable since they cannot coincide with the observed measurements.

(*c*) The class marks corresponding to columns I, II, and III of part (*a*) are given, respectively, by

I 5.195, 5.395, ..., 7.395 II 5.245, 5.545, ..., 7.345 III 5.295, 5.695, ..., 7.295

These class marks have the disadvantage of not coinciding with the observed measurements.

2.7 In answering Problem 2.6(*a*), a student chose the class intervals 5.10–5.40, 5.40–5.70, ..., 6.90–7.20, and 7.20–7.50. Was there anything wrong with this choice?

SOLUTION

These class intervals overlap at 5.40, 5.70, ..., 7.20. Thus a measurement recorded as 5.40, for example, could be placed in either of the first two class intervals. Some statisticians justify this choice by agreeing to place half of such ambiguous cases in one class and half in the other.

The ambiguity is removed by writing the class intervals as 5.10 to under 5.40, 5.40 to under 5.70, etc. In this case, the class limits coincide with the class boundaries, and the class marks can coincide with the observed data.

In general it is desirable to avoid overlapping class intervals whenever possible and to choose them so that the class boundaries are values not coinciding with actual observed data. For example, the class intervals for Problem 2.6 could have been chosen as 5.095–5.395, 5.395–5.695, etc., without ambiguity. A disadvantage of this particular choice is that the class marks do not coincide with the observed data.

2.8 In the following table the weights of 40 male students at State University are recorded to the nearest pound. Construct a frequency distribution.

138	164	150	132	144	125	149	157
146	158	140	147	136	148	152	144
168	126	138	176	163	119	154	165
146	173	142	147	135	153	140	135
161	145	135	142	150	156	145	128

SOLUTION

The largest weight is 176 lb and the smallest weight is 119 lb, so that the range is $176 - 119 = 57$ lb. If five class intervals are used, the class-interval size is $57/5 = 11$ approximately; if 20 class intervals are used, the class-interval size is $57/20 = 3$ approximately.

One convenient choice for the class-interval size is 5 lb. Also, it is convenient to choose the class marks as 120, 125, 130, 135, ... lb. Thus the class intervals can be taken as 118–122, 123–127, 128–132, ... With this choice the class boundaries are 117.5, 122.5, 127.5, ..., which do not coincide with the observed data.

The required frequency distribution is shown in Table 2.6. The center column, called a *tally*, or *score sheet*, is used to tabulate the class frequencies from the raw data and is usually omitted in the final presentation of the frequency distribution. It is unnecessary to make an array, although if it is available it can be used in tabulating the frequencies.

Another method

Of course, other possible frequency distributions exist. Table 2.7, for example, shows a frequency distribution with only seven classes in which the class interval is 9 lb.

Table 2.6

Weight (lb)	Tally	Frequency
118–122	/	1
123–127	//	2
128–132	//	2
133–137	////	4
138–142	~~////~~ /	6
143–147	~~////~~ ///	8
148–152	~~////~~	5
153–157	////	4
158–162	//	2
163–167	///	3
168–172	/	1
173–177	//	2
		Total 40

Table 2.7

Weight (lb)	Tally	Frequency
118–126	///	3
127–135	~~////~~	5
136–144	~~////~~ ////	9
145–153	~~////~~ ~~////~~ //	12
154–162	~~////~~	5
163–171	////	4
172–180	//	2
		Total 40

2.9 In the following, the heights of 45 female students at Midwestern University are recorded to the nearest inch. Use the statistical package, STATISTIX, to construct a histogram.

67	67	64	64	74	61	68	71	69	61	65	64	62	63	59
70	66	66	63	59	64	67	70	65	66	66	56	65	67	69
64	67	68	67	67	65	74	64	62	68	65	65	65	66	67

SOLUTION

After entering the data into the STATISTIX worksheet, the STATISTIX pull-down "**Statistics ⇒ Summary Statistics ⇒ Histogram**" produces the histogram of female heights in Fig. 2-5.

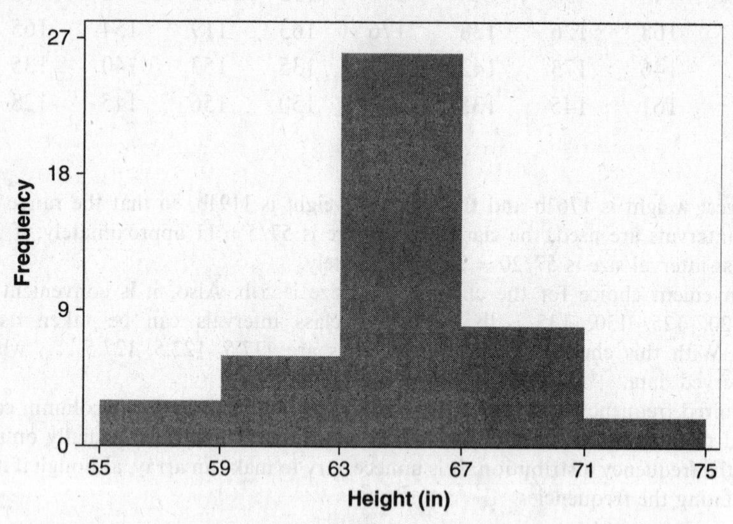

Fig. 2-5 Histogram of 45 female heights produced by STATISTIX.

2.10 Table 2.8 gives the one-way commuting distances for 50 Metropolitan College students in miles.

Table 2.8 Commuting distances from Metropolitan College (miles)

4.3	7.0	8.0	3.9	3.7	8.4	2.6	1.0	15.7	3.9
6.5	8.7	0.9	0.9	12.6	4.0	10.3	10.0	6.2	1.1
7.2	8.8	7.8	4.9	2.0	3.0	4.2	3.3	4.8	4.4
7.7	2.4	8.0	8.0	4.6	1.4	2.2	1.9	3.2	4.8
5.0	10.3	12.3	3.8	3.8	6.6	2.0	1.6	4.4	4.3

The SPSS histogram for the commuting distances in Table 2.8 is shown in Fig. 2-6. Note that the classes are 0 to 2, 2 to 4, 4 to 6, 6 to 8, 8 to 10, 10 to 12, 12 to 14, and 14 to 16. The frequencies are 7, 13, 11, 7, 6, 3, 2 and 1. A number is counted in the class if it falls on the lower class limit but in the next class if it falls on the upper class limit.

(*a*) What data values are in the first class?

(*b*) What data values are in the second class?

(*c*) What data values are in the third class?

(*d*) What data values are in the fourth class?

(*e*) What data values are in the fifth class?

(*f*) What data values are in the sixth class?

(*g*) What data values are in the seventh class?

(*h*) What data values are in the eighth class?

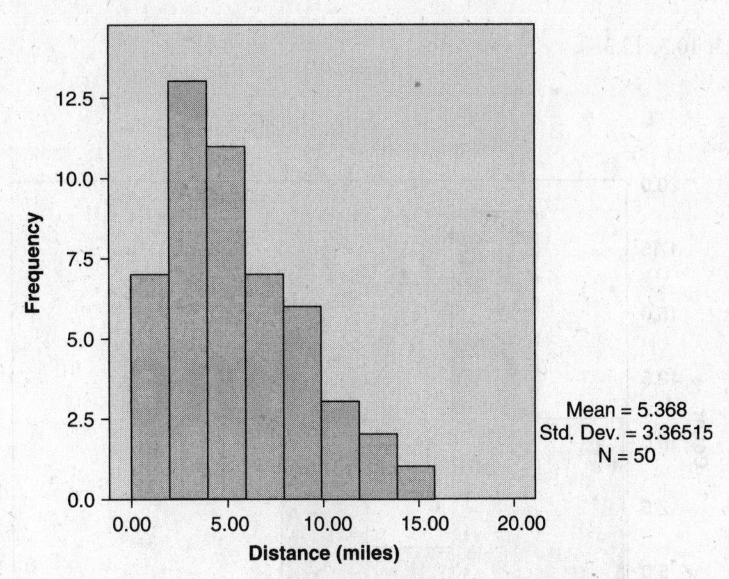

Fig. 2-6 SPSS histogram for commuting distances from Metropolitan College.

SOLUTION

(*a*) 0.9, 0.9, 1.0, 1.1, 1.4, 1.6, 1.9
(*b*) 2.0, 2.0, 2.2, 2.4, 2.6, 3.0, 3.2, 3.3, 3.7, 3.8, 3.8, 3.9, 3.9
(*c*) 4.0, 4.2, 4.3, 4.3, 4.4, 4.4, 4.6, 4.8, 4.8, 4.9, 5.0
(*d*) 6.2, 6.5, 6.6, 7.0, 7.2, 7.7, 7.8
(*e*) 8.0, 8.0, 8.0, 8.4, 8.7, 8.8
(*f*) 10.0, 10.3, 10.3
(*g*) 12.3, 12.6
(*h*) 15.7

2.11 The SAS histogram for the commuting distances in Table 2.8 is shown in Fig. 2-7. The midpoints of the class intervals are shown. The classes are 0 to 2.5, 2.5 to 5.0, 5.0 to 7.5, 7.5 to 10.0, 10.0 to 12.5, 12.5 to 15.0, 15.0 to 17.5, and 17.5 to 20.0. A number is counted in the class if it falls on the lower class limit but in the next class if it falls on the upper class limit.

(*a*) What data values (in Table 2.8) are in the first class?
(*b*) What data values are in the second class?
(*c*) What data values are in the third class?
(*d*) What data values are in the fourth class?
(*e*) What data values are in the fifth class?
(*f*) What data values are in the sixth class?
(*g*) What data values are in the seventh class?

SOLUTION

(*a*) 0.9, 0.9, 1.0, 1.1, 1.4, 1.6, 1.9, 2.0, 2.0, 2.2, 2.4
(*b*) 2.6, 3.0, 3.2, 3.3, 3.7, 3.8, 3.8, 3.9, 3.9, 4.0, 4.2, 4.3, 4.3, 4.4, 4.4, 4.6, 4.8, 4.8, 4.9
(*c*) 5.0, 6.2, 6.5, 6.6, 7.0, 7.2
(*d*) 7.7, 7.8, 8.0, 8.0, 8.0, 8.4, 8.7, 8.8

(*e*) 10.0, 10.3, 10.3, 12.3

(*f*) 12.6

(*g*) 15.7

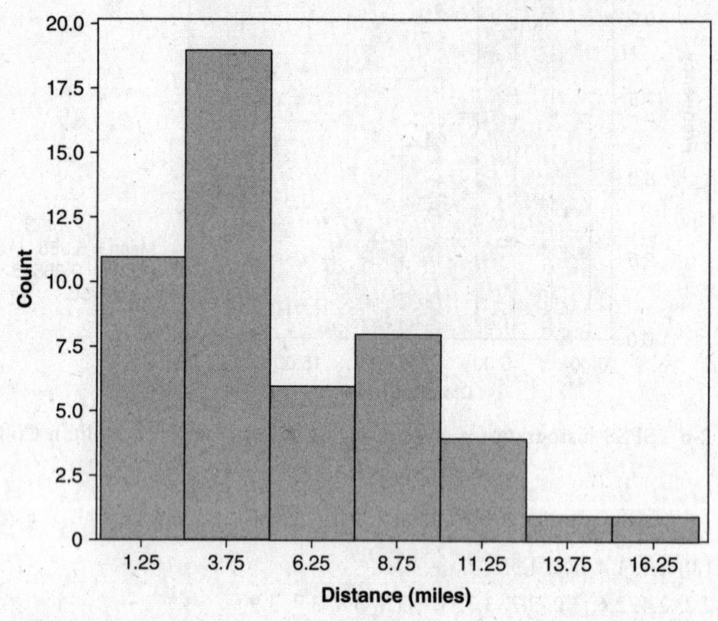

Fig. 2-7 SAS histogram for commuting distances from Metropolitan College.

2.12 At the P&R Company (Problem 2.3), five new employees were hired at weekly wages of $285.34, $316.83, $335.78, $356.21, and $374.50. Construct a frequency distribution of wages for the 70 employees.

SOLUTION

Possible frequency distributions are shown in Table 2.9.

Table 2.9(*a*)

Wages	Frequency
$250.00–$259.99	8
260.00–269.99	10
270.00–279.99	16
280.00–289.99	15
290.00–299.99	10
300.00–309.99	5
310.00–319.99	3
320.00–329.99	0
330.00–339.99	1
340.00–349.99	0
350.00–359.99	1
360.00–369.99	0
370.00–379.99	1
	Total 70

Table 2.9(*b*)

Wages	Frequency
$250.00–$259.99	8
260.00–269.99	10
270.00–279.99	16
280.00–289.99	15
290.00–299.99	10
300.00–309.99	5
310.00–319.99	3
320.00 and over	3
	Total 70

Table 2.9(*c*)

Wages	Frequency
$250.00–$269.99	18
270.00–289.99	31
290.00–309.99	15
310.00–329.99	3
330.00–349.99	1
350.00–369.99	1
370.00–389.99	1
	Total 70

Table 2.9(*d*)

Wages	Frequency
$250.00–$259.99	8
260.00–269.99	10
270.00–279.99	16
280.00–289.99	15
290.00–299.99	10
300.00–319.99	8
320.00–379.99	3
	Total 70

In Table 2.9(*a*), the same class-interval size, $10.00, has been maintained throughout the table. As a result, there are too many empty classes and the detail is much too fine at the upper end of the wage scale.

In Table 2.9(*b*), empty classes and fine detail have been avoided by use of the open class interval "$320.00 and over." A disadvantage of this is that the table becomes useless in performing certain mathematical calculations. For example, it is impossible to determine the total amount of wages paid per week, since "over $320.00" might conceivably imply that individuals could earn as high as $1400.00 per week.

In Table 2.9(*c*), a class-interval size of $20.00 has been used. A disadvantage is that much information is lost at the lower end of the wage scale and the detail is still fine at the upper end of the scale.

In Table 2.9(*d*), unequal class-interval sizes have been used. A disadvantage is that certain mathematical calculations to be made later lose a simplicity that is available when class intervals have the same size. Also, the larger the class-interval size, the greater will be the grouping error.

2.13 The EXCEL histogram for the commuting distances in Table 2.8 is shown in Fig. 2-8. The classes are 0 to 3, 3 to 6, 6 to 9, 9 to 12, 12 to 15, and 15 to 18. A number is counted in the class if it falls on the upper class limit but in the prior class if it falls on the lower class limit.

(*a*) What data values (in Table 2.8) are in the first class?

(*b*) What data values are in the second class?

(*c*) What data values are in the third class?

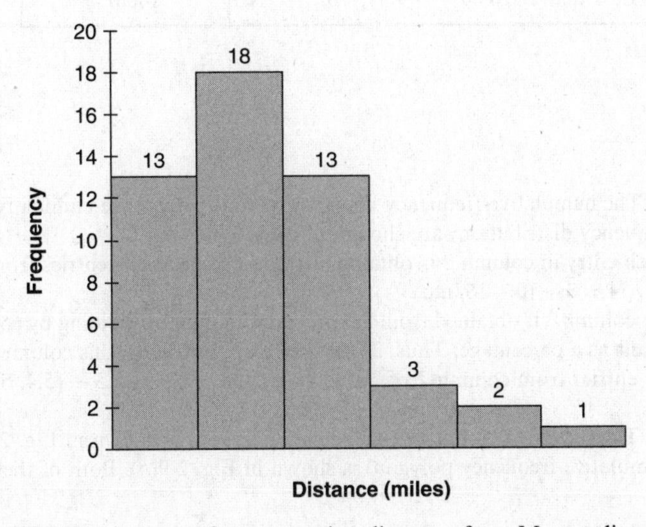

Fig. 2-8 EXCEL histogram for commuting distances from Metropolitan College.

(*d*) What data values are in the fourth class?

(*e*) What data values are in the fifth class?

(*f*) What data values are in the sixth class?

SOLUTION

(*a*) 0.9, 0.9, 1.0, 1.1, 1.4, 1.6, 1.9, 2.0, 2.0, 2.2, 2.4, 2.6, 3.0

(*b*) 3.2, 3.3, 3.7, 3.8, 3.8, 3.9, 3.9, 4.0, 4.2, 4.3, 4.3, 4.4, 4.4, 4.6, 4.8, 4.8, 4.9, 5.0

(*c*) 6.2, 6.5, 6.6, 7.0, 7.2, 7.7, 7.8, 8.0, 8.0, 8.0, 8.4, 8.7, 8.8

(*d*) 10.0, 10.3, 10.3

(*e*) 12.3, 12.6

(*f*) 15.7

CUMULATIVE-FREQUENCY DISTRIBUTIONS AND OGIVES

2.14 Construct (*a*) a cumulative-frequency distribution, (*b*) a percentage cumulative distribution, (*c*) an ogive, and (*d*) a percentage ogive from the frequency distribution in Table 2.5 of Problem 2.3.

Table 2.10

Wages	Cumulative Frequency	Percentage Cumulative Distribution
Less than $250.00	0	0.0
Less than $260.00	8	12.3
Less than $270.00	18	27.7
Less than $280.00	34	52.3
Less than $290.00	48	73.8
Less than $300.00	58	89.2
Less than $310.00	63	96.9
Less than $320.00	65	100.0

SOLUTION

(*a*) and (*b*) The cumulative-frequency distribution and percentage cumulative distribution (or cumulative relative-frequency distribution) are shown in Table 2.10.

Note that each entry in column 2 is obtained by adding successive entries from column 2 of Table 2.5, thus, $18 = 8 + 10$, $34 = 8 + 10 + 16$, etc.

Each entry in column 3 is obtained from the previous column by dividing by 65, the total frequency, and expressing the result as a percentage. Thus, $34/65 = 52.3\%$. Entries in this column can also be obtained by adding successive entries from column 2 of Table 2.8. Thus, $27.7 = 12.3 + 15.4$, $52.3 = 12.3 + 15.4 + 24.6$, etc.

(*c*) and (*d*) The ogive (or cumulative-frequency polygon) is shown in Fig. 2-9(*a*) and the percentage ogive (relative cumulative-frequency polygon) is shown in Fig. 2-9(*b*). Both of these are Minitab generated plots.

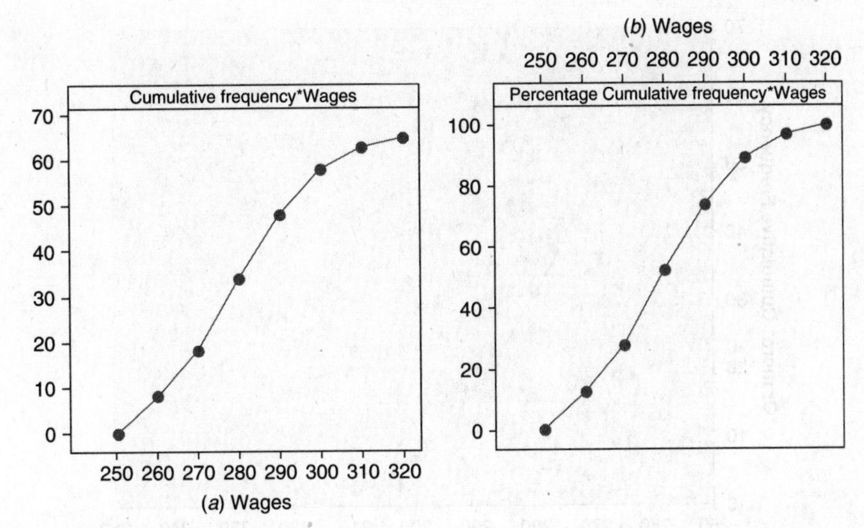

Fig. 2-9 (*a*) MINITAB cumulative frequency and (*b*) percentage cumulative frequency graphs.

2.15 From the frequency distribution in Table 2.5 of Problem 2.3, construct (*a*) an "or more" cumulative-frequency distribution and (*b*) an "or more" ogive.

SOLUTION

(*a*) Note that each entry in column 2 of Table 2.11 is obtained by adding successive entries from column 2 of Table 2.5, *starting at the bottom* of Table 2.5; thus $7 = 2 + 5$, $17 = 2 + 5 + 10$, etc. These entries can also be obtained by subtracting each entry in column 2 of Table 2.10 from the total frequency, 65; thus $57 = 65 - 8$, $47 = 65 - 18$, etc.

(*b*) Figure 2-10 shows the "or more" ogive.

Table 2.11

Wages	"Or More" Cumulative Frequency
$250.00 or more	65
260.00 or more	57
270.00 or more	47
280.00 or more	31
290.00 or more	17
300.00 or more	7
310.00 or more	2
320.00 or more	0

2.16 From the ogives in Figs. 2-9 and 2-10 (of Problems 2.14 and 2.15, respectively), estimate the number of employees earning (*a*) less than $288.00 per week, (*b*) $296.00 or more per week, and (*c*) at least $263.00 per week but less than $275.00 per week.

SOLUTION

(*a*) Referring to the "less than" ogive of Fig. 2-9, construct a vertical line intersecting the "Wages" axis at $288.00. This line meets the ogive at the point with coordinates (288, 45); hence 45 employees earn less than $288.00 per week.

(*b*) In the "or more" ogive of Fig. 2-10, construct a vertical line at $296.00. This line meets the ogive at the point (296, 11); hence 11 employees earn $296.00 or more per week.

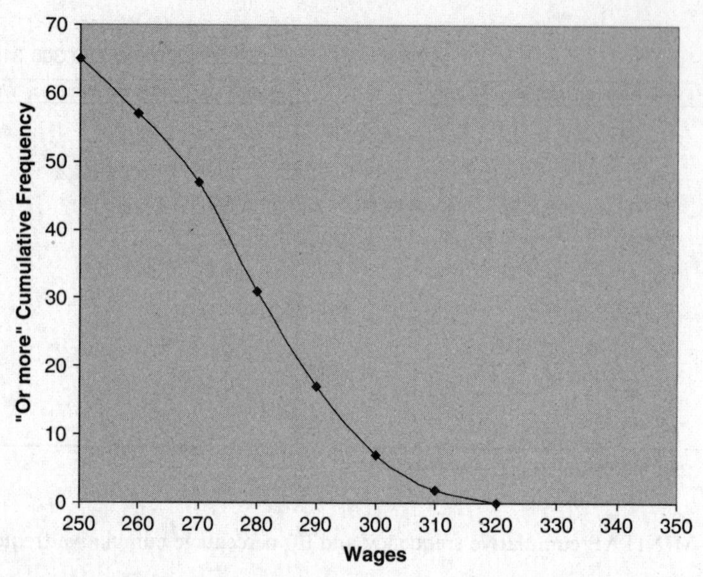

Fig. 2-10 EXCEL produced "Or more" cumulative frequency graph.

This can also be obtained from the "less than" ogive of Fig. 2.9. By constructing a line at \$296.00, we find that 54 employees earn less than \$296.00 per week; hence $65 - 54 = 11$ employees earn \$296.00 or more per week.

(*c*) Using the "less than" ogive of Fig. 2-9, we have: Required number of employees = number earning less than \$275.00 per week − number earning less than \$263.00 per week $= 26 - 11 = 15$.

Note that the above results could just as well have been obtained by the process of *interpolation* in the cumulative-frequency tables. In part (*a*), for example, since \$288.00 is 8/10, or 4/5, of the way between \$280.00 and \$290.00, the required number of employees should be 4/5 of the way between the corresponding values 34 and 48 (see Table 2.10). But 4/5 of the way between 34 and 48 is $\frac{4}{5}(48 - 34) = 11$. Thus the required number of employees is $34 + 11 = 45$.

2.17 Five pennies were tossed 1000 times, and at each toss the number of heads was observed. The number of tosses during which 0, 1, 2, 3, 4, and 5 heads were obtained is shown in Table 2.12.

(*a*) Graph the data of Table 2.12.

(*b*) Construct a table showing the percentage of tosses resulting in a number of heads less than 0, 1, 2, 3, 4, 5, or 6.

(*c*) Graph the data of the table in part (*b*).

Table 2.12

Number of Heads	Number of Tosses (frequency)
0	38
1	144
2	342
3	287
4	164
5	25
	Total 1000

SOLUTION

(*a*) The data can be shown graphically either as in Fig. 2-11 or as in Fig. 2-12.

Figure 2-11 seems to be a more natural graph to use—since, for example, the number of heads cannot be 1.5 or 3.2. The graph is called a *dotplot*. It is especially used when the data are discrete.

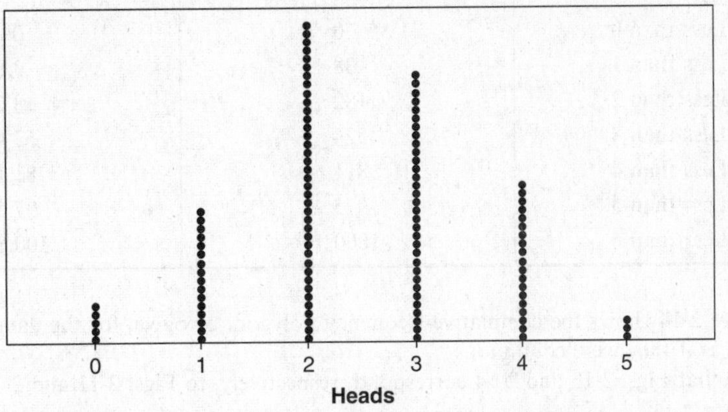

0 1 2 3 4 5
Heads

Each symbol represents up to 9 observations.

Fig. 2-11 MINITAB dotplot of the number of heads.

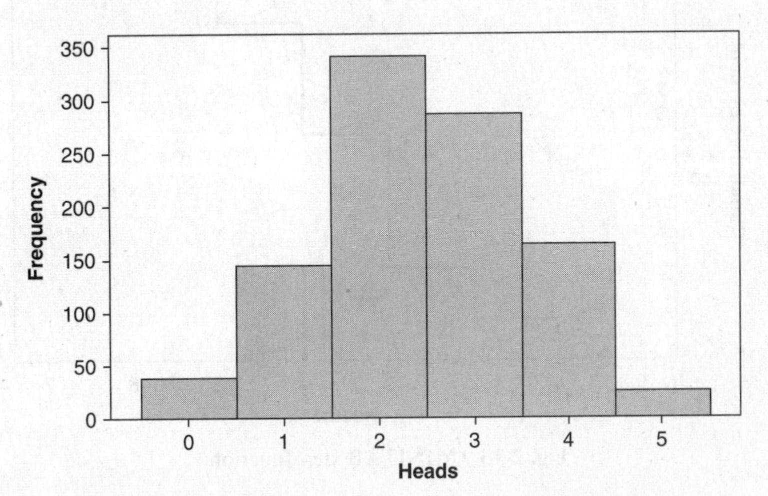

Fig. 2-12 MINITAB histogram of the number of heads.

Figure 2-12 shows a histogram of the data. Note that the total area of the histogram is the total frequency 1000, as it should be. In using the histogram representation or the corresponding frequency polygon, we are essentially treating the data *as if* they were continuous. This will later be found useful. Note that we have already used the histogram and frequency polygon for discrete data in Problem 2.10.

(*b*) Referring to the required Table 2.13, note that it shows simply a cumulative-frequency distribution and percentage cumulative distribution of the number of heads. It should be observed that the entries "Less than 1," "Less than 2," etc., could just as well have been "Less than or equal to 0," "Less than or equal to 1," etc.

(*c*) The required graph can be presented either as in Fig. 2-13 or as in Fig. 2-14.

Figure 2-13 is the most natural for presenting discrete data—since, for example, the percentage of tosses in which there will be less than 2 heads is equal to the percentage in which there will be less than 1.75, 1.56, or 1.23 heads, so that the same percentage (18.2%) should be shown for these values (indicated by the horizontal line).

Table 2.13

Number of Heads	Number of Tosses (cumulative frequency)	Percentage Number of Tosses (percentage cumulative frequency)
Less than 0	0	0.0
Less than 1	38	3.8
Less than 2	182	18.2
Less than 3	524	52.4
Less than 4	811	81.1
Less than 5	975	97.5
Less than 6	1000	100.0

Figure 2-14 shows the cumulative-frequency polygon, or ogive, for the data and essentially treats the data as if they were continuous.

Note that Figs. 2-13 and 2-14 correspond, respectively, to Figs. 2-11 and 2-12 of part (*a*).

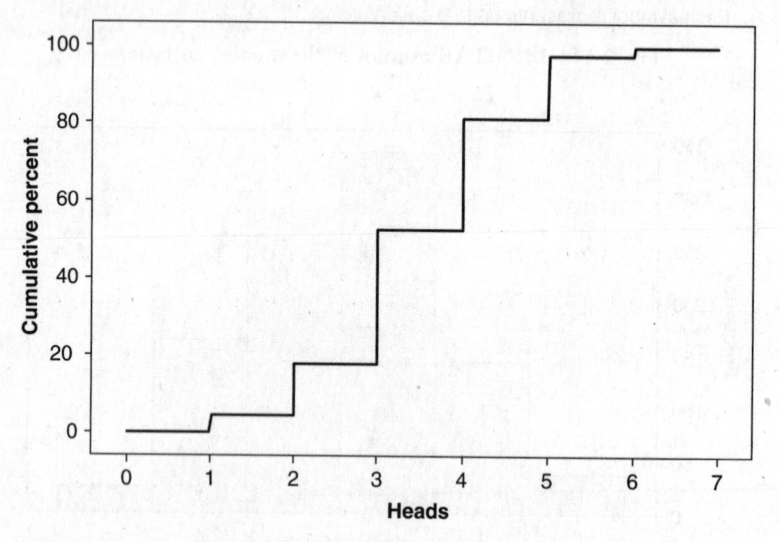

Fig. 2-13 MINITAB step function.

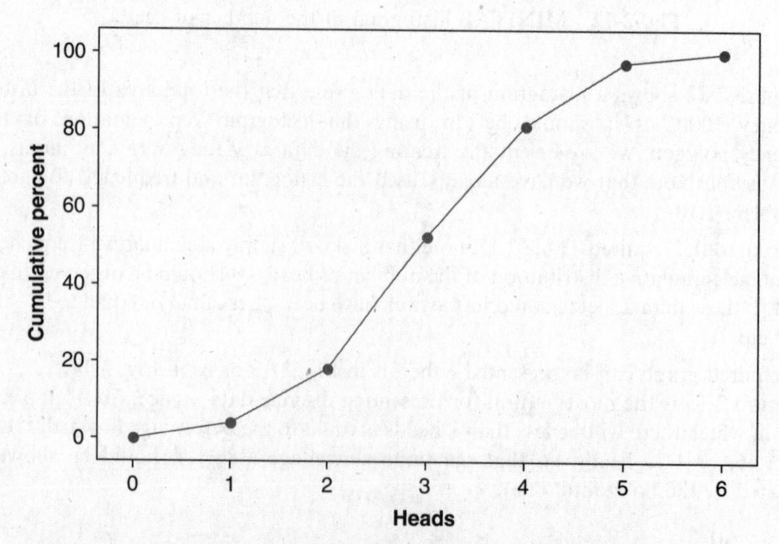

Fig. 2-14 MINITAB cumulative frequency polygon.

FREQUENCY CURVES AND SMOOTHED OGIVES

2.18 Samples from populations have histograms and frequency polygons that have certain shapes. If the samples are extremely large, the histograms and frequency polygons approach the frequency distribution of the population. Consider two such population frequency distributions. (*a*) Consider a machine that fills cola containers uniformly between 15.9 and 16.1 ounces. Construct the frequency curve and determine what percent put more than 15.95 ounces in the containers. (*b*) Consider the heights of females. They have a population frequency distribution that is symmetrical or bell shaped with an average equal to 65 in and standard deviation equal 3 in. (Standard deviation is discussed in a later chapter.) What percent of the heights are between 62 and 68 in tall, that is are within one standard deviation of the mean? What percent are within two standard deviations of the mean? What percent are within three standard deviations of the mean?

SOLUTION

Figure 2-15 shows a uniform frequency curve. The cross-hatched region corresponds to fills of more than 15.95 ounces. Note that the region covered by the frequency curve is shaped as a rectangle. The area under the frequency curve is length × width or $(16.10 - 15.90) \times 5 = 1$. The area of the cross hatched region is $(16.10 - 15.90) \times 5 = 0.75$. This is interpreted to mean that 75% of the machine fills exceed 15.95 ounces.

Figure 2-16 shows a symmetrical or bell-shaped frequency curve. It shows the heights within one standard deviation as the cross-hatched region. Calculus is used to find this area. The area within one

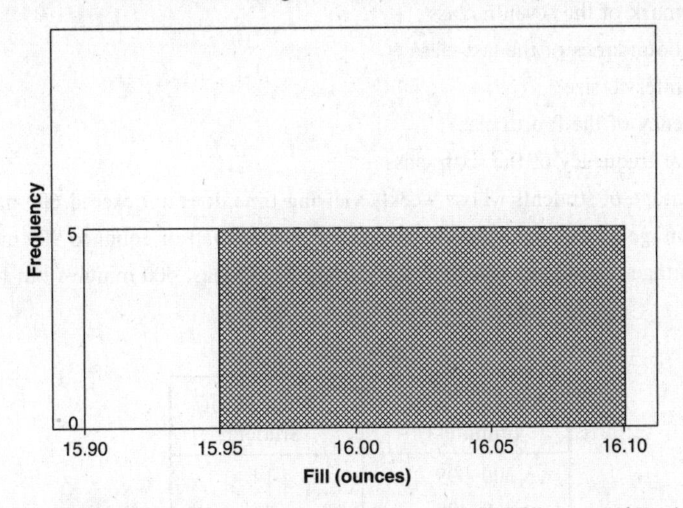

Fig. 2-15 MINITAB uniform frequency curve showing fills more than 15.95 ounces.

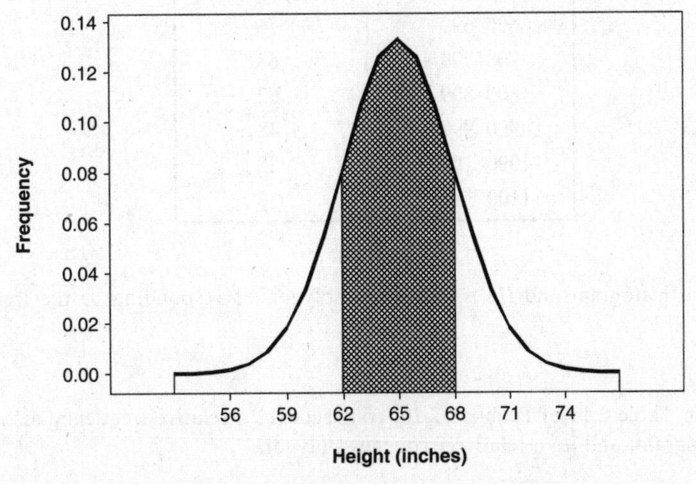

Fig. 2-16 MINITAB bell shaped frequency curve showing heights between 62 and 68 in and their frequencies.

standard deviation is approximately 68% of the total area under the curve. The area within two standard deviations is approximately 95% of the total area under the curve. The area within three standard deviations is approximately 99.7% of the total area under the curve.

We shall have more to say about finding areas under these curves in later chapters.

Supplementary Problems

2.19 (*a*) Arrange the numbers 12, 56, 42, 21, 5, 18, 10, 3, 61, 34, 65, and 24 in an array and (*b*) determine the range.

2.20 Table 2.14 shows the frequency distribution for the number of minutes per week spent watching TV by 400 junior high students. With reference to this table determine:

(*a*) The upper limit of the fifth class
(*b*) The lower limit of the eighth class
(*c*) The class mark of the seventh class
(*d*) The class boundaries of the last class
(*e*) The class-interval size
(*f*) The frequency of the fourth class
(*g*) The relative frequency of the sixth class
(*h*) The percentage of students whose weekly viewing time does not exceed 600 minutes
(*i*) The percentage of students with viewing times greater than or equal to 900 minutes
(*j*) The percentage of students whose viewing times are at least 500 minutes but less than 1000 minutes

Table 2.14

Viewing Time (minutes)	Number of Students
300–399	14
400–499	46
500–599	58
600–699	76
700–799	68
800–899	62
900–999	48
1000–1099	22
1100–1199	6

2.21 Construct (*a*) a histogram and (*b*) a frequency polygon corresponding to the frequency distribution of Table 2.14.

2.22 For the data in Table 2.14 of Problem 2.20, construct (*a*) a relative-frequency distribution, (*b*) a relative-frequency histogram, and (*c*) a relative-frequency polygon.

2.23 For the data in Table 2.14, construct (*a*) a cumulative-frequency distribution, (*b*) a percentage cumulative distribution, (*c*) an ogive, and (*d*) a percentage ogive. (Note that unless otherwise specified, a cumulative distribution refers to one made on a "less than" basis.)

2.24 Work Problem 2.23 for the case where the frequencies are cumulative on an "or more" basis.

2.25 Using the data in Table 2.14, estimate the percentage of students that have viewing times of (*a*) less than 560 minutes per week, (*b*) 970 or more minutes per week, and (*c*) between 620 and 890 minutes per week.

2.26 The inner diameters of washers produced by a company can be measured to the nearest thousandth of an inch. If the class marks of a frequency distribution of these diameters are given in inches by 0.321, 0.324, 0.327, 0.330, 0.333, and 0.336, find (*a*) the class-interval size, (*b*) the class boundaries, and (*c*) the class limits.

2.27 The following table shows the diameters in centimeters of a sample of 60 ball bearings manufactured by a company. Construct a frequency distribution of the diameters, using appropriate class intervals.

1.738	1.729	1.743	1.740	1.736	1.741	1.735	1.731	1.726	1.737
1.728	1.737	1.736	1.735	1.724	1.733	1.742	1.736	1.739	1.735
1.745	1.736	1.742	1.740	1.728	1.738	1.725	1.733	1.734	1.732
1.733	1.730	1.732	1.730	1.739	1.734	1.738	1.739	1.727	1.735
1.735	1.732	1.735	1.727	1.734	1.732	1.736	1.741	1.736	1.744
1.732	1.737	1.731	1.746	1.735	1.735	1.729	1.734	1.730	1.740

2.28 For the data of Problem 2.27 construct (*a*) a histogram, (*b*) a frequency polygon, (*c*) a relative-frequency distribution, (*d*) a relative-frequency histogram, (*e*) a relative-frequency polygon (*f*) a cumulative-frequency distribution, (*g*) a percentage cumulative distribution, (*h*) an ogive, and (*i*) a percentage ogive.

2.29 From the results in Problem 2.28, determine the percentage of ball bearings having diameters (*a*) exceeding 1.732 cm, (*b*) not more than 1.736 cm, and (*c*) between 1.730 and 1.738 cm. Compare your results with those obtained directly from the raw data of Problem 2.27.

2.30 Work Problem 2.28 for the data of Problem 2.20.

2.31 According to the U.S. Bureau of the Census, Current Population Reports, the 1996 population of the United States is 265,284,000. Table 2.15 gives the percent distribution for various age groups.

(*a*) What is the width, or size, of the second class interval? The fourth class interval?

(*b*) How many different class-interval sizes are there?

(*c*) How many open class intervals are there?

(*d*) How should the last class interval be written so that its class width will equal that of the next to last class interval?

(*e*) What is the class mark of the second class interval? The fourth class interval?

(*f*) What are the class boundaries of the fourth class interval?

(*g*) What percentage of the population is 35 years of age or older? What percentage of the population is 64 or younger?

(*h*) What percentage of the ages are between 20 and 49 inclusive?

(*i*) What percentage of the ages are over 70 years?

2.32 (*a*) Why is it impossible to construct a percentage histogram or frequency polygon for the distribution in Table 2.15?

(*b*) How would you modify the distribution so that a percentage histogram and frequency polygon could be constructed?

(*c*) Perform the construction using the modification in part (*b*).

Table 2.15

Age group in years	% of U.S.
Under 5	7.3
5–9	7.3
10–14	7.2
15–19	7.0
20–24	6.6
25–29	7.2
30–34	8.1
35–39	8.5
40–44	7.8
45–49	6.9
50–54	5.3
55–59	4.3
60–64	3.8
65–74	7.0
75–84	4.3
85 and over	1.4

Source: U.S. Bureau of the Census, Current Population Reports.

2.33 Refer to Table 2.15. Assume that the total population is 265 million and that the class "Under 5" contains babies who are not yet 1 year old. Determine the number of individuals in millions to one decimal point in each age group.

2.34 (*a*) Construct a smoothed percentage frequency polygon and smoothed percentage ogive corresponding to the data in Table 2.14.

(*b*) From the results in part (*a*), estimate the probability that a student views less than 10 hours of TV per week.

(*c*) From the results in part (*a*), estimate the probability that a student views TV for 15 or more hours per week.

(*d*) From the results in part (*a*), estimate the probability that a student views TV for less than 5 hours per week.

2.35 (*a*) Toss four coins 50 times and tabulate the number of heads at each toss.

(*b*) Construct a frequency distribution showing the number of tosses in which 0, 1, 2, 3, and 4 heads appeared.

(*c*) Construct a percentage distribution corresponding to part (*b*).

(*d*) Compare the percentage obtained in part (*c*) with the theoretical ones 6.25%, 25%, 37.5%, 25%, and 6.25% (proportional to 1, 4, 6, 4, and 1) arrived at by rules of probability.

(*e*) Graph the distributions in parts (*b*) and (*c*).

(*f*) Construct a percentage ogive for the data.

2.36 Work Problem 2.35 with 50 more tosses of the four coins and see if the experiment is more in agreement with theoretical expectation. If not, give possible reasons for the differences.

CHAPTER 3

The Mean, Median, Mode, and Other Measures of Central Tendency

INDEX, OR SUBSCRIPT, NOTATION

Let the symbol X_j (read "X sub j") denote any of the N *values* $X_1, X_2, X_3, \ldots, X_N$ assumed by a variable X. The letter j in X_j, which can stand for any of the numbers 1, 2, 3, ..., N is called a *subscript*, or *index*. Clearly any letter other than j, such as i, k, p, q, or s, could have been used as well.

SUMMATION NOTATION

The symbol $\sum_{j=1}^{N} X_j$ is used to denote the sum of all the X_j's from $j = 1$ to $j = N$; by definition,

$$\sum_{j=1}^{N} X_j = X_1 + X_2 + X_3 + \cdots + X_N$$

When no confusion can result, we often denote this sum simply by $\sum X$, $\sum X_j$, or $\sum_j X_j$. The symbol $\sum$ is the Greek capital letter *sigma*, denoting sum.

EXAMPLE 1. $\sum_{j=1}^{N} X_j Y_j = X_1 Y_1 + X_2 Y_2 + X_3 Y_3 + \cdots + X_N Y_N$

EXAMPLE 2. $\sum_{j=1}^{N} aX_j = aX_1 + aX_2 + \cdots + aX_N = a(X_1 + X_2 + \cdots + X_N) = a\sum_{j=1}^{N} X_j$

where a is a constant. More simply, $\sum aX = a\sum X$.

EXAMPLE 3. If a, b, and c are any constants, then $\sum (aX + bY - cZ) = a\sum X + b\sum Y - c\sum Z$. See Problem 3.3.

AVERAGES, OR MEASURES OF CENTRAL TENDENCY

An *average* is a value that is typical, or representative, of a set of data. Since such typical values tend to lie centrally within a set of data arranged according to magnitude, averages are also called *measures of central tendency*.

Several types of averages can be defined, the most common being the *arithmetic mean*, the *median*, the *mode*, the *geometric mean*, and the *harmonic mean*. Each has advantages and disadvantages, depending on the data and the intended purpose.

THE ARITHMETIC MEAN

The *arithmetic mean*, or briefly the *mean*, of a set of N numbers $X_1, X_2, X_3, \ldots, X_N$ is denoted by $\bar{X}$ (read "X bar") and is defined as

$$\bar{X} = \frac{X_1 + X_2 + X_3 + \cdots + X_N}{N} = \frac{\sum_{j=1}^{N} X_j}{N} = \frac{\sum X}{N} \tag{1}$$

EXAMPLE 4. The arithmetic mean of the numbers 8, 3, 5, 12, and 10 is

$$\bar{X} = \frac{8 + 3 + 5 + 12 + 10}{5} = \frac{38}{5} = 7.6$$

If the numbers $X_1, X_2, \ldots, X_K$ occur $f_1, f_2, \ldots, f_K$ times, respectively (i.e., occur with frequencies $f_1, f_2, \ldots, f_K$), the arithmetic mean is

$$\bar{X} = \frac{f_1X_1 + f_2X_2 + \cdots + f_KX_K}{f_1 + f_2 + \cdots + f_K} = \frac{\sum_{j=1}^{K} f_jX_j}{\sum_{j=1}^{K} f_j} = \frac{\sum fX}{\sum f} = \frac{\sum fX}{N} \tag{2}$$

where $N = \sum f$ is the *total frequency* (i.e., the total number of cases).

EXAMPLE 5. If 5, 8, 6, and 2 occur with frequencies 3, 2, 4, and 1, respectively, the arithmetic mean is

$$\bar{X} = \frac{(3)(5) + (2)(8) + (4)(6) + (1)(2)}{3 + 2 + 4 + 1} = \frac{15 + 16 + 24 + 2}{10} = 5.7$$

THE WEIGHTED ARITHMETIC MEAN

Sometimes we associate with the numbers $X_1, X_2, \ldots, X_K$ certain *weighting factors* (or *weights*) $w_1, w_2, \ldots, w_K$, depending on the significance or importance attached to the numbers. In this case,

$$\bar{X} = \frac{w_1X_1 + w_2X_2 + \cdots + w_KX_k}{w_1 + w_2 + \cdots + w_K} = \frac{\sum wX}{\sum w} \tag{3}$$

is called the *weighted arithmetic mean*. Note the similarity to equation (*2*), which can be considered a weighted arithmetic mean with weights $f_1, f_2, \ldots, f_K$.

EXAMPLE 6. If a final examination in a course is weighted 3 times as much as a quiz and a student has a final examination grade of 85 and quiz grades of 70 and 90, the mean grade is

$$\bar{X} = \frac{(1)(70) + (1)(90) + (3)(85)}{1 + 1 + 3} = \frac{415}{5} = 83$$

PROPERTIES OF THE ARITHMETIC MEAN

1. The algebraic sum of the deviations of a set of numbers from their arithmetic mean is zero.

EXAMPLE 7. The deviations of the numbers 8, 3, 5, 12, and 10 from their arithmetic mean 7.6 are $8-7.6$, $3-7.6$, $5-7.6$, $12-7.6$, and $10-7.6$, or 0.4, -4.6, -2.6, 4.4, and 2.4, with algebraic sum $0.4-4.6-2.6+4.4+2.4=0$.

2. The sum of the squares of the deviations of a set of numbers X_j from any number a is a minimum if and only if $a=\bar{X}$ (see Problem 4.27).
3. If f_1 numbers have mean m_1, f_2 numbers have mean $m_2, \ldots, f_K$ numbers have mean m_K, then the mean of all the numbers is

$$\bar{X}=\frac{f_1m_1+f_2m_2+\cdots+f_Km_K}{f_1+f_2+\cdots+f_K} \tag{4}$$

that is, a weighted arithmetic mean of all the means (see Problem 3.12).
4. If A is any *guessed* or *assumed arithmetic mean* (which may be any number) and if $d_j=X_j-A$ are the deviations of X_j from A, then equations (*1*) and (*2*) become, respectively,

$$\bar{X}=A+\frac{\sum_{j=1}^{N}d_j}{N}=A+\frac{\sum d}{N} \tag{5}$$

$$\bar{X}=A+\frac{\sum_{j=1}^{K}f_jd_j}{\sum_{j=1}^{K}f_j}=A+\frac{\sum fd}{N} \tag{6}$$

where $N=\sum_{j=1}^{K}f_j=\sum f$. Note that formulas (*5*) and (*6*) are summarized in the equation $\bar{X}=A+\bar{d}$ (see Problem 3.18).

THE ARITHMETIC MEAN COMPUTED FROM GROUPED DATA

When data are presented in a frequency distribution, all values falling within a given class interval are considerd to be coincident with the class mark, or midpoint, of the interval. Formulas (*2*) and (*6*) are valid for such grouped data if we interpret X_j as the class mark, f_j as its corresponding class frequency, A as any guessed or assumed class mark, and $d_j=X_j-A$ as the deviations of X_j from A.

Computations using formulas (*2*) and (*6*) are sometimes called the *long* and *short methods*, respectively (see Problems 3.15 and 3.20).

If class intervals all have equal size c, the deviations $d_j=X_j-A$ can all be expressed as cu_j, where u_j can be positive or negative integers or zero (i.e., 0, ±1, ±2, $\pm3, \ldots$), and formula (*6*) becomes

$$\bar{X}=A+\left(\frac{\sum_{j=1}^{K}f_ju_j}{N}\right)c=A+\left(\frac{\sum fu}{N}\right)c \tag{7}$$

which is equivalent to the equation $\bar{X}=A+c\bar{u}$ (see Problem 3.21). This is called the *coding method* for computing the mean. It is a very short method and should always be used for grouped data where the class-interval sizes are equal (see Problems 3.22 and 3.23). Note that in the coding method the values of the variable X are *transformed* into the values of the variable u according to $X=A+cu$.

THE MEDIAN

The *median* of a set of numbers arranged in order of magnitude (i.e., in an array) is either the middle value or the arithmetic mean of the two middle values.

EXAMPLE 8. The set of numbers 3, 4, 4, 5, 6, 8, 8, 8, and 10 has median 6.

EXAMPLE 9. The set of numbers 5, 5, 7, 9, 11, 12, 15, and 18 has median $\frac{1}{2}(9 + 11) = 10$.

For grouped data, the median, obtained by interpolation, is given by

$$\text{Median} = L_1 + \left(\frac{\frac{N}{2} - (\sum f)_1}{f_{\text{median}}}\right)c \tag{8}$$

where L_1 = lower class boundary of the median class (i.e., the class containing the median)
N = number of items in the data (i.e., total frequency)
$(\sum f)_1$ = sum of frequencies of all classes lower than the median class
f_{median} = frequency of the median class
c = size of the median class interval

Geometrically the median is the value of X (abscissa) corresponding to the vertical line which divides a histogram into two parts having equal areas. This value of X is sometimes denoted by $\tilde{X}$.

THE MODE

The *mode* of a set of numbers is that value which occurs with the greatest frequency; that is, it is the most common value. The mode may not exist, and even if it does exist it may not be unique.

EXAMPLE 10. The set 2, 2, 5, 7, 9, 9, 9, 10, 10, 11, 12, and 18 has mode 9.

EXAMPLE 11. The set 3, 5, 8, 10, 12, 15, and 16 has no mode.

EXAMPLE 12. The set 2, 3, 4, 4, 4, 5, 5, 7, 7, 7, and 9 has two modes, 4 and 7, and is called *bimodal.*

A distribution having only one mode is called *unimodal.*

In the case of grouped data where a frequency curve has been constructed to fit the data, the mode will be the value (or values) of X corresponding to the maximum point (or points) on the curve. This value of X is sometimes denoted by $\hat{X}$.

From a frequency distribution or histogram the mode can be obtained from the formula

$$\text{Mode} = L_1 + \left(\frac{\Delta_1}{\Delta_1 + \Delta_2}\right)c \tag{9}$$

where L_1 = lower class boundary of the modal class (i.e., the class containing the mode)
Δ_1 = excess of modal frequency over frequency of next-lower class
Δ_2 = excess of modal frequency over frequency of next-higher class
c = size of the modal class interval

THE EMPIRICAL RELATION BETWEEN THE MEAN, MEDIAN, AND MODE

For unimodal frequency curves that are moderately skewed (asymmetrical), we have the empirical relation

$$\text{Mean} - \text{mode} = 3(\text{mean} - \text{median}) \tag{10}$$

Figures 3-1 and 3-2 show the relative positions of the mean, median, and mode for frequency curves skewed to the right and left, respectively. For symmetrical curves, the mean, mode, and median all coincide.

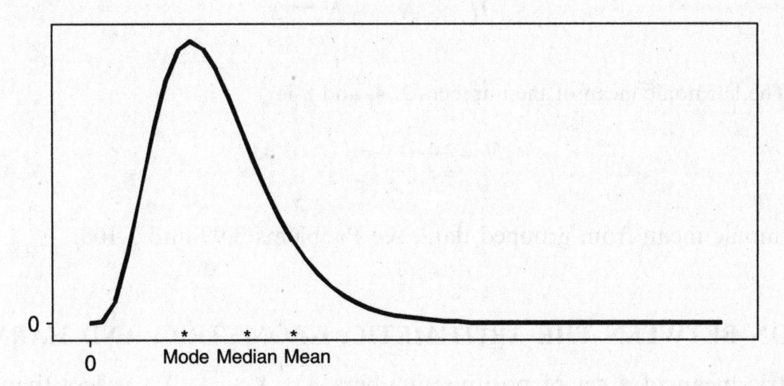

Fig. 3-1 Relative positions of mode, median, and mean for a right-skewed frequency curve.

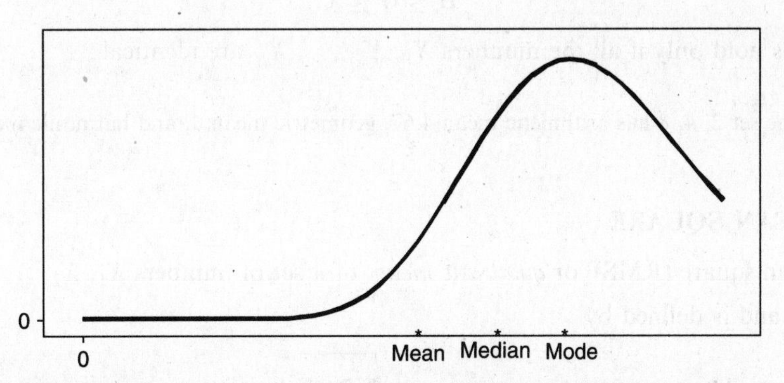

Fig. 3-2 Relative positions of mode, median, and mean for a left-skewed frequency curve.

THE GEOMETRIC MEAN G

The geometric mean G of a set of N positive numbers $X_1, X_2, X_3, \ldots, X_N$ is the Nth root of the product of the numbers:

$$G = \sqrt[N]{X_1 X_2 X_3 \cdots X_N} \tag{11}$$

EXAMPLE 13. The geometric mean of the numbers 2, 4, and 8 is $G = \sqrt[3]{(2)(4)(8)} = \sqrt[3]{64} = 4$.

We can compute G by logarithms (see Problem 3.35) or by using a calculator. For the geometric mean from grouped data, see Problems 3.36 and 3.91.

THE HARMONIC MEAN H

The harmonic mean H of a set of N numbers $X_1, X_2, X_3, \ldots, X_N$ is the reciprocal of the arithmetic mean of the reciprocals of the numbers:

$$H = \frac{1}{\dfrac{1}{N}\displaystyle\sum_{j=1}^{N}\frac{1}{X_j}} = \frac{N}{\sum \dfrac{1}{X}} \tag{12}$$

In practice it may be easier to remember that

$$\frac{1}{H} = \frac{\sum \frac{1}{X}}{N} = \frac{1}{N} \sum \frac{1}{X} \tag{13}$$

EXAMPLE 14. The harmonic mean of the numbers 2, 4, and 8 is

$$H = \frac{3}{\frac{1}{2} + \frac{1}{4} + \frac{1}{8}} = \frac{3}{\frac{7}{8}} = 3.43$$

For the harmonic mean from grouped data, see Problems 3.99 and 3.100.

THE RELATION BETWEEN THE ARITHMETIC, GEOMETRIC, AND HARMONIC MEANS

The geometric mean of a set of positive numbers $X_1, X_2, \ldots, X_N$ is less than or equal to their arithmetic mean but is greater than or equal to their harmonic mean. In symbols,

$$H \leq G \leq \bar{X} \tag{14}$$

The equality signs hold only if all the numbers $X_1, X_2, \ldots, X_N$ are identical.

EXAMPLE 15. The set 2, 4, 8 has arithmetic mean 4.67, geometric mean 4, and harmonic mean 3.43.

THE ROOT MEAN SQUARE

The root mean square (RMS), or *quadratic mean*, of a set of numbers $X_1, X_2, \ldots, X_N$ is sometimes denoted by $\sqrt{\overline{X^2}}$ and is defined by

$$\text{RMS} = \sqrt{\overline{X^2}} = \sqrt{\frac{\sum_{j=1}^{N} X_j^2}{N}} = \sqrt{\frac{\sum X^2}{N}} \tag{15}$$

This type of average is frequently used in physical applications.

EXAMPLE 16. The RMS of the set 1, 3, 4, 5, and 7 is

$$\sqrt{\frac{1^2 + 3^2 + 4^2 + 5^2 + 7^2}{5}} = \sqrt{20} = 4.47$$

QUARTILES, DECILES, AND PERCENTILES

If a set of data is arranged in order of magnitude, the middle value (or arithmetic mean of the two middle values) that divides the set into two equal parts is the median. By extending this idea, we can think of those values which divide the set into four equal parts. These values, denoted by Q_1, Q_2, and Q_3, are called the first, second, and third *quartiles*, respectively, the value Q_2 being equal to the median.

Similarly, the values that divide the data into 10 equal parts are called *deciles* and are denoted by $D_1, D_2, \ldots, D_9$, while the values dividing the data into 100 equal parts are called *percentiles* and are denoted by $P_1, P_2, \ldots, P_{99}$. The fifth decile and the 50th percentile correspond to the median. The 25th and 75th percentiles correspond to the first and third quartiles, respectively.

Collectively, quartiles, deciles, percentiles, and other values obtained by equal subdivisions of the data are called *quantiles*. For computations of these from grouped data, see Problems 3.44 to 3.46.

EXAMPLE 17. Use EXCEL to find Q_1, Q_2, Q_3, D_9, and P_{95} for the following sample of test scores:

88	45	53	86	33	86	85	30	89	53	41	96	56	38	62
71	51	86	68	29	28	47	33	37	25	36	33	94	73	46
42	34	79	72	88	99	82	62	57	42	28	55	67	62	60
96	61	57	75	93	34	75	53	32	28	73	51	69	91	35

To find the first quartile, put the data in the first 60 rows of column A of the EXCEL worksheet. Then give the command =PERCENTILE(A1:A60,0.25). EXCEL returns the value 37.75. We find that 15 out of the 60 values are less than 37.75 or 25% of the scores are less than 37.75. Similarly, we find =PERCENTILE(A1:A60,0.5) gives 57, =PERCENTILE(A1:A60,0.75) gives 76, =PERCENTILE(A1:A60,0.9) gives 89.2, and =PERCENTILE(A1:A60,0.95) gives 94.1. EXCEL gives quartiles, deciles, and percentiles all expressed as percentiles.

An algorithm that is often used to find quartiles, deciles, and percentiles by hand is described as follows. The data in example 17 is first sorted and the result is:

test	scores													
25	28	28	28	29	30	32	33	33	33	34	34	35	36	37
38	41	42	42	45	46	47	51	51	53	53	53	55	56	57
57	60	61	62	62	62	67	68	69	71	72	73	73	75	75
79	82	85	86	86	86	88	88	89	91	93	94	96	96	99

Suppose we wish to find the first quartile (25th percentile). Calculate $i = np/100 = 60(25)/100 = 15$. Since 15 is a whole number, go to the 15th and 16th numbers in the sorted array and average the 15th and 16th numbers. That is average 37 and 38 to get 37.5 as the first quartile ($Q_1 = 37.5$). To find the 93rd percentile, calculate $np/100 = 60(93)/100$ to get 55.8. Since this number is not a whole number, always round it up to get 56. The 56th number in the sorted array is 93 and $P_{93} = 93$. The EXCEL command =PERCENTILE(A1:A60,0.93) gives 92.74. Note that EXCEL does not give the same values for the percentiles, but the values are close. As the data set becomes larger, the values tend to equal each other.

SOFTWARE AND MEASURES OF CENTRAL TENDENCY

All the software packages utilized in this book give the descriptive statistics discussed in this section. The output for all five packages is now given for the test scores in Example 17.

EXCEL

If the pull-down "Tools ⇒ Data Analysis ⇒ Descriptive Statistics" is given, the measures of central tendency median, mean, and mode as well as several measures of dispersion are obtained:

Mean	59.16667
Standard Error	2.867425
Median	57
Mode	28
Standard Deviation	22.21098
Sample Variance	493.3277
Kurtosis	−1.24413
Skewness	0.167175
Range	74
Minimum	25
Maximum	99
Sum	3550
Count	60

MINITAB

If the pull-down "Stat ⇒ Basic Statistics ⇒ Display Descriptive Statistics" is given, the following output is obtained:

Descriptive Statistics: testscore

```
Variable    N  N*   Mean  SE Mean  St Dev  Minimum     Q1  Median     Q3
testscore  60   0  59.17     2.87   22.21    25.00  37.25   57.00  78.00

Variable   Maximum
testscore    99.00
```

SPSS

If the pull-down "Analyze ⇒ Descriptive Statistics ⇒ Descriptives" is given, the following output is obtained:

Descriptive Statistics

	N	Minimum	Maximum	Mean	Std. Deviation
Testscore	60	25.00	99.00	59.1667	22.21098
Valid *N* (listwise)	60				

SAS

If the pull-down "**Solutions ⇒ Analysis ⇒ Analyst**" is given and the data are read in as a file, the pull-down "**Statistics ⇒ Descriptive ⇒ Summary Statistics**" gives the following output:

```
                         The MEANS Procedure
                     Analysis Variable : Testscr

        Mean           Std Dev         N         Minimum         Maximum
ffffffffffffffffffffffffffffffffffffffffffffffffffffffffffffffffffffffff
   59.1666667      22.2109811       60      25.0000000      99.0000000
ffffffffffffffffffffffffffffffffffffffffffffffffffffffffffffffffffffffff
```

STATISTIX

If the pull-down "Statistics ⇒ Summary Statistics ⇒ Descriptive Statistics" is given in the software package STATISTIX, the following output is obtained:

```
Statistix 8.0
Descriptive Staistics

                Testscore
N                      60
Mean               59.167
SD                 22.211
Minimum            25.000
1st Quarti         37.250
3rd Quarti         78.000
Maximum            99.000
```

Solved Problems

SUMMATION NOTATION

3.1 Write out the terms in each of the following indicated sums:

(a) $\sum_{j=1}^{6} X_j$ (c) $\sum_{j=1}^{N} a$ (e) $\sum_{j=1}^{3} (X_j - a)$

(b) $\sum_{j=1}^{4} (Y_j - 3)^2$ (d) $\sum_{k=1}^{5} f_k X_k$

SOLUTION

(a) $X_1 + X_2 + X_3 + X_4 + X_5 + X_6$

(b) $(Y_1 - 3)^2 + (Y_2 - 3)^2 + (Y_3 - 3)^2 + (Y_4 - 3)^2$

(c) $a + a + a + \cdots + a = Na$

(d) $f_1X_1 + f_2X_2 + f_3X_3 + f_4X_4 + f_5X_5$

(e) $(X_1 - a) + (X_2 - a) + (X_3 - a) = X_1 + X_2 + X_3 - 3a$

3.2 Express each of the following by using the summation notation:

(a) $X_1^2 + X_2^2 + X_3^2 + \cdots + X_{10}^2$

(b) $(X_1 + Y_1) + (X_2 + Y_2) + \cdots + (X_8 + Y_8)$

(c) $f_1X_1^3 + f_2X_2^3 + \cdots + f_{20}X_{20}^3$

(d) $a_1b_1 + a_2b_2 + a_3b_3 + \cdots + a_Nb_N$

(e) $f_1X_1Y_1 + f_2X_2Y_2 + f_3X_3Y_3 + f_4X_4Y_4$

SOLUTION

(a) $\sum_{j=1}^{10} X_j^2$ (c) $\sum_{j=1}^{20} f_j X_j^3$ (e) $\sum_{j=1}^{4} f_j X_j Y_j$

(b) $\sum_{j=1}^{8} (X_j + Y_j)$ (d) $\sum_{j=1}^{N} a_j b_j$

3.3 Prove that $\sum_{j=1}^{N}(aX_j + bY_j - cZ_j) = a\sum_{j=1}^{N} X_j + b\sum_{j=1}^{N} Y_j - c\sum_{j=1}^{N} Z_j$, where a, b, and c are any constants.

SOLUTION

$$\begin{aligned}\sum_{j=1}^{N}(aX_j + bY_j - cZ_j) &= (aX_1 + bY_1 - cZ_1) + (aX_2 + bY_2 - cZ_2) + \cdots + (aX_N + bY_N - cZ_N) \\ &= (aX_1 + aX_2 + \cdots + aX_N) + (bY_1 + bY_2 + \cdots + bY_N) - (cZ_1 + cZ_2 + \cdots + cZ_N) \\ &= a(X_1 + X_2 + \cdots + X_N) + b(Y_1 + Y_2 + \cdots + Y_N) - c(Z_1 + Z_2 + \cdots + Z_N) \\ &= a\sum_{j=1}^{N} X_j + b\sum_{j=1}^{N} Y_j - c\sum_{j=1}^{N} Z_j\end{aligned}$$

or briefly, $\sum (aX + bY - cZ) = a\sum X + b\sum Y - c\sum Z$.

3.4 Two variables, X and Y, assume the values $X_1 = 2$, $X_2 = -5$, $X_3 = 4$, $X_4 = -8$ and $Y_1 = -3$, $Y_2 = -8$, $Y_3 = 10$, $Y_4 = 6$, respectively. Calculate (*a*) $\sum X$, (*b*) $\sum Y$, (*c*) $\sum XY$, (*d*) $\sum X^2$, (*e*) $\sum Y^2$, (*f*) $(\sum X)(\sum Y)$, (*g*) $\sum XY^2$, and (*h*) $\sum (X + Y)(X - Y)$.

SOLUTION

Note that in each case the subscript j on X and Y has been omitted and $\sum$ is understood as $\sum_{j=1}^{4}$. Thus $\sum X$, for example, is short for $\sum_{j=1}^{4} X_j$.

(*a*) $\sum X = (2) + (-5) + (4) + (-8) = 2 - 5 + 4 - 8 = -7$

(*b*) $\sum Y = (-3) + (-8) + (10) + (6) = -3 - 8 + 10 + 6 = 5$

(*c*) $\sum XY = (2)(-3) + (-5)(-8) + (4)(10) + (-8)(6) = -6 + 40 + 40 - 48 = 26$

(*d*) $\sum X^2 = (2)^2 + (-5)^2 + (4)^2 + (-8)^2 = 4 + 25 + 16 + 64 = 109$

(*e*) $\sum Y^2 = (-3)^2 + (-8)^2 + (10)^2 + (6)^2 = 9 + 64 + 100 + 36 = 209$

(*f*) $(\sum X)(\sum Y) = (-7)(5) = -35$, using parts (*a*) and (*b*). Note that $(\sum X)(\sum Y) \neq \sum XY$.

(*g*) $\sum XY^2 = (2)(-3)^2 + (-5)(-8)^2 + (4)(10)^2 + (-8)(6)^2 = -190$

(*h*) $\sum (X + Y)(X - Y) = \sum (X^2 - Y^2) = \sum X^2 - \sum Y^2 = 109 - 209 = -100$, using parts (*d*) and (*e*).

3.5 In a *USA Today* snapshot, it was reported that per capita state taxes collected in 2005 averaged \$2189.84 across the United States. The break down is: Sales and receipts \$1051.42, Income \$875.23, Licenses \$144.33, Others \$80.49, and Property \$38.36. Use EXCEL to show that the sum equals \$2189.84.

SOLUTION

Note that the expression = sum(A1:A5) is equivalent to $\sum_{j=1}^{5} X_j$.

1051.42	sales and receipts
875.23	income
144.33	licenses
80.49	others
38.36	property
2189.83	=sum(A1:A5)

THE ARITHMETIC MEAN

3.6 The grades of a student on six examinations were 84, 91, 72, 68, 87, and 78. Find the arithmetic mean of the grades.

SOLUTION

$$\bar{X} = \frac{\sum X}{N} = \frac{84 + 91 + 72 + 68 + 87 + 78}{6} = \frac{480}{6} = 80$$

Frequently one uses the term *average* synonymously with *arithmetic mean*. Strictly speaking, however, this is incorrect since there are averages other than the arithmetic mean.

3.7 Ten measurements of the diameter of a cylinder were recorded by a scientist as 3.88, 4.09, 3.92, 3.97, 4.02, 3.95, 4.03, 3.92, 3.98, and 4.06 centimeters (cm). Find the arithmetic mean of the measurements.

SOLUTION

$$\bar{X} = \frac{\sum X}{N} = \frac{3.88 + 4.09 + 3.92 + 3.97 + 4.02 + 3.95 + 4.03 + 3.92 + 3.98 + 4.06}{10} = \frac{39.82}{10} = 3.98\,\text{cm}$$

3.8 The following MINITAB output shows the time spent per week on line for 30 Internet users as well as the mean of the 30 times. Would you say this average is typical of the 30 times?

```
MTB > print cl
```

Data Display

```
time
3   4   4   5   5   5   5   5   5   6
6   6   6   7   7   7   7   7   8   8
9  10  10  10  10  10  10  12  55  60
```

```
MTB > mean cl
```

Column Mean

```
Mean of time = 10.400
```

SOLUTION

The mean 10.4 hours is not typical of the times. Note that 21 of the 30 times are in the single digits, but the mean is 10.4 hours. A great disadvantage of the mean is that it is strongly affected by extreme values.

3.9 Find the arithmetic mean of the numbers 5, 3, 6, 5, 4, 5, 2, 8, 6, 5, 4, 8, 3, 4, 5, 4, 8, 2, 5, and 4.

SOLUTION

First method

$$\bar{X} = \frac{\sum X}{N} = \frac{5+3+6+5+4+5+2+8+6+5+4+8+3+4+5+4+8+2+5+4}{20} = \frac{96}{20} = 4.8$$

Second method

There are six 5's, two 3's, two 6's, five 4's, two 2's and three 8's. Thus

$$\bar{X} = \frac{\sum fX}{\sum f} = \frac{\sum fX}{N} = \frac{(6)(5) + (2)(3) + (2)(6) + (5)(4) + (2)(2) + (3)(8)}{6+2+2+5+2+3} = \frac{96}{20} = 4.8$$

3.10 Out of 100 numbers, 20 were 4's, 40 were 5's, 30 were 6's and the remainder were 7's. Find the arithmetic mean of the numbers.

SOLUTION

$$\bar{X} = \frac{\sum fX}{\sum f} = \frac{\sum fX}{N} = \frac{(20)(4) + (40)(5) + (30)(6) + (10)(7)}{100} = \frac{530}{100} = 5.30$$

3.11 A student's final grades in mathematics, physics, English and hygiene are, respectively, 82, 86, 90, and 70. If the respective credits received for these courses are 3, 5, 3, and 1, determine an appropriate average grade.

SOLUTION

We use a weighted arithmetic mean, the weights associated with each grade being taken as the number of credits received. Thus

$$\bar{X} = \frac{\sum wX}{\sum w} = \frac{(3)(82) + (5)(86) + (3)(90) + (1)(70)}{3 + 5 + 3 + 1} = 85$$

3.12 In a company having 80 employees, 60 earn \$10.00 per hour and 20 earn \$13.00 per hour.

(*a*) Determine the mean earnings per hour.

(*b*) Would the answer in part (*a*) be the same if the 60 employees earn a mean hourly wage of \$10.00 per hour? Prove your answer.

(*c*) Do you believe the mean hourly wage to be typical?

SOLUTION

(*a*)

$$\bar{X} = \frac{\sum fX}{N} = \frac{(60)(\$10.00) + (20)(\$13.00)}{60 + 20} = \$10.75$$

(*b*) Yes, the result is the same. To prove this, suppose that f_1 numbers have mean m_1 and that f_2 numbers have mean m_2. We must show that the mean of all the numbers is

$$\bar{X} = \frac{f_1 m_1 + f_2 m_2}{f_1 + f_2}$$

Let the f_1 numbers add up to M_1 and the f_2 numbers add up to M_2. Then by definition of the arithmetic mean,

$$m_1 = \frac{M_1}{f_1} \qquad \text{and} \qquad m_2 = \frac{M_2}{f_2}$$

or $M_1 = f_1 m_1$ and $M_2 = f_2 m_2$. Since all $(f_1 + f_2)$ numbers add up to $(M_1 + M_2)$, the arithmetic mean of all numbers is

$$\bar{X} = \frac{M_1 + M_2}{f_1 + f_2} = \frac{f_1 m_1 + f_2 m_2}{f_1 + f_2}$$

as required. The result is easily extended.

(*c*) We can say that \$10.75 is a "typical" hourly wage in the sense that most of the employees earn \$10.00, which is not too far from \$10.75 per hour. It must be remembered that, whenever we summarize numerical data into a single number (as is true in an average), we are likely to make some error. Certainly, however, the result is not as misleading as that in Problem 3.8.

Actually, to be on safe ground, some estimate of the "spread," or "variation," of the data about the mean (or other average) should be given. This is called the *dispersion* of the data. Various measures of dispersion are given in Chapter 4.

3.13 Four groups of students, consisting of 15, 20, 10, and 18 individuals, reported mean weights of 162, 148, 153, and 140 pounds (lb), respectively. Find the mean weight of all the students.

SOLUTION

$$\bar{X} = \frac{\sum fX}{\sum f} = \frac{(15)(162) + (20)(148) + (10)(153) + (18)(140)}{15 + 20 + 10 + 18} = 150\,\text{lb}$$

3.14 If the mean annual incomes of agricultural and nonagricultural workers are \$25,000 and \$35,000, respectively, would the mean annual income of both groups together be \$30,000?

SOLUTION

It would be \$30,000 only if the numbers of agricultural and nonagricultural workers were the same. To determine the true mean annual income, we would have to know the relative numbers of workers in each group. Suppose that 10% of all workers are agricultural workers. Then the mean would be $(0.10)(25,000) + (0.90)(35,000) = \$34,000$. If there were equal numbers of both types of workers, the mean would be $(0.50)(25,000) + (0.50)(35,000) = \$30,000$.

3.15 Use the frequency distribution of heights in Table 2.1 to find the mean height of the 100 male students at XYZ university.

SOLUTION

The work is outlined in Table 3.1. Note that all students having heights 60 to 62 inches (in), 63 to 65 in, etc., are considered as having heights 61 in, 64 in, etc. The problem then reduces to finding the mean height of 100 students if 5 students have height 61 in, 18 have height 64 in, etc.

The computations involved can become tedious, especially for cases in which the numbers are large and many classes are present. Short techniques are available for lessening the labor in such cases; for example, see Problems 3.20 and 3.22.

Table 3.1

Height (in)	Class Mark (X)	Frequency (f)	fX
60–62	61	5	305
63–65	64	18	1152
66–68	67	42	2814
69–71	70	27	1890
72–74	73	8	584
		$N = \sum f = 100$	$\sum fX = 6745$

$$\bar{X} = \frac{\sum fX}{\sum f} = \frac{\sum fX}{N} = \frac{6745}{100} = 67.45 \text{ in}$$

PROPERTIES OF THE ARITHMETIC MEAN

3.16 Prove that the sum of the deviations of $X_1, X_2, \ldots, X_N$ from their mean $\bar{X}$ is equal to zero.

SOLUTION

Let $d_1 = X_1 - \bar{X}, d_2 = X_2 - \bar{X}, \ldots, d_N = X_N - \bar{X}$ be the deviations of $X_1, X_2, \ldots, X_N$ from their mean $\bar{X}$. Then

$$\text{Sum of deviations} = \sum d_j = \sum (X_j - \bar{X}) = \sum X_j - N\bar{X}$$
$$= \sum X_j - N\left(\frac{\sum X_j}{N}\right) = \sum X_j - \sum X_j = 0$$

where we have used $\sum$ in place of $\sum_{j=1}^{N}$. We could, if desired, have omitted the subscript j in X_j, provided that it is *understood*.

3.17 If $Z_1 = X_1 + Y_1, Z_2 = X_2 + Y_2, \ldots, Z_N = X_N + Y_N$, prove that $\bar{Z} = \bar{X} + \bar{Y}$.

SOLUTION

By definition,

$$\bar{X} = \frac{\sum X}{N} \qquad \bar{Y} = \frac{\sum Y}{N} \qquad \bar{Z} = \frac{\sum Z}{N}$$

Thus $$\bar{Z} = \frac{\sum Z}{N} = \frac{\sum (X+Y)}{N} = \frac{\sum X + \sum Y}{N} = \frac{\sum X}{N} + \frac{\sum Y}{N} = \bar{X} + \bar{Y}$$

where the subscripts j on X, Y, and Z have been omitted and where $\sum$ means $\sum_{j=1}^{N}$.

3.18 (*a*) If N numbers $X_1, X_2, \ldots, X_N$ have deviations from any number A given by $d_1 = X_1 - A$, $d_2 = X_2 - A, \ldots, d_N = X_N - A$, respectively, prove that

$$\bar{X} = A + \frac{\sum_{j=1}^{N} d_j}{N} = A + \frac{\sum d}{N}$$

(*b*) In case $X_1, X_2, \ldots, X_K$ have respective frequencies $f_1, f_2, \ldots, f_K$ and $d_1 = X_1 - A, \ldots, d_K = X_K - A$, show that the result in part (*a*) is replaced with

$$\bar{X} = A + \frac{\sum_{j=1}^{K} f_j d_j}{\sum_{j=1}^{K} f_j} = A + \frac{\sum fd}{N} \quad \text{where} \quad \sum_{j=1}^{K} f_j = \sum f = N$$

SOLUTION

(*a*) **First method**

Since $d_j = X_j - A$ and $X_j = A + d_j$, we have

$$\bar{X} = \frac{\sum X_j}{N} = \frac{\sum (A + d_j)}{N} = \frac{\sum A + \sum d_j}{N} = \frac{NA + \sum d_j}{N} = A + \frac{\sum d_j}{N}$$

where we have used $\sum$ in place of $\sum_{j=1}^{N}$ for brevity.

Second method

We have $d = X - A$, or $X = A + d$, omitting the subscripts on d and X. Thus, by Problem 3.17,

$$\bar{X} = \bar{A} + \bar{d} = A + \frac{\sum d}{N}$$

since the mean of a number of constants all equal to A is A.

(*b*) $$\bar{X} = \frac{\sum_{j=1}^{K} f_j X_j}{\sum_{j=1}^{K} f_j} = \frac{\sum f_j X_j}{N} = \frac{\sum f_j (A + d_j)}{N} = \frac{\sum A f_j + \sum f_j d_j}{N} = \frac{A \sum f_j + \sum f_j d_j}{N}$$

$$= \frac{AN + \sum f_j d_j}{N} = A + \frac{\sum f_j d_j}{N} = A + \frac{\sum fd}{N}$$

Note that *formally* the result is obtained from part (*a*) by replacing d_j with $f_j\, d_j$ and summing from $j = 1$ to K instead of from $j = 1$ to N. The result is equivalent to $\bar{X} = A + \bar{d}$, where $\bar{d} = (\sum fd)/N$.

THE ARITHMETIC MEAN COMPUTED FROM GROUPED DATA

3.19 Use the method of Problem 3.18(*a*) to find the arithmetic mean of the numbers 5, 8, 11, 9, 12, 6, 14, and 10, choosing as the "guessed mean" A the values (*a*) 9 and (*b*) 20.

SOLUTION

(*a*) The deviations of the given numbers from 9 are −4, −1, 2, 0, 3, −3, 5, and 1, and the sum of the deviations is $\sum d = -4 - 1 + 2 + 0 + 3 - 3 + 5 + 1 = 3$. Thus

$$\bar{X} = A + \frac{\sum d}{N} = 9 + \frac{3}{8} = 9.375$$

(*b*) The deviations of the given numbers from 20 are −15, −12, −9, −11, −8, −14, −6, and −10, and $\sum d = -85$. Thus

$$\bar{X} = A + \frac{\sum d}{N} = 20 + \frac{(-85)}{8} = 9.375$$

3.20 Use the method of Problem 3.18(*b*) to find the arithmetic mean of the heights of the 100 male students at XYZ University (see Problem 3.15).

SOLUTION

The work may be arranged as in Table 3.2. We take the guessed mean A to be the class mark 67 (which has the largest frequency), although any class mark can be used for A. Note that the computations are simpler than those in Problem 3.15. To shorten the labor even more, we can proceed as in Problem 3.22, where use is made of the fact that the deviations (column 2 of Table 3.2) are all integer multiples of the class-interval size.

Table 3.2

Class Mark (X)	Deviation $d = X - A$	Frequency (f)	fd
61	−6	5	−30
64	−3	18	−54
$A \longrightarrow 67$	0	42	0
70	3	27	81
73	6	8	48
		$N = \sum f = 100$	$\sum fd = 45$

$$\bar{X} = A + \frac{\sum fd}{N} = 67 + \frac{45}{100} = 67.45 \text{ in}$$

3.21 Let $d_j = X_j - A$ denote the deviations of any class mark X_j in a frequency distribution from a given class mark A. Show that if all class intervals have equal size c, then (*a*) the deviations are all multiples of c (i.e., $d_j = cu_j$, where $u_j = 0, \pm 1, \pm 2, \ldots$) and (*b*) the arithmetic mean can be computed from the formula

$$\bar{X} = A + \left(\frac{\sum fu}{N}\right)c$$

SOLUTION

(*a*) The result is illustrated in Table 3.2 of Problem 3.20, where it is observed that the deviations in column 2 are all multiples of the class-interval size $c = 3$ in.

To see that the result is true in general, note that if $X_1, X_2, X_3, \ldots$ are successive class marks, their common difference will for this case be equal to c, so that $X_2 = X_1 + c$, $X_3 = X_1 + 2c$, and in general $X_j = X_1 + (j - 1)c$. Then any two class marks X_p and X_q, for example, will differ by

$$X_p - X_q = [X_1 + (p - 1)c] - [X_1 + (q - 1)c] = (p - q)c$$

which is a multiple of c.

(*b*) By part (*a*), the deviations of all the class marks from any given one are multiples of c (i.e., $d_j = cu_j$). Then, using Problem 3.18(*b*), we have

$$\bar{X} = A + \frac{\sum f_j d_j}{N} = A + \frac{\sum f_j(cu_j)}{N} = A + c\frac{\sum f_j u_j}{N} = A + \left(\frac{\sum fu}{N}\right)c$$

Note that this is equivalent to the result $\bar{X} = A + c\bar{u}$, which can be obtained from $\bar{X} = A + \bar{d}$ by placing $d = cu$ and observing that $\bar{d} = c\bar{u}$ (see Problem 3.18).

3.22 Use the result of Problem 3.21(*b*) to find the mean height of the 100 male students at XYZ University (see Problem 3.20).

SOLUTION

The work may be arranged as in Table 3.3. The method is called the *coding method* and should be employed whenever possible.

Table 3.3

	X	u	f	fu
	61	−2	5	−10
	64	−1	18	−18
$A \longrightarrow$	67	0	42	0
	70	1	27	27
	73	2	8	16
			$N = 100$	$\sum fu = 15$

$$\bar{X} = A + \left(\frac{\sum fu}{N}\right)c = 67 + \left(\frac{15}{100}\right)(3) = 67.45 \text{ in}$$

3.23 Compute the mean weekly wage of the 65 employees at the P&R Company from the frequency distribution in Table 2.5, using (*a*) the long method and (*b*) the coding method.

SOLUTION

Tables 3.4 and 3.5 show the solutions to (*a*) and (*b*), respectively.

Table 3.4

X	f	fX
$255.00	8	$2040.00
265.00	10	2650.00
275.00	16	4400.00
285.00	14	3990.00
295.00	10	2950.00
305.00	5	1525.00
315.00	2	630.00
	$N = 65$	$\sum fX = \$18,185.00$

Table 3.5

	X	u	f	fu
	$255.00	−2	8	−16
	265.00	−1	10	−10
$A \longrightarrow$	275.00	0	16	0
	285.00	1	14	14
	295.00	2	10	20
	305.00	3	5	15
	315.00	4	2	8
			$N = 65$	$\sum fu = 31$

It might be supposed that error would be introduced into these tables since the class marks are actually \$254.995, \$264.995, etc., instead of \$255.00, \$265.00, etc. If in Table 3.4 these true class marks are used instead, then $\bar{X}$ turns out to be \$279.76 instead of \$279.77, and the difference is negligible.

$$\bar{X} = \frac{\sum fX}{N} = \frac{\$18{,}185.00}{65} = \$279.77 \qquad \bar{X} = A + \left(\frac{\sum fu}{N}\right)c = \$275.00 + \frac{31}{65}(\$10.00) = \$279.77$$

3.24 Using Table 2.9(*d*), find the mean wage of the 70 employees at the P&R Company.

SOLUTION

In this case the class intervals do not have equal size and we must use the long method, as shown in Table 3.6.

Table 3.6

X	f	fX
\$255.00	8	\$2040.00
265.00	10	2650.00
275.00	16	4400.00
285.00	15	4275.00
295.00	10	2950.00
310.00	8	2480.00
350.00	3	1050.00
	$N = 70$	$\sum fX = \$19{,}845.00$

$$\bar{X} = \frac{\sum fX}{N} = \frac{\$19{,}845.00}{70} = \$283.50$$

THE MEDIAN

3.25 The following MINITAB output shows the time spent per week searching on line for 30 Internet users as well as the median of the 30 times. Verify the median. Would you say this average is typical of the 30 times? Compare your results with those found in Problem 3.8.

```
MTB > print c1
```

Data Display

```
time
3   4   4   5   5   5   5   5   5   6
6   6   6   7   7   7   7   7   8   8
9  10  10  10  10  10  10  12  55  60

MTB > median c1
```

Column Median

```
Median of time = 7.0000
```

SOLUTION

Note that the two middle values are both 7 and the mean of the two middle values is 7. The mean was found to be 10.4 hours in Problem 3.8. The median is more typical of the times than the mean.

3.26 The number of ATM transactions per day were recorded at 15 locations in a large city. The data were: 35, 49, 225, 50, 30, 65, 40, 55, 52, 76, 48, 325, 47, 32, and 60. Find (*a*) the median number of transactions and (*b*) the mean number of transactions.

SOLUTION

(*a*) Arranged in order, the data are: 30, 32, 35, 40, 47, 48, 49, 50, 52, 55, 60, 65, 76, 225, and 325. Since there is an odd number of items, there is only one middle value, 50, which is the required median.

(*b*) The sum of the 15 values is 1189. The mean is $1189/15 = 79.267$.

Note that the median is not affected by the two extreme values 225 and 325, while the mean is affected by it. In this case, the median gives a better indication of the average number of daily ATM transactions.

3.27 If (*a*) 85 and (*b*) 150 numbers are arranged in an array, how would you find the median of the numbers?

SOLUTION

(*a*) Since there are 85 items, an odd number, there is only one middle value with 42 numbers below and 42 numbers above it. Thus the median is the 43rd number in the array.

(*b*) Since there are 150 items, an even number, there are two middle values with 74 numbers below them and 74 numbers above them. The two middle values are the 75th and 76th numbers in the array, and their arithmetic mean is the required median.

3.28 From Problem 2.8, find the median weight of the 40 male college students at State University by using (*a*) the frequency distribution of Table 2.7 (reproduced here as Table 3.7) and (*b*) the original data.

SOLUTION

(*a*) **First method** (using interpolation)

The weights in the frequency distribution of Table 3.7 are assumed to be continuously distributed. In such case the median is that weight for which half the total frequency ($40/2 = 20$) lies above it and half lies below it.

Table 3.7

Weight (lb)	Frequency
118–126	3
127–135	5
136–144	9
145–153	12
154–162	5
163–171	4
172–180	2
	Total 40

Now the sum of the first three class frequencies is $3 + 5 + 9 = 17$. Thus to give the desired 20, we require three more of the 12 cases in the fourth class. Since the fourth class interval, 145–153, actually corresponds to weights 144.5 to 153.5, the median must lie 3/12 of the way between 144.5 and 153.5; that is, the median is

$$144.5+\frac{3}{12}(153.5-144.5)=144.5+\frac{3}{12}(9)=146.8 \text{ lb}$$

Second method (using formula)

Since the sum of the first three and first four class frequencies are $3+5+9=17$ and $3+5+9+12=29$, respectively, it is clear that the median lies in the fourth class, which is therefore the median class. Then

$$L_1 = \text{lower class boundary of median class} = 144.5$$

$$N = \text{number of items in the data} = 40$$

$$(\sum f)_1 = \text{sum of all classes lower than the median class} = 3+5+9=17$$

$$f_{\text{median}} = \text{frequency of median class} = 12$$

$$c = \text{size of median class interval} = 9$$

and thus

$$\text{Median} = L_1+\left(\frac{N/2-(\sum f)_1}{f_{\text{median}}}\right)c = 144.5+\left(\frac{40/2-17}{12}\right)(9)=146.8 \text{ lb}$$

(*b*) Arranged in an array, the original weights are

119, 125, 126, 128, 132, 135, 135, 135, 136, 138, 138, 140, 140, 142, 142, 144, 144, 145, 145, 146,

146, 147, 147, 148, 149, 150, 150, 152, 153, 154, 156, 157, 158, 161, 163, 164, 165, 168, 173, 176

The median is the arithmetic mean of the 20th and 2lst weights in this array and is equal to 146 lb.

3.29 Figure 3-3 gives the stem-and-leaf display for the number of 2005 alcohol-related traffic deaths for the 50 states and Washington D.C.
Stem-and-Leaf Display: Deaths

Stem–and–Leaf Display: Deaths

Stem–and–leaf of Deaths N = 51
Leaf Unit = 10

14	0	22334556667889
23	1	122255778
(7)	2	0334689
21	3	124679
15	4	22669
10	5	012448
4	6	3
3	7	
3	8	
3	9	
3	10	
3	11	
3	12	
3	13	
3	14	7
2	15	6
1	16	
1	17	1

Fig. 3-3 MINITAB stem-and-leaf display for 2005 alcohol-related traffic deaths.

Find the mean, median, and mode of the alcohol-related deaths in Fig. 3-3.

SOLUTION

The number of deaths range from 20 to 1710. The distribution is bimodal. The two modes are 60 and 120, both of which occur three times.

The class `(7) 2    0334689 is the median class.` That is, the median is contained in this class. The median is the middle of the data or the 26th value in the ordered array. The 24th value is 200, the 25th value is 230, and the 26th value is also 230. Therefore the median is 230.

The sum of the 51 values is 16,660 and the mean is 16,660/51 = 326.67.

3.30 Find the median wage of the 65 employees at the P&R Company (see Problem 2.3).

SOLUTION

Here $N = 65$ and $N/2 = 32.5$. Since the sum of the first two and first three class frequencies are $8 + 10 = 18$ and $8 + 10 + 16 = 34$, respectively, the median class is the third class. Using the formula,

$$\text{Median} = L_1 + \left(\frac{N/2 - (\sum f)_1}{f_{\text{median}}}\right)c = \$269.995 + \left(\frac{32.5 - 18}{16}\right)(\$10.00) = \$279.06$$

THE MODE

3.31 Find the mean, median, and mode for the sets (*a*) 3, 5, 2, 6, 5, 9, 5, 2, 8, 6 and (*b*) 51.6, 48.7, 50.3, 49.5, 48.9.

SOLUTION

(*a*) Arranged in an array, the numbers are 2, 2, 3, 5, 5, 5, 6, 6, 8, and 9.

$$\text{Mean} = \tfrac{1}{10}(2 + 2 + 3 + 5 + 5 + 5 + 6 + 6 + 8 + 9) = 5.1$$

$$\text{Median} = \text{arithmetic mean of two middle numbers} = \tfrac{1}{2}(5 + 5) = 5$$

$$\text{Mode} = \text{number occurring most frequently} = 5$$

(*b*) Arranged in an array, the numbers are 48.7, 48.9, 49.5, 50.3, and 51.6.

$$\text{Mean} = \tfrac{1}{5}(48.7 + 48.9 + 49.5 + 50.3 + 51.6) = 49.8$$

$$\text{Median} = \text{middle number} = 49.5$$

$$\text{Mode} = \text{number occurring most frequently (nonexistent here)}$$

3.32 Suppose you desired to find the mode for the data in Problem 3.29. You could use frequencies procedure of SAS to obtain the following output. By considering the output from FREQ Procedure (Fig. 3-4), what are the modes of the alcohol-linked deaths?

The FREQ Procedure

Deaths

deaths	Frequency	Percent	Cumulative Frequency	Cumulative Percent
20	2	3.92	2	3.92
30	2	3.92	4	7.84
40	1	1.96	5	9.80
50	2	3.92	7	13.73
60	3	5.88	10	19.61
70	1	1.96	11	21.57
80	2	3.92	13	25.49
90	1	1.96	14	27.45
110	1	1.96	15	29.41
120	3	5.88	18	35.29
150	2	3.92	20	39.22
170	2	3.92	22	43.14
180	1	1.96	23	45.10
200	1	1.96	24	47.06
230	2	3.92	26	50.98
240	1	1.96	27	52.94
260	1	1.96	28	54.90
280	1	1.96	29	56.86
290	1	1.96	30	58.82
310	1	1.96	31	60.78
320	1	1.96	32	62.75
340	1	1.96	33	64.71
360	1	1.96	34	66.67
370	1	1.96	35	68.63
390	1	1.96	36	70.59
420	2	3.92	38	74.51
460	2	3.92	40	78.43
490	1	1.96	41	80.39
500	1	1.96	42	82.35
510	1	1.96	43	84.31
520	1	1.96	44	86.27
540	2	3.92	46	90.20
580	1	1.96	47	92.16
630	1	1.96	48	94.12
1470	1	1.96	49	96.08
1560	1	1.96	50	98.04
1710	1	1.96	51	100.00

Fig. 3-4 SAS output for FREQ procedure and the alcohol-related deaths.

SOLUTION

The data is bimodal and the modes are 60 and 120. This is seen by considering the SAS output and noting that both 60 and 120 have a frequency of 3 which is greater than any other of the values.

3.33 Some statistical software packages have a mode routine built in but do not return all modes in the case that the data is multimodal. Consider the output from SPSS in Fig. 3-5.

What does SPSS do when asked to find modes?

Deaths

		Frequency	Percent	Valid Percent	Cumulative Percent
Valid	20.00	2	3.9	3.9	3.9
	30.00	2	3.9	3.9	7.8
	40.00	1	2.0	2.0	9.8
	50.00	2	3.9	3.9	13.7
	60.00	3	5.9	5.9	19.6
	70.00	1	2.0	2.0	21.6
	80.00	2	3.9	3.9	25.5
	90.00	1	2.0	2.0	27.5
	110.00	1	2.0	2.0	29.4
	120.00	3	5.9	5.9	35.3
	150.00	2	3.9	3.9	39.2
	170.00	2	3.9	3.9	43.1
	180.00	1	2.0	2.0	45.1
	200.00	1	2.0	2.0	47.1
	230.00	2	3.9	3.9	51.0
	240.00	1	2.0	2.0	52.9
	260.00	1	2.0	2.0	54.9
	280.00	1	2.0	2.0	56.9
	290.00	1	2.0	2.0	58.8
	310.00	1	2.0	2.0	60.8
	320.00	1	2.0	2.0	62.7
	340.00	1	2.0	2.0	64.7
	360.00	1	2.0	2.0	66.7
	370.00	1	2.0	2.0	68.6
	390.00	1	2.0	2.0	70.6
	420.00	2	3.9	3.9	74.5
	460.00	2	3.9	3.9	78.4
	490.00	1	2.0	2.0	80.4
	500.00	1	2.0	2.0	82.4
	510.00	1	2.0	2.0	84.3
	520.00	1	2.0	2.0	86.3
	540.00	2	3.9	3.9	90.2
	580.00	1	2.0	2.0	92.2
	630.00	1	2.0	2.0	94.1
	1470.00	1	2.0	2.0	96.1
	1560.00	1	2.0	2.0	98.0
	1710.00	1	2.0	2.0	100.0
	Total	51	100.0	100.0	

Statistics

Deaths

N	Valid	51
	Missing	0
Mode		60.00[a]

[a]Multiple modes exist. The smallest value is shown.

Fig. 3-5 SPSS output for the alcohol-related deaths.

SOLUTION

SPSS prints out the smallest mode. However, it is possible to look at the frequency distribution and find all modes just as in SAS (see the output above).

THE EMPIRICAL RELATION BETWEEN THE MEAN, MEDIAN, AND MODE

3.34 (*a*) Use the empirical formula mean − mode = 3(mean − median) to find the modal wage of the 65 employees at the P&R Company.

(*b*) Compare your result with the mode obtained in Problem 3.33.

SOLUTION

(*a*) From Problems 3.23 and 3.30 we have mean = \$279.77 and median = \$279.06. Thus

$$\text{Mode} = \text{mean} - 3(\text{mean} - \text{median}) = \$279.77 - 3(\$279.77 - \$279.06) = \$277.64$$

(*b*) From Problem 3.33 the modal wage is \$277.50, so there is good agreement with the empirical result in this case.

THE GEOMETRIC MEAN

3.35 Find (*a*) the geometric mean and (*b*) the arithmetic mean of the numbers 3, 5, 6, 6, 7, 10, and 12. Assume that the numbers are exact.

SOLUTION

(*a*) Geometric mean $= G = \sqrt[7]{(3)(5)(6)(6)(7)(10)(12)} = \sqrt[7]{453{,}600}$. Using common logarithms, $\log G = \frac{1}{7}\log 453{,}600 = \frac{1}{7}(5.6567) = 0.8081$, and $G = 6.43$ (to the nearest hundredth). Alternatively, a calculator can be used.

Another method

$$\begin{aligned}\log G &= \tfrac{1}{7}(\log 3 + \log 5 + \log 6 + \log 6 + \log 7 + \log 10 + \log 12)\\ &= \tfrac{1}{7}(0.4771 + 0.6990 + 0.7782 + 0.7782 + 0.8451 + 1.0000 + 1.0792)\\ &= 0.8081\end{aligned}$$

and $\qquad G = 6.43$

(*b*) Arithmetic mean $= \bar{X} = \frac{1}{7}(3 + 5 + 6 + 6 + 7 + 10 + 12) = 7$. This illustrates the fact that the geometric mean of a set of unequal positive numbers is less than the arithmetic mean.

3.36 The numbers $X_1, X_2, \ldots, X_K$ occur with frequencies $f_1, f_2, \ldots, f_K$, where $f_1 + f_2 + \cdots + f_K = N$ is the total frequency.

(*a*) Find the geometric mean G of the numbers.

(*b*) Derive an expression for $\log G$.

(*c*) How can the results be used to find the geometric mean for data grouped into a frequency distribution?

SOLUTION

(*a*)
$$G = \sqrt[N]{\underbrace{X_1X_1\cdots X_1}_{f_1 \text{ times}}\ \underbrace{X_2X_2\cdots X_2}_{f_2 \text{ times}}\cdots\underbrace{X_KX_K\cdots X_K}_{f_K \text{ times}}} = \sqrt[N]{X_1^{f_1}X_2^{f_2}\cdots X_K^{f_K}}$$

where $N = \sum f$. This is sometimes called the *weighted geometric mean.*

(*b*)
$$\log G = \frac{1}{N}\log(X_1^{f_1} X_2^{f_2} \cdots X_K^{f_K}) = \frac{1}{N}(f_1 \log X_1 + f_2 \log X_2 + \cdots + f_K \log X_K)$$
$$= \frac{1}{N}\sum_{j=1}^{K} f_j \log X_j = \frac{\sum f \log X}{N}$$

where we assume that all the numbers are positive; otherwise, the logarithms are not defined.

Note that the logarithm of the geometric mean of a set of positive numbers is the arithmetic mean of the logarithms of the numbers.

(*c*) The result can be used to find the geometric mean for grouped data by taking $X_1, X_2, \ldots, X_K$ as class marks and $f_1, f_2, \ldots, f_K$ as the corresponding class frequencies.

3.37 During one year the ratio of milk prices per quart to bread prices per loaf was 3.00, whereas during the next year the ratio was 2.00.

(*a*) Find the arithmetic mean of these ratios for the 2-year period.

(*b*) Find the arithmetic mean of the ratios of bread prices to milk prices for the 2-year period.

(*c*) Discuss the advisability of using the arithmetic mean for averaging ratios.

(*d*) Discuss the suitability of the geometric mean for averaging ratios.

SOLUTION

(*a*)
$$\text{Mean ratio of milk to bread prices} = \tfrac{1}{2}(3.00 + 2.00) = 2.50.$$

(*b*) Since the ratio of milk to bread prices for the first year is 3.00, the ratio of bread to milk prices is $1/3.00 = 0.333$. Similarly, the ratio of bread to milk prices for the second year is $1/2.00 = 0.500$. Thus

$$\text{Mean ratio of bread to milk prices} = \tfrac{1}{2}(0.333 + 0.500) = 0.417$$

(*c*) We would expect the mean ratio of milk to bread prices to be the reciprocal of the mean ratio of bread to milk prices if the mean is an appropriate average. However, $1/0.417 = 2.40 \neq 2.50$. This shows that the arithmetic mean is a poor average to use for ratios.

(*d*)
$$\text{Geometric mean of ratios of milk to bread prices} = \sqrt{(3.00)(2.00)} = \sqrt{6.00}$$
$$\text{Geometric mean of ratios of bread to milk prices} = \sqrt{(0.333)(0.500)} = \sqrt{0.0167} = 1/\sqrt{6.00}$$

Since these averages are reciprocals, our conclusion is that the geometric mean is more suitable than the arithmetic mean for averaging ratios for this type of problem.

3.38 The bacterial count in a certain culture increased from 1000 to 4000 in 3 days. What was the average percentage increase per day?

SOLUTION

Since an increase from 1000 to 4000 is a 300% increase, one might be led to conclude that the average percentage increase per day would be $300\%/3 = 100\%$. This, however, would imply that during the first day the count went from 1000 to 2000, during the second day from 2000 to 4000, and during the third day from 4000 to 8000, which is contrary to the facts.

To determine this average percentage increase, let us denote it by r. Then

$$\text{Total bacterial count after 1 day} = 1000 + 1000r = 1000(1 + r)$$
$$\text{Total bacterial count after 2 days} = 1000(1 + r) + 1000(1 + r)r = 1000(1 + r)^2$$
$$\text{Total bacterial count after 3 days} = 1000(1 + r)^2 + 1000(1 + r)^2 r = 1000(1 + r)^3$$

This last expression must equal 4000. Thus $1000(1 + r)^3 = 4000$, $(1 + r)^3 = 4$, $1 + r = \sqrt[3]{4}$, and $r = \sqrt[3]{4} - 1 = 1.587 - 1 = 0.587$, so that $r = 58.7\%$.

In general, if we start with a quantity P and increase it at a constant rate r per unit of time, we will have after n units of time the amount

$$A = P(1+r)^n$$

This is called the *compound-interest formula* (see Problems 3.94 and 3.95).

THE HARMONIC MEAN

3.39 Find the harmonic mean H of the numbers 3, 5, 6, 6, 7, 10, and 12.

SOLUTION

$$\frac{1}{H} = \frac{1}{N}\sum \frac{1}{X} = \frac{1}{7}\left(\frac{1}{3}+\frac{1}{5}+\frac{1}{6}+\frac{1}{6}+\frac{1}{7}+\frac{1}{10}+\frac{1}{12}\right) = \frac{1}{7}\left(\frac{140+84+70+70+60+42+35}{420}\right)$$
$$= \frac{501}{2940}$$

and $$H = \frac{2940}{501} = 5.87$$

It is often convenient to express the fractions in decimal form first. Thus

$$\frac{1}{H} = \tfrac{1}{7}(0.3333+0.2000+0.1667+0.1667+0.1429+0.1000+0.0833)$$
$$= \frac{1.1929}{7}$$

and $$H = \frac{7}{1.1929} = 5.87$$

Comparison with Problem 3.35 illustrates the fact that the harmonic mean of several positive numbers not all equal is less than their geometric mean, which is in turn less than their arithmetic mean.

3.40 During four successive years, a home owner purchased oil for her furnace at respective costs of \$0.80, \$0.90, \$1.05, and \$1.25 per gallon (gal). What was the average cost of oil over the 4-year period?

SOLUTION

Case 1

Suppose that the home owner purchased the same quantity of oil each year, say 1000 gal. Then

$$\text{Average cost} = \frac{\text{total cost}}{\text{total quantity purchased}} = \frac{\$800+\$900+\$1050+\$1250}{4000\text{ gal}} = \$1.00/\text{gal}$$

This is the same as the arithmetic mean of the costs per gallon; that is, $\frac{1}{4}(\$0.80+\$0.90+\$1.05+\$1.25) =$ 1.00/gal. This result would be the same even if x gallons were used each year.

Case 2

Suppose that the home owner spends the same amount of money each year, say \$1000. Then

$$\text{Average cost} = \frac{\text{total cost}}{\text{total quantity purchased}} = \frac{\$4000}{(1250+1111+952+800)\text{gal}} = \$0.975/\text{gal}$$

This is the same as the harmonic mean of the costs per gallon:

$$\frac{4}{\dfrac{1}{0.80}+\dfrac{1}{0.90}+\dfrac{1}{1.05}+\dfrac{1}{1.25}} = 0.975$$

This result would be the same even if y dollars were spent each year.

Both averaging processes are correct, each average being computed under different prevailing conditions.

It should be noted that in case the number of gallons used changes from one year to another instead of remaining the same, the ordinary arithmetic mean of Case 1 is replaced by a weighted arithmetic mean. Similarly, if the total amount spent changes from one year to another, the ordinary harmonic mean of Case 2 is replaced by a weighted harmonic mean.

3.41 A car travels 25 miles at 25 miles per hour (mi/h), 25 miles at 50 mph, and 25 miles at 75 mph. Find the arithmetic mean of the three velocities and the harmonic mean of the three velocities. Which is correct?

SOLUTION

The average velocity is equal to the distance traveled divided by the total time and is equal to the following:

$$\frac{75}{\left(1+\dfrac{1}{2}+\dfrac{1}{3}\right)} = 40.9 \text{ mi/h}$$

The arithmetic mean of the three velocities is:

$$\frac{25+50+75}{3} = 50 \text{ mi/h}$$

The harmonic mean is found as follows:

$$\frac{1}{H} = \frac{1}{N}\sum\frac{1}{X} = \frac{1}{3}\left(\frac{1}{25}+\frac{1}{50}+\frac{1}{75}\right) = \frac{11}{450} \quad \text{and} \quad H = \frac{450}{11} = 40.9$$

The harmonic mean is the correct measure of the average velocity.

THE ROOT MEAN SQUARE, OR QUADRATIC MEAN

3.42 Find the quadratic mean of the numbers 3, 5, 6, 6, 7, 10, and 12.

SOLUTION

$$\text{Quadratic mean} = \text{RMS} = \sqrt{\frac{3^2+5^2+6^2+6^2+7^2+10^2+12^2}{7}} = \sqrt{57} = 7.55$$

3.43 Prove that the quadratic mean of two positive unequal numbers, a and b, is greater than their geometric mean.

SOLUTION

We are required to show that $\sqrt{\frac{1}{2}(a^2+b^2)} > \sqrt{ab}$. If this is true, then by squaring both sides, $\frac{1}{2}(a^2+b^2) > ab$, so that $a^2+b^2 > 2ab$, $a^2-2ab+b^2>0$, or $(a-b)^2>0$. But this last inequality is true, since the square of any real number not equal to zero must be positive.

The proof consists in establishing the reversal of the above steps. Thus starting with $(a-b)^2 > 0$, which we know to be true, we can show that $a^2 + b^2 > 2ab$, $\frac{1}{2}(a^2 + b^2) > ab$, and finally $\sqrt{\frac{1}{2}(a^2 + b^2)} > \sqrt{ab}$, as required.

Note that $\sqrt{\frac{1}{2}(a^2 + b^2)} = \sqrt{ab}$ if and only if $a = b$.

QUARTILES, DECILES, AND PERCENTILES

3.44 Find (*a*) the quartiles Q_1, Q_2, and Q_3 and (*b*) the deciles $D_1, D_2, \ldots, D_9$ for the wages of the 65 employees at the P&R Company (see Problem 2.3).

SOLUTION

(*a*) The first quartile Q_1 is that wage obtained by counting $N/4 = 65/4 = 16.25$ of the cases, beginning with the first (lowest) class. Since the first class contains 8 cases, we must take 8.25 (16.25 − 8) of the 10 cases from the second class. Using the method of linear interpolation, we have

$$Q_1 = \$259.995 + \frac{8.25}{10}(\$10.00) = \$268.25$$

The second quartile Q_2 is obtained by counting the first $2N/4 = N/2 = 65/2 = 32.5$ of the cases. Since the first two classes comprise 18 cases, we must take $32.5 - 18 = 14.5$ of the 16 cases from the third class; thus

$$Q_2 = \$269.995 + \frac{14.5}{16}(\$10.00) = \$279.06$$

Note that Q_2 is actually the median.

The third quartile Q_3 is obtained by counting the first $3N/4 = \frac{3}{4}(65) = 48.75$ of the cases. Since the first four classes comprise 48 cases, we must take $48.75 - 48 = 0.75$ of the 10 cases from the fifth class; thus

$$Q_3 = \$289.995 + \frac{0.75}{10}(\$10.00) = \$290.75$$

Hence 25% of the employees earn $268.25 or less, 50% earn $279.06 or less, and 75% earn $290.75 or less.

(*b*) The first, second, ..., ninth deciles are obtained by counting $N/10, 2N/10, \ldots, 9N/10$ of the cases, beginning with the first (lowest) class. Thus

$$D_1 = \$249.995 + \frac{6.5}{8}(\$10.00) = \$258.12 \qquad D_6 = \$279.995 + \frac{5}{14}(\$10.00) = \$283.57$$

$$D_2 = \$259.995 + \frac{5}{10}(\$10.00) = \$265.00 \qquad D_7 = \$279.995 + \frac{11.5}{14}(\$10.00) = \$288.21$$

$$D_3 = \$269.995 + \frac{1.5}{16}(\$10.00) = \$270.94 \qquad D_8 = \$289.995 + \frac{4}{10}(\$10.00) = \$294.00$$

$$D_4 = \$269.995 + \frac{8}{16}(\$10.00) = \$275.00 \qquad D_9 = \$299.995 + \frac{0.5}{5}(\$10.00) = \$301.00$$

$$D_5 = \$269.995 + \frac{14.5}{16}(\$10.00) = \$279.06$$

Hence 10% of the employees earn $258.12 or less, 20% earn $265.00 or less, ..., 90% earn $301.00 or less.

Note that the fifth decile is the median. The second, fourth, sixth, and eighth deciles, which divide the distribution into five equal parts and which are called *quintiles*, are sometimes used in practice.

3.45 Determine (*a*) the 35th and (*b*) the 60th percentiles for the distribution in Problem 3.44.

SOLUTION

(*a*) The 35th percentile, denoted by P_{35}, is obtained by counting the first $35N/100 = 35(65)/100 = 22.75$ of the cases, beginning with the first (lowest) class. Then, as in Problem 3.44,

$$P_{35} = \$269.995 + \frac{4.75}{16}(\$10.00) = \$272.97$$

This means that 35% of the employees earn $272.97 or less.

(*b*) The 60th percentile is $P_{60} = \$279.995 + \frac{5}{14}(\$10.00) = \$283.57$. Note that this is the same as the sixth decile or third quintile.

3.46 The following EXCEL worksheet is contained in A1:D26. It contains the per capita income for the 50 states. Give the EXCEL commands to find Q_1, Q_2, Q_3, and P_{95}. Also, give the states that are on both sides those quartiles or percentiles.

State	Per capita income	State	Per capita income
Wyoming	36778	Pennsylvania	34897
Montana	29387	Wisconsin	33565
North Dakota	31395	Massachusetts	44289
New Mexico	27664	Missouri	31899
West Virginia	27215	Idaho	28158
Rhode Island	36153	Kentucky	28513
Virginia	38390	Minnesota	37373
South Dakota	31614	Florida	33219
Alabama	29136	South Carolina	28352
Arkansas	26874	New York	40507
Maryland	41760	Indiana	31276
Iowa	32315	Connecticut	47819
Nebraska	33616	Ohio	32478
Hawaii	34539	New Hampshire	38408
Mississippi	25318	Texas	32462
Vermont	33327	Oregon	32103
Maine	31252	New Jersey	43771
Oklahoma	29330	California	37036
Delaware	37065	Colorado	37946
Alaska	35612	North Carolina	30553
Tennessee	31107	Illinois	36120
Kansas	32836	Michigan	33116
Arizona	30267	Washington	35409
Nevada	35883	Georgia	31121
Utah	28061	Louisiana	24820

SOLUTION

		Nearest states
=PERCENTILE(A2:D26,0.25)	$30338.5	Arizona and North Carolina
=PERCENTILE(A2:D26,0.50)	$32657	Ohio and Kansas
=PERCENTILE(A2:D26,0.75)	$36144.75	Illinois and Rhode Island
=PERCENTILE(A2:D26,0.95)	$42866.05	Maryland and New Jersey

Supplementary Problems

SUMMATION NOTATION

3.47 Write the terms in each of the following indicated sums:

(a) $\sum_{j=1}^{4}(X_j+2)$ (c) $\sum_{j=1}^{3}U_j(U_j+6)$ (e) $\sum_{j=1}^{4}4X_jY_j$

(b) $\sum_{j=1}^{5}f_jX_j^2$ (d) $\sum_{k=1}^{N}(Y_k^2-4)$

3.48 Express each of the following by using the summation notation:

(a) $(X_1+3)^3+(X_2+3)^3+(X_3+3)^3$

(b) $f_1(Y_1-a)^2+f_2(Y_2-a)^2+\cdots+f_{15}(Y_{15}-a)^2$

(c) $(2X_1-3Y_1)+(2X_2-3Y_2)+\cdots+(2X_N-3Y_N)$

(d) $(X_1/Y_1-1)^2+(X_2/Y_2-1)^2+\cdots+(X_8/Y_8-1)^2$

(e) $\dfrac{f_1a_1^2+f_2a_2^2+\cdots+f_{12}a_{12}^2}{f_1+f_2+\cdots+f_{12}}$

3.49 Prove that $\sum_{j=1}^{N}(X_j-1)^2=\sum_{j=1}^{N}X_j^2-2\sum_{j=1}^{N}X_j+N$.

3.50 Prove that $\sum(X+a)(Y+b)=\sum XY+a\sum Y+b\sum X+Nab$, where a and b are constants. What subscript notation is implied?

3.51 Two variables, U and V, assume the values $U_1=3$, $U_2=-2$, $U_3=5$, and $V_1=-4$, $V_2=-1$, $V_3=6$, respectively. Calculate (a) $\sum UV$, (b) $\sum(U+3)(V-4)$, (c) $\sum V^2$, (d) $(\sum U)(\sum V)^2$, (e) $\sum UV^2$, (f) $\sum(U^2-2V^2+2)$, and (g) $\sum(U/V)$.

3.52 Given $\sum_{j=1}^{4}X_j=7$, $\sum_{j=1}^{4}Y_j=-3$, and $\sum_{j=1}^{4}X_jY_j=5$, find (a) $\sum_{j=1}^{4}(2X_j+5Y_j)$ and (b) $\sum_{j=1}^{4}(X_j-3)(2Y_j+1)$.

THE ARITHMETIC MEAN

3.53 A student received grades of 85, 76, 93, 82, and 96 in five subjects. Determine the arithmetic mean of the grades.

3.54 The reaction times of an individual to certain stimuli were measured by a psychologist to be 0.53, 0.46, 0.50, 0.49, 0.52, 0.53, 0.44, and 0.55 seconds respectively. Determine the mean reaction time of the individual to the stimuli.

3.55 A set of numbers consists of six 6's, seven 7's, eight 8's, nine 9's and ten 10's. What is the arithmetic mean of the numbers?

3.56 A student's grades in the laboratory, lecture, and recitation parts of a physics course were 71, 78, and 89, respectively.

(a) If the weights accorded these grades are 2, 4, and 5, respectively, what is an appropriate average grade?

(b) What is the average grade if equal weights are used?

3.57 Three teachers of economics reported mean examination grades of 79, 74, and 82 in their classes, which consisted of 32, 25, and 17 students, respectively. Determine the mean grade for all the classes.

3.58 The mean annual salary paid to all employees in a company is \$36,000. The mean annual salaries paid to male and female employees of the company is \$34,000 and \$40,000 respectively. Determine the percentages of males and females employed by the company.

3.59 Table 3.8 shows the distribution of the maximum loads in short tons (1 short ton = 2000 lb) supported by certain cables produced by a company. Determine the mean maximum loading, using (*a*) the long method and (*b*) the coding method.

Table 3.8

Maximum Load (short tons)	Number of Cables
9.3–9.7	2
9.8–10.2	5
10.3–10.7	12
10.8–11.2	17
11.3–11.7	14
11.8–12.2	6
12.3–12.7	3
12.8–13.2	1
	Total 60

3.60 Find $\bar{X}$ for the data in Table 3.9, using (*a*) the long method and (*b*) the coding method.

Table 3.9

X	462	480	498	516	534	552	570	588	606	624
f	98	75	56	42	30	21	15	11	6	2

3.61 Table 3.10 shows the distribution of the diameters of the heads of rivets manufactured by a company. Compute the mean diameter.

3.62 Compute the mean for the data in Table 3.11.

Table 3.10

Diameter (cm)	Frequency
0.7247–0.7249	2
0.7250–0.7252	6
0.7253–0.7255	8
0.7256–0.7258	15
0.7259–0.7261	42
0.7262–0.7264	68
0.7265–0.7267	49
0.7268–0.7270	25
0.7271–0.7273	18
0.7274–0.7276	12
0.7277–0.7279	4
0.7280–0.7282	1
	Total 250

Table 3.11

Class	Frequency
10 to under 15	3
15 to under 20	7
20 to under 25	16
25 to under 30	12
30 to under 35	9
35 to under 40	5
40 to under 45	2
	Total 54

3.63 Compute the mean TV viewing time for the 400 junior high students per week in Problem 2.20.

3.64 (*a*) Use the frequency distribution obtained in Problem 2.27 to compute the mean diameter of the ball bearings.

(*b*) Compute the mean directly from the raw data and compare it with part (*a*), explaining any discrepancy.

THE MEDIAN

3.65 Find the mean and median of these sets of numbers: (*a*) 5, 4, 8, 3, 7, 2, 9 and (*b*) 18.3, 20.6, 19.3, 22.4, 20.2, 18.8, 19.7, 20.0.

3.66 Find the median grade of Problem 3.53.

3.67 Find the median reaction time of Problem 3.54.

3.68 Find the median of the set of numbers in Problem 3.55.

3.69 Find the median of the maximum loads of the cables in Table 3.8 of Problem 3.59.

3.70 Find the median $\tilde{X}$ for the distribution in Table 3.9 of Problem 3.60.

3.71 Find the median diameter of the rivet heads in Table 3.10 of Problem 3.61.

3.72 Find the median of the distribution in Table 3.11 of Problem 3.62.

3.73 Table 3.12 gives the number of deaths in thousands due to heart disease in 1993. Find the median age for individuals dying from heart disease in 1993.

Table 3.12

Age Group	Thousands of Deaths
Total	743.3
Under 1	0.7
1 to 4	0.3
5 to 14	0.3
15 to 24	1.0
25 to 34	3.5
35 to 44	13.1
45 to 54	32.7
55 to 64	72.0
65 to 74	158.1
75 to 84	234.0
85 and over	227.6

Source: U.S. National Center for Health Statistics, Vital Statistics of the U.S., annual.

3.74 Find the median age for the U.S. using the data for Problem 2.31.

3.75 Find the median TV viewing time for the 400 junior high students in Problem 2.20.

THE MODE

3.76 Find the mean, median, and mode for each set of numbers: (*a*) 7, 4, 10, 9, 15, 12, 7, 9, 7 and (*b*) 8, 11, 4, 3, 2, 5, 10, 6, 4, 1, 10, 8, 12, 6, 5, 7.

3.77 Find the modal grade in Problem 3.53.

3.78 Find the modal reaction time in Problem 3.54.

3.79 Find the mode of the set of numbers in Problem 3.55.

3.80 Find the mode of the maximum loads of the cables of Problem 3.59.

3.81 Find the mode $\hat{X}$ for the distribution in Table 3.9 of Problem 3.60.

3.82 Find the modal diameter of the rivet heads in Table 3.10 of Problem 3.61.

3.83 Find the mode of the distribution in Problem 3.62.

3.84 Find the modal TV viewing time for the 400 junior high students in Problem 2.20.

3.85 (*a*) What is the modal age group in Table 2.15?
(*b*) What is the modal age group in Table 3.12?

3.86 Using formulas (*9*) and (*10*) in this chapter, find the mode of the distributions given in the following Problems. Compare your answers obtained in using the two formulas.

(*a*) Problem 3.59 (*b*) Problem 3.61 (*c*) Problem 3.62 (*d*) Problem 2.20.

3.87 A continuous random variable has probability associated with it that is described by the following probability density function. $f(x) = -0.75x^2 + 1.5x$ for $0 < x < 2$ and $f(x) = 0$ for all other x values. The mode occurs where the function attains a maximum. Use your knowledge of quadratic functions to show that the mode occurs when $x = 1$.

THE GEOMETRIC MEAN

3.88 Find the geometric mean of the numbers (*a*) 4.2 and 16.8 and (*b*) 3.00 and 6.00.

3.89 Find (*a*) the geometric mean G and (*b*) the arithmetic mean $\bar{X}$ of the set 2, 4, 8, 16, 32.

3.90 Find the geometric mean of the sets (*a*) 3, 5, 8, 3, 7, 2 and (*b*) 28.5, 73.6, 47.2, 31.5, 64.8.

3.91 Find the geometric mean for the distributions in (*a*) Problem 3.59 and (*b*) Problem 3.60. Verify that the geometric mean is less than or equal to the arithmetic mean for these cases.

3.92 If the price of a commodity doubles in a period of 4 years, what is the average percentage increase per year?

3.93 In 1980 and 1996 the population of the United States was 226.5 million and 266.0 million, respectively. Using the formula given in Problem 3.38, answer the following:

(*a*) What was the average percentage increase per year?

(*b*) Estimate the population in 1985.

(*c*) If the average percentage increase of population per year from 1996 to 2000 is the same as in part (*a*), what would the population be in 2000?

3.94 A principal of \$1000 is invested at an 8% annual rate of interest. What will the total amount be after 6 years if the original principal is not withdrawn?

3.95 If in Problem 3.94 the interest is compounded quarterly (i.e., there is a 2% increase in the money every 3 months), what will the total amount be after 6 years?

3.96 Find two numbers whose arithmetic mean is 9.0 and whose geometric mean is 7.2.

THE HARMONIC MEAN

3.97 Find the harmonic mean of the numbers (*a*) 2, 3, and 6 and (*b*) 3.2, 5.2, 4.8, 6.1, and 4.2.

3.98 Find the (*a*) arithmetic mean, (*b*) geometric mean, and (*c*) harmonic mean of the numbers 0, 2, 4, and 6.

3.99 If $X_1, X_2, X_3, \ldots$ represent the class marks in a frequency distribution with corresponding class frequencies $f_1, f_2, f_3, \ldots$, respectively, prove that the harmonic mean H of the distribution is given by

$$\frac{1}{H} = \frac{1}{N}\left(\frac{f_1}{X_1} + \frac{f_2}{X_2} + \frac{f_3}{X_3} + \cdots\right) = \frac{1}{N}\sum \frac{f}{X}$$

where $N = f_1 + f_2 + \cdots = \sum f$.

3.100 Use Problem 3.99 to find the harmonic mean of the distributions in (*a*) Problem 3.59 and (*b*) Problem 3.60. Compare with Problem 3.91.

3.101 Cities A, B, and C are equidistant from each other. A motorist travels from A to B at 30 mi/h, from B to C at 40 mi/h, and from C to A at 50 mi/h. Determine his average speed for the entire trip.

3.102 (*a*) An airplane travels distances of d_1, d_2, and d_3 miles at speeds v_1, v_2, and v_3 mi/h, respectively. Show that the average speed is given by V, where

$$\frac{d_1 + d_2 + d_3}{V} = \frac{d_1}{v_1} + \frac{d_2}{v_2} + \frac{d_3}{v_3}$$

This is a weighted harmonic mean.

(*b*) Find V if $d_1 = 2500$, $d_2 = 1200$, $d_3 = 500$, $v_1 = 500$, $v_2 = 400$, and $v_3 = 250$.

3.103 Prove that the geometric mean of two positive numbers a and b is (*a*) less than or equal to the arithmetic mean and (*b*) greater than or equal to the harmonic mean of the numbers. Can you extend the proof to more than two numbers?

THE ROOT MEAN SQUARE, OR QUADRATIC MEAN

3.104 Find the RMS (or quadratic mean) of the numbers (*a*) 11, 23, and 35 and (*b*) 2.7, 3.8, 3.2, and 4.3.

3.105 Show that the RMS of two positive numbers, *a* and *b*, is (*a*) greater than or equal to the arithmetic mean and (*b*) greater than or equal to the harmonic mean. Can you extend the proof to more than two numbers?

3.106 Derive a formula that can be used to find the RMS for grouped data and apply it to one of the frequency distributions already considered.

QUARTILES, DECILES, AND PERCENTILES

3.107 Table 3.13 shows a frequency distribution of grades on a final examination in college algebra. (*a*) Find the quartiles of the distribution, and (*b*) interpret clearly the significance of each.

Table 3.13

Grade	Number of Students
90–100	9
80–89	32
70–79	43
60–69	21
50–59	11
40–49	3
30–39	1
	Total 120

3.108 Find the quartiles Q_1, Q_2, and Q_3 for the distributions in (*a*) Problem 3.59 and (*b*) Problem 3.60. Interpret clearly the significance of each.

3.109 Give six different statistical terms for the balance point or central value of a bell-shaped frequency curve.

3.110 Find (*a*) P_{10}, (*b*) P_{90}, (*c*) P_{25}, and (*d*) P_{75} for the data of Problem 3.59, interpreting clearly the significance of each.

3.111 (*a*) Can all quartiles and deciles be expressed as percentiles? Explain.

(*b*) Can all quantiles be expressed as percentiles? Explain.

3.112 For the data of Problem 3.107, determine (*a*) the lowest grade scored by the top 25% of the class and (*b*) the highest grade scored by the lowest 20% of the class. Interpret your answers in terms of percentiles.

3.113 Interpret the results of Problem 3.107 graphically by using (*a*) a percentage histogram, (*b*) a percentage frequency polygon, and (*c*) a percentage ogive.

3.114 Answer Problem 3.113 for the results of Problem 3.108.

3.115 (*a*) Develop a formula, similar to that of equation (*8*) in this chapter, for computing any percentile from a frequency distribution.

(*b*) Illustrate the use of the formula by applying it to obtain the results of Problem 3.110.

The Standard Deviation and Other Measures of Dispersion

DISPERSION, OR VARIATION

The degree to which numerical data tend to spread about an average value is called the *dispersion*, or *variation*, of the data. Various measures of this dispersion (or variation) are available, the most common being the range, mean deviation, semi-interquartile range, 10–90 percentile range, and standard deviation.

THE RANGE

The *range* of a set of numbers is the difference between the largest and smallest numbers in the set.

EXAMPLE 1. The range of the set 2, 3, 3, 5, 5, 5, 8, 10, 12 is 12 − 2 = 10. Sometimes the range is given by simply quoting the smallest and largest numbers; in the above set, for instance, the range could be indicated as 2 to 12, or 2–12.

THE MEAN DEVIATION

The *mean deviation*, or *average deviation*, of a set of N numbers $X_1, X_2, \ldots, X_N$ is abbreviated MD and is defined by

$$\text{Mean deviation (MD)} = \frac{\sum_{j=1}^{N} |X_j - \bar{X}|}{N} = \frac{\sum |X - \bar{X}|}{N} = \overline{|X - \bar{X}|} \qquad (1)$$

where $\bar{X}$ is the arithmetic mean of the numbers and $|X_j - \bar{X}|$ is the absolute value of the deviation of X_j from $\bar{X}$. (The *absolute value* of a number is the number without the associated sign and is indicated by two vertical lines placed around the number; thus $|-4| = 4$, $|+3| = 3$, $|6| = 6$, and $|-0.84| = 0.84$.)

EXAMPLE 2. Find the mean deviation of the set 2, 3, 6, 8, 11.

$$\text{Arithmetic mean } (\bar{X}) = \frac{2+3+6+8+11}{5} = 6$$

$$\text{MD} = \frac{|2-6|+|3-6|+|6-6|+|8-6|+|11-6|}{5} = \frac{|-4|+|-3|+|0|+|2|+|5|}{5} = \frac{4+3+0+2+5}{5} = 2.8$$

If $X_1, X_2, \ldots, X_K$ occur with frequencies $f_1, f_2, \ldots, f_K$, respectively, the mean deviation can be written as

$$\text{MD} = \frac{\sum_{j=1}^{K} f_j |X_j - \bar{X}|}{N} = \frac{\sum f |X - \bar{X}|}{N} = \overline{|X - \bar{X}|} \tag{2}$$

where $N = \sum_{j=1}^{K} f_j = \sum f$. This form is useful for grouped data, where the X_j's represent class marks and the f_j's are the corresponding class frequencies.

Occasionally the mean deviation is defined in terms of absolute deviations from the median or other average instead of from the mean. An interesting property of the sum $\sum_{j=1}^{N} |X_j - a|$ is that it is a minimum when a is the median (i.e., the mean deviation about the median is a minimum).

Note that it would be more appropriate to use the terminology *mean absolute deviation* than *mean deviation*.

THE SEMI-INTERQUARTILE RANGE

The *semi-interquartile range*, or *quartile deviation*, of a set of data is denoted by Q and is defined by

$$Q = \frac{Q_3 - Q_1}{2} \tag{3}$$

where Q_1 and Q_3 are the first and third quartiles for the data (see Problems 4.6 and 4.7). The interquartile range $Q_3 - Q_1$ is sometimes used, but the semi-interquartile range is more common as a measure of dispersion.

THE 10–90 PERCENTILE RANGE

The *10–90 percentile range* of a set of data is defined by

$$\text{10–90 percentile range} = P_{90} - P_{10} \tag{4}$$

where P_{10} and P_{90} are the 10th and 90th percentiles for the data (see Problem 4.8). The semi-10–90 percentile range, $\frac{1}{2}(P_{90} - P_{10})$, can also be used but is not commonly employed.

THE STANDARD DEVIATION

The *standard deviation* of a set of N numbers $X_1, X_2, \ldots, X_N$ is denoted by s and is defined by

$$s = \sqrt{\frac{\sum_{j=1}^{N} (X_j - \bar{X})^2}{N}} = \sqrt{\frac{\sum (X - \bar{X})^2}{N}} = \sqrt{\frac{\sum x^2}{N}} = \sqrt{\overline{(X - \bar{X})^2}} \tag{5}$$

where x represents the deviations of each of the numbers X_j from the mean $\bar{X}$. Thus s is the root mean square (RMS) of the deviations from the mean, or, as it is sometimes called, the *root-mean-square deviation*.

If $X_1, X_2, \ldots, X_K$ occur with frequencies $f_1, f_2, \ldots, f_K$, respectively, the standard deviation can be written

$$s = \sqrt{\frac{\sum_{j=1}^{K} f_j(X_j - \bar{X})^2}{N}} = \sqrt{\frac{\sum f(X - \bar{X})^2}{N}} = \sqrt{\frac{\sum fx^2}{N}} = \sqrt{\overline{(X - \bar{X})^2}} \tag{6}$$

where $N = \sum_{j=1}^{K} f_j = \sum f$. In this form it is useful for grouped data.

Sometimes the standard deviation of a sample's data is defined with $(N - 1)$ replacing N in the denominators of the expressions in equations (5) and (6) because the resulting value represents a better estimate of the standard deviation of a population from which the sample is taken. For large values of N (certainly $N > 30$), there is practically no difference betwen the two definitions. Also, when the better estimate is needed we can always obtain it by multiplying the standard deviation computed according to the first definition by $\sqrt{N/(N-1)}$. Hence we shall adhere to the form (5) and (6).

THE VARIANCE

The *variance* of a set of data is defined as the square of the standard deviation and is thus given by s^2 in equations (5) and (6).

When it is necessary to distinguish the standard deviation of a population from the standard deviation of a sample drawn from this population, we often use the symbol s for the latter and σ (lowercase Greek *sigma*) for the former. Thus s^2 and σ^2 would represent the *sample variance* and *population variance*, respectively.

SHORT METHODS FOR COMPUTING THE STANDARD DEVIATION

Equations (5) and (6) can be written, respectively, in the equivalent forms

$$s = \sqrt{\frac{\sum_{j=1}^{N} X_j^2}{N} - \left(\frac{\sum_{j=1}^{N} X_j}{N}\right)^2} = \sqrt{\frac{\sum X^2}{N} - \left(\frac{\sum X}{N}\right)^2} = \sqrt{\overline{X^2} - \bar{X}^2} \tag{7}$$

$$s = \sqrt{\frac{\sum_{j=1}^{K} f_j X_j^2}{N} - \left(\frac{\sum_{j=1}^{K} f_j X_j}{N}\right)^2} = \sqrt{\frac{\sum fX^2}{N} - \left(\frac{\sum fX}{N}\right)^2} = \sqrt{\overline{X^2} - \bar{X}^2} \tag{8}$$

where $\overline{X^2}$ denotes the mean of the squares of the various values of X, while $\bar{X}^2$ denotes the square of the mean of the various values of X (see Problems 4.12 to 4.14).

If $d_j = X_j - A$ are the deviations of X_j from some arbitrary constant A, results (7) and (8) become, respectively,

$$s = \sqrt{\frac{\sum_{j=1}^{N} d_j^2}{N} - \left(\frac{\sum_{j=1}^{N} d_j}{N}\right)^2} = \sqrt{\frac{\sum d^2}{N} - \left(\frac{\sum d}{N}\right)^2} = \sqrt{\overline{d^2} - \bar{d}^2} \tag{9}$$

$$s = \sqrt{\frac{\sum_{j=1}^{K} f_j d_j^2}{N} - \left(\frac{\sum_{j=1}^{K} f_j d_j}{N}\right)^2} = \sqrt{\frac{\sum fd^2}{N} - \left(\frac{\sum fd}{N}\right)^2} = \sqrt{\overline{d^2} - \bar{d}^2} \tag{10}$$

(See Problems 4.15 and 4.17.)

When data are grouped into a frequency distribution whose class intervals have equal size c, we have $d_j = cu_j$ or $X_j = A + cu_j$ and result *(10)* becomes

$$s = c\sqrt{\frac{\sum_{j=1}^{K} f_j u_j^2}{N} - \left(\frac{\sum_{j=1}^{K} f_j u_j}{N}\right)^2} = c\sqrt{\frac{\sum fu^2}{N} - \left(\frac{\sum fu}{N}\right)^2} = c\sqrt{\overline{u^2} - \bar{u}^2} \qquad (11)$$

This last formula provides a very short method for computing the standard deviation and should always be used for grouped data when the class-interval sizes are equal. It is called the *coding method* and is exactly analogous to that used in Chapter 3 for computing the arithmetic mean of grouped data. (See Problems 4.16 to 4.19.)

PROPERTIES OF THE STANDARD DEVIATION

1. The standard deviation can be defined as

$$s = \sqrt{\frac{\sum_{j=1}^{N} (X_j - a)^2}{N}}$$

where a is an average besides the arithmetic mean. Of all such standard deviations, the minimum is that for which $a = \bar{X}$, because of Property 2 in Chapter 3. This property provides an important reason for defining the standard deviation as above. For a proof of this property, see Problem 4.27.

2. For normal distributions (see Chapter 7), it turns out that (as shown in Fig. 4-1):

   (*a*) 68.27% of the cases are included between $\bar{X} - s$ and $\bar{X} + s$ (i.e., one standard deviation on either side of the mean).

   (*b*) 95.45% of the cases are included between $\bar{X} - 2s$ and $\bar{X} + 2s$ (i.e., two standard deviations on either side of the mean).

   (*c*) 99.73% of the cases are included between $\bar{X} - 3s$ and $\bar{X} + 3s$ (i.e., three standard deviations on either side of the mean).

   For moderately skewed distributions, the above percentages may hold approximately (see Problem 4.24).

3. Suppose that two sets consisting of N_1 and N_2 numbers (or two frequency distributions with total frequencies N_1 and N_2) have variances given by s_1^2 and s_2^2, respectively, and have the *same* mean $\bar{X}$. Then the *combined*, or *pooled*, *variance* of both sets (or both frequency distributions) is given by

$$s^2 = \frac{N_1 s_1^2 + N_2 s_2^2}{N_1 + N_2} \qquad (12)$$

   Note that this is a weighted arithmetic mean of the variances. This result can be generalized to three or more sets.

4. Chebyshev's theorem states that for $k > 1$, there is at least $(1 - (1/k^2)) \times 100\%$ of the probability distribution for any variable within k standard deviations of the mean. In particular, when $k = 2$, there is at least $(1 - (1/2^2)) \times 100\%$ or 75% of the data in the interval $(\bar{x} - 2S, \bar{x} + 2S)$, when $k = 3$ there is at least $(1 - (1/3^2)) \times 100\%$ or 89% of the data in the interval $(\bar{x} - 3S, \bar{x} + 3S)$, and when $k = 4$ there is at least $(1 - (1/4^2)) \times 100\%$ or 93.75% of the data in the interval $(\bar{x} - 4S, \bar{x} + 4S)$.

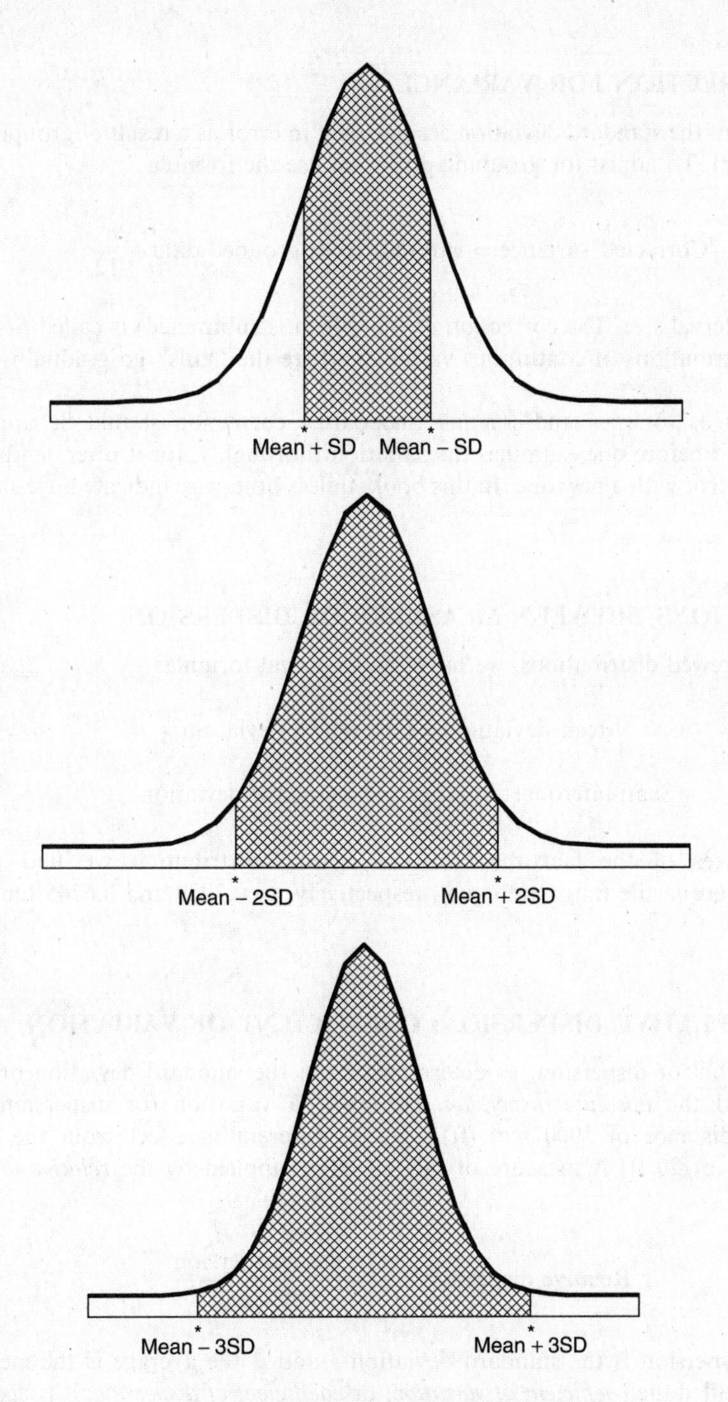

Fig. 4-1 Illustration of the empirical rule.

CHARLIER'S CHECK

Charlier's check in computations of the mean and standard deviation by the coding method makes use of the identities

$$\sum f(u+1) = \sum fu + \sum f = \sum fu + N$$

$$\sum f(u+1)^2 = \sum f(u^2+2u+1) = \sum fu^2 + 2\sum fu + \sum f = \sum fu^2 + 2\sum fu + N$$

(See Problem 4.20.)

SHEPPARD'S CORRECTION FOR VARIANCE

The computation of the standard deviation is somewhat in error as a result of grouping the data into classes (grouping error). To adjust for grouping error, we use the formula

$$\text{Corrected variance} = \text{variance from grouped data} - \frac{c^2}{12} \tag{13}$$

where c is the class-interval size. The correction $c^2/12$ (which is subtracted) is called *Sheppard's correction.* It is used for distributions of continuous variables where the "tails" go gradually to zero in both directions.

Statisticians differ as to *when* and *whether* Sheppard's correction should be applied. It should certainly not be applied before one examines the situation thoroughly, for it often tends to *overcorrect*, thus replacing an old error with a new one. In this book, unless otherwise indicated, we shall not be using Sheppard's correction.

EMPIRICAL RELATIONS BETWEEN MEASURES OF DISPERSION

For moderately skewed distributions, we have the empirical formulas

$$\text{Mean deviation} = \tfrac{4}{5}(\text{standard deviation})$$

$$\text{Semi-interquartile range} = \tfrac{2}{3}(\text{standard deviation})$$

These are consequences of the fact that for the normal distribution we find that the mean deviation and semi-interquartile range are equal, respectively, to 0.7979 and 0.6745 times the standard deviation.

ABSOLUTE AND RELATIVE DISPERSION; COEFFICIENT OF VARIATION

The actual variation, or dispersion, as determined from the standard deviation or other measure of dispersion is called the *absolute dispersion.* However, a variation (or dispersion) of 10 inches (in) in measuring a distance of 1000 feet (ft) is quite different in effect from the same variation of 10 in in a distance of 20 ft. A measure of this effect is supplied by the *relative dispersion*, which is defined by

$$\text{Relative dispersion} = \frac{\text{absolute dispersion}}{\text{average}} \tag{14}$$

If the absolute dispersion is the standard deviation s and if the average is the mean $\bar{X}$, then the relative dispersion is called the *coefficient of variation*, or *coefficient of dispersion*; it is denoted by V and is given by

$$\text{Coefficient of variation } (V) = \frac{s}{\bar{X}} \tag{15}$$

and is generally expressed as a percentage. Other possibilities also occur (see Problem 4.30).

Note that the coefficient of variation is independent of the units used. For this reason, it is useful in comparing distributions where the units may be different. A disadvantage of the coefficient of variation is that it fails to be useful when $\bar{X}$ is close to zero.

STANDARDIZED VARIABLE; STANDARD SCORES

The variable that measures the deviation from the mean in units of the standard deviation is called a *standardized variable*, is a dimensionless quantity (i.e., is independent of the units used), and is given by

$$z = \frac{X - \bar{X}}{s} \tag{16}$$

If the deviations from the mean are given in units of the standard deviation, they are said to be expressed in *standard units*, or *standard scores*. These are of great value in the comparison of distributions (see Problem 4.31).

SOFTWARE AND MEASURES OF DISPERSION

The statistical software gives a variety of measures for dispersion. The dispersion measures are usually given in descriptive statistics. EXCEL allows for the computation of all the measures discussed in this book. MINITAB and EXCEL are discussed here and outputs of other packages are given in the solved problems.

EXAMPLE 3.

(*a*) EXCEL provides calculations for several measures of dispersion. The following example illustrates several. A survey was taken at a large company and the question asked was how many e-mails do you send per week? The results for 75 employees are shown in A1:E15 of an EXCEL worksheet.

32	113	70	60	84
114	31	58	86	102
113	79	86	24	40
44	42	54	71	25
42	116	68	30	63
121	74	77	77	100
51	31	61	28	26
47	54	74	57	35
77	80	125	105	61
102	45	115	36	52
58	24	24	39	40
95	99	54	35	31
77	29	69	58	32
49	118	44	95	65
71	65	74	122	99

The range is given by =MAX(A1:E15)-MIN(A1:E15) or $125 - 24 = 101$. The mean deviation or average deviation is given by =AVEDEV(A1:E15) or 24.42. The semi-interquartile range is given by the expression =(PERCENTILE(A1:E15,0.75)-PERCENTILE(A1:E15,0.25))/2 or 22. The 10–90 percentile range is given by PERCENTILE(A1:E15,0.9)-PERCENTILE (A1:E15,0.1) or 82.6.

The standard deviation and variance is given by =STDEV(A1:E15) or 29.2563, and =VAR(A1:E15) or 855.932 for samples and =STDEVP(A1:E15) or 29.0606 and =VARP(A1:E15) or 844.52 for populations.

(*b*)

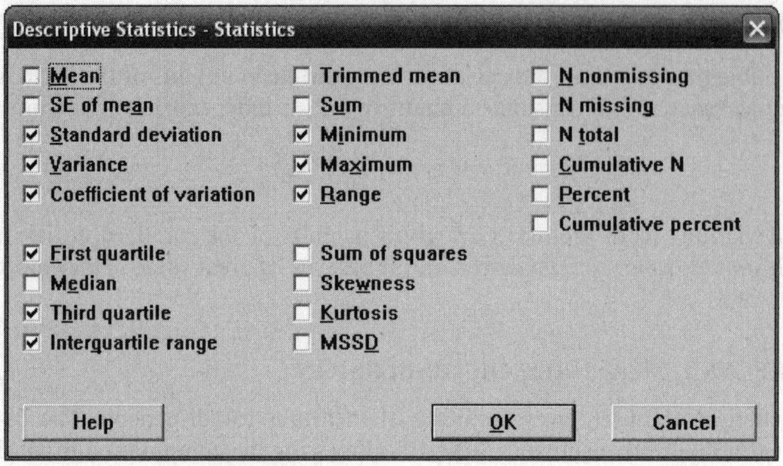

Descriptive Statistics - Statistics

☐ Mean ☐ Trimmed mean ☐ N nonmissing
☐ SE of mean ☐ Sum ☐ N missing
☑ Standard deviation ☑ Minimum ☐ N total
☑ Variance ☑ Maximum ☐ Cumulative N
☑ Coefficient of variation ☑ Range ☐ Percent
☐ Cumulative percent
☑ First quartile ☐ Sum of squares
☐ Median ☐ Skewness
☑ Third quartile ☐ Kurtosis
☑ Interquartile range ☐ MSSD

Help OK Cancel

Fig. 4-2 Dialog box for MINITAB.

The dialog box in Fig. 4-2 shows MINITAB choices for measures of dispersion and measures of central tendency. Output is as follows:

Descriptive Statistics: e-mails

Variable	StDev	Variance	CoefVar	Minimum	Q1	Q3	Maximum	Range	IQR
e-mails	29.26	855.93	44.56	24.00	40.00	86.00	125.00	101.00	46.00

Solved Problems

THE RANGE

4.1 Find the range of the sets (*a*) 12, 6, 7, 3, 15, 10, 18, 5 and (*b*) 9, 3, 8, 8, 9, 8, 9, 18.

SOLUTION

In both cases, range = largest number − smallest number = 18 − 3 = 15. However, as seen from the arrays of sets (*a*) and (*b*),

(*a*) 3, 5, 6, 7, 10, 12, 15, 18 (*b*) 3, 8, 8, 8, 9, 9, 9, 18

there is much more variation, or dispersion, in (*a*) than in (*b*). In fact, (*b*) consists mainly of 8's and 9's.

Since the range indicates no difference between the sets, it is not a very good measure of dispersion in this case. Where extreme values are present, the range is generally a poor measure of dispersion.

An improvement is achieved by throwing out the extreme cases, 3 and 18. Then for set (*a*) the range is (15 − 5) = 10, while for set (*b*) the range is (9 − 8) = 1, clearly showing that (*a*) has greater dispersion than (*b*). However, this is not the way the range is defined. The semi-interquartile range and the 10–90 percentile range were designed to improve on the range by eliminating extreme cases.

4.2 Find the range of heights of the students at XYZ University as given in Table 2.1.

SOLUTION

There are two ways of defining the range for grouped data.

First method

$$\text{Range} = \text{class mark of highest class} - \text{class mark of lowest class}$$
$$= 73 - 61 = 12 \text{ in}$$

Second method

$$\text{Range} = \text{upper class boundary of highest class} - \text{lower class boundary of lowest class}$$
$$= 74.5 - 59.5 = 15 \text{ in}$$

The first method tends to eliminate extreme cases to some extent.

THE MEAN DEVIATION

4.3 Find the mean deviation of the sets of numbers in Problem 4.1.

SOLUTION

(*a*) The arithmetic mean is

$$\bar{X} = \frac{12 + 6 + 7 + 3 + 15 + 10 + 18 + 5}{8} = \frac{76}{8} = 9.5$$

The mean deviation is

$$\text{MD} = \frac{\sum |X - \bar{X}|}{N}$$
$$= \frac{|12 - 9.5| + |6 - 9.5| + |7 - 9.5| + |3 - 9.5| + |15 - 9.5| + |10 - 9.5| + |18 - 9.5| + |5 - 9.5|}{8}$$
$$= \frac{2.5 + 3.5 + 2.5 + 6.5 + 5.5 + 0.5 + 8.5 + 4.5}{8} = \frac{34}{8} = 4.25$$

(*b*)
$$\bar{X} = \frac{9 + 3 + 8 + 8 + 9 + 8 + 9 + 18}{8} = \frac{72}{8} = 9$$

$$\text{MD} = \frac{\sum |X - \bar{X}|}{N}$$
$$= \frac{|9 - 9| + |3 - 9| + |8 - 9| + |8 - 9| + |9 - 9| + |8 - 9| + |9 - 9| + |18 - 9|}{8}$$
$$= \frac{0 + 6 + 1 + 1 + 0 + 1 + 0 + 9}{8} = 2.25$$

The mean deviation indicates that set (*b*) shows less dispersion than set (*a*), as it should.

4.4 Find the mean deviation of the heights of the 100 male students at XYZ University (see Table 3.2 of Problem 3.20).

SOLUTION

From Problem 3.20, $\bar{X} = 67.45$ in. The work can be arranged as in Table 4.1. It is also possible to devise a coding method for computing the mean deviation (see Problem 4.47).

Table 4.1

Height (in)	Class Mark (X)	$\lvert X-\bar{X}\rvert = \lvert X-67.45\rvert$	Frequency (f)	$f\lvert X-\bar{X}\rvert$
60–62	61	6.45	5	32.25
63–65	64	3.45	18	62.10
66–68	67	0.45	42	18.90
69–71	70	2.55	27	68.85
72–74	73	5.55	8	44.40
			$N=\sum f=100$	$\sum f\lvert X-\bar{X}\rvert = 226.50$

$$\text{MD} = \frac{\sum f|X-\bar{X}|}{N} = \frac{226.50}{100} = 2.26 \text{ in}$$

4.5 Determine the percentage of the students' heights in Problem 4.4 that fall within the ranges (*a*) $\bar{X} \pm \text{MD}$, (*b*) $\bar{X} \pm 2\,\text{MD}$, and (*c*) $\bar{X} \pm 3\,\text{MD}$.

SOLUTION

(*a*) The range from 65.19 to 69.71 in is $\bar{X} \pm \text{MD} = 67.45 \pm 2.26$. This range includes all individuals in the third class $+\frac{1}{3}(65.5 - 65.19)$ of the students in the second class $+\frac{1}{3}(69.71 - 68.5)$ of the students in the fourth class (since the class-interval size is 3 in, the upper class boundary of the second class is 65.5 in, and the lower class boundary of the fourth class is 68.5 in). The number of students in the range $\bar{X} \pm \text{MD}$ is

$$42 + \frac{0.31}{3}(18) + \frac{1.21}{3}(27) = 42 + 1.86 + 10.89 = 54.75 \qquad \text{or} \qquad 55$$

which is 55% of the total.

(*b*) The range from 62.93 to 71.97 in is $\bar{X} \pm 2\,\text{MD} = 67.45 \pm 2(2.26) = 67.45 \pm 4.52$. The number of students in the range $\bar{X} \pm 2\,\text{MD}$ is

$$18 - \left(\frac{62.93 - 62.5}{3}\right)(18) + 42 + 27 + \left(\frac{71.97 - 71.5}{3}\right)(8) = 85.67 \qquad \text{or} \qquad 86$$

which is 86% of the total.

(*c*) The range from 60.67 to 74.23 in is $\bar{X} \pm 3\,\text{MD} = 67.45 \pm 3(2.26) = 67.45 \pm 6.78$. The number of students in the range $\bar{X} \pm 3\,\text{MD}$ is

$$5 - \left(\frac{60.67 - 59.5}{3}\right)(5) + 18 + 42 + 27 + \left(\frac{74.5 - 74.23}{3}\right)(8) = 97.33 \qquad \text{or} \qquad 97$$

which is 97% of the total.

THE SEMI-INTERQUARTILE RANGE

4.6 Find the semi-interquartile range for the height distribution of the students at XYZ University (see Table 4.1 of Problem 4.4).

SOLUTION

The lower and upper quartiles are $Q_1 = 65.5 + \frac{2}{42}(3) = 65.64$ in and $Q_3 = 68.5 + \frac{10}{27}(3) = 69.61$ in, respectively, and the semi-interquartile range (or quartile deviation) is $Q = \frac{1}{2}(Q_3 - Q_1) = \frac{1}{2}(69.61 - 65.64) = 1.98$ in. Note that 50% of the cases lie between Q_1 and Q_3 (i.e., 50 students have heights between 65.64 and 69.61 in).

We can consider $\frac{1}{2}(Q_1 + Q_3) = 67.63$ in to be a measure of central tendency (i.e., average height). It follows that 50% of the heights lie in the range 67.63 ± 1.98 in.

4.7 Find the semi-interquartile range for the wages of the 65 employees at the P&R Company (see Table 2.5 of Problem 2.3).

SOLUTION

From Problem 3.44, $Q_1 = \$268.25$ and $Q_3 = \$290.75$. Thus the semi-interquartile range $Q = \frac{1}{2}(Q_3 - Q_1) = \frac{1}{2}(\$290.75 - \$268.25) = \$11.25$. Since $\frac{1}{2}(Q_1 + Q_3) = \$279.50$, we can conclude that 50% of the employees earn wages lying in the range $\$279.50 \pm \$11.25$.

THE 10–90 PERCENTILE RANGE

4.8 Find the 10–90 percentile range of the heights of the students at XYZ University (see Table 2.1).

SOLUTION

Here $P_{10} = 62.5 + \frac{5}{18}(3) = 63.33$ in, and $P_{90} = 68.5 + \frac{25}{27}(3) = 71.27$ in. Thus the 10–90 percentile range is $P_{90} - P_{10} = 71.27 - 63.33 = 7.94$ in. Since $\frac{1}{2}(P_{10} + P_{90}) = 67.30$ in and $\frac{1}{2}(P_{90} - P_{10}) = 3.97$ in, we can conclude that 80% of the students have heights in the range 67.30 ± 3.97 in.

THE STANDARD DEVIATION

4.9 Find the standard deviation s of each set of numbers in Problem 4.1.

SOLUTION

(*a*)

$$\bar{X} = \frac{\sum X}{N} = \frac{12 + 6 + 7 + 3 + 15 + 10 + 18 + 5}{8} = \frac{76}{8} = 9.5$$

$$s = \sqrt{\frac{\sum(X - \bar{X})^2}{N}}$$

$$= \sqrt{\frac{(12-9.5)^2+(6-9.5)^2+(7-9.5)^2+(3-9.5)^2+(15-9.5)^2+(10-9.5)^2+(18-9.5)^2+(5-9.5)^2}{8}}$$

$$= \sqrt{23.75} = 4.87$$

(*b*)

$$\bar{X} = \frac{9 + 3 + 8 + 8 + 9 + 8 + 9 + 18}{8} = \frac{72}{8} = 9$$

$$s = \sqrt{\frac{\sum(X - \bar{X})^2}{N}}$$

$$= \sqrt{\frac{(9-9)^2 + (3-9)^2 + (8-9)^2 + (8-9)^2 + (9-9)^2 + (8-9)^2 + (9-9)^2 + (18-9)^2}{8}}$$

$$= \sqrt{15} = 3.87$$

The above results should be compared with those of Problem 4.3. It will be noted that the standard deviation does indicate that set (*b*) shows less dispersion than set (*a*). However, the effect is masked by the fact that extreme values affect the standard deviation much more than they affect the mean deviation. This is to be expected, of course, since the deviations are squared in computing the standard deviation.

4.10 The standard deviations of the two data sets given in Problem 4.1 were found using MINITAB and the results are shown below. Compare the answers with those obtained in Problem 4.9.

```
MTB > print c1
set1
      12     6     7     3    15    10    18     5
MTB > print c2
set2
       9     3     8     8     9     8     9    18
MTB > standard deviation c1
```

Column Standard Deviation

```
    Standard deviation of set1 = 5.21
MTB > standard deviation c2
```

Column Standard Deviation

```
    Standard deviation of set2 = 4.14
```

SOLUTION

The MINITAB package uses the formula

$$s = \sqrt{\frac{\sum(X - \bar{X})^2}{N - 1}}$$

and therefore the standard deviations are not the same in Problems 4.9 and 4.10. The answers in Problem 4.10 are obtainable from those in Problem 4.9 if we multiply those in Problem 4.9 by $\sqrt{N/(N-1)}$. Since $N = 8$ for both sets $\sqrt{N/(N-1)} = 1.069045$, and for set 1, we have $(1.069045)(4.87) = 5.21$, the standard deviation given by MINITAB. Similarly, $(1.069045)(3.87) = 4.14$, the standard deviation given for set 2 by MINITAB.

4.11 Find the standard deviation of the heights of the 100 male students at XYZ University (see Table 2.1).

SOLUTION

From Problem 3.15, 3.20, or 3.22, $\bar{X} = 67.45$ in. The work can be arranged as in Table 4.2.

$$s = \sqrt{\frac{\sum f(X - \bar{X})^2}{N}} = \sqrt{\frac{852.7500}{100}} = \sqrt{8.5275} = 2.92 \text{ in}$$

Table 4.2

Height (in)	Class Mark (X)	$X - \bar{X} = X - 67.45$	$(X - \bar{X})^2$	Frequency (f)	$f(X - \bar{X})^2$
60–62	61	−6.45	41.6025	5	208.0125
63–65	64	−3.45	11.9025	18	214.2450
66–68	67	−0.45	0.2025	42	8.5050
69–71	70	2.55	6.5025	27	175.5675
72–74	73	5.55	30.8025	8	246.4200
				$N = \sum f = 100$	$\sum f(X - \bar{X})^2 = 852.7500$

COMPUTING THE STANDARD DEVIATIONS FROM GROUPED DATA

4.12 (*a*) Prove that

$$s = \sqrt{\frac{\sum X^2}{N} - \left(\frac{\sum X}{N}\right)^2} = \sqrt{\overline{X^2} - \bar{X}^2}$$

(*b*) Use the formula in part (*a*) to find the standard deviation of the set 12, 6, 7, 3, 15, 10, 18, 5.

SOLUTION

(*a*) By definition,

$$s = \sqrt{\frac{\sum (X - \bar{X})^2}{N}}$$

Then

$$s^2 = \frac{\sum (X - \bar{X})^2}{N} = \frac{\sum (X^2 - 2\bar{X}X + \bar{X}^2)}{N} = \frac{\sum X^2 - 2\bar{X}\sum X + N\bar{X}^2}{N}$$

$$= \frac{\sum X^2}{N} - 2\bar{X}\frac{\sum X}{N} + \bar{X}^2 = \frac{\sum X^2}{N} - 2\bar{X}^2 + \bar{X}^2 = \frac{\sum X^2}{N} - \bar{X}^2$$

$$= \overline{X^2} = \bar{X}^2 = \frac{\sum X^2}{N} - \left(\frac{\sum X}{N}\right)^2$$

or

$$s = \sqrt{\frac{\sum X^2}{N} - \left(\frac{\sum X}{N}\right)^2} = \sqrt{\overline{X^2} - \bar{X}^2}$$

Note that in the above summations we have used the abbreviated form, with X replacing X_j and with $\sum$ replacing $\sum_{j=1}^{N}$.

Another method

$$s^2 = \overline{(X - \bar{X})^2} = \overline{X^2 - 2X\bar{X} + \bar{X}^2} = \overline{X^2} - \overline{2X\bar{X}} + \overline{\bar{X}^2} = \overline{X^2} - 2\bar{X}\bar{X} + \bar{X}^2 = \overline{X^2} - \bar{X}^2$$

(*b*)

$$\overline{X^2} = \frac{\sum X^2}{N} = \frac{(12)^2 + (6)^2 + (7)^2 + (3)^2 + (15)^2 + (10)^2 + (18)^2 + (5)^2}{8} = \frac{912}{8} = 114$$

$$\bar{X} = \frac{\sum X}{N} = \frac{12 + 6 + 7 + 3 + 15 + 10 + 18 + 5}{8} = \frac{76}{8} = 9.5$$

Thus

$$s = \sqrt{\overline{X^2} - \bar{X}^2} = \sqrt{114 - 90.25} = \sqrt{23.75} = 4.87$$

This method should be compared with that of Problem 4.9(*a*).

4.13 Modify the formula of Problem 4.12(*a*) to allow for frequencies corresponding to the various values of X.

SOLUTION

The appropriate modification is

$$s = \sqrt{\frac{\sum fX^2}{N} - \left(\frac{\sum fX}{N}\right)^2} = \sqrt{\overline{X^2} - \bar{X}^2}$$

As in Problem 4.12(*a*), this can be established by starting with

$$s = \sqrt{\frac{\sum f(X - \bar{X})^2}{N}}$$

Then
$$s^2 = \frac{\sum f(X-\bar{X})^2}{N} = \frac{\sum f(X^2 - 2\bar{X}X + \bar{X}^2)}{N} = \frac{\sum fX^2 - 2\bar{X}\sum fX + \bar{X}^2 \sum f}{N}$$
$$= \frac{\sum fX^2}{N} - 2\bar{X}\frac{\sum fX}{N} + \bar{X}^2 = \frac{\sum fX^2}{N} - 2\bar{X}^2 + \bar{X}^2 = \frac{\sum fX^2}{N} - \bar{X}^2$$
$$= \frac{\sum fX^2}{N} - \left(\frac{\sum fX}{N}\right)^2$$

or
$$s = \sqrt{\frac{\sum fX^2}{N} - \left(\frac{\sum fX}{N}\right)^2}$$

Note that in the above summations we have used the abbreviated form, with X and f replacing X_j and f_j, $\sum$ replacing $\sum_{j=1}^{K}$, and $\sum_{j=1}^{K} f_j = N$.

4.14 Using the formula of Problem 4.13, find the standard deviation for the data in Table 4.2 of Problem 4.11.

SOLUTION

The work can be arranged as in Table 4.3, where $\bar{X} = (\sum fX)/N = 67.45$ in, as obtained in Problem 3.15. Note that this method, like that of Problem 4.11, entails much tedious computation. Problem 4.17 shows how the coding method simplifies the calculations immensely.

Table 4.3

Height (in)	Class Mark (X)	X^2	Frequency (f)	fX^2
60–62	61	3721	5	18,605
63–65	64	4096	18	73,728
66–68	67	4489	42	188,538
69–71	70	4900	27	132,300
72–74	73	5329	8	42,632
			$N = \sum f = 100$	$\sum fX^2 = 455{,}803$

$$s = \sqrt{\frac{\sum fX^2}{N} - \left(\frac{\sum fX}{N}\right)^2} = \sqrt{\frac{455{,}803}{100} - (67.45)^2} = \sqrt{8.5275} = 2.92 \text{ in}$$

4.15 If $d = X - A$ are the deviations of X from an arbitrary constant A, prove that
$$s = \sqrt{\frac{\sum fd^2}{N} - \left(\frac{\sum fd}{N}\right)^2}$$

SOLUTION

Since $d = X - A$, $X = A + d$, and $\bar{X} = A + \bar{d}$ (see Problem 3.18), then
$$X - \bar{X} = (A + d) - (A + \bar{d}) = d - \bar{d}$$

so that
$$s = \sqrt{\frac{\sum f(X-\bar{X})^2}{N}} = \sqrt{\frac{\sum f(d-\bar{d})^2}{N}} = \sqrt{\frac{\sum fd^2}{N} - \left(\frac{\sum fd}{N}\right)^2}$$

using the result of Problem 4.13 and replacing X and $\bar{X}$ with d and $\bar{d}$, respectively.

Another method

$$s^2 = \overline{(X - \bar{X})^2} = \overline{(d - \bar{d})^2} = \overline{d^2 - 2\bar{d}d + \bar{d}^2}$$

$$= \overline{d^2} - 2\bar{d}^2 + \bar{d}^2 = \overline{d^2} - \bar{d}^2 = \frac{\sum fd^2}{N} - \left(\frac{\sum fd}{N}\right)^2$$

and the result follows on taking the positive square root.

4.16 Show that if each class mark X in a frequency distribution having class intervals of equal size c is coded into a corresponding value u according to the relation $X = A + cu$, where A is a given class mark, then the standard deviation can be written as

$$s = c\sqrt{\frac{\sum fu^2}{N} - \left(\frac{\sum fu}{N}\right)^2} = c\sqrt{\overline{u^2} - \bar{u}^2}$$

SOLUTION

This follows at once from Problem 4.15 since $d = X - A = cu$. Thus, since c is a constant,

$$s = \sqrt{\frac{\sum f(cu)^2}{N} - \left(\frac{\sum f(cu)}{N}\right)^2} = \sqrt{c^2\frac{\sum fu^2}{N} - c^2\left(\frac{\sum fu}{N}\right)^2} = c\sqrt{\frac{\sum fu^2}{N} - \left(\frac{\sum fu}{N}\right)^2}$$

Another method

We can also prove the result directly without using Problem 4.15. Since $X = A + cu$, $\bar{X} = A + c\bar{u}$, and $X - \bar{X} = c(u - \bar{u})$, then

$$s^2 = \overline{(X - \bar{X})^2} = \overline{c^2(u - \bar{u})^2} = \overline{c^2(u^2 - 2\bar{u}u + \bar{u}^2)} = c^2(\overline{u^2} - 2\bar{u}^2 + \bar{u}^2) = c^2(\overline{u^2} - \bar{u}^2)$$

and

$$s = c\sqrt{\overline{u^2} - \bar{u}^2} = c\sqrt{\frac{\sum fu^2}{N} - \left(\frac{\sum fu}{N}\right)^2}$$

4.17 Find the standard deviation of the heights of the students at XYZ University (see Table 2.1) by using (*a*) the formula derived in Problem 4.15 and (*b*) the coding method of Problem 4.16.

SOLUTION

In Tables 4.4 and 4.5, A is arbitrarily chosen as being equal to the class mark 67. Note that in Table 4.4 the deviations $d = X - A$ are all multiples of the class-interval size $c = 3$. This factor is removed in Table 4.5. As a result, the computations in Table 4.5 are greatly simplified (compare them with those of Problems 4.11 and 4.14). For this reason, the coding method should be used wherever possible.

(*a*) See Table 4.4.

Table 4.4

Class Mark (X)	$d = X - A$	Frequency (f)	fd	fd^2
61	−6	5	−30	180
64	−3	18	−54	162
$A \rightarrow$ 67	0	42	0	0
70	3	27	81	243
73	6	8	48	288
		$N = \sum f = 100$	$\sum fd = 45$	$\sum fd^2 = 873$

$$s = \sqrt{\frac{\sum fd^2}{N} - \left(\frac{\sum fd}{N}\right)^2} = \sqrt{\frac{873}{100} - \left(\frac{45}{100}\right)^2} = \sqrt{8.5275} = 2.92 \text{ in}$$

(*b*) See Table 4.5.

Table 4.5

Class Mark (X)	$u = \frac{X - A}{c}$	Frequency (f)	fu	fu^2
61	−2	5	−10	20
64	−1	18	−18	18
$A \rightarrow 67$	0	42	0	0
70	1	27	27	27
73	2	8	16	32
		$N = \sum f = 100$	$\sum fu = 15$	$\sum fu^2 = 97$

$$s = c\sqrt{\frac{\sum fu^2}{N} - \left(\frac{\sum fu}{N}\right)^2} = 3\sqrt{\frac{97}{100} - \left(\frac{15}{100}\right)^2} = 3\sqrt{0.9475} = 2.92 \text{ in}$$

4.18 Using coding methods, find (*a*) the mean and (*b*) the standard deviation for the wage distribution of the 65 employees at the P&R Company (see Table 2.5 of Problem 2.3).

SOLUTION

The work can be arranged simply, as shown in Table 4.6.

(*a*)
$$\bar{X} = A + c\bar{u} = A + c\frac{\sum fu}{N} = \$275.00 + (\$10.00)\left(\frac{31}{65}\right) = \$279.77$$

(*b*)
$$s = c\sqrt{\overline{u^2} - \bar{u}^2} = c\sqrt{\frac{\sum fu^2}{N} - \left(\frac{\sum fu}{N}\right)^2} = (\$10.00)\sqrt{\frac{173}{65} - \left(\frac{31}{65}\right)^2} = (\$10.00)\sqrt{2.4341} = \$15.60$$

Table 4.6

X	u	f	fu	fu^2
\$255.00	−2	8	−16	32
265.00	−1	10	−10	10
$A \rightarrow$ 275.00	0	16	0	0
285.00	1	14	14	14
295.00	2	10	20	40
305.00	3	5	15	45
315.00	4	2	8	32
		$N = \sum f = 65$	$\sum fu = 31$	$\sum fu^2 = 173$

4.19 Table 4.7 shows the IQ's of 480 school children at a certain elementary school. Using the coding method, find (*a*) the mean and (*b*) the standard deviation.

Table 4.7

Class mark (X)	70	74	78	82	86	90	94	98	102	106	110	114	118	122	126
Frequency (f)	4	9	16	28	45	66	85	72	54	38	27	18	11	5	2

SOLUTION

The intelligence quotient is

$$\text{IQ} = \frac{\text{mental age}}{\text{chronological age}}$$

expressed as a percentage. For example, an 8-year-old child who (according to certain educational procedures) has a mentality equivalent to that of a 10-year-old child would have an IQ of $10/8 = 1.25 = 125\%$, or simply 125, the % sign being understood.

To find the mean and standard deviation of the IQ's in Table 4.7, we can arrange the work as in Table 4.8.

(*a*)
$$\bar{X} = A + c\bar{u} = A + c\frac{\sum fu}{N} = 94 + 4\left(\frac{236}{480}\right) = 95.97$$

(*b*)
$$s = c\sqrt{\overline{u^2} - \bar{u}^2} = c\sqrt{\frac{\sum fu^2}{N} - \left(\frac{\sum fu}{N}\right)^2} = 4\sqrt{\frac{3404}{480} - \left(\frac{236}{480}\right)^2} = 4\sqrt{6.8499} = 10.47$$

CHARLIER'S CHECK

4.20 Use Charlier's check to help verify the computations of (*a*) the mean and (*b*) the standard deviation performed in Problem 4.19.

SOLUTION

To supply the required check, the columns of Table 4.9 are added to those of Table 4.8 (with the exception of column 2, which is repeated in Table 4.9 for convenience).

(*a*) From Table 4.9, $\sum f(u+1) = 716$; from Table 4.8, $\sum fu + N = 236 + 480 = 716$. This provides the required check on the mean.

Table 4.8

X	u	f	fu	fu^2
70	−6	4	−24	144
74	−5	9	−45	225
78	−4	16	−64	256
82	−3	28	−84	252
86	−2	45	−90	180
90	−1	66	−66	66
$A \rightarrow$ 94	0	85	0	0
98	1	72	72	72
102	2	54	108	216
106	3	38	114	342
110	4	27	108	432
114	5	18	90	450
118	6	11	66	396
122	7	5	35	245
126	8	2	16	128
		$N = \sum f = 480$	$\sum fu = 236$	$\sum fu^2 = 3404$

Table 4.9

$u+1$	f	$f(u+1)$	$f(u+1)^2$
−5	4	−20	100
−4	9	−36	144
−3	16	−48	144
−2	28	−56	112
−1	45	−45	45
0	66	0	0
1	85	85	85
2	72	144	288
3	54	162	486
4	38	152	608
5	27	135	675
6	18	108	648
7	11	77	539
8	5	40	320
9	2	18	162
	$N = \sum f = 480$	$\sum f(u+1) = 716$	$\sum f(u+1)^2 = 4356$

(*b*) From Table 4.9, $\sum f(u+1)^2 = 4356$; from Table 4.8, $\sum fu^2 + 2\sum fu + N = 3404 + 2(236) + 480 = 4356$. This provides the required check on the standard deviation.

SHEPPARD'S CORRECTION FOR VARIANCE

4.21 Apply Sheppard's correction to determine the standard deviation of the data in (*a*) Problem 4.17, (*b*) Problem 4.18, and (*c*) Problem 4.19.

SOLUTION

(*a*) $s^2 = 8.5275$, and $c = 3$. Corrected variance $= s^2 - c^2/12 = 8.5275 - 3^2/12 = 7.7775$. Corrected standard deviation $= \sqrt{\text{correct variance}} = \sqrt{7.7775} = 2.79$ in.

(*b*) $s^2 = 243.41$, and $c = 10$. Corrected variance $= s^2 - c^2/12 = 243.41 - 10^2/12 = 235.08$. Corrected standard deviation $= \sqrt{235.08} = \$15.33$.

(*c*) $s^2 = 109.60$, and $c = 4$. Corrected variance $= s^2 - c^2/12 = 109.60 - 4^2/12 = 108.27$. Corrected standard deviation $= \sqrt{108.27} = 10.41$.

4.22 For the second frequency distribution of Problem 2.8, find (*a*) the mean, (*b*) the standard deviation, (*c*) the standard deviation using Sheppard's correction, and (*d*) the actual standard deviation from the ungrouped data.

SOLUTION

The work is arranged in Table 4.10.

(*a*)
$$\bar{X} = A + c\bar{u} = A + c\frac{\sum fu}{N} = 149 + 9\left(\frac{-9}{40}\right) = 147.0 \text{ lb}$$

(*b*)
$$s = c\sqrt{\overline{u^2} - \bar{u}^2} = c\sqrt{\frac{\sum fu^2}{N} - \left(\frac{\sum fu}{N}\right)^2} = 9\sqrt{\frac{95}{40} - \left(\frac{-9}{40}\right)^2} = 9\sqrt{2.324375} = 13.7 \text{ lb}$$

(*c*) Corrected variance $= s^2 - c^2/12 = 188.27 - 9^2/12 = 181.52$. Corrected standard deviation $= 13.5$ lb.

Table 4.10

X	u	f	fu	fu^2
122	−3	3	−9	27
131	−2	5	−10	20
140	−1	9	−9	9
$A \rightarrow$ 149	0	12	0	0
158	1	5	5	5
167	2	4	8	16
176	3	2	6	18
		$N = \sum f = 40$	$\sum fu = -9$	$\sum fu^2 = 95$

(*d*) To compute the standard deviation from the actual weights of the students given in the problem, it is convenient first to subtract a suitable number, say $A = 150$ lb, from each weight and then use the method of Problem 4.15. The deviations $d = X - A = X - 150$ are then given in the following table:

−12	14	0	−18	−6	−25	−1	7
−4	8	−10	−3	−14	−2	2	−6
18	−24	−12	26	13	−31	4	15
−4	23	−8	−3	−15	3	−10	−15
11	−5	−15	−8	0	6	−5	−22

from which we find that $\sum d = -128$ and $\sum d^2 = 7052$. Then

$$s = \sqrt{\overline{d^2} - \bar{d}^2} = \sqrt{\frac{\sum d^2}{N} - \left(\frac{\sum d}{N}\right)^2} = \sqrt{\frac{7052}{40} - \left(\frac{-128}{40}\right)^2} = \sqrt{166.06} = 12.9 \text{ lb}$$

Hence Sheppard's correction supplied some improvement in this case.

EMPIRICAL RELATIONS BETWEEN MEASURES OF DISPERSION

4.23 For the distribution of the heights of the students at XYZ University, discuss the validity of the empirical formulas (*a*) mean deviation $= \frac{4}{5}$(standard deviation) and (*b*) semi-interquartile range $= \frac{2}{3}$(standard deviation).

SOLUTION

(*a*) From Problems 4.4 and 4.11, mean deviation ÷ standard deviation = 2.26/2.92 = 0.77, which is close to $\frac{4}{5}$.

(*b*) From Problems 4.6 and 4.11, semi-interquartile range ÷ standard deviation = 1.98/2.92 = 0.68, which is close to $\frac{2}{3}$.

Thus the empirical formulas are valid in this case.

Note that in the above we have not used the standard deviation with Sheppard's correction for grouping, since no corresponding correction has been made for the mean deviation or semi-interquartile range.

PROPERTIES OF THE STANDARD DEVIATION

4.24 Determine the percentage of the students' IQ's in Problem 4.19 that fall within the ranges (*a*) $\bar{X} \pm s$, (*b*) $\bar{X} \pm 2s$, and (*c*) $\bar{X} \pm 3s$.

SOLUTION

(*a*) The range of IQ's from 85.5 to 106.4 is $\bar{X} \pm s = 95.97 \pm 10.47$. The number of IQ's in the range $\bar{X} \pm s$ is

$$\left(\frac{88 - 85.5}{4}\right)(45) + 66 + 85 + 72 + 54 + \left(\frac{106.4 - 104}{4}\right)(38) = 339$$

The percentage of IQ's in the range $\bar{X} \pm s$ is $339/480 = 70.6\%$.

(*b*) The range of IQ's from 75.0 to 116.9 is $\bar{X} \pm 2s = 95.97 \pm 2(10.47)$. The number of IQ's in the range $\bar{X} \pm 2s$ is

$$\left(\frac{76 - 75.0}{4}\right)(9) + 16 + 28 + 45 + 66 + 85 + 72 + 54 + 38 + 27 + 18 + \left(\frac{116.9 - 116}{4}\right)(11) = 451$$

The percentage of IQ's in the range $\bar{X} \pm 2s$ is $451/480 = 94.0\%$.

(*c*) The range of IQ's from 64.6 to 127.4 is $\bar{X} \pm 3s = 95.97 \pm 3(10.47)$. The number of IQ's in the range $\bar{X} \pm 3s$ is

$$480 - \left(\frac{128 - 127.4}{4}\right)(2) = 479.7 \quad \text{or} \quad 480$$

The percentage of IQ's in the range $\bar{X} \pm 3s$ is $479.7/480 = 99.9\%$, or practically 100%.

The percentages in parts (*a*), (*b*), and (*c*) agree favorably with those to be expected for a normal distribution: 68.27%, 95.45%, and 99.73%, respectively.

Note that we have not used Sheppard's correction for the standard deviation. If this is used, the results in this case agree closely with the above. Note also that the above results can also be obtained by using Table 4.11 of Problem 4.32.

4.25 Given the sets 2, 5, 8, 11, 14, and 2, 8, 14, find (*a*) the mean of each set, (*b*) the variance of each set, (*c*) the mean of the combined (or pooled) sets, and (*d*) the variance of the combined sets.

SOLUTION

(*a*) Mean of first set $= \frac{1}{5}(2 + 5 + 8 + 11 + 14) = 8$. Mean of second set $= \frac{1}{3}(2 + 8 + 14) = 8$.

(*b*) Variance of first set $= s_1^2 = \frac{1}{5}[(2-8)^2 + (5-8)^2 + (8-8)^2 + (11-8)^2 + (14-8)^2] = 18$. Variance of second set $= s_2^2 = \frac{1}{3}[(2-8)^2 + (8-8)^2 + (14-8)^2] = 24$.

(*c*) The mean of the combined sets is

$$\frac{2 + 5 + 8 + 11 + 14 + 2 + 8 + 14}{5 + 3} = 8$$

(*d*) The variance of the combined sets is

$$s^2 = \frac{(2-8)^2 + (5-8)^2 + (8-8)^2 + (11-8)^2 + (14-8)^2 + (2-8)^2 + (8-8)^2 + (14-8)^2}{5+3} = 20.25$$

Another method (by formula)

$$s^2 = \frac{N_1 s_1^2 + N_2 s_2^2}{N_1 + N_2} = \frac{(5)(18) + (3)(24)}{5 + 3} = 20.25$$

4.26 Work Problem 4.25 for the sets 2, 5, 8, 11, 14 and 10, 16, 22.

SOLUTION

Here the means of the two sets are 8 and 16, respectively, while the variances are the *same* as the sets of the preceding problem, namely, $s_1^2 = 18$ and $s_2^2 = 24$.

$$\text{Mean of combined sets} = \frac{2+5+8+11+14+10+16+22}{5+3} = 11$$

$$s^2 = \frac{(2-11)^2 + (5-11)^2 + (8-11)^2 + (11-11)^2 + (14-11)^2 + (10-11)^2 + (16-11)^2 + (22-11)^2}{5+3}$$

$$= 35.25$$

Note that the formula

$$s^2 = \frac{N_1 s_1^2 + N_2 s_2^2}{N_1 + N_2}$$

which gives the value 20.25, is *not* applicable in this case since the means of the two sets are *not* the same.

4.27 (*a*) Prove that $w^2 + pw + q$, where p and q are given constants, is a minimum if and only if $w = -\frac{1}{2}p$.

(*b*) Using part (*a*), prove that

$$\frac{\sum_{j=1}^{N}(X_j - a)^2}{N} \quad \text{or briefly} \quad \frac{\sum(X-a)^2}{N}$$

is a minimum if and only if $a = \bar{X}$.

SOLUTION

(*a*) We have $w^2 + pw + q = (w + \frac{1}{2}p)^2 + q - \frac{1}{4}p^2$. Since $(q - \frac{1}{4}p^2)$ is a constant, the expression has the least value (i.e., is a minimum) if and only if $w + \frac{1}{2}p = 0$ (i.e., $w = -\frac{1}{2}p$).

(*b*) $$\frac{\sum(X-a)^2}{N} = \frac{\sum(X^2 - 2aX + a^2)}{N} = \frac{\sum X^2 - 2a\sum X + Na^2}{N} = a^2 - 2a\frac{\sum X}{N} + \frac{\sum X^2}{N}$$

Comparing this last expression with $(w^2 + pw + q)$, we have

$$w = a \qquad p = -2\frac{\sum X}{N} \qquad q = \frac{\sum X^2}{N}$$

Thus the expression is a minimum when $a = -\frac{1}{2}p = (\sum X)/N = \bar{X}$, using the result of part (*a*).

ABSOLUTE AND RELATIVE DISPERSION; COEFFICIENT OF VARIATION

4.28 A manufacturer of television tubes has two types of tubes, A and B. Respectively, the tubes have mean lifetimes of $\bar{X}_A = 1495$ hours and $\bar{X}_B = 1875$ hours, and standard deviations of $s_A = 280$ hours and $s_B = 310$ hours. Which tube has the greater (*a*) absolute dispersion and (*b*) relative dispersion?

SOLUTION

(*a*) The absolute dispersion of A is $s_A = 280$ hours, and of B is $s_B = 310$ hours. Thus tube B has the greater absolute dispersion.

(*b*) The coefficients of variation are

$$A = \frac{s_A}{\bar{X}_A} = \frac{280}{1495} = 18.7\% \qquad B = \frac{s_B}{\bar{X}_B} = \frac{310}{1875} = 16.5\%$$

Thus tube A has the greater relative variation, or dispersion.

4.29 Find the coefficients of variation, V, for the data of (*a*) Problem 4.14 and (*b*) Problem 4.18, using both uncorrected and corrected standard deviations.

SOLUTION

(*a*)
$$V(\text{uncorrected}) = \frac{s(\text{uncorrected})}{\bar{X}} = \frac{2.92}{67.45} = 0.0433 = 4.3\%$$

$$V(\text{corrected}) = \frac{s(\text{corrected})}{\bar{X}} = \frac{2.79}{67.45} = 0.0413 = 4.1\% \qquad \text{by Problem 4.21}(a)$$

(*b*)
$$V(\text{uncorrected}) = \frac{s(\text{uncorrected})}{\bar{X}} = \frac{15.60}{79.77} = 0.196 = 19.6\%$$

$$V(\text{corrected}) = \frac{s(\text{corrected})}{\bar{X}} = \frac{15.33}{79.77} = 0.192 = 19.2\% \qquad \text{by Problem 4.21}(b)$$

4.30 (*a*) Define a measure of relative dispersion that could be used for a set of data for which the quartiles are known.

(*b*) Illustrate the calculation of the measure defined in part (*a*) by using the data of Problem 4.6.

SOLUTION

(*a*) If Q_1 and Q_3 are given for a set of data, then $\frac{1}{2}(Q_1 + Q_3)$ is a measure of the data's central tendency, or average, while $Q = \frac{1}{2}(Q_3 - Q_1)$, the semi-interquartile range, is a measure of the data's dispersion. We can thus define a measure of relative dispersion as

$$V_Q = \frac{\frac{1}{2}(Q_3 - Q_1)}{\frac{1}{2}(Q_1 + Q_3)} = \frac{Q_3 - Q_1}{Q_3 + Q_1}$$

which we call the *quartile coefficient of variation*, or *quartile coefficient of relative dispersion*.

(*b*)
$$V_Q = \frac{Q_3 - Q_1}{Q_3 + Q_1} = \frac{69.61 - 65.64}{69.61 + 65.64} = \frac{3.97}{135.25} = 0.0293 = 2.9\%$$

STANDARDIZED VARIABLE; STANDARD SCORES

4.31 A student received a grade of 84 on a final examination in mathematics for which the mean grade was 76 and the standard deviation was 10. On the final examination in physics, for which the mean grade was 82 and the standard deviation was 16, she received a grade of 90. In which subject was her relative standing higher?

SOLUTION

The standardized variable $z = (X - \bar{X})/s$ measures the deviation of X from the mean $\bar{X}$ in terms of standard deviation s. For mathematics, $z = (84 - 76)/10 = 0.8$; for physics, $z = (90 - 82)/16 = 0.5$. Thus the student had a grade 0.8 of a standard deviation above the mean in mathematics, but only 0.5 of a standard deviation above the mean in physics. Thus her relative standing was higher in mathematics.

The variable $z = (X - \bar{X})/s$ is often used in educational testing, where it is known as a *standard score*.

SOFTWARE AND MEASURES OF DISPERSION

4.32 The STATISTIX analysis of the data in Example 3 of this chapter gave the following output.

```
Statistix 8.0
Descriptive Statistics

Variable      SD      Variance      C.V.       MAD
e-mails    29.256      855.93     44.562    21.000
```

The MAD value is the *median absolute deviation.* It is the median value of the absolute differences among the individual values and the sample median. Confirm that the MAD value for this data equals 21.

SOLUTION

The sorted original data are:

24	24	24	25	26	28	29	30	31	31	31	32	32
35	35	36	39	40	40	42	42	44	44	45	47	49
51	52	54	54	54	57	58	58	58	60	61	61	63
65	65	68	69	70	71	71	74	74	74	77	77	77
77	79	80	84	86	86	95	95	99	99	100	102	102
105	113	113	114	115	116	118	121	122	125			

The median of the original data is 61.
If 61 is subtracted from each value, the data become:

−37	−37	−37	−36	−35	−33	−32	−31	−30	−30	−30	−29	−29
−26	−26	−25	−22	−21	−21	−19	−19	−17	−17	−16	−14	−12
−10	−9	−7	−7	−7	−4	−3	−3	−3	−1	0	0	2
4	4	7	8	9	10	10	13	13	13	16	16	16
16	18	19	23	25	25	34	34	38	38	39	41	41
44	52	52	53	54	55	57	60	61	64			

Now, take the absolute value of each of these values:

37	37	37	36	35	33	32	31	30	30	30	29	29	26	26
25	22	21	21	19	19	17	17	16	14	12	10	9	7	7
7	4	3	3	3	1	0	0	2	4	4	7	8	9	10
10	13	13	13	16	16	16	16	18	19	23	25	25	34	34
38	38	39	41	41	44	52	52	53	54	55	57	60	61	64

The median of this last set of data is 21. Therefore MAD = 21.

Supplementary Problems

THE RANGE

4.33 Find the range of the sets (*a*) 5, 3, 8, 4, 7, 6, 12, 4, 3 and (*b*) 8.772, 6.453, 10.624, 8.628, 9.434, 6.351.

4.34 Find the range of the maximum loads given in Table 3.8 of Problem 3.59.

4.35 Find the range of the rivet diameters in Table 3.10 of Problem 3.61.

4.36 The largest of 50 measurements is 8.34 kilograms (kg). If the range is 0.46 kg, find the smallest measurement.

4.37 The following table gives the number of weeks needed to find a job for 25 older workers that lost their jobs as a result of corporation downsizing. Find the range of the data.

13	13	17	7	22
22	26	17	13	14
16	7	6	18	20
10	17	11	10	15
16	8	16	21	11

THE MEAN DEVIATION

4.38 Find the absolute values of (*a*) -18.2, (*b*) $+3.58$, (*c*) 6.21, (*d*) 0, (*e*) $-\sqrt{2}$, and (*f*) $4.00 - 2.36 - 3.52$.

4.39 Find the mean deviation of the set (*a*) 3, 7, 9, 5 and (*b*) 2.4, 1.6, 3.8, 4.1, 3.4.

4.40 Find the mean deviation of the sets of numbers in Problem 4.33.

4.41 Find the mean deviation of the maximum loads in Table 3.8 of Problem 3.59.

4.42 (*a*) Find the mean deviation (MD) of the rivet diameters in Table 3.10 of Problem 3.61.
(*b*) What percentage of the rivet diameters lie between ($\bar{X} \pm \text{MD}$), ($\bar{X} \pm 2\text{MD}$), and ($\bar{X} \pm 3\,\text{MD}$)?

4.43 For the set 8, 10, 9, 12, 4, 8, 2, find the mean deviation (*a*) from the mean and (*b*) from the median. Verify that the mean deviation from the median is not greater than the mean deviation from the mean.

4.44 For the distribution in Table 3.9 of Problem 3.60, find the mean deviation (*a*) about the mean and (*b*) about the median. Use the results of Problems 3.60 and 3.70.

4.45 For the distribution in Table 3.11 of Problem 3.62, find the mean deviation (*a*) about the mean and (*b*) about the median. Use the results of Problems 3.62 and 3.72.

4.46 Find the mean deviation for the data given in Problem 4.37.

4.47 Derive coding formulas for computing the mean deviation (*a*) about the mean and (*b*) about the median from a frequency distribution. Apply these formulas to verify the results of Problems 4.44 and 4.45.

THE SEMI-INTERQUARTILE RANGE

4.48 Find the semi-interquartile range for the distributions of (*a*) Problem 3.59, (*b*) Problem 3.60, and (*c*) Problem 3.107. Interpret the results clearly in each case.

4.49 Find the semi-interquartile range for the data given in Problem 4.37.

4.50 Prove that for any frequency distribution the total percentage of cases falling in the interval $\frac{1}{2}(Q_1 + Q_3) \pm \frac{1}{2}(Q_3 - Q_1)$ is 50%. Is the same true for the interval $Q_2 \pm \frac{1}{2}(Q_3 - Q_1)$? Explain your answer.

4.51 (*a*) How would you graph the semi-interquartile range corresponding to a given frequency distribution?
(*b*) What is the relationship of the semi-interquartile range to the ogive of the distribution?

THE 10–90 PERCENTILE RANGE

4.52 Find the 10–90 percentile range for the distributions of (*a*) Problem 3.59 and (*b*) Problem 3.107. Interpret the results clearly in each case.

4.53 The tenth percentile for home selling prices in a city is \$35,500 and the ninetieth percentile for home selling prices in the same city is \$225,000. Find the 10–90 percentile range and give a range within which 80% of the selling prices fall.

4.54 What advantages or disadvantages would a 20–80 percentile range have in comparison to a 10–90 percentile range?

4.55 Answer Problem 4.51 with reference to the (*a*) 10–90 percentile range, (*b*) 20–80 percentile range, and (*c*) 25–75 percentile range. What is the relationship between (*c*) and the semi-interquartile range?

THE STANDARD DEVIATION

4.56 Find the standard deviation of the sets (*a*) 3, 6, 2, 1, 7, 5; (*b*) 3.2, 4.6, 2.8, 5.2, 4.4; and (*c*) 0, 0, 0, 0, 0, 1, 1, 1.

4.57 (*a*) By adding 5 to each of the numbers in the set 3, 6, 2, 1, 7, 5, we obtain the set 8, 11, 7, 6, 12, 10. Show that the two sets have the same standard deviation but different means. How are the means related?

(*b*) By multiplying each of the numbers 3, 6, 2, 1, 7, and 5 by 2 and then adding 5, we obtain the set 11, 17, 9, 7, 19, 15. What is the relationship between the standard deviations and the means for the two sets?

(*c*) What properties of the mean and standard deviation are illustrated by the particular sets of numbers in parts (*a*) and (*b*)?

4.58 Find the standard deviation of the set of numbers in the arithmetic progression 4, 10, 16, 22, ..., 154.

4.59 Find the standard deviation for the distributions of (*a*) Problem 3.59, (*b*) Problem 3.60, and (*c*) Problem 3.107.

4.60 Demonstrate the use of Charlier's check in each part of Problem 4.59.

4.61 Find (*a*) the mean and (*b*) the standard deviation for the distribution of Problem 2.17, and explain the significance of the results obtained.

4.62 When data have a bell-shaped distribution, the standard deviation may be approximated by dividing the range by 4. For the data given in Problem 4.37, compute the standard deviation and compare it with the range divided by 4.

4.63 (*a*) Find the standard deviation s of the rivet diameters in Table 3.10 of Problem 3.61.
(*b*) What percentage of the rivet diameters lies between $\bar{X} \pm s$, $\bar{X} \pm 2s$, and $\bar{X} \pm 3s$?
(*c*) Compare the percentages in part (*b*) with those which would theoretically be expected if the distribution were normal, and account for any observed differences.

4.64 Apply Sheppard's correction to each standard deviation in Problem 4.59. In each case, discuss whether such application is or is not justified.

4.65 What modifications occur in Problem 4.63 when Sheppard's correction is applied?

4.66 (*a*) Find the mean and standard deviation for the data of Problem 2.8.
(*b*) Construct a frequency distribution for the data and find the standard deviation.
(*c*) Compare the results of part (*b*) with that of part (*a*). Determine whether an application of Sheppard's correction produces better results.

4.67 Work Problem 4.66 for the data of Problem 2.27.

4.68 (*a*) Of a total of N numbers, the fraction p are 1's, while the fraction $q = 1 - p$ are 0's. Prove that the standard deviation of the set of numbers is $\sqrt{pq}$.
(*b*) Apply the result of part (*a*) to Problem 4.56(*c*).

4.69 (*a*) Prove that the variance of the set of n numbers $a, a + d, a + 2d, \ldots, a + (n - 1)d$ (i.e., an arithmetic progression with the first term a and common difference d) is given by $\frac{1}{12}(n^2 - 1)d^2$.
(*b*) Use part (*a*) for Problem 4.58. [*Hint:* Use $1 + 2 + 3 \cdots + (n - 1) = \frac{1}{2}n(n - 1)$, $1^2 + 2^2 + 3^2 + \cdots + (n - 1)^2 = \frac{1}{6}n(n - 1)(2n - 1)$.]

4.70 Generalize and prove Property 3 of this chapter.

EMPIRICAL RELATIONS BETWEEN MEASURES OF DISPERSION

4.71 By comparing the standard deviations obtained in Problem 4.59 with the corresponding mean deviations of Problems 4.41, 4.42, and 4.44, determine whether the following empirical relation holds: Mean deviation $= \frac{4}{5}$(standard deviation). Account for any differences that may occur.

4.72 By comparing the standard deviations obtained in Problem 4.59 with the corresponding semi-interquartile ranges of Problem 4.48, determine whether the following empirical relation holds: Semi-interquartile range $= \frac{2}{3}$(standard deviation). Account for any differences that may occur.

4.73 What empirical relation would you expect to exist between the semi-interquartile range and the mean deviation for bell-shaped distributions that are moderately skewed?

4.74 A frequency distribution that is approximately normal has a semi-interquartile range equal to 10. What values would you expect for (*a*) the standard deviation and (*b*) the mean deviation?

ABSOLUTE AND RELATIVE DISPERSION; COEFFICIENT OF VARIATION

4.75 On a final examination in statistics, the mean grade of a group of 150 students was 78 and the standard deviation was 8.0. In algebra, however, the mean final grade of the group was 73 and the standard deviation was 7.6. In which subject was there the greater (*a*) absolute dispersion and (*b*) relative dispersion?

4.76 Find the coefficient of variation for the data of (*a*) Problem 3.59 and (*b*) Problem 3.107.

4.77 The distribution of SAT scores for a group of high school students has a first quartile score equal to 825 and a third quartile score equal to 1125. Calculate the quartile coefficient of variation for the distribution of SAT scores for this group of high school students.

4.78 For the age group 15–24 years, the first quartile of household incomes is equal to \$16,500 and the third quartile of household incomes for this same age group is \$25,000. Calculate the quartile coefficient of variation for the distribution of incomes for this age group.

STANDARDIZED VARIABLES; STANDARD SCORES

4.79 On the examinations referred to in Problem 4.75, a student scored 75 in statistics and 71 in algebra. In which examination was his relative standing higher?

4.80 Convert the set 6, 2, 8, 7, 5 into standard scores.

4.81 Prove that the mean and standard deviation of a set of standard scores are equal to 0 and 1, respectively. Use Problem 4.80 to illustrate this.

4.82 (*a*) Convert the grades of Problem 3.107 into standard scores, and (*b*) construct a graph of relative frequency versus standard score.

SOFTWARE AND MEASURES OF DISPERSION

4.83 Table 4.11 gives the per capita income for the 50 states in 2005.

Table 4.11 Per Capita Income for the 50 States

State	Per Capita Income	State	Per Capita Income
Wyoming	36,778	Pennsylvania	34,897
Montana	29,387	Wisconsin	33,565
North Dakota	31,395	Massachusetts	44,289
New Mexico	27,664	Missouri	31,899
West Virginia	27,215	Idaho	28,158
Rhode Island	36,153	Kentucky	28,513
Virginia	38,390	Minnesota	37,373
South Dakota	31,614	Florida	33,219
Alabama	29,136	South Carolina	28,352
Arkansas	26,874	New York	40,507
Maryland	41,760	Indiana	31,276
Iowa	32,315	Connecticut	47,819
Nebraska	33,616	Ohio	32,478
Hawaii	34,539	New Hampshire	38,408
Mississippi	25,318	Texas	32,462
Vermont	33,327	Oregon	32,103
Maine	31,252	New Jersey	43,771
Oklahoma	29,330	California	37,036
Delaware	37,065	Colorado	37,946
Alaska	35,612	North Carolina	30,553
Tennessee	31,107	Illinois	36,120
Kansas	32,836	Michigan	33,116
Arizona	30,267	Washington	35,409
Nevada	35,883	Georgia	31,121
Utah	28,061	Louisiana	24,820

The SPSS analysis of the data is as follows:

Descriptive Statistics

	N	Range	Std. Deviation	Variance
Income	50	22999.00	4893.54160	2E+007
Valid N (listwise)	50			

Verify the range, standard deviation, and variance.

Moments, Skewness, and Kurtosis

MOMENTS

If $X_1, X_2, \ldots, X_N$ are the N values assumed by the variable X, we define the quantity

$$\overline{X^r} = \frac{X_1^r + X_2^r + \cdots + X_N^r}{N} = \frac{\sum_{j=1}^{N} X_j^r}{N} = \frac{\sum X^r}{N} \tag{1}$$

called the rth *moment*. The first moment with $r = 1$ is the arithmetic mean $\bar{X}$.

The rth *moment about the mean* $\bar{X}$ is defined as

$$m_r = \frac{\sum_{j=1}^{N} (X_j - \bar{X})^r}{N} = \frac{\sum (X - \bar{X})^r}{N} = \overline{(X - \bar{X})^r} \tag{2}$$

If $r = 1$, then $m_1 = 0$ (see Problem 3.16). If $r = 2$, then $m_2 = s^2$, the variance.

The rth *moment about any origin* A is defined as

$$m'_r = \frac{\sum_{j=1}^{N} (X_j - A)^r}{N} = \frac{\sum (X - A)^r}{N} = \frac{\sum d^r}{N} = \overline{(X - A)^r} \tag{3}$$

where $d = X - A$ are the deviations of X from A. If $A = 0$, equation (*3*) reduces to equation (*1*). For this reason, equation (*1*) is often called the rth *moment about zero*.

MOMENTS FOR GROUPED DATA

If $X_1, X_2, \ldots, X_K$ occur with frequencies $f_1, f_2, \ldots, f_K$, respectively, the above moments are given by

$$\overline{X^r} = \frac{f_1 X_1^r + f_2 X_2^r + \cdots + f_K X_K^r}{N} = \frac{\sum_{j=1}^{K} f_j X_j^r}{N} = \frac{\sum f X^r}{N} \tag{4}$$

$$m_r = \frac{\sum_{j=1}^{K} f_j(X_j - \bar{X})^r}{N} = \frac{\sum f(X - \bar{X})^r}{N} = \overline{(X - \bar{X})^r} \tag{5}$$

$$m'_r = \frac{\sum_{j=1}^{K} f_j(X_j - A)^r}{N} = \frac{\sum f\,(X - A)^r}{N} = \overline{(X - A)^r} \tag{6}$$

where $N = \sum_{j=1}^{K} f_j = \sum f$. The formulas are suitable for calculating moments from grouped data.

RELATIONS BETWEEN MOMENTS

The following relations exist between moments about the mean m_r and moments about an arbitrary origin m'_r:

$$\begin{aligned} m_2 &= m'_2 - m'^2_1 \\ m_3 &= m'_3 - 3m'_1 m'_2 + 2m'^3_1 \\ m_4 &= m'_4 - 4m'_1 m'_3 + 6m'^2_1 m'_2 - 3m'^4_1 \end{aligned} \tag{7}$$

etc. (see Problem 5.5). Note that $m'_1 = \bar{X} - A$.

COMPUTATION OF MOMENTS FOR GROUPED DATA

The coding method given in previous chapters for computing the mean and standard deviation can also be used to provide a short method for computing moments. This method uses the fact that $X_j = A + cu_j$ (or briefly, $X = A + cu$), so that from equation (6) we have

$$m'_r = c^r \frac{\sum fu^r}{N} = c^r \overline{u^r} \tag{8}$$

which can be used to find m_r by applying equations (7).

CHARLIER'S CHECK AND SHEPPARD'S CORRECTIONS

Charlier's check in computing moments by the coding method uses the identities:

$$\begin{aligned} \sum f(u+1) &= \sum fu + N \\ \sum f(u+1)^2 &= \sum fu^2 + 2\sum fu + N \\ \sum f(u+1)^3 &= \sum fu^3 + 3\sum fu^2 + 3\sum fu + N \\ \sum f(u+1)^4 &= \sum fu^4 + 4\sum fu^3 + 6\sum fu^2 + 4\sum fu + N \end{aligned} \tag{9}$$

Sheppard's corrections for moments are as follows:

$$\text{Corrected } m_2 = m_2 - \tfrac{1}{12}c^2 \qquad \text{Corrected } m_4 = m_4 - \tfrac{1}{2}c^2 m_2 + \tfrac{7}{240}c^4$$

The moments m_1 and m_3 need no correction.

MOMENTS IN DIMENSIONLESS FORM

To avoid particular units, we can define the *dimensionless moments* about the mean as

$$a_r = \frac{m_r}{s^r} = \frac{m_r}{(\sqrt{m_2})^r} = \frac{m_r}{\sqrt{m_2^r}} \tag{10}$$

where $s = \sqrt{m_2}$ is the standard deviation. Since $m_1 = 0$ and $m_2 = s^2$, we have $a_1 = 0$ and $a_2 = 1$.

SKEWNESS

Skewness is the degree of asymmetry, or departure from symmetry, of a distribution. If the frequency curve (smoothed frequency polygon) of a distribution has a longer tail to the right of the central maximum than to the left, the distribution is said to be *skewed to the right*, or to have *positive skewness*. If the reverse is true, it is said to be *skewed to the left*, or to have *negative skewness*.

For skewed distributions, the mean tends to lie on the same side of the mode as the longer tail (see Figs. 3-1 and 3-2). Thus a measure of the asymmetry is supplied by the difference: mean–mode. This can be made dimensionless if we divide it by a measure of dispersion, such as the standard deviation, leading to the definition

$$\text{Skewness} = \frac{\text{mean} - \text{mode}}{\text{standard deviation}} = \frac{\bar{X} - \text{mode}}{s} \qquad (11)$$

To avoid using the mode, we can employ the empirical formula (*10*) of Chapter 3 and define

$$\text{Skewness} = \frac{3(\text{mean} - \text{median})}{\text{standard deviation}} = \frac{3(\bar{X} - \text{median})}{s} \qquad (12)$$

Equations (*11*) and (*12*) are called, respectively, *Pearson's first and second coefficients of skewness*.

Other measures of skewness, defined in terms of quartiles and percentiles, are as follows:

$$\text{Quartile coefficient of skewness} = \frac{(Q_3 - Q_2) - (Q_2 - Q_1)}{Q_3 - Q_1} = \frac{Q_3 - 2Q_2 + Q_1}{Q_3 - Q_1} \qquad (13)$$

$$\text{10–90 percentile coefficient of skewness} = \frac{(P_{90} - P_{50}) - (P_{50} - P_{10})}{P_{90} - P_{10}} = \frac{P_{90} - 2P_{50} + P_{10}}{P_{90} - P_{10}} \qquad (14)$$

An important measure of skewness uses the third moment about the mean expressed in dimensionless form and is given by

$$\text{Moment coefficient of skewness} = a_3 = \frac{m_3}{s^3} = \frac{m_3}{(\sqrt{m_2})^3} = \frac{m_3}{\sqrt{m_2^3}} \qquad (15)$$

Another measure of skewness is sometimes given by $b_1 = a_3^2$. For perfectly symmetrical curves, such as the normal curve, a_3 and b_1 are zero.

KURTOSIS

Kurtosis is the degree of peakedness of a distribution, usually taken relative to a normal distribution. A distribution having a relatively high peak is called *leptokurtic*, while one which is flat-topped is called *platykurtic*. A normal distribution, which is not very peaked or very flat-topped, is called *mesokurtic*.

One measure of kurtosis uses the fourth moment about the mean expressed in dimensionless form and is given by

$$\text{Moment coefficient of kurtosis} = a_4 = \frac{m_4}{s^4} = \frac{m_4}{m_2^2} \qquad (16)$$

which is often denoted by b_2. For the normal distribution, $b_2 = a_4 = 3$. For this reason, the kurtosis is sometimes defined by $(b_2 - 3)$, which is positive for a leptokurtic distribution, negative for a platykurtic distribution, and zero for the normal distribution.

Another measure of kurtosis is based on both quartiles and percentiles and is given by

$$\kappa = \frac{Q}{P_{90} - P_{10}} \qquad (17)$$

where $Q = \frac{1}{2}(Q_3 - Q_1)$ is the semi-interquartile range. We refer to κ (the lowercase Greek letter *kappa*) as the *percentile coefficient of kurtosis*; for the normal distribution, κ has the value 0.263.

POPULATION MOMENTS, SKEWNESS, AND KURTOSIS

When it is necessary to distinguish a sample's moments, measures of skewness, and measures of kurtosis from those corresponding to a population of which the sample is a part, it is often the custom to use Latin symbols for the former and Greek symbols for the latter. Thus if the sample's moments are denoted by m_r and m'_r, the corresponding Greek symbols would be μ_r and μ'_r (μ is the Greek letter *mu*). Subscripts are always denoted by Latin symbols.

Similarly, if the sample's measures of skewness and kurtosis are denoted by a_3 and a_4, respectively, the population's skewness and kurtosis would be α_3 and α_4 (α is the Greek letter *alpha*).

We already know from Chapter 4 that the standard deviation of a sample and of a population are denoted by s and σ, respectively.

SOFTWARE COMPUTATION OF SKEWNESS AND KURTOSIS

The software that we have discussed in the text so far may be used to compute skewness and kurtosis measures for sample data. The data in Table 5.1 samples of size 50 from a normal distribution, a skewed-right distribution, a skewed-left distribution, and a uniform distribution.

The normal data are female height measurements, the skewed-right data are age at marriage for females, the skewed-left data are obituary data that give the age at death for females, and the uniform data are the amount of cola put into a 12 ounce container by a soft drinks machine. The distribution of

Table 5.1

Normal		Skewed-right		Skewed-left		Uniform	
67	69	31	40	102	87	12.1	11.6
70	62	43	24	55	104	12.1	11.6
63	67	30	29	70	75	12.4	12.0
65	59	30	24	95	80	12.1	11.6
68	66	38	27	73	66	12.1	11.6
60	65	26	35	79	93	12.2	11.7
70	63	29	33	60	90	12.2	12.3
64	65	55	75	73	84	12.2	11.7
69	60	46	38	89	73	11.9	11.7
61	67	26	34	85	98	12.2	11.7
66	64	29	85	72	79	12.3	11.8
65	68	57	29	92	35	12.3	12.5
71	61	34	40	76	71	11.7	11.8
62	69	34	41	93	90	12.3	11.8
66	65	36	35	76	71	12.3	11.8
68	62	40	26	97	63	12.4	11.9
64	67	28	34	10	58	12.4	11.9
67	70	26	19	70	82	12.1	11.9
62	64	66	23	85	72	12.4	12.2
66	63	63	28	25	93	12.4	11.9
65	68	30	26	83	44	12.5	12.0
63	64	33	31	58	65	11.8	11.9
66	65	24	25	10	77	12.5	12.0
65	61	35	22	92	81	12.5	12.0
63	66	34	28	82	77	12.5	12.0

the four sets of sample data is shown in Fig. 5-1. The distributions of the four samples are illustrated by dotplots.

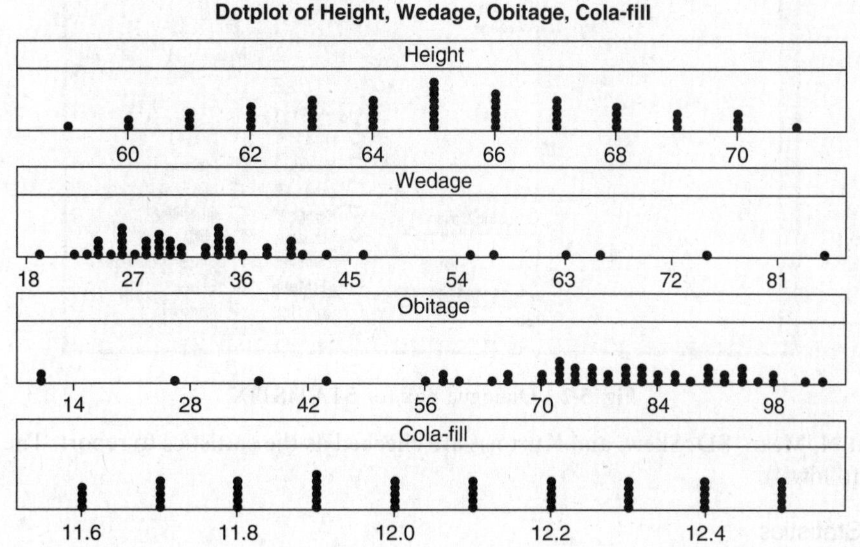

Fig. 5-1 MINITAB plot of four distributions: normal, right-skewed, left-skewed, and uniform.

The variable Height is the height of 50 adult females, the variable Wedage is the age at the wedding of 50 females, the variable Obitage is the age at death of 50 females, and the variable Cola-fill is the amount of cola put into 12 ounce containers. Each sample is of size 50. Using the terminology that we have learned in this chapter: the height distribution is mesokurtic, the cola-fill distribution is platykurtic, the wedage distribution is skewed to the right, and the obitage distribution is skewed to the left.

EXAMPLE 1. If MINITAB is used to find skewness and kurtosis values for the four variables, the following results are obtained by using the pull-down menu "**Stat ⇒ Basic statistics ⇒ Display descriptive statistics.**"

Descriptive Statistics: Height, Wedage, Obitage, Cola-fill

Variable	N	N*	Mean	StDev	Skewness	Kurtosis
Height	50	0	65.120	2.911	−0.02	−0.61
Wedage	50	0	35.48	13.51	1.98	4.10
Obitage	50	0	74.20	20.70	−1.50	2.64
Cola-fill	50	0	12.056	0.284	0.02	−1.19

The skewness values for the uniform and normal distributions are seen to be near 0. The skewness measure is positive for a distribution that is skewed to the right and negative for a distribution that is skewed to the left.

EXAMPLE 2. Use EXCEL to find the skewness and kurtosis values for the data in Fig. 5-1. If the variable names are entered into A1:D1 and the sample data are entered into A2:D51 and any open cell is used to enter =SKEW(A2:A51) the value −0.0203 is returned. The function =SKEW(B2:B51) returns 1.9774, the function =SKEW(C2:C51) returns −1.4986, and =SKEW(D2:D51) returns 0.0156. The kurtosis values are obtained by =KURT(A2:A51) which gives −0.6083, =KURT(B2:B51) which gives 4.0985, =KURT(C2:C51) which gives 2.6368, and =KURT(D2:D51) which gives −1.1889. It can be seen that MINITAB and EXCEL give the same values for kurtosis and skewness.

EXAMPLE 3. When STATISTIX is used to analyze the data shown in Fig. 5-1, the pull-down "**Statistics ⇒ Summary Statistics ⇒ Descriptive Statistics**" gives the dialog box in Fig. 5-2.

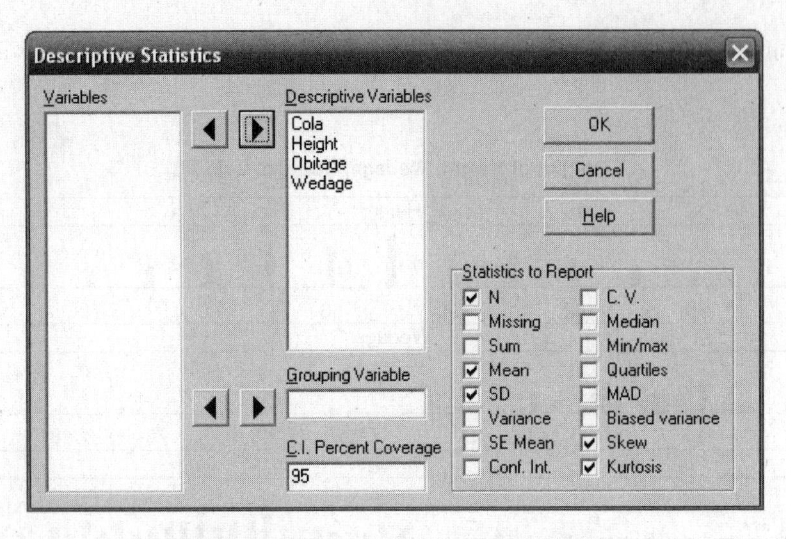

Fig. 5-2 Dialogue box for STATISTIX.

Note that N, Mean, SD, Skew, and Kurtosis are checked as the statistics to report. The STATISTIX output is as follows.

Descriptive Statistics

Variable	N	Mean	SD	Skew	Kurtosis
Cola	50	12.056	0.2837	0.0151	−1.1910
Height	50	65.120	2.9112	−0.0197	−0.6668
Obitage	50	74.200	20.696	−1.4533	2.2628
Wedage	50	35.480	13.511	1.9176	3.5823

Since the numerical values differ slightly from MINITAB and EXCEL, it is obvious that slightly different measures of skewness and kurtosis are used by the software.

EXAMPLE 4. The SPSS pull-down menu "**Analyze ⇒ Descriptive Statistics ⇒ Descriptives**" gives the dialog box in Fig. 5-3 and the routines Mean, Std. deviation, Kurtosis, and Skewness are chosen. SPSS gives the same measures of skewness and kurtosis as EXCEL and MINITAB.

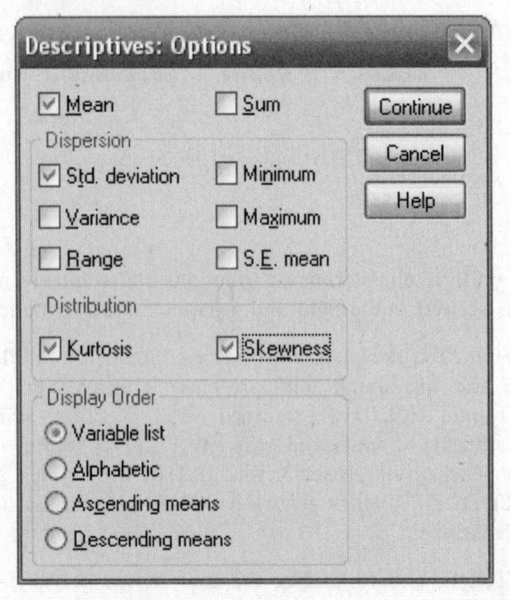

Fig. 5-3 Dialogue box for SPSS.

The following SPSS output is given.

Descriptive Statistics

	N	Mean	Std.	Skewness		Kurtosis	
	Statistic	Statistic	Statistic	Statistic	Std. Error	Statistic	Std. Error
Height	50	65.1200	2.91120	−.020	.337	−.608	.662
Wedage	50	35.4800	13.51075	1.977	.337	4.098	.662
Obitage	50	74.2000	20.69605	−1.499	.337	2.637	.662
Colafill	50	12.0560	.28368	.016	.337	−1.189	.662
Valid N (listwise)	50						

EXAMPLE 5. When SAS is used to compute the skewness and kurtosis values, the following output results. It is basically the same as that given by EXCEL, MINITAB, and SPSS.

```
                         The MEANS Procedure
Variable        Mean         Std Dev      N    Skewness       Kurtosis
ffffffffffffffffffffffffffffffffffffffffffffffffffffffffffffffffffffffff
Height       65.1200000    2.9112029    50   -0.0203232    -0.6083437
Wedage       35.4800000   13.5107516    50    1.9774237     4.0984607
Obitage      74.2000000   20.6960511    50   -1.4986145     2.6368045
Cola_fill    12.0560000    0.2836785    50    0.0156088    -1.1889600
ffffffffffffffffffffffffffffffffffffffffffffffffffffffffffffffffffffffff
```

Solved Problems

MOMENTS

5.1 Find the (*a*) first, (*b*) second, (*c*) third, and (*d*) fourth moments of the set 2, 3, 7, 8, 10.

SOLUTION

(*a*) The first moment, or arithmetic mean, is

$$\bar{X} = \frac{\sum X}{N} = \frac{2+3+7+8+10}{5} = \frac{30}{5} = 6$$

(*b*) The second moment is

$$\overline{X^2} = \frac{\sum X^2}{N} = \frac{2^2+3^2+7^2+8^2+10^2}{5} = \frac{226}{5} = 45.2$$

(*c*) The third moment is

$$\overline{X^3} = \frac{\sum X^3}{N} = \frac{2^3+3^3+7^3+8^3+10^3}{5} = \frac{1890}{5} = 378$$

(*d*) The fourth moment is

$$\overline{X^4} = \frac{\sum X^4}{N} = \frac{2^4+3^4+7^4+8^4+10^4}{5} = \frac{16{,}594}{5} = 3318.8$$

5.2 Find the (*a*) first, (*b*) second, (*c*) third, and (*d*) fourth moments about the mean for the set of numbers in Problem 5.1.

SOLUTION

$$(a)\quad m_1 = \overline{(X-\bar{X})} = \frac{\sum(X-\bar{X})}{N} = \frac{(2-6)+(3-6)+(7-6)+(8-6)+(10-6)}{5} = \frac{0}{5} = 0$$

m_1 is always equal to zero since $\overline{X-\bar{X}} = \bar{X}-\bar{X} = 0$ (see Problem 3.16).

$$(b)\quad m_2 = \overline{(X-\bar{X})^2} = \frac{\sum(X-\bar{X})^2}{N} = \frac{(2-6)^2+(3-6)^2+(7-6)^2+(8-6)^2+(10-6)^2}{6} = \frac{46}{5} = 9.2$$

Note that m_2 is the variance s^2.

$$(c)\quad m_3 = \overline{(X-\bar{X})^3} = \frac{\sum(X-\bar{X})^3}{N} = \frac{(2-6)^3+(3-6)^3+(7-6)^3+(8-6)^3+(10-6)^3}{5} = \frac{-18}{5} = -3.6$$

$$(d)\quad m_4 = \overline{(X-\bar{X})^4} = \frac{\sum(X-\bar{X})^4}{N} = \frac{(2-6)^4+(3-6)^4+(7-6)^4+(8-6)^4+(10-6)^4}{5} = \frac{610}{5} = 122$$

5.3 Find the (*a*) first, (*b*) second, (*c*) third, and (*d*) fourth moments about the origin 4 for the set of numbers in Problem 5.1.

SOLUTION

$$(a)\quad m_1' = \overline{(X-4)} = \frac{\sum(X-4)}{N} = \frac{(2-4)+(3-4)+(7-4)+(8-4)+(10-4)}{5} = 2$$

$$(b)\quad m_2' = \overline{(X-4)^2} = \frac{\sum(X-4)^2}{N} = \frac{(2-4)^2+(3-4)^2+(7-4)^2+(8-4)^2+(10-4)^2}{5} = \frac{66}{5} = 13.2$$

$$(c)\quad m_3' = \overline{(X-4)^3} = \frac{\sum(X-4)^3}{N} = \frac{(2-4)^3+(3-4)^3+(7-4)^3+(8-4)^3+(10-4)^3}{5} = \frac{298}{5} = 59.6$$

$$(d)\quad m_4' = \overline{(X-4)^4} = \frac{\sum(X-4)^4}{N} = \frac{(2-4)^4+(3-4)^4+(7-4)^4+(8-4)^4+(10-4)^4}{5} = \frac{1650}{5} = 330$$

5.4 Using the results of Problems 5.2 and 5.3, verify the relations between the moments (*a*) $m_2 = m_2' - m_1'^2$, (*b*) $m_3 = m_3' - 3m_1'm_2' + 2m_1'^3$, and (*c*) $m_4 = m_4' - 4m_1'm_3' + 6m_1'^2m_2' - 3m_1'^4$.

SOLUTION

From Problem 5.3 we have $m_1' = 2$, $m_2' = 13.2$, $m_3' = 59.6$, and $m_4' = 330$. Thus:

(*a*) $m_2 = m_2' - m_1'^2 = 13.2 - (2)^2 = 13.2 - 4 = 9.2$

(*b*) $m_3 = m_3' - 3m_1'm_2' + 2m_1'^3 = 59.6 - (3)(2)(13.2) + 2(2)^3 = 59.6 - 79.2 + 16 = -3.6$

(*c*) $m_4 = m_4' - 4m_1'm_3' + 6m_1'^2m_2' - 3m_1'^4 = 330 - 4(2)(59.6) + 6(2)^2(13.2) - 3(2)^4 = 122$

in agreement with Problem 5.2.

5.5 Prove that (*a*) $m_2 = m_2' - m_1'^2$, (*b*) $m_3 = m_3' - 3m_1'm_2' + 2m_1'^3$, and (*c*) $m_4 = m_4' - 4m_1'm_3' + 6m_1'^2m_2' - 3m_1'^4$.

SOLUTION

If $d = X - A$, then $X = A + d$, $\bar{X} = A + \bar{d}$, and $X - \bar{X} = d - \bar{d}$. Thus:

$$(a)\quad \begin{aligned} m_2 &= \overline{(X-\bar{X})^2} = \overline{(d-\bar{d})^2} = \overline{d^2 - 2\bar{d}d + \bar{d}^2} \\ &= \overline{d^2} - 2\bar{d}^2 + \bar{d}^2 = \overline{d^2} - \bar{d}^2 = m_2' - m_1'^2 \end{aligned}$$

(*b*)
$$m_3 = \overline{(X-\bar{X})^3} = \overline{(d-\bar{d})^3} = \overline{(d^3 - 3d^2\bar{d} + 3d\bar{d}^2 - \bar{d}^3)}$$
$$= \overline{d^3} - 3\bar{d}\overline{d^2} + 3\bar{d}^3 - \bar{d}^3 = \overline{d^3} - 3\bar{d}\overline{d^2} + 2\bar{d}^3 = m_3' - 3m_1'm_2' + 2m_1'^3$$

(*c*)
$$m_4 = \overline{(X-\bar{X})^4} = \overline{(d-\bar{d})^4} = \overline{(d^4 - 4d^3\bar{d} + 6d^2\bar{d}^2 - 4d\bar{d}^3 + \bar{d}^4)}$$
$$= \overline{d^4} - 4\bar{d}\overline{d^3} + 6\bar{d}^2\overline{d^2} - 4\bar{d}^4 + \bar{d}^4 = \overline{d^4} - 4\bar{d}\overline{d^3} + 6\bar{d}^2\overline{d^2} - 3\bar{d}^4$$
$$= m_4' - 4m_1'm_3' + 6m_1'^2m_2' - 3m_1'^4$$

By extension of this method, we can derive similar results for m_5, m_6, etc.

COMPUTATION OF MOMENTS FROM GROUPED DATA

5.6 Find the first four moments about the mean for the height distribution of Problem 3.22.

SOLUTION

The work can be arranged as in Table 5.2, from which we have

$$m_1' = c\frac{\sum fu}{N} = (3)\left(\frac{15}{100}\right) = 0.45 \qquad m_3' = c^3\frac{\sum fu^3}{N} = (3)^3\left(\frac{33}{100}\right) = 8.91$$

$$m_2' = c^2\frac{\sum fu^2}{N} = (3)^2\left(\frac{97}{100}\right) = 8.73 \qquad m_4' = c^4\frac{\sum fu^4}{N} = (3)^4\left(\frac{253}{100}\right) = 204.93$$

Thus
$$m_1 = 0$$
$$m_2 = m_2' - m_1'^2 = 8.73 - (0.45)^2 = 8.5275$$
$$m_3 = m_3' - 3m_1'm_2' + m_1'^3 = 8.91 - 3(0.45)(8.73) + 2(0.45)^3 = -2.6932$$
$$m_4 = m_4' - 4m_1'm_3' + 6m_1'^2m_2' - 3m_1'^4$$
$$= 204.93 - 4(0.45)(8.91) + 6(0.45)^2(8.73) - 3(0.45)^4 = 199.3759$$

Table 5.2

X	u	f	fu	fu^2	fu^3	fu^4
61	−2	5	−10	20	−40	80
64	−1	18	−18	18	−18	18
67	0	42	0	0	0	0
70	1	27	27	27	27	27
73	2	8	16	32	64	128
		$N = \sum f = 10$	$\sum fu = 15$	$\sum fu^2 = 97$	$\sum fu^3 = 33$	$\sum fu^4 = 253$

5.7 Find (*a*) m_1', (*b*) m_2', (*c*) m_3', (*d*) m_4', (*e*) m_1, (*f*) m_2, (*g*) m_3, (*h*) m_4, (*i*) $\bar{X}$, (*j*) s, (*k*) $\overline{X^2}$, and (*l*) $\overline{X^3}$ for the distribution in Table 4.7 of Problem 4.19.

SOLUTION

The work can be arranged as in Table 5.3.

Table 5.3

X	u	f	fu	fu^2	fu^3	fu^4
70	−6	4	−24	144	−864	5184
74	−5	9	−45	225	−1125	5625
78	−4	16	−64	256	−1024	4096
82	−3	28	−84	252	−756	2268
86	−2	45	−90	180	−360	720
90	−1	66	−66	66	−66	66
$A \longrightarrow$ 94	0	85	0	0	0	0
98	1	72	72	72	72	72
102	2	54	108	216	432	864
106	3	38	114	342	1026	3078
110	4	27	108	432	1728	6912
114	5	18	90	450	2250	11250
118	6	11	66	396	2376	14256
122	7	5	34	245	1715	12005
126	8	2	16	128	1024	8192
		$N = \sum f = 480$	$\sum fu = 236$	$\sum fu^2 = 3404$	$\sum fu^3 = 6428$	$\sum fu^4 = 74,588$

(*a*) $m_1' = c\frac{\sum fu}{N} = (4)\left(\frac{236}{480}\right) = 1.9667$

(*b*) $m_2' = c^2\frac{\sum fu^2}{N} = (4)^2\left(\frac{3404}{480}\right) = 113.4667$

(*c*) $m_3' = c^3\frac{\sum fu^3}{N} = (4)^3\left(\frac{6428}{480}\right) = 857.0667$

(*d*) $m_4' = c^4\frac{\sum fu^4}{N} = (4)^4\left(\frac{74,588}{480}\right) = 39,780.2667$

(*e*) $m_1 = 0$

(*f*) $m_2 = m_2' - m_1'^2 = 113.4667 - (1.9667)^2 = 109.5988$

(*g*) $m_3 = m_3' - 3m_1'm_2' + 2m_1'^3 = 857.0667 - 3(1.9667)(113.4667) + 2(1.9667)^3 = 202.8158$

(*h*) $m_4 = m_4' - 4m_1'm_3' + 6m_1'^2m_2' - 3m_1'^4 = 35,627.2853$

(*i*) $\bar{X} = \overline{(A + d)} = A + m_1' = A + c\frac{\sum fu}{N} = 94 + 1.9667 = 95.97$

(*j*) $s = \sqrt{m_2} = \sqrt{109.5988} = 10.47$

(*k*) $\overline{X^2} = \overline{(A+d)^2} = \overline{(A^2 + 2Ad + d^2)} = A^2 + 2A\bar{d} + \overline{d^2} = A^2 + 2Am_1' + m_2'$
$= (94)^2 + 2(94)(1.9667) + 113.4667 = 9319.2063$, or 9319 to four significant figures

(*l*) $\overline{X^3} = \overline{(A+d)^3} = \overline{(A^3 + 3A^2d + 3Ad^2 + d^3)} = A^3 + 3A^2\bar{d} + 3A\overline{d^2} + \overline{d^3}$
$= A^3 + 3A^2m_1' + 3Am_2' + m_3' = 915,571.9597$, or 915,600 to four significant figures

CHARLIER'S CHECK

5.8 Illustrate the use of Charlier's check for the computations in Problem 5.7.

SOLUTION

To supply the required check, we add to Table 5.3 the columns shown in Table 5.4 (with the exception of column 2, which is repeated in Table 5.3 for convenience).

In each of the following groupings, the first is taken from Table 5.4 and the second is taken from Table 5.2. Equality of results in each grouping provides the required check.

Table 5.4

$u+1$	f	$f(u+1)$	$f(u+1)^2$	$f(u+1)^3$	$f(u+1)^4$
−5	4	−20	100	−500	2500
−4	9	−36	144	−576	2304
−3	16	−48	144	−432	1296
−2	28	−56	112	−224	448
−1	45	−45	45	−45	45
0	66	0	0	0	0
1	85	85	85	85	85
2	72	144	288	576	1152
3	54	162	486	1458	4374
4	38	152	608	2432	9728
5	27	135	675	3375	16875
6	18	108	648	3888	23328
7	11	77	539	3773	26411
8	5	40	320	2560	20480
9	2	18	162	1458	13122
	$N=\sum f$ = 480	$\sum f(u+1)$ = 716	$\sum f(u+1)^2$ = 4356	$\sum f(u+1)^3$ = 17,828	$\sum f(u+1)^4$ = 122,148

$$\sum f(u+1) = 716$$

$$\sum fu + N = 236 + 480 = 716$$

$$\sum f(u+1)^2 = 4356$$

$$\sum fu^2 + 2\sum fu + N = 3404 + 2(236) + 480 = 4356$$

$$\sum f(u+1)^3 = 17{,}828$$

$$\sum fu^3 + 3\sum fu^2 + 3\sum fu + N = 6428 + 3(3404) + 3(236) + 480 = 17{,}828$$

$$\sum f(u+1)^4 = 122{,}148$$

$$\sum fu^4 + 4\sum fu^3 + 6\sum fu^2 + 4\sum fu + N = 74{,}588 + 4(6428) + 6(3404) + 4(236) + 480 = 122{,}148$$

SHEPPARD'S CORRECTIONS FOR MOMENTS

5.9 Apply Sheppard's corrections to determine the moments about the mean for the data in (*a*) Problem 5.6 and (*b*) Problem 5.7.

SOLUTION

(a) Corrected $m_2 = m_2 - c^2/12 = 8.5275 - 3^2/12 = 7.7775$

Corrected $m_4 = m_4 - \frac{1}{2}c^2 m_2 + \frac{7}{240}c^4$

$= 199.3759 - \frac{1}{2}(3)^2(8.5275) + \frac{7}{240}(3)^4$

$= 163.3646$

m_1 and m_2 need no correction.

(b) Corrected $m_2 = m_2 - c^2/12 = 109.5988 - 4^2/12 = 108.2655$

Corrected $m_4 = m_4 - \frac{1}{2}c^2 m_2 + \frac{7}{240}c^4$

$= 35{,}627.2853 - \frac{1}{2}(4)^2(109.5988) + \frac{7}{240}(4)^4$

$= 34{,}757.9616$

SKEWNESS

5.10 Find Pearson's (a) first and (b) second coefficients of skewness for the wage distribution of the 65 employees at the P&R Company (see Problems 3.44 and 4.18).

SOLUTION

Mean = \$279.76, median = \$279.06, mode = \$277.50, and standard deviation s = \$15.60. Thus:

(a) $$\text{First coefficient of skewness} = \frac{\text{mean} - \text{mode}}{s} = \frac{\$279.76 - \$277.50}{\$15.60} = 0.1448, \text{ or } 0.14$$

(b) $$\text{Second coefficient of skewness} = \frac{3(\text{mean} - \text{median})}{s} = \frac{3(\$279.76 - \$279.06)}{\$15.60} = 0.1346, \text{ or } 0.13$$

If the corrected standard deviation is used [see Problem 4.21(b)], these coefficients become, respectively:

(a) $$\frac{\text{Mean} - \text{mode}}{\text{Corrected } s} = \frac{\$279.76 - \$277.50}{\$15.33} = 0.1474, \text{ or } 0.15$$

(b) $$\frac{3(\text{mean} - \text{median})}{\text{Corrected } s} = \frac{3(\$279.76 - \$279.06)}{\$15.33} = 0.1370, \text{ or } 0.14$$

Since the coefficients are positive, the distribution is skewed positively (i.e., to the right).

5.11 Find the (a) quartile and (b) percentile coefficients of skewness for the distribution of Problem 5.10 (see Problem 3.44).

SOLUTION

$Q_1 = \$268.25$, $Q_2 = P_{50} = \$279.06$, $Q_3 = \$290.75$, $P_{10} = D_1 = \$258.12$, and $P_{90} = D_9 = \$301.00$. Thus:

(a) $$\text{Quartile coefficient of skewness} = \frac{Q_3 - 2Q_2 + Q_1}{Q_3 - Q_1} = \frac{\$290.75 - 2(\$279.06) + \$268.25}{\$290.75 - \$268.25} = 0.0391$$

(b) $$\text{Percentile coefficient of skewness} = \frac{P_{90} - 2P_{50} + P_{10}}{P_{90} - P_{10}} = \frac{\$301.00 - 2(\$279.06) + \$258.12}{\$301.00 - \$258.12} = 0.0233$$

5.12 Find the moment coefficient of skewness, a_3, for (a) the height distribution of students at XYZ University (see Problem 5.6) and (b) the IQ's of elementary school children (see Problem 5.7).

SOLUTION

(*a*) $m_2 = s^2 = 8.5275$, and $m_3 = -2.6932$. Thus:

$$a_3 = \frac{m_3}{s^3} = \frac{m_3}{(\sqrt{m_2})^3} = \frac{-2.6932}{(\sqrt{8.5275})^3} = -0.1081 \quad \text{or} \quad -0.11$$

If Sheppard's corrections for grouping are used [see Problem 5.9(*a*)], then

$$\text{Corrected } a_3 = \frac{m_3}{(\sqrt{\text{corrected } m_2})^3} = \frac{-2.6932}{(\sqrt{7.7775})^3} = -0.1242 \quad \text{or} \quad -0.12$$

(*b*)
$$a_3 = \frac{m_3}{s^3} = \frac{m_3}{(\sqrt{m_2})^3} = \frac{202.8158}{(\sqrt{109.5988})^3} = 0.1768 \quad \text{or} \quad 0.18$$

If Sheppard's corrections for grouping are used [see Problem 5.9(*b*)], then

$$\text{Corrected } a_3 = \frac{m_3}{(\sqrt{\text{corrected } m_2})^3} = \frac{202.8158}{(\sqrt{108.2655})^3} = 0.1800 \quad \text{or} \quad 0.18$$

Note that both distributions are moderately skewed, distribution (*a*) to the left (negatively) and distribution (*b*) to the right (positively). Distribution (*b*) is more skewed than (*a*); that is, (*a*) is more symmetrical than (*b*), as is evidenced by the fact that the numerical value (or absolute value) of the skewness coefficient for (*b*) is greater than that for (*a*).

KURTOSIS

5.13 Find the moment coefficient of kurtosis, a_4, for the data of (*a*) Problem 5.6 and (*b*) Problem 5.7.

SOLUTION

(*a*)
$$a_4 = \frac{m_4}{s^4} = \frac{m_4}{m_2^2} = \frac{199.3759}{(8.5275)^2} = 2.7418 \quad \text{or} \quad 2.74$$

If Sheppard's corrections are used [see Problem 5.9(*a*)], then

$$\text{Corrected } a_4 = \frac{\text{corrected } m_4}{(\text{corrected } m_2)^2} = \frac{163.36346}{(7.7775)^5} = 2.7007 \quad \text{or} \quad 2.70$$

(*b*)
$$a_4 = \frac{m_4}{s^4} = \frac{m_4}{m_2^2} = \frac{35{,}627.2853}{(109.5988)^2} = 2.9660 \quad \text{or} \quad 2.97$$

If Sheppard's corrections are used [see Problem 5.9(*b*)], then

$$\text{Corrected } a_4 = \frac{\text{corrected } m_4}{(\text{corrected } m_2)^2} = \frac{34{,}757.9616}{(108.2655)^2} = 2.9653 \quad \text{or} \quad 2.97$$

Since for a normal distribution $a_4 = 3$, it follows that both distributions (*a*) and (*b*) are *platykurtic* with respect to the normal distribution (i.e., less peaked than the normal distribution).

Insofar as peakedness is concerned, distribution (*b*) approximates the normal distribution much better than does distribution (*a*). However, from Problem 5.12 distribution (*a*) is more symmetrical than (*b*), so that as far as symmetry is concerned, (*a*) approximates the normal distribution better than (*b*) does.

SOFTWARE COMPUTATION OF SKEWNESS AND KURTOSIS

5.14 Sometimes the scores on a test do not follow the normal distribution, although they usually do. We sometimes find that students score low or high with very few in the middle. The distribution shown in Fig. 5-4 is such a distribution. This distribution is referred to as the U distribution. Find the mean, standard deviation, skewness, and kurtosis for the data using EXCEL.

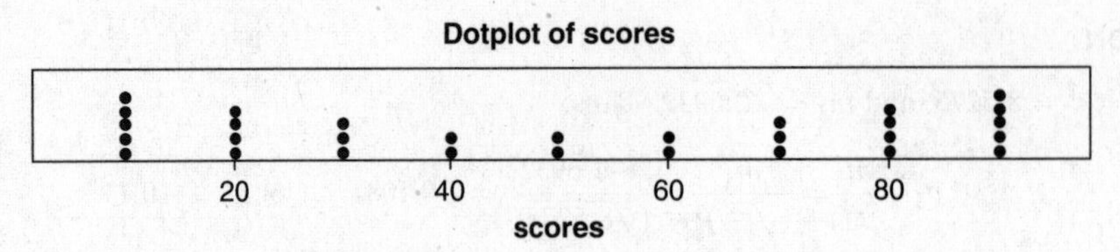

Fig. 5-4 .A MINITAB plot of data that follows a U distribution.

SOLUTION

The data is entered into A1:A30 of an EXCEL worksheet. The command "**=AVERAGE(A1:A30)**" gives 50. The command "**=STDEV(A1:A30)**" gives 29.94. The command "**=SKEW(A1:A30)**" gives 0. The command "**=KURT(A1:A30)**" gives −1.59.

Supplementary Problems

MOMENTS

5.15 Find the (*a*) first, (*b*) second, (*c*) third, and (*d*) fourth moments for the set 4, 7, 5, 9, 8, 3, 6.

5.16 Find the (*a*) first, (*b*) second, (*c*) third, and (*d*) fourth moments about the mean for the set of numbers in Problem 5.15.

5.17 Find the (*a*) first, (*b*) second, (*c*) third, and (*d*) fourth moments about the number 7 for the set of numbers in Problem 5.15.

5.18 Using the results of Problems 5.16 and 5.17, verify the relations between the moments (*a*) $m_2 = m_2' - m_1'^2$, (*b*) $m_3 = m_3' - 3m_1'm_2' + 2m_1'^3$, and (*c*) $m_4 = m_4' - 4m_1'm_3' + 6m_1'^2m_2' - 3m_1'^4$.

5.19 Find the first four moments about the mean of the set of numbers in the arithmetic progression 2, 5, 8, 11, 14, 17.

5.20 Prove that (*a*) $m_2' = m_2 + h^2$, (*b*) $m_3' = m_3 + 3hm_2 + h^3$, and (*c*) $m_4' = m_4 + 4hm_3 + 6h^2m_2 + h^4$, where $h = m_1'$.

5.21 If the first moment about the number 2 is equal to 5, what is the mean?

5.22 If the first four moments of a set of numbers about the number 3 are equal to −2, 10, −25, and 50, determine the corresponding moments (*a*) about the mean, (*b*) about the number 5, and (*c*) about zero.

5.23 Find the first four moments about the mean of the numbers 0, 0, 0, 1, 1, 1, 1, and 1.

5.24 (*a*) Prove that $m_5 = m_5' - 5m_1'm_4' + 10m_1'^2m_3' - 10m_1'^3m_2' + 4m_1'^5$.

(*b*) Derive a similar formula for m_6.

5.25 Of a total of N numbers, the fraction p are 1's, while the fraction $q = 1 - p$ are 0's. Find (*a*) m_1, (*b*) m_2, (*c*) m_3, and (*d*) m_4 for the set of numbers. Compare with Problem 5.23.

5.26 Prove that the first four moments about the mean of the arithmetic progression $a, a+d, a+2d, \ldots, a+(n-1)d$ are $m_1 = 0$, $m_2 = \frac{1}{12}(n^2-1)d^2$, $m_3 = 0$, and $m_4 = \frac{1}{240}(n^2-1)(3n^2-7)d^4$. Compare with Problem 5.19 (see also Problem 4.69). [*Hint*: $1^4 + 2^4 + 3^4 + \cdots + (n-1)^4 = \frac{1}{30}n(n-1)(2n-1)(3n^2-3n-1)$.]

MOMENTS FOR GROUPED DATA

5.27 Calculate the first four moments about the mean for the distribution of Table 5.5.

Table 5.5

X	f
12	1
14	4
16	6
18	10
20	7
22	2
	Total 30

5.28 Illustrate the use of Charlier's check for the computations in Problem 5.27.

5.29 Apply Sheppard's corrections to the moments obtained in Problem 5.27.

5.30 Calculate the first four moments about the mean for the distribution of Problem 3.59(*a*) without Sheppard's corrections and (*b*) with Sheppard's corrections.

5.31 Find (*a*) m_1, (*b*) m_2, (*c*) m_3, (*d*) m_4, (*e*) $\bar{X}$, (*f*) s, (*g*) $\overline{X^2}$, (*h*) $\overline{X^3}$, (*i*) $\overline{X^4}$, and (*j*) $\overline{(X+1)^3}$ for the distribution of Problem 3.62.

SKEWNESS

5.32 Find the moment coefficient of skewness, a_3, for the distribution of Problem 5.27 (*a*) without and (*b*) with Sheppard's corrections.

5.33 Find the moment coefficient of skewness, a_3, for the distribution of Problem 3.59 (see Problem 5.30).

5.34 The second moments about the mean of two distributions are 9 and 16, while the third moments about the mean are -8.1 and -12.8, respectively. Which distribution is more skewed to the left?

5.35 Find Pearson's (*a*) first and (*b*) second coefficients of skewness for the distribution of Problem 3.59, and account for the difference.

5.36 Find the (*a*) quartile and (*b*) percentile coefficients of skewness for the distribution of Problem 3.59. Compare your results with those of Problem 5.35 and explain.

5.37 Table 5.6 gives three different distributions for the variable X. The frequencies for the three distributions are given by f_1, f_2, and f_3. Find Pearson's first and second coefficients of skewness for the three distributions. Use the corrected standard deviation when computing the coefficients.

Table 5.6

X	f_1	f_2	f_3
0	10	1	1
1	5	2	2
2	2	14	2
3	2	2	5
4	1	1	10

KURTOSIS

5.38 Find the moment coefficient of kurtosis, a_4, for the distribution of Problem 5.27 (*a*) without and (*b*) with Sheppard's corrections.

5.39 Find the moment coefficient of kurtosis for the distribution of Problem 3.59 (*a*) without and (*b*) with Sheppard's corrections (see Problem 5.30).

5.40 The fourth moments about the mean of the two distributions of Problem 5.34 are 230 and 780, respectively. Which distribution more nearly approximates the normal distribution from the viewpoint of (*a*) peakedness and (*b*) skewness?

5.41 Which of the distributions in Problem 5.40 is (*a*) leptokurtic, (*b*) mesokurtic, and (*c*) platykurtic?

5.42 The standard deviation of a symmetrical distribution is 5. What must be the value of the fourth moment about the mean in order that the distribution be (*a*) leptokurtic, (*b*) mesokurtic, and (*c*) platykurtic?

5.43 (*a*) Calculate the percentile coefficient of kurtosis, κ, for the distribution of Problem 3.59.

(*b*) Compare your result with the theoretical value 0.263 for the normal distribution, and interpret.

(*c*) How do you reconcile this result with that of Problem 5.39?

SOFTWARE COMPUTATION OF SKEWNESS AND KURTOSIS

5.44 The data in Fig. 5-5 show a sharp peak at 50. This should show up in the kurtosis measure of the data. Using EXCEL, show that the skewness is basically zero and that the kurtosis is 2.0134.

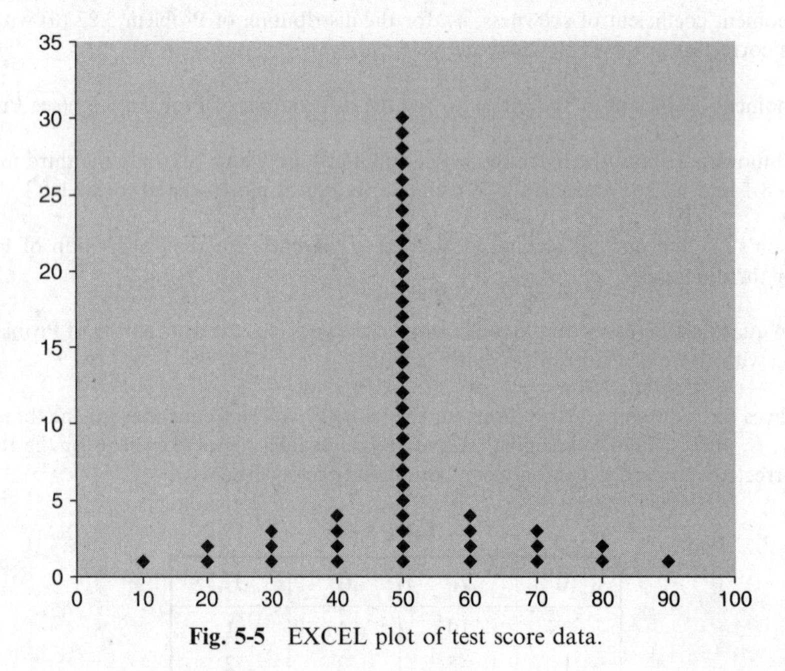

Fig. 5-5 EXCEL plot of test score data.

Elementary Probability Theory

DEFINITIONS OF PROBABILITY

Classic Definition

Suppose that an event E can happen in h ways out of a total of n possible equally likely ways. Then the probability of occurrence of the event (called its *success*) is denoted by

$$p = \Pr\{E\} = \frac{h}{n}$$

The probability of nonoccurrence of the event (called its *failure*) is denoted by

$$q = \Pr\{\text{not } E\} = \frac{n-h}{n} = 1 - \frac{h}{n} = 1 - p = 1 - \Pr\{E\}$$

Thus $p + q = 1$, or $\Pr\{E\} + \Pr\{\text{not } E\} = 1$. The event "not E" is sometimes denoted by $\bar{E}$, $\tilde{E}$, or $\sim E$.

EXAMPLE 1. When a die is tossed, there are 6 equally possible ways in which the die can fall:

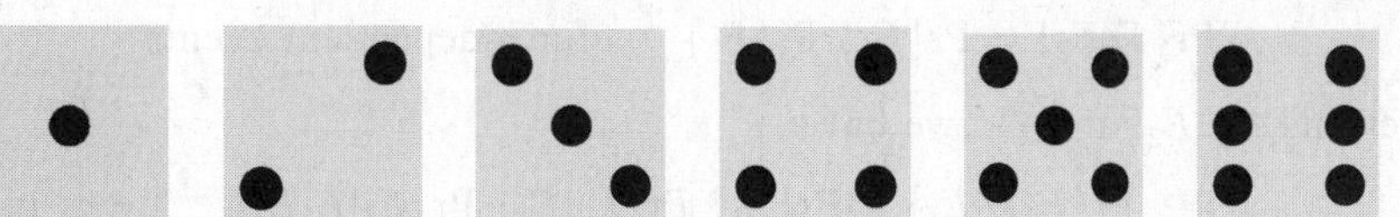

The event E, that a 3 or 4 turns up, is:

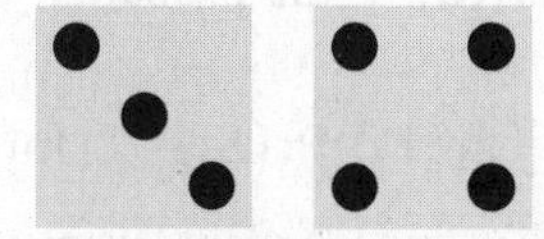

and the probability of E is $\Pr\{E\} = 2/6$ or $1/3$. The probability of not getting a 3 or 4 (i.e., getting a 1, 2, 5, or 6) is $\Pr\{\bar{E}\} = 1 - \Pr\{E\} = 2/3$.

Note that the probability of an event is a number between 0 and 1. If the event cannot occur, its probability is 0. If it must occur (i.e., its occurrence is *certain*), its probability is 1.

If p is the probability that an event will occur, the *odds* in favor of its happening are $p : q$ (read "p to q"); the odds against its happening are $q : p$. Thus the odds against a 3 or 4 in a single toss of a fair die are $q : p = \frac{2}{3} : \frac{1}{3} = 2 : 1$ (i.e., 2 to 1).

Relative-Frequency Definition

The classic definition of probability has a disadvantage in that the words "equally likely" are vague. In fact, since these words seem to be synonymous with "equally probable," the definition is *circular* because we are essentially defining probability in terms of itself. For this reason, a statistical definition of probability has been advocated by some people. According to this the estimated probability, or *empirical probability*, of an event is taken to be the *relative frequency* of occurrence of the event when the number of observations is very large. The probability itself is the *limit* of the relative frequency as the number of observations increases indefinitely.

EXAMPLE 2. If 1000 tosses of a coin result in 529 heads, the relative frequency of heads is $529/1000 = 0.529$. If another 1000 tosses results in 493 heads, the relative frequency in the total of 2000 tosses is $(529 + 493)/2000 = 0.511$. According to the statistical definition, by continuing in this manner we should ultimately get closer and closer to a number that represents the probability of a head in a single toss of the coin. From the results so far presented, this should be 0.5 to one significant figure. To obtain more significant figures, further observations must be made.

The statistical definition, although useful in practice, has difficulties from a mathematical point of view, since an actual limiting number may not really exist. For this reason, modern probability theory has been developed *axiomatically*; that is, the theory leaves the concept of probability undefined, much the same as *point* and *line* are undefined in geometry.

CONDITIONAL PROBABILITY; INDEPENDENT AND DEPENDENT EVENTS

If E_1 and E_2 are two events, the probability that E_2 occurs given that E_1 has occurred is denoted by $\Pr\{E_2|E_1\}$, or $\Pr\{E_2 \text{ given } E_1\}$, and is called the *conditional probability* of E_2 given that E_1 has occurred.

If the occurrence or nonoccurrence of E_1 does not affect the probability of occurrence of E_2, then $\Pr\{E_2|E_1\} = \Pr\{E_2\}$ and we say that E_1 and E_2 are *independent events*; otherwise, they are *dependent events*.

If we denote by E_1E_2 the event that "both E_1 and E_2 occur," sometimes called a *compound event*, then

$$\Pr\{E_1E_2\} = \Pr\{E_1\}\Pr\{E_2|E_1\} \tag{1}$$

In particular,

$$\Pr\{E_1E_2\} = \Pr\{E_1\}\Pr\{E_2\} \qquad \text{for independent events} \tag{2}$$

For three events E_1, E_2, and E_3, we have

$$\Pr\{E_1E_2E_3\} = \Pr\{E_1\}\Pr\{E_2|E_1\}\Pr\{E_3|E_1E_2\} \tag{3}$$

That is, the probability of occurrence of E_1, E_2, and E_3 is equal to (the probability of E_1) $\times$ (the probability of E_2 given that E_1 has occurred) $\times$ (the probability of E_3 given that both E_1 and E_2 have occurred). In particular,

$$\Pr\{E_1E_2E_3\} = \Pr\{E_1\}\Pr\{E_2\}\Pr\{E_3\} \qquad \text{for independent events} \tag{4}$$

In general, if $E_1, E_2, E_3, \ldots, E_n$ are n independent events having respective probabilities $p_1, p_2, p_3, \ldots, p_n$, then the probability of occurrence of E_1 and E_2 and E_3 and $\cdots E_n$ is $p_1p_2p_3 \cdots p_n$.

EXAMPLE 3. Let E_1 and E_2 be the events "heads on fifth toss" and "heads on sixth toss" of a coin, respectively. Then E_1 and E_2 are independent events, and thus the probability of heads on both the fifth and sixth tosses is (assuming the coin to be fair)

$$\Pr\{E_1E_2\} = \Pr\{E_1\}\Pr\{E_2\} = \left(\frac{1}{2}\right)\left(\frac{1}{2}\right) = \frac{1}{4}$$

EXAMPLE 4. If the probability that A will be alive in 20 years is 0.7 and the probability that B will be alive in 20 years is 0.5, then the probability that they will both be alive in 20 years is $(0.7)(0.5) = 0.35$.

EXAMPLE 5. Suppose that a box contains 3 white balls and 2 black balls. Let E_1 be the event "first ball drawn is black" and E_2 the event "second ball drawn is black," where the balls are not replaced after being drawn. Here E_1 and E_2 are dependent events.

The probability that the first ball drawn is black is $\Pr\{E_1\} = 2/(3+2) = \frac{2}{5}$. The probability that the second ball drawn is black, given that the first ball drawn was black, is $\Pr\{E_2|E_1\} = 1/(3+1) = \frac{1}{4}$. Thus the probability that both balls drawn are black is

$$\Pr\{E_1E_2\} = \Pr\{E_1\}\Pr\{E_2|E_1\} = \frac{2}{5}\cdot\frac{1}{4} = \frac{1}{10}$$

MUTUALLY EXCLUSIVE EVENTS

Two or more events are called *mutually exclusive* if the occurrence of any one of them excludes the occurrence of the others. Thus if E_1 and E_2 are mutually exclusive events, then $\Pr\{E_1E_2\} = 0$.

If $E_1 + E_2$ denotes the event that "either E_1 or E_2 or both occur," then

$$\Pr\{E_1 + E_2\} = \Pr\{E_1\} + \Pr\{E_2\} - \Pr\{E_1E_2\} \tag{5}$$

In particular,

$$\Pr\{E_1 + E_2\} = \Pr\{E_1\} + \Pr\{E_2\} \qquad \text{for mutually exclusive events} \tag{6}$$

As an extension of this, if $E_1, E_2, \ldots, E_n$ are n mutually exclusive events having respective probabilities of occurrence $p_1, p_2, \ldots, p_n$, then the probability of occurrence of either E_1 or E_2 or $\cdots E_n$ is $p_1 + p_2 + \cdots + p_n$.

Result (*5*) can also be generalized to three or more mutually exclusive events.

EXAMPLE 6. If E_1 is the event "drawing an ace from a deck of cards" and E_2 is the event "drawing a king," then $\Pr\{E_1\} = \frac{4}{52} = \frac{1}{13}$ and $\Pr\{E_2\} = \frac{4}{52} = \frac{1}{13}$. The probability of drawing either an ace or a king in a single draw is

$$\Pr\{E_1 + E_2\} = \Pr\{E_1\} + \Pr\{E_2\} = \frac{1}{13} + \frac{1}{13} = \frac{2}{13}$$

since both an ace and a king cannot be drawn in a single draw and are thus mutually exclusive events (Fig. 6-1).

A♣	2♣	3♣	4♣	5♣	6♣	7♣	8♣	9♣	10♣	J♣	Q♣	K♣
A♦	2♦	3♦	4♦	5♦	6♦	7♦	8♦	9♦	10♦	J♦	Q♦	K♦
A♥	2♥	3♥	4♥	5♥	6♥	7♥	8♥	9♥	10♥	J♥	Q♥	K♥
A♠	2♠	3♠	4♠	5♠	6♠	7♠	8♠	9♠	10♠	J♠	Q♠	K♠

Fig. 6-1 E_1 is the event "drawing an ace" and E_2 is the event "drawing a king."

Note that E_1 and E_2 have no outcomes in common. They are mutually exclusive.

EXAMPLE 7. If E_1 is the event "drawing an ace" from a deck of cards and E_2 is the event "drawing a spade," then E_1 and E_2 are not mutually exclusive since the ace of spades can be drawn (Fig. 6-2). Thus the probability of drawing either an ace or a spade or both is

$$\Pr\{E_1 + E_2\} = \Pr\{E_1\} + \Pr\{E_2\} - \Pr\{E_1E_2\} = \frac{4}{52} + \frac{13}{52} - \frac{1}{52} = \frac{16}{52} = \frac{4}{13}$$

A♣	2♣	3♣	4♣	5♣	6♣	7♣	8♣	9♣	10♣	J♣	Q♣	K♣
A♦	2♦	3♦	4♦	5♦	6♦	7♦	8♦	9♦	10♦	J♦	Q♦	K♦
A♥	2♥	3♥	4♥	5♥	6♥	7♥	8♥	9♥	10♥	J♥	Q♥	K♥
A♠	2♠	3♠	4♠	5♠	6♠	7♠	8♠	9♠	10♠	J♠	Q♠	K♠

Fig. 6-2 E_1 is the event "drawing an ace" and E_2 is the event "drawing a spade."

Note that the event "E_1 and E_2" consisting of those outcomes in both events is the ace of spades.

PROBABILITY DISTRIBUTIONS

Discrete

If a variable X can assume a discrete set of values $X_1, X_2, \ldots, X_K$ with respective probabilities $p_1, p_2, \ldots, p_K$, where $p_1 + p_2 + \cdots + p_K = 1$, we say that a *discrete probability distribution* for X has been defined. The function $p(X)$, which has the respective values $p_1, p_2, \ldots, p_K$ for $X = X_1, X_2, \ldots, X_K$, is called the *probability function*, or *frequency function*, of X. Because X can assume certain values with given probabilities, it is often called a *discrete random variable*. A random variable is also known as a *chance variable* or *stochastic variable*.

EXAMPLE 8. Let a pair of fair dice be tossed and let X denote the sum of the points obtained. Then the probability distribution is as shown in Table 6.1. For example, the probability of getting sum 5 is $\frac{4}{36} = \frac{1}{9}$; thus in 900 tosses of the dice we would expect 100 tosses to give the sum 5.

Note that this is analogous to a relative-frequency distribution with probabilities replacing the relative frequencies. Thus we can think of probability distributions as theoretical or ideal limiting

Table 6.1

X	2	3	4	5	6	7	8	9	10	11	12
$p(X)$	1/36	2/36	3/36	4/36	5/36	6/36	5/36	4/36	3/36	2/36	1/36

forms of relative-frequency distributions when the number of observations made is very large. For this reason, we can think of probability distributions as being distributions of *populations*, whereas relative-frequency distributions are distributions of *samples* drawn from this population.

The probability distribution can be represented graphically by plotting $p(X)$ against X, just as for relative-frequency distributions (see Problem 6.11).

By cumulating probabilities, we obtain *cumulative probability distributions*, which are analogous to cumulative relative-frequency distributions. The function associated with this distribution is sometimes called a *distribution function*.

The distribution in Table 6.1 can be built using EXCEL. The following portion of an EXCEL worksheet is built by entering Diel into A1, Die2 into B1, and Sum into C1. The 36 outcomes on the dice are entered into A2:B37. In C2, =SUM(A2:B2) is entered and a click-and-drag is performed from C2 to C37. By noting that the sum 2 occurs once, the sum 3 occurs twice, etc., the probability distribution in Table 6.1 is formed.

Die1	Die2	Sum
1	1	2
1	2	3
1	3	4
1	4	5
1	5	6
1	6	7
2	1	3
2	2	4
2	3	5
2	4	6
2	5	7

Die1	Die2	Sum
2	6	8
3	1	4
3	2	5
3	3	6
3	4	7
3	5	8
3	6	9
4	1	5
4	2	6
4	3	7
4	4	8
4	5	9
4	6	10
5	1	6
5	2	7
5	3	8
5	4	9
5	5	10
5	6	11
6	1	7
6	2	8
6	3	9
6	4	10
6	5	11
6	6	12

Continuous

The above ideas can be extended to the case where the variable X may assume a continuous set of values. The relative-frequency polygon of a sample becomes, in the theoretical or limiting case of a population, a continuous curve (such as shown in Fig. 6-3) whose equation is $Y = p(X)$. The total area under this curve bounded by the X axis is equal to 1, and the area under the curve between lines $X = a$ and $X = b$ (shaded in Fig. 6-3) gives the probability that X lies between a and b, which can be denoted by $\Pr\{a < X < b\}$.

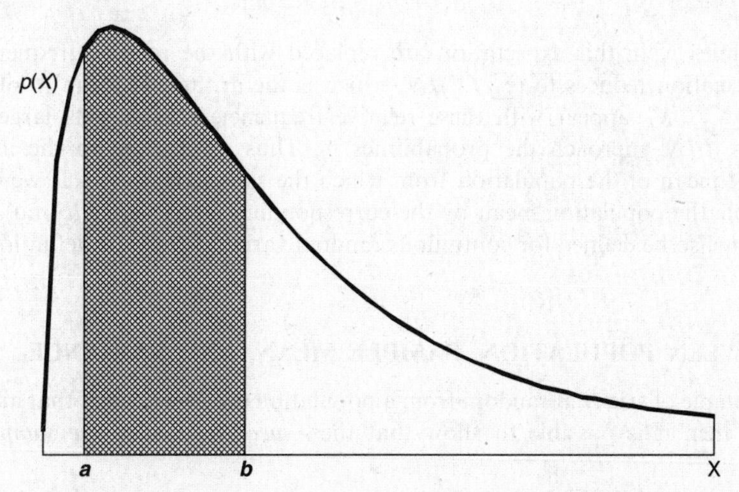

Fig. 6-3 $\Pr\{a<X<b\}$ is shown as the cross-hatched area under the density function.

We call $p(X)$ a *probability density function*, or briefly a *density function*, and when such a function is given we say that a *continuous probability distribution* for X has been defined. The variable X is then often called a *continuous random variable*.

As in the discrete case, we can define cumulative probability distributions and the associated distribution functions.

MATHEMATICAL EXPECTATION

If p is the probability that a person will receive a sum of money S, the *mathematical expectation* (or simply the *expectation*) is defined as pS.

EXAMPLE 9. Find $E(X)$ for the distribution of the sum of the dice given in Table 6.1. The distribution is given in the following EXCEL printout. The distribution is given in A2:B12 where the $p(X)$ values have been converted to their decimal equivalents. In C2, the expression =A2*B2 is entered and a click-and-drag is performed from C2 to C12. In C13, the expression =Sum(C2:C12) gives the mathematical expectation which equals 7.

X	p(X)	XP(X)
2	0.027778	0.055556
3	0.055556	0.166667
4	0.083333	0.333333
5	0.111111	0.555556
6	0.138889	0.833333
7	0.166667	1.166667
8	0.138889	1.111111
9	0.111111	1
10	0.083333	0.833333
11	0.055556	0.611111
12	0.027778	0.333333
		7

The concept of expectation is easily extended. If X denotes a discrete random variable that can assume the values $X_1, X_2, \ldots, X_K$ with respective probabilities $p_1, p_2, \ldots, p_K$, where $p_1 + p_2 + \cdots + p_K = 1$, the *mathematical expectation* of X (or simply the *expectation* of X), denoted by $E(X)$, is defined as

$$E(X) = p_1X_1 + p_2X_2 + \cdots + p_KX_K = \sum_{j=1}^{K} p_jX_j = \sum pX \qquad (7)$$

If the probabilities p_j in this expectation are replaced with the relative frequencies f_j/N, where $N = \sum f_j$, the expectation reduces to $(\sum fX)/N$, which is the arithmetic mean $\bar{X}$ of a sample of size N in which $X_1, X_2, \ldots, X_K$ appear with these relative frequencies. As N gets larger and larger, the relative frequencies f_j/N approach the probabilities p_j. Thus we are led to the interpretation that $E(X)$ represents the mean of the population from which the sample is drawn. If we call m the sample mean, we can denote the population mean by the corresponding Greek letter μ (mu).

Expectation can also be defined for continuous random variables, but the definition requires the use of calculus.

RELATION BETWEEN POPULATION, SAMPLE MEAN, AND VARIANCE

If we select a sample of size N at random from a population (i.e., we assume that all such samples are equally probable), then it is possible to show that the *expected value of the sample mean m is the population mean μ*.

It does not follow, however, that the expected value of any quantity computed from a sample is the corresponding population quantity. For example, the expected value of the sample variance as we have defined it is not the population variance, but $(N-1)/N$ times this variance. This is why some statisticians choose to define the sample variance as our variance multiplied by $N/(N-1)$.

COMBINATORIAL ANALYSIS

In obtaining probabilities of complex events, an enumeration of cases is often difficult, tedious, or both. To facilitate the labor involved, use is made of basic principles studied in a subject called *combinatorial analysis.*

Fundamental Principle

If an event can happen in any one of n_1 ways, and if when this has occurred another event can happen in any one of n_2 ways, then the number of ways in which both events can happen in the specified order is $n_1 n_2$.

EXAMPLE 10. The numbers 0 through 5 are entered into A1 through A6 in EXCEL and =FACT(A1) is entered into B1 and a click-and-drag is performed from B1 through B6. Then the chart wizard is used to plot the points. The function =FACT(n) is the same as $n!$. For $n = 0, 1, 2, 3, 4$, and 5, =FACT(n) equals 1, 1, 2, 6, 24, and 120. Figure 6.4 was generated by the chart wizard in EXCEL.

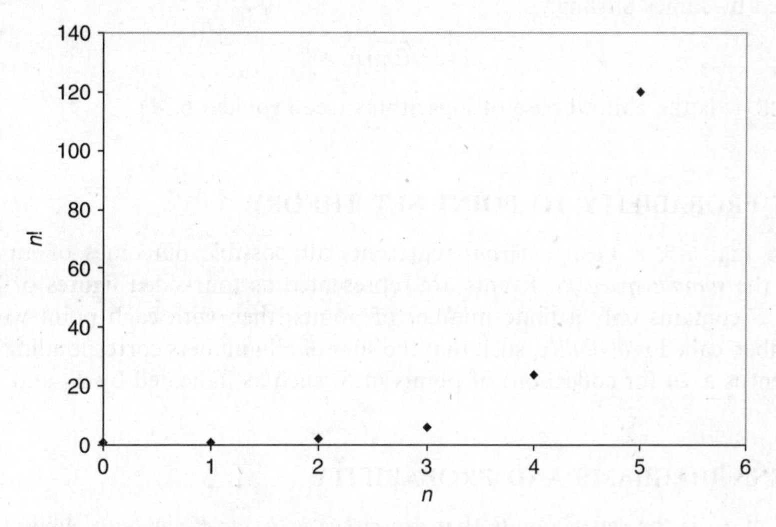

Fig. 6-4 Chart for $n!$ generated by EXCEL.

EXAMPLE 11. The number of permutations of the letters *a*, *b*, and *c* taken two at a time is ${}_3P_2 = 3 \cdot 2 = 6$. These are *ab*, *ba*, *ac*, *ca*, *bc*, and *cb*.

The number of permutations of n objects consisting of groups of which n_1 are alike, n_2 are alike, $\cdots$ is

$$\frac{n!}{n_1!\, n_2! \cdots} \qquad \text{where } n = n_1 + n_2 + \cdots \tag{10}$$

EXAMPLE 12. The number of permutations of letters in the word *statistics* is

$$\frac{10!}{3!\,3!\,1!\,2!\,1!} = 50,400$$

since there are 3*s*'s, 3*t*'s, 1*a*, 2*i*'s, and 1*c*.

COMBINATIONS

A combination of n different objects taken r at a time is a selection of r out of the n objects, with no attention given to the order of arrangement. The number of combinations of n objects taken r at a time is denoted by the symbol $\binom{n}{r}$ and is given by

$$\binom{n}{r} = \frac{n(n-1)\cdots(n-r+1)}{r!} = \frac{n!}{r!(n-r)!} \tag{11}$$

EXAMPLE 13. The number of combinations of the letters a, b, and c taken two at a time is

$$\binom{3}{2} = \frac{3 \cdot 2}{2!} = 3$$

These are ab, ac, and bc. Note that ab is the same combination as ba, but not the same permutation.

The number of combinations of 3 things taken 2 at a time is given by the EXCEL command = COMBIN(3,2) which gives the number 3.

STIRLING'S APPROXIMATION TO *n*!

When n is large, a direct evaluation of $n!$ is impractical. In such cases, use is made of an approximate formula developed by James Stirling:

$$n! \approx \sqrt{2\pi n}\, n^n\, e^{-n} \tag{12}$$

where $e = 2.71828\cdots$ is the natural base of logarithms (see Problem 6.31).

RELATION OF PROBABILITY TO POINT SET THEORY

As shown in Fig. 6-5, a *Venn diagram* represents all possible outcomes of an *experiment* as a rectangle, called the *sample space*, S. Events are represented as four-sided figures or circles inside the sample space. If S contains only a finite number of points, then with each point we can associate a nonnegative number, called *probability*, such that the sum of all numbers corresponding to all points in S add to 1. An event is a set (or collection) of points in S, such as indicated by E_1 and E_2 in Fig. 6-5.

EULER OR VENN DIAGRAMS AND PROBABILITY

The event $E_1 + E_2$ is the set of points that are *either in* E_1 *or* E_2 *or both*, while the event E_1E_2 is the set of points *common to both* E_1 *and* E_2. Thus the probability of an event such as E_1 is the sum of the probabilities associated with all points contained in the set E_1. Similarly, the probability of $E_1 + E_2$, denoted by $\Pr\{E_1 + E_2\}$, is the sum of the probabilities associated with all points contained in the set $E_1 + E_2$. If E_1 and E_2 have no points in common (i.e., the events are mutually exclusive), then $\Pr\{E_1 + E_2\} = \Pr\{E_1\} + \Pr\{E_2\}$. If they have points in common, then $\Pr\{E_1 + E_2\} = \Pr\{E_1\} + \Pr\{E_2\} - \Pr\{E_1E_2\}$.

The set $E_1 + E_2$ is sometimes denoted by $E_1 \cup E_2$ and is called the *union* of the two sets. The set E_1E_2 is sometimes denoted by $E_1 \cap E_2$ and is called the *intersection* of the two sets. Extensions to more than two sets can be made; thus instead of $E_1 + E_2 + E_3$ and $E_1E_2E_3$, we could use the notations $E_1 \cup E_2 \cup E_3$ and $E_1 \cap E_2 \cap E_3$, respectively.

The symbol ϕ (the Greek letter *phi*) is sometimes used to denote a set with no points in it, called the *null set*. The probability associated with an event corresponding to this set is zero (i.e., $\Pr\{\phi\} = 0$). If E_1 and E_2 have no points in common, we can write $E_1E_2 = \phi$, which means that the corresponding

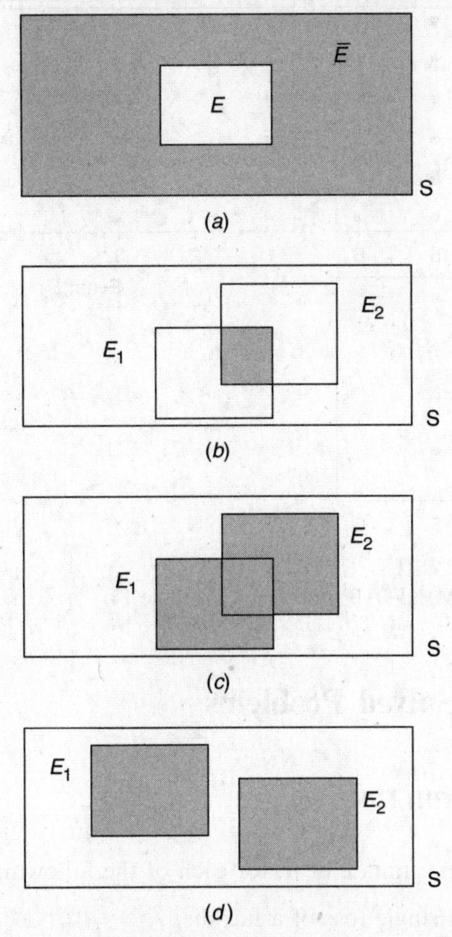

Fig. 6-5 Operations on events. (*a*) Complement of event E shown in gray, written as $\bar{E}$; (*b*) Intersection of events E_1 and E_2 shown in gray, written as $E_1 \cap E_2$; (*c*) Union of events E_1 and E_2 shown in gray, written as $E_1 \cup E_2$; (*d*) Mutually exclusive events E_1 and E_2, that is $E_1 \cap E_2 = \phi$.

events are mutually exclusive, whereby $\Pr\{E_1E_2\} = 0$.

With this modern approach, a random variable is a function defined at each point of the sample space. For example, in Problem 6.37 the random variable is the sum of the coordinates of each point.

In the case where S has an infinite number of points, the above ideas can be extended by using concepts of calculus.

EXAMPLE 14. An experiment consists of rolling a pair of dice. Event E_1 is the event that a 7 occurs, i. e., the sum on the dice is 7. Event E_2 is the event that an odd number occurs on die 1. Sample space S, events E_1 and E_2 are shown. Find $\Pr\{E_1\}$, $\Pr\{E_2\}$, $\Pr\{E_1 \cap E_2\}$ and $\Pr(E_1 \cup E_2)$. E_1, E_2, and S are shown in separate panels of the MINITAB output shown in Fig. 6-6.

$$P(E_1) = 6/36 = 1/6 \quad P(E_2) = 18/36 = 1/2 \quad P(E_1 \cap E_2) = 3/36 = 1/12$$
$$P(E_1 \cup E_2) = 6/36 + 18/36 - 3/36 = 21/36.$$

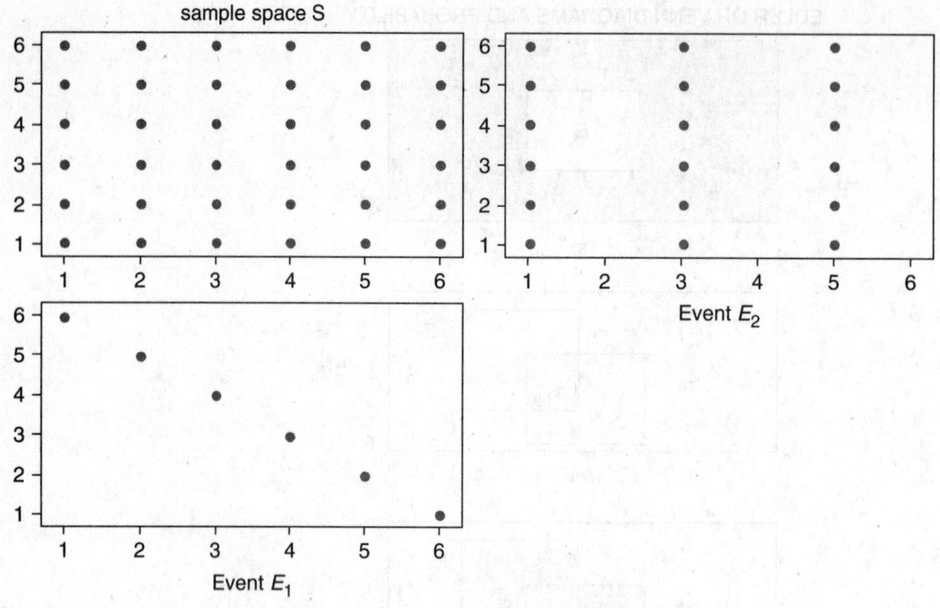

Fig. 6-6 MINITAB output for Example 14.

Solved Problems

FUNDAMENTAL RULES OF PROBABILITY

6.1 Determine the probability p, or an estimate of it, for each of the following events:

(*a*) An odd number appears in a single toss of a fair die.

(*b*) At least one head appears in two tosses of a fair coin.

(*c*) An ace, 10 of diamonds, or 2 of spades appears in drawing a single card from a well-shuffled ordinary deck of 52 cards.

(*d*) The sum 7 appears in a single toss of a pair of fair dice.

(*e*) A tail appears in the next toss of a coin if out of 100 tosses 56 were heads.

SOLUTION

(*a*) Out of six possible equally likely cases, three cases (where the die comes up 1, 3, or 5) are favorable to the event. Thus $p = \frac{3}{6} = \frac{1}{2}$.

(*b*) If H denotes "head" and T denotes "tail," the two tosses can lead to four cases: HH, HT, TH, and TT, all equally likely. Only the first three cases are favorable to the event. Thus $p = \frac{3}{4}$.

(*c*) The event can occur in six ways (ace of spades, ace of hearts, ace of clubs, ace of diamonds, 10 of diamonds, and 2 of spades) out of 52 equally likely cases. Thus $p = \frac{6}{52} = \frac{3}{26}$.

(*d*) Each of the six faces of one die can be associated with each of the six faces of the other die, so that the total number of cases that can arise, all equally likely, is $6 \cdot 6 = 36$. These can be denoted by $(1, 1)$, $(2, 1)$, $(3, 1), \ldots, (6, 6)$.

There are six ways of obtaining the sum 7, denoted by $(1, 6)$, $(2, 5)$, $(3, 4)$, $(4, 3)$, $(5, 2)$, and $(6, 1)$. Thus $p = \frac{6}{36} = \frac{1}{6}$.

(*e*) Since $100 - 56 = 44$ tails were obtained in 100 tosses, the *estimated* (or *empirical*) *probability* of a tail is the relative frequency $44/100 = 0.44$.

6.2 An experiment consists of tossing a coin and a die. If E_1 is the event that "head" comes up in tossing the coin and E_2 is the event that "3 or 6" comes up in tossing the die, state in words the meaning of each of the following:

(*a*) $\bar{E}_1$ (*c*) E_1E_2 (*e*) $\Pr\{E_1|E_2\}$
(*b*) $\bar{E}_2$ (*d*) $\Pr\{E_1\bar{E}_2\}$ (*f*) $\Pr\{\bar{E}_1+\bar{E}_2\}$

SOLUTION

(*a*) Tails on the coin and anything on the die
(*b*) 1, 2, 4, or 5 on the die and anything on the coin
(*c*) Heads on the coin and 3 or 6 on the die
(*d*) Probability of heads on the coin and 1, 2, 4, or 5 on the die
(*e*) Probability of heads on the coin, given that a 3 or 6 has come up on the die
(*f*) Probability of tails on the coin or 1, 2, 4, or 5 on the die, or both

6.3 A ball is drawn at random from a box containing 6 red balls, 4 white balls, and 5 blue balls. Determine the probability that the ball drawn is (*a*) red, (*b*) white, (*c*) blue, (*d*) not red, and (*e*) red or white.

SOLUTION

Let *R*, *W*, and *B* denote the events of drawing a red ball, white ball, and blue ball, respectively. Then:

(*a*) $$\Pr\{R\}=\frac{\text{ways of choosing a red ball}}{\text{total ways of choosing a ball}}=\frac{6}{6+4+5}=\frac{6}{15}=\frac{2}{5}$$

(*b*) $$\Pr\{W\}=\frac{4}{6+4+5}=\frac{4}{15}$$

(*c*) $$\Pr\{B\}=\frac{5}{6+4+5}=\frac{5}{15}=\frac{1}{3}$$

(*d*) $$\Pr\{\bar{R}\}=1-\Pr\{R\}=1-\frac{2}{5}=\frac{3}{5} \quad \text{by part } (a)$$

(*e*) $$\Pr\{R+W\}=\frac{\text{ways of choosing a red or white ball}}{\text{total ways of choosing a ball}}=\frac{6+4}{6+4+5}=\frac{10}{15}=\frac{2}{3}$$

Another method

$$\Pr\{R+W\}=\Pr\{\bar{B}\}=1-\Pr\{B\}=1-\frac{1}{3}=\frac{2}{3} \quad \text{by part } (c)$$

Note that $\Pr\{R+W\}=\Pr\{R\}+\Pr\{W\}$ (i.e., $\frac{2}{3}=\frac{2}{5}+\frac{4}{15}$). This is an illustration of the general rule $\Pr\{E_1+E_2\}=\Pr\{E_1\}+\Pr\{E_2\}$ that is true for *mutually exclusive* events E_1 and E_2.

6.4 A fair die is tossed twice. Find the probability of getting a 4, 5, or 6 on the first toss and a 1, 2, 3, or 4 on the second toss.

SOLUTION

Let E_1 = event "4, 5, or 6" on the first toss, and let E_2 = event "1, 2, 3, or 4" on the second toss. Each of the six ways in which the die can fall on the first toss can be associated with each of the six ways in which it can fall on the second toss, a total of $6 \cdot 6 = 36$ ways, all equally likely. Each of the three ways in which E_1 can occur can be associated with each of the four ways in which E_2 can occur, to give $3 \cdot 4 = 12$ ways in which both E_1 and E_2, or E_1E_2 occur. Thus $\Pr\{E_1E_2\}=12/36=1/3$.

Note that $\Pr\{E_1E_2\}=\Pr\{E_1\}\Pr\{E_2\}$ (i.e., $\frac{1}{3}=\frac{3}{6}\cdot\frac{4}{6}$) is valid for the *independent events* E_1 and E_2.

6.5 Two cards are drawn from a well-shuffled ordinary deck of 52 cards. Find the probability that they are both aces if the first card is (*a*) replaced and (*b*) not replaced.

SOLUTION

Let E_1 = event "ace" on the first draw, and let E_2 = event "ace" on the second draw.

(*a*) If the first card is replaced, E_1 and E_2 are independent events. Thus Pr{both cards drawn are aces} $= \Pr\{E_1E_2\} = \Pr\{E_1\}\Pr\{E_2\} = (\frac{4}{52})(\frac{4}{52}) = \frac{1}{169}$.

(*b*) The first card can be drawn in any one of 52 ways, and the second card can be drawn in any one of 51 ways since the first card is not replaced. Thus both cards can be drawn in $52 \cdot 51$ ways, all equally likely.

There are four ways in which E_1 can occur and three ways in which E_2 can occur, so that both E_1 and E_2, or E_1E_2, can occur in $4 \cdot 3$ ways. Thus $\Pr\{E_1E_2\} = (4 \cdot 3)/(52 \cdot 51) = \frac{1}{221}$.

Note that $\Pr\{E_2|E_1\}$ = Pr{second card is an ace given that first card is an ace} $= \frac{3}{51}$. Thus our result is an illustration of the general rule that $\Pr\{E_1E_2\} = \Pr\{E_1\}\Pr\{E_2|E_1\}$ when E_1 and E_2 are dependent events.

6.6 Three balls are drawn successively from the box of Problem 6.3. Find the probability that they are drawn in the order red, white, and blue if each ball is (*a*) replaced and (*b*) not replaced.

SOLUTION

Let R = event "red" on the first draw, W = event "white" on the second draw, and B = event "blue" on the third draw. We require $\Pr\{RWB\}$.

(*a*) If each ball is replaced, then R, W, and B are independent events and

$$\Pr\{RWB\} = \Pr\{R\}\Pr\{W\}\Pr\{B\} = \left(\frac{6}{6+4+5}\right)\left(\frac{4}{6+4+5}\right)\left(\frac{5}{6+4+5}\right) = \left(\frac{6}{15}\right)\left(\frac{4}{15}\right)\left(\frac{5}{15}\right) = \frac{8}{225}$$

(*b*) If each ball is not replaced, then R, W, and B are dependent events and

$$\Pr\{RWB\} = \Pr\{R\}\Pr\{W|R\}\Pr\{B|WR\} = \left(\frac{6}{6+4+5}\right)\left(\frac{4}{5+4+5}\right)\left(\frac{5}{5+3+5}\right)$$
$$= \left(\frac{6}{15}\right)\left(\frac{4}{14}\right)\left(\frac{5}{13}\right) = \frac{4}{91}$$

where $\Pr\{B|WR\}$ is the conditional probability of getting a blue ball if a white and red ball have already been chosen.

6.7 Find the probability of a 4 turning up at least once in two tosses of a fair die.

SOLUTION

Let E_1 = event "4" on the first toss, E_2 = event "4" on the second toss, and $E_1 + E_2$ = event "4" on the first toss or "4" on the second toss or both = event that at least one 4 turns up. We require $\Pr\{E_1 + E_2\}$.

First method

The total number or equally likely ways in which both dice can fall is $6 \cdot 6 = 36$. Also,

Number of ways in which E_1 occurs but not E_2 = 5
Number of ways in which E_2 occurs but not E_1 = 5
Number of ways in which both E_1 and E_2 occur = 1

Thus the number of ways in which at least one of the events E_1 or E_2 occurs is $5 + 5 + 1 = 11$, and thus $\Pr\{E_1 + E_2\} = \frac{11}{36}$.

Second method

Since E_1 and E_2 are not mutually exclusive, $\Pr\{E_1 + E_2\} = \Pr\{E_1\} + \Pr\{E_2\} - \Pr\{E_1E_2\}$. Also, since E_1 and E_2 are independent, $\Pr\{E_1E_2\} = \Pr\{E_1\}\Pr\{E_2\}$. Thus $\Pr\{E_1 + E_2\} = \Pr\{E_1\} + \Pr\{E_2\} - \Pr\{E_1\}\Pr\{E_2\} = \frac{1}{6} + \frac{1}{6} - (\frac{1}{6})(\frac{1}{6}) = \frac{11}{36}$.

Third method

$$\Pr\{\text{at least one 4 comes up}\} + \Pr\{\text{no 4 comes up}\} = 1$$

Thus
$$\begin{aligned}\Pr\{\text{at least one 4 comes up}\} &= 1 - \Pr\{\text{no 4 comes up}\}\\ &= 1 - \Pr\{\text{no 4 on first toss and no 4 on second toss}\}\\ &= 1 - \Pr\{\bar{E}_1\bar{E}_2\} = 1 - \Pr\{\bar{E}_1\}\Pr\{\bar{E}_2\}\\ &= 1 - (\tfrac{5}{6})(\tfrac{5}{6}) = \tfrac{11}{36}\end{aligned}$$

6.8 One bag contains 4 white balls and 2 black balls; another contains 3 white balls and 5 black balls. If one ball is drawn from each bag, find the probability that (*a*) both are white, (*b*) both are black, and (*c*) one is white and one is black.

SOLUTION

Let W_1 = event "white" ball from the first bag, and let W_2 = event "white" ball from the second bag.

(*a*)
$$\Pr\{W_1W_2\} = \Pr\{W_1\}\Pr\{W_2\} = \left(\frac{4}{4+2}\right)\left(\frac{3}{3+5}\right) = \frac{1}{4}$$

(*b*)
$$\Pr\{\bar{W}_1\bar{W}_2\} = \Pr\{\bar{W}_1\}\Pr\{\bar{W}_2\} = \left(\frac{2}{4+2}\right)\left(\frac{5}{3+5}\right) = \frac{5}{24}$$

(*c*) The event "one is white and one is black" is the same as the event "either the first is white and the second is black *or* the first is black and the second is white"; that is, $W_1\bar{W}_2 + \bar{W}_1W_2$. Since events $W_1\bar{W}_2$ and $\bar{W}_1W_2$ are mutually exclusive, we have

$$\begin{aligned}\Pr\{W_1\bar{W}_2 + \bar{W}_1W_2\} &= \Pr\{W_1\bar{W}_2\} + \Pr\{\bar{W}_1W_2\}\\ &= \Pr\{W_1\}\Pr\{\bar{W}_2\} + \Pr\{\bar{W}_1\}\Pr\{W_2\}\\ &= \left(\frac{4}{4+2}\right)\left(\frac{5}{3+5}\right) + \left(\frac{2}{4+2}\right)\left(\frac{3}{3+5}\right) = \frac{13}{24}\end{aligned}$$

Another method

The required probability is $1 - \Pr\{W_1W_2\} - \Pr\{\bar{W}_1\bar{W}_2\} = 1 - \frac{1}{4} - \frac{5}{24} = \frac{13}{24}$.

6.9 *A* and *B* play 12 games of chess, of which 6 are won by *A*, 4 are won by *B*, and 2 end in a draw. They agree to play a match consisting of 3 games. Find the probability that (*a*) *A* wins all 3 games, (*b*) 2 games end in a draw, (*c*) *A* and *B* win alternately, and (*d*) *B* wins at least 1 game.

SOLUTION

Let A_1, A_2, and A_3 denote the events "*A* wins" in the first, second, and third games, respectively; let B_1, B_2, and B_3 denote the events "*B* wins" in the first, second, and third games, respectively; and let, D_1, D_2, and D_3 denote the events "there is a draw" in the first, second, and third games, respectively.

On the basis of their past experience (empirical probability), we shall assume that Pr{*A* wins any one game} $= \frac{6}{12} = \frac{1}{2}$, that Pr{*B* wins any one game} $= \frac{4}{12} = \frac{1}{3}$, and that Pr{any one game ends in a draw} $= \frac{2}{12} = \frac{1}{6}$.

(*a*) $\Pr\{A \text{ wins all 3 games}\} = \Pr\{A_1A_2A_3\} = \Pr\{A_1\}\Pr\{A_2\}\Pr\{A_3\} = \left(\frac{1}{2}\right)\left(\frac{1}{2}\right)\left(\frac{1}{2}\right) = \frac{1}{8}$

assuming that the results of each game are independent of the results of any others, which appears to be justifiable (unless, of course, the players happen to be *psychologically influenced* by the other one's winning or losing).

(*b*) $$\begin{aligned}\Pr\{2 \text{ games end in a draw}\} &= \Pr\{\text{1st and 2nd } or \text{ 1st and 3rd } or \text{ 2nd and 3rd games end in a draw}\}\\ &= \Pr\{D_1D_2\bar{D}_3\} + \Pr\{D_1\bar{D}_2D_3\} + \Pr\{\bar{D}_1D_2D_3\}\\ &= \Pr\{D_1\}\Pr\{D_2\}\Pr\{\bar{D}_3\} + \Pr\{D_1\}\Pr\{\bar{D}_2\}\Pr\{D_3\}\\ &\quad + \Pr\{\bar{D}_1\}\Pr\{D_2\}\Pr\{D_3\}\\ &= \left(\frac{1}{6}\right)\left(\frac{1}{6}\right)\left(\frac{5}{6}\right) + \left(\frac{1}{6}\right)\left(\frac{5}{6}\right)\left(\frac{1}{6}\right) + \left(\frac{5}{6}\right)\left(\frac{1}{6}\right)\left(\frac{1}{6}\right) = \frac{15}{216} = \frac{5}{72}\end{aligned}$$

(*c*) $$\begin{aligned}\Pr\{A \text{ and } B \text{ win alternately}\} &= \Pr\{A \text{ wins then } B \text{ wins then } A \text{ wins } or\ B \text{ wins then } A \text{ wins then } B \text{ wins}\}\\ &= \Pr\{A_1B_2A_3 + B_1A_2B_3\} = \Pr\{A_1B_2A_3\} + \Pr\{B_1A_2B_3\}\\ &= \Pr\{A_1\}\Pr\{B_2\}\Pr\{A_3\} + \Pr\{B_1\}\Pr\{A_2\}\Pr\{B_3\}\\ &= \left(\frac{1}{2}\right)\left(\frac{1}{3}\right)\left(\frac{1}{2}\right) + \left(\frac{1}{3}\right)\left(\frac{1}{2}\right)\left(\frac{1}{3}\right) = \frac{5}{36}\end{aligned}$$

(*d*) $$\begin{aligned}\Pr\{B \text{ wins at least 1 game}\} &= 1 - \Pr\{B \text{ wins no game}\}\\ &= 1 - \Pr\{\bar{B}_1\bar{B}_2\bar{B}_3\} = 1 - \Pr\{\bar{B}_1\}\Pr\{\bar{B}_2\}\Pr\{\bar{B}_3\}\\ &= 1 - \left(\frac{2}{3}\right)\left(\frac{2}{3}\right)\left(\frac{2}{3}\right) = \frac{19}{27}\end{aligned}$$

PROBABILITY DISTRIBUTIONS

6.10 Find the probability of boys and girls in families with three children, assuming equal probabilities for boys and girls.

SOLUTION

Let B = event "boy in the family," and let G = event "girl in the family." Thus according to the assumption of equal probabilities, $\Pr\{B\} = \Pr\{G\} = \frac{1}{2}$. In families of three children the following mutually exclusive events can occur with the corresponding indicated probabilities:

(*a*) Three boys (*BBB*):

$$\Pr\{BBB\} = \Pr\{B\}\Pr\{B\}\Pr\{B\} = \frac{1}{8}$$

Here we assume that the birth of a boy is not influenced in any manner by the fact that a previous child was also a boy, that is, we assume that the events are independent.

(*b*) Three girls (*GGG*): As in part (*a*) or by symmetry,

$$\Pr\{GGG\} = \frac{1}{8}$$

(*c*) Two boys and one girl ($BBG + BGB + GBB$):

$$\begin{aligned}\Pr\{BBG + BGB + GBB\} &= \Pr\{BBG\} + \Pr\{BGB\} + \Pr\{GBB\}\\ &= \Pr\{B\}\Pr\{B\}\Pr\{G\} + \Pr\{B\}\Pr\{G\}\Pr\{B\} + \Pr\{G\}\Pr\{B\}\Pr\{B\}\\ &= \frac{1}{8} + \frac{1}{8} + \frac{1}{8} = \frac{3}{8}\end{aligned}$$

(*d*) Two girls and one boy ($GGB + GBG + BGG$): As in part (*c*) or by symmetry, the probability is $\frac{3}{8}$.

If we call X the *random variable* showing the number of boys in families with three children, the probability distribution is as shown in Table 6.2.

Table 6.2

Number of boys X	0	1	2	3
Probability $p(X)$	1/8	3/8	3/8	1/8

6.11 Graph the distribution of Problem 6.10.

SOLUTION

The graph can be represented either as in Fig. 6-7 or Fig. 6-8. Note that the sum of the areas of the rectangles in Fig. 6-8 is 1; in this figure, called a *probability histogram*, we are considering X as a continuous variable even though it is actually discrete, a procedure that is often found useful. Figure 6-7, on the other hand, is used when one does not wish to consider the variable as continuous.

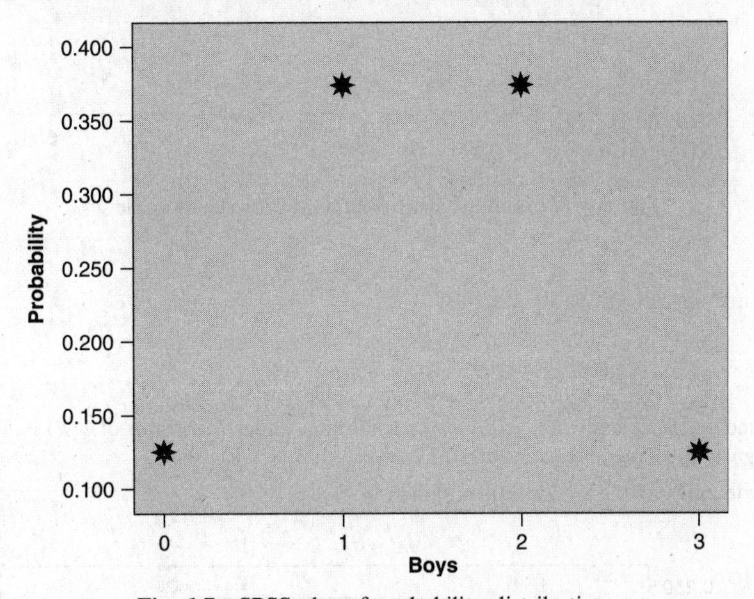

Fig. 6-7 SPSS plot of probability distribution.

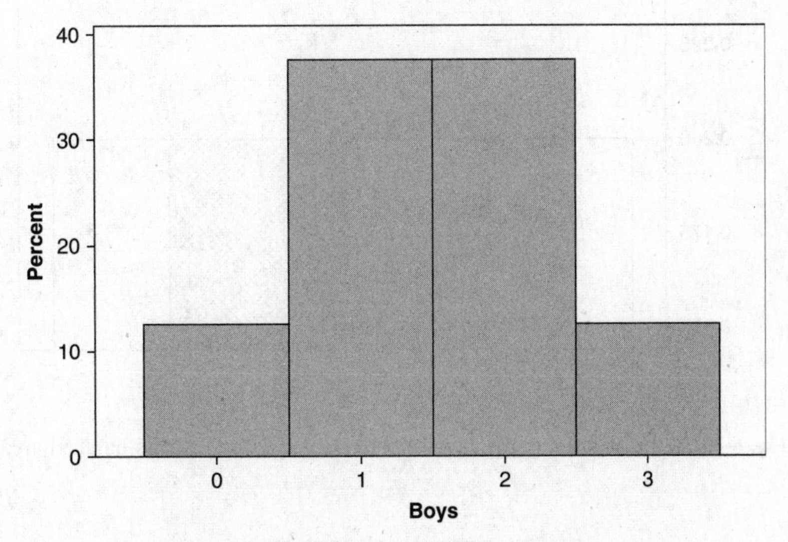

Fig. 6-8 MINITAB probability histogram.

6.12 A continuous random variable X, having values only between 0 and 5, has a density function given by $p(X) = \begin{cases} 0.2, & 0 < X < 5 \\ 0, & \text{otherwise} \end{cases}$. The graph is shown in Fig. 6-9.

(*a*) Verify that it is a density function.

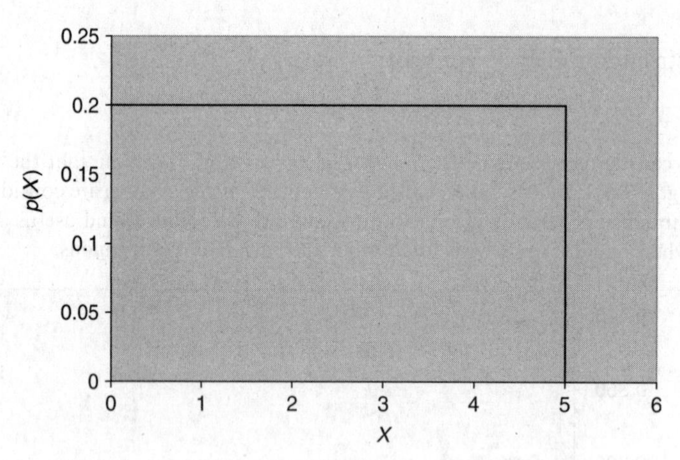

Fig. 6-9 Probability density function for the variable X.

(*b*) Find and graph $\Pr\{2.5 < X < 4.0\}$.

SOLUTION

(*a*) The function $p(X)$ is always ≥ 0 and the total area under the graph of $p(X)$ is $5 \times 0.2 = 1$ since it is rectangular in shape and has width 0.2 and length 5 (see Fig. 6-9).

(*b*) The probability $\Pr\{2.5 < X < 4.0\}$ is shown in Fig. 6-10.

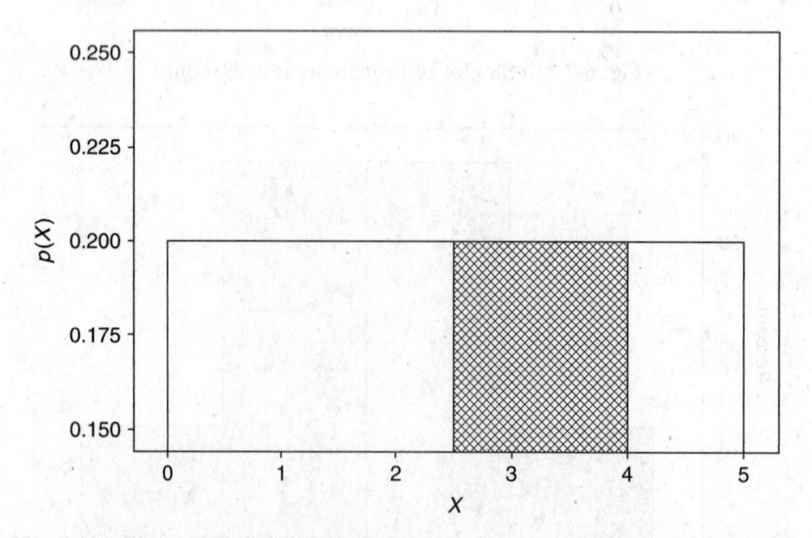

Fig. 6-10 The probability $\Pr(2.5 < X < 4.0)$ is shown as the cross-hatched area.

The rectangular area, $\Pr\{2.5 < X < 4.0\}$, is $(4 - 2.5) \times 0.2 = 0.3$.

MATHEMATICAL EXPECTATION

6.13 If a man purchases a raffle ticket, he can win a first prize of \$5000 or a second prize of \$2000 with probabilities 0.001 and 0.003. What should be a fair price to pay for the ticket?

SOLUTION

His expectation is (\$5000)(0.001) + (\$2000)(0.003) = \$5 + \$6 = \$11, which is a fair price to pay.

6.14 In a given business venture a lady can make a profit of \$300 with probability 0.6 or take a loss of \$100 with probability 0.4. Determine her expectation.

SOLUTION

Her expectation is (\$300)(0.6) + (−\$100)(0.4) = \$180 − \$40 = \$140.

6.15 Find (*a*) $E(X)$, (*b*) $E(X^2)$, and (*c*) $E[(X-\bar{X})^2)]$ for the probability distribution shown in Table 6.3.
(*d*) Give the EXCEL solution to parts (*a*), (*b*), and (*c*).

Table 6.3

X	8	12	16	20	24
$p(X)$	1/8	1/6	3/8	1/4	1/12

SOLUTION

(*a*) $E(X) = \sum Xp(X) = (8)(\frac{1}{8}) + (12)(\frac{1}{6}) + (16)(\frac{3}{8}) + (20)(\frac{1}{4}) + (24)(\frac{1}{12}) = 16$; this represents the *mean* of the distribution.

(*b*) $E(X^2) = \sum X^2p(X) = (8)^2(\frac{1}{8}) + (12)^2(\frac{1}{6}) + (16)^2(\frac{3}{8}) + (20)^2(\frac{1}{4}) + (24)^2(\frac{1}{12}) = 276$; this represents the *second moment* about the origin zero.

(*c*) $E[(X-\bar{X})^2] = \sum(X-\bar{X})^2p(X) = (8-16)^2(\frac{1}{8}) + (12-16)^2(\frac{1}{6}) + (16-16)^2(\frac{3}{8}) + (20-16)^2(\frac{1}{4}) + (24-16)^2(\frac{1}{12}) = 20$; this represents the *variance* of the distribution.

(*d*) The labels are entered into A1:E1 as shown. The X values and the probability values are entered into A2:B6. The expected values of X are computed in C2:C7. The expected value is given in C7. The second moment about the origin is computed in D2:D7. The second moment is given in D7. The variance is computed in E2:E7. The variance is given in E7.

A	B	C	D	E
X	P(X)	Xp(X)	X^2p(X)	(X − E(X))^2*p(X)
8	0.125	1	8	8
12	0.166667	2	24	2.666666667
16	0.375	6	96	0
20	0.25	5	100	4
24	0.083333	2	48	5.333333333
		16	276	20

6.16 A bag contains 2 white balls and 3 black balls. Each of four persons, A, B, C, and D, in the order named, draws one ball and does not replace it. The first to draw a white ball receives \$10. Determine the expectations of A, B, C, and D.

SOLUTION

Since only 3 black balls are present, one person must win on his or her first attempt. Denote by A, B, C, and D the events "A wins," "B wins," "C wins," and "D wins," respectively.

$$\Pr\{A \text{ wins}\} = \Pr\{A\} = \frac{2}{3+2} = \frac{2}{5}$$

Thus A's expectation $= \frac{2}{5}(\$10) = \$4$.

$$\Pr\{A \text{ loses and } B \text{ wins}\} = \Pr\{\bar{A}B\} = \Pr\{\bar{A}\}\Pr\{B|\bar{A}\} = \left(\frac{3}{5}\right)\left(\frac{2}{4}\right) = \frac{3}{10}$$

Thus B's expectation $= \$3$.

$$\Pr\{A \text{ and } B \text{ lose and } C \text{ wins}\} = \Pr\{\bar{A}\bar{B}C\} = \Pr\{\bar{A}\}\Pr\{\bar{B}|\bar{A}\}\Pr\{C\bar{A}\bar{B}\} = \left(\frac{3}{5}\right)\left(\frac{2}{4}\right)\left(\frac{2}{3}\right) = \frac{1}{5}$$

Thus C's expectation $= \$2$.

$$\begin{aligned}\Pr\{A, B, \text{ and } C \text{ lose and } D \text{ wins}\} &= \Pr\{\bar{A}\bar{B}\bar{C}D\} \\ &= \Pr\{\bar{A}\}\Pr\{\bar{B}|\bar{A}\}\Pr\{\bar{C}|\bar{A}\bar{B}\}\Pr\{D|\bar{A}\bar{B}\bar{C}\} \\ &= \left(\frac{3}{5}\right)\left(\frac{2}{4}\right)\left(\frac{1}{3}\right)\left(\frac{1}{1}\right) = \frac{1}{10}\end{aligned}$$

Thus D's expectation $= \$1$.

Check: $\$4 + \$3 + \$2 + \$1 = \$10$, and $\frac{2}{5} + \frac{3}{10} + \frac{1}{5} + \frac{1}{10} = 1$.

PERMUTATIONS

6.17 In how many ways can 5 differently colored marbles be arranged in a row?

SOLUTION

We must arrange the 5 marbles in 5 positions: – – – – –. The first position can be occupied by any one of 5 marbles (i.e., there are 5 ways of filling the first position). When this has been done, there are 4 ways of filling the second position. Then there are 3 ways of filling the third position, 2 ways of filling the fourth position, and finally only 1 way of filling the last position. Therefore:

$$\text{Number of arrangements of 5 marbles in a row} = 5 \cdot 4 \cdot 3 \cdot 2 \cdot 1 = 5! = 120$$

In general,

$$\text{Number of arrangements of } n \text{ different objects in a row} = n(n-1)(n-2)\cdots 1 = n!$$

This is also called the number of *permutations* of n different objects taken n at a time and is denoted by ${}_nP_n$.

6.18 In how many ways can 10 people be seated on a bench if only 4 seats are available?

SOLUTION

The first seat can be filled in any one of 10 ways, and when this has been done there are 9 ways of filling the second seat, 8 ways of filling the third seat, and 7 ways of filling the fourth seat. Therefore:

$$\text{Number of arrangements of 10 people taken 4 at a time} = 10 \cdot 9 \cdot 8 \cdot 7 = 5040$$

In general,

$$\text{Number of arrangements of } n \text{ different objects taken } r \text{ at a time} = n(n-1)\cdots(n-r+1)$$

This is also called the number of *permutations* of n different objects taken r at a time and is denoted by $_nP_r$, $P(n, r)$ or $P_{n,r}$. Note that when $r = n$, $_nP_n = n!$, as in Problem 6.17.

6.19 Evaluate (*a*) $_8P_3$, (*b*) $_6P_4$, (*c*) $_{15}P_1$, and $_3P_3$, and (*e*) parts (*a*) through (*d*) using EXCEL.

SOLUTION

(*a*) $_8P_3 = 8 \cdot 7 \cdot 6 = 336$, (*b*) $_6P_4 = 6 \cdot 5 \cdot 4 \cdot 3 = 360$, (*c*) $_{15}P_1 = 15$, and (*d*) $_3P_3 = 3 \cdot 2 \cdot 1 = 6$,

(*e*) =PERMUT(8, 3) = 336 = PERMUT(6, 4) = 360
=PERMUT(15, 1) = 15 = PERMUT(15, 1) = 6

6.20 It is required to seat 5 men and 4 women in a row so that the women occupy the even places. How many such arrangements are possible?

SOLUTION

The men may be seated in $_5P_5$ ways and the women in $_4P_4$ ways; each arrangement of the men may be associated with each arrangement of the women. Hence the required number of arrangements is $_5P_5 \cdot {_4P_4} = 5!4! = (120)(24) = 2880$.

6.21 How many four-digit numbers can be formed with the 10 digits 0, 1, 2, 3, ..., 9, if (*a*) repetitions are allowed, (*b*) repetitions are not allowed, and (*c*) the last digit must be zero and repetitions are not allowed?

SOLUTION

(*a*) The first digit can be any one of 9 (since 0 is not allowed). The second, third and fourth digits can be any one of 10. Then $9 \cdot 10 \cdot 10 \cdot 10 = 9000$ numbers can be formed.

(*b*) The first digit can be any one of 9 (any one but 0).
The second digit can be any one of 9 (any but that used for the first digit).
The third digit can be any one of 8 (any but those used for the first two digits).
The fourth digit can be any one of 7 (any but those used for the first three digits).
Thus $9 \cdot 9 \cdot 8 \cdot 7 = 4536$ numbers can be formed.

Another method

The first digit can be any one of 9 and the remaining three can be chosen in $_9P_3$ ways. Thus $9 \cdot {_9P_3} = 9 \cdot 9 \cdot 8 \cdot 7 = 4536$ numbers can be formed.

(*c*) The first digit can be chosen in 9 ways, the second in 8 ways, and the third in 7 ways. Thus $9 \cdot 8 \cdot 7 = 504$ numbers can be formed.

Another method

The first digit can be chosen in 9 ways and the next two digits in $_9P_2$ ways. Thus $9 \cdot {_8P_2} = 9 \cdot 8 \cdot 7 = 504$ numbers can be found.

6.22 Four different mathematics books, 6 different physics books, and 2 different chemistry books are to be arranged on a shelf. How many different arrangements are possible if (*a*) the books in each particular subject must all stand together and (*b*) only the mathematics books must stand together?

SOLUTION

(*a*) The mathematics books can be arranged among themselves in $_4P_4 = 4!$ ways, the physics books in $_6P_6 = 6!$ ways, the chemistry books in $_2P_2 = 2!$ ways, and the three groups in $_3P_3 = 3!$ ways. Thus the required number of arrangements $= 4!\,6!\,2!\,3! = 207{,}360$.

(*b*) Consider the 4 mathematics books as one big book. Then we have 9 books that can be arranged in ${}_9P_9 = 9!$ ways. In all of these ways the mathematics books are together. But the mathematics books can be arranged among themselves in ${}_4P_4 = 4!$ ways. Thus the required number of arrangements $= 9!\,4! = 8{,}709{,}120$.

6.23 Five red marbles, 2 white marbles, and 3 blue marbles are arranged in a row. If all the marbles of the same color are not distinguishable from each other, how many different arrangements are possible? Use the EXCEL function =MULTINOMIAL to evaluate the expression.

SOLUTION

Assume that there are P different arrangements. Multiplying P by the numbers of ways of arranging (*a*) the 5 red marbles among themselves, (*b*) the 2 white marbles among themselves, and (*c*) the 3 blue marbles among themselves (i.e., multiplying P by $5!\,2!\,3!$), we obtain the number of ways of arranging the 10 marbles if they are distinguishable (i.e., 10!). Thus

$$(5!2!3!)P = 10! \qquad \text{and} \qquad P = \frac{10!}{5!2!3!}$$

In general, the number of different arrangements of n objects of which n_1 are alike, n_2 are alike, $\ldots, n_k$ are alike is

$$\frac{n!}{n_1!n_2!\cdots n_k!}$$

where $n_1 + n_2 + \cdots + n_k = n$.

The EXCEL function =MULTINOMIAL(5,2,3) gives the value 2520.

6.24 In how many ways can 7 people be seated at a round table if (*a*) they can sit anywhere and (*b*) 2 particular people must not sit next to each other?

SOLUTION

(*a*) Let 1 of them be seated anywhere. Then the remaining 6 people can be seated in $6! = 720$ ways, which is the total number of ways of arranging the 7 people in a circle.

(*b*) Consider the 2 particular people as 1 person. Then there are 6 people altogether and they can be arranged in 5! ways. But the 2 people considered as 1 can be arranged among themselves in 2! ways. Thus the number of ways of arranging 6 people at a round table with 2 particular people sitting together $= 5!2! = 240$.

Then using part (*a*), the total number of ways in which 6 people can be seated at a round table so that the 2 particular people do not sit together $= 720 - 240 = 480$ ways.

COMBINATIONS

6.25 In how many ways can 10 objects be split into two groups containing 4 and 6 objects, respectively.

SOLUTION

This is the same as the number of arrangements of 10 objects of which 4 objects are alike and 6 other objects are alike. By Problem 6.23, this is

$$\frac{10!}{4!6!} = \frac{10 \cdot 9 \cdot 8 \cdot 7}{4!} = 210$$

The problem is equivalent to finding the number of selections of 4 out of 10 objects (or 6 out of 10 objects), the order of selection being immaterial.

In general the number of selections of r out of n objects, called the number of *combinations* of n things taken r at a time, is denoted by $\binom{n}{r}$ and is given by

$$\binom{n}{r} = \frac{n!}{r!(n-r)!} = \frac{n(n-1)\cdots(n-r+1)}{r!} = \frac{{}_nP_r}{r!}$$

6.26 Evaluate (*a*) $\binom{7}{4}$, (*b*) $\binom{6}{5}$, (*c*) $\binom{4}{4}$, and (*d*) parts (*a*) through (*c*) using EXCEL.

SOLUTION

(*a*)
$$\binom{7}{4} = \frac{7!}{4!\,3!} = \frac{7 \cdot 6 \cdot 5 \cdot 4}{4!} = \frac{7 \cdot 6 \cdot 5}{3 \cdot 2 \cdot 1} = 35$$

(*b*)
$$\binom{6}{5} = \frac{6!}{5!\,1!} = \frac{6 \cdot 5 \cdot 4 \cdot 3 \cdot 2}{5!} = 6 \quad \text{or} \quad \binom{6}{5} = \binom{6}{1} = 6$$

(*c*) $\binom{4}{4}$ is the number of selections of 4 objects taken all at a time, and there is only one such selection; thus $\binom{4}{4} = 1$, Note that formally

$$\binom{4}{4} = \frac{4!}{4!\,0!} = 1$$

if we *define* $0! = 1$.

(*d*) = COMBIN(7, 4) gives 35, = COMBIN(6, 5) gives 6, and = COMBIN(4, 4) gives 1.

6.27 In how many ways can a committee of 5 people be chosen out of 9 people?

SOLUTION

$$\binom{9}{5} = \frac{9!}{5!\,4!} = \frac{9 \cdot 8 \cdot 7 \cdot 6 \cdot 5}{5!} = 126$$

6.28 Out of 5 mathematicians and 7 physicists, a committee consisting of 2 mathematicians and 3 physicists is to be formed. In how many ways can this be done if (*a*) any mathematician and any physicist can be included, (*b*) one particular physicist must be on the committee, and (*c*) two particular mathematicians cannot be on the committee?

SOLUTION

(*a*) Two mathematicians out of 5 can be selected in $\binom{5}{2}$ ways, and 3 physicists out of 7 can be selected in $\binom{7}{3}$ ways. The total number of possible selections is

$$\binom{5}{2} \cdot \binom{7}{3} = 10 \cdot 35 = 350$$

(*b*) Two mathematicians out of 5 can be selected in $\binom{5}{2}$ ways, and 2 additional physicists out of 6 can be selected in $\binom{6}{2}$ ways. The total number of possible selections is

$$\binom{5}{2} \cdot \binom{6}{2} = 10 \cdot 15 = 150$$

(*c*) Two mathematicians out of 3 can be selected in $\binom{3}{2}$ ways, and 3 physicists out of 7 can be selected in $\binom{7}{3}$ ways. The total number possible selections is

$$\binom{3}{2} \cdot \binom{7}{3} = 3 \cdot 35 = 105$$

6.29 A girl has 5 flowers, each of a different variety. How many different bouquets can she form?

SOLUTION

Each flower can be dealt with in 2 ways: It can be chosen or not chosen. Since each of the 2 ways of dealing with a flower is associated with 2 ways of dealing with each of the other flowers, the number of ways of dealing with the 5 flowers = 2^5. But these 2^5 ways include the case in which no flower is chosen. Hence the required number of bouquets = $2^5 - 1 = 31$.

Another method

She can select either 1 out of 5 flowers, 2 out of 5 flowers, ..., 5 out of 5 flowers. Thus the required number of bouquets is

$$\binom{5}{1} + \binom{5}{2} + \binom{5}{3} + \binom{5}{4} + \binom{5}{5} = 5 + 10 + 10 + 5 + 1 = 31$$

In general, for any positive integer n,

$$\binom{n}{1} + \binom{n}{2} + \binom{n}{3} + \cdots + \binom{n}{n} = 2^n - 1$$

6.30 From 7 consonants and 5 vowels, how many words can be formed consisting of 4 different consonants and 3 different vowels? The words need not have meaning.

SOLUTION

The 4 different consonants can be selected in $\binom{7}{4}$ ways, the 3 different vowels can be selected in $\binom{5}{3}$ ways, and the resulting 7 different letters (4 consonants and 3 vowels) can then be arranged among themselves in ${}_7P_7 = 7!$ ways. Thus the number of words is

$$\binom{7}{4} \cdot \binom{5}{3} \cdot 7! = 35 \cdot 10 \cdot 5040 = 1{,}764{,}000$$

STIRLING'S APPROXIMATION TO *n*!

6.31 Evaluate 50!.

SOLUTION

For large n, we have $n! \approx \sqrt{2\pi n}\, n^n e^{-n}$; thus

$$50! \approx \sqrt{2\pi(50)}\, 50^{50} e^{-50} = S$$

To evaluate S, use logarithms to the base 10. Thus

$$\begin{aligned}\log S = \log(\sqrt{100\pi}\, 50^{50} e^{-50}) &= \tfrac{1}{2}\log 100 + \tfrac{1}{2}\log \pi + 50 \log 50 - 50 \log e \\ &= \tfrac{1}{2}\log 100 + \tfrac{1}{2}\log 3.142 + 50 \log 50 - 50 \log 2.718 \\ &= \tfrac{1}{2}(2) + \tfrac{1}{2}(0.4972) + 50(1.6990) - 50(0.4343) = 64.4846\end{aligned}$$

from which $S = 3.05 \times 10^{64}$, a number that has 65 digits.

PROBABILITY AND COMBINATORIAL ANALYSIS

6.32 A box contains 8 red, 3 white, and 9 blue balls. If 3 balls are drawn at random, determine the probability that (*a*) all 3 are red, (*b*) all 3 are white, (*c*) 2 are red and 1 is white, (*d*) at least 1 is white, (*e*) 1 of each color is drawn, and (*f*) the balls are drawn in the order red, white, blue.

SOLUTION

(*a*) **First method**

Let R_1, R_2, and R_3 denote the events "red ball on first draw," "red ball on second draw," and "red ball on third draw," respectively. Then $R_1 R_2 R_3$ denotes the event that all 3 balls drawn are red.

$$\Pr\{R_1 R_2 R_3\} = \Pr\{R_1\} \Pr\{R_2|R_1\} \Pr\{R_3|R_1 R_2\} = \left(\frac{8}{20}\right)\left(\frac{7}{19}\right)\left(\frac{6}{18}\right) = \frac{14}{285}$$

Second method

$$\text{Required probability} = \frac{\text{number of selections of 3 out of 8 red balls}}{\text{number of selections of 3 out of 20 balls}} = \frac{\binom{8}{3}}{\binom{20}{3}} = \frac{14}{285}$$

(*b*) Using the second method of part (*a*),

$$\Pr\{\text{all 3 are white}\} = \frac{\binom{3}{3}}{\binom{20}{3}} = \frac{1}{1140}$$

The first method of part (*a*) can also be used.

(*c*) $$\Pr\{\text{2 are red and 1 is white}\} = \frac{\begin{pmatrix}\text{selections of 2 out}\\ \text{of 8 red balls}\end{pmatrix}\begin{pmatrix}\text{selections of 1 out}\\ \text{of 3 white balls}\end{pmatrix}}{\text{number of selections of 3 out of 20 balls}} = \frac{\binom{8}{2}\binom{3}{1}}{\binom{20}{3}} = \frac{7}{95}$$

(*d*) $$\Pr\{\text{none is white}\} = \frac{\binom{17}{3}}{\binom{20}{3}} = \frac{34}{57} \quad \text{so} \quad \Pr\{\text{at least 1 is white}\} = 1 - \frac{34}{57} = \frac{23}{57}$$

(*e*) $$\Pr\{\text{1 of each color is drawn}\} = \frac{\binom{8}{1}\binom{3}{1}\binom{9}{1}}{\binom{20}{3}} = \frac{18}{95}$$

(*f*) Using part (*e*),

$$\Pr\{\text{balls drawn in order red, white, blue}\} = \frac{1}{3!}\Pr\{\text{1 of each color is drawn}\} = \frac{1}{6}\left(\frac{18}{95}\right) = \frac{3}{95}$$

Another method

$$\Pr\{R_1 W_2 B_2\} = \Pr\{R_1\}\Pr\{W_2|R_1\}\ \Pr\{B_3|R_1 W_2\} = \left(\frac{8}{20}\right)\left(\frac{3}{19}\right)\left(\frac{9}{18}\right) = \frac{3}{95}$$

6.33 Five cards are drawn from a pack of 52 well-shuffled cards. Find the probability that (*a*) 4 are aces; (*b*) 4 are aces and 1 is a king; (*c*) 3 are 10's and 2 are jacks; (*d*) a 9, 10, jack, queen, and king are obtained in any order; (*e*) 3 are of any one suit and 2 are of another; and (*f*) at least 1 ace is obtained.

SOLUTION

(*a*) $$\Pr\{\text{4 aces}\} = \binom{4}{4}\cdot\frac{\binom{48}{1}}{\binom{52}{5}} = \frac{1}{54{,}145}$$

(*b*) $$\Pr\{\text{4 aces and 1 king}\} = \binom{4}{4}\cdot\frac{\binom{4}{1}}{\binom{52}{5}} = \frac{1}{649{,}740}$$

(*c*) $$\Pr\{3 \text{ are 10's and 2 are jacks}\} = \binom{4}{3} \cdot \frac{\binom{4}{2}}{\binom{52}{5}} = \frac{1}{108{,}290}$$

(*d*) $$\Pr\{9,\ 10,\ \text{jack, queen, king in any order}\} = \frac{\binom{4}{1} \cdot \binom{4}{1} \cdot \binom{4}{1} \cdot \binom{4}{1} \cdot \binom{4}{1}}{\binom{52}{5}} = \frac{64}{162{,}435}$$

(*e*) Since there are 4 ways of choosing the first suit and 3 ways of choosing the second suit,

$$\Pr\{3 \text{ of any one suit, 2 of another}\} = \frac{4\binom{13}{3} \cdot 3\binom{13}{2}}{\binom{52}{5}} = \frac{429}{4165}$$

(*f*) $$\Pr\{\text{no ace}\} = \frac{\binom{48}{5}}{\binom{52}{5}} = \frac{35{,}673}{54{,}145} \quad \text{and} \quad \Pr\{\text{at least 1 ace}\} = 1 - \frac{35{,}673}{54{,}145} = \frac{18{,}482}{54{,}145}$$

6.34 Determine the probability of three 6's in five tosses of a fair die.

SOLUTION

Let the tosses of the die be represented by the 5 spaces – – – – –. In each space we will have either the events 6 or non-6 ($\bar{6}$); for example, three 6's and two non-6's can occur as $6\ 6\ \bar{6}\ 6\ \bar{6}$ or as $6\ \bar{6}\ 6\ \bar{6}\ 6$, etc.

Now the probability of an event such as $6\ 6\ \bar{6}\ 6\ \bar{6}$ is

$$\Pr\{6 6 \bar{6} 6 \bar{6}\} = \Pr\{6\}\Pr\{6\}\ \Pr\{\bar{6}\}\ \Pr\{6\}\ \Pr\{\bar{6}\} = \frac{1}{6} \cdot \frac{1}{6} \cdot \frac{5}{6} \cdot \frac{1}{6} \cdot \frac{5}{6} = \left(\frac{1}{6}\right)^3 \left(\frac{5}{6}\right)^2$$

Similarly, $\Pr\{6 \bar{6} 6 \bar{6} 6\} = (\frac{1}{6})^3 (\frac{5}{6})^2$, etc., for all events in which three 6's and two non-6's occur. But there are $\binom{5}{3} = 10$ such events, and these events are mutually exclusive; hence the required probability is

$$\Pr\{6 6 \bar{6} 6 \bar{6} \text{ or } 6 \bar{6} 6 \bar{6} 6 \text{ or etc.}\} = \binom{5}{3} \left(\frac{1}{6}\right)^3 \left(\frac{5}{6}\right)^2 = \frac{125}{3888}$$

In general, if $p = \Pr\{E\}$ and $q = \Pr\{\bar{E}\}$, then by using the same reasoning as given above, the probability of getting exactly X E's in N trials is $\binom{N}{X} p^X q^{N-X}$.

6.35 A factory finds that, on average, 20% of the bolts produced by a given machine will be defective for certain specified requirements. If 10 bolts are selected at random from the day's production of this machine, find the probability (*a*) that exactly 2 will be defective, (*b*) that 2 or more will be defective, and (*c*) that more than 5 will be defective.

SOLUTION

(*a*) Using reasoning similar to that of Problem 6.34,

$$\Pr\{2 \text{ defective bolts}\} = \binom{10}{2} (0.2)^2 (0.8)^8 = 45(0.04)(0.1678) = 0.3020$$

(*b*) $$\begin{aligned} \Pr\{2 \text{ or more defective bolts}\} &= 1 - \Pr\{0 \text{ defective bolts}\} - \Pr\{1 \text{ defective bolt}\} \\ &= 1 - \binom{10}{0}(0.2)^0(0.8)^{10} - \binom{10}{1}(0.2)^1(0.8)^9 \\ &= 1 - (0.8)^{10} - 10(0.2)(0.8)^9 \\ &= 1 - 0.1074 - 0.2684 = 0.6242 \end{aligned}$$

(c)
$$\begin{aligned}\Pr\{\text{more than 5 defective bolts}\} &= \Pr\{6 \text{ defective bolts}\} + \Pr\{7 \text{ defective bolts}\} \\ &\quad + \Pr\{8 \text{ defective bolts}\} + \Pr\{9 \text{ defective bolts}\} \\ &\quad + \Pr\{10 \text{ defective bolts}\} \\ &= \binom{10}{6}(0.2)^6(0.8)^4 + \binom{10}{7}(0.2)^7(0.8)^3 + \binom{10}{8}(0.2)^8(0.8)^2 \\ &\quad + \binom{10}{9}(0.2)^9(0.8) + \binom{10}{10}(0.2)^{10} \\ &= 0.00637\end{aligned}$$

6.36 If 1000 samples of 10 bolts each were taken in Problem 6.35, in how many of these samples would we expect to find (a) exactly 2 defective bolts, (b) 2 or more defective bolts, and (c) more than 5 defective bolts?

SOLUTION

(a) Expected number = (1000)(0.3020) = 302, by Problem 6.35(a).

(b) Expected number = (1000)(0.6242) = 624, by Problem 6.35(b).

(c) Expected number = (1000)(0.00637) = 6, by Problem 6.35(c).

EULER OR VENN DIAGRAMS AND PROBABILITY

6.37 Figure 6-11 shows how to represent the sample space for tossing a fair coin 4 times and event E_1 that exactly two heads and two tails occurred and event E_2 that the coin showed the same thing on the first and last toss. This is one way to represent Venn diagrams and events in a computer worksheet.

	sample	space		event E1	event E2
h	h	h	h		Y
h	h	h	t		
h	h	t	h		Y
h	h	t	t	X	
h	t	h	h		Y
h	t	h	t	X	
h	t	t	h	X	Y
h	t	t	t		
t	h	h	h		
t	h	h	t	X	Y
t	h	t	h	X	
t	h	t	t		Y
t	t	h	h	X	
t	t	h	t		Y
t	t	t	h		
t	t	t	t		Y

Fig. 6-11 EXCEL display of sample space and events E_1 and E_2.

The outcomes in E_1 have an X beside them under event E_1 and the outcomes in E_2 have a Y beside them under event E_2.

(a) Give the outcomes in $E_1 \cap E_2$ and $E_1 \cup E_2$.

(b) Give the probabilities $\Pr\{E_1 \cap E_2\}$ and $\Pr\{E_1 \cup E_2\}$.

SOLUTION

(a) The outcomes in $E_1 \cap E_2$ have an X and a Y beside them. Therefore $E_1 \cap E_2$ consists of outcomes htth and thht. The outcomes in $E_1 \cup E_2$ have an X, Y, or X and Y beside them. The outcomes in $E_1 \cup E_2$ are: hhhh, hhth, hhtt, hthh, htth, thht, thth, thtt, tthh, ttht, and tttt.

(b) $\Pr\{E_1 \cap E_2\} = 2/16 = 1/8$ or 0.125. $\Pr\{E_1 \cup E_2\} = 12/16 = 3/4$ or 0.75.

6.38 Using a sample space and Venn diagrams, show that:

(*a*) $\Pr\{A \cup B\} = \Pr\{A\} + \Pr\{B\} - \Pr\{A \cap B\}$

(*b*) $\Pr\{A \cup B \cup C\} = \Pr\{A\} + \Pr\{B\} + \Pr\{C\} - \Pr\{A \cap B\} - \Pr\{B \cap C\} - \Pr\{A \cap C\} + \Pr\{A \cap B \cap C\}$

SOLUTION

(*a*) The nonmutually exclusive union $A \cup B$ is expressible as the mutually exclusive union of $A \cap \bar{B}$, $B \cap \bar{A}$, and $A \cap B$.

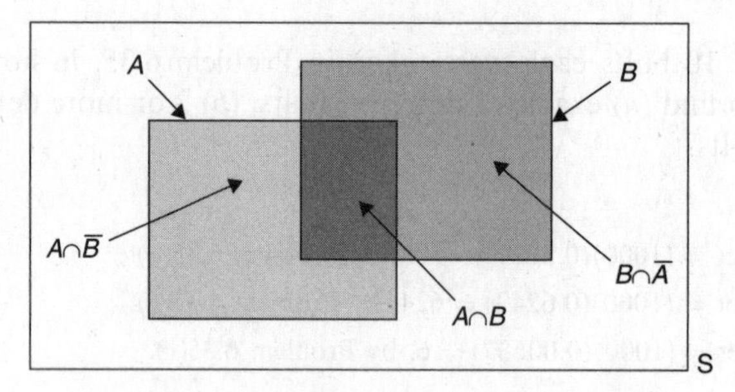

Fig. 6-12 A union expressed as a disjoint union.

$$\Pr\{A \cup B\} = \Pr\{A \cap \bar{B}\} + \Pr\{B \cap \bar{A}\} + \Pr\{A \cap B\}$$

Now add and subtract $\Pr\{A \cap B\}$ from the right-hand side of this equation.

$$\Pr\{A \cup B\} = \Pr\{A \cap \bar{B}) + \Pr\{B \cap \bar{A}\} + \Pr\{A \cap B\} + [\Pr\{A \cap B\} - \Pr\{A \cap B\}]$$

Rearrange this equation as follows:

$$\Pr\{A \cup B\} = [\Pr\{A \cap \bar{B}) + \Pr\{A \cap B\}] + [\Pr\{B \cap \bar{A}\} + Pr\{A \cap B\}] - \Pr\{A \cap B\}$$
$$\Pr\{A \cup B\} = \Pr\{A\} + \Pr\{B\} - \Pr\{A \cap B\}$$

(*b*) In Fig. 6-13, event A is composed of regions 1, 2, 3, and 6, event B is composed of 1, 3, 4, and 7, and

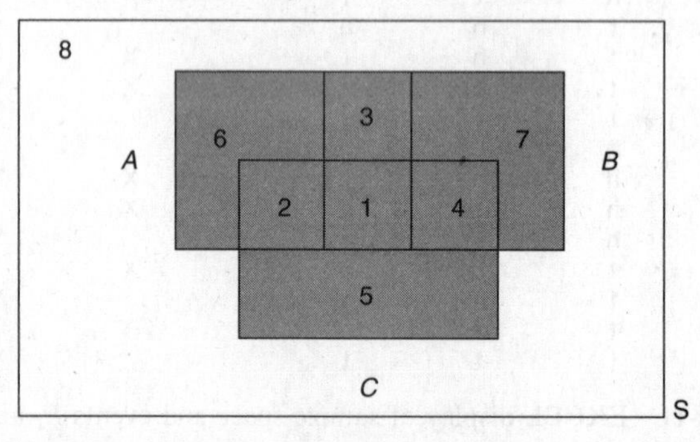

Fig. 6-13 The union of three nonmutually exclusive events, $A \cup B \cup C$.

event C is composed of regions 1, 2, 4, and 5.

The sample space in Fig. 6-13 is made up of 8 mutually exclusive regions. The eight regions are described as follows: region 1 is $A \cap B \cap C$, region 2 is $A \cap C \cap \bar{B}$, region 3 is $A \cap B \cap \bar{C}$, region 4 is $\bar{A} \cap C \cap B$, region 5 is $\bar{A} \cap C \cap \bar{B}$, region 6 is $A \cap \bar{C} \cap \bar{B}$, region 7 is $\bar{A} \cap \bar{C} \cap B$, and region 8 is $\bar{A} \cap \bar{C} \cap \bar{B}$.

The probability $\Pr\{A \cup B \cup C\}$ may be expressed as the probability of the 7 mutually exclusive regions that make up $A \cup B \cup C$ as follows:

$$\Pr\{A \cap B \cap C\} + \Pr\{A \cap C \cap \bar{B}\} + \Pr\{A \cap B \cap \bar{C}\} + \Pr\{\bar{A} \cap C \cap B\}$$
$$+ \Pr\{\bar{A} \cap C \cap \bar{B}\} + \Pr\{A \cap \bar{C} \cap \bar{B}\} + \Pr\{\bar{A} \cap \bar{C} \cap B\}$$

Each part of this equation may be re-expressed and the whole simplified to the result:

$$\Pr\{A \cup B \cup C\} = \Pr\{A\} + \Pr\{B\} + \Pr\{C\} - \Pr\{A \cap B\} - \Pr\{B \cap C\} - \Pr\{A \cap C\} + \Pr\{A \cap B \cap C\}$$

For example, $\Pr\{\bar{A} \cap C \cap \bar{B}\}$ is expressible as:

$$\Pr\{C\} - \Pr\{A \cap C\} - Pr\{B \cap C\} + \Pr\{A \cap B \cap C\}$$

6.39 In a survey of 500 adults were asked the three-part question (1) Do you own a cell phone, (2) Do you own an ipod, and (3) Do you have an internet connection? The results of the survey were as follows (no one answered no to all three parts):

cell phone	329	cell phone and ipod	83
ipod	186	cell phone and internet connection	217
internet connection	295	ipod and internet connection	63

Give the probabilities of the following events:

(*a*) answered yes to all three parts, (*b*) had a cell phone but not an internet connection, (*c*) had an ipod but not a cell phone, (*d*) had an internet connection but not an ipod, (*e*) had a cell phone or an internet connection but not an ipod and, (*f*) had a cell phone but not an ipod or an internet connection.

SOLUTION

Event A is that the respondent had a cell phone, event B is that the respondent had an ipod, and event C is that the respondent had an internet connection.

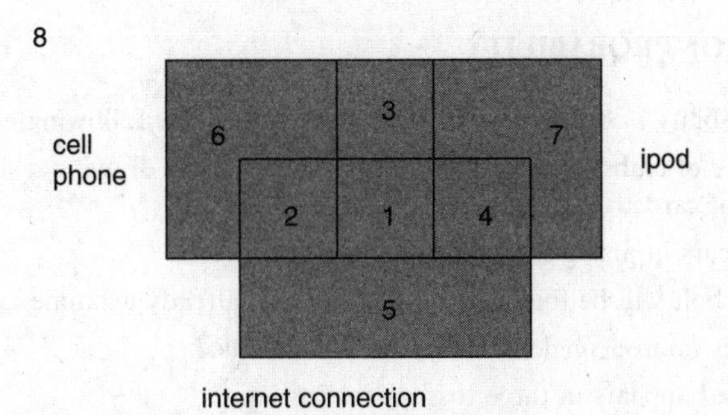

Fig. 6-14 Venn diagram for Problem 6.39.

(*a*) The probability that everyone is in the union is 1 since no one answered no to all three parts. $\Pr\{A \cup B \cup C\}$ is given by the following expression:

$$\Pr\{A\} + \Pr\{B\} + \Pr\{C\} - \Pr\{A \cap B\} - \Pr\{B \cap C\} - \Pr\{A \cap C\} + \Pr\{A \cap B \cap C\}$$

$$1 = 329/500 + 186/500 + 295/500 - 83/500 - 63/500 - 217/500 + \Pr\{A \cap B \cap C\}$$

Solving for $\Pr\{A \cap B \cap C\}$, we obtain $1 - 447/500$ or $53/500 = 0.106$.

Before answering the other parts, it is convenient to fill in the regions in Fig. 6-14 as shown in Fig. 6-15. The number in region 2 is the number in region $A \cap C$ minus the number in region 1 or $217 - 53 = 164$. The number in region 3 is the number in region $A \cap B$ minus the number in region 1 or $83 - 53 = 30$. The number in region 4 is the number in region $B \cap C$ minus the number in region 1 or $63 - 53 = 10$. The number in region 5 is the number in region C minus the number in regions 1, 2, and 4 or $295 - 53 - 164 - 10 = 68$. The number in region 6 is the number in region A minus the number in regions 1, 2, and 3 or $329 - 53 - 164 - 30 = 82$. The number in region 7 is $186 - 53 - 30 - 10 = 93$.

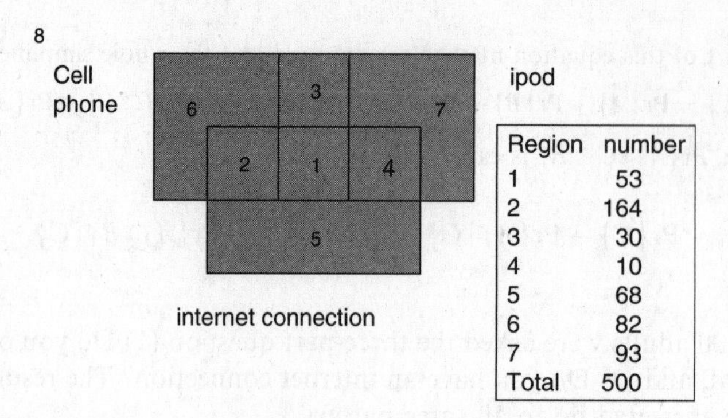

Fig. 6-15 *A*, *B*, and *C* broken into mutually exclusive regions.

(*b*) regions 3 and 6 or $30+82=112$ and the probability is $112/500=0.224$.

(*c*) regions 4 and 7 or $10+93=103$ and the probability is $103/500=0.206$

(*d*) regions 2 and 5 or $164+68=232$ and the probability is $232/500=0.464$.

(*e*) regions 2, 5, or 6 or $164+68+82=314$ and the probability is $314/500=0.628$.

(*f*) region 6 or 82 and the probability is $82/500=0.164$.

Supplementary Problems

FUNDAMENTAL RULES OF PROBABILITY

6.40 Determine the probability p, or an estimate of it, for each of the following events:

(*a*) A king, ace, jack of clubs, or queen of diamonds appears in drawing a single card from a well-shuffled ordinary deck of cards.

(*b*) The sum 8 appears in a single toss of a pair of fair dice.

(*c*) A nondefective bolt will be found if out of 600 bolts already examined, 12 were defective.

(*d*) A 7 or 11 comes up in a single toss of a pair of fair dice.

(*e*) At least one head appears in three tosses of a fair coin.

6.41 An experiment consists of drawing three cards in succession from a well-shuffled ordinary deck of cards. Let E_1 be the event "king" on the first draw, E_2 the event "king" on the second draw, and E_3 the event "king" on the third draw. State in words the meaning of each of the following:

(*a*) $\Pr\{E_1\bar{E}_2\}$ (*c*) $\bar{E}_1+\bar{E}_2$ (*e*) $\bar{E}_1\bar{E}_2\bar{E}_3$

(*b*) $\Pr\{E_1+E_2\}$ (*d*) $\Pr\{E_3|E_1\bar{E}_2\}$ (*f*) $\Pr\{E_1E_2+\bar{E}_2E_3\}$

6.42 A ball is drawn at random from a box containing 10 red, 30 white, 20 blue, and 15 orange marbles. Find the probability that the ball drawn is (*a*) orange or red, (*b*) not red or blue, (*c*) not blue, (*d*) white, and (*e*) red, white, or blue.

6.43 Two marbles are drawn in succession from the box of Problem 6.42, replacement being made after each drawing. Find the probability that (*a*) both are white, (*b*) the first is red and the second is white, (*c*) neither is orange, (*d*) they are either red or white or both (red and white), (*e*) the second is not blue, (*f*) the first is orange, (*g*) at least one is blue, (*h*) at most one is red, (*i*) the first is white but the second is not, and (*j*) only one is red.

6.44 Work Problem 6.43 if there is no replacement after each drawing.

6.45 Find the probability of scoring a total of 7 points (*a*) once, (*b*) at least once, and (*c*) twice in two tosses of a pair of fair dice.

6.46 Two cards are drawn successively from an ordinary deck of 52 well-shuffled cards. Find the probability that (*a*) the first card is not a 10 of clubs or an ace, (*b*) the first card is an ace but the second is not, (*c*) at least one card is a diamond, (*d*) the cards are not of the same suit, (*e*) not more than one card is a picture card (jack, queen, king), (*f*) the second card is not a picture card, (*g*) the second card is not a picture card given that the first was a picture card, and (*h*) the cards are picture cards or spades or both.

6.47 A box contains 9 tickets numbered from 1 to 9 inclusive. If 3 tickets are drawn from the box one at a time, find the probability that they are alternately either (1) odd, even, odd or (2) even, odd, even.

6.48 The odds in favor of A winning a game of chess against B are 3:2. If three games are to be played, what are the odds (*a*) in favor of A's winning at least two games out of the three and (*b*) against A losing the first two games to B?

6.49 A purse contains 2 silver coins and 4 copper coins, and a second purse contains 4 silver coins and 3 copper coins. If a coin is selected at random from one of the two purses, what is the probability that it is a silver coin?

6.50 The probability that a man will be alive in 25 years is $\frac{3}{5}$, and the probability that his wife will be alive in 25 years is $\frac{2}{3}$. Find the probability that (*a*) both will be alive, (*b*) only the man will be alive, (*c*) only the wife will be alive, and (*d*) at least one will be alive.

6.51 Out of 800 families with 4 children each, what percentage would be expected to have (*a*) 2 boys and 2 girls, (*b*) at least 1 boy, (*c*) no girls, and (*d*) at most 2 girls? Assume equal probabilities for boys and girls.

PROBABILITY DISTRIBUTIONS

6.52 If X is the random variable showing the number of boys in families with 4 children (see Problem 6.51), (*a*) construct a table showing the probability distribution of X and (*b*) represent the distribution in part (*a*) graphically.

6.53 A continuous random variable X that can assume values only between $X = 2$ and 8 inclusive has a density function given by $a(X + 3)$, where a is a constant. (*a*) Calculate a. Find (*b*) $\Pr\{3 < X < 5\}$, (*c*) $\Pr\{X \geq 4\}$, and (*d*) $\Pr\{|X - 5| < 0.5\}$.

6.54 Three marbles are drawn without replacement from an urn containing 4 red and 6 white marbles. If X is a random variable that denotes the total number of red marbles drawn, (*a*) construct a table showing the probability distribution of X and (*b*) graph the distribution.

6.55 (*a*) Three dice are rolled and $X =$ the sum on the three upturned faces. Give the probability distribution for X. (*b*) Find $\Pr\{7 \leq X \leq 11\}$.

MATHEMATICAL EXPECTATION

6.56 What is a fair price to pay to enter a game in which one can win \$25 with probability 0.2 and \$10 with probability 0.4?

6.57 If it rains, an umbrella salesman can earn \$30 per day. If it is fair, he can lose \$6 per day. What is his expectation if the probability of rain is 0.3?

6.58 A and B play a game in which they toss a fair coin 3 times. The one obtaining heads first wins the game. If A tosses the coin first and if the total value of the stakes is \$20, how much should be contributed by each in order that the game be considered fair?

6.59 Find (*a*) $E(X)$, (*b*) $E(X^2)$, (*c*) $E[(X-\bar{X})^2]$, and (*d*) $E(X^3)$ for the probability distribution of Table 6.4.

Table 6.4

X	-10	-20	30
$p(X)$	1/5	3/10	1/2

6.60 Referring to Problem 6.54, find the (*a*) mean, (*b*) variance, and (*c*) standard deviation of the distribution of X, and interpret your results.

6.61 A random variable assumes the value 1 with probability p, and 0 with probability $q = 1 - p$. Prove that (*a*) $E(X) = p$ and (*b*) $E[(X-\bar{X})^2] = pq$.

6.62 Prove that (*a*) $E(2X + 3) = 2E(X) + 3$ and (*b*) $E[(X-\bar{X})^2] = E(X^2) - [E(X)]^2$.

6.63 In Problem 6.55, find the expected value of X.

PERMUTATIONS

6.64 Evaluate (*a*) ${}_4P_2$, (*b*) ${}_7P_5$, and (*c*) ${}_{10}P_3$. Give the EXCEL function for parts (*a*), (*b*), and (*c*).

6.65 For what value of n is ${}_{n+1}P_3 = {}_nP_4$?

6.66 In how many ways can 5 people be seated on a sofa if there are only 3 seats available?

6.67 In how many ways can 7 books be arranged on a shelf if (*a*) any arrangement is possible, (*b*) 3 particular books must always stand together, and (*c*) 2 particular books must occupy the ends?

6.68 How many numbers consisting of five different digits each can be made from the digits 1, 2, 3, ..., 9 if (*a*) the numbers must be odd and (*b*) the first two digits of each number are even?

6.69 Solve Problem 6.68 if repetitions of the digits are allowed.

6.70 How many different three-digit numbers can be made with three 4's, four 2's, and two 3's?

6.71 In how many ways can 3 men and 3 women be seated at a round table if (*a*) no restriction is imposed, (*b*) 2 particular women must not sit together, and (*c*) each woman is to be between 2 men?

COMBINATIONS

6.72 Evaluate (*a*) $\binom{7}{3}$, (*b*) $\binom{8}{4}$, and (*c*) $\binom{10}{8}$. Give the EXCEL function for parts (*a*), (*b*), and (*c*).

6.73 For what value of n does $3\binom{n+1}{3} = 7\binom{n}{2}$?

6.74 In how many ways can 6 questions be selected out of 10?

6.75 How many different committees of 3 men and 4 women can be formed from 8 men and 6 women?

6.76 In how many ways can 2 men, 4 women, 3 boys, and 3 girls be selected from 6 men, 8 women, 4 boys, and 5 girls if (*a*) no restrictions are imposed and (*b*) a particular man and woman must be selected?

6.77 In how many ways can a group of 10 people be divided into (*a*) two groups consisting of 7 and 3 people and (*b*) three groups consisting of 4, 3, and 2 people?

6.78 From 5 statisticians and 6 economists a committee consisting of 3 statisticians and 2 economists is to be formed. How many different committees can be formed if (*a*) no restrictions are imposed, (*b*) 2 particular statisticians must be on the committee, and (*c*) 1 particular economist cannot be on the committee?

6.79 Find the number of (*a*) combinations and (*b*) permutations of four letters each that can be made from the letters of the word *Tennessee*?

6.80 Prove that $1 - \binom{n}{1} + \binom{n}{2} - \binom{n}{3} + \cdots + (-1)^n\binom{n}{n} = 0$.

STIRLING'S APPROXIMATION TO *n*!

6.81 In how many ways can 30 individuals be selected out of 100?

6.82 Show that $\binom{2n}{n} = 2^{2n}/\sqrt{\pi n}$ approximately, for large values of n.

MISCELLANEOUS PROBLEMS

6.83 Three cards are drawn from a deck of 52 cards. Find the probability that (*a*) two are jacks and one is a king, (*b*) all cards are of one suit, (*c*) all cards are of different suits, and (*d*) at least two aces are drawn.

6.84 Find the probability of at least two 7's in four tosses of a pair of dice.

6.85 If 10% of the rivets produced by a machine are defective, what is the probability that out of 5 rivets chosen at random (*a*) none will be defective, (*b*) 1 will be defective, and (*c*) at least 2 will be defective?

6.86 (*a*) Set up a sample space for the outcomes of 2 tosses of a fair coin, using 1 to represent "heads" and 0 to represent "tails."

(*b*) From this sample space determine the probability of at least one head.

(*c*) Can you set up a sample space for the outcomes of 3 tosses of a coin? If so, determine with the aid of it the probability of at most two heads?

6.87 A sample poll of 200 voters revealed the following information concerning three candidates (A, B, and C) of a certain party who were running for three different offices:

28 in favor of both A and B	122 in favor of B or C but not A
98 in favor of A or B but not C	64 in favor of C but not A or B
42 in favor of B but not A or C	14 in favor of A and C but not B

How many of the voters were in favor of (*a*) all three candidates, (*b*) A irrespective of B or C, (*c*) B irrespective of A or C, (*d*) C irrespective of A or B, (*e*) A and B but not C, and (*f*) only one of the candidates?

6.88 (*a*) Prove that for any events E_1 and E_2, $\Pr\{E_1 + E_2\} \leq \Pr\{E_1\} + \Pr\{E_2\}$.

(*b*) Generalize the result in part (*a*).

6.89 Let E_1, E_2, and E_3 be three different events, at least one of which is known to have occurred. Suppose that any of these events can result in another event A, which is also known to have occurred. If all the probabilities $\Pr\{E_1\}$, $\Pr\{E_2\}$, $\Pr\{E_3\}$ and $\Pr\{A|E_1\}$, $\Pr\{A|E_2\}$, $\Pr\{A|E_3\}$ are assumed known, prove that

$$\Pr\{E_1|A\} = \frac{\Pr\{E_1\}\,\Pr\{A|E_1\}}{\Pr\{E_1\}\,\Pr\{A|E_1\} + \Pr\{E_2\}\,\Pr\{A|E_2\} + \Pr\{E_3\}\,\Pr\{A|E_3\}}$$

with similar results for $\Pr\{E_2|A\}$ and $\Pr\{E_3|A\}$. This is known as *Bayes' rule* or *theorem*. It is useful in computing probabilities of various *hypotheses* E_1, E_2, or E_3 that have resulted in the event A. The result can be generalized.

6.90 Each of three identical jewelry boxes has two drawers. In each drawer of the first box there is a gold watch. In each drawer of the second box there is a silver watch. In one drawer of the third box there is a gold watch, while in the other drawer there is a silver watch. If we select a box at random, open one of the drawers, and find it to contain a silver watch, what is the probability that the other drawer has the gold watch? [*Hint:* Apply Problem 6.89.]

6.91 Find the probability of winning a state lottery in which one is required to choose six of the numbers 1, 2, 3, ..., 40 in any order.

6.92 Work Problem 6.91 if one is required to choose (*a*) five, (*b*) four, and (*c*) three of the numbers.

6.93 In the game of poker, five cards from a standard deck of 52 cards are dealt to each player. Determine the odds against the player receiving:

(*a*) A royal flush (the ace, king, queen, jack, and 10 of the same suit)

(*b*) A straight flush (any five cards in sequence and of the same suit, such as the 3, 4, 5, 6, and 7 of spades)

(*c*) Four of a kind (such as four 7's)

(*d*) A full house (three of one kind and two of another, such as three kings and two 10's)

6.94 A and B decide to meet between 3 and 4 P.M. but agree that each should wait no longer than 10 minutes for the other. Determine the probability that they meet.

6.95 Two points are chosen at random on a line segment whose length is $a > 0$. Find the probability that the three line segments thus formed can be the sides of a triangle.

6.96 A regular tetrahedron consists of four sides. Each is equally likely to be the one facing down when the tetrahedron is tossed and comes to rest. The numbers 1, 2, 3, and 4 are on the four faces. Three regular tetrahedrons are tossed upon a table. Let X = the sum of the three faces that are facing down. Give the probability distribution for X.

6.97 In Problem 6.96, find the expected value of X.

6.98 In a survey of a group of people it was found that 25% were smokers and drinkers, 10% were smokers but not drinkers, and 35% were drinkers but not smokers. What percent in the survey were either smokers or drinkers or both?

6.99 Acme electronics manufactures MP3 players at three locations. The plant at Omaha manufactures 50% of the MP3 players and 1% are defective. The plant at Memphis manufactures 30% and 2% at that plant are defective. The plant at Fort Meyers manufactures 20% and 3% at that plant are defective. If an MP3 player is selected at random, what is the probability that it is defective?

6.100 Refer to Problem 6.99. An MP3 player is found to be defective. What is the probability that it was manufactured in Fort Meyers?

The Binomial, Normal, and Poisson Distributions

THE BINOMIAL DISTRIBUTION

If p is the probability that an event will happen in any single trial (called the probability of a *success*) and $q = 1 - p$ is the probability that it will fail to happen in any single trial (called the probability of a *failure*), then the probability that the event will happen exactly X times in N trials (i.e., X successes and $N - X$ failures will occur) is given by

$$p(X) = \binom{N}{X} p^X q^{N-X} = \frac{N!}{X!\,(N-X)!} p^X q^{N-X} \tag{1}$$

where $X = 0, 1, 2, \ldots, N$; $N! = N(N-1)(N-2)\cdots 1$; and $0! = 1$ by definition (see Problem 6.34).

EXAMPLE 1. The probability of getting exactly 2 heads in 6 tosses of a fair coin is

$$\binom{6}{2}\left(\frac{1}{2}\right)^2\left(\frac{1}{2}\right)^{6-2} = \frac{6}{2!\,4!}\left(\frac{1}{2}\right)^6 = \frac{15}{64}$$

using formula (*1*) with $N = 6$, $X = 2$, and $p = q = \frac{1}{2}$.

Using EXCEL, the evaluation of the probability of 2 heads in 6 tosses is given by the following: =BINOMDIST(2,6,0.5,0), where the function BINOMDIST has 4 parameters.

The first parameter is the number of successes, the second is the number of trials, the third is the probability of success, and the fourth is a 0 or 1. A zero gives the probability of the number of successes and a 1 gives the cumulative probability. The function =BINOMDIST(2,6,0.5,0) gives 0.234375 which is the same as 15/64.

EXAMPLE 2. The probability of getting at least 4 heads in 6 tosses of a fair coin is

$$\binom{6}{4}\left(\frac{1}{2}\right)^4\left(\frac{1}{2}\right)^{6-4} + \binom{6}{5}\left(\frac{1}{2}\right)^5\left(\frac{1}{2}\right)^{6-5} + \binom{6}{6}\left(\frac{1}{2}\right)^6\left(\frac{1}{2}\right)^{6-6} = \frac{15}{64} + \frac{6}{64} + \frac{1}{64} = \frac{11}{32}$$

The discrete probability distribution (*1*) is often called the *binomial distribution* since for $X = 0, 1, 2, \ldots, N$ it corresponds to successive terms of the *binomial formula*, or *binomial expansion*,

$$(q+p)^N = q^N + \binom{N}{1} q^{N-1} p + \binom{N}{2} q^{N-2} p^2 + \cdots + p^N \tag{2}$$

where 1, $\binom{N}{1}$, $\binom{N}{2}$, ... are called the *binomial coefficients*.

Using EXCEL, the solution is =1-BINOMDIST(3,6,0.5,1) or 0.34375 which is the same as 11/32. Since $\Pr\{X \geq 4\} = 1 - \Pr\{X \leq 3\}$, and BINOMDIST(3,6,0.5,1) $= \Pr\{X \leq 3\}$, this computation will give the probability of at least 4 heads.

EXAMPLE 3.

$$\begin{aligned}(q+p)^4 &= q^4 + \binom{4}{1} q^3 p + \binom{4}{2} q^2 p^2 + \binom{4}{3} q p^3 + p^4 \\ &= q^4 + 4q^3 p + 6q^2 p^2 + 4qp^3 + p^4\end{aligned}$$

Some properties of the binomial distribution are listed in Table 7.1.

Table 7.1 Binomial Distribution

Mean	$\mu = Np$
Variance	$\sigma^2 = Npq$
Standard deviation	$\sigma = \sqrt{Npq}$
Moment coefficient of skewness	$\alpha_3 = \dfrac{q-p}{\sqrt{Npq}}$
Moment coefficient of kurtosis	$\alpha_4 = 3 + \dfrac{1-6pq}{Npq}$

EXAMPLE 4. In 100 tosses of a fair coin the mean number of heads is $\mu = Np = (100)(\frac{1}{2}) = 50$; this is the *expected* number of heads in 100 tosses of the coin. The standard deviation is $\sigma = \sqrt{Npq} = \sqrt{(100)(\frac{1}{2})(\frac{1}{2})} = 5$.

THE NORMAL DISTRIBUTION

One of the most important examples of a continuous probability distribution is the *normal distribution, normal curve*, or *gaussian distribution*. It is defined by the equation

$$Y = \frac{1}{\sigma\sqrt{2\pi}} e^{-1/2(X-\mu)^2/\sigma^2} \tag{3}$$

where μ = mean, σ = standard deviation, $\pi = 3.14159\cdots$, and $e = 2.71828\cdots$. The total area bounded by curve (*3*) and the X axis is 1; hence the area under the curve between two ordinates $X = a$ and $X = b$, where $a < b$, represents the probability that X lies between a and b. This probability is denoted by $\Pr\{a < X < b\}$.

When the variable X is expressed in terms of standard units $[z = (X - \mu)/\sigma]$, equation (*3*) is replaced by the so-called *standard form*:

$$Y = \frac{1}{\sqrt{2\pi}} e^{-1/2z^2} \tag{4}$$

In such cases we say that z is *normally distributed with mean 0 and variance 1*. Figure 7-1 is a graph of this standardized normal curve. It shows that the areas included between $z = -1$ and $+1$, $z = -2$ and $+2$, and $z = -3$ and $+3$ are equal, respectively, to 68.27%, 95.45%, and 99.73% of the total area, which is 1. The table in Appendix II shows the areas under this curve bounded by the ordinates at $z = 0$ and any positive value of z. From this table the area between any two ordinates can be found by using the symmetry of the curve about $z = 0$.

Some properties of the normal distribution given by equation (3) are listed in Table 7.2.

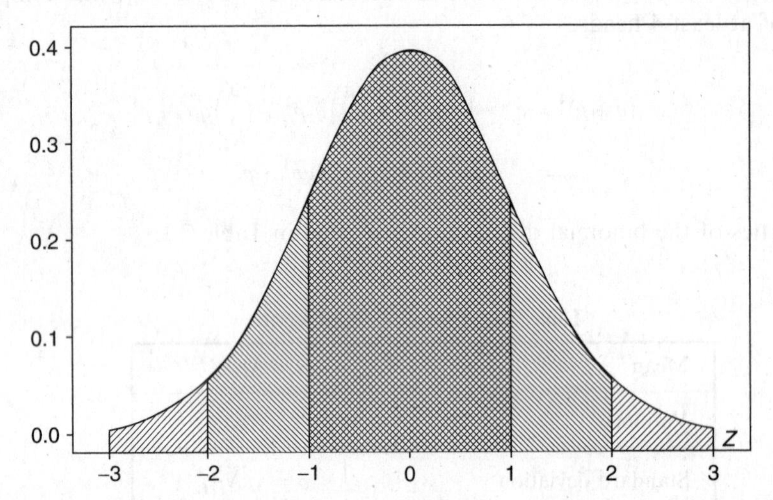

Fig. 7-1 Standard normal curve: 68.27% of the area is between $z = -1$ and $z = 1$, 95.45% is between $z = -2$ and $z = 2$, and 99.73% is between $z = -3$ and $z = 3$.

Table 7.2 Normal Distribution

Mean	μ
Variance	σ^2
Standard deviation	σ
Moment coefficient of skewness	$\alpha_3 = 0$
Moment coefficient of kurtosis	$\alpha_4 = 3$
Mean deviation	$\sigma\sqrt{2/\pi} = 0.7979\sigma$

RELATION BETWEEN THE BINOMIAL AND NORMAL DISTRIBUTIONS

If N is large and if neither p nor q is too close to zero, the binomial distribution can be closely approximated by a normal distribution with standardized variable given by

$$z = \frac{X - Np}{\sqrt{Npq}}$$

The approximation becomes better with increasing N, and in the limiting case it is exact; this is shown in Tables 7.1 and 7.2, where it is clear that as N increases, the skewness and kurtosis for the binomial distribution approach that of the normal distribution. In practice the approximation is very good if both Np and Nq are greater than 5.

EXAMPLE 5. Figure 7-2 shows the binomial distribution with $N=16$ and $p=0.5$, reflecting the probabilities of getting X heads in 16 flips of a coin and the normal distribution with mean equal to 8 and standard deviation 2. Notice how similar and close the two are. X is binomial with mean $= Np = 16(0.5) = 8$ and standard deviation $\sqrt{Npq} = \sqrt{16(0.5)(0.5)} = 2$. Y is a normal curve with mean $= 8$ and standard deviation 2.

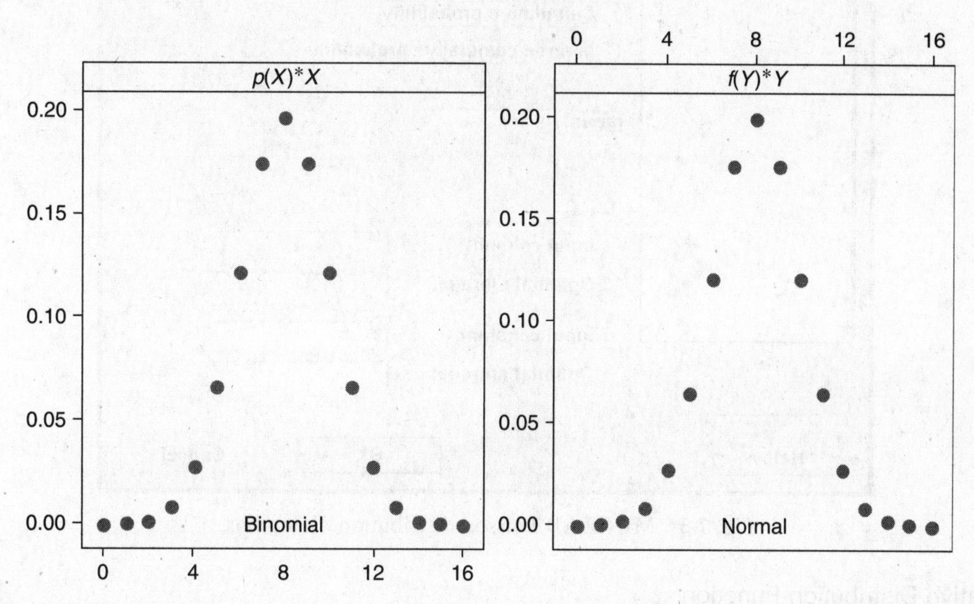

Fig. 7-2 Plot of binomial for $N=16$ and $p=0.5$ and normal curve with mean $=8$ and standard deviation $=2$.

THE POISSON DISTRIBUTION

The discrete probability distribution

$$p(X) = \frac{\lambda^X e^{-\lambda}}{X!} \qquad X = 0, 1, 2, \ldots \tag{5}$$

where $e = 2.71828\cdots$ and λ is a given constant, is called the *Poisson distribution* after Siméon-Denis Poisson, who discovered it in the early part of the nineteenth century. The values of $p(X)$ can be computed by using the table in Appendix VIII (which gives values of $e^{-\lambda}$ for various values of λ) or by using logarithms.

EXAMPLE 6. The number of admissions per day at an emergency room has a Poisson distribution and the mean is 5. Find the probability of at most 3 admissions per day and the probability of at least 8 admissions per day. The probability of at most 3 is $\Pr\{X \leq 3\} = e^{-5}\{5^0/0! + 5^1/1! + 5^2/2! + 5^3/3!\}$. From Appendix VIII, $e^{-5} = 0.006738$, and $\Pr\{X \leq 3\} = 0.006738\{1 + 5 + 12.5 + 20.8333\} = 0.265$. Using MINITAB, the pull-down "**Calc ⇒ Probability distribution ⇒ Poisson**" gives the Poisson distribution dialog box which is filled out as shown in Fig. 7-3.

The following output is given:

Cumulative Distribution Function

```
Poisson with mean = 5

x    P(X<=x)
3    0.265026
```

The same answer as found using Appendix VIII is given.

The probability of at least 8 admissions is $\Pr\{X \geq 8\} = 1 - \Pr\{X \leq 7\}$. Using MINITAB, we find:

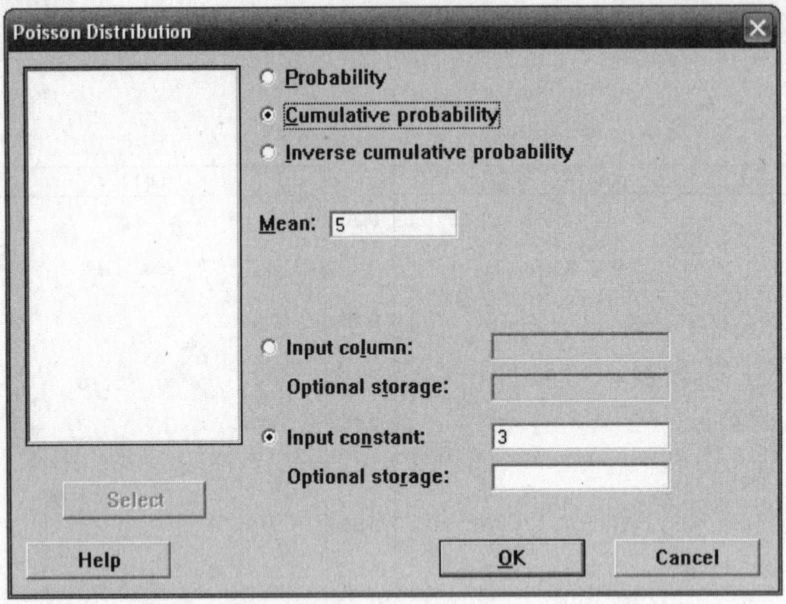

Poisson Distribution

Probability
Cumulative probability
Inverse cumulative probability

Mean: 5

Input column:
Optional storage:
Input constant: 3
Optional storage:

Select
Help
OK
Cancel

Fig. 7-3 MINITAB Poisson distribution dialog box.

```
Cumulative Distribution Function
Poisson with mean = 5
x     P(X <= x)
7     0.866628
```

$\Pr\{X \geq 8\} = 1 - 0.867 = 0.133$.

Some properties of the Poisson distribution are listed in Table 7.3.

Table 7.3 Poisson's Distribution

Mean	$\mu = \lambda$
Variance	$\sigma^2 = \lambda$
Standard deviation	$\sigma = \sqrt{\lambda}$
Moment coefficient of skewness	$\alpha_3 = 1/\sqrt{\lambda}$
Moment coefficient of kurtosis	$\alpha_4 = 3 + 1/\lambda$

RELATION BETWEEN THE BINOMIAL AND POISSON DISTRIBUTIONS

In the binomial distribution (*1*), if N is large while the probability p of the occurrence of an event is close to 0, so that $q = 1 - p$ is close to 1, the event is called a *rare event*. In practice we shall consider an event to be rare if the number of trials is at least 50 ($N \geq 50$) while Np is less than 5. In such case the binomial distribution (*1*) is very closely approximated by the Poisson distribution (*5*) with $\lambda = Np$. This is indicated by comparing Tables 7.1 and 7.3—for by placing $\lambda = Np$, $q \approx 1$, and $p \approx 0$ in Table 7.1, we get the results in Table 7.3.

Since there is a relation between the binomial and normal distributions, it follows that there also is a relation between the Poisson and normal distributions. It can in fact be shown that the Poisson distribution approaches a normal distribution with standardized variable $(X - \lambda)/\sqrt{\lambda}$ as λ increases indefinitely.

THE MULTINOMIAL DISTRIBUTION

If events $E_1, E_2, \ldots, E_K$ can occur with probabilities $p_1, p_2, \ldots, p_K$, respectively, then the probability that $E_1, E_2, \ldots, E_K$ will occur $X_1, X_2, \ldots, X_K$ times, respectively, is

$$\frac{N!}{X_1!X_2!\cdots X_K!} p_1^{X_1} p_2^{X_2} \cdots p_K^{X_K} \tag{6}$$

where $X_1 + X_2 + \cdots + X_K = N$. This distribution, which is a generalization of the binomial distribution, is called the *multinomial distribution* since equation (*6*) is the general term in the *multinomial expansion* $(p_1 + p_2 + \cdots + p_K)^N$.

EXAMPLE 7. If a fair die is tossed 12 times, the probability of getting 1, 2, 3, 4, 5, and 6 points exactly twice each is

$$\frac{12!}{2!2!2!2!2!2!}\left(\frac{1}{6}\right)^2\left(\frac{1}{6}\right)^2\left(\frac{1}{6}\right)^2\left(\frac{1}{6}\right)^2\left(\frac{1}{6}\right)^2\left(\frac{1}{6}\right)^2 = \frac{1925}{559{,}872} = 0.00344$$

The *expected* numbers of times that $E_1, E_2, \ldots, E_K$ will occur in N trials are $Np_1, Np_2, \ldots, Np_K$, respectively.

EXCEL can be used to find the answer as follows. =MULTINOMIAL(2,2,2,2,2,2) is used to evaluate $\frac{12!}{2!2!2!2!2!2!}$ which gives 7,484,400. This is then divided by 6^{12} which is 2,176,782,336. The quotient is 0.00344.

FITTING THEORETICAL DISTRIBUTIONS TO SAMPLE FREQUENCY DISTRIBUTIONS

When one has some indication of the distribution of a population by probabilistic reasoning or otherwise, it is often possible to fit such theoretical distributions (also called *model* or *expected* distributions) to frequency distributions obtained from a sample of the population. The method used consists in general of employing the mean and standard deviation of the sample to estimate the mean and sample of the population (see Problems 7.31, 7.33, and 7.34).

In order to test the *goodness of fit* of the theoretical distributions, we use the *chi-square test* (which is given in Chapter 12). In attempting to determine whether a normal distribution represents a good fit for given data, it is convenient to use *normal-curve graph paper*, or *probability graph paper* as it is sometimes called (see Problem 7.32).

Solved Problems

THE BINOMIAL DISTRIBUTION

7.1 Evaluate the following:

(*a*) 5! (*c*) $\binom{8}{3}$ (*e*) $\binom{4}{4}$

(*b*) $\frac{6!}{2!\,4!}$ (*d*) $\binom{7}{5}$ (*f*) $\binom{4}{0}$

SOLUTION

(*a*) $5! = 5 \cdot 4 \cdot 3 \cdot 2 \cdot 1 = 120$

(*b*) $\frac{6!}{2!\,4!} = \frac{6 \cdot 5 \cdot 4 \cdot 3 \cdot 2 \cdot 1}{(2 \cdot 1)(4 \cdot 3 \cdot 2 \cdot 1)} = \frac{6 \cdot 5}{2 \cdot 1} = 15$

(*c*) $\binom{8}{3} = \frac{8!}{3!\,(8-3)!} = \frac{8!}{3!\,5!} = \frac{8 \cdot 7 \cdot 6 \cdot 5 \cdot 4 \cdot 3 \cdot 2 \cdot 1}{(3 \cdot 2 \cdot 1)(5 \cdot 4 \cdot 3 \cdot 2 \cdot 1)} = \frac{8 \cdot 7 \cdot 6}{3 \cdot 2 \cdot 1} = 56$

(*d*) $\binom{7}{5} = \frac{7!}{5!\,2!} = \frac{7 \cdot 6 \cdot 5 \cdot 4 \cdot 3 \cdot 2 \cdot 1}{(5 \cdot 4 \cdot 3 \cdot 2 \cdot 1)(2 \cdot 1)} = \frac{7 \cdot 6}{2 \cdot 1} = 21$

(*e*) $\binom{4}{4} = \frac{4!}{4!\,0!} = 1$ since $0! = 1$ by definition

(*f*) $\binom{4}{0} = \frac{4!}{0!\,4!} = 1$

7.2 Suppose that 15% of the population is left-handed. Find the probability that in group of 50 individuals there will be (*a*) at most 10 left-handers, (*b*) at least 5 left-handers, (*c*) between 3 and 6 left-handers inclusive, and (*d*) exactly 5 left-handers. Use EXCEL to find the solutions.

SOLUTION

(*a*) The EXCEL expression =BINOMDIST(10,50,0.15,1) gives $\Pr\{X \le 10\}$ or 0.8801.

(*b*) We are asked to find $\Pr\{X \ge 5\}$ which equals $1 - \Pr\{X \le 4\}$ since $X \ge 5$ and $X \le 4$ are complementary events. The EXCEL expression that gives the desired result is =1-BINOMDIST(4,50,0.15,1) or 0.8879.

(*c*) We are asked to find $\Pr\{3 \le X \le 6\}$ which equals $\Pr\{X \le 6\} - \Pr\{X \le 2\}$. The EXCEL expression to give the result is =BINOMDIST(6,50,0.15,1)-BINOMDIST(2,50,0.15,1) or 0.3471.

(*d*) The EXCEL expression =BINOMDIST(5,50,0.15,0) gives $\Pr\{X = 5\}$ or 0.1072.

7.3 Find the probability that in five tosses of a fair die a 3 appears (*a*) at no time, (*b*) once, (*c*) twice, (*d*) three times, (*e*) four times, (*f*) five times, and (*g*) give the MINITAB solution.

SOLUTION

The probability of 3 in a single toss $= p = \frac{1}{6}$, and the probability of no 3 in a single toss $= q = 1 - p = \frac{5}{6}$; thus:

(*a*) $\Pr\{3 \text{ occurs zero times}\} = \binom{5}{0}\left(\frac{1}{6}\right)^0\left(\frac{5}{6}\right)^5 = (1)(1)\left(\frac{5}{6}\right)^5 = \frac{3125}{7776}$

(*b*) $\Pr\{3 \text{ occurs one time}\} = \binom{5}{1}\left(\frac{1}{6}\right)^1\left(\frac{5}{6}\right)^4 = (5)\left(\frac{1}{6}\right)\left(\frac{5}{6}\right)^4 = \frac{3125}{7776}$

(*c*) $\Pr\{3 \text{ occurs two times}\} = \binom{5}{2}\left(\frac{1}{6}\right)^2\left(\frac{5}{6}\right)^3 = (10)\left(\frac{1}{36}\right)\left(\frac{125}{216}\right) = \frac{625}{3888}$

(*d*) $\Pr\{3 \text{ occurs three times}\} = \binom{5}{3}\left(\frac{1}{6}\right)^3\left(\frac{5}{6}\right)^2 = (10)\left(\frac{1}{216}\right)\left(\frac{25}{36}\right) = \frac{125}{3888}$

(*e*) $\Pr\{3 \text{ occurs four times}\} = \binom{5}{4}\left(\frac{1}{6}\right)^4\left(\frac{5}{6}\right)^1 = (5)\left(\frac{1}{1296}\right)\left(\frac{5}{6}\right) = \frac{25}{7776}$

(*f*) $\Pr\{3 \text{ occurs five times}\} = \binom{5}{5}\left(\frac{1}{6}\right)^5\left(\frac{5}{6}\right)^0 = (1)\left(\frac{1}{7776}\right)(1) = \frac{1}{7776}$

Note that these probabilities represent the terms in the binomial expansion

$$\left(\frac{5}{6}+\frac{1}{6}\right)^5 = \left(\frac{5}{6}\right)^5 + \binom{5}{1}\left(\frac{5}{6}\right)^4\left(\frac{1}{6}\right) + \binom{5}{2}\left(\frac{5}{6}\right)^3\left(\frac{1}{6}\right)^2 + \binom{5}{3}\left(\frac{5}{6}\right)^2\left(\frac{1}{6}\right)^3 + \binom{5}{4}\left(\frac{5}{6}\right)\left(\frac{1}{6}\right)^4 + \left(\frac{1}{6}\right)^5 = 1$$

(*g*) enter the integers 0 through 5 in column C1 and then fill in the dialog box for the binomial distribution as in Fig. 7-4.

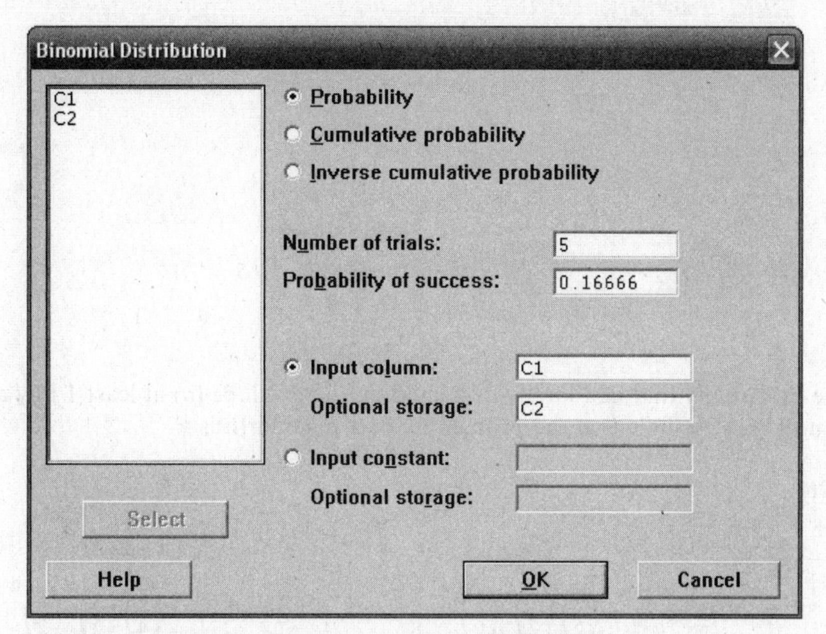

Fig. 7-4 MINITAB dialog box for Problem 7.3 (*g*).

The following output is produced in the worksheet:

C1	C2
0	0.401894
1	0.401874
2	0.160742
3	0.032147
4	0.003215
5	0.000129

Show that the fractions given in parts (*a*) through (*f*) convert to the decimals that MINITAB gives.

7.4 Write the binomial expansion for $(a)(q+p)^4$ and $(b)(q+p)^6$.

SOLUTION

(*a*)
$$(q+p)^4 = q^4 + \binom{4}{1}q^3p + \binom{4}{2}q^2p^2 + \binom{4}{3}qp^3 + p^4$$
$$= q^4 + 4q^3p + 6q^2p^2 + 4qp^3 + p^4$$

(*b*)
$$(q+p)^6 = q^6 + \binom{6}{1}q^5p + \binom{6}{2}q^4p^2 + \binom{6}{3}q^3p^3 + \binom{6}{4}q^2p^4 + \binom{6}{5}qp^5 + p^6$$
$$= q^6 + 6q^5p + 15q^4p^2 + 20q^3p^3 + 15q^2p^4 + 6qp^5 + p^6$$

The coefficients 1, 4, 6, 4, 1 and 1, 6, 15, 20, 15, 6, 1 are called the *binomial coefficients* corresponding to $N = 4$ and $N = 6$, respectively. By writing these coefficients for $N = 0, 1, 2, 3, \ldots$, as shown in the following array, we obtain an arrangement called *Pascal's triangle*. Note that the first and last numbers in each row are 1 and that any other number can be obtained by adding the two numbers to the right and left of it in the preceding row.

1

1 1

1 2 1

1 3 3 1

1 4 6 4 1

1 5 10 10 5 1

1 6 15 20 15 6 1

7.5 Find the probability that in a family of 4 children there will be (*a*) at least 1 boy and (*b*) at least 1 boy and 1 girl. Assume that the probability of a male birth is $\frac{1}{2}$.

SOLUTION

(*a*)
$$\Pr\{1 \text{ boy}\} = \binom{4}{1}\left(\frac{1}{2}\right)^1\left(\frac{1}{2}\right)^3 = \frac{1}{4} \qquad \Pr\{3 \text{ boys}\} = \binom{4}{3}\left(\frac{1}{2}\right)^3\left(\frac{1}{2}\right) = \frac{1}{4}$$
$$\Pr\{2 \text{ boys}\} = \binom{4}{2}\left(\frac{1}{2}\right)^2\left(\frac{1}{2}\right)^2 = \frac{3}{8} \qquad \Pr\{4 \text{ boys}\} = \binom{4}{4}\left(\frac{1}{2}\right)^4\left(\frac{1}{2}\right)^0 = \frac{1}{16}$$

Thus
$$\Pr\{\text{at least 1 boy}\} = \Pr\{1 \text{ boy}\} + \Pr\{2 \text{ boys}\} + \Pr\{3 \text{ boys}\} + \Pr\{4 \text{ boys}\}$$
$$= \frac{1}{4} + \frac{3}{8} + \frac{1}{4} + \frac{1}{16} = \frac{15}{16}$$

Another method

$$\Pr\{\text{at least 1 boy}\} = 1 - \Pr\{\text{no boy}\} = 1 - \left(\frac{1}{2}\right)^4 = 1 - \frac{1}{16} = \frac{15}{16}$$

(*b*)
$$\Pr\{\text{at least 1 boy and 1 girl}\} = 1 - \Pr\{\text{no boy}\} - \Pr\{\text{no girl}\} = 1 - \frac{1}{16} - \frac{1}{16} = \frac{7}{8}$$

7.6 Out of 2000 families with 4 children each, how many would you expect to have (*a*) at least 1 boy, (*b*) 2 boys, (*c*) 1 or 2 girls, and (*d*) no girls? Refer to Problem 7.5(*a*).

SOLUTION

(*a*) Expected number of families with at least 1 boy $= 2000(\frac{15}{16}) = 1875$

(*b*) Expected number of families with 2 boys $= 2000 \cdot \Pr\{2 \text{ boys}\} = 2000(\frac{3}{8}) = 750$

(*c*) $\Pr\{1 \text{ or } 2 \text{ girls}\} = \Pr\{1 \text{ girl}\} + \Pr\{2 \text{ girls}\} = \Pr\{1 \text{ boy}\} + \Pr\{2 \text{ boys}\} = \frac{1}{4} + \frac{3}{8} = \frac{5}{8}$. Expected number of families with 1 or 2 girls $= 2000(\frac{5}{8}) = 1250$

(*d*) Expected number of families with no girls $= 2000(\frac{1}{16}) = 125$

7.7 If 20% of the bolts produced by a machine are defective, determine the probability that, out of 4 bolts chosen at random, (*a*) 1, (*b*) 0, and (*c*) at most 2 bolts will be defective.

SOLUTION

The probability of a defective bolt is $p = 0.2$, and the probability of a nondefective bolt is $q = 1 - p = 0.8$.

(*a*)
$$\Pr\{1 \text{ defective bolt out of } 4\} = \binom{4}{1}(0.2)^1(0.8)^3 = 0.4096$$

(*b*) $\Pr\{0 \text{ defective bolts}\} = \binom{4}{0}(0.2)^0(0.8)^4 = 0.4096$

(*c*) $\Pr\{2 \text{ defective bolts}\} = \binom{4}{2}(0.2)^2(0.8)^2 = 0.1536$

Thus

$$\Pr\{\text{at most 2 defective bolts}\} = \Pr\{0 \text{ defective bolts}\} + \Pr\{1 \text{ defective bolt}\} + \Pr\{2 \text{ defective bolts}\}$$
$$= 0.4096 + 0.4096 + 0.1536 = 0.9728$$

7.8 The probability that an entering student will graduate is 0.4. Determine the probability that out of 5 students, (*a*) none will graduate, (*b*) 1 will graduate, (*c*) at least 1 will graduate, (*d*) all will graduate, and (*e*) use STATISTIX to answer parts (*a*) through (*d*).

SOLUTION

(*a*) $\Pr\{\text{none will graduate}\} = \binom{5}{0}(0.4)^0(0.6)^5 = 0.07776$ or about 0.08

(*b*) $\Pr\{1 \text{ will graduate}\} = \binom{5}{1}(0.4)^1(0.6)^4 = 0.2592$ or about 0.26

(*c*) $\Pr\{\text{at least 1 will graduate}\} = 1 - \Pr\{\text{none will graduate}\} = 0.92224$ or about 0.92

(*d*) $\Pr\{\text{all will graduate}\} = \binom{5}{5}(0.4)^5(0.6)^0 = 0.01024$ or about 0.01

(*e*) STATISTIX evaluates only the cumulative binomial. The dialog box in Fig. 7-5 gives the cumulative binomial distribution for $N = 5$, $p = 0.4$, $q = 0.6$, and $x = 0$, 1, 4, and 5.

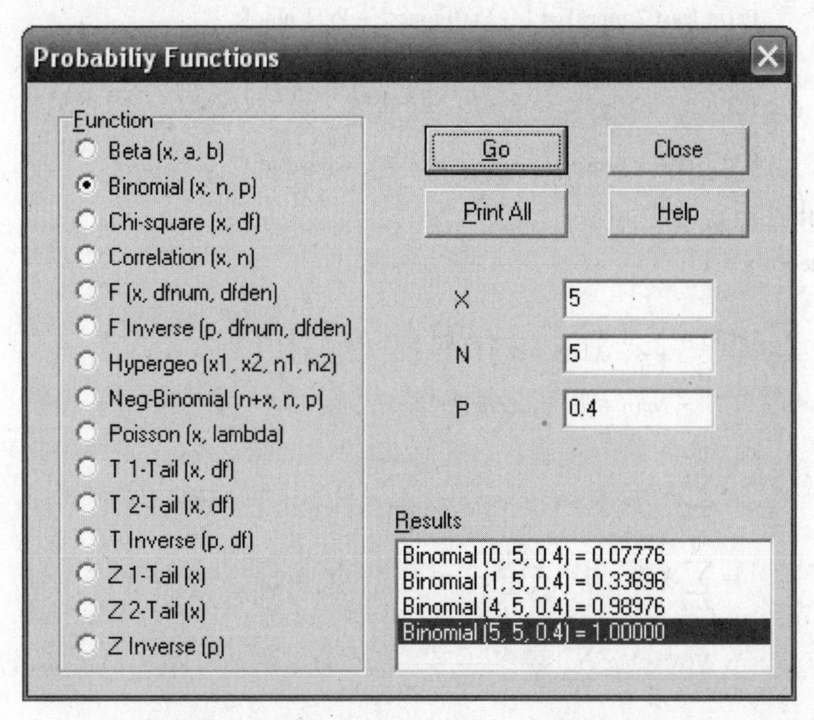

Fig. 7-5 STATISTIX dialog box for Problem 7.8 (*e*).

Referring to the dialog box, the probability none will graduate is $\Pr\{X=0\} = \text{Binomial}(0,5,0.4) = 0.07776$. The probability 1 will graduate is $\Pr\{X=1\} = \Pr\{X \le 1\} - \Pr\{X \le 0\} = \text{Binomial}(1,5,0.4) - \text{Binomial}(0,5,0.4) = 0.33696 - 0.07776 = 0.2592$. The probability at least 1 will graduate is $\Pr\{X \ge 1\} = 1 - \Pr\{X=0\} = 1 - \text{Binomial}(0,5,0.4) = 0.92224$. The probability all will graduate is $\Pr\{X=5\} = \Pr\{X \le 5\} - \Pr\{X \le 4\} = \text{Binomial}(5,5,0.4) - \text{Binomial}(4,5,0.4) = 1.00000 - 0.98976 = 0.01024$. Note that STATISTIX gives only the cumulative binomial and some textbook tables also give only cumulative binomial probabilities.

7.9 What is the probability of getting a total of 9 (*a*) twice and (*b*) at least twice in 6 tosses of a pair of dice?

SOLUTION

Each of the 6 ways in which the first die can fall can be associated with each of the 6 ways in which the second die can fall; thus there are $6 \cdot 6 = 36$ ways in which both dice can fall. These are: 1 on the first die and 1 on the second die, 1 on the first die and 2 on the second die, etc., denoted by (1, 1), (1, 2), etc.

Of these 36 ways (all equally likely if the dice are fair), a total of 9 occurs in 4 cases: (3, 6), (4, 5), (5, 4), and (6, 3). Thus the probability of a total of 9 in a single toss of a pair of dice is $p = \frac{4}{36} = \frac{1}{9}$, and the probability of not getting a total of 9 in a single toss of a pair of dice is $q = 1 - p = \frac{8}{9}$.

(*a*) $$\Pr\{2 \text{ nines in 6 tosses}\} = \binom{6}{2}\left(\frac{1}{9}\right)^2\left(\frac{8}{9}\right)^{6-2} = \frac{61{,}440}{531{,}441}$$

(*b*) $\Pr\{\text{at least 2 nines}\} = \Pr\{2 \text{ nines}\} + \Pr\{3 \text{ nines}\} + \Pr\{4 \text{ nines}\} + \Pr\{5 \text{ nines}\} + \Pr\{6 \text{ nines}\}$

$$= \binom{6}{2}\left(\frac{1}{9}\right)^2\left(\frac{8}{9}\right)^4 + \binom{6}{3}\left(\frac{1}{9}\right)^3\left(\frac{8}{9}\right)^3 + \binom{6}{4}\left(\frac{1}{9}\right)^4\left(\frac{8}{9}\right)^2 + \binom{6}{5}\left(\frac{1}{9}\right)^5\left(\frac{8}{9}\right)^1 + \binom{6}{6}\left(\frac{1}{9}\right)^6\left(\frac{8}{9}\right)^0$$

$$= \frac{61{,}440}{531{,}441} + \frac{10{,}240}{531{,}441} + \frac{960}{531{,}441} + \frac{48}{531{,}441} + \frac{1}{531{,}441} = \frac{72{,}689}{531{,}441}$$

Another method

$$\Pr\{\text{at least 2 nines}\} = 1 - \Pr\{0 \text{ nines}\} - \Pr\{1 \text{ nine}\}$$

$$= 1 - \binom{6}{0}\left(\frac{1}{9}\right)^0\left(\frac{8}{9}\right)^6 - \binom{6}{1}\left(\frac{1}{9}\right)^1\left(\frac{8}{9}\right)^5 = \frac{72{,}689}{531{,}441}$$

7.10 Evaluate (*a*) $\sum_{X=0}^{N} Xp(X)$ and (*b*) $\sum_{X=0}^{N} X^2p(X)$, where $p(X) = \binom{N}{X}p^X q^{N-X}$.

SOLUTION

(*a*) Since $q + p = 1$,

$$\sum_{X=0}^{N} Xp(X) = \sum_{X=1}^{N} X \frac{N!}{X!\,(N-X)!} p^X q^{N-X} = Np \sum_{X=1}^{N} \frac{(N-1)!}{(X-1)!(N-X)!} p^{X-1} q^{N-X}$$

$$= Np(q+p)^{N-1} = Np$$

(*b*) $$\sum_{X=0}^{N} X^2 p(X) = \sum_{X=1}^{N} X^2 \frac{N!}{X!(N-X)!} p^X q^{N-X} = \sum_{X=1}^{N} [X(X-1) + X] \frac{N!}{X!(N-X)!} p^X q^{N-X}$$

$$= \sum_{X=2}^{N} X(X-1) \frac{N!}{X!(N-X)!} p^X q^{N-X} + \sum_{X=1}^{N} X \frac{N!}{X!(N-X)!} p^X q^{N-X}$$

$$= N(N-1)p^2 \sum_{X=2}^{N} \frac{(N-2)!}{(X-2)!(N-X)!} p^{X-2} q^{N-X} + Np = N(N-1)p^2(q+p)^{N-2} + Np$$

$$= N(N-1)p^2 + Np$$

Note: The results in parts (*a*) and (*b*) are the *expectations* of X and X^2, denoted by $E(X)$ and $E(X^2)$, respectively (see Chapter 6).

7.11 If a variable is binomially distributed, determine its (*a*) mean μ and (*b*) variance σ^2.

SOLUTION

(*a*) By Problem 7.10(*a*),

$$\mu = \text{expectation of variable} = \sum_{X=0}^{N} Xp(X) = Np$$

(*b*) Using $\mu = Np$ and the results of Problem 7.10,

$$\sigma^2 = \sum_{X=0}^{N}(X-\mu)^2 p(X) = \sum_{X=0}^{N}(X^2 - 2\mu X + \mu^2)p(X) = \sum_{X=0}^{N} X^2 p(X) - 2\mu \sum_{X=0}^{N} Xp(X) + \mu^2 \sum_{X=0}^{N} p(X)$$

$$= N(N-1)p^2 + Np - 2(Np)(Np) + (Np)^2(1) = Np - Np^2 = Np(1-p) = Npq$$

It follows that the standard deviation of a binomially distributed variable is $\sigma = \sqrt{Npq}$.

Another method

By Problem 6.62(*b*),

$$E[(X-\bar{X})]^2 = E(X^2) - [E(X)]^2 = N(N-1)p^2 + Np - N^2p^2 = Np - Np^2 = Npq$$

7.12 If the probability of a defective bolt is 0.1, find (*a*) the mean and (*b*) the standard deviation for the distribution of defective bolts in a total of 400.

SOLUTION

(*a*) The mean is $Np = 400(0.1) = 40$; that is, we can *expect* 40 bolts to be defective.

(*b*) The variance is $Npq = 400(0.1)(0.9) = 36$. Hence the standard deviation is $\sqrt{36} = 6$.

7.13 Find the moment coefficients of (*a*) skewness and (*b*) kurtosis of the distribution in Problem 7.12.

SOLUTION

(*a*) $$\text{Moment coefficient of skewness} = \frac{q-p}{\sqrt{Npq}} = \frac{0.9-0.1}{6} = 0.133$$

Since this is positive, the distribution is skewed to the right.

(*b*) $$\text{Moment coefficient of kurtosis} = 3 + \frac{1-6pq}{Npq} = 3 + \frac{1-6(0.1)(0.9)}{36} = 3.01$$

The distribution is slightly *leptokurtic* with respect to the normal distribution (i.e., slightly more peaked; see Chapter 5).

THE NORMAL DISTRIBUTION

7.14 On a final examination in mathematics, the mean was 72 and the standard deviation was 15. Determine the standard scores (i.e., grades in standard-deviation units) of students receiving the grades (*a*) 60, (*b*) 93, and (*c*) 72.

SOLUTION

(*a*) $$z = \frac{X-\bar{X}}{s} = \frac{60-72}{15} = -0.8$$

(*c*) $$z = \frac{X-\bar{X}}{s} = \frac{72-72}{15} = 0$$

(*b*) $$z = \frac{X-\bar{X}}{s} = \frac{93-72}{15} = 1.4$$

7.15 Referring to Problem 7.14, find the grades corresponding to the standard scores (*a*) −1 and (*b*) 1.6.

SOLUTION

(*a*) $X = \bar{X} + zs = 72 + (-1)(15) = 57$ (*b*) $X = \bar{X} + zs = 72 + (1.6)(15) = 96$

7.16 Suppose the number of games in which major league baseball players play during their careers is normally distributed with mean equal to 1500 games and standard deviation equal to 350 games. Use EXCEL to solve the following problems. (*a*) What percentage play in fewer than 750 games? (*b*) What percentage play in more than 2000 games? (*c*) Find the 90th percentile for the number of games played during a career.

SOLUTION

(*a*) The EXCEL statement =NORMDIST(750, 1500, 350, 1) requests the area to the left of 750 for a normal curve having mean equal to 1500 and standard deviation equal to 350. The answer is $\Pr\{X < 1500\} = 0.0161$ or 1.61% play in less than 750 games.

(*b*) The EXCEL statement =1-NORMDIST(2000, 1500, 350, 1) requests the area to the right of 2000 for a normal curve having mean equal to 1500 and standard deviation equal to 350. The answer is $\Pr\{X > 2000\} = 0.0766$ or 7.66 % play in more than 2000 games.

(*c*) The EXCEL statement =NORMINV(0.9, 1500, 350) requests the value on the horizontal axis which is such that there is 90% of the area to its left under the normal curve having mean equal to 1500 and standard deviation equal to 350. In the notation of Chapter 3, $P_{90} = 1948.5$.

7.17 Find the area under the standard normal curve in the following case:

(*a*) Between $z = 0.81$ and $z = 1.94$.

(*b*) To the right of $z = -1.28$.

(*c*) To the right of $z = 2.05$ or to the left of $z = -1.44$.

Use Appendix II as well as EXCEL to solve (*a*) through (*c*) [see Figs. 7-6 (*a*), (*b*), and (*c*)].

SOLUTION

(*a*) In Appendix II go down the z column until you reach 1.9; then proceed right to the column marked 4. The result 0.4738 is $\Pr\{0 \le z \le 1.94\}$. Next, go down the z column until you reach 0.8; then proceed right to the column marked 1. The result 0.2910 is $\Pr\{0 \le z \le 0.81\}$. The area $\Pr\{0.81 \le z \le 1.94\}$ is the difference of the two, $\Pr\{0 \le z \le 1.94\} - \Pr\{0 \le z \le 0.81\} = 0.4738 - 0.2910 = 0.1828$. Using EXCEL, the answer is =NORMSDIST(1.94)-NORMSDIST(0.81) = 0.1828. When EXCEL is used, the area $\Pr\{0.81 \le z \le 1.94\}$ is the difference $\Pr\{-\text{infinity} \le z \le 1.94\} - \Pr\{-\text{infinity} \le z \le 0.81\}$. Note that the table in Appendix II gives areas from 0 to a positive z-value and EXCEL gives the area from –infinity to the same z-value.

(*b*) The area to the right of $z = -1.28$ is the same as the area to the left of $z = 1.28$. Using Appendix II, the area to the left of $z = 1.28$ is $\Pr\{z \le 0\} + \Pr\{0 < z \le 1.28\}$ or $0.5 + 0.3997 = 0.8997$. Using EXCEL, $\Pr\{z \ge -1.28\} = \Pr\{z \le 1.28\}$ and $\Pr\{z \le 1.28\}$ is given by =NORMSDIST(1.28) or 0.8997.

(*c*) Using Appendix II, the area to the right of 2.05 is $0.5 - \Pr\{z \le 2.05\}$ or $0.5 - 0.4798 = 0.0202$. The area to the left of −1.44 is the same as the area to the right of 1.44. The area to the right of 1.44 is $0.5 - \Pr\{z \le 1.44\} = 0.5 - 0.4251 = 0.0749$. The two tail areas are equal to $0.0202 + 0.0749 = 0.0951$. Using EXCEL, the area is given by the following: =NORMSDIST(−1.44) + (1 − NORMSDIST(2.05)) or 0.0951.

Note in EXCEL, =NORMSDIST(z) gives the area under the standard normal curve to the left of z whereas =NORMDIST(z, μ, σ, 1) gives the area under the normal curve having mean μ and standard deviation σ to the left of z.

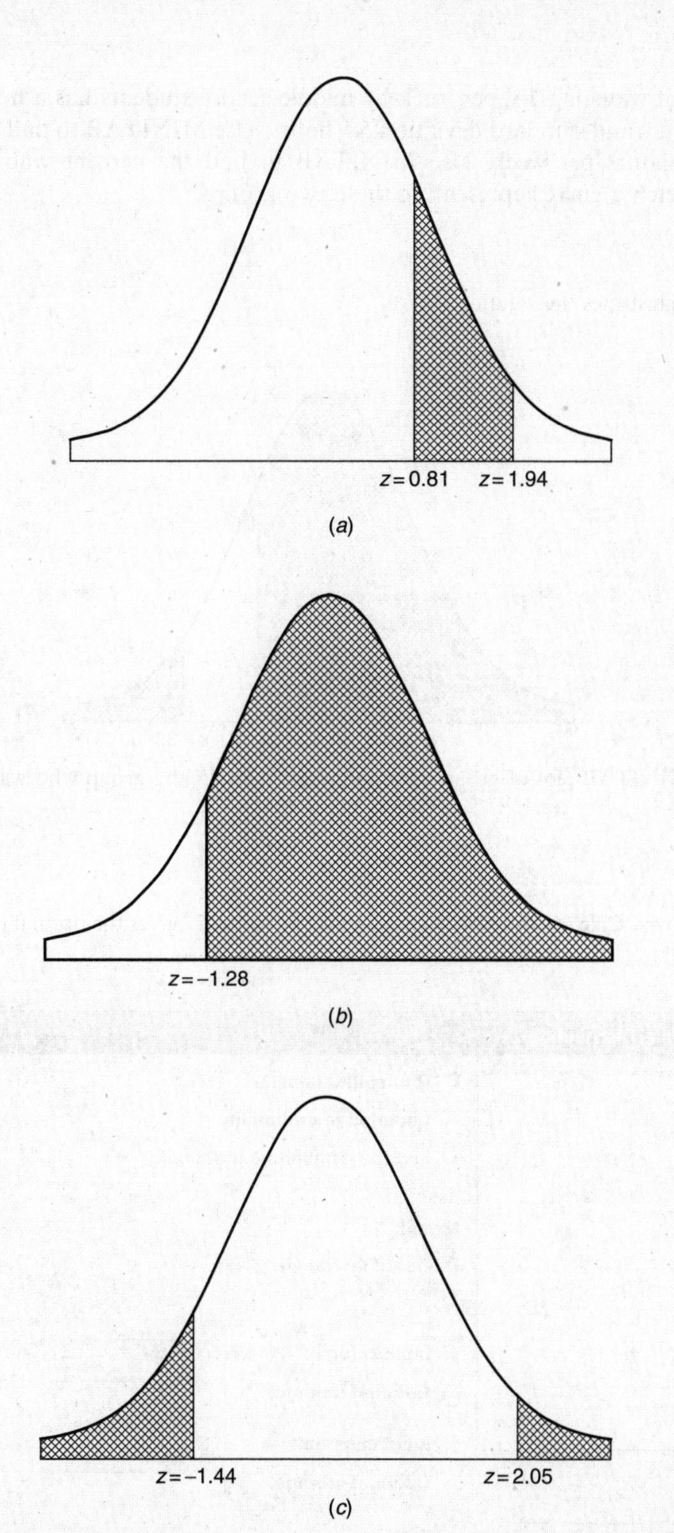

Fig. 7-6 Areas under the standard normal curve. (a) Area between $z=0.81$ and $z=1.94$; (b) Area to the right of $z=-1.28$; (c) Area to the left of $z=-1.44$ plus the area to the right of $z=2.05$.

7.18 The time spent watching TV per week by middle-school students has a normal distribution with mean 20.5 hours and standard deviation 5.5 hours. Use MINITAB to find the percent who watch less than 25 hours per week. Use MINITAB to find the percent who watch over 30 hours per week. Sketch a curve representing these two groups.

SOLUTION

Figure 7-7 illustrates the solution.

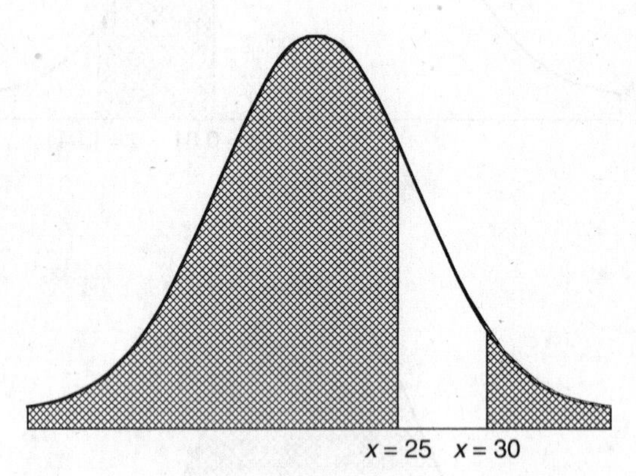

Fig. 7-7 MINITAB plot of group who watch less than 25 and group who watch more than 30 hours per week of TV.

The pull down "**Calc** ⇒ **Probability distributions** ⇒ **Normal**" gives the normal distribution dialog box as in Fig. 7-8.

Normal Distribution

Probability density
Cumulative probability
Inverse cumulative probability

Mean: 20.5
Standard deviation: 5.5

Input column:
Optional storage:
Input constant: 25
Optional storage:

Select
Help
OK
Cancel

Fig. 7-8 Normal distribution MINITAB dialog box.

When the dialog box is completed as shown in Fig. 7.8 and executed, the following output results:

Cumulative Distribution Function

```
Normal with mean = 20.5 and standard deviation = 5.5

x     P(X<=x)
25    0.793373
```

79.3% of the middle-school students watch 25 hours or less per week.

When 30 is entered as the input constant, the following output results:

Cumulative Distribution Function

```
Normal with mean = 20.5 and standard deviation = 5.5

x     P(X<=x)
30    0.957941
```

The percent who watch more than 30 hours per week is $1 - 0.958 = 0.042$ or 4.2%.

7.19 Find the ordinates of the normal curve at (*a*) $z = 0.84$, (*b*) $z = -1.27$, and (*c*) $z = -0.05$.

SOLUTION

(*a*) In Appendix I, proceed downward under the column headed z until reaching the entry 0.8; then proceed right to the column headed 4. The entry 0.2803 is the required ordinate.

(*b*) By symmetry: (ordinate at $z = -1.27$) = (ordinate at $z = 1.27$) = 0.1781.

(*c*) (Ordinate at $z = -0.05$) = (ordinate at $z = 0.05$) = 0.3984.

7.20 Use EXCEL to evaluate some suitably chosen ordinates of the normal curve with mean equal 13.5 and standard deviation 2.5. Then use the chart wizard to plot the resulting points. This plot represents the normal distribution for variable X, where X represents the time spent on the internet per week for college students.

SOLUTION

Abscissas are chosen from 6 to 21 at 0.5 intervals and entered into the EXCEL worksheet in A1:A31. The expression =NORMDIST(A1,13.5,3.5,0) is entered into cell B1 and a click-and-drag is performed. These are points on the normal curve with mean equal to 13.5 and standard deviation 3.5:

6	0.001773
6.5	0.003166
7	0.005433
7.5	0.008958
8	0.01419
8.5	0.021596
9	0.03158
9.5	0.044368
10	0.059891
10.5	0.077674
11	0.096788
11.5	0.115877
12	0.13329
12.5	0.147308
13	0.156417
13.5	0.159577

14	0.156417
14.5	0.147308
15	0.13329
15.5	0.115877
16	0.096788
16.5	0.077674
17	0.059891
17.5	0.044368
18	0.03158
18.5	0.021596
19	0.01419
19.5	0.008958
20	0.005433
20.5	0.003166
21	0.001773

The chart wizard is used to plot these points. The result is shown in Fig. 7-9.

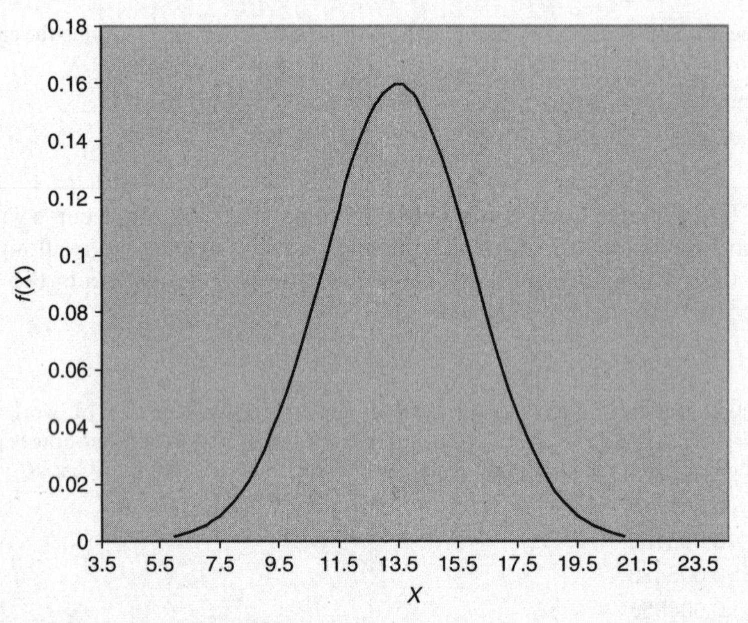

Fig. 7-9 The EXCEL plot of the normal curve having mean = 13.5 and standard deviation = 2.5.

7.21 Determine the second quartile (Q_2), the third quartile (Q_3), and the 90th percentile (P_{90}) for the times that college students spend on the internet. Use the normal distribution given in Problem 7.20.

SOLUTION

The second quartile or 50th percentile for the normal distribution will occur at the center of the curve. Because of the symmetry of the distribution it will occur at the same point as the mean. In the case of the internet usage, that will be 13.5 hours per week. The EXCEL function =NORMINV(0.5, 13.5, 2.5) is used to find the 50th percentile. The 50th percentile means that 0.5 of the area is to the left of the second quartile, and the mean is 13.5 and the standard deviation is 2.5. The result that EXCEL gives is 13.5. Using MINITAB, we find:

Inverse Cumulative Distribution Function

```
Normal with mean = 13.5 and standard deviation = 2.5

p(X<=x)    x
0.5        13.5
```

Figure 7-10(a) illustrates this discussion. Referring to Fig. 7-10(b), we see that 75% of the area under the curve is to the left of Q_3. The EXCEL function =NORMINV(0.75,13.5,2.5) gives $Q_3 = 15.19$.

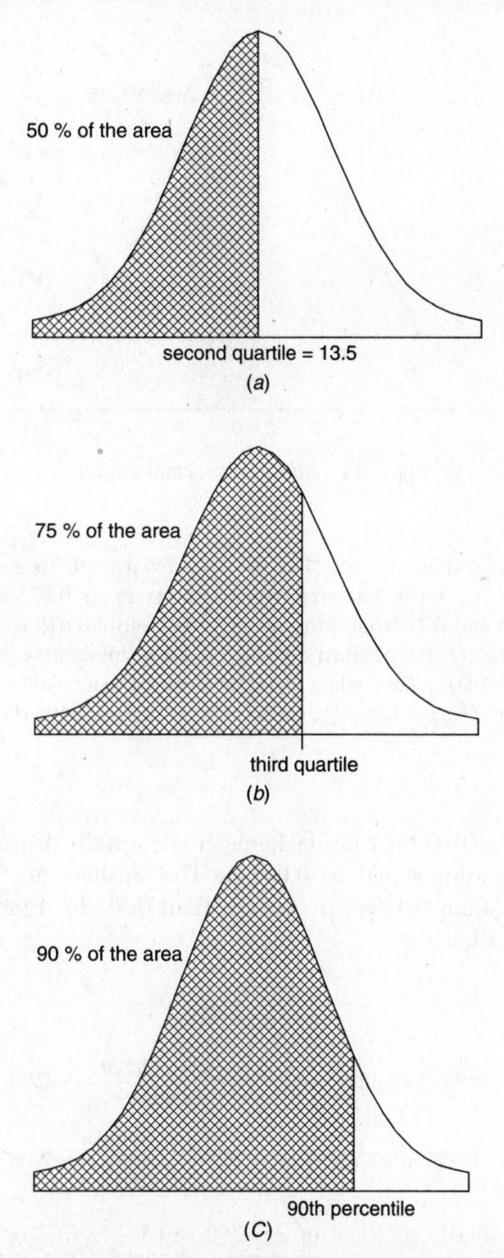

Fig. 7-10 Finding percentiles and quartiles using MINITAB and EXCEL. (a) Q_2 is the value such that 50% of the times are less than that value; (b) Q_3 is the value such that 75% of the times are less than that value; (c) P_{90} is the value such that 90% of the times are less than that value.

Figure 7-10 (*c*) shows that 90% of the area is to the left of P_{90}. The EXCEL function =NORMINV (0.90,13.5,2.5) gives $P_{90} = 16.70$.

7.22 Using Appendix II, find Q_3, in Problem 7.21.

SOLUTION

When software such as EXCEL or MINITAB is not available you must work with the standard normal distribution since it is the only one available in tables.

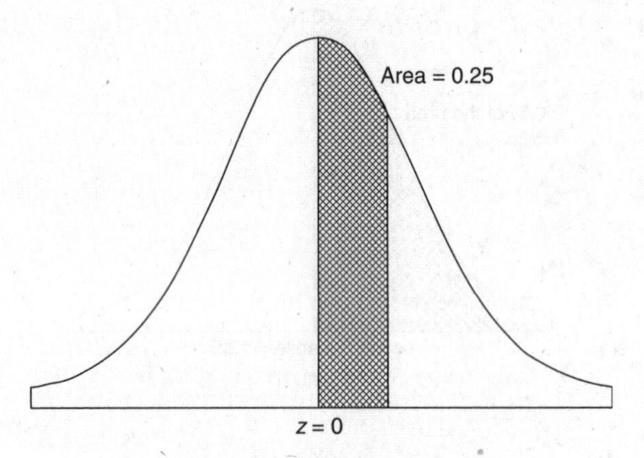

Fig. 7-11 Standard normal curve.

Using Appendix II in reverse, we see that the area from $z = 0$ to $z = 0.67$ is 0.2486 and from $z = 0$ to $z = 0.68$ is 0.2518 (see Fig. 7-11). The area from −infinity to $z = 0.675$ is approximately 0.75 since there is 0.5 from −infinity to 0 and 0.25 from 0 to $z = 0.675$. Therefore 0.675 is roughly the third quartile on the standard normal curve. Let Q_3 be the third quartile on the normal curve having mean = 13.5 and standard deviation = 2.5 [see Fig. 7-10 (*b*)]. Now when Q_3 is transformed to a z-value, we have $0.675 = (Q_3 - 13.5)/2.5$. Solving this equation for Q_3, we have $Q_3 = 2.5(0.675) + 13.5 = 15.19$, the same answer EXCEL gave in Problem 7.21.

7.23 Washers are produced so that their inside diameter is normally distributed with mean 0.500 inches (in) and standard deviation equal to 0.005 in. The washers are considered defective if their inside diameter is less than 0.490 in or greater than 0.510 in. Find the percent defective using Appendix II and EXCEL.

SOLUTION

$$0.490 \text{ in standard units is } \frac{0.490 - 0.500}{0.005} = -2.00$$

$$0.510 \text{ in standard units is } \frac{0.510 - 0.500}{0.005} = 2.00$$

From Appendix II, the area to the right of $Z = 2.00$ is 0.5 − 0.4772 or 0.0228. The area to the left of $Z = -2.00$ is 0.0228. The percent defective is $(0.0228 + 0.0228) \times 100 = 4.56\%$. When finding areas under a normal curve using Appendix II, we convert to the standard normal curve to find the answer.

If EXCEL is used, the answer is = 2*NORMDIST(0.490, 0.500, 0.005, 1) which also gives 4.56%.

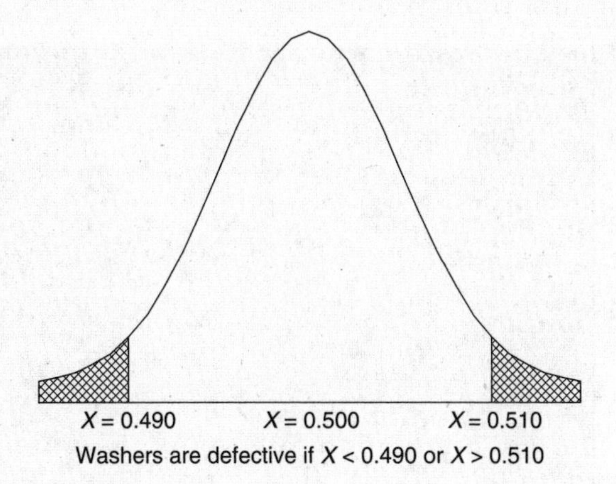

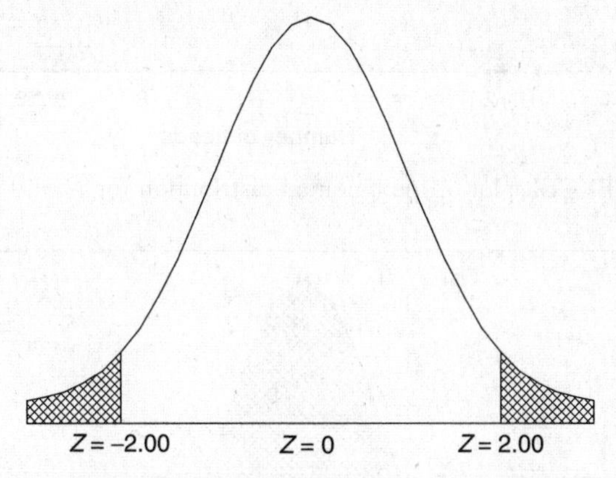

Fig. 7-12 Area to the right of $X = 0.510$ equals the area to the right of $Z = 2.000$ and area to the left of $X = 0.490$ equals the area to the left of $Z = -2.00$.

NORMAL APPROXIMATION TO THE BINOMIAL DISTRIBUTION

7.24 Find the probability of getting between 3 and 6 heads inclusive in 10 tosses of a fair coin by using (*a*) the binomial distribution and (*b*) the normal approximation to the binomial distribution.

SOLUTION

(*a*)

$$\Pr\{3\text{ heads}\} = \binom{10}{3}\left(\frac{1}{2}\right)^3\left(\frac{1}{2}\right)^7 = \frac{15}{128} \qquad \Pr\{5\text{ heads}\} = \binom{10}{5}\left(\frac{1}{2}\right)^5\left(\frac{1}{2}\right)^5 = \frac{63}{256}$$

$$\Pr\{4\text{ heads}\} = \binom{10}{4}\left(\frac{1}{2}\right)^4\left(\frac{1}{2}\right)^6 = \frac{105}{512} \qquad \Pr\{6\text{ heads}\} = \binom{10}{6}\left(\frac{1}{2}\right)^6\left(\frac{1}{2}\right)^4 = \frac{105}{512}$$

Thus

$$\Pr\{\text{between 3 and 6 heads inclusive}\} = \frac{15}{128} + \frac{105}{512} + \frac{63}{256} + \frac{105}{512} = \frac{99}{128} = 0.7734$$

(*b*) The EXCEL plot of the binomial distribution for $N = 10$ tosses of a fair coin is shown in Fig. 7-13.

Note that even though the binomial distribution is discrete, it has the shape of the continuous normal distribution. When approximating the binomial probability at 3, 4, 5, and 6 heads by the area under the normal curve, find the normal curve area from $X = 2.5$ to $X = 6.5$. The 0.5 that you go on either side of $X = 3$ and $X = 6$ is called the *continuity correction*. The following are the steps to follow when approximating the binomial with the normal. Choose the normal curve with mean $Np = 10(0.5) = 5$ and standard deviation $= \sqrt{Npq} = \sqrt{10(0.5)(0.5)} = 1.58$. You are choosing the normal curve with the same center and variation as the binomial distribution. Then find the area under the

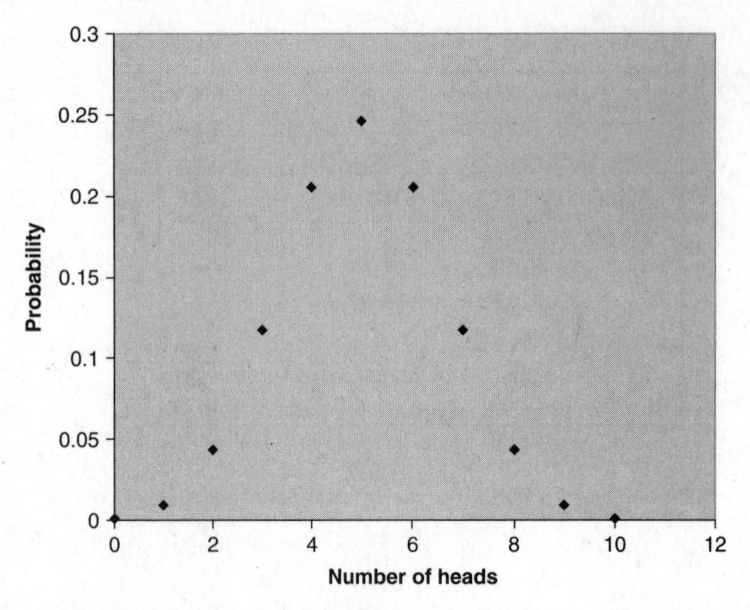

Fig. 7-13 EXCEL plot of the binomial distribution for $N=10$ and $p=0.5$.

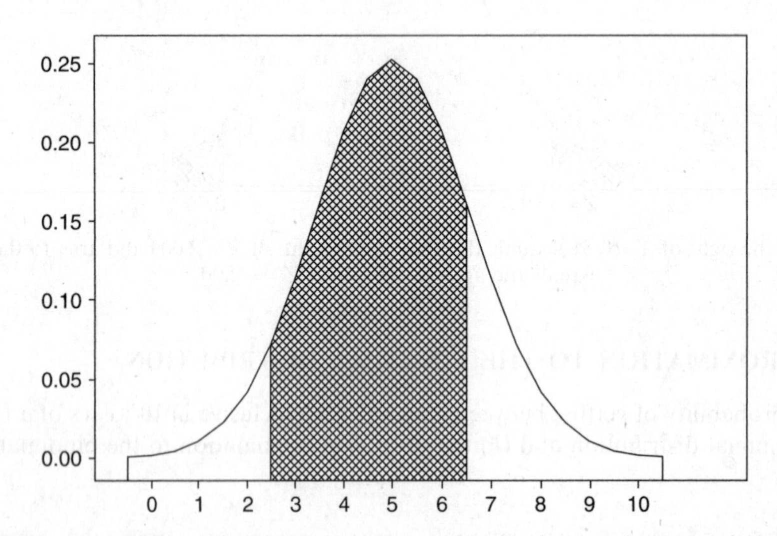

Fig. 7-14 Normal approximation to 3, 4, 5, or 6 heads when a coin is tossed 10 times.

curve from 2.5 to 6.5 as shown in Fig. 7-14. This is the normal approximation to the binomial distribution.

The solution when using an EXCEL worksheet is given by =NORMDIST(6.5, 5, 1.58, 1) – NORMDIST (2.5, 5, 1.58, 1) which equals 0.7720.

If the technique using Appendix II is applied, the normal values 6.5 and 2.5 are first converted to standard normal values. (2.5 in standard units is −1.58 and 6.5 in standard units is 0.95.) The area between −1.58 and 0.95 from Appendix II is 0.7718. Whichever method is used the answer is very close to the binomial answer of 0.7734.

7.25 A fair coin is tossed 500 times. Find the probability that the number of heads will not differ from 250 by (*a*) more than 10 and (*b*) more than 30.

SOLUTION

$$\mu = Np = (500)(\tfrac{1}{2}) = 250 \qquad \sigma = \sqrt{Npq} = \sqrt{(500)(\tfrac{1}{2})(\tfrac{1}{2})} = 11.18$$

(*a*) We require the probability that the number of heads will lie between 240 and 260 or, considering the data to be continuous, between 239.5 and 260.5. Since 239.5 in standard units is $(239.5 - 250)/11.18 = -0.94$, and 260.5 in standard units is 0.94, we have

$$\text{Required probability} = (\text{area under normal curve between } z = -0.94 \text{ and } z = 0.94)$$
$$= (\text{twice area between } z = 0 \text{ and } z = 0.94) = 2(0.3264) = 0.6528$$

(*b*) We require the probability that the number of heads will lie between 220 and 280 or, considering the data to be continuous, between 219.5 and 280.5. Since 219.5 in standard units is $(219.5 - 250)/11.18 = -2.73$, and 280.5 in standard units is 2.73, we have

$$\text{Required probability} = (\text{twice area under normal curve between } z = 0 \text{ and } z = -2.73)$$
$$= 2(0.4968) = 0.9936$$

It follows that we can be very confident that the number of heads will not differ from that expected (250) by more than 30. Thus if it turned out that the *actual* number of heads was 280, we would strongly believe that the coin was not fair (i.e., was loaded).

7.26 Suppose 75% of the age group 1 through 4 years regularly utilize seat belts. Find the probability that in a random stop of 100 automobiles containing 1 through 4 year olds, 70 or fewer are found to be wearing a seat belt. Find the solution using the binomial distribution as well as the normal approximation to the binomial distribution. Use MINITAB to find the solutions.

SOLUTION

The MINITAB output given below shows that the probability that 70 or fewer will be found to be wearing a seat belt is equal to 0.1495.

```
MTB > cdf 70;
SUBC> binomial 100 .75.

Cumulative Distribution Function

Binomial with n = 100 and p = 0.750000
      x      P( X <= x)
70.00         0.1495
```

The solution using the normal approximation to the binomial distribution is found as follows: The mean of the binomial distribution is $\mu = Np = 100(0.75) = 75$ and the standard deviation is $\sigma = \sqrt{Npq} = \sqrt{100(0.75)(0.25)} = 4.33$. The MINITAB output given below shows the normal approximation to equal 0.1493. The approximation is very close to the true value.

```
MTB > cdf 70.5;
SUBC> normal mean = 75 sd = 4.33.

Cumulative Distribution Function

Normal with mean = 75.0000 and standard deviation = 4.33000
      x        P( X <= x)
70.5000          0.1493
```

THE POISSON DISTRIBUTION

7.27 Ten percent of the tools produced in a certain manufacturing process turn out to be defective. Find the probability that in a sample of 10 tools chosen at random exactly 2 will be defective by using (*a*) the binomial distribution and (*b*) the Poisson approximation to the binomial distribution.

SOLUTION

The probability of a defective tool is $p = 0.1$.

(*a*) $$\Pr\{2 \text{ defective tools in } 10\} = \binom{10}{2}(0.1)^2(0.9)^8 = 0.1937 \quad \text{or} \quad 0.19$$

(*b*) With $\lambda = Np = 10(0.1) = 1$ and using $e = 2.718$,

$$\Pr\{2 \text{ defective tools in } 10\} = \frac{\lambda^X e^{-\lambda}}{X!} = \frac{(1)^2 e^{-1}}{2!} = \frac{e^{-1}}{2} = \frac{1}{2e} = 0.1839 \quad \text{or} \quad 0.18$$

In general, the approximation is good if $p \leq 0.1$ and $\lambda = Np \leq 5$.

7.28 If the probability that an individual suffers a bad reaction from injection of a given serum is 0.001, determine the probability that out of 2000 individuals (*a*) exactly 3 and (*b*) more than 2 individuals will suffer a bad reaction. Use MINITAB and find the answers using both the Poisson and the binomial distributions.

SOLUTION

(*a*) The following MINITAB output gives first the binomial probability that exactly 3 suffer a bad reaction. Using $\lambda = Np = (2000)(0.001) = 2$, the Poisson probability is shown following the binomial probability. The Poisson approximation is seen to be extremely close to the binomial probability.

```
MTB > pdf 3;
SUBC> binomial 2000.001.

Probability Density Function

Binomial with n = 2000 and p = 0.001
     x          P( X = x)
   3.0           0.1805
MTB > pdf 3;
SUBC> poisson 2.

Probability Density Function

Poisson with mu = 2
     x          P( X = x)
  3.00           0.1804
```

(*b*) The probability that more than 2 individuals suffer a bad reaction is given by $1 - P(X \leq 2)$. The following MINITAB output gives the probability that $X \leq 2$ as 0.6767 using both the binomial and the Poisson distribution. The probability that more than 2 suffer a bad reaction is $1 - 0.6767 = 0.3233$.

```
MTB > cdf 2;
SUBC> binomial 2000 .001.

Cumulative Distribution Function

Binomial with n = 2000 and p = 0.001
      x      P ( X <= x)
    2.0         0.6767
MTB > cdf 2;
SUBC> poisson 2.

Cumulative Distribution Function

Poisson with mu = 2
      x          P( X <= x)
   2.00            0.6767
```

7.29 A Poisson distribution is given by

$$p(X) = \frac{(0.72)^X e^{-0.72}}{X!}$$

Find (*a*) $p(0)$, (*b*) $p(1)$, (*c*) $p(2)$, and (*d*) $p(3)$.

SOLUTION

(*a*) $$p(0) = \frac{(0.72)^0 e^{-0.72}}{0!} = \frac{(1)\,e^{-0.72}}{1} = e^{-0.72} = 0.4868 \qquad \text{using Appendix VIII}$$

(*b*) $$p(1) = \frac{(0.72)^1 e^{-0.72}}{1!} = (0.72)e^{-0.72} = (0.72)(0.4868) = 0.3505$$

(*c*) $$p(2) = \frac{(0.72)^2 e^{-0.72}}{2!} = \frac{(0.5184)e^{-0.72}}{2} = (0.2592)(0.4868) = 0.1262$$

Another method

$$p(2) = \frac{0.72}{2}\,p(1) = (0.36)(0.3505) = 0.1262$$

(*d*) $$p(3) = \frac{(0.72)^3 e^{-0.72}}{3!} = \frac{0.72}{3}\,p(2) = (0.24)(0.1262) = 0.0303$$

THE MULTINOMIAL DISTRIBUTION

7.30 A box contains 5 red balls, 4 white balls, and 3 blue balls. A ball is selected at random from the box, its color is noted, and then the ball is replaced. Find the probability that out of 6 balls selected in this manner, 3 are red, 2 are white, and 1 is blue.

SOLUTION

Pr{red at any drawing} = $\frac{5}{12}$, Pr{white at any drawing} = $\frac{4}{12}$, and Pr{blue at any drawing} = $\frac{3}{12}$; thus

$$\Pr\{3 \text{ are red, } 2 \text{ are white, } 1 \text{ is blue}\} = \frac{6!}{3!2!1!}\left(\frac{5}{12}\right)^3\left(\frac{4}{12}\right)^2\left(\frac{3}{12}\right)^1 = \frac{625}{5184}$$

FITTING OF DATA BY THEORETICAL DISTRIBUTIONS

7.31 Fit a binomial distribution to the data of Problem 2.17.

SOLUTION

We have Pr{X heads in a toss of 5 pennies} = $p(X) = \binom{5}{X}p^X q^{5-X}$, where p and q are the respective probabilities of a head and a tail on a single toss of a penny. By Problem 7.11(*a*), the mean number of heads is $\mu = Np = 5p$. For the actual (or observed) frequency distribution, the mean number of heads is

$$\frac{\sum fX}{\sum f} = \frac{(38)(0) + (144)(1) + (342)(2) + (287)(3) + (164)(4) + (25)(5)}{1000} = \frac{2470}{1000} = 2.47$$

Equating the theoretical and actual means, $5p = 2.47$, or $p = 0.494$. Thus the fitted binomial distribution is given by $p(X) = \binom{5}{X}(0.494)^X(0.506)^{5-X}$.

Table 7.4 lists these probabilities as well as the expected (theoretical) and actual frequencies. The fit is seen to be fair.

Table 7.4

Number of Heads (X)	Pr{X heads}	Expected Frequency	Observed Frequency
0	0.0332	33.2, or 33	38
1	0.1619	161.9, or 162	144
2	0.3162	316.2, or 316	342
3	0.3087	308.7, or 309	287
4	0.1507	150.7, or 151	164
5	0.0294	29.4, or 29	25

7.32 Use the Kolmogorov–Smirnov test in MINITAB to test the data in Table 7.5 for normality. The data represent the time spent on a cell phone per week for 30 college students.

Table 7.5

16	17	15
14	14	16
12	16	12
13	11	14
10	15	14
13	15	16
17	13	15
14	18	14
11	12	13
13	15	12

SOLUTION

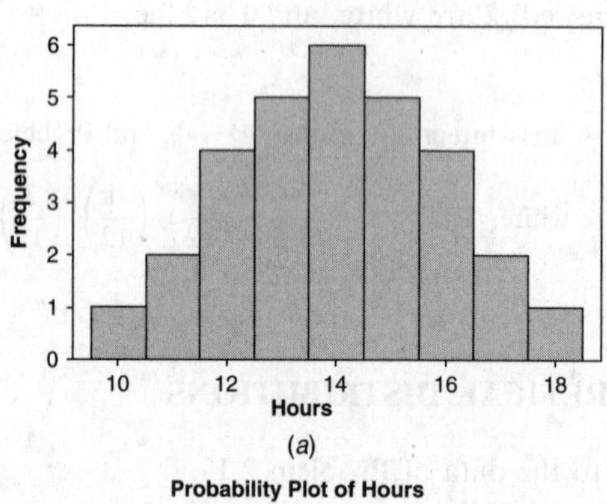

(a)

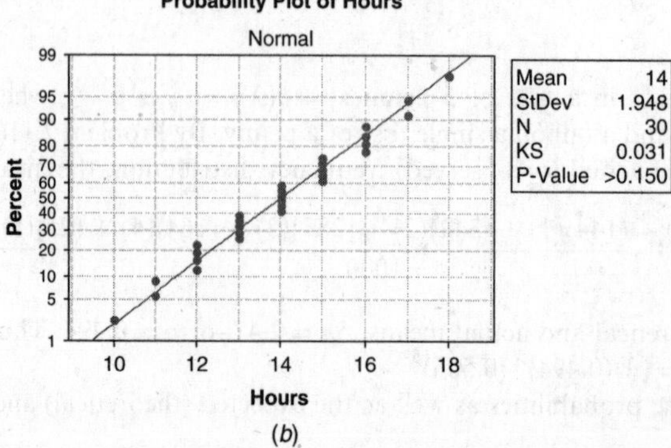

(b)

Fig. 7-15 Kolmogorov–Smirnov test for normality: normal data. (*a*) Histogram reveals a set of data that is normally distributed; (*b*) Kolmogorov–Smirnov test of normality indicates normality p-value > 0.150.

The histogram in Fig. 7-15(*a*) indicates that the data in this survey are normally distributed. The Kolmogorov–Smirnov test also indicates that the sample data came from a population that is normally distributed. Most statisticians recommend that if the *p*-value is less than 0.05, then reject normality. The *p*-value here is > 0.15.

7.33 Use the Kolmogorov–Smirnov test in MINITAB to test the data in Table 7.6 for normality. The data represent the time spent on a cell phone per week for 30 college students.

Table 7.6

18	16	11
17	12	13
17	17	17
16	18	17
16	17	17
16	18	15
18	18	16
16	16	17
14	18	15
11	18	10

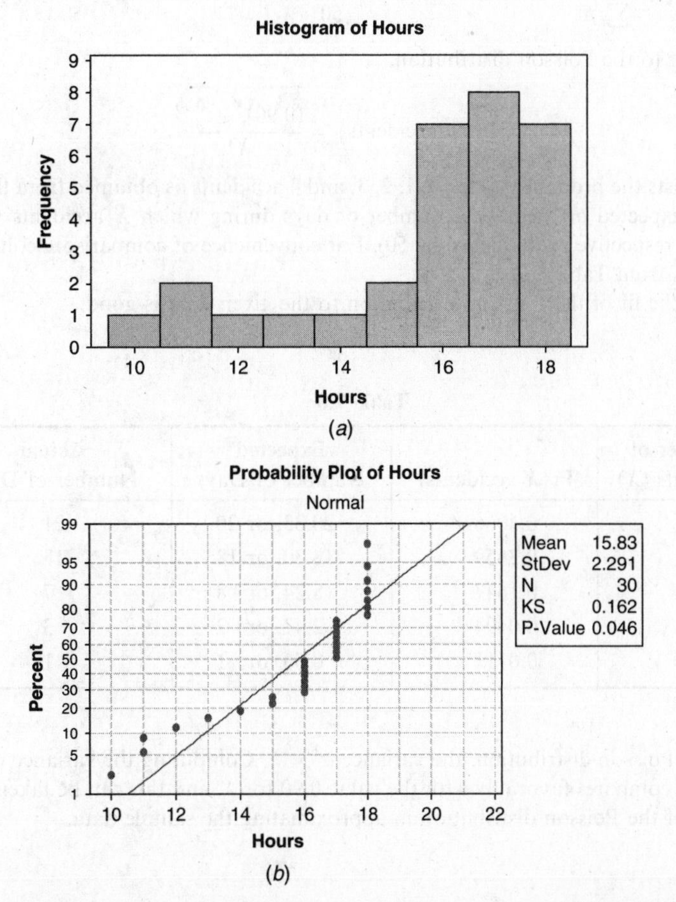

Fig. 7-16 Kolmogorov–Smirnov test for normality: non-normal data *p*-value = 0.046. (*a*) Histogram reveals a set of data that is skewed to the left; (*b*) Kolmogorov–Smirnov test of normality indicates lack of normality.

Most statisticians recommend that if the p-value is less than 0.05, then reject normality. The p-value here is less than 0.05.

7.34 Table 7.7 shows the number of days, f, in a 50-day period during which X automobile accidents occurred in a city. Fit a Poisson distribution to the data.

Table 7.7

Number of Accidents (X)	Number of Days (f)
0	21
1	18
2	7
3	3
4	1
	Total 50

SOLUTION

The mean number of accidents is

$$\lambda = \frac{\sum fX}{\sum f} = \frac{(21)(0) + (18)(1) + (7)(2) + (3)(3) + (1)(4)}{50} = \frac{45}{50} = 0.90$$

Thus, according to the Poisson distribution,

$$\Pr\{X \text{ accidents}\} = \frac{(0.90)^X e^{-0.90}}{X!}$$

Table 7.8 lists the probabilities for 0, 1, 2, 3, and 4 accidents as obtained from this Poisson distribution, as well as the expected or theoretical number of days during which X accidents take place (obtained by multiplying the respective probabilities by 50). For convenience of comparison, column 4 repeats the actual number of days from Table 7.7.

Note that the fit of the Poisson distribution to the given data is good.

Table 7.8

Number of Accidents (X)	Pr{X accidents}	Expected Number of Days	Actual Number of Days
0	0.4066	20.33, or 20	21
1	0.3659	18.30, or 18	18
2	0.1647	8.24, or 8	7
3	0.0494	2.47, or 2	3
4	0.0111	0.56, or 1	1

For a true Poisson distribution, the variance $\sigma^2 = \lambda$. Computing the variance of the given distribution gives 0.97. This compares favorably with the value 0.90 for λ, and this can be taken as further evidence for the suitability of the Poisson distribution in approximating the sample data.

Supplementary Problems

THE BINOMIAL DISTRIBUTION

7.35 Evaluate (*a*) 7!, (*b*) 10!/(6!4!), (*c*) $\binom{9}{5}$, (*d*) $\binom{11}{8}$, and (*e*) $\binom{6}{1}$.

7.36 Expand (*a*) $(q+p)^7$ and (*b*) $(q+p)^{10}$.

7.37 Find the probability that in tossing a fair coin six times there will appear (*a*) 0, (*b*) 1, (*c*) 2, (*d*), 3, (*e*) 4, (*f*) 5, and (*g*) 6 heads, (*h*) Use MINITAB to build the distribution of X = the number of heads in 6 tosses of a coin.

7.38 Find the probability of (*a*) 2 or more heads and (*b*) fewer than 4 heads in a single toss of 6 fair coins, (*c*) Use EXCEL to find the answers to (*a*) and (*b*).

7.39 If X denotes the number of heads in a single toss of 4 fair coins, find (*a*) $\Pr\{X = 3\}$, (*b*) $\Pr\{X < 2\}$, (*c*) $\Pr\{X \leq 2\}$, and (*d*) $\Pr\{1 < X \leq 3\}$.

7.40 Out of 800 families with 5 children each, how many would you expect to have (*a*) 3 boys, (*b*) 5 girls, and (*c*) either 2 or 3 boys? Assume equal probabilities for boys and girls.

7.41 Find the probability of getting a total of 11 (*a*) once and (*b*) twice in two tosses of a pair of fair dice.

7.42 What is the probability of getting a 9 exactly once in 3 throws with a pair of dice?

7.43 Find the probability of guessing correctly at least 6 of the 10 answers on a true-false examination.

7.44 An insurance salesperson sells policies to 5 men, all of identical age and in good health. According to the actuarial tables, the probability that a man of this particular age will be alive 30 years hence is $\frac{2}{3}$. Find the probability that in 30 years (*a*) all 5 men, (*b*) at least 3 men, (*c*) only 2 men, and (*d*) at least 1 man will be alive, (*e*) Use EXCEL to answer (*a*) through (*d*).

7.45 Compute the (*a*) mean, (*b*) standard deviation, (*c*) moment coefficient of skewness, and (*d*) moment coefficient of kurtosis for a binomial distribution in which $p = 0.7$ and $N = 60$. Interpret the results.

7.46 Show that if a binomial distribution with $N = 100$ is symmetrical, its moment coefficient of kurtosis is 2.98.

7.47 Evaluate (*a*) $\sum (X-\mu)^3 p(X)$ and (*b*) $\sum (X-\mu)^4 p(X)$ for the binomial distribution.

7.48 Prove formulas (*1*) and (*2*) at the beginning of this chapter for the moment coefficients of skewness and kurtosis.

THE NORMAL DISTRIBUTION

7.49 On a statistics examination the mean was 78 and the standard deviation was 10.

(*a*) Determine the standard scores of two students whose grades were 93 and 62, respectively.

(*b*) Determine the grades of two students whose standard scores were -0.6 and 1.2, respectively.

7.50 Find (*a*) the mean and (*b*) the standard deviation on an examination in which grades of 70 and 88 correspond to standard scores of -0.6 and 1.4, respectively.

7.51 Find the area under the normal curve between (*a*) $z = -1.20$ and $z = 2.40$, (*b*) $z = 1.23$ and $z = 1.87$, and (*c*) $z = -2.35$ and $z = -0.50$, (*d*) Work parts (*a*) through (*c*) using EXCEL.

7.52 Find the area under the normal curve (*a*) to the left of $z = -1.78$, (*b*) to the left of $z = 0.56$, (*c*) to the right of $z = -1.45$, (*d*) corresponding to $z \geq 2.16$, (*e*) corresponding to $-0.80 \leq z \leq 1.53$, and (*f*) to the left of $z = -2.52$ and to the right of $z = 1.83$, (*g*) Solve parts (*a*) through (*f*) using EXCEL.

7.53 If z is normally distributed with mean 0 and variance 1, find (*a*) $\Pr\{z \geq -1.64\}$, (*b*) $\Pr\{-1.96 \leq z \leq 1.96\}$, and (*c*) $\Pr\{|z| \geq 1\}$.

7.54 Find the value of z such that (*a*) the area to the right of z is 0.2266, (*b*) the area to the left of z is 0.0314, (*c*) the area between -0.23 and z is 0.5722, (*d*) the area between 1.15 and z is 0.0730, and (*e*) the area between $-z$ and z is 0.9000.

7.55 Find z_1 if $\Pr\{z \geq z_1\} = 0.84$, where z is normally distributed with mean 0 and variance 1.

7.56 Find the ordinates of the normal curve at (*a*) $z = 2.25$, (*b*) $z = -0.32$, and (*c*) $z = -1.18$ using Appendix I. (*d*) Find the answers to (*a*) through (*c*) using EXCEL.

7.57 Adult males have normally distributed heights with mean equal to 70 in and standard deviation equal to 3 in. (*a*) What percent are shorter than 65 in? (*b*) What percent are taller than 72 in? (*c*) What percent are between 68 and 73 in?

7.58 The amount spent for goods online is normally distributed with mean equal to \$125 and standard deviation equal to \$25 for a certain age group. (*a*) What percent spend more than \$175? (*b*) What percent spend between \$100 and \$150? (*c*) What percent spend less than \$50?

7.59 The mean grade on a final examination was 72 and the standard deviation was 9. The top 10% of the students are to receive A's. What is the minimum grade that a student must get in order to receive an A?

7.60 If a set of measurements is normally distributed, what percentage of the measurements differ from the mean by (*a*) more than half the standard deviation and (*b*) less than three-quarters of the standard deviation?

7.61 If $\bar{X}$ is the mean and s is the standard deviation of a set of normally distributed measurements, what percentage of the measurements are (*a*) within the range $\bar{X} \pm 2s$, (*b*) outside the range $\bar{X} \pm 1.2s$, and (*c*) greater than $\bar{X} - 1.5s$?

7.62 In Problem 7.61, find the constant a such that the percentage of the cases (*a*) within the range $\bar{X} \pm as$ is 75% and (*b*) less than $\bar{X} - as$ is 22%.

NORMAL APPROXIMATION TO THE BINOMIAL DISTRIBUTION

7.63 Find the probability that 200 tosses of a coin will result in (*a*) between 80 and 120 heads inclusive, (*b*) less than 90 heads, (*c*) less than 85 or more than 115 heads, and (*d*) exactly 100 heads.

7.64 Find the probability that on a true-false examination a student can guess correctly the answers to (*a*) 12 or more out of 20 questions and (*b*) 24 or more out of 40 questions.

7.65 Ten percent of the bolts that a machine produces are defective. Find the probability that in a random sample of 400 bolts produced by this machine, (*a*) at most 30, (*b*) between 30 and 50, (*c*) between 35 and 45, and (*d*) 55 or more of the bolts will be defective.

7.66 Find the probability of getting more than 25 sevens in 100 tosses of a pair of fair dice.

THE POISSON DISTRIBUTION

7.67 If 3% of the electric bulbs manufactured by a company are defective, find the probability that in a sample of 100 bulbs (*a*) 0, (*b*) 1, (*c*) 2, (*d*) 3, (*e*) 4, and (*f*) 5 bulbs will be defective.

7.68 In Problem 7.67, find the probability that (*a*) more than 5, (*b*) between 1 and 3, and (*c*) less than or equal to 2 bulbs will be defective.

7.69 A bag contains 1 red and 7 white marbles. A marble is drawn from the bag and its color is observed. Then the marble is put back into the bag and the contents are thoroughly mixed. Using (*a*) the binomial distribution and (*b*) the Poisson approximation to the binomial distribution, find the probability that in 8 such drawings a red ball is selected exactly 3 times.

7.70 According to the National Office of Vital Statistics of the U.S. Department of Health, Education, and Welfare, the average number of accidental drownings per year in the United States is 3.0 per 100,000 population. Find the probability that in a city of population 200,000 there will be (*a*) 0, (*b*) 2, (*c*) 6, (*d*) 8, (*e*) between 4 and 8, and (*f*) fewer than 3 accidental drownings per year.

7.71 Between the hours of 2 and 4 P.M. the average number of phone calls per minute coming into the switchboard of a company is 2.5. Find the probability that during one particular minute there will be (*a*) 0, (*b*) 1, (*c*) 2, (*d*) 3, (*e*) 4 or fewer, and (*f*) more than 6 phone calls.

THE MULTINOMIAL DISTRIBUTION

7.72 A fair die is tossed six times. Find the probability (*a*) that one 1, two 2's, and three 3's turn up and (*b*) that each side turns up only once.

7.73 A box contains a very large number of red, white, blue, and yellow marbles in the ratio 4:3:2:1, respectively. Find the probability that in 10 drawings (*a*) 4 red, 3 white, 2 blue, and 1 yellow marble will be drawn and (*b*) 8 red and 2 yellow marbles will be drawn.

7.74 Find the probability of not getting a 1, 2, or 3 in four tosses of a fair die.

FITTING OF DATA BY THEORETICAL DISTRIBUTIONS

7.75 Fit a binomial distribution to the data in Table 7.9.

Table 7.9

X	0	1	2	3	4
f	30	62	46	10	2

7.76 A survey of middle-school students and their number of hours of exercise per week was determined. Construct a histogram of the data using STATISTIX. Use the Shapiro–Wilk test of STATISTIX to determine if the data were taken from a normal distribution. The data are shown in Table 7.10.

Table 7.10

5	10	2	3	2
5	5	1	3	15
1	2	20	3	1
4	4	4	3	5

7.77 Using the data in Table 7.5 of Problem 7.32, construct a histogram of the data using STATISTIX. Use the Shapiro–Wilk test of STATISTIX to determine if the data were taken from a normal distribution.

7.78 The test scores in Table 7.11 follow a U distribution. This is just the opposite of a normal distribution. Using the data in Table 7.11, construct a histogram of the data using STATISTIX. Use the Shapiro–Wilk test of STATISTIX to determine if the data were taken from a normal distribution.

Table 7.11

20	90	10
40	90	20
80	70	50
70	40	90
90	70	10
60	30	80
10	20	30
30	10	20
10	80	90
60	50	80

7.79 Use the test scores in Table 7.11, the Anderson–Darling test in MINITAB, and the Ryan–Joiner in MINITAB to test that the data came from a normal population.

7.80 For 10 Prussian army corps units over a period of 20 years (1875 to 1894), Table 7.12 shows the number of deaths per army corps per year resulting from the kick of a horse. Fit a Poisson distribution to the data.

Table 7.12

X	0	1	2	3	4
f	109	65	22	3	1

Elementary Sampling Theory

SAMPLING THEORY

Sampling theory is a study of relationships existing between a population and samples drawn from the population. It is of great value in many connections. For example, it is useful in *estimating* unknown population quantities (such as population mean and variance), often called *population parameters* or briefly *parameters*, from a knowledge of corresponding sample quantities (such as sample mean and variance), often called *sample statistics* or briefly *statistics*. Estimation problems are considered in Chapter 9.

Sampling theory is also useful in determining whether the observed differences between two samples are due to chance variation or whether they are really significant. Such questions arise, for example, in testing a new serum for use in treatment of a disease or in deciding whether one production process is better than another. Their answers involve the use of so-called *tests of significance and hypotheses* that are important in the *theory of decisions*. These are considered in Chapter 10.

In general, a study of the inferences made concerning a population by using samples drawn from it, together with indications of the accuracy of such inferences by using probability theory, is called *statistical inference*.

RANDOM SAMPLES AND RANDOM NUMBERS

In order that the conclusions of sampling theory and statistical inference be valid, samples must be chosen so as to be *representative* of a population. A study of sampling methods and of the related problems that arise is called the *design of the experiment*.

One way in which a representative sample may be obtained is by a process called *random sampling*, according to which each member of a population has an equal chance of being included in the sample. One technique for obtaining a random sample is to assign numbers to each member of the population, write these numbers on small pieces of paper, place them in an urn, and then draw numbers from the urn, being careful to mix thoroughly before each drawing. An alternative method is to use a table of *random numbers* (see Appendix IX) specially constructed for such purposes. See Problem 8.6.

SAMPLING WITH AND WITHOUT REPLACEMENT

If we draw a number from an urn, we have the choice of replacing or not replacing the number into the urn before a second drawing. In the first case the number can come up again and again, whereas in the second it can only come up once. Sampling where each member of the population may be chosen more than once is called *sampling with replacement*, while if each member cannot be chosen more than once it is called *sampling without replacement*.

Populations are either finite or infinite. If, for example, we draw 10 balls successively without replacement from an urn containing 100 balls, we are sampling from a finite population; while if we toss a coin 50 times and count the number of heads, we are sampling from an infinite population.

A finite population in which sampling is with replacement can theoretically be considered infinite, since any number of samples can be drawn without exhausting the population. For many practical purposes, sampling from a finite population that is very large can be considered to be sampling from an infinite population.

SAMPLING DISTRIBUTIONS

Consider all possible samples of size N that can be drawn from a given population (either with or without replacement). For each sample, we can compute a statistic (such as the mean and the standard deviation) that will vary from sample to sample. In this manner we obtain a distribution of the statistic that is called its *sampling distribution*.

If, for example, the particular statistic used is the sample mean, then the distribution is called the *sampling distribution of means*, or the *sampling distribution of the mean*. Similarly, we could have sampling distributions of standard deviations, variances, medians, proportions, etc.

For each sampling distribution, we can compute the mean, standard deviation, etc. Thus we can speak of the mean and standard deviation of the sampling distribution of means, etc.

SAMPLING DISTRIBUTION OF MEANS

Suppose that all possible samples of size N are drawn without replacement from a finite population of size $N_p > N$. If we denote the mean and standard deviation of the sampling distribution of means by $\mu_{\bar{X}}$ and $\sigma_{\bar{X}}$ and the population mean and standard deviation by μ and σ, respectively, then

$$\mu_{\bar{X}} = \mu \quad \text{and} \quad \sigma_{\bar{X}} = \frac{\sigma}{\sqrt{N}}\sqrt{\frac{N_p - N}{N_p - 1}} \tag{1}$$

If the population is infinite or if sampling is with replacement, the above results reduce to

$$\mu_{\bar{X}} = \mu \quad \text{and} \quad \sigma_{\bar{X}} = \frac{\sigma}{\sqrt{N}} \tag{2}$$

For large values of N ($N \geq 30$), the sampling distribution of means is approximately a normal distribution with mean $\mu_{\bar{X}}$ and standard deviation $\sigma_{\bar{X}}$, irrespective of the population (so long as the population mean and variance are finite and the population size is at least twice the sample size). This result for an infinite population is a special case of the *central limit theorem* of advanced probability theory, which shows that the accuracy of the approximation improves as N gets larger. This is sometimes indicated by saying that the sampling distribution is *asymptotically normal*.

In case the population is normally distributed, the sampling distribution of means is also normally distributed even for small values of N (i.e., $N < 30$).

SAMPLING DISTRIBUTION OF PROPORTIONS

Suppose that a population is infinite and that the probability of occurrence of an event (called its success) is p, while the probability of nonoccurrence of the event is $q = 1 - p$. For example, the population may be all possible tosses of a fair coin in which the probability of the event "heads" is $p = \frac{1}{2}$. Consider all possible samples of size N drawn from this population, and for each sample determine the proportion P of successes. In the case of the coin, P would be the proportion of heads turning up in N tosses. We thus obtain a *sampling distribution of proportions* whose mean μ_P and standard deviation σ_P are given by

$$\mu_P = p \quad \text{and} \quad \sigma_P = \sqrt{\frac{pq}{N}} = \sqrt{\frac{p(1-p)}{N}} \tag{3}$$

which can be obtained from equations (*2*) by placing $\mu = p$ and $\sigma = \sqrt{pq}$. For large values of N ($N \geq 30$), the sampling distribution is very closely normally distributed. Note that the population is *binomially distributed.*

Equations (*3*) are also valid for a finite population in which sampling is with replacement. For finite populations in which sampling is without replacement, equations (*3*) are replaced by equations (*1*) with $\mu = p$ and $\sigma = \sqrt{pq}$.

Note that equations (*3*) are obtained most easily by dividing the mean and standard deviation (Np and $\sqrt{Npq}$) of the binomial distribution by N (see Chapter 7).

SAMPLING DISTRIBUTIONS OF DIFFERENCES AND SUMS

Suppose that we are given two populations. For each sample of size N_1 drawn from the first population, let us compute a statistic S_1; this yields a sampling distribution for the statistic S_1, whose mean and standard deviation we denote by μ_{S1} and σ_{S1}, respectively. Similarly, for each sample of size N_2 drawn from the second population, let us compute a statistic S_2; this yields a sampling distribution for the statistic S_2, whose mean and standard deviation are denoted by μ_{S2} and σ_{S2}. From all possible combinations of these samples from the two populations we can obtain a distribution of the differences, $S_1 - S_2$, which is called the *sampling distribution of differences of the statistics*. The mean and standard deviation of this sampling distribution, denoted respectively by μ_{S1-S2} and σ_{S1-S2}, are given by

$$\mu_{S1-S2} = \mu_{S1} - \mu_{S2} \quad \text{and} \quad \sigma_{S1-S2} = \sqrt{\sigma_{S1}^2 + \sigma_{S2}^2} \tag{4}$$

provided that the samples chosen do not in any way depend on each other (i.e., the samples are *independent*).

If S_1, and S_2 are the sample means from the two populations—which means we denote by $\bar{X}_1$ and $\bar{X}_2$, respectively—then the sampling distribution of the differences of means is given for infinite populations with means and standard deviations (μ_1, σ_1) and (μ_2, σ_2), respectively, by

$$\mu_{\bar{X}1-\bar{X}2} = \mu_{\bar{X}1} - \mu_{\bar{X}2} = \mu_1 - \mu_2 \quad \text{and} \quad \sigma_{\bar{X}1-X2} = \sqrt{\sigma_{\bar{X}1}^2 + \sigma_{\bar{X}2}^2} = \sqrt{\frac{\sigma_1^2}{N_1} + \frac{\sigma_2^2}{N_2}} \tag{5}$$

using equations (*2*). The result also holds for finite populations if sampling is with replacement. Similar results can be obtained for finite populations in which sampling is without replacement by using equations (*1*).

Table 8.1 Standard Error for Sampling Distributions

Sampling Distribution	Standard Error	Special Remarks
Means	$\sigma_{\bar{X}} = \dfrac{\sigma}{\sqrt{N}}$	This is true for large or small samples. The sampling distribution of means is very nearly normal for $N \geq 30$ even when the population is nonnormal. $\mu_{\bar{X}} = \mu$, the population mean, in all cases.
Proportions	$\sigma_P = \sqrt{\dfrac{p(1-p)}{N}} = \sqrt{\dfrac{pq}{N}}$	The remarks made for means apply here as well. $\mu_P = p$ in all cases.
Standard deviations	(*1*) $\sigma_s = \dfrac{\sigma}{\sqrt{2N}}$ (*2*) $\sigma_s = \sqrt{\dfrac{\mu_4 - \mu_2^2}{4N\mu_2}}$	For $N \geq 100$, the sampling distribution of s is very nearly normal. σ_s is given by (*1*) only if the population is normal (or approximately normal). If the population is nonnormal, (*2*) can be used. Note that (*2*) reduces to (*1*) when $\mu_2 = \sigma^2$ and $\mu_4 = 3\sigma^4$, which is true for normal populations. For $N \geq 100$, $\mu_s = \sigma$ very nearly
Medians	$\sigma_{\text{med}} = \sigma\sqrt{\dfrac{\pi}{2N}} = \dfrac{1.2533\sigma}{\sqrt{N}}$	For $N \geq 30$, the sampling distribution of the median is very nearly normal. The given result holds only if the population is normal (or approximately normal). $\mu_{\text{med}} = \mu$
First and third quartiles	$\sigma_{Q1} = \sigma_{Q3} = \dfrac{1.3626\sigma}{\sqrt{N}}$	The remarks made for medians apply here as well. μ_{Q1} and μ_{Q3} are very nearly equal to the first and third quartiles of the population. Note that $\sigma_{Q2} = \sigma_{\text{med}}$
Deciles	$\sigma_{D1} = \sigma_{D9} = \dfrac{1.7094\sigma}{\sqrt{N}}$ $\sigma_{D2} = \sigma_{D8} = \dfrac{1.4288\sigma}{\sqrt{N}}$ $\sigma_{D3} = \sigma_{D7} = \dfrac{1.3180\sigma}{\sqrt{N}}$ $\sigma_{D4} = \sigma_{D6} = \dfrac{1.2680\sigma}{\sqrt{N}}$	The remarks made for medians apply here as well. $\mu_{D1}, \mu_{D2}, \ldots$ are very nearly equal to the first, second, ... deciles of the population. Note that $\sigma_{D5} = \sigma_{\text{med}}$.
Semi-interquartile ranges	$\sigma_Q = \dfrac{0.7867\sigma}{\sqrt{N}}$	The remarks made for medians apply here as well. μ_Q is very nearly equal to the population semi-interquartile range
Variances	(*1*) $\sigma_{S^2} = \sigma^2\sqrt{\dfrac{2}{N}}$ (*2*) $\sigma_{S^2} = \sqrt{\dfrac{\mu_4 - \dfrac{N-3}{N-1}\mu_2^2}{N}}$	The remarks made for standard deviation apply here as well. Note that (*2*) yields (*1*) in the case that the population is normal. $\mu_{S^2} = \sigma^2(N-1)/N$, which is very nearly σ^2 for large N.
Coefficients of variation	$\sigma_V = \dfrac{v}{\sqrt{2N}}\sqrt{1 + 2v^2}$	Here $v = \sigma/\mu$ is the population coefficient of variation. The given result holds for normal (or nearly normal) populations and $N \geq 100$.

Corresponding results can be obtained for the sampling distributions of differences of proportions from two binomially distributed populations with parameters (p_1, q_1) and (p_2, q_2), respectively. In this case S_1 and S_2 correspond to the proportion of successes, P_1 and P_2, and equations (4) yield the results

$$\mu_{P1-P2} = \mu_{P1} - \mu_{P2} = p_1 - p_2 \qquad \text{and} \qquad \sigma_{P1-P2} = \sqrt{\sigma_{P1}^2 + \sigma_{P2}^2} = \sqrt{\frac{p_1 q_1}{N_1} + \frac{p_2 q_2}{N_2}} \tag{6}$$

If N_1, and N_2 are large ($N_1, N_2 \geq 30$), the sampling distributions of differences of means or proportions are very closely normally distributed.

It is sometimes useful to speak of the *sampling distribution of the sum of statistics*. The mean and standard deviation of this distribution are given by

$$\mu_{S1+S2} = \mu_{S1} + \mu_{S2} \qquad \text{and} \qquad \sigma_{S1+S2} = \sqrt{\sigma_{S1}^2 + \sigma_{S2}^2} \tag{7}$$

assuming that the samples are *independent*.

STANDARD ERRORS

The standard deviation of a sampling distribution of a statistic is often called its *standard error*. Table 8.1 lists standard errors of sampling distributions for various statistics under the conditions of random sampling from an infinite (or very large) population or of sampling with replacement from a finite population. Also listed are special remarks giving conditions under which results are valid and other pertinent statements.

The quantities μ, σ, p, μ_r and $\bar{X}$, s, P, m_r denote, respectively, the population and sample means, standard deviations, proportions, and rth moments about the mean.

It is noted that if the sample size N is large enough, the sampling distributions are normal or nearly normal. For this reason, the methods are known as *large sampling methods.* When $N < 30$, samples are called *small*. The theory of *small* samples, or *exact sampling theory* as it is sometimes called, is treated in Chapter 11.

When population parameters such as σ, p, or μ_r, are unknown, they may be estimated closely by their corresponding sample statistics namely, s (or $\hat{s} = \sqrt{N/(N-1)}s$), P, and m_r—if the samples are large enough.

SOFTWARE DEMONSTRATION OF ELEMENTARY SAMPLING THEORY

EXAMPLE 1.

A large population has the following random variable defined on it. X represents the number of computers per household and X is uniformly distributed, that is, $p(x) = 0.25$ for $x = 1$, 2, 3, and 4. In other words, 25% of the households have 1 computer, 25% have 2 computers, 25% have 3 computers, and 25% have 4 computers. The mean value of X is $\mu = \Sigma x p(x) = 0.25 + 0.5 + 0.75 + 1 = 2.5$. The variance of X is $\sigma^2 = \Sigma x^2 p(x) - \mu^2 = 0.25 + 1 + 2.25 + 4 - 6.25 = 1.25$. We say that the mean number of computers per household is 2.5 and the variance of the number of computers is 1.25 per household.

EXAMPLE 2.

MINITAB may be used to list all samples of two households taken with replacement. The worksheet would be as in Table 8.2. The 16 samples are shown in C1 and C2 and the mean for each sample in C3. Because the population is uniformly distributed, each sample mean has probability 1/16. Summarizing, the probability distribution is given in C4 and C5.

Note that $\mu_{\bar{x}} = \Sigma \bar{x} p(\bar{x}) = 1(0.0625) + 1.5(0.1250) + \cdots + 4(0.0625) = 2.5$. We see that $\mu_{\bar{x}} = \mu$. Also, $\sigma_{\bar{x}}^2 = \Sigma \bar{x}^2 p(\bar{x}) - \mu_{\bar{x}}^2 = 1(0.0625) + 2.25(0.1250) + \cdots + 16(0.0625) - 6.25 = 0.625$ which gives $\sigma_{\bar{x}}^2 = (\sigma^2/2)$. If MINITAB is used to draw the graph of the probability distribution of xbar, the result shown in Fig. 8-1 is obtained. (Note that $\bar{X}$ and xbar are used interchangeably.)

Table 8.2

C1 household1	C2 household2	C3 mean	C4 xbar	C5 p(xbar)
1	1	1.0	1.0	0.0625
1	2	1.5	1.5	0.1250
1	3	2.0	2.0	0.1875
1	4	2.5	2.5	0.2500
2	1	1.5	3.0	0.1875
2	2	2.0	3.5	0.1250
2	3	2.5	4.0	0.0625
2	4	3.0		
3	1	2.0		
3	2	2.5		
3	3	3.0		
3	4	3.5		
4	1	2.5		
4	2	3.0		
4	3	3.5		
4	4	4.0		

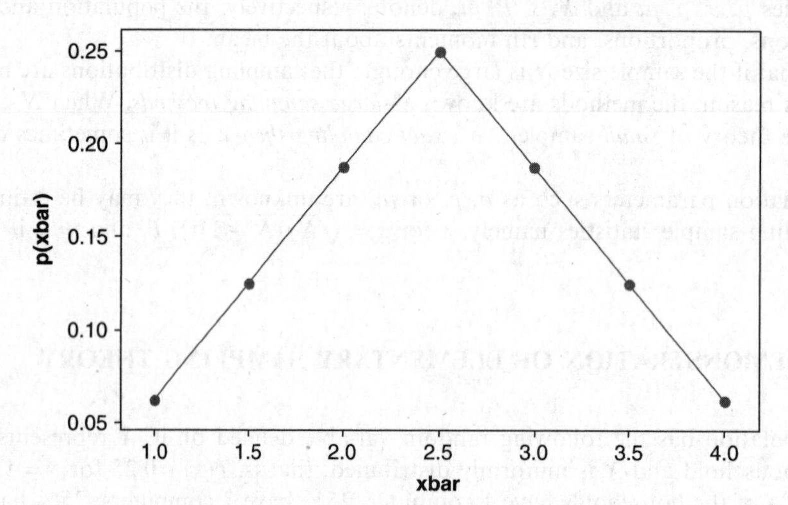

Fig. 8-1 Scatterplot of *p*(xbar) vs xbar.

Solved Problems

SAMPLING DISTRIBUTION OF MEANS

8.1 A population consists of the five numbers 2, 3, 6, 8, and 11. Consider all possible samples of size 2 that can be drawn with replacement from this population. Find (*a*) the mean of the population, (*b*) the standard deviation of the population, (*c*) the mean of the sampling distribution of means, and (*d*) the standard deviation of the sampling distribution of means (i.e., the standard error of means).

SOLUTION

(*a*)
$$\mu = \frac{2+3+6+8+11}{5} = \frac{30}{5} = 6.0$$

(*b*)
$$\sigma^2 = \frac{(2-6)^2+(3-6)^2+(6-6)^2+(8-6)^2+(11-6)^2}{5} = \frac{16+9+0+4+25}{5} = 10.8$$

and $\sigma = 3.29$.

(*c*) There are 5(5) = 25 samples of size 2 that can be drawn with replacement (since any one of the five numbers on the first draw can be associated with any one of the five numbers on the second draw). These are

(2, 2)	(2, 3)	(2, 6)	(2, 8)	(2, 11)
(3, 2)	(3, 3)	(3, 6)	(3, 8)	(3, 11)
(6, 2)	(6, 3)	(6, 6)	(6, 8)	(6, 11)
(8, 2)	(8, 3)	(8, 6)	(8, 8)	(8, 11)
(11, 2)	(11, 3)	(11, 6)	(11, 8)	(11, 11)

The corresponding sample means are

2.0	2.5	4.0	5.0	6.5	
2.5	3.0	4.5	5.5	7.0	
4.0	4.5	6.0	7.0	8.5	(8)
5.0	5.5	7.0	8.0	9.5	
6.5	7.0	8.5	9.5	11.0	

and the mean of sampling distribution of means is

$$\mu_{\bar{X}} = \frac{\text{sum of all sample means in } (8)}{25} = \frac{150}{25} = 6.0$$

illustrating the fact that $\mu_{\bar{X}} = \mu$.

(*d*) The variance $\sigma^2_{\bar{X}}$ of the sampling distribution of means is obtained by subtracting the mean 6 from each number in (*8*), squaring the result, adding all 25 numbers thus obtained, and dividing by 25. The final result is $\sigma^2_{\bar{X}} = 135/25 = 5.40$, and thus $\sigma_{\bar{X}} = \sqrt{5.40} = 2.32$. This illustrates the fact that for finite populations involving sampling with replacement (or infinite populations), $\sigma^2_{\bar{X}} = \sigma^2/N$ since the right-hand side is $10.8/2 = 5.40$, agreeing with the above value.

8.2 Solve Problem 8.1 for the case that the sampling is without replacement.

SOLUTION

As in parts (*a*) and (*b*) of Problem 8.1, $\mu = 6$ and $\sigma = 3.29$.

(*c*) There are $\binom{5}{2} = 10$ samples of size 2 that can be drawn without replacement (this means that we draw one number and then another number different from the first) from the population: (2, 3), (2, 6), (2, 8), (2, 11), (3, 6), (3, 8), (3, 11), (6, 8), (6, 11), and (8, 11). The selection (2, 3), for example, is considered the same as (3, 2).

The corresponding sample means are 2.5, 4.0, 5.0, 6.5, 4.5, 5.5, 7.0, 7.0, 8.5, and 9.5, and the mean of sampling distribution of means is

$$\mu_{\bar{X}} = \frac{2.5+4.0+5.0+6.5+4.5+5.5+7.0+7.0+8.5+9.5}{10} = 6.0$$

illustrating the fact that $\mu_{\bar{X}} = \mu$.

(*d*) The variance of sampling distribution of means is

$$\sigma_{\bar{X}}^2 = \frac{(2.5-6.0)^2 + (4.0-6.0)^2 + (5.0-6.0)^2 + \cdots + (9.5-6.0)^2}{10} = 4.05$$

and $\sigma_{\bar{X}} = 2.01$. This illustrates

$$\sigma_{\bar{X}}^2 = \frac{\sigma^2}{N}\left(\frac{N_p - N}{N_p - 1}\right)$$

since the right side equals

$$\frac{10.8}{2}\left(\frac{5-2}{5-1}\right) = 4.05$$

as obtained above.

8.3 Assume that the heights of 3000 male students at a university are normally distributed with mean 68.0 inches (in) and standard deviation 3.0 in. If 80 samples consisting of 25 students each are obtained, what would be the expected mean and standard deviation of the resulting sampling distribution of means if the sampling were done (*a*) with replacement and (*b*) without replacement?

SOLUTION

The numbers of samples of size 25 that could be obtained theoretically from a group of 3000 students with and without replacement are $(3000)^{25}$ and $\binom{3000}{25}$, which are much larger than 80. Hence we do not get a true sampling distribution of means, but only an *experimental* sampling distribution. Nevertheless, since the number of samples is large, there should be close agreement between the two sampling distributions. Hence the expected mean and standard deviation would be close to those of the theoretical distribution. Thus we have:

(*a*) $$\mu_{\bar{X}} = \mu = 68.0 \text{ in} \quad \text{and} \quad \sigma_{\bar{X}} = \frac{\sigma}{\sqrt{N}} = \frac{3}{\sqrt{25}} = 0.6 \text{ in}$$

(*b*) $$\mu_{\bar{X}} = 68.0 \text{ in} \quad \text{and} \quad \sigma_{\bar{X}} = \frac{\sigma}{\sqrt{N}}\sqrt{\frac{N_p - N}{N_p - 1}} = \frac{3}{\sqrt{25}}\sqrt{\frac{3000-25}{3000-1}}$$

which is only very slightly less than 0.6 in and can therefore, for all practical purposes, be considered the same as in sampling with replacement.

Thus we would expect the experimental sampling distribution of means to be approximately normally distributed with mean 68.0 in and standard deviation 0.6 in.

8.4 In how many samples of Problem 8.3 would you expect to find the mean (*a*) between 66.8 and 68.3 in and (*b*) less than 66.4 in?

SOLUTION

The mean $\bar{X}$ of a sample in standard units is here given by

$$z = \frac{\bar{X} - \mu_{\bar{X}}}{\sigma_{\bar{X}}} = \frac{\bar{X} - 68.0}{0.6}$$

(*a*) $$66.8 \text{ in standard units} = \frac{66.8 - 68.0}{0.6} = -2.0$$

$$68.3 \text{ in standard units} = \frac{68.3 - 68.0}{0.6} = 0.5$$

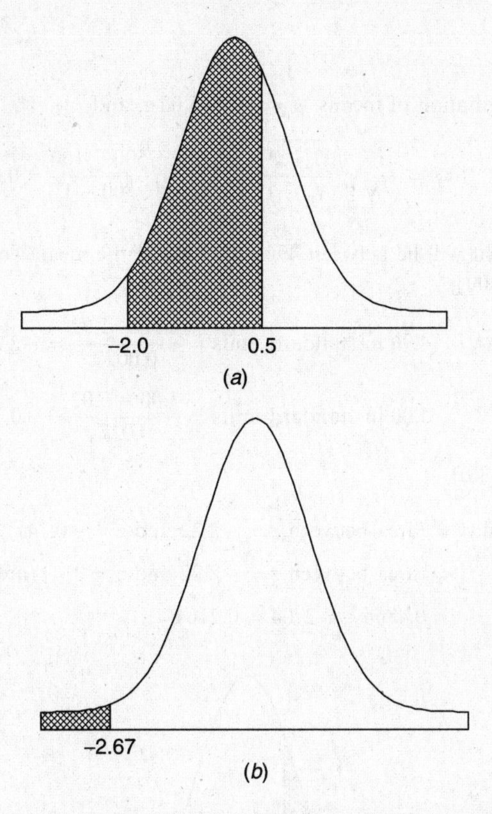

Fig. 8-2 Areas under the standard normal curve. (*a*) Standard normal curve showing the area between $z = -2$ and $z = 0.5$; (*b*) Standard normal curve showing the area to the left of $z = -2.67$.

As shown in Fig. 8-2(*a*),

Proportion of samples with means between 66.8 and 68.3 in

$$= (\text{area under normal curve between } z = -2.0 \text{ and } z = 0.5)$$
$$= (\text{area between } z = -2 \text{ and } z = 0) + (\text{area between } z = 0 \text{ and } z = 0.5)$$
$$= 0.4772 + 0.1915 = 0.6687$$

Thus the expected number of samples is $(80)(0.6687) = 53.496$, or 53.

(*b*)
$$66.4 \text{ in standard units} = \frac{66.4 - 68.0}{0.6} = -2.67$$

As shown in Fig. 8.2(*b*),

$$\text{Proportion of samples with means less than 66.4 in} = (\text{area under normal curve to left of } z = -2.67)$$
$$= (\text{area to left of } z = 0)$$
$$-(\text{area between } z = -2.67 \text{ and } z = 0)$$
$$= 0.5 - 0.4962 = 0.0038$$

Thus the expected number of samples is $(80)(0.0038) = 0.304$, or zero.

8.5 Five hundred ball bearings have a mean weight of 5.02 grams (g) and a standard deviation of 0.30 g. Find the probability that a random sample of 100 ball bearings chosen from this group will have a combined weight of (*a*) between 496 and 500 g and (*b*) more than 510 g.

SOLUTION

For the sample distribution of means, $\mu_{\bar{X}} = \mu = 5.02\,\text{g}$, and

$$\sigma_{\bar{X}} = \frac{\sigma}{\sqrt{N}}\sqrt{\frac{N_p - N}{N_p - 1}} = \frac{0.30}{\sqrt{100}}\sqrt{\frac{500-100}{500-1}} = 0.027\,\text{g}$$

(*a*) The combined weight will lie between 496 and 500 g if the mean weight of the 100 ball bearings lies between 4.96 and 5.00 g.

$$4.96 \text{ in standard units} = \frac{4.96 - 5.02}{0.0027} = -2.22$$

$$5.00 \text{ in standard units} = \frac{5.00 - 5.02}{0.027} = -0.74$$

As shown in Fig. 8-3(*a*),

$$\begin{aligned}\text{Required probability} &= (\text{area between } z = -2.22 \text{ and } z = -0.74)\\ &= (\text{area between } z = -2.22 \text{ and } z = 0) - (\text{area between } z = -0.74 \text{ and } z = 0)\\ &= 0.4868 - 0.2704 = 0.2164\end{aligned}$$

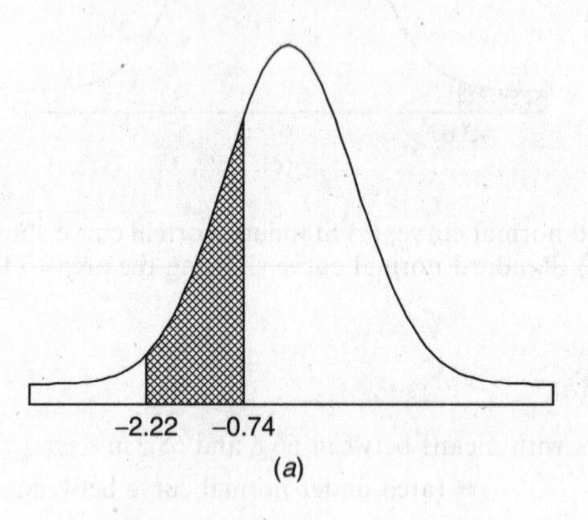

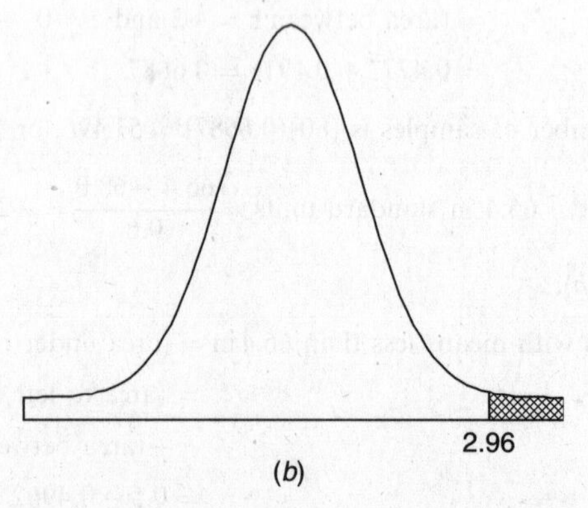

Fig. 8-3 Sample probabilities are found as areas under the standard normal curve. (*a*) Standard normal curve showing the area between $z = -2.22$ and $z = -0.74$; (*b*) Standard normal curve showing the area to the right of $z = 2.96$.

(*b*) The combined weight will exceed 510 g if the mean weight of the 100 bearings exceeds 5.10 g.

$$5.10 \text{ in standard units} = \frac{5.10 - 5.02}{0.027} = 2.96$$

As shown in Fig. 8-3(*b*),

$$\begin{aligned}\text{Required probability} &= (\text{area to right of } z = 2.96)\\ &= (\text{area to right of } z = 0) - (\text{area between } z = 0 \text{ and } z = 2.96)\\ &= 0.5 - 0.4985 = 0.0015\end{aligned}$$

Thus there are only 3 chances in 2000 of picking a sample of 100 ball bearings with a combined weight exceeding 510 g.

8.6 (*a*) Show how to select 30 random samples of 4 students each (with replacement) from Table 2.1 by using random numbers.

(*b*) Find the mean and standard deviation of the sampling distribution of means in part (*a*).

(*c*) Compare the results of part (*b*) with theoretical values, explaining any discrepancies.

SOLUTION

(*a*) Use two digits to number each of the 100 students: 00, 01, 02, ..., 99 (see Table 8.3). Thus the 5 students with heights 60–62 in are numbered 00–04, the 18 students with heights 63–65 in are numbered 05–22, etc. Each student number is called a *sampling number*.

Table 8.3

Height (in)	Frequency	Sampling Number
60–62	5	00–04
63–65	18	05–22
66–68	42	23–64
69–71	27	65–91
72–74	8	92–99

We now draw sampling numbers from the random-number table (Appendix IX). From the first line we find the sequence 51, 77, 27, 46, 40, etc., which we take as random sampling numbers, each of which yields the height of a particular student. Thus 51 corresponds to a student having height 66–68 in, which we take as 67 in (the class mark). Similarly, 77, 27, and 46 yield heights of 70, 67, and 67 in, respectively.

By this process we obtain Table 8.4, which shows the sample numbers drawn, the corresponding heights, and the mean height for each of 30 samples. It should be mentioned that although we have entered the random-number table on the first line, we could have started *anywhere* and chosen any specified pattern.

(*b*) Table 8.5 gives the frequency distribution of the sample mean heights obtained in part (*a*). This is a *sampling distribution of means*. The mean and the standard deviation are obtained as usual by the coding methods of Chapters 3 and 4:

$$\text{Mean} = A + c\bar{u} = A + \frac{c\sum fu}{N} = 67.00 + \frac{(0.75)(23)}{30} = 67.58 \text{ in}$$

$$\text{Standard deviation} = c\sqrt{\overline{u^2} - \bar{u}^2} = c\sqrt{\frac{\sum fu^2}{N} - \left(\frac{\sum fu}{N}\right)^2} = 0.75\sqrt{\frac{123}{30} - \left(\frac{23}{30}\right)^2} = 1.41 \text{ in}$$

(*c*) The theoretical mean of the sampling distribution of means, given by $\mu_{\bar{X}}$, should equal the population mean μ, which is 67.45 in (see Problem 3.22), in agreement with the value 67.58 in of part (*b*).

The theoretical standard deviation (standard error) of the sampling distribution of means, given by $\sigma_{\bar{X}}$, should equal $\sigma/\sqrt{N}$, where the population standard deviation $\sigma = 2.92$ in (see Problem 4.17) and the sample size $N = 4$. Since $\sigma/\sqrt{N} = 2.92/\sqrt{4} = 1.46$ in, we have agreement with the value 1.41 in of part (*b*). The discrepancies result from the fact that only 30 samples were selected and the sample size was small.

Table 8.4

Sample Number Drawn	Corresponding Height	Mean Height	Sample Number Drawn	Corresponding Height	Mean Height
1. 51, 77, 27, 46	67, 70, 67, 67	67.75	**16.** 11, 64, 55, 58	64, 67, 67, 67	66.25
2. 40, 42, 33, 12	67, 67, 67, 64	66.25	**17.** 70, 56, 97, 43	70, 67, 73, 67	69.25
3. 90, 44, 46, 62	70, 67, 67, 67	67.75	**18.** 74, 28, 93, 50	70, 67, 73, 67	69.25
4. 16, 28, 98, 93	64, 67, 73, 73	69.25	**19.** 79, 42, 71, 30	70, 67, 70, 67	68.50
5. 58, 20, 41, 86	67, 64, 67, 70	67.00	**20.** 58, 60, 21, 33	67, 67, 64, 67	66.25
6. 19, 64, 08, 70	64, 67, 64, 70	66.25	**21.** 75, 79, 74, 54	70, 70, 70, 67	69.25
7. 56, 24, 03, 32	67, 67, 61, 67	65.50	**22.** 06, 31, 04, 18	64, 67, 61, 64	64.00
8. 34, 91, 83, 58	67, 70, 70, 67	68.50	**23.** 67, 07, 12, 97	70, 64, 64, 73	67.75
9. 70, 65, 68, 21	70, 70, 70, 64	68.50	**24.** 31, 71, 69, 88	67, 70, 70, 70	69.25
10. 96, 02, 13, 87	73, 61, 64, 70	67.00	**25.** 11, 64, 21, 87	64, 67, 64, 70	66.25
11. 76, 10, 51, 08	70, 64, 67, 64	66.25	**26.** 03, 58, 57, 93	61, 67, 67, 73	67.00
12. 63, 97, 45, 39	67, 73, 67, 67	68.50	**27.** 53, 81, 93, 88	67, 70, 73, 70	70.00
13. 05, 81, 45, 93	64, 70, 67, 73	68.50	**28.** 23, 22, 96, 79	67, 64, 73, 70	68.50
14. 96, 01, 73, 52	73, 61, 70, 67	67.75	**29.** 98, 56, 59, 36	73, 67, 67, 67	68.50
15. 07, 82, 54, 24	64, 70, 67, 67	67.00	**30.** 08, 15, 08, 84	64, 64, 64, 70	65.50

Table 8.5

Sample Mean	Tally	f	u	fu	fu^2
64.00	/	1	−4	−4	16
64.75		0	−3	0	0
65.50	//	2	−2	−4	8
66.25	~~////~~ /	6	−1	−6	6
$A \rightarrow$ 67.00	////	4	0	0	0
67.75	////	4	1	4	4
68.50	~~////~~ //	7	2	14	28
69.25	~~////~~	5	3	15	45
70.00	/	1	4	4	16
		$\sum f = N = 30$		$\sum fu = 23$	$\sum fu^2 = 123$

SAMPLING DISTRIBUTION OF PROPORTIONS

8.7 Find the probability that in 120 tosses of a fair coin (*a*) less than 40% or more than 60% will be heads and (*b*) $\frac{5}{8}$ or more will be heads.

SOLUTION

First method

We consider the 120 tosses of the coin to be a sample from the infinite population of all possible tosses of the coin. In this population the probability of heads is $p = \frac{1}{2}$ and the probability of tails is $q = 1 - p = \frac{1}{2}$.

(*a*) We require the probability that the number of heads in 120 tosses will be less than 48 or more than 72. We proceed as in Chapter 7, using the normal approximation to the binomial. Since the number of heads is a discrete variable, we ask for the probability that the number of heads is less than 47.5 or greater than 72.5.

$$\mu = \text{expected number of heads} = Np = 120(\tfrac{1}{2}) = 60 \qquad \text{and} \qquad \sigma = \sqrt{Npq} = \sqrt{(120)(\tfrac{1}{2})(\tfrac{1}{2})} = 5.48$$

$$47.5 \text{ in standard units} = \frac{47.5 - 60}{5.48} = -2.28$$

$$72.5 \text{ in standard units} = \frac{72.5 - 60}{5.48} = 2.28$$

As shown in Fig. 8-4,

$$\text{Required probability} = (\text{area to the left of } -2.28 \text{ plus area to the right of } 2.28)$$
$$= (2(0.0113) = 0.0226)$$

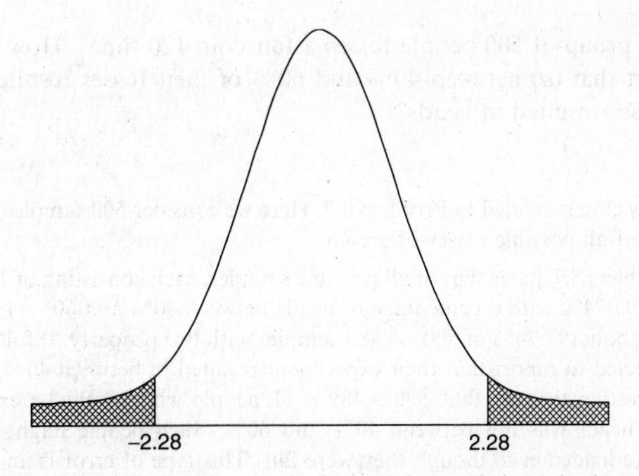

Fig. 8-4 Normal approximation to the binomial uses the standard normal curve.

Second method

$$\mu_P = p = \tfrac{1}{2} = 0.50 \qquad \sigma_P = \sqrt{\frac{pq}{N}} = \sqrt{\frac{(\frac{1}{2})(\frac{1}{2})}{120}} = 0.0456$$

$$40\% \text{ in standard units} = \frac{0.40 - 0.50}{0.0456} = -2.19$$

$$60\% \text{ in standard units} = \frac{0.60 - 0.50}{0.0456} = 2.19$$

$$\text{Required probability} = (\text{area to the left of } -2.19 \text{ plus area to the right of } 2.19)$$
$$= (2(0.0143) = 0.0286)$$

Although this result is accurate to two significant figures, it does not agree exactly since we have not used the fact that the proportion is actually a discrete variable. To account for this, we subtract

$1/2N = 1/2(120)$ from 0.40 and add $1/2N = 1/2(120)$ to 0.60; thus, since $1/240 = 0.00417$, the required proportions in standard units are

$$\frac{0.40 - 0.00417 - 0.50}{0.0456} = -2.28 \quad \text{and} \quad \frac{0.60 + 0.00417 - 0.50}{0.0456} = 2.28$$

so that agreement with the first method is obtained.

Note that $(0.40 - 0.00417)$ and $(0.60 + 0.00417)$ correspond to the proportions $47.5/120$ and $72.5/120$ in the first method.

(*b*) Using the second method of part (*a*), we find that since $\frac{5}{8} = 0.6250$,

$$(0.6250 - 0.00417) \text{ in standard units} = \frac{0.6250 - 0.00417 - 0.50}{0.0456} = 2.65$$

$$\begin{aligned}\text{Required probability} &= (\text{area under normal curve to right of } z = 2.65)\\ &= (\text{area to right of } z = 0) - (\text{area between } z = 0 \text{ and } z = 2.65)\\ &= 0.5 - 0.4960 = 0.0040\end{aligned}$$

8.8 Each person of a group of 500 people tosses a fair coin 120 times. How many people should be expected to report that (*a*) between 40% and 60% of their tosses resulted in heads and (*b*) $\frac{5}{8}$ or more of their tosses resulted in heads?

SOLUTION

This problem is closely related to Problem 8.7. Here we consider 500 samples, of size 120 each, from the infinite population of all possible tosses of a coin.

(*a*) Part (*a*) of Problem 8.7 states that of all possible samples, each consisting of 120 tosses of a coin, we can expect to find 97.74% with a percentage of heads between 40% and 60%. In 500 samples we can thus expect to find about (97.74% of 500) = 489 samples with this property. It follows that about 489 people would be expected to report that their experiment resulted in between 40% and 60% heads.

It is interesting to note that $500 - 489 = 11$ people who would be expected to report that the percentage of heads was not between 40% and 60%. Such people might reasonably conclude that their coins were loaded even though they were fair. This type of error is an everpresent *risk* whenever we deal with probability.

(*b*) By reasoning as in part (*a*), we conclude that about $(500)(0.0040) = 2$ persons would report that $\frac{5}{8}$ or more of their tosses resulted in heads.

8.9 It has been found that 2% of the tools produced by a certain machine are defective. What is the probability that in a shipment of 400 such tools (*a*) 3% or more and (*b*) 2% or less will prove defective?

SOLUTION

$$\mu_P = p = 0.02 \quad \text{and} \quad \sigma_P = \sqrt{\frac{pq}{N}} = \sqrt{\frac{(0.02)(0.98)}{400}} = \frac{0.14}{20} = 0.007$$

(*a*) **First method**

Using the correction for discrete variables, $1/2N = 1/800 = 0.00125$, we have

$$(0.03 - 0.00125) \text{ in standard units} = \frac{0.03 - 0.00125 - 0.02}{0.007} = 1.25$$

$$\text{Required probability} = (\text{area under normal curve to right of } z = 1.25) = 0.1056$$

If we had not used the correction, we would have obtained 0.0764.

Another method

(3% of 400) = 12 defective tools. On a continuous basis 12 or more tools means 11.5 or more.

$$\bar{X} = (2\% \text{ of } 400) = 8 \quad \text{and} \quad \sigma = \sqrt{Npq} = \sqrt{(400)(0.02)(0.98)} = 2.8$$

Then, 11.5 in standard units = (11.5 − 8)/2.8 = 1.25, and as before the required probability is 0.1056.

(*b*)
$$(0.02 + 0.00125) \text{ in standard units} = \frac{0.02 + 0.00125 - 0.02}{0.007} = 0.18$$

$$\text{Required probability} = (\text{area under normal curve to left of } z = 0.18)$$
$$= 0.5000 + 0.0714 = 0.5714$$

If we had not used the correction, we would have obtained 0.5000. The second method of part (*a*) can also be used.

8.10 The election returns showed that a certain candidate received 46% of the votes. Determine the probability that a poll of (*a*) 200 and (*b*) 1000 people selected at random from the voting population would have shown a majority of votes in favor of the candidate.

SOLUTION

(*a*)
$$\mu_P = p = 0.46 \quad \text{and} \quad \sigma_P = \sqrt{\frac{pq}{N}} = \sqrt{\frac{(0.46)(0.54)}{200}} = 0.0352$$

Since $1/2N = 1/400 = 0.0025$, a majority is indicated in the sample if the proportion in favor of the candidate is $0.50 + 0.0025 = 0.5025$ or more. (This proportion can also be obtained by realizing that 101 or more indicates a majority, but as a continuous variable this is 100.5, and so the proportion is $100.5/200 = 0.5025$.)

$$0.5025 \text{ in standard units} = \frac{0.5025 - 0.46}{0.0352} = 1.21$$

$$\text{Required probability} = (\text{area under normal curve to right of } z = 1.21)$$
$$= 0.5000 - 0.3869 = 0.1131$$

(*b*)
$$\mu_p = p = 0.46 \quad \text{and} \quad \sigma_P = \sqrt{\frac{pq}{N}} = \sqrt{\frac{(0.46)(0.54)}{1000}} = 0.0158$$

$$0.5025 \text{ in standard units} = \frac{0.5025 - 0.46}{0.0158} = 2.69$$

$$\text{Required probability} = (\text{area under normal curve to right of } z = 2.69)$$
$$= 0.5000 - 0.4964 = 0.0036$$

SAMPLING DISTRIBUTIONS OF DIFFERENCES AND SUMS

8.11 Let U_1 be a variable that stands for any of the elements of the population 3, 7, 8 and U_2 be a variable that stands for any of the elements of the population 2, 4. Compute (*a*) μ_{U1}, (*b*) μ_{U2}, (*c*) μ_{U1-U2}, (*d*) σ_{U1}, (*e*) σ_{U2}, and (*f*) σ_{U1-U2}.

SOLUTION

(*a*) μ_{U1} = mean of population $U_1 = \frac{1}{3}(3 + 7 + 8) = 6$

(*b*) μ_{U2} = mean of population $U_2 = \frac{1}{2}(2 + 4) = 3$

(*c*) The population consisting of the differences of any member of U_1 and any member of U_2 is

$$\begin{matrix} 3-2 & 7-2 & 8-2 \\ 3-4 & 7-4 & 8-4 \end{matrix} \quad \text{or} \quad \begin{matrix} 1 & 5 & 6 \\ -1 & 3 & 4 \end{matrix}$$

Thus
$$\mu_{U1-U2} = \text{mean of } (U_1 - U_2) = \frac{1+5+6+(-1)+3+4}{6} = 3$$

This illustrates the general result $\mu_{U1-U2} = \mu_{U1} - \mu_{U2}$, as seen from parts (*a*) and (*b*).

(*d*)
$$\sigma^2_{U1} = \text{variance of population } U_1 = \frac{(3-6)^2 + (7-6)^2 + (8-6)^2}{3} = \frac{14}{3}$$

or
$$\sigma_{U1} = \sqrt{\frac{14}{3}}$$

(*e*)
$$\sigma^2_{U2} = \text{variance of population } U_2 = \frac{(2-3)^2 + (4-3)^2}{2} = 1 \quad \text{or} \quad \sigma_{U2} = 1$$

(*f*)
$$\sigma^2_{U1-U2} = \text{variance of population } (U_1 - U_2)$$
$$= \frac{(1-3)^2 + (5-3)^2 + (6-3)^2 + (-1-3)^2 + (3-3)^2 + (4-3)^2}{6} = \frac{17}{3}$$

or
$$\sigma_{U1-U2} = \sqrt{\frac{17}{3}}$$

This illustrates the general result for independent samples, $\sigma_{U1-U2} = \sqrt{\sigma^2_{U1} + \sigma^2_{U2}}$, as seen from parts (*d*) and (*e*).

8.12 The electric light bulbs of manufacturer *A* have a mean lifetime of 1400 hours (h) with a standard deviation of 200 h, while those of manufacturer *B* have a mean lifetime of 1200 h with a standard deviation of 100 h. If random samples of 125 bulbs of each brand are tested, what is the probability that the brand *A* bulbs will have a mean lifetime that is at least (*a*) 160 h and (*b*) 250 h more than the brand *B* bulbs?

SOLUTION

Let $\bar{X}_A$ and $\bar{X}_B$ denote the mean lifetimes of samples *A* and *B*, respectively. Then

$$\mu_{\bar{X}_A - \bar{X}_B} = \mu_{\bar{X}_A} - \mu_{\bar{X}_B} = 1400 - 1200 = 200\text{ h}$$

and
$$\sigma_{\bar{X}_A - \bar{X}_B} = \sqrt{\frac{\sigma^2_A}{N_A} + \frac{\sigma^2_B}{N_B}} = \sqrt{\frac{(100)^2}{125} + \frac{(200)^2}{125}} = 20\text{ h}$$

The standardized variable for the difference in means is

$$z = \frac{(\bar{X}_A - \bar{X}_B) - (\mu_{\bar{X}_A - \bar{X}_B})}{\sigma_{\bar{X}_A - \bar{X}_B}} = \frac{(\bar{X}_A - \bar{X}_B) - 200}{20}$$

and is very closely normally distributed.

(*a*) The difference 160 h in standard units is $(160 - 200)/20 = -2$. Thus

$$\text{Required probability} = (\text{area under normal curve to right of } z = -2)$$
$$= 0.5000 + 0.4772 = 0.9772$$

(*b*) The difference 250 h in standard units is $(250 - 200)/20 = 2.50$. Thus

$$\text{Required probability} = (\text{area under normal curve to right of } z = 2.50)$$
$$= 0.5000 - 0.4938 = 0.0062$$

8.13 Ball bearings of a given brand weigh 0.50 g with a standard deviation of 0.02 g. What is the probability that two lots of 1000 ball bearings each will differ in weight by more than 2 g?

SOLUTION

Let $\bar{X}_1$ and $\bar{X}_2$ denote the mean weights of ball bearings in the two lots. Then

$$\mu_{\bar{X}_1-\bar{X}_2} = \mu_{\bar{X}_1} - \mu_{\bar{X}_2} = 0.50 - 0.50 = 0$$

and

$$\sigma_{\bar{X}_1-X_2} = \sqrt{\frac{\sigma_1^2}{N_1} + \frac{\sigma_2^2}{N_2}} = \sqrt{\frac{(0.02)^2}{1000} + \frac{(0.02)^2}{1000}} = 0.000895$$

The standardized variable for the difference in means is

$$z = \frac{(\bar{X}_1 - \bar{X}_2) - 0}{0.000895}$$

and is very closely normally distributed.

A difference of 2 g in the lots is equivalent to a difference of $2/1000 = 0.002$ g in the means. This can occur either if $\bar{X}_1 - \bar{X}_2 \geq 0.002$ or $\bar{X}_1 - X_2 \leq -0.002$; that is,

$$z \geq \frac{0.002 - 0}{0.000895} = 2.23 \qquad \text{or} \qquad z \leq \frac{-0.002 - 0}{0.000895} = -2.23$$

Then $\Pr\{z \geq 2.23 \text{ or } z \leq -2.23\} = \Pr\{z \geq 2.23\} + \Pr\{z \leq -2.23\} = 2(0.5000 - 0.4871) = 0.0258$.

8.14 *A* and *B* play a game of "heads and tails," each tossing 50 coins. *A* will win the game if she tosses 5 or more heads than *B*; otherwise, *B* wins. Determine the odds against *A* winning any particular game.

SOLUTION

Let P_A and P_B denote the proportion of heads obtained by *A* and *B*. If we assume that the coins are all fair, the probability p of heads is $\frac{1}{2}$. Then

$$\mu_{P_A-P_B} = \mu_{P_A} - \mu_{P_B} = 0$$

and

$$\sigma_{P_A-P_B} = \sqrt{\sigma_{P_A}^2 + \sigma_{P_B}^2} = \sqrt{\frac{pq}{N_A} + \frac{pq}{N_B}} = \sqrt{\frac{2(\frac{1}{2})(\frac{1}{2})}{50}} = 0.10$$

The standardized variable for the difference in proportions is $z = (P_A - P_B - 0)/0.10$.

On a continuous-variable basis, 5 or more heads means 4.5 or more heads, so that the difference in proportions should be $4.5/50 = 0.09$ or more; that is, z is greater than or equal to $(0.09 - 0)/0.10 = 0.9$ (or $z \geq 0.9$). The probability of this is the area under the normal curve to the right of $z = 0.9$, which is $(0.5000 - 0.3159) = 0.1841$.

Thus the odds against *A* winning are $(1 - 0.1841){:}0.1841 = 0.8159{:}0.1841$, or 4.43 to 1.

8.15 Two distances are measured as 27.3 centimeters (cm) and 15.6 cm with standard deviations (standard errors) of 0.16 cm and 0.08 cm, respectively. Determine the mean and standard deviation of (*a*) the sum and (*b*) the difference of the distances.

SOLUTION

If the distances are denoted by D_1 and D_2, then:

(*a*)

$$\mu_{D1+D2} = \mu_{D1} + \mu_{D2} = 27.3 + 15.6 = 42.9 \text{ cm}$$

$$\sigma_{D1+D2} = \sqrt{\sigma_{D1}^2 + \sigma_{D2}^2} = \sqrt{(0.16)^2 + (0.08)^2} = 0.18 \text{ cm}$$

(*b*)

$$\mu_{D1-D2} = \mu_{D1} - \mu_{D2} = 27.3 - 15.6 = 11.7 \text{ cm}$$

$$\sigma_{D1-D2} = \sqrt{\sigma_{D1}^2 + \sigma_{D2}^2} = \sqrt{(0.16)^2 + (0.08)^2} = 0.18 \text{ cm}$$

8.16 A certain type of electric light bulb has a mean lifetime of 1500 h and a standard deviation of 150 h. Three bulbs are connected so that when one burns out, another will go on. Assuming that the lifetimes are normally distributed, what is the probability that lighting will take place for (*a*) at least 5000 h and (*b*) at most 4200 h?

SOLUTION

Assume the lifetimes to be L_1, L_2, and L_3. Then

$$\mu_{L1+L2+L3} = \mu_{L1} + \mu_{L2} + \mu_{L3} = 1500 + 1500 + 1500 = 4500\,\text{h}$$

$$\sigma_{L1+L2+L3} = \sqrt{\sigma_{L1}^2 + \sigma_{L2}^2 + \sigma_{L3}^2} = \sqrt{3(150)^2} = 260\,\text{h}$$

(*a*)

$$50000\,\text{h in standard units} = \frac{5000 - 4500}{260} = 1.92$$

$$\text{Required probability} = (\text{area under normal curve to right of } z = 1.92)$$

$$= 0.5000 - 0.4726 = 0.0274$$

(*b*)

$$4200\,\text{h in standard units} = \frac{4200 - 4500}{260} = -1.15$$

$$\text{Required probability} = (\text{area under normal curve to left of } z = -1.15)$$

$$= 0.5000 - 0.3749 = 0.1251$$

SOFTWARE DEMONSTRATION OF ELEMENTARY SAMPLING THEORY

8.17 Midwestern University has 1/3 of its students taking 9 credit hours, 1/3 taking 12 credit hours, and 1/3 taking 15 credit hours. If X represents the credit hours a student is taking, the distribution of X is $p(x) = 1/3$ for $x = 9$, 12, and 15. Find the mean and variance of X. What type of distribution does X have?

SOLUTION

The mean of X is $\mu = \sum xp(x) = 9(1/3) + 12(1/3) + 15(1/3) = 12$. The variance of X is $\sigma^2 = \sum x^2 p(x) - \mu^2 = 81(1/3) + 144(1/3) + 225(1/3) - 144 = 150 - 144 = 6$. The distribution of X is uniform.

8.18 List all samples of size $n = 2$ that are possible (with replacement) from the population in Problem 8.17. Use the chart wizard of EXCEL to plot the sampling distribution of the mean to show that $\mu_{\bar{x}} = \mu$, and show that $\sigma_{\bar{x}}^2 = \sigma^2/2$.

SOLUTION

A	B	C	D	E	F	G
		mean	xbar	p(xbar)	xbar × p(xbar)	xbar² × p(xbar)
9	9	9	9	0.111111	1	9
9	12	10.5	10.5	0.222222	2.333333333	24.5
9	15	12	12	0.333333	4	48
12	9	10.5	13.5	0.222222	3	40.5
12	12	12	15	0.111111	1.666666667	25
12	15	13.5			12	147
15	9	12				
15	12	13.5				
15	15	15				

The EXCEL worksheet shows the possible sample values in A and B and the mean in C. The sampling distribution of xbar is built and displayed in D and E. In C2, the function =AVERAGE(A2:B2) is entered and a click-and-drag is performed from C2 to C10. Because the population is uniform each sample has probability 1/9 of being selected. The sample mean is represented by xbar. The mean of the sample mean is $\mu_{\bar{x}} = \Sigma \bar{x} p(\bar{x})$and is computed in F2 through F6. The function =SUM(F2:F6) is in F7 and is equal to 12 showing that $\mu_{\bar{x}} = \mu$. The variance of the sample mean is $\sigma_{\bar{x}}^2 = \Sigma \bar{x}^2 p(\bar{x}) - \mu_{\bar{x}}^2$ and is computed as follows. $\Sigma \bar{x}^2 p(\bar{x})$ is computed in G2 through G6. The function =SUM(G2:G6) is in G7 and is equal to 147. When 12^2 or 144 is subtracted from 147, we get 3 or $\sigma_{\bar{x}}^2 = \sigma^2/2$. Figure 8-5 shows that even with sample size 2, the sampling distribution of $\bar{x}$ is somewhat like a normal distribution. The larger probabilities are near 12 and they tail off to the right and left of 12.

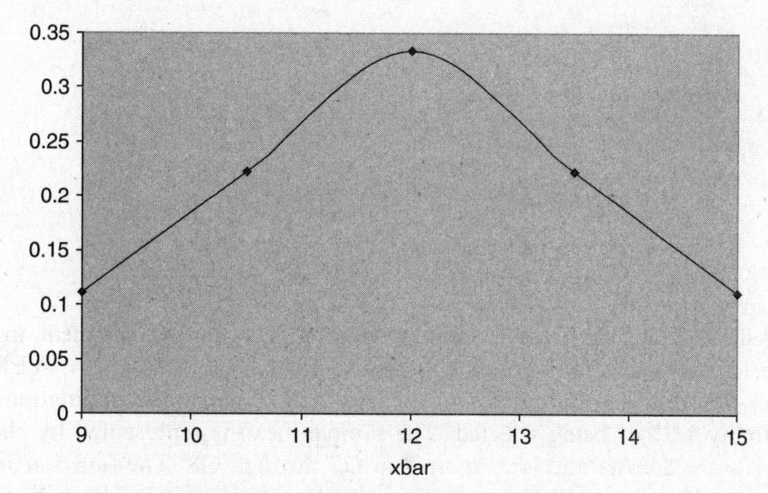

Fig. 8-5 Distribution of xbar for $n = 2$.

8.19 List all samples of size $n = 3$ that are possible (with replacement) from the population in Problem 8.17. Use EXCEL to construct the sampling distribution of the mean. Use the chart wizard of EXCEL to plot the sampling distribution of the mean. Show $\mu_{\bar{x}} = \mu$, and $\sigma_{\bar{x}}^2 = \frac{\sigma^2}{3}$.

SOLUTION

A	B	C	D	E	F	G	H
			mean	xbar	p(xbar)	xbar*p(xbar)	xbar^2p(xbar)
9	9	9	9	9	0.037037037	0.333333333	3
9	9	12	10	10	0.111111111	1.111111111	11.11111111
9	9	15	11	11	0.222222222	2.444444444	26.88888889
9	12	9	10	12	0.259259259	3.111111111	37.33333333
9	12	12	11	13	0.222222222	2.888888889	37.55555556
9	12	15	12	14	0.111111111	1.555555556	21.77777778
9	15	9	11	15	0.037037037	0.555555556	8.333333333
9	15	12	12			12	146
9	15	15	13				
12	9	9	10				

Continued

A	B	C	D mean	E xbar	F p(xbar)	G xbar*p(xbar)	H xbar^2p(xbar)
12	9	12	11				
12	9	15	12				
12	12	9	11				
12	12	12	12				
12	12	15	13				
12	15	9	12				
12	15	12	13				
12	15	15	14				
15	9	9	11				
15	9	12	12				
15	9	15	13				
15	12	9	12				
15	12	12	13				
15	12	15	14				
15	15	9	13				
15	15	12	14				
15	15	15	15				

The EXCEL worksheet shows the possible sample values in A, B, and C, the mean in D, the sampling distribution of xbar is computed and given in E and F. In D2, the function =AVERAGE(A2:C2) is entered and a click-and-drag is performed from D2 to D28. Because the population is uniform each sample has probability 1/27 of being selected. The sample mean is represented by xbar. The mean of the sample mean is $\mu_{\bar{x}} = \Sigma \bar{x} p(\bar{x})$ and is computed in G2 through G8. The function =SUM(G2:G8) is in G9 and is equal to 12 showing that $\mu_{\bar{x}} = \mu$ for samples of size $n = 3$. The variance of the sample mean is $\sigma_{\bar{x}}^2 = \Sigma \bar{x}^2 p(\bar{x}) - \mu_{\bar{x}}^2$ and is computed as follows. $\Sigma \bar{x}^2 p(\bar{x})$ is computed in H2 through H8. The function =SUM(H2:H8) is in H9 and is equal to 146. When 12^2 or 144 is subtracted from 146 we get 2. Note that $\sigma_{\bar{x}}^2 = \sigma^2/3$. Figure 8-6 shows the normal distribution tendency of the distribution of xbar.

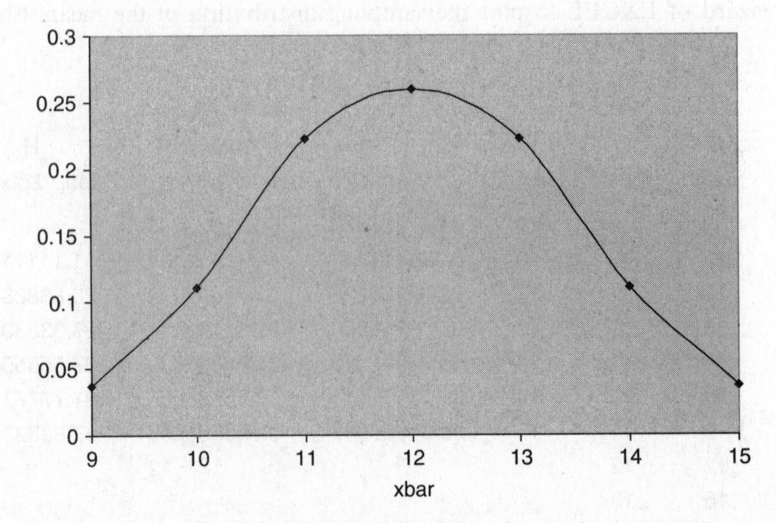

Fig. 8-6 Distribution of xbar for $n = 3$.

8.20 List all 81 samples of size $n=4$ that are possible (with replacement) from the population in Problem 8.17. Use EXCEL to construct the sampling distribution of the mean. Use the chart wizard of EXCEL to plot the sampling distribution of the mean, show that $\mu_{\bar{x}} = \mu$, and show that $\sigma_{\bar{x}}^2 = \sigma^2/4$.

SOLUTION

The method used in Problems 8.18 and 8.19 is extended to samples of size 4. From the EXCEL worksheet, the following distribution for xbar is obtained. In addition, it can be shown that $\mu_{\bar{x}} = \mu$ and $\sigma_{\bar{x}}^2 = \sigma^2/4$.

xbar	p(xbar)	xbar*p(xbar)	xbar^2p(xbar)
9	0.012345679	0.111111111	1
9.75	0.049382716	0.481481481	4.694444444
10.5	0.12345679	1.296296296	13.61111111
11.25	0.197530864	2.222222222	25
12	0.234567901	2.814814815	33.77777778
12.75	0.197530864	2.518518519	32.11111111
13.5	0.12345679	1.666666667	22.5
14.25	0.049382716	0.703703704	10.02777778
15	0.012345679	0.185185185	2.777777778
	1	12	145.5

The EXCEL plot of the distribution of xbar for samples of size 4 is shown in Fig. 8-7.

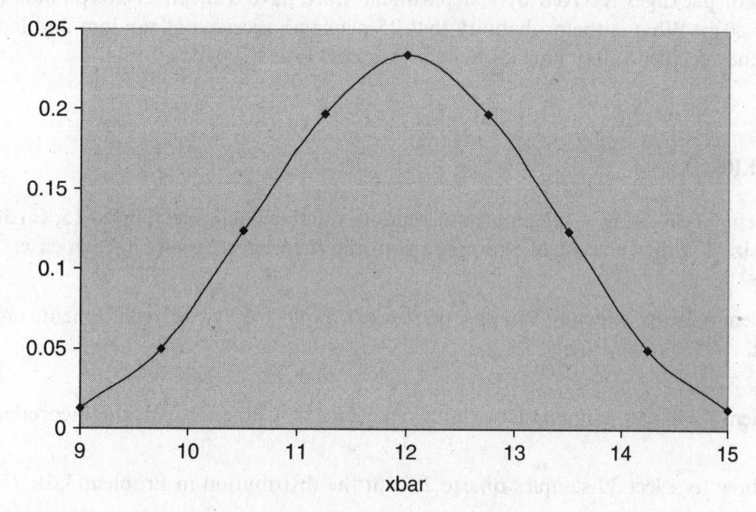

Fig. 8-7 Distribution of xbar for $n=4$.

Supplementary Problems

SAMPLING DISTRIBUTION OF MEANS

8.21 A population consists of the four numbers 3, 7, 11, and 15. Consider all possible samples of size 2 that can be drawn with replacement from this population. Find (*a*) the population mean, (*b*) the population standard deviation, (*c*) the mean of the sampling distribution of means, and (*d*) the standard deviation of the sampling distribution of means. Verify parts (*c*) and (*d*) directly from (*a*) and (*b*) by using suitable formulas.

8.22 Solve Problem 8.21 if the sampling is without replacement.

8.23 The masses of 1500 ball bearings are normally distributed, with a mean of 22.40 g and a standard deviation of 0.048 g. If 300 random samples of size 36 are drawn from this population, determine the expected mean and standard deviation of the sampling distribution of means if the sampling is done (*a*) with replacement and (*b*) without replacement.

8.24 Solve Problem 8.23 if the population consists of 72 ball bearings.

8.25 How many of the random samples in Problem 8.23 would have their means (*a*) between 22.39 and 22.41 g, (*b*) greater than 22.42 g, (*c*) less than 22.37 g, and (*d*) less than 22.38 g or more than 22.41 g?

8.26 Certain tubes manufactured by a company have a mean lifetime of 800 h and a standard deviation of 60 h. Find the probability that a random sample of 16 tubes taken from the group will have a mean lifetime of (*a*) between 790 and 810 h, (*b*) less than 785 h, (*c*) more than 820 h, and (*d*) between 770 and 830 h.

8.27 Work Problem 8.26 if a random sample of 64 tubes is taken. Explain the difference.

8.28 The weights of packages received by a department store have a mean of 300 pounds (lb) and a standard deviation of 50 lb. What is the probability that 25 packages received at random and loaded on an elevator will exceed the specified safety limit of the elevator, listed as 8200 lb?

RANDOM NUMBERS

8.29 Work Problem 8.6 by using a different set of random numbers and selecting (*a*) 15, (*b*) 30, (*c*) 45, and (*d*) 60 samples of size 4 with replacement. Compare with the theoretical results in each case.

8.30 Work Problem 8.29 by selecting samples of size (*a*) 2 and (*b*) 8 with replacement, instead of size 4 with replacement.

8.31 Work Problem 8.6 if the sampling is without replacement. Compare with the theoretical results.

8.32 (*a*) Show how to select 30 samples of size 2 from the distribution in Problem 3.61.

(*b*) Compute the mean and standard deviation of the resulting sampling distribution of means, and compare with theoretical results.

8.33 Work Problem 8.32 by using samples of size 4.

SAMPLING DISTRIBUTION OF PROPORTIONS

8.34 Find the probability that of the next 200 children born, (*a*) less than 40% will be boys, (*b*) between 43% and 57% will be girls, and (*c*) more than 54% will be boys. Assume equal probabilities for the births of boys and girls.

8.35 Out of 1000 samples of 200 children each, in how many would you expect to find that (*a*) less than 40% are boys, (*b*) between 40% and 60% are girls, and (*c*) 53% or more are girls?

8.36 Work Problem 8.34 if 100 instead of 200 children are considered, and explain the differences in results.

8.37 An urn contains 80 marbles, of which 60% are red and 40% are white. Out of 50 samples of 20 marbles, each selected with replacement from the urn, how many samples can be expected to consist of (*a*) equal numbers of red and white marbles, (*b*) 12 red and 8 white marbles, (*c*) 8 red and 12 white marbles, and (*d*) 10 or more white marbles?

8.38 Design an experiment intended to illustrate the results of Problem 8.37. Instead of red and white marbles, you may use slips of paper on which R and W are written in the correct proportions. What errors might you introduce by using two different sets of coins?

8.39 A manufacturer sends out 1000 lots, each consisting of 100 electric bulbs. If 5% of the bulbs are normally defective, in how many of the lots should we expect (*a*) fewer than 90 good bulbs and (*b*) 98 or more good bulbs?

SAMPLING DISTRIBUTIONS OF DIFFERENCES AND SUMS

8.40 A and B manufacture two types of cables that have mean breaking strengths of 4000 lb and 4500 lb and standard deviations of 300 lb and 200 lb, respectively. If 100 cables of brand A and 50 cables of brand B are tested, what is the probability that the mean breaking strength of B will be (*a*) at least 600 lb more than A and (*b*) at least 450 lb more than A?

8.41 What are the probabilities in Problem 8.40 if 100 cables of both brands are tested? Account for the differences.

8.42 The mean score of students on an aptitude test is 72 points with a standard deviation of 8 points. What is the probability that two groups of students, consisting of 28 and 36 students, respectively, will differ in their mean scores by (*a*) 3 or more points, (*b*) 6 or more points, and (*c*) between 2 and 5 points?

8.43 An urn contains 60 red marbles and 40 white marbles. Two sets of 30 marbles each are drawn with replacement from the urn and their colors are noted. What is the probability that the two sets differ by 8 or more red marbles?

8.44 Solve Problem 8.43 if the sampling is without replacement in obtaining each set.

8.45 Election returns showed that a certain candidate received 65% of the votes. Find the probability that two random samples, each consisting of 200 voters, indicated more than a 10% difference in the proportions who voted for the candidate.

8.46 If U_1 and U_2 are the sets of numbers in Problem 8.11, verify that (*a*) $\mu_{U1+U2} = \mu_{U1} + \mu_{U2}$ and (*b*) $\sigma_{U1+U2} = \sqrt{\sigma_{U1}^2 + \sigma_{U2}^2}$.

8.47 Three masses are measured as 20.48, 35.97, and 62.34 g, with standard deviations of 0.21, 0.46, and 0.54 g, respectively. Find the (*a*) mean and (*b*) standard deviation of the sum of the masses.

8.48 The mean voltage of a battery is 15.0 volts (V) and the standard deviation is 0.2 V. What is the probability that four such batteries connected in series will have a combined voltage of 60.8 V or more?

SOFTWARE DEMONSTRATION OF ELEMENTARY SAMPLING THEORY

8.49 The credit hour distribution at Metropolitan Technological College is as follows:

x	6	9	12	15	18
$p(x)$	0.1	0.2	0.4	0.2	0.1

Find μ and σ^2. Give the 25 (with replacement) possible samples of size 2, their means, and their probabilities.

8.50 Refer to problem 8.49. Give and plot the probability distribution of xbar for $n = 2$.

8.51 Refer to problem 8.50. Show that $\mu_{\bar{x}} = \mu$ and $\sigma_{\bar{x}}^2 = \frac{\sigma^2}{2}$.

8.52 Refer to problem 8.49. Give and plot the probability distribution of xbar for $n = 3$.

Statistical Estimation Theory

ESTIMATION OF PARAMETERS

In the Chapter 8 we saw how sampling theory can be employed to obtain information about samples drawn at random from a known population. From a practical viewpoint, however, it is often more important to be able to infer information about a population from samples drawn from it. Such problems are dealt with in *statistical inference*, which uses principles of sampling theory.

One important problem of statistical inference is the estimation of *population parameters*, or briefly *parameters* (such as population mean and variance), from the corresponding *sample statistics*, or briefly *statistics* (such as sample mean and variance). We consider this problem in this chapter.

UNBIASED ESTIMATES

If the mean of the sampling distribution of a statistic equals the corresponding population parameter, the statistic is called an *unbiased estimator* of the parameter; otherwise, it is called a *biased estimator*. The corresponding values of such statistics are called *unbiased* or *biased* estimates, respectively.

EXAMPLE 1. The mean of the sampling distribution of means $\mu_{\bar{X}}$ is μ, the population mean. Hence the sample mean $\bar{X}$ is an unbiased estimate of the population mean μ.

EXAMPLE 2. The mean of the sampling distribution of variances is

$$\mu_{s^2} = \frac{N-1}{N}\sigma^2$$

where σ^2 is the population variance and N is the sample size (see Table 8.1). Thus the sample variance s^2 is a biased estimate of the population variance σ^2. By using the modified variance

$$\hat{s}^2 = \frac{N}{N-1}s^2$$

we find $\mu_{\hat{s}^2} = \sigma^2$, so that $\hat{s}^2$ is an unbiased estimate of σ^2. However, $\hat{s}$ is a biased estimate of σ.

In the language of expectation (see Chapter 6) we could say that a statistic is unbiased if its expectation equals the corresponding population parameter. Thus $\bar{X}$ and $\hat{s}^2$ are unbiased since $E\{\bar{X}\} = \mu$ and $E\{\hat{s}^2\} = \sigma^2$.

EFFICIENT ESTIMATES

If the sampling distributions of two statistics have the same mean (or expectation), then the statistic with the smaller variance is called an *efficient estimator* of the mean, while the other statistic is called an *inefficient estimator*. The corresponding values of the statistics are called *efficient and unefficient estimates*.

If we consider all possible statistics whose sampling distributions have the same mean, the one with the smallest variance is sometimes called the *most efficient*, or *best*, *estimator* of this mean.

EXAMPLE 3. The sampling distributions of the mean and median both have the same mean, namely, the population mean. However, the variance of the sampling distribution of means is smaller than the variance of the sampling distribution of medians (see Table 8.1). Hence the sample mean gives an efficient estimate of the population mean, while the sample median gives an inefficient estimate of it.

Of all statistics estimating the population mean, the sample mean provides the best (or most efficient) estimate.

In practice, inefficient estimates are often used because of the relative ease with which some of them can be obtained.

POINT ESTIMATES AND INTERVAL ESTIMATES; THEIR RELIABILITY

An estimate of a population parameter given by a single number is called a *point estimate* of the parameter. An estimate of a population parameter given by two numbers between which the parameter may be considered to lie is called an *interval estimate* of the parameter.

Interval estimates indicate the precision, or accuracy, of an estimate and are therefore preferable to point estimates.

EXAMPLE 4. If we say that a distance is measured as 5.28 meters (m), we are giving a point estimate. If, on the other hand, we say that the distance is 5.28 ± 0.03 m (i.e., the distance lies between 5.25 and 5.31 m), we are giving an interval estimate.

A statement of the error (or precision) of an estimate is often called its *reliability*.

CONFIDENCE-INTERVAL ESTIMATES OF POPULATION PARAMETERS

Let μ_S and σ_S be the mean and standard deviation (standard error), respectively, of the sampling distribution of a statistic S. Then if the sampling distribution of S is approximately normal (which as we have seen is true for many statistics if the sample size $N \geq 30$), we can expect to find an actual sample statistic S lying in the intervals $\mu_S - \sigma_S$ to $\mu_S + \sigma_S$, $\mu_S - 2\sigma_S$ to $\mu_S + 2\sigma_S$, or $\mu_S - 3\sigma_S$ to $\mu_S + 3\sigma_S$ about 68.27%, 95.45%, and 99.73% of the time, respectively.

Equivalently, we can expect to find (or we can be *confident* of finding) μ_S in the intervals $S - \sigma_S$ to $S + \sigma_S$, $S - 2\sigma_S$ to $S + 2\sigma_S$, or $S - 3\sigma_S$ to $S + 3\sigma_S$ about 68.27%, 95.45%, and 99.73% of the time, respectively. Because of this, we call these respective intervals the 68.27%, 95.45%, and 99.73% *confidence intervals* for estimating μ_S. The end numbers of these intervals ($S \pm \sigma_S$, $S \pm 2\sigma_S$, and $S \pm 3\sigma_S$) are then called the 68.27%, 95.45%, and 99.73% *confidence limits*, or *fiducial limits*.

Similarly, $S \pm 1.96\sigma_S$ and $S \pm 2.58\sigma_S$ are the 95% and 99% (or 0.95 and 0.99) confidence limits for S. The percentage confidence is often called the *confidence level*. The numbers 1.96, 2.58, etc., in the confidence limits are called *confidence coefficients*, or *critical values*, and are denoted by z_c. From confidence levels we can find confidence coefficients, and vice versa.

Table 9.1 shows the values of z_c corresponding to various confidence levels used in practice. For confidence levels not presented in the table, the values of z_c can be found from the normal-curve area tables (see Appendix II).

Table 9.1

Confidence level	99.73%	99%	98%	96%	95.45%	95%	90%	80%	68.27%	50%
z_c	3.00	2.58	2.33	2.05	2.00	1.96	1.645	1.28	1.00	0.6745

Confidence Intervals for Means

If the statistic S is the sample mean $\bar{X}$, then the 95% and 99% confidence limits for estimating the population mean μ, are given by $\bar{X} \pm 1.96\sigma_{\bar{X}}$ and $\bar{X} \pm 2.58\sigma_{\bar{X}}$, respectively. More generally, the confidence limits are given by $\bar{X} \pm z_c\sigma_{\bar{X}}$, where z_c (which depends on the particular level of confidence desired) can be read from Table 9.1. Using the values of $\sigma_{\bar{X}}$ obtained in Chapter 8, we see that the confidence limits for the population mean are given by

$$\bar{X} \pm z_c \frac{\sigma}{\sqrt{N}} \tag{1}$$

if the sampling is either from an infinite population or with replacement from a finite population, and are given by

$$\bar{X} \pm z_c \frac{\sigma}{\sqrt{N}} \sqrt{\frac{N_p - N}{N_p - 1}} \tag{2}$$

if the sampling is without replacement from a population of finite size N_p.

Generally, the population standard deviation σ is unknown; thus, to obtain the above confidence limits, we use the sample estimate $\hat{s}$ or s. This will prove satisfactory when $N \geq 30$. For $N < 30$, the approximation is poor and small sampling theory must be employed (see Chapter 11).

Confidence Intervals for Proportions

If the statistic S is the proportion of "successes" in a sample of size N drawn from a binomial population in which p is the proportion of successes (i.e., the probability of success), then the confidence limits for p are given by $P \pm z_c\sigma_P$, where P is the proportion of successes in the sample of size N. Using the values of σ_P obtained in Chapter 8, we see that the confidence limits for the population proportion are given by

$$P \pm z_c \sqrt{\frac{pq}{N}} = P \pm z_c \sqrt{\frac{p(1-p)}{N}} \tag{3}$$

if the sampling is either from an infinite population or with replacement from a finite population and are given by

$$P \pm z_c \sqrt{\frac{pq}{N}} \sqrt{\frac{N_p - N}{N_p - 1}} \tag{4}$$

if the sampling is without replacement from a population of finite size N_p.

To compute these confidence limits, we can use the sample estimate P for p, which will generally prove satisfactory if $N \geq 30$. A more exact method for obtaining these confidence limits is given in Problem 9.12.

Confidence Intervals for Differences and Sums

If S_1 and S_2 are two sample statistics with approximately normal sampling distributions, confidence limits for the difference of the population parameters corresponding to S_1 and S_2 are given by

$$S_1 - S_2 \pm z_c\sigma_{S_1-S_2} = S_1 - S_2 \pm z_c\sqrt{\sigma_{S_1}^2 + \sigma_{S_2}^2} \tag{5}$$

while confidence limits for the sum of the population parameters are given by

$$S_1 + S_2 \pm z_c\sigma_{S_1+S_2} = S_1 + S_2 \pm z_c\sqrt{\sigma_{S_1}^2 + \sigma_{S_2}^2} \tag{6}$$

provided that the samples are independent (see Chapter 8).

For example, confidence limits for the difference of two population means, in the case where the populations are infinite, are given by

$$\bar{X}_1 - \bar{X}_2 \pm z_c\sigma_{\bar{X}_1-\bar{X}_2} = \bar{X}_1 - \bar{X}_2 \pm z_c\sqrt{\frac{\sigma_1^2}{N_1} + \frac{\sigma_2^2}{N_2}} \tag{7}$$

where $\bar{X}_1$, σ_1, N_1 and $\bar{X}_2$, σ_2, N_2 are the respective means, standard deviations, and sizes of the two samples drawn from the populations.

Similarly, confidence limits for the difference of two population proportions, where the populations are infinite, are given by

$$P_1 - P_2 \pm z_c\sigma_{P_1-P_2} = P_1 - P_2 \pm z_c\sqrt{\frac{p_1(1-p_1)}{N_1} + \frac{p_2(1-p_2)}{N_2}} \tag{8}$$

where P_1 and P_2 are the two sample proportions, N_1 and N_2 are the sizes of the two samples drawn from the populations, and p_1 and p_2 are the proportions in the two populations (estimated by P_1 and P_2).

Confidence Intervals for Standard Deviations

The confidence limits for the standard deviation σ of a normally distributed population, as estimated from a sample with standard deviation s, are given by

$$s \pm z_c\sigma_s = s \pm z_c\frac{\sigma}{\sqrt{2N}} \tag{9}$$

using Table 8.1. In computing these confidence limits, we use s or $\hat{s}$ to estimate σ.

PROBABLE ERROR

The 50% confidence limits of the population parameters corresponding to a statistic S are given by $S \pm 0.6745\sigma_S$. The quantity $0.6745\sigma_S$ is known as the *probable error* of the estimate.

Solved Problems

UNBIASED AND EFFICIENT ESTIMATES

9.1 Give an example of estimators (or estimates) that are (*a*) unbiased and efficient, (*b*) unbiased and inefficient, and (*c*) biased and inefficient.

SOLUTION

(*a*) The sample mean $\bar{X}$ and the modified sample variance

$$\hat{s}^2 = \frac{N}{N-1} s^2$$

are two such examples.

(*b*) The sample median and the sample statistic $\frac{1}{2}(Q_1 + Q_3)$, where Q_1 and Q_3 are the lower and upper sample quartiles, are two such examples. Both statistics are unbiased estimates of the population mean, since the mean of their sampling distributions is the population mean.

(*c*) The sample standard deviation s, the modified standard deviation $\hat{s}$, the mean deviation, and the semi-interquartile range are four such examples.

9.2 In a sample of five measurements, the diameter of a sphere was recorded by a scientist as 6.33, 6.37, 6.36, 6.32, and 6.37 centimeters (cm). Determine unbiased and efficient estimates of (*a*) the true mean and (*b*) the true variance.

SOLUTION

(*a*) The unbiased and efficient estimate of the true mean (i.e., the populations mean) is

$$\bar{X} = \frac{\sum X}{N} = \frac{6.33 + 6.37 + 6.36 + 6.32 + 6.37}{5} = 6.35\,\text{cm}$$

(*b*) The unbiased and efficient estimate of the true variance (i.e., the population variance) is

$$\hat{s}^2 = \frac{N}{N-1} s^2 = \frac{\sum (X - \bar{X})^2}{N-1}$$

$$= \frac{(6.33 - 6.35)^2 + (6.37 - 6.35)^2 + (6.36 - 6.35)^2 + (6.32 - 6.35)^2 + (6.37 - 6.35)^2}{5-1}$$

$$= 0.00055\,\text{cm}^2$$

Note that although $\hat{s} = \sqrt{0.00055} = 0.023\,\text{cm}$ is an estimate of the true standard deviation, this estimate is neither unbiased nor efficient.

9.3 Suppose that the heights of 100 male students at XYZ University represent a random sample of the heights of all 1546 students at the university. Determine unbiased and efficient estimates of (*a*) the true mean and (*b*) the true variance.

SOLUTION

(*a*) From Problem 3.22, the unbiased and efficient estimate of the true mean height is $\bar{X} = 67.45$ inches (in).

(*b*) From Problem 4.17, the unbiased and efficient estimate of the true variance is

$$\hat{s}^2 = \frac{N}{N-1} s^2 = \frac{100}{99}(8.5275) = 8.6136$$

Thus $\hat{s} = \sqrt{8.6136} = 2.93$ in. Note that since N is large, there is essentially no difference between s^2 and $\hat{s}^2$ or between s and $\hat{s}$.

Note that we have not used Sheppard's correction for grouping. To take this into account, we would use $s = 2.79$ in (see Problem 4.21).

9.4 Give an unbiased and inefficient estimate of the true mean diameter of the sphere of Problem 9.2.

SOLUTION

The median is one example of an unbiased and inefficient estimate of the population mean. For the five measurements arranged in order of magnitude, the median is 6.36 cm.

CONFIDENCE INTERVALS FOR MEANS

9.5 Find the (*a*) 95% and (*b*) 99% confidence intervals for estimating the mean height of the XYZ University students in Problem 9.3.

SOLUTION

(*a*) The 95% confidence limits are $\bar{X} \pm 1.96\sigma/\sqrt{N}$. Using $\bar{X} = 67.45$ in, and $\hat{s} = 2.93$ in as an estimate of σ (see Problem 9.3), the confidence limits are $67.45 \pm 1.96(2.93/\sqrt{100})$, or 67.45 ± 0.57 in. Thus the 95% confidence interval for the population mean μ is 66.88 to 68.02 in, which can be denoted by $66.88 < \mu < 68.02$.

We can therefore say that the probability that the population mean height lies between 66.88 and 68.02 in is about 95%, or 0.95. In symbols we write $\Pr\{66.88 < \mu < 68.02\} = 0.95$. This is equivalent to saying that we are 95% *confident* that the population mean (or true mean) lies between 66.88 and 68.02 in.

(*b*) The 99% confidence limits are $\bar{X} \pm 2.58\sigma/\sqrt{N} = \bar{X} \pm 2.58\hat{s}/\sqrt{N} = 67.45 \pm 2.58(2.93/\sqrt{100}) = 67.45 \pm 0.76$ in. Thus the 99% confidence interval for the population mean μ is 66.69 to 68.21 in, which can be denoted by $66.69 < \mu < 68.21$.

In obtaining the above confidence intervals, we assumed that the population was infinite or so large that we could consider conditions to be the same as sampling with replacement. For finite populations where sampling is without replacement, we should use

$$\frac{\sigma}{\sqrt{N}}\sqrt{\frac{N_p - N}{N_p - 1}} \quad \text{in place of} \quad \frac{\sigma}{\sqrt{N}}$$

However, we can consider the factor

$$\sqrt{\frac{N_p - N}{N_p - 1}} = \sqrt{\frac{1546 - 100}{1546 - 1}} = 0.967$$

to be essentially 1.0, and thus it need not be used. If it is used, the above confidence limits become 67.45 ± 0.56 in and 67.45 ± 0.73 in, respectively.

9.6 Blaises' Christmas Tree Farm has 5000 trees that are mature and ready to be cut and sold. One-hundred of the trees are randomly selected and their heights measured. The heights in inches are given in Table 9.2. Use MINITAB to set a 95% confidence interval on the mean height of all 5000 trees. If the trees sell for \$2.40 per foot, give a lower and an upper bound on the value of the 5000 trees.

SOLUTION

The MINITAB confidence interval given below indicates that the mean height for the 5000 trees could be as small as 57.24 or as large as 61.20 inches. The total number of inches for all 5000 trees ranges between (57.24)(5000) = 286,2000 and (61.20)(5000) = 306,000. If the trees sell for \$2.40 per foot, then cost

Table 9.2

56	61	52	62	63	34	47	35	44	59
70	61	65	51	65	72	55	71	57	75
53	48	55	67	60	60	73	74	43	74
71	53	78	59	56	62	48	65	68	51
73	62	80	53	64	44	67	45	58	48
50	57	72	55	56	62	72	57	49	62
46	61	52	46	72	56	46	48	57	52
54	73	71	70	66	67	58	71	75	50
44	59	56	54	63	43	68	69	55	63
48	49	70	60	67	47	49	69	66	73

per inch is \$0.2. The value of the trees ranges between (286,000)(0.2) = \$57,200 and (306,000)(0.2) = \$61,200 with 95% confidence.

Data Display

```
height
         56   70   53   71   73   50   46   54   44
         48   61   61   48   53   62   57   61   73
         59   49   52   65   55   78   80   72   52
         71   56   70   62   51   67   59   53   55
         46   70   54   60   63   65   60   56   64
         56   72   66   63   67   34   72   60   62
         44   62   56   67   43   47   47   55   73
         48   67   72   46   58   68   49   35   71
         74   65   45   57   48   71   69   69   44
         57   43   68   58   49   57   75   55   66
         59   75   74   51   48   62   52   50   63
         73

MTB > standard deviation cl
```

Column Standard Deviation

```
Standard deviation of height = 10.111
MTB > zinterval 95 percent confidence sd = 10.111 data in cl
```

Confidence Intervals

```
The assumed sigma = 10.1

Variable     N      Mean      StDev    SE Mean      95.0 % CI
height     100     59.22      10.11       1.01   (57.24, 61.20)
```

9.7 In a survey of Catholic priests each priest reported the total number of baptisms, marriages, and funerals conducted during the past calendar year. The responses are given in Table 9.3. Use this data to construct a 95% confidence interval on μ, the mean number of baptisms, marriages,

Table 9.3

32	44	48	35	34	29	31	61	37	41
31	40	44	43	41	40	41	31	42	45
29	40	42	51	16	24	40	52	62	41
32	41	45	24	41	30	42	47	30	46
38	42	26	34	45	58	57	35	62	46

and funerals conducted during the past calendar year per priest for all priests. Construct the interval by use of the confidence interval formula, and also use the `Zinterval` command of MINITAB to find the interval.

SOLUTION

After entering the data from Table 9.3 into column 1 of MINITAB'S worksheet, and naming the column 'number', the mean and standard deviation commands were given.

```
MTB > mean c1
```

Column Mean

```
Mean of Number = 40.261
MTB > standard deviation c1
```

Column Standard Deviation

```
Standard deviation of Number = 9.9895
```

The standard error of the mean is equal to $9.9895/\sqrt{50} = 1.413$, the critical value is 1.96, and the 95% margin of error is $1.96(1.413) = 2.769$. The confidence interval extends from $40.261 - 2.769 = 37.492$ to $40.261 + 2.769 = 43.030$.

The `Zinterval` command produces the following output.

```
MTB > Zinterval 95% confidence sd = 9.9895 data in c1
```

Z Confidence Intervals

```
The assumed sigma = 9.99

Variable     N     Mean      StDev    SE Mean    95.00 % CI
Number      50    40.26      9.99        1.41   (37.49, 43.03)
```

We are 95% confident that the true mean for all priests is between 37.49 and 43.03.

9.8 In measuring reaction time, a psychologist estimates that the standard deviation is 0.05 seconds (s). How large a sample of measurements must he take in order to be (*a*) 95% and (*b*) 99% confident that the error of his estimate will not exceed 0.01 s?

SOLUTION

(*a*) The 95% confidence limits are $\bar{X} \pm 1.96\sigma/\sqrt{N}$, the error of the estimate being $1.96\sigma/\sqrt{N}$. Taking $\sigma = s = 0.05$ s, we see that this error will be equal to 0.01 s if $(1.96)(0.05)/\sqrt{N} = 0.01$; that is, $\sqrt{N} = (1.96)(0.05)/0.01 = 9.8$, or $N = 96.04$. Thus we can be 95% confident that the error of the estimate will be less than 0.01 s if N is 97 or larger.

Another method

$$\frac{(1.96)(0.05)}{\sqrt{N}} \leq 0.01 \quad \text{if} \quad \frac{\sqrt{N}}{(1.96)(0.05)} \geq \frac{1}{0.01} \quad \text{or} \quad \sqrt{N} \geq \frac{(1.96)(0.05)}{0.01} = 9.8$$

Then $N \geq 96.04$, or $N \geq 97$.

(*b*) The 99% confidence limits are $\bar{X} \pm 2.58\sigma/\sqrt{N}$. Then $(2.58)(0.05)/\sqrt{N} = 0.01$, or $N = 166.4$. Thus we can be 99% confident that the error of the estimate will be less than 0.01 s only if N is 167 or larger.

9.9 A random sample of 50 mathematics grades out of a total of 200 showed a mean of 75 and a standard deviation of 10.

(*a*) What are the 95% confidence limits for estimates of the mean of the 200 grades?

(*b*) With what degree of confidence could we say that the mean of all 200 grades is 75 ± 1?

SOLUTION

(*a*) Since the population size is not very large compared with the sample size, we must adjust for it. Then the 95% confidence limits are

$$\bar{X} \pm 1.96\sigma_{\bar{X}} = \bar{X} \pm 1.96 \frac{\sigma}{\sqrt{N}} \sqrt{\frac{N_p - N}{N_p - 1}} = 75 \pm 1.96 \frac{10}{\sqrt{50}} \sqrt{\frac{200 - 50}{200 - 1}} = 75 \pm 2.4$$

(*b*) The confidence limits can be represented by

$$\bar{X} \pm z_c \sigma_{\bar{X}} = \bar{X} \pm z_c \frac{\sigma}{\sqrt{N}} \sqrt{\frac{N_p - N}{N_p - 1}} = 75 \pm z_c \frac{10}{\sqrt{50}} \sqrt{\frac{200 - 50}{200 - 1}} = 75 \pm 1.23 z_c$$

Since this must equal 75 ± 1, we have $1.23 z_c = 1$, or $z_c = 0.81$. The area under the normal curve from $z = 0$ to $z = 0.81$ is 0.2910; hence the required degree of confidence is $2(0.2910) = 0.582$, or 58.2%.

CONFIDENCE INTERVALS FOR PROPORTIONS

9.10 A sample poll of 100 voters chosen at random from all voters in a given district indicated that 55% of them were in favor of a particular candidate. Find the (*a*) 95%, (*b*) 99%, and (*c*) 99.73% confidence limits for the proportion of all the voters in favor of this candidate.

SOLUTION

(*a*) The 95% confidence limits for the population p are $P \pm 1.96\sigma_P = P \pm 1.96\sqrt{p(1-p)/N} = 0.55 \pm 1.96\sqrt{(0.55)(0.45)/100} = 0.55 \pm 0.10$, where we have used the sample proportion P to estimate p.

(*b*) The 99% confidence limits for p are $0.55 \pm 2.58\sqrt{(0.55)(0.45)/100} = 0.55 \pm 0.13$.

(*c*) The 99.73% confidence limits for p are $0.55 \pm 3\sqrt{(0.55)(0.45)/100} = 0.55 \pm 0.15$.

9.11 How large a sample of voters should we take in Problem 9.10 in order to be (*a*) 95% and (*b*) 99.73% confident that the candidate will be elected?

SOLUTION

The confidence limits for p are $P \pm z_c\sqrt{p(1-p)/N} = 0.55 \pm z_c\sqrt{(0.55)(0.45)/N} = 0.55 \pm 0.50 z_c/\sqrt{N}$, where we have used the estimate $P = p = 0.55$ on the basis of Problem 9.10. Since the candidate will win only if she receives more than 50% of the population's votes, we require that $0.50 z_c/\sqrt{N}$ be less than 0.05.

(a) For 95% confidence, $0.50z_c/\sqrt{N} = 0.50(1.96)/\sqrt{N} = 0.05$ when $N = 384.2$. Thus N should be at least 385.

(b) For 99.73% confidence, $0.50z_c/\sqrt{N} = 0.50(3)/\sqrt{N} = 0.05$ when $N = 900$. Thus N should be at least 901.

Another method

$1.50/\sqrt{N} < 0.05$ when $\sqrt{N}/1.50 > 1/0.05$ or $\sqrt{N} > 1.50/0.05$. Then $\sqrt{N} > 30$ or $N > 900$, so that N should be at least 901.

9.12 A survey is conducted and it is found that 156 out of 500 adult males are smokers. Use the software package STATISTIX to set a 99% confidence interval on p, the proportion in the population of adult males who are smokers. Check the confidence interval by computing it by hand.

SOLUTION

The STATISTIX output is as follows. The 99% confidence interval is shown in bold.

```
One-Sample Proportion Test

Sample Size              500
Successes                156
Proportion           0.31200

Null Hypothesis:     P = 0.5
Alternative Hyp:   P < > 0.5

Difference          -0.18800
Standard Error       0.02072
Z (uncorrected)        -8.41      P 0.0000
Z (corrected)          -8.36      P 0.0000

                     99% Confidence Interval
Uncorrected          (0.25863, 0.36537)
Corrected            (0.25763, 0.36637)
```

We are 99% confident that the true percent of adult male smokers is between 25.9% and 36.5%.

Check:

$$P = 0.312,\ z_c = 2.58,\ \sqrt{\frac{0.312(0.688)}{500}} = 0.0207$$

$P \pm z_c\sqrt{\dfrac{p(1-p)}{N}}$ or $0.312 \pm 2.58(0.0207)$ or (**0.258**, **0.365**) This is the same as given above by the software package STATISTIX.

9.13 Refer to Problem 9.12. Set a 99% confidence interval on p using the software package MINITAB.

SOLUTION

The 99% confidence interval is shown in bold below. It is the same as the STATISTIX confidence interval shown in Problem 9.12.

```
Sample    X    N    Sample P    99% CI                  z-Value    P-Value
1       156  500    0.312000    (0.258629, 0.365371)      -8.41      0.000
```

CONFIDENCE INTERVALS FOR DIFFERENCES AND SUMS

9.14 A study was undertaken to compare the mean time spent on cell phones by male and female college students per week. Fifty male and 50 female students were selected from Midwestern University and the number of hours per week spent talking on their cell phones determined. The results in hours are shown in Table 9.4. Set a 95% confidence interval on $\mu_1 - \mu_2$ using MINITAB. Check the results by calculating the interval by hand.

Table 9.4

Males					Females				
12	4	11	13	11	11	9	7	10	9
7	9	10	10	7	10	10	7	9	10
7	12	6	9	15	11	8	9	6	11
10	11	12	7	8	10	7	9	12	14
8	9	11	10	9	11	12	12	8	12
10	9	9	7	9	12	9	10	11	7
11	7	10	10	11	12	7	9	8	11
9	12	12	8	13	10	8	13	8	10
9	10	8	11	10	9	9	9	11	9
13	13	9	10	13	9	8	9	12	11

SOLUTION

When both samples exceed 30, the two sample t test and the z test may be used interchangeably since the t distribution and the z distribution are very similar.

```
Two-sample T for males vs females
          N     Mean     StDev     SE Mean
males     50    9.82      2.15       0.30
females   50    9.70      1.78       0.25
Difference = mu (males) - mu (females)
Estimate for difference: 0.120000
95% CI for difference: (-0.663474, 0.903474)
T-Test of difference = 0 (vs not =): T-Value = 0.30 P-Value = 0.762
DF = 98
Both use Pooled StDev = 1.9740
```

According to the MINITAB output, the difference in the population means is between −0.66 and 0.90. There is a good chance that there is no difference in the population means.

Check:

The formula for a 95% confidence interval is $(\bar{x}_1 - \bar{x}_2) \pm z_c\left(\sqrt{(s_1^2/n_1) + (s_2^2/n_2)}\right)$. Substituting, we obtain $0.12 \pm 1.96(0.395)$ and get the answer given by MINITAB.

9.15 Use STATISTIX and SPSS to solve Problem 9.14.

SOLUTION

The STATISTIX solution is given below. Note that the 95% confidence interval is the same as that in Problem 9.14. We shall comment later on why we assume equal variances.

```
Two-Sample T Tests for males vs females

Variable       Mean      N       SD        SE
males        9.8200     50   2.1542    0.3046
females      9.7000     50   1.7757    0.2511
Difference   0.1200

Null Hypothesis: difference = 0
Alternative Hyp: difference < > 0

                                                     95% CI for Difference
Assumption                T       DF        P        Lower        Upper
Equal Variances        0.30       98   0.7618      -0.6635       0.9035
Unequal Variances      0.30     94.6   0.7618      -0.6638       0.9038

Test for Equality         F         DF           P
   of Variances        1.47      49,49      0.0899
```

The SPSS solution is as follows:

Group Statistics

	Sex	N	Mean	Std. Deviation	Std. Error Mean
time	1.00	50	9.7000	1.77569	.25112
	2.00	50	9.8200	2.15416	.30464

Independent Samples Test

	Levene's Test for Equality of Variances		t-test for Equality of Means						
								95% Confidence Interval of the Difference	
	F	Sig.	t	df	Sig. (2-tailed)	Mean Difference	Std. Error Difference	Lower	Upper
time Equal variances assumed	.898	.346	−.304	98	.762	−.12000	.39480	−.90347	.66347
Equal variances not assumed			−.304	94.556	.762	−.12000	.39480	−.90383	.66383

9.16 Use SAS to work Problem 9.14. Give the forms of the data files that SAS allows for the analysis.

SOLUTION

The SAS analysis is as follows. The confidence interval is shown in bold at the bottom of the following output.

```
Two Sample t-test for the Means of males and females
Sample Statistics
 Group      N      Mean       Std. Dev.       Std. Error
 -----------------------------------------------------
 males     50      9.82        2.1542          0.3046
 females   50      9.7         1.7757          0.2511

 Null hypothesis: Mean 1- Mean 2 = 0
 Alternative:     Mean 1- Mean 2 ^= 0

 If Variances Are        t statistic    Df       Pr > t
 ------------------------------------------------------
 Equal                     0.304        98       0.7618
 Not Equal                 0.304     94.56       0.7618
 95% Confidence Interval for the Difference between Two Means

          Lower Limit      Upper Limit
          -----------      -----------
             -0.66            0.90
```

The data file used in the SAS analysis may have the data for males and females in separate columns or the data may consist of hours spent on the cell phone in one column and the sex of the person (Male or Female) in the other column. Male and female may be coded as 1 or 2. The first form consists of 2 columns and 50 rows. The second form consists of 2 columns and 100 rows.

CONFIDENCE INTERVALS FOR STANDARD DEVIATIONS

9.17 A confidence interval for the variance of a population utilizes the chi-square distribution. The $(1-\alpha) \times 100\%$ confidence interval is $\frac{(n-1)S^2}{(\chi^2_{\alpha/2})} < \sigma^2 < \frac{(n-1)S^2}{(\chi^2_{1-\alpha/2})}$, where n is the sample size, S^2 is the sample variance, $\chi^2_{\alpha/2}$ and $\chi^2_{1-\alpha/2}$ come from the chi-square distribution with $(n-1)$ degrees of freedom. Use EXCEL software to find a 99% confidence interval for the variance of twenty 180 ounce containers. The data from twenty containers is shown in Table 9.5.

Table 9.5

181.5	180.8
179.7	182.4
178.7	178.5
183.9	182.2
179.7	180.9
180.6	181.4
180.4	181.4
178.5	180.6
178.8	180.1
181.3	182.2

SOLUTION

The EXCEL worksheet is as follows. The data are in A1:B10. Column D shows the function in column C which gives the values shown.

A	B	C	D
181.5	180.8	2.154211	=VAR(A1:B10)
179.1	182.4	40.93	=19*C1
178.7	178.5	38.58226	=CHIINV(0.005,19)
183.9	182.2	6.843971	=CHIINV(0.995,19)
179.7	180.9		
180.6	181.4	1.06085	=C2/C3
180.4	181.4	5.980446	=C2/C4
178.5	180.6		
178.8	180.1		
181.3	182.2		

The following is a 99% confidence interval for σ^2 : $(1.06085 < \sigma^2 < 5.980446)$.The following is a 99% confidence interval for σ : (1.03, 2.45).

Note that =VAR(A1:B10) gives S^2, =CHIINV(0.005,19) is the chi-square value with area 0.005 to its right, and =CHIINV(0.995,19) is the chi-square value with area 0.995 to its right. In all cases, the chi-square distribution has 19 degrees of freedom.

9.18 When comparing the variance of one population with the variance of another population, the following $(1 - \alpha) \times 100\%$ confidence interval may be used:

$$\frac{S_1^2}{S_2^2} \cdot \frac{1}{F_{\alpha/2(\nu_1, \nu_2)}} < \frac{\sigma_1^2}{\sigma_2^2} < \frac{S_1^2}{S_2^2} F_{\alpha/2(\nu_2,\nu_1)},$$

where n_1 and n_2 are the two sample sizes, S_1^2 and S_2^2 are the two sample variances, $v_1 = n_1 - 1$ and $v_2 = n_2 - 1$ are the numerator and denominator degrees of freedom for the F distribution and the F values are from the F distribution. Table 9.6 gives the number of e-mails sent per week by employees at two different companies.

Set a 95% confidence interval for $\dfrac{\sigma_1}{\sigma_2}$.

Table 9.6

Company 1	Company 2
81	99
104	100
115	104
111	98
85	103
121	113
95	95
112	107
100	98
117	95
113	101
109	109
101	99
	93
	105

SOLUTION

The EXCEL worksheet is shown below. Column D shows the function in column C which gives the values shown. The two sample variances are computed in C1 and C2. The F values are computed in C3 and C4. The end points of the confidence interval for the ratio of the variances are computed in C5 and C6. We see that a 95% confidence interval for $\frac{\sigma_1^2}{\sigma_2^2}$ is (1.568, 15.334). A 95% confidence interval for $\frac{\sigma_1}{\sigma_2}$ is (1.252, 3.916). Note that =FINV(0.025,12,14) is the point associated with the F distribution with $\nu_1 = 12$ and $\nu_2 = 14$ degrees of freedom so that 0.025 of the area is to the right of that point.

A	B	C	D
Company 1	Company 2	148.5769231	=VAR(A2:A14)
81	99	31.06666667	=VAR(B2:B16)
104	100	3.050154789	=FINV(0.025,12,14)
115	104	3.2062117	=FINV(0.025,14,12)
111	98	1.567959436	=(C1/C2)/C3
85	103	15.33376832	=(C1/C2)*C4
121	113		
95	95	1.25218187	=SQRT(C5)
112	107	3.915835584	=SQRT(C6)
100	98		
117	95		
113	101		
109	109		
101	99		
	93		
	105		

PROBABLE ERROR

9.19 The voltages of 50 batteries of the same type have a mean of 18.2 volts (V) and a standard deviation of 0.5 V. Find (*a*) the probable error of the mean and (*b*) the 50% confidence limits.

SOLUTION

(*a*)
$$\text{Probable error of the mean} = 0.674\sigma_{\bar{X}} = 0.6745\frac{\sigma}{\sqrt{N}} = 0.6745\frac{\hat{s}}{\sqrt{N}}$$
$$= 0.6745\frac{s}{\sqrt{N-1}} = 0.6745\frac{0.5}{\sqrt{49}} = 0.048 \text{ V}$$

Note that if the standard deviation of 0.5 V is computed as $\hat{s}$, the probable error is $0.6745(0.5/\sqrt{50}) = 0.048$ also, so that either estimate can be used if N is large enough.

(*b*) The 50% confidence limits are 18 ± 0.048 V.

9.20 A measurement was recorded as 216.480 grams (g) with a probable error of 0.272 g. What are the 95% confidence limits for the measurement?

SOLUTION

The probable error is $0.272 = 0.6745\sigma_{\bar{X}}$, or $\sigma_{\bar{X}} = 0.272/0.6745$. Thus the 95% confidence limits are $\bar{X} \pm 1.96\sigma_{\bar{X}} = 216.480 \pm 1.96(0.272/0.6745) = 216.480 \pm 0.790$ g.

Supplementary Problems

UNBIASED AND EFFICIENT ESTIMATES

9.21 Measurements of a sample of masses were determined to be 8.3, 10.6, 9.7, 8.8, 10.2, and 9.4 kilograms (kg), respectively. Determine unbiased and efficient estimates of (*a*) the population mean and (*b*) the population variance, and (*c*) compare the sample standard deviation with the estimated population standard deviation.

9.22 A sample of 10 television tubes produced by a company showed a mean lifetime of 1200 hours (h) and a standard deviation of 100 h. Estimate (*a*) the mean and (*b*) the standard deviation of the population of all television tubes produced by this company.

9.23 (*a*) Work Problem 9.22 if the same results are obtained for 30, 50, and 100 television tubes.

(*b*) What can you conclude about the relation between sample standard deviations and estimates of population standard deviations for different sample sizes?

CONFIDENCE INTERVALS FOR MEANS

9.24 The mean and standard deviation of the maximum loads supported by 60 cables (see Problem 3.59) are given by 11.09 tons and 0.73 ton, respectively. Find the (*a*) 95% and (*b*) 99% confidence limits for the mean of the maximum loads of all cables produced by the company.

9.25 The mean and standard deviation of the diameters of a sample of 250 rivet heads manufactured by a company are 0.72642 in and 0.00058 in, respectively (see Problem 3.61). Find the (*a*) 99%, (*b*) 98%, (*c*) 95%, and (*d*) 90% confidence limits for the mean diameter of all the rivet heads manufactured by the company.

9.26 Find (*a*) the 50% confidence limits and (*b*) the probable error for the mean diameters in Problem 9.25.

9.27 If the standard deviation of the lifetimes of television tubes is estimated to be 100 h, how large a sample must we take in order to be (*a*) 95%, (*b*) 90%, (*c*) 99%, and (*d*) 99.73% confident that the error in the estimated mean lifetime will not exceed 20 h?

9.28 A group of 50 internet shoppers were asked how much they spent per year on the Internet. Their responses are shown in Table 9.7.

Find an 80% confidence interval for μ, the mean amount spent by all internet shoppers, using the equations found in Chapter 9 as well as statistical software.

Table 9.7

418	379	77	212	378
363	434	348	245	341
331	356	423	330	247
351	151	220	383	257
307	297	448	391	210
158	310	331	348	124
523	356	210	364	406
331	364	352	299	221
466	150	282	221	432
366	195	96	219	202

9.29 A company has 500 cables. A test of 40 cables selected at random showed a mean breaking strength of 2400 pounds (lb) and a standard deviation of 150 lb.

(*a*) What are the 95% and 99% confidence limits for estimating the mean breaking strength of the remaining 460 cables?

(*b*) With what degree of confidence could we say that the mean breaking strength of the remaining 460 cables is 2400 ± 35 lb?

CONFIDENCE INTERVALS FOR PROPORTIONS

9.30 An urn contains an unknown proportion of red and white marbles. A random sample of 60 marbles selected with replacement from the urn showed that 70% were red. Find the (*a*) 95%, (*b*) 99%, and (*c*) 99.73% confidence limits for the actual proportion of red marbles in the urn.

9.31 A poll of 1000 individuals over the age of 65 years was taken to determine the percent of the population in this age group who had an Internet connection. It was found that 387 of the 1000 had an internet connection. Using the equations in the book as well as statistical software, find a 97.5% confidence interval for p.

9.32 It is believed that an election will result in a very close vote between two candidates. What is the least number of voters that one should poll in order to be (*a*) 80%, (*b*) 90%, (*c*) 95%, and (*d*) 99% confident of a decision in favor of either one of the candidates?

CONFIDENCE INTERVALS FOR DIFFERENCES AND SUMS

9.33 Of two similar groups of patients, A and B, consisting of 50 and 100 individuals, respectively, the first was given a new type of sleeping pill and the second was given a conventional type. For the patients in group A, the mean number of hours of sleep was 7.82 with a standard deviation of 0.24 h. For the patients in group B, the mean number of hours of sleep was 6.75 with a standard deviation of 0.30 h. Find the (*a*) 95% and (*b*) 99% confidence limits for the difference in the mean number of hours of sleep induced by the two types of sleeping pills.

9.34 A study was conducted to compare the mean lifetimes of males with the mean lifetimes of females. Random samples were collected from the obituary pages and the data in Table 9.8 were collected.

Table 9.8

Males					Females				
85	53	100	49	65	64	93	82	71	77
60	51	61	83	65	64	60	75	87	60
55	99	56	55	55	61	84	91	61	85
90	72	62	69	59	105	90	59	86	62
49	72	58	60	68	71	99	98	54	94
90	74	85	80	77	98	61	108	79	50
62	65	81	55	71	66	74	60	90	95
78	49	78	80	75	81	86	65	86	81
53	82	109	87	78	92	77	82	86	79
72	104	70	31	50	91	93	63	93	53

Using the above data, the equations in the book, and statistical software, set an 85% confidence interval on $\mu_{\text{MALE}} - \mu_{\text{FEMALE}}$.

9.35 Two areas of the country are compared as to the percent of teenagers who have at least one cavity. One area fluoridates its water and the other doesn't. The sample from the non-fluoridated area finds that 425 out of 1000 have at least one cavity. The sample from the fluoridated area finds that 376 out of 1000 have at least one cavity. Set a 99% confidence interval on the difference in percents using the equations in the book as well as statistical software.

CONFIDENCE INTERVALS FOR STANDARD DEVIATIONS

9.36 The standard deviation of the breaking strengths of 100 cables tested by a company was 180 lb. Find the (*a*) 95%, (*b*) 99%, and (*c*) 99.73% confidence limits for the standard deviation of all cables produced by the company.

9.37 Solve Problem 9.17 using SAS.

9.38 Solve Problem 9.18 using SAS.

Statistical Decision Theory

STATISTICAL DECISIONS

Very often in practice we are called upon to make decisions about populations on the basis of sample information. Such decisions are called *statistical decisions*. For example, we may wish to decide on the basis of sample data whether a new serum is really effective in curing a disease, whether one educational procedure is better than another, or whether a given coin is loaded.

STATISTICAL HYPOTHESES

In attempting to reach decisions, it is useful to make assumptions (or guesses) about the populations involved. Such assumptions, which may or may not be true, are called *statistical hypotheses*. They are generally statements about the probability distributions of the populations.

Null Hypotheses

In many instances we formulate a statistical hypothesis for the sole purpose of rejecting or nullifying it. For example, if we want to decide whether a given coin is loaded, we formulate the hypothesis that the coin is fair (i.e., $p = 0.5$, where p is the probability of heads). Similarly, if we want to decide whether one procedure is better than another, we formulate the hypothesis that there is *no difference* between the procedures (i.e., any observed differences are due merely to fluctuations in sampling from the *same* population). Such hypotheses are often called *null hypotheses* and are denoted by H_0.

Alternative Hypotheses

Any hypothesis that differs from a given hypothesis is called an *alternative hypothesis*. For example, if one hypothesis is $p = 0.5$, alternative hypotheses might be $p = 0.7$, $p \neq 0.5$, or $p > 0.5$. A hypothesis alternative to the null hypothesis is denoted by H_1.

TESTS OF HYPOTHESES AND SIGNIFICANCE, OR DECISION RULES

If we suppose that a particular hypothesis is true but find that the results observed in a random sample differ markedly from the results expected under the hypothesis (i.e., expected on the basis of pure chance, using sampling theory), then we would say that the observed differences are *significant* and would thus be inclined to reject the hypothesis (or at least not accept it on the basis of the evidence obtained). For example, if 20 tosses of a coin yield 16 heads, we would be inclined to reject the hypothesis that the coin is fair, although it is conceivable that we might be wrong.

Procedures that enable us to determine whether observed samples differ significantly from the results expected, and thus help us decide whether to accept or reject hypotheses, are called *tests of hypotheses, tests of significance, rules of decision*, or simply *decision rules*.

TYPE I AND TYPE II ERRORS

If we reject a hypothesis when it should be accepted, we say that a *Type I error* has been made. If, on the other hand, we accept a hypothesis when it should be rejected, we say that a *Type II error* has been made. In either case, a wrong decision or error in judgment has occurred.

In order for decision rules (or tests of hypotheses) to be good, they must be designed so as to minimize errors of decision. This is not a simple matter, because for any given sample size, an attempt to decrease one type of error is generally accompanied by an increase in the other type of error. In practice, one type of error may be more serious than the other, and so a compromise should be reached in favor of limiting the more serious error. The only way to reduce both types of error is to increase the sample size, which may or may not be possible.

LEVEL OF SIGNIFICANCE

In testing a given hypothesis, the maximum probability with which we would be willing to risk a Type I error is called the *level of significance*, or *significance level*, of the test. This probability, often denoted by α, is generally specified before any samples are drawn so that the results obtained will not influence our choice.

In practice, a significance level of 0.05 or 0.01 is customary, although other values are used. If, for example, the 0.05 (or 5%) significance level is chosen in designing a decision rule, then there are about 5 chances in 100 that we would reject the hypothesis when it should be accepted; that is, we are about 95% *confident* that we have made the right decision. In such case we say that the hypothesis has been rejected at the 0.05 significance level, which means that the hypothesis has a 0.05 probability of being wrong.

TESTS INVOLVING NORMAL DISTRIBUTIONS

To illustrate the ideas presented above, suppose that under a given hypothesis the sampling distribution of a statistic S is a normal distribution with mean μ_S and standard deviation σ_S. Thus the distribution of the standardized variable (or z score), given by $z = (S - \mu_S)/\sigma_S$, is the standardized normal distribution (mean 0, variance 1), as shown in Fig. 10-1.

As indicated in Fig. 10-1, we can be 95% confident that if the hypothesis is true, then the z score of an actual sample statistic S will lie between -1.96 and 1.96 (since the area under the normal curve between these values is 0.95). However, if on choosing a single sample at random we find that the z score of its statistic lies *outside* the range -1.96 to 1.96, we would conclude that such an event could happen with a probability of only 0.05 (the total shaded area in the figure) if the given hypothesis were true. We would then say that this z score differed *significantly* from what would be expected under the hypothesis, and we would then be inclined to reject the hypothesis.

The total shaded area 0.05 is the significance level of the test. It represents the probability of our being wrong in rejecting the hypothesis (i.e., the probability of making a Type I error). Thus we say that

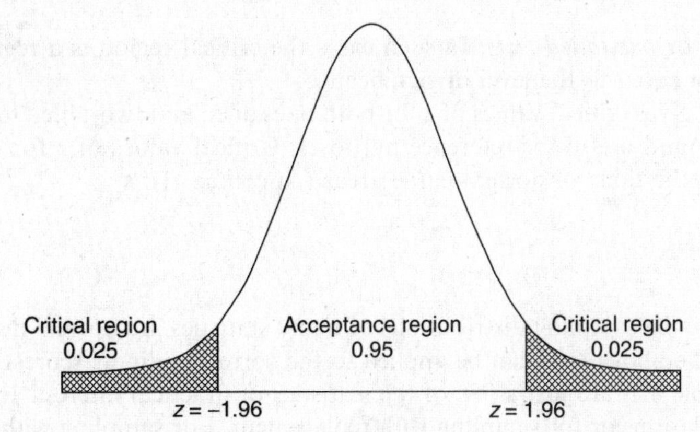

Fig. 10-1 Standard normal curve with the critical region (0.05) and acceptance region (0.95).

the hypothesis is *rejected at 0.05 the significance level* or that the z score of the given sample statistic is *significant at the 0.05 level.*

The set of z scores outside the range -1.96 to 1.96 constitutes what is called the *critical region of the hypothesis, the region of rejection of the hypothesis,* or *the region of significance.* The set of z scores inside the range -1.96 to 1.96 is thus called the *region of acceptance of the hypothesis,* or the *region of non-significance.*

On the basis of the above remarks, we can formulate the following decision rule (or test of hypothesis or significance):

Reject the hypothesis at the 0.05 significance level if the z score of the statistic S lies outside the range -1.96 to 1.96 (i.e., either $z > 1.96$ or $z < -1.96$). This is equivalent to saying that the observed sample statistic is significant at the 0.05 level.

Accept the hypothesis otherwise (or, if desired, make no decision at all).

Because the z score plays such an important part in tests of hypotheses, it is also called a *test statistic.*

It should be noted that other significance levels could have been used. For example, if the 0.01 level were used, we would replace 1.96 everywhere above with 2.58 (see Table 10.1). Table 9.1 can also be used, since the sum of the significance and confidence levels is 100%.

Table 10.1

Level of significance, α	0.10	0.05	0.01	0.005	0.002
Critical values of z for one-tailed tests	-1.28 *or* 1.28	-1.645 *or* 1.645	-2.33 *or* 2.33	-2.58 *or* 2.58	-2.88 *or* 2.88
Critical values of z for two-tailed tests	-1.645 *and* 1.645	-1.96 *and* 1.96	-2.58 *and* 2.58	-2.81 *and* 2.81	-3.08 *and* 3.08

TWO-TAILED AND ONE-TAILED TESTS

In the above test we were interested in extreme values of the statistic S or its corresponding z score on *both* sides of the mean (i.e., in both tails of the distribution). Such tests are thus called *two-sided tests,* or *two-tailed tests.*

Often, however, we may be interested only in extreme values to one side of the mean (i.e., in one tail of the distribution), such as when we are testing the hypothesis that one process is better than another (which is different from testing whether one process is better or worse than the other). Such tests are called *one-sided tests*, or *one-tailed tests*. In such cases the critical region is a region to one side of the distribution, with area equal to the level of significance.

Table 10.1, which gives critical values of z for both one-tailed and two-tailed tests at various levels of significance, will be found useful for reference purposes. Critical values of z for other levels of significance are found from the table of normal-curve areas (Appendix II).

SPECIAL TESTS

For large samples, the sampling distributions of many statistics are normal distributions (or at least nearly normal), and the above tests can be applied to the corresponding z scores. The following special cases, taken from Table 8.1, are just a few of the statistics of practical interest. In each case the results hold for infinite populations or for sampling with replacement. For sampling without replacement from finite populations, the results must be modified. See page 182.

1. **Means.** Here $S = \bar{X}$, the sample mean; $\mu_S = \mu_{\bar{X}} = \mu$, the population mean; and $\sigma_S = \sigma_{\bar{X}} = \sigma/\sqrt{N}$, where σ is the population standard deviation and N is the sample size. The z score is given by

$$z = \frac{\bar{X} - \mu}{\sigma/\sqrt{N}}$$

   When necessary, the sample deviation s or $\hat{s}$ is used to estimate σ.

2. **Proportions.** Here $S = P$, the proportion of "successes" in a sample; $\mu_S = \mu_P = p$, where p is the population proportion of successes and N is the sample size; and $\sigma_S = \sigma_P = \sqrt{pq/N}$, where $q = 1 - p$.

   The z score is given by

$$z = \frac{P - p}{\sqrt{pq/N}}$$

   In case $P = X/N$, where X is the actual number of successes in a sample, the z score becomes

$$z = \frac{X - Np}{\sqrt{Npq}}$$

   That is, $\mu_X = \mu = Np$, $\sigma_X = \sigma = \sqrt{Npq}$, and $S = X$.

The results for other statistics can be obtained similarly.

OPERATING-CHARACTERISTIC CURVES; THE POWER OF A TEST

We have seen how the Type I error can be limited by choosing the significance level properly. It is possible to avoid risking Type II errors altogether simply by not making them, which amounts to never accepting hypotheses. In many practical cases, however, this cannot be done. In such cases, use is often made of *operating-characteristic curves*, or *OC curves*, which are graphs showing the probabilities of Type II errors under various hypotheses. These provide indications of how well a given test will enable us to minimize Type II errors; that is, they indicate the *power of a test* to prevent us from making wrong decisions. They are useful in designing experiments because they show such things as what sample sizes to use.

p-VALUES FOR HYPOTHESES TESTS

The *p*-value is the probability of observing a sample statistic as extreme or more extreme than the one observed under the assumption that the null hypothesis is true. When testing a hypothesis, state the value of α. Calculate your p-value and if the p-value $\leq \alpha$, then reject H_0. Otherwise, do not reject H_0. For testing means, using large samples ($n > 30$), calculate the p-value as follows:

1. For $H_0: \mu = \mu_0$ and $H_1: \mu < \mu_0$, p-value $= P(Z <$ computed test statistic),
2. For $H_0: \mu = \mu_0$ and $H_1: \mu > \mu_0$, p-value $= P(Z >$ computed test statistic), and
3. For $H_0: \mu = \mu_0$ and $H_1: \mu \neq \mu_0$, p-value $= P(Z < -|$ computed test statistic $|) + P(Z > |$ computed test statistic $|)$.

The computed test statistic is $\dfrac{\bar{x} - \mu_0}{(s/\sqrt{n})}$, where $\bar{x}$ is the mean of the sample, s is the standard deviation of the sample, and μ_0 is the value specified for μ in the null hypothesis. Note that if σ is unknown, it is estimated from the sample by using s. This method of testing hypothesis is equivalent to the method of finding a critical value or values and if the computed test statistic falls in the rejection region, reject the null hypothesis. The same decision will be reached using either method.

CONTROL CHARTS

It is often important in practice to know when a process has changed sufficiently that steps should be taken to remedy the situation. Such problems arise, for example, in quality control. Quality control supervisors must often decide whether observed changes are due simply to chance fluctuations or are due to actual changes in a manufacturing process because of deteriorating machine parts, employees' mistakes, etc. *Control charts* provide a useful and simple method for dealing with such problems (see Problem 10.16).

TESTS INVOLVING SAMPLE DIFFERENCES

Differences of Means

Let $\bar{X}_1$ and $\bar{X}_2$ be the sample means obtained in large samples of sizes N_1 and N_2 drawn from respective populations having means μ_1 and μ_2 and standard deviations σ_1 and σ_2. Consider the null hypothesis that there is *no difference* between the population means (i.e., $\mu_1 = \mu_2$), which is to say that the samples are drawn from two populations having the same mean.

Placing $\mu_1 = \mu_2$ in equation (*5*) of Chapter 8, we see that the sampling distribution of differences in means is approximately normally distributed, with its mean and standard deviation given by

$$\mu_{\bar{X}1-\bar{X}2} = 0 \qquad \text{and} \qquad \sigma_{\bar{X}1-\bar{X}2} = \sqrt{\frac{\sigma_1^2}{N_1} + \frac{\sigma_2^2}{N_2}} \tag{1}$$

where we can, if necessary, use the sample standard deviations s_1 and s_2 (or $\hat{s}_1$ and $\hat{s}_2$) as estimates of σ_1 and σ_2.

By using the standardized variable, or z score, given by

$$z = \frac{\bar{X}_1 - \bar{X}_2 - 0}{\sigma_{\bar{X}1-\bar{X}2}} = \frac{\bar{X}_1 - \bar{X}_2}{\sigma_{\bar{X}1-\bar{X}2}} \tag{2}$$

we can test the null hypothesis against alternative hypotheses (or the significance of an observed difference) at an appropriate level of significance.

Differences of Proportions

Let P_1 and P_2 be the sample proportions obtained in large samples of sizes N_1 and N_2 drawn from respective populations having proportions p_1 and p_2. Consider the null hypothesis that there is *no difference* between the population parameters (i.e., $p_1 = p_2$) and thus that the samples are really drawn from the same population.

Placing $p_1 = p_2 = p$ in equation (6) of Chapter 8, we see that the sampling distribution of differences in proportions is approximately normally distributed, with its mean and standard deviation given by

$$\mu_{P1-P2} = 0 \quad \text{and} \quad \sigma_{P1-P2} = \sqrt{pq\left(\frac{1}{N_1} + \frac{1}{N_2}\right)} \tag{3}$$

where

$$p = \frac{N_1P_1 + N_2P_2}{N_1 + N_2}$$

is used as an estimate of the population proportion and where $q = 1 - p$.

By using the standardized variable

$$z = \frac{P_1 - P_2 - 0}{\sigma_{P1-P2}} = \frac{P_1 - P_2}{\sigma_{P1-P2}} \tag{4}$$

we can test observed differences at an appropriate level of significance and thereby test the null hypothesis.

Tests involving other statistics can be designed similarly.

TESTS INVOLVING BINOMIAL DISTRIBUTIONS

Tests involving binomial distributions (as well as other distributions) can be designed in a manner analogous to those using normal distributions; the basic principles are essentially the same. See Problems 10.23 to 10.28.

Solved Problems

TESTS OF MEANS AND PROPORTIONS, USING NORMAL DISTRIBUTIONS

10.1 Find the probability of getting between 40 and 60 heads inclusive in 100 tosses of a fair coin.

SOLUTION

According to the binomial distribution, the required probability is

$$\binom{100}{40}\left(\frac{1}{2}\right)^{40}\left(\frac{1}{2}\right)^{60} + \binom{100}{41}\left(\frac{1}{2}\right)^{41}\left(\frac{1}{2}\right)^{59} + \cdots + \binom{100}{60}\left(\frac{1}{2}\right)^{60}\left(\frac{1}{2}\right)^{40}$$

Since $Np = 100(\frac{1}{2})$ and $Nq = 100(\frac{1}{2})$ are both greater than 5, the normal approximation to the binomial distribution can be used in evaluating this sum. The mean and standard deviation of the number of heads in 100 tosses are given by

$$\mu = Np = 100(\tfrac{1}{2}) = 50 \quad \text{and} \quad \sigma = \sqrt{Npq} = \sqrt{(100)(\tfrac{1}{2})(\tfrac{1}{2})} = 5$$

On a continuous scale, between 40 and 60 heads inclusive is the same as between 39.5 and 60.5 heads. We thus have

$$39.5 \text{ in standard units} = \frac{39.5 - 50}{5} = -2.10 \qquad 60.5 \text{ in standard units} = \frac{60.5 - 50}{5} = 2.10$$

$$\begin{aligned}\text{Required probability} &= \text{area under normal curve between } z = -2.10 \text{ and } z = 2.10\\ &= 2(\text{area between } z = 0 \text{ and } z = 2.10) = 2(0.4821) = 0.9642\end{aligned}$$

10.2 To test the hypothesis that a coin is fair, adopt the following decision rule:

Accept the hypothesis if the number of heads in a single sample of 100 tosses is between 40 and 60 inclusive.

Reject the hypothesis otherwise.

(*a*) Find the probability of rejecting the hypothesis when it is actually correct.

(*b*) Graph the decision rule and the result of part (*a*).

(*c*) What conclusions would you draw if the sample of 100 tosses yielded 53 heads? And if it yielded 60 heads?

(*d*) Could you be wrong in your conclusions about part (*c*)? Explain.

SOLUTION

(*a*) From Problem 10.1, the probability of not getting between 40 and 60 heads inclusive if the coin is fair is $1 - 0.9642 = 0.0358$. Thus the probability of rejecting the hypothesis when it is correct is 0.0358.

(*b*) The decision rule is illustrated in Fig. 10-2, which shows the probability distribution of heads in 100 tosses of a fair coin. If a single sample of 100 tosses yields a z score between -2.10 and 2.10, we accept the hypothesis; otherwise, we reject the hypothesis and decide that the coin is not fair.

The error made in rejecting the hypothesis when it should be accepted is the *Type 1 error* of the decision rule; and the probability of making this error, equal to 0.0358 from part (*a*), is represented by the total shaded area of the figure. If a single sample of 100 tosses yields a number of heads whose z score (or z statistic) lies in the shaded regions, we would say that this z score differed *significantly* from what would be expected if the hypothesis were true. For this reason, the total shaded area (i.e., the probability of a Type I error) is called the *significance level* of the decision rule and equals 0.0358 in this case. Thus we speak of rejecting the hypothesis at the 0.0358 (or 3.58%) significance level.

(*c*) According to the decision rule, we would have to accept the hypothesis that the coin is fair in both cases. One might argue that if only one more head had been obtained, we would have rejected the hypothesis. This is what one must face when any sharp line of division is used in making decisions.

(*d*) Yes. We could accept the hypothesis when it actually should be rejected—as would be the case, for example, when the probability of heads is really 0.7 instead of 0.5. The error made in accepting the hypothesis when it should be rejected is the *Type II error* of the decision.

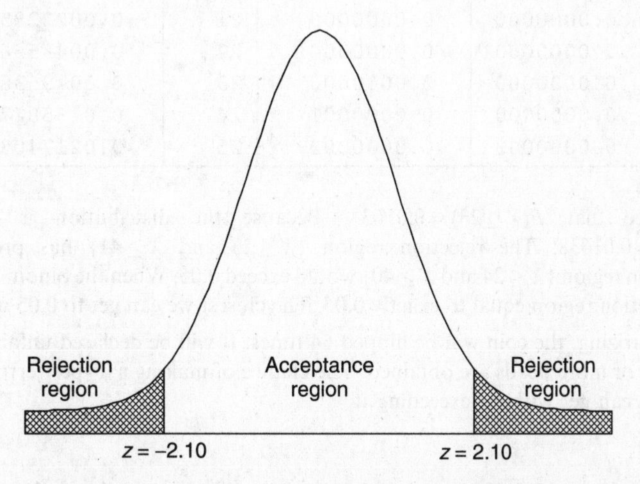

Fig. 10-2 Standard normal curve showing the acceptance and rejection regions for testing that the coin is fair.

10.3 Using the binomial distribution and not the normal approximation to the binomial distribution, design a decision rule to test the hypothesis that a coin is fair if a sample of 64 tosses of the coin is taken and a significance level of 0.05 is used. Use MINITAB to assist with the solution.

SOLUTION

The binomial plot of probabilities when a fair coin is tossed 64 times is given in Fig. 10.3. Partial cumulative probabilities generated by MINITAB are shown below the Fig. 10-3.

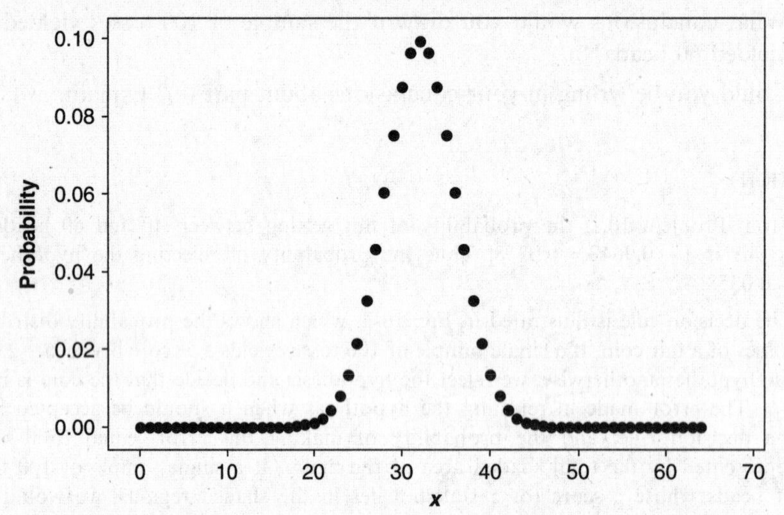

Fig. 10-3 MINITAB plot of the binomial distribution for $n=64$ and $p=0.5$.

x	Probability	Cumulative	x	Probability	Cumulative
0	0.0000000	0.0000000	13	0.0000007	0.0000009
1	0.0000000	0.0000000	14	0.0000026	0.0000035
2	0.0000000	0.0000000	15	0.0000086	0.0000122
3	0.0000000	0.0000000	16	0.0000265	0.0000387
4	0.0000000	0.0000000	17	0.0000748	0.0001134
5	0.0000000	0.0000000	18	0.0001952	0.0003087
6	0.0000000	0.0000000	19	0.0004727	0.0007814
7	0.0000000	0.0000000	20	0.0010636	0.0018450
8	0.0000000	0.0000000	21	0.0022285	0.0040735
9	0.0000000	0.0000000	22	0.0043556	0.0084291
10	0.0000000	0.0000000	23	0.0079538	0.0163829
11	0.0000000	0.0000001	24	0.0135877	0.0299706
12	0.0000002	0.0000002	25	0.0217403	0.0517109

We see that $P(X \leq 23) = 0.01638$. Because the distribution is symmetrical, we also know $P(X \geq 41) = 0.01638$. The rejection region $\{X \leq 23 \text{ and } X \geq 41\}$ has probability $2(0.01638) = 0.03276$. The rejection region $\{X \leq 24 \text{ and } X \geq 40\}$ would exceed 0.05. When the binomial distribution is used we cannot have a rejection region equal to exactly 0.05. The closest we can get to 0.05 without exceeding it is 0.03276.

Summarizing, the coin will be flipped 64 times. It will be declared unfair or not balanced if 23 or fewer heads or 41 or more heads are obtained. The chance of making a Type 1 error is 0.03276 which is as close to 0.05 as you can get without exceeding it.

10.4 Refer to Problem 10.3. Using the binomial distribution and not the normal approximation to the binomial distribution, design a decision rule to test the hypothesis that a coin is fair if a sample of 64 tosses of the coin is taken and a significance level of 0.05 is used. Use EXCEL to assist with the solution.

SOLUTION

The outcomes 0 through 64 are entered into column A of the EXCEL worksheet. The expressions = BINOMDIST(A1, 64, 0.5, 0) and = BINOMDIST(A1, 64, 0.5, 1) are used to obtain the binomial and cumulative binomial distributions. The 0 for the fourth parameter requests individual probabilities and the 1 requests cumulative probabilities. A click-and-drag in column B gives the individual probabilities, and in column C a click-and-drag gives the cumulative probabilities.

A	B	C	A	B	C
X	Probability	Cumulative	x	Probability	Cumulative
0	5.42101E-20	5.42101E-20	13	7.12151E-07	9.40481E-07
1	3.46945E-18	3.52366E-18	14	2.59426E-06	3.53474E-06
2	1.09288E-16	1.12811E-16	15	8.64754E-06	1.21823E-05
3	2.25861E-15	2.37142E-15	16	2.64831E-05	3.86654E-05
4	3.44438E-14	3.68152E-14	17	7.47758E-05	0.000113441
5	4.13326E-13	4.50141E-13	18	0.000195248	0.000308689
6	4.06437E-12	4.51451E-12	19	0.000472706	0.000781395
7	3.36762E-11	3.81907E-11	20	0.001063587	0.001844982
8	2.39943E-10	2.78134E-10	21	0.002228469	0.004073451
9	1.49298E-09	1.77111E-09	22	0.004355644	0.008429095
10	8.21138E-09	9.98249E-09	23	0.007953785	0.01638288
11	4.03104E-08	5.02929E-08	24	0.013587715	0.029970595
12	1.78038E-07	2.28331E-07	25	0.021740344	0.051710939

It is found, as in Problem 10.3, that $P(X \le 23) = 0.01638$ and because of symmetry, $P(X \ge 41) = 0.01638$ and that the rejection region is $\{X \le 23 \text{ or } X \ge 41\}$ and the significance level is $0.01638 + 0.01638$ or 0.03276.

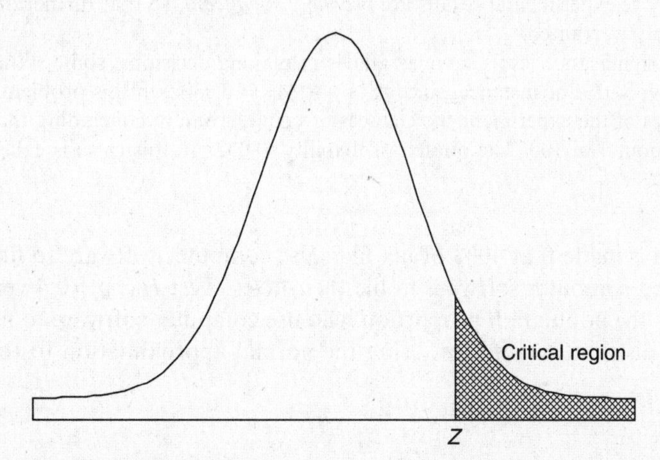

Fig. 10-4 Determining the Z value that will give a critical region equal to 0.05.

10.5 In an experiment on extrasensory perception (ESP), an individual (subject) in one room is asked to state the color (red or blue) of a card chosen from a deck of 50 well-shuffled cards by an individual in another room. It is unknown to the subject how many red or blue cards are in the deck. If the subject identifies 32 cards correctly, determine whether the results are significant at the (*a*) 0.05 and (*b*) 0.01 levels.

SOLUTION

If p is the probability of the subject choosing the color of a card correctly, then we have to decide between two hypotheses:

$H_0 : p = 0.5$, and the subject is simply guessing (i.e., the results are due to chance).

$H_1 : p > 0.5$, and the subject has powers of ESP.

Since we are not interested in the subject's ability to obtain extremely low scores, but only in the ability to obtain high scores, we choose a one-tailed test. If hypothesis H_0 is true, then the mean and standard deviation of the number of cards identified correctly are given by

$$\mu = Np = 50(0.5) = 25 \quad \text{and} \quad \sigma = \sqrt{Npq} = \sqrt{50(0.5)(0.5)} = \sqrt{12.5} = 3.54$$

(*a*) For a one-tailed test at the 0.05 significance level, we must choose z in Fig. 10-4 so that the shaded area in the critical region of high scores is 0.05. The area between 0 and z is 0.4500, and $z = 1.645$; this can also be read from Table 10.1. Thus our decision rule (or test of significance) is:

If the z score observed is greater than 1.645, the results are significant at the 0.05 level and the individual has powers of ESP.

If the z score is less than 1.645, the results are due to chance (i.e., not significant at the 0.05 level).

Since 32 in standard units is $(32 - 25)/3.54 = 1.98$, which is greater than 1.645, we conclude at the 0.05 level that the individual has powers of ESP.

Note that we should really apply a continuity correction, since 32 on a continuous scale is between 31.5 and 32.5. However, 31.5 has a standard score of $(31.5 - 25)/3.54 = 1.84$, and so the same conclusion is reached.

(*b*) If the significance level is 0.01, then the area between 0 and z is 0.4900, from which we conclude that $z = 2.33$.

Since 32 (or 31.5) in standard units is 1.98 (or 1.84), which is less than 2.33, we conclude that the results are *not significant* at the 0.01 level.

Some statisticians adopt the terminology that results significant at the 0.01 level are *highly significant*, that results significant at the 0.05 level but not at the 0.01 level are *probably significant*, and that results significant at levels larger than 0.05 are *not significant*. According to this terminology, we would conclude that the above experimental results are *probably significant*, so that further investigations of the phenomena are probably warranted.

Since significance levels serve as guides in making decisions, some statisticians quote the actual probabilities involved. For instance, since $pr\{z \geq 1.84\} = 0.0322$, in this problem, the statistician could say that on the basis of the experiment the chances of being wrong in concluding that the individual has powers of ESP are about 3 in 100. The quoted probability (0.0322 in this case) is called the p-value for the test.

10.6 The claim is made that 40% of tax filers use computer software to file their taxes. In a sample of 50, 14 used computer software to file their taxes. Test $H_0: p = 0.4$ versus $H_a: p < 0.4$ at $\alpha = 0.05$ where p is the population proportion who use computer software to file their taxes. Test using the binomial distribution and test using the normal approximation to the binomial distribution.

SOLUTION

If the exact test of $H_0: p = 0.4$ versus $H_a: p < 0.4$ at $\alpha = 0.05$ is used, the null is rejected if $X \leq 15$. This is called the rejection region. If the test based on the normal approximation to the binomial is used, the null is rejected if $Z < -1.645$ and this is called the rejection region. $X = 14$ is called the test statistic. The binomial test statistic is in the rejection region and the null is rejected. Using the normal approximation, the test statistic is $z = \frac{14-20}{3.46} = -1.73$. The actual value of α is 0.054 and the rejection region is $X \leq 15$ and the cumulative binomial probability $P(X \leq 15)$ is used. If the normal approximation is used, you would also reject since $z = -1.73$ is in the rejection region which is $Z < -1.645$. Note that if the binomial distribution is used to perform the test, the test statistic has a binomial distribution. If the normal distribution is used to test the hypothesis, the test statistic, Z, has a standard normal distribution.

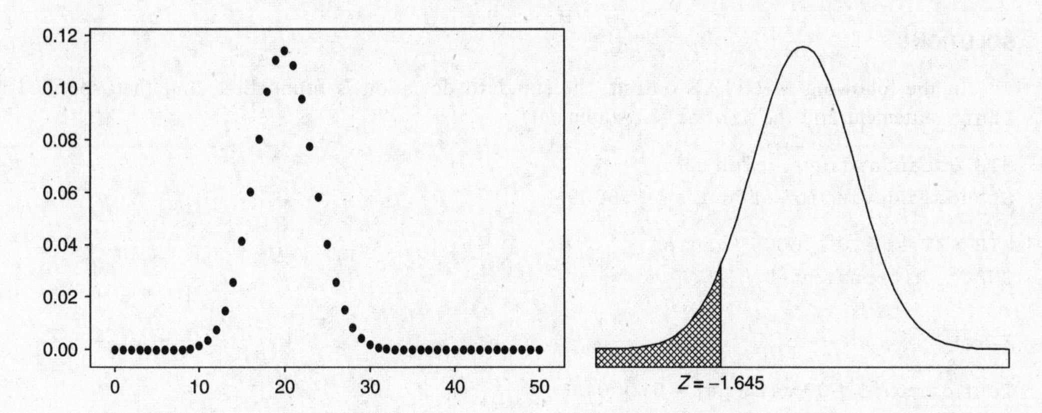

Fig. 10-5 Comparison of the exact test on the left (Binomial) and the approximate test on the right (standard normal).

10.7 The *p*-value for a test of hypothesis is defined to be the smallest level of significance at which the null hypothesis is rejected. This problem illustrates the computation of the *p*-value for a statistical test. Use the data in Problem 9.6 to test the null hypothesis that the mean height of all the trees on the farm equals 5 feet (ft) versus the alternative hypothesis that the mean height is less than 5 ft. Find the *p*-value for this test.

SOLUTION

The computed value for z is $z = (59.22 - 60)/1.01 = -0.77$. The smallest level of significance at which the null hypothesis would be rejected is *p*-value $= P(z < -0.77) = 0.5 - 0.2794 = 0.2206$. The null hypothesis is rejected if the *p*-value is less than the pre-set level of significance. In this problem, if the level of significance is pre-set at 0.05, then the null hypothesis is not rejected. The MINITAB solution is as follows where the subcommand `Alternative-1` indicates a lower-tail test.

```
MTB > ZTest mean = 60 sd = 10.111 data in cl ;
SUBC> Alternative -1.
```

Z-Test

```
Test of mu = 60.00 vs mu < 60.00
The assumed sigma = 10.1

Variable     N     Mean    StDev   SE Mean      Z       P
height     100    59.22    10.11      1.01   -0.77    0.22
```

10.8 A random sample of 33 individuals who listen to talk radio was selected and the hours per week that each listens to talk radio was determined. The data are as follows.

9 8 7 4 8 6 8 8 7 10 8 10 6 7 7 8 9

6 5 8 5 6 8 7 8 5 5 8 7 6 6 4 5

Test the null hypothesis that $\mu = 5$ hours (h) versus the alternative hypothesis that $\mu \neq 5$ at level of significance $\alpha = 0.05$ in the following three equivalent ways:

(*a*) Compute the value of the test statistic and compare it with the critical value for $\alpha = 0.05$.

(*b*) Compute the *p*-value corresponding to the computed test statistic and compare the *p*-value with $\alpha = 0.05$.

(*c*) Compute the $1 - \alpha = 0.95$ confidence interval for μ and determine whether 5 falls in this interval.

SOLUTION

In the following MINITAB output, the standard deviation is found first, and then specified in the `ztest` statement and the `Zinterval` statement.

```
MTB > standard deviation c1
Standard deviation of hours = 1.6005

MTB > ZTest 5.0 1.6005 'hours';
SUBC>  Alternative 0.
```

Z-Test

```
Test of mu = 5.000 vs mu not = 5.000
The assumed sigma = 1.60
Variable      N     Mean    StDev    SE Mean      Z        P
hours        33    6.897    1.600     0.279     6.81   0.0000

MTB > ZInterval 95.0 1.6005 'hours'.
Variable      N    Mean    StDev    SE Mean        95.0 % CI
hours        33   6.897    1.600     0.279     ( 6.351, 7.443)
```

(*a*) The computed value of the test statistic is $Z = \dfrac{6.897 - 5}{0.279} = 6.81$, the critical values are ± 1.96, and the null hypothesis is rejected. Note that this is the computed value shown in the MINITAB output.

(*b*) The computed *p*-value from the MINITAB output is 0.0000 and since the *p*-value $< \alpha = 0.05$, the null hypothesis is rejected.

(*c*) Since the value specified by the null hypothesis, 5, is not contained in the 95% confidence interval for μ, the null hypothesis is rejected.

These three procedures for testing a null hypothesis against a two-tailed alternative are equivalent.

10.9 The breaking strengths of cables produced by a manufacturer have a mean of 1800 pounds (lb) and a standard deviation of 100 lb. By a new technique in the manufacturing process, it is claimed that the breaking strength can be increased. To test this claim, a sample of 50 cables is tested and it is found that the mean breaking strength is 1850 lb. Can we support the claim at the 0.01 significance level?

SOLUTION

We have to decide between the two hypotheses:

$H_0 : \mu = 1800$ lb, and there is really no change in breaking strength.

$H_1 : \mu > 1800$ lb, and there is a change in breaking strength.

A one-tailed test should be used here; the diagram associated with this test is identical with Fig. 10-4 of Problem 10.5(*a*). At the 0.01 significance level, the decision rule is:

If the *z* score observed is greater than 2.33, the results are significant at the 0.01 level and H_0 is rejected.

Otherwise, H_0 is accepted (or the decision is withheld).

Under the hypothesis that H_0 is true, we find that

$$z = \frac{\bar{X} - \mu}{\sigma/\sqrt{N}} = \frac{1850 - 1800}{100/\sqrt{50}} = 3.55$$

which is greater than 2.33. Hence we conclude that the results are *highly significant* and that the claim should thus be supported.

p-VALUES FOR HYPOTHESES TESTS

10.10 A group of 50 Internet shoppers were asked how much they spent per year on the Internet. Their responses are shown in Table 10.2. It is desired to test that they spend $325 per year versus it is different from $325. Find the *p*-value for the test of hypothesis. What is your conclusion for $\alpha = 0.05$?

Table 10.2

418	379	77	212	378
363	434	348	245	341
331	356	423	330	247
351	151	220	383	257
307	297	448	391	210
158	310	331	348	124
523	356	210	364	406
331	364	352	299	221
466	150	282	221	432
366	195	96	219	202

SOLUTION

The mean of the data is 304.60, the standard deviation is 101.51, the computed test statistic is $z = \dfrac{304.60 - 325}{101.50/\sqrt{50}} = -1.43$. The Z statistic has an approximate standard normal distribution. The computed *p*-value is the following $P(Z < -|\text{computed test statistic}|)$ or $Z > |\text{computed test statistic}|)$ or $P(Z < -1.43) + P(Z > 1.43)$. The answer may be found using Appendix II, or using EXCEL. Using EXCEL, the *p*-value =2* NORMSDIST(−1.43) = 0.1527, since the normal curve is symmetrical and the areas to the left of −1.43 and to the right of 1.43 are the same, we may simply double the area to the left of −1.43. Because the *p*-value is not less than 0.05, we do not reject the null hypothesis. Refer to Fig. 10-6 to see the graphically the computed *p*-value for this problem.

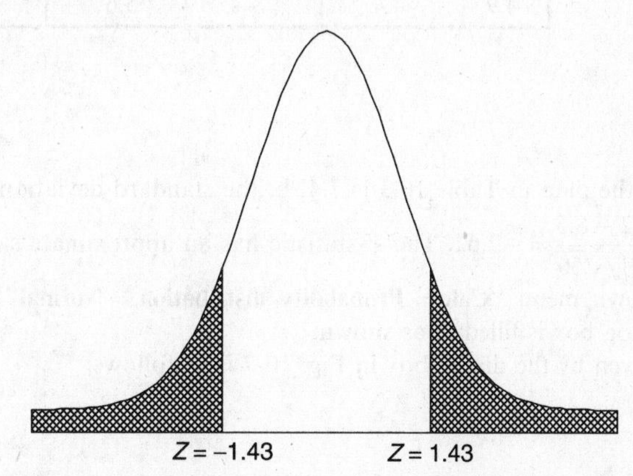

Fig. 10-6 The *p*-value is the sum of the area to the left of $Z = -1.43$ and the area to the right of $Z = 1.43$.

10.11 Refer to Problem 10.10. Use the statistical software MINITAB to analyze the data. Note that the software gives the *p*-value and you, the user, are left to make a decision about the hypothesis based on whatever value you may have assigned to α.

SOLUTION

The pull-down "**Stat ⇒ Basic statistics ⇒ 1 sample Z**" gives the following analysis. The test statistic and the *p*-value are computed for you.

One-Sample Z: Amount

```
Test of mu = 325 vs not = 325
The assumed standard deviation = 101.51

Variable    N      Mean     StDev    SE Mean        Z        P
Amount     50   304.460   101.508     14.356    -1.43    0.152
```

Note that the software gives the value of the test statistic (−1.43) and the *p*-value (0.152).

10.12 A survey of individuals who use computer software to file their tax returns is shown in Table 10.3. The recorded response is the time spent in filing their tax returns. The null hypothesis is $H_0 : \mu = 8.5$ hours versus the alternative hypothesis $H_1 : \mu < 8.5$. Find the *p*-value for the test of hypothesis. What is your conclusion for $\alpha = 0.05$?

Table 10.3

6.2	4.8	8.9	5.6	6.5
11.5	8.6	6.2	8.5	5.2
2.7	14.9	11.2	6.9	7.9
4.8	9.5	12.4	9.7	10.7
8.0	11.8	7.4	9.1	4.9
9.1	6.4	9.5	7.6	6.7
2.6	3.5	6.4	4.3	7.9
3.3	10.3	3.2	11.5	1.7
10.4	8.5	10.8	6.9	5.3
4.9	4.4	9.4	5.6	7.0

SOLUTION

The mean of the data in Table 10.3 is 7.42 h, the standard deviation is 2.91 h and the computed test statistic is $Z = \dfrac{7.42 - 8.5}{2.91/\sqrt{50}} = -2.62$. The Z statistic has an approximate standard normal distribution. The MINITAB pull-down menu "**Calc ⇒ Probability distribution ⇒ Normal**" gives the dialog box shown in Fig. 10-7. The dialog box is filled in as shown.

The output given by the dialog box in Fig. 10-7 is as follows.

Cumulative Distribution Function

```
Normal with mean = 0 and standard deviation = 1

x       P( X <= x )
-2.62   0.0043965
```

The *p*-value is 0.0044 and since the *p*-value < α, the null hypothesis is rejected. Refer to Fig. 10-8 to see graphically the computed value of the *p*-value for this problem.

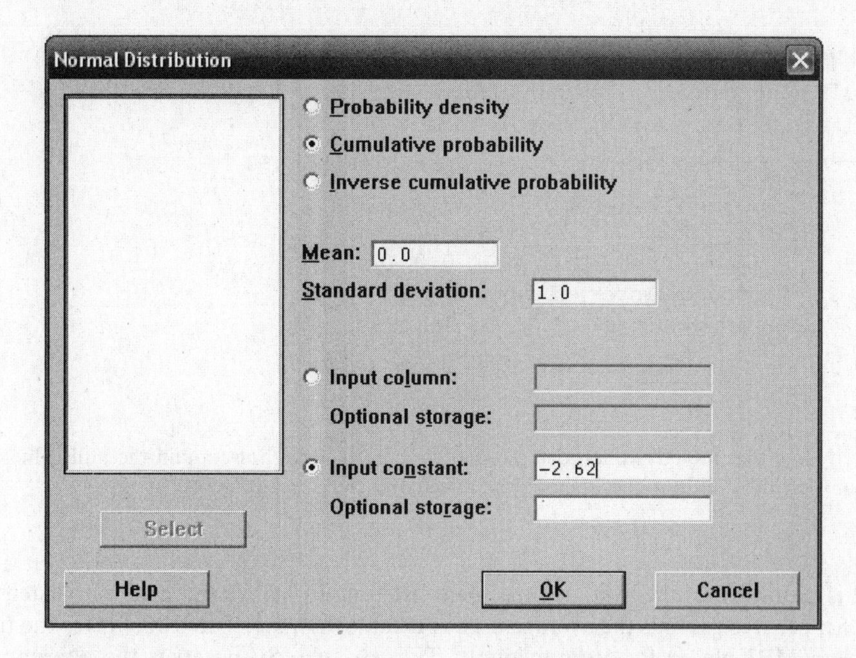

Fig. 10-7 Dialog box for figuring the p-value when the test statistic equals −2.62.

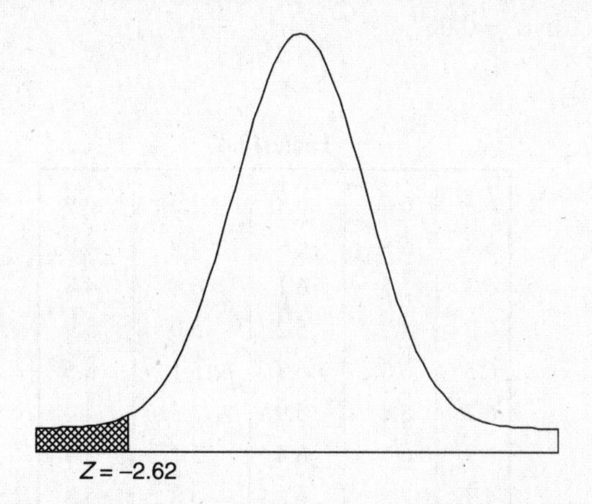

Fig. 10-8 The p-value is the area to the left of $Z = -2.62$.

10.13 Refer to Problem 10.12. Use the statistical software SAS to analyze the data. Note that the software gives the p-value and you, the user, are left to make a decision about the hypothesis based on whatever value you may have assigned to α.

SOLUTION

The SAS output is shown below. The p-value is shown as Prob $> z = 0.0044$, the same value as was obtained in Problem 10.12. It is the area under the standard normal curve to the left of −2.62. Compare the other quantities in the SAS output with those in Problem 10.12.

SAS OUTPUT:

```
One Sample Z Test for a Mean
Sample Statistics for time
        N           Mean           Std. Dev.   Std. Error
-------------------------------------------------------------
        50          7.42             2.91        0.41
```

```
Hypothesis Test
Null hypothesis:        Mean of time => 8.5
Alternative:            Mean of time <  8.5
with a specified known standard deviation of 2.91
           Z Statistic       Prob > Z
           --------------------
             -2.619          0.0044
  95% Confidence Interval for the Mean
  (Upper Bound Only)
           Lower Limit       Upper Limit
           --------------------
             -infinity       8.10
```

Note that the 95% one-sided interval $(-\infty, 8.10)$ does not contain the null value, 8.5. This is another sign that the null hypothesis should be rejected at the $\alpha = 0.05$ level.

10.14 It is claimed that the average time spent listening to MP3's by those who listen to these devices is 5.5 h per week versus the average time is greater than 5.5. Table 10.4 gives the time spent listening to an MP3 player for 50 individuals. Test $H_0 : \mu = 5.5$ h versus the alternative $H_1 : \mu > 5.5$ h. Find the *p*-value for the test of hypothesis using the computer software STATISTIX. What is your conclusion for $\alpha = 0.05$?

Table 10.4

6.4	6.4	6.8	7.6	6.9
5.8	5.9	6.9	5.9	6.0
6.3	5.5	6.1	6.4	4.8
6.3	4.2	6.2	5.0	5.9
6.5	6.8	6.8	5.1	6.5
6.7	5.4	5.9	3.5	4.4
6.9	6.7	6.4	5.1	5.4
4.7	7.0	6.0	5.8	5.8
5.7	5.2	4.9	6.6	8.2
6.9	5.5	5.2	3.3	8.3

SOLUTION

The STATISTIX package gives

```
Statistix 8.0
Descriptive Statistics
Variable     N         Mean          SD
MP3          50        5.9700        1.0158
```

The computed test statistic is $Z = \dfrac{5.97 - 5.5}{1.0158/\sqrt{50}} = 3.27$. The *p*-value is computed in Fig. 10-9.

Figure 10-10 shows graphically the computed value of the *p*-value for this problem.

The computed *p*-value is 0.00054 and since this is less than 0.05 the null hypothesis is rejected.

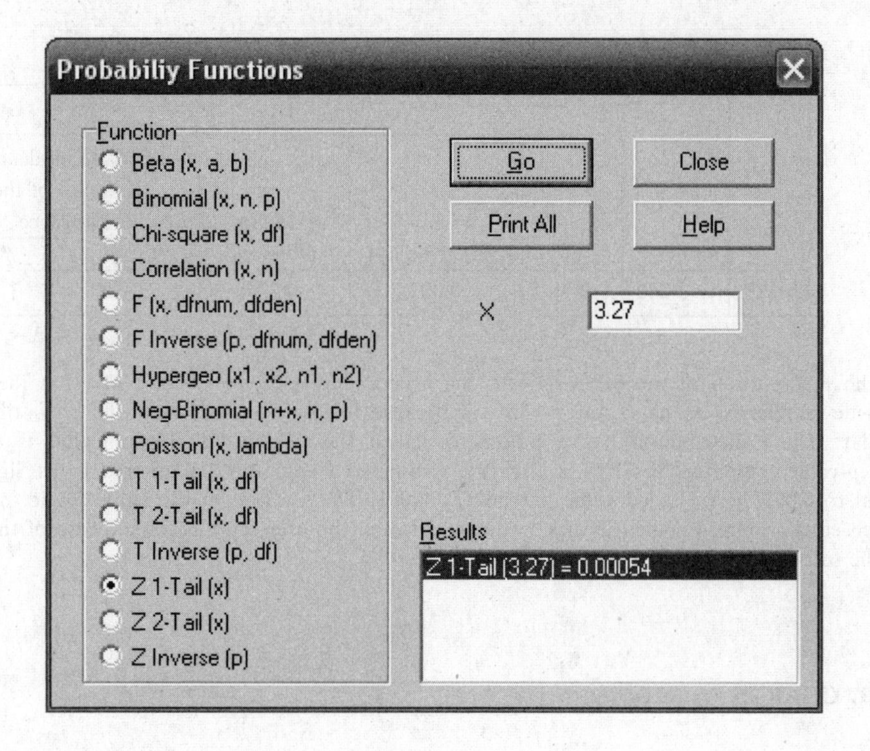

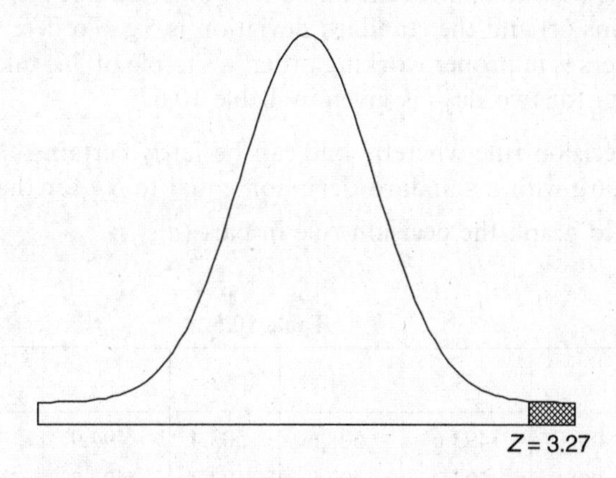

Fig. 10-10 The p-value is the area to the right of $Z = 3.27$

10.15 Using SPSS and the pull-down "**Analyze ⇒ Compare means ⇒ one-sample t test**" and the data in Problem 10.14, test $H_0 : \mu = 5.5$ h versus the alternative $H_0 : \mu > 5.5$ h at $\alpha = 0.05$ by finding the p-value and comparing it with α.

SOLUTION

The SPSS output is as follows:

One-Sample Statistics

	N	Mean	Std. Deviation	Std. Error Mean
MPE	50	5.9700	1.0184	.14366

One-Sample Test

	Test Value=5.5					
					95% Confidence Interval of the Diffrence	
	t	df	Sig. (2-tailed)	Mean Difference	Lower	Upper
MPE	3.272	49	0.002	.47000	.1813	.7587

In the first portion of the SPSS output, the needed statistics are given. Note that the computed test statistic is referred to as t, not z. This is because for $n > 30$, the t and the z distribution are very similar. The t distribution has a parameter called the degrees of freedom that is equal to $n - 1$. The p-value computed by SPSS is always a two-tailed p-value and is referred to as Sig.(2-tailed). It is equal to 0.002. The 1-tailed value is $0.002/2 = 0.001$. This is close to the value found in Problem 10.14 which equals 0.00054. When using computer software, the user must become aware of the idiosyncrasies of the software.

CONTROL CHARTS

10.16 A control chart is used to control the amount of mustard put into containers. The mean amount of fill is 496 grams (g) and the standard deviation is 5 g. To determine if the machine filling the mustard containers is in proper working order, a sample of 5 is taken every hour for all eight h in the day. The data for two days is given in Table 10.5.

(*a*) Design a decision rule whereby one can be fairly certain that the mean fill is being maintained at 496 g with a standard deviation equal to 5 g for the two days.

(*b*) Show how to graph the decision rule in part (*a*).

Table 10.5

1	2	3	4	5	6	7	8
492.2	486.2	493.6	508.6	503.4	494.9	497.5	490.5
487.9	489.5	503.2	497.8	493.4	492.3	497.0	503.0
493.8	495.9	486.0	493.4	493.9	502.9	493.8	496.4
495.4	494.1	498.4	495.8	493.8	502.8	497.1	489.7
491.7	494.0	496.5	508.0	501.3	498.9	488.3	492.6

9	10	11	12	13	14	15	16
492.2	486.2	493.6	508.6	503.4	494.9	497.5	490.5
487.9	489.5	503.2	497.8	493.4	492.3	497.0	503.0
493.8	495.9	486.0	493.4	493.9	502.9	493.8	496.4
495.4	494.1	498.4	495.8	493.8	502.8	497.1	489.7
491.7	494.0	496.5	508.0	501.3	498.9	488.3	492.6

SOLUTION

(*a*) With 99.73% confidence we can say that the sample mean $\bar{x}$ must lie in the range $\mu_{\bar{x}} - 3\sigma_{\bar{x}}$ to $\mu_{\bar{x}} + 3\sigma_{\bar{x}}$ or $\mu - 3\frac{\sigma}{\sqrt{n}}$ to $\mu + 3\frac{\sigma}{\sqrt{n}}$. Since $\mu = 496$, $\sigma = 5$, and $n = 5$, it follows that with 99.73% confidence, the sample mean should fall in the interval $496 - 3\frac{5}{\sqrt{5}}$ to $496 + 3\frac{5}{\sqrt{5}}$ or between 489.29 and 502.71. Hence our decision rule is as follows:

If a sample mean falls inside the range 489.29 g to 502.71 g, assume the machine fill is correct. Otherwise, conclude that the filling machine is not in proper working order and seek to determine the reasons for the incorrect fills.

(*b*) A record of the sample means can be kept by the use of a chart such as shown in Fig. 10-11, called a *quality control chart*. Each time a sample mean is computed, it is represented by a particular point. As long as the points lie between the lower limit and the upper limit, the process is under control. When a point goes outside these control limits, there is a possibility that something is wrong and investigation is warranted.

The 80 observations are entered into column C1. The pull down menu **"Stat ⇒ Control Charts ⇒ Variable charts for subgroups ⇒ Xbar"** gives a dialog box which when filled in gives the control chart shown in Fig. 10-11.

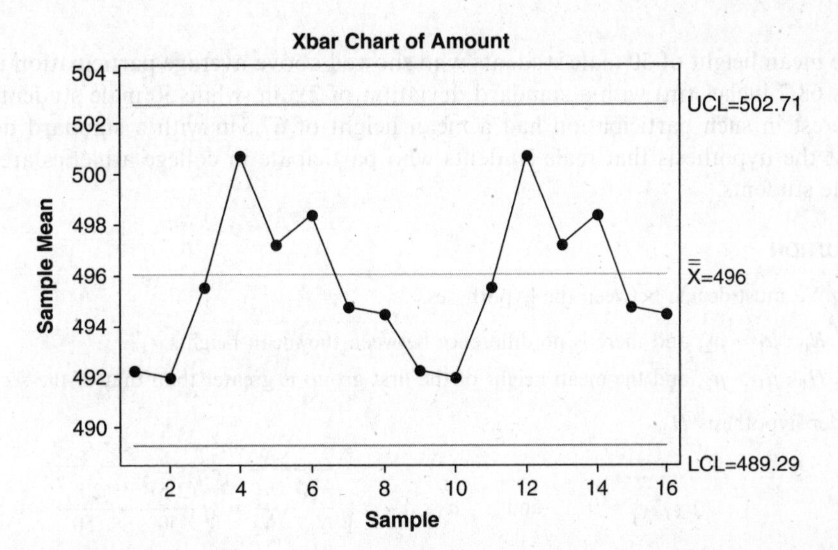

Fig. 10-11 Control chart with 3σ limits for controlling the mean fill of mustard containers.

The control limits specified above are called the 99.73% confidence limits, or briefly, the 3σ limits. Other confidence limits (such as 99% or 95% limits) could be determined as well. The choice in each case depends on the particular circumstances.

TESTS INVOLVING DIFFERENCES OF MEANS AND PROPORTIONS

10.17 An examination was given to two classes consisting of 40 and 50 students, respectively. In the first class the mean grade was 74 with a standard deviation of 8, while in the second class the mean grade was 78 with a standard deviation of 7. Is there a significant difference between the performance of the two classes at the (*a*) 0.05 and (*b*) 0.01 levels?

SOLUTION

Suppose that the two classes come from two populations having the respective means μ_1 and μ_2. We thus need to decide between the hypotheses:

$H_0 : \mu_1 = \mu_2$, and the difference is due merely to chance.

$H_1 : \mu_1 \neq \mu_2$, and there is a significant difference between the classes.

Under hypothesis H_0, both classes come from the same population. The mean and standard deviation of the difference in means are given by

$$\mu_{\bar{X}1-\bar{X}2} = 0 \qquad \text{and} \qquad \sigma_{\bar{X}1-\bar{X}2} = \sqrt{\frac{\sigma_1^2}{N_1} + \frac{\sigma_2^2}{N_2}} = \sqrt{\frac{8^2}{40} + \frac{7^2}{50}} = 1.606$$

where we have used the sample standard deviations as estimates of σ_1 and σ_2. Thus

$$z = \frac{\bar{X}_1 - \bar{X}}{\sigma_{\bar{X}1-\bar{X}2}} = \frac{74 - 78}{1.606} = -2.49$$

(*a*) For a two-tailed test, the results are significant at the 0.05 level if z lies outside the range -1.96 to 1.96. Hence we conclude that at the 0.05 level there is a significant difference in performance between the two classes and that the second class is probably better.

(*b*) For a two-tailed test, the results are significant at the 0.01 level if z lies outside the range -2.58 and 2.58. Hence we conclude that at the 0.01 level there is no significant difference between the classes.

Since the results are significant at the 0.05 level but not at the 0.01 level, we conclude that the results are *probably significant* (according to the terminology used at the end of Problem 10.5).

10.18 The mean height of 50 male students who showed above-average participation in college athletics was 68.2 inches (in) with a standard deviation of 2.5 in, while 50 male students who showed no interest in such participation had a mean height of 67.5 in with a standard deviation of 2.8 in. Test the hypothesis that male students who participate in college athletics are taller than other male students.

SOLUTION

We must decide between the hypotheses:

$H_0 : \mu_1 = \mu_2$, and there is no difference between the mean heights.

$H_1 : \mu_1 > \mu_2$, and the mean height of the first group is greater than that of the second group.

Under hypothesis H_0,

$$\mu_{\bar{X}1-\bar{X}2} = 0 \qquad \text{and} \qquad \sigma_{\bar{X}1-\bar{X}2} = \sqrt{\frac{\sigma_1^2}{N_1} + \frac{\sigma_2^2}{N_2}} = \sqrt{\frac{(2.5)^2}{50} + \frac{(2.8)^2}{50}} = 0.53$$

where we have used the sample standard deviations as estimates of σ_1 and σ_2. Thus

$$z = \frac{\bar{X}_1 - \bar{X}_2}{\sigma_{\bar{X}1-X2}} = \frac{68.2 - 67.5}{0.53} = 1.32$$

Using a one-tailed test at the 0.05 significance level, we would reject hypothesis H_0 if the z score were greater than 1.645. Thus we cannot reject the hypothesis at this level of significance.

It should be noted, however, that the hypothesis can be rejected at the 0.10 level if we are willing to take the risk of being wrong with a probability of 0.10 (i.e., 1 chance in 10).

10.19 A study was undertaken to compare the mean time spent on cell phones by male and female college students per week. Fifty male and 50 female students were selected from Midwestern University and the number of hours per week spent talking on their cell phones determined. The results in hours are shown in Table 10.6. It is desired to test $H_0 : \mu_1 - \mu_2 = 0$ versus $H_a : \mu_1 - \mu_2 \neq 0$ based on these samples. Use EXCEL to find the p-value and reach a decision about the null hypothesis.

Table 10.6 Hours spent talking on cell phone for males and females at Midwestern University

Males					Females				
12	4	11	13	11	11	9	7	10	9
7	9	10	10	7	10	10	7	9	10
7	12	6	9	15	11	8	9	6	11
10	11	12	7	8	10	7	9	12	14
8	9	11	10	9	11	12	12	8	12
10	9	9	7	9	12	9	10	11	7
11	7	10	10	11	12	7	9	8	11
9	12	12	8	13	10	8	13	8	10
9	10	8	11	10	9	9	9	11	9
13	13	9	10	13	9	8	9	12	11

SOLUTION

The data in Table 10.6 are entered into an EXCEL worksheet as shown in Fig. 10-12. The male data are entered into A2 : E11 and the female data are entered into F2:J11. The variance of the male data is computed by entering =VAR(A2 : E11) into A14. The variance of female data is computed by entering =VAR(F2 : J11) into A15. The mean of male data is computed by entering =AVERAGE(A2 : E11)

Microsoft Excel - Book1

File Edit View Insert Format Tools Data Window Help

Arial

M8 fx

	A	B	C	D	E	F	G	H	I	J	K
1			males					Females			
2	12	4	11	13	11	11	9	7	10	9	
3	7	9	10	10	7	10	10	7	9	10	
4	7	12	6	9	15	11	8	9	6	11	
5	10	11	12	7	8	10	7	9	12	14	
6	8	9	11	10	9	11	12	12	8	12	
7	10	9	9	7	9	12	9	10	11	7	
8	11	7	10	10	11	12	7	9	8	11	
9	9	12	12	8	13	10	8	13	8	10	
10	9	10	8	11	10	9	9	9	11	9	
11	13	13	9	10	13	9	8	9	12	11	
12											
13											
14	4.640408	VAR(A2:E11)									
15	3.153061	VAR(F2:J11)									
16	9.82	AVERAGE(A2:E11)									
17	9.7	AVERAGE(F2:J11)									
18											
19	0.303949	(A16-A17)/SQRT(A14/50+A15/50)									
20											
21	0.761167	2*(1-NORMSDIST(A19))									

Fig. 10-12 EXCEL worksheet for computing the p-value in problem 10-19.

into A16. The mean of female data is computed by entering =AVERAGE(F2:J11) into A17. The test statistic is =(A16-A17)/SQRT(A14/50+A15/50) and is shown in A19. The test statistic has a standard normal distribution and a value of 0.304. The expression =2*(1-NORMSDIST(A19)) computes the area to the right of 0.304 and doubles it. This gives a p-value of 0.761.

Since the p-value is not smaller than any of the usual α values such as 0.01 or 0.05, the null hypothesis is not rejected. The probability of obtaining samples like the one we obtained is 0.761, assuming the null hypothesis to be true. Therefore, there is no evidence to suggest that the null hypothesis is false and that it should be rejected.

10.20 Two groups, A and B, consist of 100 people each who have a disease. A serum is given to group A but not to group B (which is called the *control*); otherwise, the two groups are treated identically. It is found that in groups A and B, 75 and 65 people, respectively, recover from the disease. At significance levels of (*a*) 0.01, (*b*) 0.05, and (*c*) 0.10, test the hypothesis that the serum helps cure the disease. Compute the p-value and show that p-value>0.01, p-value>0.05, but p-value<0.10.

SOLUTION

Let p_1 and p_2 denote the population proportions cured by (1) using the serum and (2) not using the serum, respectively. We must decide between two hypotheses:

$H_0 : p_1 = p_2$, and the observed differences are due to chance (i.e. the serum is ineffective).

$H_1 : p_1 > p_2$, and the serum is effective.

Under hypothesis H_0,

$$\mu_{P1-P2} = 0 \quad \text{and} \quad \sigma_{P1-P2} = \sqrt{pq\left(\frac{1}{N_1}+\frac{1}{N_2}\right)} = \sqrt{(0.70)(0.30)\left(\frac{1}{100}+\frac{1}{100}\right)} = 0.0648$$

where we have used as an estimate of p the average proportion of cures in the two sample groups given by $(75+65)/200 = 0.70$, and where $q = 1 - p = 0.30$. Thus

$$z = \frac{P_1 - P_2}{\sigma_{P1-P2}} = \frac{0.750 - 0.650}{0.0648} = 1.54$$

(*a*) Using a one-tailed test at the 0.01 significance level, we would reject hypothesis H_0 only if the z score were greater than 2.33. Since the z score is only 1.54, we must conclude that the results are due to chance at this level of significance.

(*b*) Using a one-tailed test at the 0.05 significance level, we would reject H_0 only if the z score were greater than 1.645. Hence we must conclude that the results are due to chance at this level also.

(*c*) If a one-tailed test at the 0.10 significance level were used, we would reject H_0 only if the z score were greater than 1.28. Since this condition is satisfied, we conclude that the serum is effective at the 0.10 level.

(*d*) Using EXCEL, the p-value is given by=1 - NORMSDIST(1.54) which equals 0.06178. This is the area to the right of 1.54. Note that the p-value is greater than 0.01, 0.05, but is less than 0.10.

Note that these conclusions depend how much we are willing to risk being wrong. If the results are actually due to chance, but we conclude that they are due to the serum (Type I error), we might proceed to give the serum to large groups of people—only to find that it is actually ineffective. This is a risk that we are not always willing to assume.

On the other hand, we could conclude that the serum does not help, whereas it actually does help (Type II error). Such a conclusion is very dangerous, especially if human lives are at stake.

10.21 Work Problem 10.20 if each group consists of 300 people and if 225 people in group A and 195 people in group B are cured. Find the p-value using EXCEL and comment on your decision.

SOLUTION

Note that in this case the proportions of people cured in the two groups are $225/300 = 0.750$ and $195/300 = 0.650$, respectively, which are the same as in Problem 10.20. Under hypothesis H_0,

$$\mu_{P1-P2} = 0 \quad \text{and} \quad \sigma_{P1-P2} = \sqrt{pq\left(\frac{1}{N_1} + \frac{1}{N_2}\right)} = \sqrt{(0.70)(0.30)\left(\frac{1}{300} + \frac{1}{300}\right)} = 0.0374$$

where $(225 + 195)/600 = 0.70$ is used as an estimate of p. Thus

$$z = \frac{P_1 - P_2}{\sigma_{P1-P2}} = \frac{0.750 - 0.650}{0.0374} = 2.67$$

Since this value of z is greater than 2.33, we can reject the hypothesis at the 0.01 significance level; that is, we can conclude that the serum is effective with only a 0.01 probability of being wrong.

This shows how increasing the sample size can increase the reliability of decisions. In many cases, however, it may be impractical to increase sample sizes. In such cases we are forced to make decisions on the basis of available information and must therefore contend with greater risks of incorrect decisions.

p-value = 1 - NORMSDIST(2.67) = 0.003793. This is less than 0.01.

10.22 A sample poll of 300 voters from district A and 200 voters from district B showed that 56% and 48%, respectively, were in favor of a given candidate. At a significance level of 0.05, test the hypotheses (*a*) that there is a difference between the districts and (*b*) that the candidate is preferred in district A (*c*) calculate the p-value for parts (*a*) and (*b*).

SOLUTION

Let p_1 and p_2 denote the proportions of all voters from districts A and B, respectively, who are in favor of the candidate. Under the hypothesis $H_0 : p_1 = p_2$, we have

$$\mu_{P1-P2} = 0 \quad \text{and} \quad \sigma_{P1-P2} = \sqrt{pq\left(\frac{1}{N_1} + \frac{1}{N_2}\right)} = \sqrt{(0.528)(0.472)\left(\frac{1}{300} + \frac{1}{200}\right)} = 0.0456$$

where we have used as estimates of p and q the values $[(0.56)(300) + (0.48)(200)]/500 = 0.528$ and $(1 - 0.528) = 0.472$, respectively. Thus

$$z = \frac{P_1 - P_2}{\sigma_{P1-P2}} = \frac{0.560 - 0.480}{0.0456} = 1.75$$

(*a*) If we wish only to determine whether there is a difference between the districts, we must decide between the hypotheses $H_0 : p_1 = p_2$ and $H_1 : p_1 \neq p_2$, which involves a two-tailed test. Using a two-tailed test at the 0.05 significance level, we would reject H_0 if z were outside the interval -1.96 to 1.96. Since $z = 1.75$ lies inside this interval, we cannot reject H_0 at this level; that is, there is no significant difference between the districts.

(*b*) If we wish to determine whether the candidate is preferred in district A, we must decide between the hypotheses $H_0 : p_1 = p_2$ and $H_1 : p_1 > p_2$, which involves a one-tailed test. Using a one-tailed test at the 0.05 significance level, we would reject H_0 if z were greater than 1.645. Since this is the case, we can reject H_0 at this level and conclude that the candidate is preferred in district A.

(*c*) For the two-tailed alternative, the p-value = 2*(1-NORMSDIST(1.75)) = 0.0801. You cannot reject the null hypothesis at $\alpha = 0.05$. For the one-tailed alternative, p-value = 1-NORMSDIST(1.75) = 0.04006. You can reject the null hypothesis at alpha = 0.05.

TESTS INVOLVING BINOMIAL DISTRIBUTIONS

10.23 An instructor gives a short quiz involving 10 true-false questions. To test the hypothesis that students are guessing, the instructor adopts the following decision rule:

If seven or more answers are correct, the student is not guessing.
If less than seven answers are correct, the student is guessing.

Find the following probability of rejecting the hypothesis when it is correct using (*a*) the binomial probability formula and (*b*) EXCEL.

SOLUTION

(*a*) Let p be the probability that a question is answered correctly. The probability of getting X problems out of 10 correct is $\binom{10}{X}p^X q^{10-X}$, where $q = 1 - p$. Then under the hypothesis $p = 0.5$ (i.e., the student is guessing),

$$\Pr\{7 \text{ or more correct}\} = \Pr\{7 \text{ correct}\} + \Pr\{8 \text{ correct}\} + \Pr\{9 \text{ correct}\} + \Pr\{10 \text{ correct}\}$$

$$= \binom{10}{7}\left(\frac{1}{2}\right)^7\left(\frac{1}{2}\right)^3 + \binom{10}{8}\left(\frac{1}{2}\right)^8\left(\frac{1}{2}\right)^2 + \binom{10}{9}\left(\frac{1}{2}\right)^9\left(\frac{1}{2}\right) + \binom{10}{10}\left(\frac{1}{2}\right)^{10} = 0.1719$$

Thus the probability of concluding that students are not guessing when in fact they are guessing is 0.1719. Note that this is the probability of a Type I error.

(*b*) Enter the numbers 7, 8, 9, and 10 into A1:A4 of the EXCEL worksheet. Next enter = BINOMDIST(A1, 10, 0.5, 0). Next perform a click-and-drag from B1 to B4. In B5 enter = SUM(B1 : B4). The answer appears in B5.

A	B
7	0.117188
8	0.043945
9	0.009766
10	0.000977
	0.171875

10.24 In Problem 10.23, find the probability of accepting the hypothesis $p = 0.5$ when actually $p = 0.7$. Find the answer (*a*) using the binomial probability formula and (*b*) using EXCEL.

SOLUTION

(*a*) Under the hypothesis $p = 0.7$,

$$\Pr\{\text{less than 7 correct}\} = 1 - \Pr\{7 \text{ or more correct}\}$$

$$= 1 - \left[\binom{10}{7}(0.7)^7(0.3)^3 + \binom{10}{8}(0.7)^8(0.3)^2 + \binom{10}{9}(0.7)^9(0.3) + \binom{10}{10}(0.3)^{10}\right]$$

$$= 0.3504$$

(*b*) The EXCEL solution is:

Pr{less than 7 correct when $p = 0.7$} is given by = BINOMDIST(6,10,0.7,1) which equals 0.350389. The 1 in the function BINOMDIST says to accumulate from 0 to 6 the binomial probabilities with $n = 10$ and $p = 0.7$.

10.25 In Problem 10.23, find the probability of accepting the hypothesis $p = 0.5$ when actually (*a*) $p = 0.6$, (*b*) $p = 0.8$, (*c*) $p = 0.9$, (*d*) $p = 0.4$, (*e*) $p = 0.3$, (*f*) $p = 0.2$, and (*g*) $p = 0.1$.

SOLUTION

(*a*) If $p = 0.6$,

$$\text{Required probability} = 1 - [\Pr\{7 \text{ correct}\} + \Pr\{8 \text{ correct}\} + \Pr\{9 \text{ correct}\} + \Pr\{10 \text{ correct}\}]$$

$$= 1 - \left[\binom{10}{7}(0.6)^7(0.4)^3 + \binom{10}{8}(0.6)^8(0.4)^2 + \binom{10}{9}(0.6)^9(0.4) + \binom{10}{10}(0.6)^{10}\right] = 0.618$$

The results for parts (*b*) through (*g*) can be found similarly and are shown in Table 10.7, together with the values corresponding to $p = 0.5$ and to $p = 0.7$. Note that the probability is denoted in Table 10.7 by β (probability of a Type II error); the β entry for $p = 0.5$ is given by $\beta = 1 - 0.1719 = 0.828$ (from Problem 10.23), and the β entry for $p = 0.7$ is from Problem 10.24.

Table 10.7

p	0.1	0.2	0.3	0.4	0.5	0.6	0.7	0.8	0.9
β	1.000	0.999	0.989	0.945	0.828	0.618	0.350	0.121	0.013

10.26 Use Problem 10.25 to construct the graph of β versus p.

SOLUTION

The required graph is shown in Fig. 10-13.

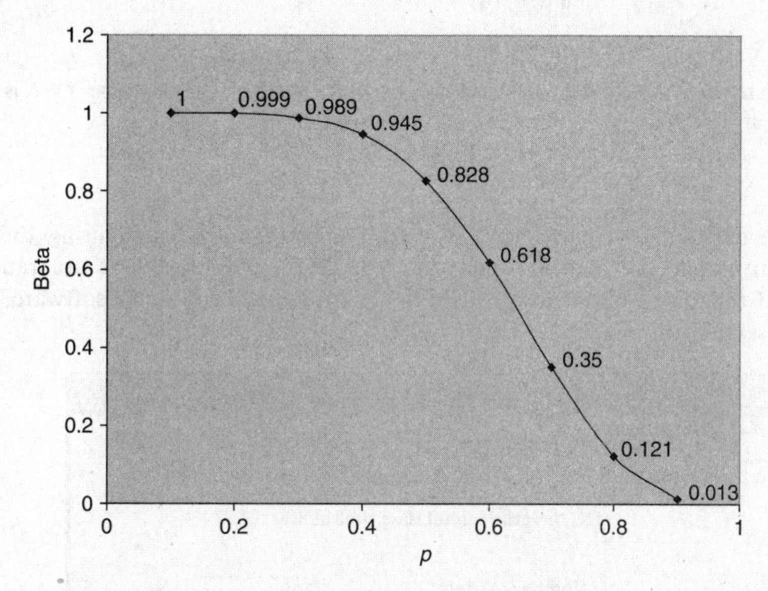

Fig. 10-13 Graph of Type II errors for Problem 10.25.

10.27 The null hypothesis is that a die is fair and the alternative hypothesis is that the die is biased such that the face six is more likely to occur than it would be if the die were fair. The hypothesis is tested by rolling it 18 times and observing how many times the face six occurs. Find the p-value if the face six occurs 7 times in the 18 rolls of the die.

SOLUTION

Zero through 18 are entered into A1:A19 in the EXCEL worksheet. =BINOMDIST(A1, 18, 0.16666, 0) is entered in B1 and a click-and-drag is performed from B1 to B19 to give the individual binomial probabilities. =BINOMDIST(A1, 18, 0.16666, 1) is entered into C1 and a click-and-drag is performed from C1 to C19 to give the cumulative binomial probabilities.

A	B	C
0	0.037566	0.037566446
1	0.135233	0.17279916
2	0.229885	0.402683738
3	0.245198	0.647882186
4	0.18389	0.831772194
5	0.102973	0.934745656
6	0.04462	0.979365347
7	0.015297	0.994662793
8	0.004207	0.998869389
9	0.000935	0.999804143
10	0.000168	0.999972391
11	2.45E-05	0.999996862
12	2.85E-06	0.999999717
13	2.64E-07	0.99999998
14	1.88E-08	0.999999999
15	1E-09	1
16	3.76E-11	1
17	8.86E-13	1
18	9.84E-15	1

The p-value is $p\{x \geq 7\} = 1 - P\{X \leq 6\} = 1 - 0.979 = 0.021$.The outcome $X = 6$ is significant at $\alpha = 0.05$ but not at $\alpha = 0.01$.

10.28 To test that 40% of taxpayers use computer software when figuring their taxes against the alternative that the percent is greater than 40%, 300 taxpayers are randomly selected and asked if they use computer software. If 131 of 300 use computer software, find the p-value for this observation.

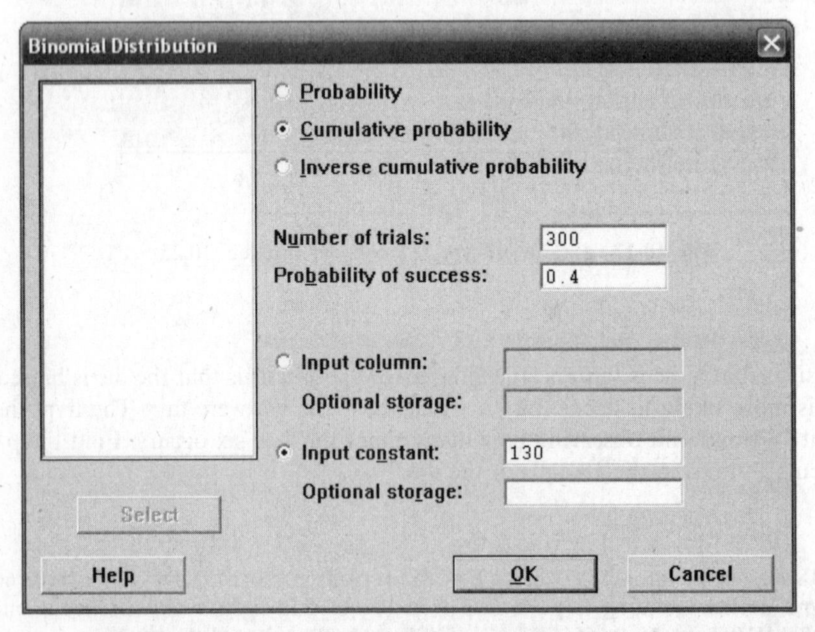

Fig. 10-14 Binomial dialog box for computing 130 or fewer computer software users in 300 when 40% of all taxpayers use computer software.

SOLUTION

The null hypothesis is $H_0: p = 0.4$ versus the alternative $H_a: p > 0.4$. The observed valued of X is 131, where X is the number who use computer software. The p-value $= P\{X \geq 131$ when $p = 0.4\}$. The p – value = 1 – P{X ≤ 130 when $p = 0.4$} Using MINITAB, the pull down "**Calc ⇒ Probability Distribution ⇒ Binomial**" gives the dialog box shown in Fig 10-14.

The dialog box in Fig. 10-14 produces the following output.

Cumulative Distribution Function

```
Binomial with n=300 and p=0.4

x      P(X<=x)
130    0.891693
```

The p-value is $1 - P\{X \leq 130 \text{ when } p = 0.4\} = 1 - 0.8971 = 0.1083$. The outcome $X = 131$ is not significant at 0.01, 0.05, or 0.10.

Supplementary Problems

TESTS OF MEANS AND PROPORTIONS, USING NORMAL DISTRIBUTIONS

10.29 An urn contains marbles that are either red or blue. To test the null hypothesis of equal proportions of these colors, we agree to sample 64 marbles with replacement, noting the colors drawn, and to adopt the following decision rule:

Accept the null hypothesis if $28 \leq X \leq 36$, where X = number of red marbles in the 64.
Reject the null hypothesis if $X \leq 27$ or if $X \geq 37$.

(*a*) Find the probability of rejecting the null hypothesis when it is correct.

(*b*) Graph the decision rule and the result obtained in part (*a*).

10.30 (*a*) What decision rule would you adopt in Problem 10.29 if you require that the probability of rejecting the hypothesis when it is actually correct be no more than 0.01 (i.e., you want a 0.01 significance level)?

(*b*) At what level of confidence would you accept the hypothesis?

(*c*) What would the decision rule be if the 0.05 significance level were adopted?

10.31 Suppose that in Problem 10.29 we wish to test the hypothesis that there is a *greater proportion* of red than blue marbles.

(*a*) What would you take as the null hypothesis, and what would be the alternative hypothesis?

(*b*) Should you use a one- or a two-tailed test? Why?

(*c*) What decision rule should you adopt if the significance level is 0.05?

(*d*) What is the decision rule if the significance level is 0.01?

10.32 A pair of dice is tossed 100 times and it is observed that 7's appear 23 times. Test the hypothesis that the dice are fair (i.e., not loaded) at the 0.05 significance level by using (*a*) a two-tailed test and (*b*) a one-tailed test. Discuss your reasons, if any, for preferring one of these tests over the other.

10.33 Work Problem 10.32 if the significance level is 0.01.

10.34 A manufacturer claimed that at least 95% of the equipment that she supplied to a factory conformed to specifications. An examination of a sample of 200 pieces of equipment revealed that 18 were faulty. Test her claim at significance levels of (*a*) 0.01 and (*b*) 0.05.

10.35 The claim is made that Internet shoppers spend on the average \$335 per year. It is desired to test that this figure is not correct at $\alpha = 0.075$. Three hundred Internet shoppers are surveyed and it is found that the sample mean = \$354 and the standard deviation = \$125. Find the value of the test statistic, the critical values, and give your conclusion.

10.36 It has been found from experience that the mean breaking strength of a particular brand of thread is 9.72 ounces (oz) with a standard deviation of 1.40 oz. A recent sample of 36 pieces of this thread showed a mean breaking strength of 8.93 oz. Test the null hypothesis $H_0: \mu = 9.72$ versus the alternative $H_0: \mu < 9.72$ by giving the value of the test statistic and the critical value for $\alpha = 0.10$ and $\alpha = 0.025$. Is the result significant at $\alpha = 0.10$. Is the result significant at $\alpha = 0.025$?

10.37 A study was designed to test the null hypothesis that the average number of e-mails sent weekly by employees in a large city equals 25.5 versus the mean is greater than 25.5. Two hundred employees were surveyed across the city and it was found that $\bar{x} = 30.3$ and $s = 10.5$. Give the value of the test statistic, the critical value for $\alpha = 0.03$, and your conclusion.

10.38 For large n ($n > 30$) and known standard deviation the standard normal distribution is used to perform a test concerning the mean of the population from which the sample was selected. The alternative hypothesis $H_a : \mu < \mu_0$ is called a *lower-tailed alternative* and the alternative hypothesis $H_a : \mu > \mu_0$ is called a *upper-tailed alternative*. For an upper-tailed alternative, give the EXCEL expression for the critical value if $\alpha = 0.1$, $\alpha = 0.01$, and $\alpha = 0.001$.

p-VALUES FOR HYPOTHESES TESTS

10.39 To test a coin for its balance, it is flipped 15 times. The number of heads obtained is 12. Give the *p*-value corresponding to this outcome. Use BINOMDIST of EXCEL to find the *p*-value.

10.40 Give the *p*-value for the outcome in Problem 10.35.

10.41 Give the *p*-value for the outcome in Problem 10.36.

10.42 Give the *p*-value for the outcome in Problem 10.37.

QUALITY CONTROL CHARTS

10.43 In the past a certain type of thread produced by a manufacturer has had a mean breaking strength of 8.64 oz and a standard deviation of 1.28 oz. To determine whether the product is conforming to standards, a sample of 16 pieces of thread is taken every 3 hours and the mean breaking strength is determined. Record the (*a*) 99.73% (or 3σ), (*b*) 99%, and (*c*) 95% control limits on a quality control chart and explain their applications.

10.44 On average, about 3% of the bolts produced by a company are defective. To maintain this quality of performance, a sample of 200 bolts produced is examined every 4 hours. Determine the (*a*) 99% and (*b*) 95% control limits for the number of defective bolts in each sample. Note that only *upper control limits* are needed in this case.

TESTS INVOLVING DIFFERENCES OF MEANS AND PROPORTIONS

10.45 A study compared the mean lifetimes in hours of two types of types of light bulbs. The results of the study are shown in Table 10.8.

Table 10.8

	Environmental Bulb	Traditional Bulb
n	75	75
Mean	1250	1305
Std. dev	55	65

Test $H_0 : \mu_1 - \mu_2 = 0$ versus $H_a : \mu_1 - \mu_2 \neq 0$ at $\alpha = 0.05$. Give the value of the test statistic and compute the p-value and compare the p-value with $\alpha = 0.05$. Give your conclusion.

10.46 A study compared the grade point averages (GPAS) of 50 high school seniors with a TV in their bedroom with the GPAS of 50 high school seniors without a TV in their bedrooms. The results are shown in Table 10.9. The alternative is that the mean GPA is greater for the group with no TV in their bedroom. Give the value of the test statistic assuming no difference in mean GPAs. Give the p-value and your conclusion for $\alpha = 0.05$ and for $\alpha = 0.10$.

Table 10.9

	TV in Bedroom	No TV in Bedroom
n	50	50
Mean	2.58	2.77
Std. dev	0.55	0.65

10.47 On an elementary school spelling examination, the mean grade of 32 boys was 72 with a standard deviation of 8, while the mean grade of 36 girls was 75 with a standard deviation of 6. The alternative is that the girls are better at spelling than the boys. Give the value of the test statistic assuming no difference in boys and girls at spelling. Give the p-value and your conclusion for $\alpha = 0.05$ and for $\alpha = 0.10$.

10.48 To test the effects of a new fertilizer on wheat production, a tract of land was divided into 60 squares of equal areas, all portions having identical qualities in terms of soil, exposure to sunlight, etc. The new fertilizer was applied to 30 squares and the old fertilizer was applied to the remaining squares. The mean number of bushels (bu) of wheat harvested per square of land using the new fertilizer was 18.2 bu with a standard deviation of 0.63 bu. The corresponding mean and standard deviation for the squares using the old fertilizer were 17.8 and 0.54 bu, respectively. Using significance levels of (*a*) 0.05 and (*b*) 0.01, test the hypothesis that the new fertilizer is better than the old one.

10.49 Random samples of 200 bolts manufactured by machine A and of 100 bolts manufactured by machine B showed 19 and 5 defective bolts, respectively.

(*a*) Give the test statistic, the p-value, and your conclusion at $\alpha = 0.05$ for testing that the two machines show different qualities of performance

(*b*) Give the test statistic, the p-value, and your conclusion at $\alpha = 0.05$ for testing that machine B is performing better than machine A.

10.50 Two urns, A and B, contain equal numbers of marbles, but the proportion of red and white marbles in each of the urns is unknown. A sample of 50 marbles from each urn is selected from each urn with replacement. There are 32 red in the 50 from urn A and 23 red in the 50 from urn B.

(*a*) Test, at $\alpha = 0.05$, that the proportion of red is the same versus the proportion is different by giving the computed test statistic, the computed *p*-value and your conclusion.
(b) Test, at $\alpha = 0.05$, that A has a greater proportion than B by giving the computed test statistic, the *p*-value, and your conclusion.

10.51 A coin is tossed 15 times in an attempt to determine whether it is biased so that heads are more likely to occur than tails. Let X = the number of heads to occur in 15 tosses. The coin is declared biased in favor of heads if $X \geq 11$. Use EXCEL to find α.

10.52 A coin is tossed 20 times to determine if it is unfair. It is declared unfair if $X = 0, 1, 2, 18, 19, 20$ where X = the number of tails to occur. Use EXCEL to find α.

10.53 A coin is tossed 15 times in an attempt to determine whether it is biased so that heads are more likely to occur than tails. Let X = the number of heads to occur in 15 tosses. The coin is declared biased in favor of heads if $X \geq 11$. Use EXCEL find β if $p = 0.6$.

10.54 A coin is tossed 20 times to determine if it is unfair. It is declared unfair if $X = 0, 1, 2, 18, 19, 20$ where X = the number of tails to occur. Use EXCEL to find β if $p = 0.9$.

10.55 A coin is tossed 15 times in an attempt to determine whether it is biased so that heads are more likely to occur than tails. Let X = the number of heads to occur in 15 tosses. The coin is declared biased in favor of heads if $X \geq 11$. Find the *p*-value for the outcome $X = 10$. Compare the *p*-value with the value of α in this problem.

10.56 A coin is tossed 20 times to determine if it is unfair. It is declared unfair if $X = 0, 1, 2, 3, 4, 16, 17, 18, 19, 20$ where X = the number of tails to occur. Find the *p*-value for the outcome $X = 17$. Compare the *p*-value with the value of α in this problem.

10.57 A production line manufactures cell phones. Three percent defective is considered acceptable. A sample of size 50 is selected from a day's production. If more than 3 are found defective in the sample, it is concluded that the defective percent exceeds the 3% figure and the line is stopped until it meets the 3% figure. Use EXCEL to determine α?

10.58 In Problem 10.57, find the probability that a 4% defective line will not be shut down.

10.59 To determine if it is balanced, a die is rolled 20 times. It is declared to be unbalanced so that the face six occurs more often than 1/6 of the time if more than 5 sixes occur in the 20 rolls. Find the value for α. If the die is rolled 20 times and a six occurs 6 times, find the *p*-value for this outcome.

Small Sampling Theory

SMALL SAMPLES

In previous chapters we often made use of the fact that for samples of size $N > 30$, called *large samples*, the sampling distributions of many statistics are approximately normal, the approximation becoming better with increasing N. For samples of size $N < 30$, called *small samples*, this approximation is not good and becomes worse with decreasing N, so that appropriate modifications must be made.

A study of sampling distributions of statistics for small samples is called *small sampling theory*. However, a more suitable name would be *exact sampling theory*, since the results obtained hold for large as well as for small samples. In this chapter we study three important distributions: Student's t distribution, the chi-square distribution, and the F distribution.

STUDENT'S t DISTRIBUTION

Let us define the statistic

$$t = \frac{\bar{X} - \mu}{s}\sqrt{N-1} = \frac{\bar{X} - \mu}{\hat{s}/\sqrt{N}} \tag{1}$$

which is analogous to the z statistic given by

$$z = \frac{\bar{X} - \mu}{\sigma/\sqrt{N}}.$$

If we consider samples of size N drawn from a normal (or approximately normal) population with mean μ and if for each sample we compute t, using the sample mean $\bar{X}$ and sample standard deviation s or $\hat{s}$, the sampling distribution for t can be obtained. This distribution (see Fig. 11-1) is given by

$$Y = \frac{Y_0}{\left(1 + \dfrac{t^2}{N-1}\right)^{N/2}} = \frac{Y_0}{\left(1 + \dfrac{t^2}{\nu}\right)^{(\nu+1)/2}} \tag{2}$$

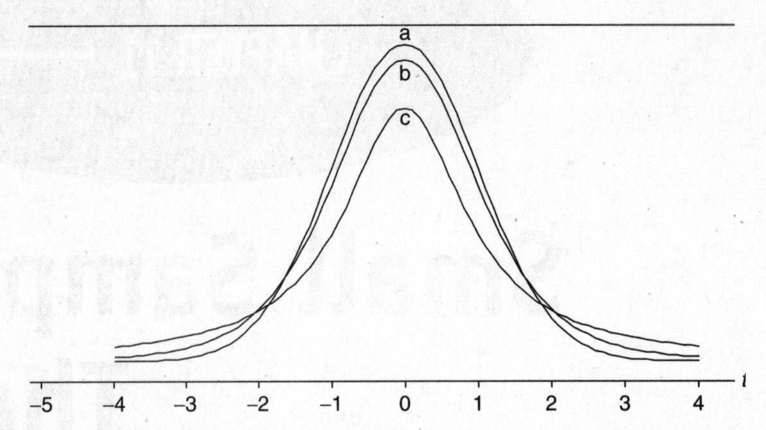

Fig. 11-1 (*a*) Standard normal, (*b*) Student t with $\nu = 5$, (*c*) Student t with $\nu = 1$.

where Y_0 is a constant depending on N such that the total area under the curve is 1, and where the constant $\nu = (N - 1)$ is called the *number of degrees of freedom* (ν is the Greek letter *nu*).

Distribution (*2*) is called *Student's t distribution* after its discoverer, W. S. Gossett, who published his works under the pseudonym "Student" during the early part of the twentieth century.

For large values of ν or N (certainly $N \geq 30$) the curves (*2*) closely approximate the standardized normal curve

$$Y = \frac{1}{\sqrt{2\pi}} e^{-(1/2)t^2}$$

as shown in Fig. 11-1.

CONFIDENCE INTERVALS

As done with normal distributions in Chapter 9, we can define 95%, 99%, or other confidence intervals by using the table of the t distribution in Appendix III. In this manner we can estimate within specified limits of confidence the population mean μ.

For example, if $-t_{.975}$ and $t_{.975}$ are the values of t for which 2.5% of the area lies in each tail of the t distribution, then the 95% confidence interval for t is

$$-t_{.975} < \frac{\bar{X} - \mu}{s}\sqrt{N - 1} < t_{.975} \tag{3}$$

from which we see that μ is estimated to lie in the interval

$$\bar{X} - t_{.975}\frac{s}{\sqrt{N - 1}} < \mu < \bar{X} + t_{.975}\frac{s}{\sqrt{N - 1}} \tag{4}$$

with 95% confidence (i.e. probability 0.95). Note that $t_{.975}$ represents the 97.5 percentile value, while $t_{.025} = -t_{.975}$ represents the 2.5 percentile value.

In general, we can represent confidence limits for population means by

$$\bar{X} \pm t_c \frac{s}{\sqrt{N - 1}} \tag{5}$$

where the values $\pm t_c$, called *critical values* or *confidence coefficients*, depend on the level of confidence desired and on the sample size. They can be read from Appendix III.

The sample is assumed to be taken from a normal population. This assumption may be checked out using the Komogorov–Smirnov test for normality.

A comparison of equation (*5*) with the confidence limits $(\bar{X} \pm z_c\sigma/\sqrt{N})$ of Chapter 9, shows that for small samples we replace z_c (obtained from the normal distribution) with t_c (obtained from the t distribution) and that we replace σ with $\sqrt{N/(N-1)}s = \hat{s}$, which is the sample estimate of σ. As N increases, both methods tend toward agreement.

TESTS OF HYPOTHESES AND SIGNIFICANCE

Tests of hypotheses and significance, or decision rules (as discussed in Chapter 10), are easily extended to problems involving small samples, the only difference being that the *z score*, or *z statistic*, is replaced by a suitable *t score*, or *t statistic*.

1. **Means.** To test the hypothesis H_0 that a normal population has mean μ, we use the t score (or t statistic)

$$t = \frac{\bar{X} - \mu}{s}\sqrt{N-1} = \frac{\bar{X} - \mu}{\hat{s}}\sqrt{N} \tag{6}$$

where $\bar{X}$ is the mean of a sample of size N. This is analogous to using the z score

$$z = \frac{\bar{X} - \mu}{\sigma/\sqrt{N}}$$

for large N, except that $\hat{s} = \sqrt{N/(N-1)}s$ is used in place of σ. The difference is that while z is normally distributed, t follows Student's distribution. As N increases, these tend toward agreement.

2. **Differences of Means.** Suppose that two random samples of sizes N_1 and N_2 are drawn from normal populations whose standard deviations are equal $(\sigma_1 = \sigma_2)$. Suppose further that these two samples have means given by $\bar{X}_1$ and $\bar{X}_2$ and standard deviations given by s_1 and s_2, respectively. To test the hypothesis H_0 that the samples come from the same population (i.e., $\mu_1 = \mu_2$ as well as $\sigma_1 = \sigma_2$), we use the t score given by

$$t = \frac{\bar{X}_1 - \bar{X}_2}{\sigma\sqrt{1/N_1 + 1/N_2}} \qquad \text{where} \qquad \sigma = \sqrt{\frac{N_1 s_1^2 + N_2 s_2^2}{N_1 + N_2 - 2}} \tag{7}$$

The distribution of t is Student's distribution with $\nu = N_1 + N_2 - 2$ degrees of freedom. The use of equation (*7*) is made plausible on placing $\sigma_1 = \sigma_2 = \sigma$ in the z score of equation (*2*) of Chapter 10 and then using as an estimate of σ^2 the weighted mean

$$\frac{(N_1 - 1)\hat{s}_1^2 + (N_2 - 1)\hat{s}_2^2}{(N_1 - 1) + (N_2 - 1)} = \frac{N_1 s_1^2 + N_2 s_2^2}{N_1 + N_2 - 2}$$

where $\hat{s}_1^2$ and $\hat{s}_2^2$ are the unbiased estimates of σ_1^2 and σ_2^2.

THE CHI-SQUARE DISTRIBUTION

Let us define the statistic

$$\chi^2 = \frac{Ns^2}{\sigma^2} = \frac{(X_1 - \bar{X})^2 + (X_2 - \bar{X})^2 + \cdots + (X_N - \bar{X})^2}{\sigma^2} \tag{8}$$

where χ is the Greek letter *chi* and χ^2 is read "chi-square."

If we consider samples of size N drawn from a normal population with standard deviation σ, and if for each sample we compute χ^2, a sampling distribution for χ^2 can be obtained. This distribution, called the *chi-square distribution*, is given by

$$Y = Y_0(\chi^2)^{(1/2)(\nu-2)}\, e^{-(1/2)\chi^2} = Y_0\chi^{\nu-2}\, e^{-(1/2)\chi^2} \tag{9}$$

where $\nu = N - 1$ is the *number of degrees of freedom*, and Y_0 is a constant depending on ν such that the total area under the curve is 1. The chi-square distributions corresponding to various values of ν are shown in Fig. 11-2. The maximum value of Y occurs at $\chi^2 = \nu - 2$ for $\nu \geq 2$.

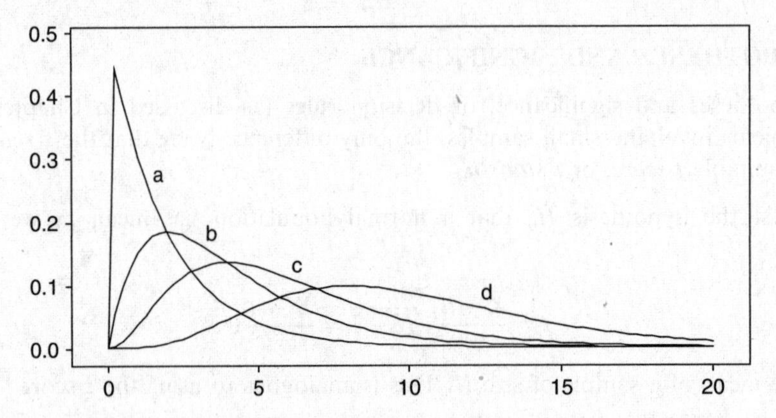

Fig. 11-2 Chi-square distributions with (a) 2, (b) 4, (c) 6, and (d) 10 degrees of freedom.

CONFIDENCE INTERVALS FOR σ

As done with the normal and t distribution, we can define 95%, 99%, or other confidence limits by using the table of the χ^2 distribution in Appendix IV. In this manner we can estimate within specified limits of confidence the population standard deviation σ in terms of a sample standard deviation s.

For example, if $\chi^2_{.025}$ and $X^2_{.975}$ are the values of χ^2 (called *critical values*) for which 2.5% of the area lies in each tail of the distribution, then the 95% confidence interval is

$$\chi^2_{.025} < \frac{Ns^2}{\sigma^2} < \chi^2_{.975} \tag{10}$$

from which we see that σ is estimated to lie in the interval

$$\frac{s\sqrt{N}}{\chi_{.975}} < \sigma < \frac{s\sqrt{N}}{\chi_{.025}} \tag{11}$$

with 95% confidence. Other confidence intervals can be found similarly. The values $\chi_{.025}$ and $\chi_{.095}$ represent, respectively, the 2.5 and 97.5 percentile values.

Appendix IV gives percentile values corresponding to the number of degrees of freedom ν. For large values of ν ($\nu \geq 30$), we can use the fact that $(\sqrt{2\chi^2} - \sqrt{2\nu - 1})$ is very nearly normally distributed with mean 0 and standard deviation 1; thus normal distribution tables can be used if $\nu \geq 30$. Then if χ^2_p and z_p are the pth percentiles of the chi-square and normal distributions, respectively, we have

$$\chi^2_p = \tfrac{1}{2}(z_p + \sqrt{2\nu - 1})^2 \tag{12}$$

In these cases, agreement is close to the results obtained in Chapters 8 and 9.

For further applications of the chi-square distribution, see Chapter 12.

DEGREES OF FREEDOM

In order to compute a statistic such as (*1*) or (*8*), it is necessary to use observations obtained from a sample as well as certain population parameters. If these parameters are unknown, they must be estimated from the sample.

The *number of degrees of freedom* of a statistic, generally denoted by ν, is defined as the number N of independent observations in the sample (i.e., the sample size) minus the number k of population parameters, which must be estimated from sample observations. In symbols, $\nu = N - k$.

In the case of statistic *(1)*, the number of independent observations in the sample is N, from which we can compute $\bar{X}$ and s. However, since we must estimate μ, $k = 1$ and so $\nu = N - 1$.

In the case of statistic *(8)*, the number of independent observations in the sample is N, from which we can compute s. However, since we must estimate σ, $k = 1$ and so $\nu = N - 1$.

THE *F* DISTRIBUTION

As we have seen, it is important in some applications to know the sampling distribution of the difference in means $(\bar{X}_1 - \bar{X}_2)$ of two samples. Similarly, we may need the sampling distribution of the difference in variances $(S_1^2 - S_2^2)$. It turns out, however, that this distribution is rather complicated. Because of this, we consider instead the statistic S_1^2/S_2^2, since a large or small ratio would indicate a large difference, while a ratio nearly equal to 1 would indicate a small difference. The sampling distribution in such a case can be found and is called the *F distribution*, named after R. A. Fisher.

More precisely, suppose that we have two samples, 1 and 2, of sizes N_1 and N_2, respectively, drawn from two normal (or nearly normal) populations having variances σ_1^2 and σ_2^2. Let us define the statistic

$$F = \frac{\hat{S}_1^2/\sigma_1^2}{\hat{S}_2^2/\sigma_2^2} = \frac{N_1 S_1^2/(N_1 - 1)\sigma_1^2}{N_2 S_2^2/(N_2 - 1)\sigma_2^2} \tag{13}$$

where

$$\hat{S}_1^2 = \frac{N_1 S_1^2}{N_1 - 1} \qquad \hat{S}_2^2 = \frac{N_2 S_2^2}{N_2 - 1}. \tag{14}$$

Then the sampling distribution of F is called Fisher's F distribution, or briefly the F distribution, with $\nu_1 = N_1 - 1$ and $\nu_2 = N_2 - 1$ degrees of freedom. This distribution is given by

$$Y = \frac{CF^{(\nu_1/2)-1}}{(\nu_1 F + \nu_2)^{(\nu_1+\nu_2)/2}} \tag{15}$$

where C is a constant depending on ν_1 and ν_2 such that the total area under the curve is 1. The curve has a shape similar to that shown in Fig. 11-3, although this shape can vary considerably for different values of ν_1 and ν_2.

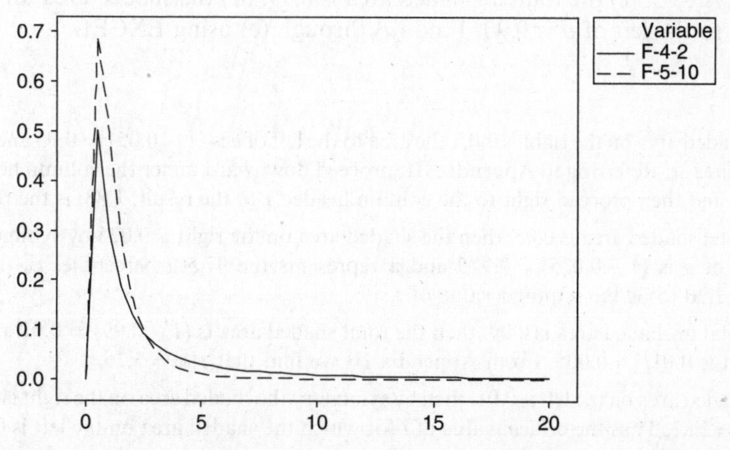

Fig. 11-3 The solid curve is the F distribution with 4 and 2 degrees of freedom and the dashed curve is the F distribution with 5 and 10 degrees of freedom.

Appendixes V and VI give percentile values of F for which the areas in the right-hand tail are 0.05 and 0.01, denoted by $F_{.95}$ and $F_{.99}$, respectively. Representing the 5% and 1% significance levels, these can be used to determine whether or not the variance S_1^2 is significantly larger than S_2^2. In practice, the sample with the larger variance is chosen as sample 1.

Statistical software has added to the ability to find areas under the Student t distribution, the chi-square distribution, and the F distribution. It has also added to our ability to sketch the various distributions. We will illustrate this in the solved problems section in this chapter.

Solved Problems

STUDENT'S t DISTRIBUTION

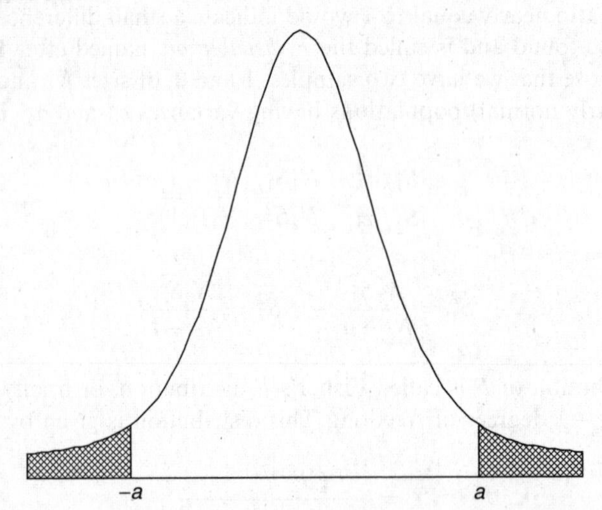

Fig. 11-4 Student's t distribution with 9 degrees of freedom.

11.1 The graph of Student's t distribution with nine degrees of freedom is shown in Fig. 11-4. Use Appendix III to find the values of a for which (a) the area to the right of a is 0.05, (b) the total shaded area is 0.05, (c) the total unshaded area is 0.99, (d) the shaded area on the left is 0.01, and (e) the area to the left of a is 0.90. Find (a) through (e) using EXCEL.

SOLUTION

(a) If the shaded area on the right is 0.05, the area to the left of a is $(1 - 0.05) = 0.95$ and a represents the 95th percentile, $t_{.95}$. Referring to Appendix III, proceed downward under the column headed v until reaching entry 9, and then proceed right to the column headed $t_{.95}$; the result, 1.83, is the required value of t.

(b) If the total shaded area is 0.05, then the shaded area on the right is 0.025 by symmetry. Thus the area to the left of a is $(1 - 0.025) = 0.975$ and a represents the 97.5th percentile, $t_{.975}$. From Appendix III we find 2.26 to be the required value of t.

(c) If the total unshaded area is 0.99, then the total shaded area is $(1 - 0.99) = 0.01$ and the shaded area to the right is $0.01/2 = 0.005$. From Appendix III we find that $t_{.995} = 3.25$.

(d) If the shaded area on the left is 0.01, then by symmetry the shaded area on the right is 0.01. From Appendix III, $t_{.99} = 2.82$. Thus the critical value of t for which the shaded area on the left is 0.01 equals -2.82.

(e) If the area to the left of a is 0.90, the a corresponds to the 90th percentile, $t_{.90}$, which from Appendix III equals 1.38.

Using EXCEL, the expression = TINV(0.1,9) gives 1.833113. EXCEL requires the sum of the areas in the two tails and the degrees of freedom. Similarly, =TINV(0.05,9) gives 2.262157, =TINV(0.01,9) gives 3.249836, =TINV(0.02,9) gives 2.821438, and =TINV(0.2,9) gives 1.383029.

11.2 Find the critical values of t for which the area of the right-hand tail of the t distribution is 0.05 if the number of degrees of freedom, ν, is equal to (*a*) 16, (*b*) 27, and (*c*) 200.

SOLUTION

Using Appendix III, we find in the column headed $t_{.95}$ the values (*a*) 1.75, corresponding to $\nu = 16$; (*b*) 1.70, corresponding to $\nu = 27$; and (*c*) 1.645, corresponding to $\nu = 200$. (The latter is the value that would be obtained by using the normal curve; in Appendix III it corresponds to the entry in the last row marked ∞, or infinity.)

11.3 The 95% confidence coefficients (two-tailed) for the normal distribution are given by ± 1.96. What are the corresponding coefficients for the t distribution if (*a*) $\nu = 9$, (*b*) $\nu = 20$, (*c*) $\nu = 30$, and (*d*) $\nu = 60$?

SOLUTION

For the 95% confidence coefficients (two-tailed), the total shaded area in Fig. 11-4 must be 0.05. Thus the shaded area in the right tail is 0.025 and the corresponding critical value of t is $t_{.975}$. Then the required confidence coefficients are $\pm t_{.975}$; for the given values of ν, these are (*a*) ± 2.26, (*b*) ± 2.09, (*c*) ± 2.04, and (*d*) ± 2.00.

11.4 A sample of 10 measurements of the diameter of a sphere gave a mean $\bar{X} = 438$ centimeters (cm) and a standard deviation $s = 0.06$ cm. Find the (*a*) 95% and (*b*) 99% confidence limits for the actual diameter.

SOLUTION

(*a*) The 95% confidence limits are given by $\bar{X} \pm t_{.975}(s/\sqrt{N-1})$.

Since $\nu = N - 1 = 10 - 1 = 9$, we find $t_{.975} = 2.26$ [see also Problem 11.3(*a*)]. Then, using $\bar{X} = 4.38$ and $s = 0.06$, the required 95% confidence limits are $4.38 \pm 2.26(0.06/\sqrt{10-1}) = 4.38 \pm 0.0452$ cm. Thus we can be 95% confident that the true mean lies between $(438 - 0.045) = 4.335$ cm and $(4.38 + 0.045) = 4.425$ cm.

(*b*) The 99% confidence limits are given by $\bar{X} \pm t_{.995}(s/\sqrt{N-1})$.

For $\nu = 9$, $t_{.995} = 3.25$. Then the 99% confidence limits are $4.38 \pm 3.25(0.06/\sqrt{10-1}) = 4.38 \pm 0.0650$ cm, and the 99% confidence interval is 4.315 to 4.445 cm.

11.5 The number of days absent from work last year due to job-related cases of carpal tunnel syndrome were recorded for 25 randomly selected workers. The results are given in Table 11.1. When the data are used to set a confidence interval on the mean of the population of all job-related cases of carpal tunnel syndrome, a basic assumption underlying the procedure is that the number of days absent are normally distributed for the population. Use the data to test the normality assumption and if you are willing to assume normality, then set a 95% confidence interval on μ.

Table 11.1

21	23	33	32	37
40	37	29	23	29
24	32	24	46	32
17	29	26	46	27
36	38	28	33	18

SOLUTION

The normal probability plot from MINITAB (Fig. 11-5) indicates that it would be reasonable to assume normality since the p-value exceeds 0.15. This p-value is used to test the null hypothesis that the data were selected from a normally distributed population. If the conventional level of significance, 0.05, is used then normality of the population distribution would be rejected only if the p-value is less than 0.05. Since the p-value associated with the Kolmogorov–Smirnov test for normality is reported as p-value > 0.15, we do not reject the assumption of normality.

The confidence interval is found using MINITAB as follows. The 95% confidence interval for the population mean extends from 27.21 to 33.59 days per year.

```
MTB > tinterval 95% confidence for data in c1
```

Confidence Intervals

Variable	N	Mean	StDev	SE Mean	95.0 % CI
days	25	30.40	7.72	1.54	(27.21, 33.59)

11.6 In the past, a machine has produced washers having a thickness of 0.050 inch (in). To determine whether the machine is in proper working order, a sample of 10 washers is chosen, for which the mean thickness is 0.053 in and the standard deviation is 0.003 in. Test the hypothesis that the machine is in proper working order, using significance levels of (*a*) 0.05 and (*b*) 0.01.

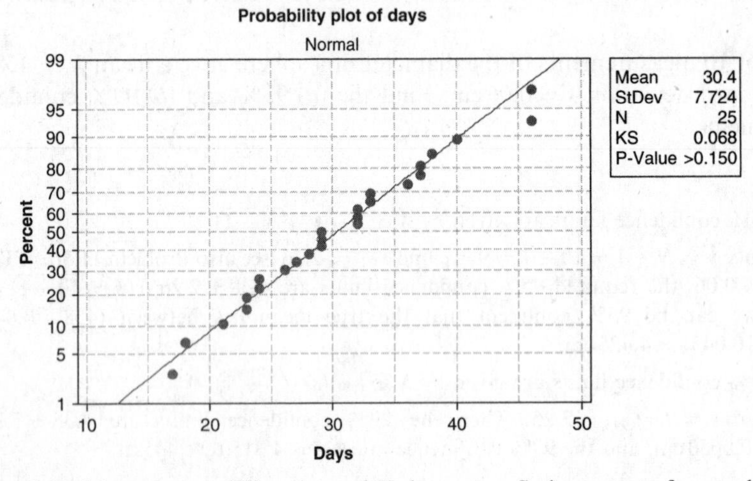

Fig. 11-5 Normal probability plot and Kolmogorov–Smirnov test of normality.

SOLUTION

We wish to decide between the hypotheses:

$H_0 : \mu = 0.050$, and the machine is in proper working order

$H_1 : \mu \neq 0.050$, and the machine is not in proper working order

Thus a two-tailed test is required. Under hypothesis H_0, we have

$$t = \frac{\bar{X} - \mu}{s}\sqrt{N-1} = \frac{0.053 - 0.050}{0.003}\sqrt{10-1} = 3.00$$

(*a*) For a two-tailed test at the 0.05 significance level, we adopt the decision rule:

Accept H_0 if t lies inside the interval $-t_{.975}$ to $t_{.975}$, which for $10 - 1 = 9$ degrees of freedom is the interval -2.26 to 2.26.

Reject H_0 otherwise

Since $t = 3.00$, we reject H_0 at the 0.05 level.

(*b*) For a two-tailed test at the 0.01 significance level, we adopt the decision rule:

Accept H_0 if t lies inside the interval $-t_{.995}$ to $t_{.995}$, which for $10 - 1 = 9$ degrees of freedom is the interval -3.25 to 3.25.

Reject H_0 otherwise

Since $t = 3.00$, we accept H_0 at the 0.01 level.

Because we can reject H_0 at the 0.05 level but not at the 0.01 level, we say that the sample result is *probably significant* (see the terminology at the end of Problem 10.5). It would thus be advisable to check the machine or at least to take another sample.

11.7 A mall manager conducts a test of the null hypothesis that $\mu = \$50$ versus the alternative hypothesis that $\mu \neq \$50$, where μ represents the mean amount spent by all shoppers making purchases at the mall. The data shown in Table 11.2 give the dollar amount spent for 28 shoppers. The test of hypothesis using the Student's t distribution assumes that the data used in the test are selected from a normally distributed population. This normality assumption may be checked out using anyone of several different *tests of normality*. MINITAB gives 3 different choices for a normality test. Test for normality at the conventional level of significance equal to $\alpha = 0.05$. If the normality assumption is not rejected, then proceed to test the hypothesis that $\mu = \$50$ versus the alternative hypothesis that $\mu \neq \$50$ at $\alpha = 0.05$.

Table 11.2

68	49	45	76	65	50
54	92	24	36	60	66
57	74	52	75	36	40
62	56	94	57	64	
72	65	59	45	33	

SOLUTION

The Anderson–Darling normality test from MINITAB gives a p-value = 0.922, the Ryan–Joyner normality test gives the p-value as greater than 0.10, and the Kolmogorov–Smirnov normality test gives the p-value as greater than 0.15. In all 3 cases, the null hypothesis that the data were selected from a normally distributed population would not be rejected at the conventional 5% level of significance. Recall that the null hypothesis is rejected only if the p-value is less than the preset level of significance. The MINITAB analysis for the test of the mean amount spent per customer is shown below. If the classical method of testing hypothesis is used then the null hypothesis is rejected if the computed value of the test statistic exceeds 2.05 in absolute value. The critical value, 2.05, is found by using Student's t distribution with 27 degrees of freedom. Since the computed value of the test statistic equals 18.50, we would reject the null hypothesis and conclude that the mean amount spent exceeds \$50. If the p-value approach is used to test the hypothesis, then since the computed p-value = 0.0000 is less than the level of significance (0.05), we also reject the null hypothesis.

Data Display

```
Amount
68     54   57   62   72   49   92   74   56
65     45   24   52   94   59   76   36   75
57     45   65   60   36   64   33   50   66
40

MTB > TTest 0.0 'Amount';
SUBC > Alternative 0.
```

T-Test of the Mean

```
Test of mu = 0.00 vs mu not = 0.00

Variable    N      Mean     StDev    SE Mean        T         P
Amount     28     58.07     16.61      3.14     18.50    0.0000
```

11.8 The intelligence quotients (IQs) of 16 students from one area of a city showed a mean of 107 and a standard deviation of 10, while the IQs of 14 students from another area of the city showed a mean of 112 and a standard deviation of 8. Is there a significant difference between the IQs of the two groups at significance levels of (*a*) 0.01 and (*b*) 0.05?

SOLUTION

If μ_1 and μ_2 denote the population mean IQs of students from the two areas, respectively, we have to decide between the hypotheses:

$H_0 : \mu_1 = \mu_2$, and there is essentially no difference between the groups.
$H_1 : \mu_1 \neq \mu_2$, and there is a significant difference between the groups.

Under hypothesis H_0,

$$t = \frac{\bar{X}_1 - \bar{X}_2}{\sigma\sqrt{1/N_1 + 1/N_2}} \qquad \text{where} \qquad \sigma = \sqrt{\frac{N_1 s_1^2 + N_2 s_2^2}{N_1 + N_2 - 2}}$$

Thus
$$\sigma = \sqrt{\frac{16(10)^2 + 14(8)^2}{16 + 14 - 2}} = 9.44 \qquad \text{and} \qquad t = \frac{112 - 107}{9.44\sqrt{1/16 + 1/14}} = 1.45$$

(*a*) Using a two-tailed test at the 0.01 significance level, we would reject H_0 if t were outside the range $-t_{.995}$ to $t_{.995}$, which for $(N_1 + N_2 - 2) = (16 + 14 - 2) = 28$ degrees of freedom is the range -2.76 to 2.76. Thus we cannot reject H_0 at the 0.01 significance level.

(*b*) Using a two-tailed test at the 0.05 significance level, we would reject H_0 if t were outside the range $-t_{.975}$ to $t_{.975}$, which for 28 degrees of freedom is the range -2.05 to 2.05. Thus we cannot reject H_0 at the 0.05 significance levels.

We conclude that there is no significant difference between the IQs of the two groups.

11.9 The costs (in thousands of dollars) for tuition, room, and board per year at 15 randomly selected public colleges and 10 randomly selected private colleges are shown in Table 11.3. Test the null hypothesis that the mean yearly cost at private colleges exceeds the mean yearly cost at public colleges by 10 thousand dollars versus the alternative hypothesis that the difference is not 10 thousand dollars. Use level of significance 0.05. Test the assumptions of normality and equal variances at level of significance 0.05 before performing the test concerning the means.

Table 11.3

Public Colleges			Private Colleges	
4.2	9.1	11.6	13.0	17.7
6.1	7.7	10.4	18.8	17.6
4.9	6.5	5.0	13.2	19.8
8.5	6.2	10.4	14.4	16.8
4.6	10.2	8.1	17.7	16.1

SOLUTION

The Anderson–Darling normality test from MINITAB for the public colleges data is shown in Fig. 11-6. Since the *p*-value (0.432) is not less than 0.05, the normality assumption is not rejected. A similar test for private colleges indicates that the normality assumption is valid for private colleges.

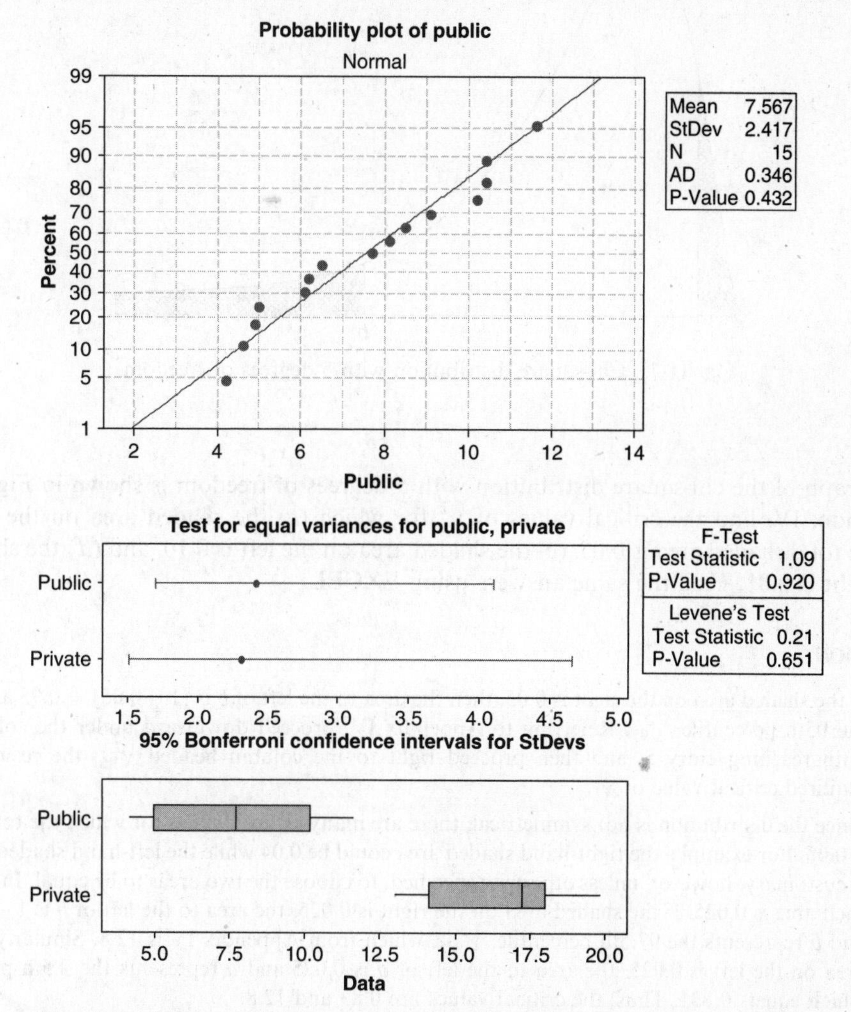

Fig. 11-6 Anderson–Darling test of normality and *F*-test of equal variances.

The *F* test shown in the lower part of Fig. 11-6 indicates that equal variances may be assumed. The pull-down menu "**Stat ⇒ Basic Statistics ⇒ 2-sample t**" gives the following output. The output indicates that we cannot reject that the cost at private colleges exceeds the cost at public colleges by 10,000 dollars.

Two-Sample T-Test and CI: Public, Private

```
Two-sample T for Public vs Private

          N      Mean      StDev    SE Mean
Public   15      7.57       2.42       0.62
Private  10     16.51       2.31       0.73

Difference = mu(Public) - mu(Private)
Estimate for difference: -8.9433
95% CI for difference: (-10.9499, -6.9367)
T-Test of difference = -10 (vs not =): T-Value = 1.09 P-Value = 0.287
DF = 23
Both use Pooled StDev = 2.3760
```

THE CHI-SQUARE DISTRIBUTION

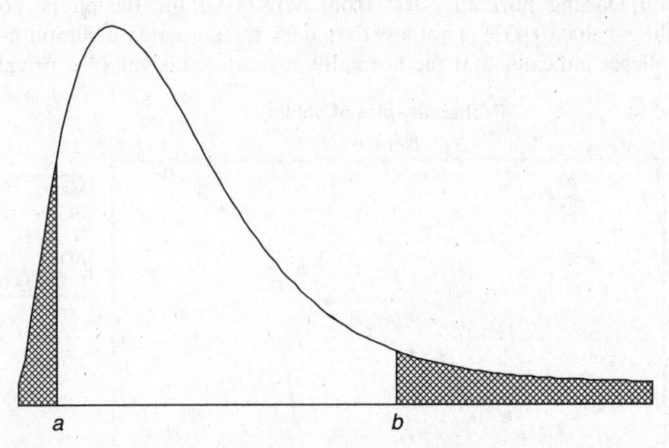

Fig. 11-7 Chi-square distribution with 5 degrees of freedom.

11.10 The graph of the chi-square distribution with 5 degrees of freedom is shown in Fig. 11-7. Using Appendix IV, find the critical values of χ^2 for which (*a*) the shaded area on the right is 0.05, (*b*) the total shaded area is 0.05, (*c*) the shaded area on the left is 0.10, and (*d*) the shaded area on the right is 0.01. Find the same answers using EXCEL.

SOLUTION

(*a*) If the shaded area on the right is 0.05, then the area to the left of b is $(1 - 0.05) = 0.95$ and b represents the 95th percentile, $\chi^2_{.95}$. Referring to Appendix IV, proceed downward under the column headed v until reaching entry 5, and then proceed right to the column headed $\chi^2_{.95}$; the result, 11.1, is the required critical value of χ^2.

(*b*) Since the distribution is not symmetrical, there are many critical values for which the total shaded area is 0.05. For example, the right-hand shaded area could be 0.04 while the left-hand shaded area is 0.01. It is customary, however, unless otherwise specified, to choose the two areas to be equal. In this case, then, each area is 0.025. If the shaded area on the right is 0.025, the area to the left of b is $1 - 0.025 = 0.975$ and b represents the 97.5th percentile, $\chi^2_{.975}$, which from Appendix IV is 12.8. Similarly, if the shaded area on the left is 0.025, the area to the left of a is 0.025 and a represents the 2.5th percentile, $\chi^2_{.025}$, which equals 0.831. Thus, the critical values are 0.83 and 12.8.

(*c*) If the shaded area on the left is 0.10, a represents the 10th percentile, $\chi^2_{.10}$, which equals 1.61.

(*d*) If the shaded area on the right is 0.01, the area to the left of b is 0.99 and b represents the 99th percentile, $\chi^2_{.99}$, which equals 15.1.

The EXCEL answer to (*a*) is =CHIINV(0.05,5) or 11.0705. The first parameter in CHIINV is the area to the right of the point and the second is the degrees of freedom. The answer to (*b*) is =CHIINV(0.975,5) or 0.8312 and =CHIINV(0.025,5) or 12.8325. The answer to (*c*) is =CHIINV(0.9,5) or 1.6103. The answer to (*d*) is =CHIINV(0.01,5) or 15.0863.

11.11 Find the critical values of χ^2 for which the area of the right-hand tail of the χ^2 distribution is 0.05, if the number of degrees of freedom, ν, is equal to (*a*) 15, (*b*) 21, and (*c*) 50.

SOLUTION

Using Appendix IV, we find in the column headed $\chi^2_{.95}$ the values (*a*) 25.0, corresponding to $\nu = 15$; (*b*) 32.7, corresponding to $\nu = 21$; and (*c*) 67.5, corresponding to $\nu = 50$.

11.12 Find the median value of χ^2 corresponding to (*a*) 9, (*b*) 28, and (*c*) 40 degrees of freedom.

SOLUTION

Using Appendix IV, we find in the column headed $\chi^2_{.50}$ (since the median is the 50th percentile) the values (*a*) 8.34, corresponding to $\nu = 9$; (*b*) 27.3, corresponding to $\nu = 28$; and (*c*) 39.3, corresponding to $\nu = 40$.

It is of interest to note that the median values are very nearly equal to the number of degrees of freedom. In fact, for $\nu > 10$ the median values are equal to $(\nu - 0.7)$, as can be seen from the table.

11.13 The standard deviation of the heights of 16 male students chosen at random in a school of 1000 male students is 2.40 in. Find the (*a*) 95% and (*b*) 99% confidence limits of the standard deviation for all male students at the school.

SOLUTION

(*a*) The 95% confidence limits are given by $s\sqrt{N}/\chi_{.975}$ and $s\sqrt{N}/\chi_{.025}$.

For $\nu = 16 - 1 = 15$ degrees of freedom, $\chi^2_{.975} = 27.5$ (or $\chi_{.975} = 5.24$) and $\chi^2_{.025} = 6.26$ (or $\chi_{.025} = 2.50$). Then the 95% confidence limits are $2.40\sqrt{16}/5.24$ and $2.40\sqrt{16}/2.50$ (i.e., 1.83 and 3.84 in). Thus we can be 95% confident that the population standard deviation lies between 1.83 and 3.84 in.

(*b*) The 99% confidence limits are given by $s\sqrt{N}/\chi_{.995}$ and $s/\sqrt{N}/\chi_{.005}$.

For $\nu = 16 - 1 = 15$ degrees of freedom, $\chi^2_{.995} = 32.8$ (or $\chi_{.995} = 5.73$) and $\chi^2_{.005} = 4.60$ (or $\chi_{.005} = 2.14$). Then the 99% confidence limits are $2.40\sqrt{16}/5.73$ and $2.40\sqrt{16}/2.14$ (i.e., 1.68 and 4.49 in). Thus we can be 99% confident that the population standard deviation lies between 1.68 and 4.49 in.

11.14 Find $\chi^2_{.95}$ for (*a*) $\nu = 50$ and (*b*) $\nu = 100$ degrees of freedom.

SOLUTION

For ν greater than 30, we can use the fact that $\sqrt{2\chi^2} - \sqrt{2\nu - 1}$ is very closely normally distributed with mean 0 and standard deviation 1. Then if z_p is the *z*-score percentile of the standardized normal distribution, we can write, to a high degree of approximation,

$$\sqrt{2\chi^2_p} - \sqrt{2\nu - 1} = z_p \qquad \text{or} \qquad \sqrt{2\chi^2_p} = z_p + \sqrt{2\nu - 1}$$

from which $\chi^2_p = \frac{1}{2}(z_p + \sqrt{2\nu - 1})^2$.

(*a*) If $\nu = 50$, $\chi^2_{.95} = \frac{1}{2}(z_{.95} + \sqrt{2(50) - 1})^2 = \frac{1}{2}(1.64 + \sqrt{99})^2 = 67.2$, which agrees very well with the value of 67.5 given in Appendix IV.

(*b*) If $\nu = 100$, $\chi^2_{.95} = \frac{1}{2}(z_{.95} + \sqrt{2(100) - 1})^2 = \frac{1}{2}(1.64 + \sqrt{199})^2 = 124.0$ (actual value = 124.3).

11.15 The standard deviation of the lifetimes of a sample of 200 electric light bulbs is 100 hours (h). Find the (*a*) 95% and (*b*) 99% confidence limits for the standard deviation of all such electric light bulbs.

SOLUTION

(*a*) The 95% confidence limits are given by $s\sqrt{N}/\chi_{.975}$ and $s\sqrt{N}/\chi_{.025}$.

For $\nu = 200 - 1 = 199$ degrees of freedom, we find (as in Problem 11.14)

$$\chi^2_{.975} = \tfrac{1}{2}(z_{.975} + \sqrt{2(199) - 1})^2 = \tfrac{1}{2}(1.96 + 19.92)^2 = 239$$

$$\chi^2_{.025} = \tfrac{1}{2}(z_{.025} + \sqrt{2(199) - 1})^2 = \tfrac{1}{2}(-1.96 + 19.92)^2 = 161$$

from which $\chi_{.975} = 15.5$ and $X_{0.025} = 12.7$. Then the 95% confidence limits are $100\sqrt{200}/15.5 = 91.2$ h and $100\sqrt{200}/12.7 = 111.3$ h, respectively. Thus we can be 95% confident that the population standard deviation will lie between 91.2 and 111.3 h.

(*b*) The 99% confidence limits are given by $s\sqrt{N}/\chi_{.995}$ and $s\sqrt{N}/\chi_{.005}$.
For $\nu = 200 - 1 = 199$ degrees of freedom,

$$\chi^2_{.995} = \tfrac{1}{2}(z_{.995} + \sqrt{2(199) - 1})^2 = \tfrac{1}{2}(2.58 + 19.92)^2 = 253$$
$$\chi^2_{.005} = \tfrac{1}{2}(z_{.005} + \sqrt{2(199) - 1})^2 = \tfrac{1}{2}(-2.58 + 19.92)^2 = 150$$

from which $\chi_{.995} = 15.9$ and $\chi_{.005} = 12.2$. Then the 99% confidence limits are $100\sqrt{200}/15.9 = 88.9$ h and $100\sqrt{200}/12.2 = 115.9$ h, respectively. Thus we can be 99% confident that the population standard deviation will lie between 88.9 and 115.9 h.

11.16 A manufacturer of axles must maintain a mean diameter of 5.000 cm in the manufacturing process. In addition, in order to insure that the wheels fit on the axle properly, it is necessary that the standard deviation of the diameters equal 0.005 cm or less. A sample of 20 axles is obtained and the diameters are given in Table 11.4.

Table 11.4

4.996	4.998	5.002	4.999
5.010	4.997	5.003	4.998
5.006	5.004	5.000	4.993
5.002	4.996	5.005	4.992
5.007	5.003	5.000	5.000

The manufacturer wishes to test the null hypothesis that the population standard deviation is 0.005 cm versus the alternative hypothesis that the population standard deviation exceeds 0.005 cm. If the alternative hypothesis is supported, then the manufacturing process must be stopped and repairs to the machinery must be made. The test procedure assumes that the axle diameters are normally distributed. Test this assumption at the 0.05 level of significance. If you are willing to assume normality, then test the hypothesis concerning the population standard deviation at the 0.05 level of significance.

SOLUTION

The Shapiro–Wilk test of normality is given in Fig. 11-8. The large *p*-value (0.9966) would indicate that normality would not be rejected. This probability plot and Shapiro–Wilk analysis was produced by the statistical software package STATISTIX.

We have to decide between the hypothesis:

$H_0: \sigma = 0.005$ cm, and the observed value is due to chance.

$H_1: \sigma > 0.005$ cm, and the variability is too large.

The SAS analysis is as follows:

```
One Sample Chi-square Test for a Variance
Sample Statistics for diameter
N                    Mean          Std. Dev.           Variance
-----------------------------------------------------------------
20                  5.0006         0.0046              215E-7


Hypothesis Test
Null hypothesis: Variance of diameter <= 0.000025
Alternative: Variance of diameter > 0.000025
Chi-square            Df            Prob
-------------------------------------------
16.358                19            0.6333
```

The large *p*-value (0.6333) would indicate that the null hypothesis would not be rejected.

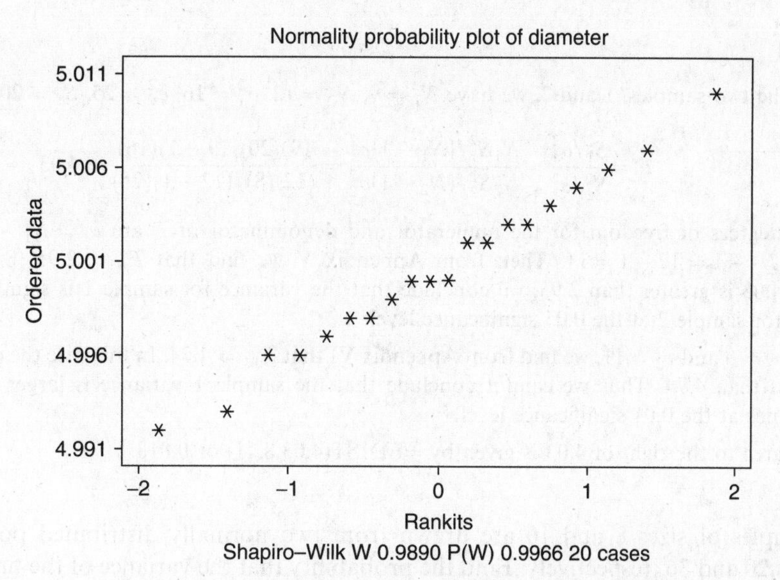

Fig. 11-8 Shapiro–Wilk test of normality from STATISTIX.

11.17 In the past, the standard deviation of weights of certain 40.0-ounce packages filled by a machine was 0.25 ounces (oz). A random sample of 20 packages showed a standard deviation of 0.32 oz. Is the apparent increase in variability significant at the (*a*) 0.05 and (*b*) 0.01 levels?

SOLUTION

We have to decide between the hypotheses:

$H_0 : \sigma = 0.25$ oz, and the observed result is due to chance.

$H_1 : a > 0.25$ oz, and the variability has increased.

The value of χ^2 for the sample is

$$\chi^2 = \frac{Ns^2}{\sigma^2} = \frac{20(0.32)^2}{(0.25)^2} = 32.8$$

(*a*) Using a one-tailed test, we would reject H_0 at the 0.05 significance level if the sample value of χ^2 were greater than $\chi^2_{.95}$, which equals 30.1 for $\nu = 20 - 1 = 19$ degrees of freedom. Thus we would reject H_0 at the 0.05 significance level.

(*b*) Using a one-tailed test, we would reject H_0 at the 0.01 significance level if the sample value of χ^2 were greater than $\chi^2_{.99}$, which equals 36.2 for 19 degrees of freedom. Thus we would not reject H_0 at the 0.01 significance level.

We conclude that the variability has probably increased. An examination of the machine should be made.

THE *F* DISTRIBUTION

11.18 Two samples of sizes 9 and 12 are drawn from two normally distributed populations having variances 16 and 25, respectively. If the sample variances are 20 and 8, determine whether the first sample has a significantly larger variance than the second sample at significance levels of (*a*) 0.05, (*b*) 0.01, and (*c*) Use EXCEL to show that the area to the right of 4.03 is between 0.01 and 0.05.

SOLUTION

For the two samples, 1 and 2, we have $N_1 = 9$, $N_2 = 12$, $\sigma_1^2 = 16$, $\sigma_2^2 = 25$, $S_1^2 = 20$, and $S_2^2 = 8$. Thus

$$F = \frac{\hat{S}_1^2/\sigma_1^2}{\hat{S}_2^2/\sigma_2^2} = \frac{N_1S_1^2/(N_1-1)\sigma_1^2}{N_2S_2^2/(N_2-1)\sigma_2^2} = \frac{(9)(20)/(9-1)(16)}{(12)(8)/(12-1)(25)} = 4.03$$

(*a*) The degrees of freedom for the numerator and denominator of F are $\nu_1 = N_1 - 1 = 9 - 1 = 8$ and $\nu_2 = N_2 - 1 = 12 - 1 = 11$. Then from Appendix V we find that $F_{.95} = 2.95$. Since the calculated $F = 4.03$ is greater than 2.95, we conclude that the variance for sample 1 is significantly larger than that for sample 2 at the 0.05 significance level.

(*b*) For $\nu_1 = 8$ and $\upsilon_2 = 11$, we find from Appendix VI that $F_{.01} = 4.74$. In this case the calculated $F = 4.03$ is less than 4.74. Thus we cannot conclude that the sample 1 variance is larger than the sample 2 variance at the 0.01 significance level.

(*c*) The area to the right of 4.03 is given by =FDIST(4.03,8,11) or 0.018.

11.19 Two samples of sizes 8 and 10 are drawn from two normally distributed populations having variances 20 and 36, respectively. Find the probability that the variance of the first sample is more than twice the variance of the second sample.
Use EXCEL to find the exact probability that F with 7 and 9 degrees of freedom exceeds 3.70.

SOLUTION

We have $N_1 = 8$, $N_2 = 10$, $\sigma_1^2 = 20$, and $\sigma_2^2 = 36$. Thus

$$F = \frac{8S_1^2/(7)(20)}{10S_2^2/(9)(36)} = 1.85\frac{S_1^2}{S_2^2}$$

The number of degrees of freedom for the numerator and denominator are $\nu_1 = N_1 - 1 = 8 - 1 = 7$ and $\nu_2 = N_2 - 1 = 10 - 1 = 9$. Now if S_1^2 is more than twice S_2^2, then

$$F = 1.85\frac{S_1^2}{S_2^2} > (1.85)(2) = 3.70$$

Looking up 3.70 in Appendixes V and VI, we find that the probability is less than 0.05 but greater than 0.01. For exact values, we need a more extensive tabulation of the F distribution.

The EXCEL answer is =FDIST(3.7,7,9) or 0.036 is the probability that F with 7 and 9 degrees of freedom exceeds 3.70.

Supplementary Problems

STUDENT'S *t* DISTRIBUTION

11.20 For a Student's distribution with 15 degrees of freedom, find the value of t_1 such that (*a*) the area to the right of t_1 is 0.01, (*b*) the area to the left of t_1 is 0.95, (*c*) the area to the right of t_1 is 0.10, (*d*) the combined area to the right of t_1 and to the left of $-t_1$ is 0.01, and (*e*) the area between $-t_1$ and t_1 is 0.95.

11.21 Find the critical values of t for which the area of the right-hand tail of the t distribution is 0.01 if the number of degrees of freedom, ν, is equal to (*a*) 4, (*b*) 12, (*c*) 25, (*d*) 60, and (*e*) 150 using Appendix III. Give the solutions to (*a*) through (*e*) using EXCEL.

11.22 Find the values of t_1 for Student's distribution that satisfy each of the following conditions:

(a) The area between $-t_1$ and t_1 is 0.90 and $\nu = 25$.

(b) The area to the left of $-t_1$ is 0.025 and $\nu = 20$.

(c) The combined area to the right of t_1 and left of $-t_1$ is 0.01 and $\nu = 5$.

(d) The area to the right of t_1 is 0.55 and $\nu = 16$.

11.23 If a variable U has a Student's distribution with $\nu = 10$, find the constant C such that (a) $\Pr\{U > C\} = 0.05$, (b) $\Pr\{-C \le U \le C\} = 0.98$, (c) $\Pr\{U \le C\} = 0.20$, and (d) $\Pr\{U \ge C\} = 0.90$.

11.24 The 99% confidence coefficients (two-tailed) for the normal distribution are given by ± 2.58. What are the corresponding coefficients for the t distribution if (a) $\nu = 4$, (b) $\nu = 12$, (c) $\nu = 25$, (d) $\nu = 30$, and (e) $\nu = 40$?

11.25 A sample of 12 measurements of the breaking strength of cotton threads gave a mean of 7.38 grams (g) and a standard deviation of 1.24 g. Find the (a) 95%, (b) 99% confidence limits for the actual breaking strength, and (c) the MINITAB solutions using the summary statistics.

11.26 Work Problem 11.25 by assuming that the methods of large sampling theory are applicable, and compare the results obtained.

11.27 Five measurements of the reaction time of an individual to certain stimuli were recorded as 0.28, 0.30, 0.27, 0.33, and 0.31 seconds. Find the (a) 95% and (b) 99% confidence limits for the actual reaction time.

11.28 The mean lifetime of electric light bulbs produced by a company has in the past been 1120 h with a standard deviation of 125 h. A sample of eight electric light bulbs recently chosen from a supply of newly produced bulbs showed a mean lifetime of 1070 h. Test the hypothesis that the mean lifetime of the bulbs has not changed, using significance levels of (a) 0.05 and (b) 0.01.

11.29 In Problem 11.28, test the hypothesis $\mu = 1120$ h against the alternative hypothesis $\mu < 1120$ h, using significance levels of (a) 0.05 and (b) 0.01.

11.30 The specifications for the production of a certain alloy call for 23.2% copper. A sample of 10 analyses of the product showed a mean copper content of 23.5% and a standard deviation of 0.24%. Can we conclude at (a) 0.01 and (b) 0.05 significance levels that the product meets the required specifications?

11.31 In Problem 11.30, test the hypothesis that the mean copper content is higher than in the required specifications, using significance levels of (a) 0.01 and (b) 0.05.

11.32 An efficiency expert claims that by introducing a new type of machinery into a production process, he can substantially decrease the time required for production. Because of the expense involved in maintenance of the machines, management feels that unless the production time can be decreased by at least 8.0%, it cannot afford to introduce the process. Six resulting experiments show that the time for production is decreased by 8.4% with a standard deviation of 0.32%. Using significance levels of (a) 0.01 and (b) 0.05, test the hypothesis that the process should be introduced.

11.33 Using brand A gasoline, the mean number of miles per gallon traveled by five similar automobiles under identical conditions was 22.6 with a standard deviation of 0.48. Using brand B, the mean number was 21.4 with a standard deviation of 0.54. Using a significance level of 0.05, investigate whether brand A is really better than brand B in providing more mileage to the gallon.

11.34 Two types of chemical solutions, A and B, were tested for their pH (degree of acidity of the solution). Analysis of six samples of A showed a mean pH of 7.52 with a standard deviation of 0.024. Analysis of five samples of B showed a mean pH of 7.49 with a standard deviation of 0.032. Using the 0.05 significance level, determine whether the two types of solutions have different pH values.

11.35 On an examination in psychology, 12 students in one class had a mean grade of 78 with a standard deviation of 6, while 15 students in another class had a mean grade of 74 with a standard deviation of 8. Using a significance level of 0.05, determine whether the first group is superior to the second group.

THE CHI-SQUARE DISTRIBUTION

11.36 For a chi-square distribution with 12 degrees of freedom, use Appendix IV to find the value of χ_c^2 such that (*a*) the area to the right of χ_c^2 is 0.05, (*b*) the area to the left of χ_c^2 is 0.99, (*c*) the area to the right of χ_c^2 is 0.025, and (*d*) find (*a*) through (*c*) using EXCEL.

11.37 Find the critical values of χ^2 for which the area of the right-hand tail of the χ^2 distribution is 0.05 if the number of degrees of freedom, ν, is equal to (*a*) 8, (*b*) 19, (*c*) 28, and (*d*) 40.

11.38 Work Problem 11.37 if the area of the right-hand tail is 0.01.

11.39 (*a*) Find χ_1^2 and χ_2^2 such that the area under the χ^2 distribution corresponding to $\nu = 20$ between χ_1^2 and χ_2^2 is 0.95, assuming equal areas to the right of χ_2^2 and left of χ_1^2.

(*b*) Show that if the assumption of equal areas in part (*a*) is not made, the values χ_1^2 and χ_2^2 are not unique.

11.40 If the variable U is chi-square distributed with $\nu = 7$, find χ_1^2 and χ_2^2 such that (*a*) $\Pr\{U > \chi_2^2\} = 0.025$, (*b*) $\Pr\{U < \chi_1^2\} = 0.50$, and (*c*) $\Pr\{\chi_1^2 \le U \le \chi_2^2\} = 0.90$.

11.41 The standard deviation of the lifetimes of 10 electric light bulbs manufactured by a company is 120 h. Find the (*a*) 95% and (*b*) 99% confidence limits for the standard deviation of all bulbs manufactured by the company.

11.42 Work Problem 11.41 if 25 electric light bulbs show the same standard deviation of 120 h.

11.43 Find (*a*)$\chi_{.05}^2$ and (*b*) $\chi_{.95}^2$ for $\nu = 150$ using $\chi_p^2 = \frac{1}{2}(z_p + \sqrt{2v - 1})^2$ and (*c*) compare the answers when EXCEL is used.

11.44 Find (*a*) $\chi_{.025}^2$ and (*b*) $\chi_{.975}^2$ for $\nu = 250$ using $\chi_p^2 = \frac{1}{2}(z_p + \sqrt{2v - 1})^2$ and (*c*) compare the answers when EXCEL is used.

11.45 Show that for large values of ν, a good approximation to χ^2 is given by $(v + z_p\sqrt{2\nu})$, where z_p is the pth percentile of the standard normal distribution.

11.46 Work Problem 11.39 by using the χ^2 distributions if a sample of 100 electric bulbs shows the same standard deviation of 120 h. Compare the results with those obtained by the methods of Chapter 9.

11.47 What is the 95% confidence interval of Problem 11.44 that has the least width?

11.48 The standard deviation of the breaking strengths of certain cables produced by a company is given as 240 pounds (lb). After a change was introduced in the process of manufacture of these cables, the breaking strengths of a sample of eight cables showed a standard deviation of 300 lb. Investigate the significance of the apparent increase in variability, using significance levels of (*a*) 0.05 and (*b*) 0.01.

11.49 The standard deviation of the annual temperatures of a city over a period of 100 years was 16° Fahrenheit. Using the mean temperature on the 15th day of each month during the last 15 years, a standard deviation of annual temperatures was computed as 10° Fahrenheit. Test the hypothesis that the temperatures in the city have become less variable than in the past, using significance levels of (*a*) 0.05 and (*b*) 0.01.

THE *F* DISTRIBUTION

11.50 Find the values of F in parts *a*, *b*, *c*, and *d* using Appendix V and VI.

(*a*) $F_{0.95}$ with $V_1 = 8$ and $V_2 = 10$.

(*b*) $F_{0.99}$ with $V_1 = 24$ and $V_2 = 11$.

(*c*) $F_{0.85}$ with $N_1 = 16$ and $N_2 = 25$.

(*d*) $F_{0.90}$ with $N_1 = 21$ and $N_2 = 23$.

11.51 Solve Problem 11.50 using EXCEL.

11.52 Two samples of sizes 10 and 15 are drawn from two normally distributed populations having variances 40 and 60, respectively. If the sample variances are 90 and 50, determine whether the sample 1 variance is significantly greater than the sample 2 variance at significance levels of (*a*) 0.05 and (*b*) 0.01.

11.53 Two companies, *A* and *B*, manufacture electric light bulbs. The lifetimes for the *A* and *B* bulbs are very nearly normally distributed, with standard deviations of 20 h and 27 h, respectively. If we select 16 bulbs from company *A* and 20 bulbs from company *B* and determine the standard deviations of their lifetimes to be 15 h and 40 h, respectively, can we conclude at significance levels of (*a*) 0.05 and (*b*) 0.01 that the variability of the *A* bulbs is significantly less than that of the *B* bulbs?

The Chi-Square Test

OBSERVED AND THEORETICAL FREQUENCIES

As we have already seen many times, the results obtained in samples do not always agree exactly with the theoretical results expected according to the rules of probability. For example, although theoretical considerations lead us to expect 50 heads and 50 tails when we toss a fair coin 100 times, it is rare that these results are obtained exactly.

Suppose that in a particular sample a set of possible events $E_1, E_2, E_3, \ldots, E_k$ (see Table 12.1) are observed to occur with frequencies $o_1, o_2, o_3, \ldots, o_k$, called *observed frequencies*, and that according to probability rules they are expected to occur with frequencies $e_1, e_2, e_3, \ldots, e_k$, called *expected*, or *theoretical, frequencies*. Often we wish to know whether the observed frequencies differ significantly from the expected frequencies.

Table 12.1

Event	E_1	E_2	E_3	$\cdots$	E_k
Observed frequency	o_1	o_2	o_3	$\cdots$	o_k
Expected frequency	e_1	e_2	e_3	$\cdots$	e_k

DEFINITION OF χ^2

A measure of the discrepancy existing between the observed and expected frequencies is supplied by the statistic χ^2 (read chi-square) given by

$$\chi^2 = \frac{(o_1 - e_1)^2}{e_1} + \frac{(o_2 - e_2)^2}{e_2} + \cdots + \frac{(o_k - e_k)^2}{e_k} = \sum_{j=1}^{k} \frac{(o_j - e_j)^2}{e_j} \tag{1}$$

where if the total frequency is N,

$$\sum o_j = \sum e_j = N \tag{2}$$

An expression equivalent to formula (*1*) is (see Problem 12.11)

$$\chi^2 = \sum \frac{o_j^2}{e_j} - N \tag{3}$$

If $\chi^2 = 0$, the observed and theoretical frequencies agree exactly; while if $\chi^2 > 0$, they do not agree exactly. The larger the value of χ^2, the greater is the discrepancy between the observed and expected frequencies.

The sampling distribution of χ^2 is approximated very closely by the chi-square distribution

$$Y = Y_0(\chi^2)^{1/2(\nu-2)}e^{-1/2\chi^2} = Y_0\chi^{\nu-2}e^{-1/2\chi^2} \tag{4}$$

(already considered in Chapter 11) if the expected frequencies are at least equal to 5. The approximation improves for larger values.

The number of degrees of freedom, ν, is given by

(1) $\nu = k - 1$ if the expected frequencies can be computed without having to estimate the population parameters from sample statistics. Note that we subtract 1 from k because of constraint condition (*2*), which states that if we know $k - 1$ of the expected frequencies, the remaining frequency can be determined.

(2) $\nu = k - 1 - m$ if the expected frequencies can be computed only by estimating m population parameters from sample statistics.

SIGNIFICANCE TESTS

In practice, expected frequencies are computed on the basis of a hypothesis H_0. If under this hypothesis the computed value of χ^2 given by equation (*1*) or (*3*) is greater than some critical value (such as $\chi^2_{.95}$ or $\chi^2_{.99}$, which are the critical values of the 0.05 and 0.01 significance levels, respectively), we would conclude that the observed frequencies differ *significantly* from the expected frequencies and would reject H_0 at the corresponding level of significance; otherwise, we would accept it (or at least not reject it). This procedure is called *the chi-square test* of hypothesis or significance.

It should be noted that we must look with suspicion upon circumstances where χ^2 is *too close to zero*, since it is rare that observed frequencies agree *too well* with expected frequencies. To examine such situations, we can determine whether the computed value of χ^2 is less than $\chi^2_{.05}$ or $\chi^2_{.01}$, in which cases we would decide that the agreement is *too good* at the 0.05 or 0.01 significance levels, respectively.

THE CHI-SQUARE TEST FOR GOODNESS OF FIT

The chi-square test can be used to determine how well theoretical distributions (such as the normal and binomial distributions) fit empirical distributions (i.e., those obtained from sample data). See Problems 12.12 and 12.13.

EXAMPLE 1. A pair of dice is rolled 500 times with the sums in Table 12.2 showing on the dice:

Table 12.2

Sum	2	3	4	5	6	7	8	9	10	11	12
Observed	15	35	49	58	65	76	72	60	35	29	6

The expected number, if the dice are fair, are determined from the distribution of x as in Table 12.3.

Table 12.3

x	2	3	4	5	6	7	8	9	10	11	12
$p(x)$	1/36	2/36	3/36	4/36	5/36	6/36	5/36	4/36	3/36	2/36	1/36

We have the observed and expected frequencies in Table 12.4.

Table 12.4

Observed	15	35	49	58	65	76	72	60	35	29	6
Expected	13.9	27.8	41.7	55.6	69.5	83.4	69.5	55.6	41.7	27.8	13.9

If the observed and expected are entered into B1:L2 in the EXCEL worksheet, the expression =(B1-B2)^2/B2 is entered into B4, a click-and-drag is executed from B4 to L4, and then the quantities in B4:L4 are summed we obtain 10.34 for $\chi^2 = \sum_j ((o_j - e_j)^2/e_j)$.

The *p*-value corresponding to 10.34 is given by the EXCEL expression =CHIDIST(10.34,10). The *p*-value is 0.411. Because of this large *p*-value, we have no reason to doubt the fairness of the dice.

CONTINGENCY TABLES

Table 12.1, in which the observed frequencies occupy a single row, is called a *one-way classification table*. Since the number of columns is k, this is also called a $1 \times k$ (read "1 by k") *table*. By extending these ideas, we can arrive at *two-way classification tables*, or $h \times k$ *tables*, in which the observed frequencies occupy h rows and k columns. Such tables are often called *contingency tables.*

Corresponding to each observed frequency in an $h \times k$ contingency table, there is an *expected* (or *theoretical*) *frequency* that is computed subject to some hypothesis according to rules of probability. These frequencies, which occupy the *cells* of a contingency table, are called *cell frequencies*. The total frequency in each row or each column is called the *marginal frequency*.

To investigate agreement between the observed and expected frequencies, we compute the statistic

$$\chi^2 = \sum_j \frac{(o_j - e_j)^2}{e_j} \tag{5}$$

where the sum is taken over all cells in the contingency table and where the symbols o_j and e_j represent, respectively, the observed and expected frequencies in the jth cell. This sum, which is analogous to equation (*1*), contains hk terms. The sum of all observed frequencies is denoted by N and is equal to the sum of all expected frequencies [compare with equation (*2*)].

As before, statistic (*5*) has a sampling distribution given very closely by (*4*), provided the expected frequencies are not too small. The number of degrees of freedom, ν, of this chi-square distribution is given for $h > 1$ and $k > 1$ by

1. $\nu = (h-1)(k-1)$ if the expected frequencies can be computed without having to estimate population parameters from sample statistics. For a proof of this, see Problem 12.18.
2. $\nu = (h-1)(k-1) - m$ if the expected frequencies can be computed only by estimating m population parameters from sample statistics.

Significance tests for $h \times k$ tables are similar to those for $1 \times k$ tables. The expected frequencies are found subject to a particular hypothesis H_0. A hypothesis commonly assumed is that the two classifications are independent of each other.

Contingency tables can be extended to higher dimensions. Thus, for example, we can have $h \times k \times l$ tables, where three classifications are present.

EXAMPLE 2. The data in Table 12.5 were collected on how individuals prepared their taxes and their education level. The null hypothesis is that the way people prepare their taxes (computer software or pen and paper) is independent of their education level. Table 12.5 is a contingency table.

Table 12.5

	Education Level		
Tax prep.	High school	Bachelors	Masters
computer	23	35	42
Pen and paper	45	30	25

If MINITAB is used to analyze this data, the following results are obtained.

```
Chi-Square Test: highschool, bachelors, masters

Expected counts are printed below observed counts
Chi-Square contributions are printed below expected counts

      highschool  bachelors    masters    Total
1             23         35         42      100
           34.00      32.50      33.50
           3.559      0.192      2.157

2             45         30         25      100
           34.00      32.50      33.50
           3.559      0.192      2.157

Total         68         65         67      200

Chi-Sq = 11.816, DF = 2, P-Value = 0.003
```

Because of the small p-value, the hypothesis of independence would be rejected and we would conclude that tax preparation would be contingent upon education level.

YATES' CORRECTION FOR CONTINUITY

When results for continuous distributions are applied to discrete data, certain corrections for continuity can be made, as we have seen in previous chapters. A similar correction is available when the chi-square distribution is used. The correction consists in rewriting equation (*1*) as

$$\chi^2(\text{corrected}) = \frac{(|o_1 - e_1| - 0.5)^2}{e_1} + \frac{(|o_2 - e_2| - 0.5)^2}{e_2} + \cdots + \frac{(|o_k - e_k| - 0.5)^2}{e_k} \tag{6}$$

and is often referred to as *Yates' correction*. An analogous modification of equation (*5*) also exists.

In general, the correction is made only when the number of degrees of freedom is $\nu = 1$. For large samples, this yields practically the same results as the uncorrected χ^2, but difficulties can arise near critical values (see Problem 12.8). For small samples where each expected frequency is between 5 and 10, it is perhaps best to compare both the corrected and uncorrected values of χ^2. If both values lead to the same conclusion regarding a hypothesis, such as rejection at the 0.05 level, difficulties are rarely encountered. If they lead to different conclusions, one can resort to increasing the sample sizes or, if this proves impractical, one can employ methods of probability involving the *multinomial distribution* of Chapter 6.

SIMPLE FORMULAS FOR COMPUTING χ^2

Simple formulas for computing χ^2 that involve only the observed frequencies can be derived. The following gives the results for 2×2 and 2×3 contingency tables (see Tables 12.6 and 12.7, respectively).

2 × 2 Tables

$$\chi^2 = \frac{N(a_1 b_2 - a_2 b_1)^2}{(a_1 + b_1)(a_2 + b_2)(a_1 + a_2)(b_1 + b_2)} = \frac{N\Delta^2}{N_1 N_2 N_A N_B} \tag{7}$$

Table 12.6

	I	II	Total
A	a_1	a_2	N_A
B	b_1	b_2	N_B
Total	N_1	N_2	N

Table 12.7

	I	II	III	Total
A	a_1	a_2	a_3	N_A
B	b_1	b_2	b_3	N_B
Total	N_1	N_2	N_3	N

where $\Delta = a_1b_2 - a_2b_1$, $N = a_1 + a_2 + b_1 + b_2$, $N_1 = a_1 + b_1$, $N_2 = a_2 + b_2$, $N_A = a_1 + a_2$, and $N_B = b_1 + b_2$ (see Problem 12.19). With Yates' correction, this becomes

$$\chi^2 \text{ (corrected)} = \frac{N(|a_1b_2 - a_2b_1| - \frac{1}{2}N)^2}{(a_1 + b_1)(a_2 + b_2)(a_1 + a_2)(b_1 + b_2)} = \frac{N(|\Delta| - \frac{1}{2}N)^2}{N_1N_2N_AN_B} \tag{8}$$

2 × 3 Tables

$$\chi^2 = \frac{N}{N_A}\left[\frac{a_1^2}{N_1} + \frac{a_2^2}{N_2} + \frac{a_3^2}{N_3}\right] + \frac{N}{N_B}\left[\frac{b_1^2}{N_1} + \frac{b_2^2}{N_2} + \frac{b_3^2}{N_3}\right] - N \tag{9}$$

where we have used the general result valid for all contingency tables (see Problem 12.43):

$$\chi^2 = \sum \frac{o_j^2}{e_j} - N \tag{10}$$

Result (*9*) for $2 \times k$ tables where $k > 3$ can be generalized (see Problem 12.46).

COEFFICIENT OF CONTINGENCY

A measure of the degree of relationship, association, or dependence of the classifications in a contingency table is given by

$$C = \sqrt{\frac{\chi^2}{\chi^2 + N}} \tag{11}$$

which is called the *coefficient of contingency*. The larger the value of C, the greater is the degree of association. The number of rows and columns in the contingency table determines the maximum value of C, which is never greater than 1. If the number of rows and columns of a contingency table is equal to k, the maximum value of C is given by $\sqrt{(k-1)/k}$ (see Problems 12.22, 12.52, and 12.53).

EXAMPLE 3. Find the coefficient of contingency for Example 2.

$$C = \sqrt{\frac{\chi^2}{\chi^2 + N}} = \sqrt{\frac{11.816}{11.816 + 200}} = 0.236$$

CORRELATION OF ATTRIBUTES

Because classifications in a contingency table often describe characteristics of individuals or objects, they are often referred to as *attributes*, and the degree of dependence, association, or relationship is called the *correlation* of attributes. For $k \times k$ tables, we define

$$r = \sqrt{\frac{\chi^2}{N(k-1)}} \tag{12}$$

as the correlation coefficient between attributes (or classifications). This coefficient lies between 0 and 1 (see Problem 12.24). For 2×2 tables in which $k = 2$, the correlation is often called *tetrachoric correlation.*

The general problem of correlation of numerical variables is considered in Chapter 14.

ADDITIVE PROPERTY OF χ^2

Suppose that the results of repeated experiments yield sample values of χ^2 given by $\chi_1^2, \chi_2^2, \chi_3^2, \ldots$ with $\nu_1, \nu_2, \nu_3, \ldots$ degrees of freedom, respectively. Then the result of all these experiments can be considered equivalent to a χ^2 value given by $\chi_1^2 + \chi_2^2 + \chi_3^2 + \cdots$ with $\nu_1 + \nu_2 + \nu_3 + \cdots$ degrees of freedom (see Problem 12.25).

Solved Problems

THE CHI-SQUARE TEST

12.1 In 200 tosses of a coin, 115 heads and 85 tails were observed. Test the hypothesis that the coin is fair, using Appendix IV and significance levels of (*a*) 0.05 and (*b*) 0.01. Test the hypothesis by computing the *p*-value and (*c*) comparing it to levels 0.05 and 0.01.

SOLUTION

The observed frequencies of heads and tails are $o_1 = 115$ and $o_2 = 85$, respectively, and the expected frequencies of heads and tails (if the coin is fair) are $e_1 = 100$ and $e_2 = 100$, respectively. Thus

$$\chi^2 = \frac{(o_1 - e_1)^2}{e_1} + \frac{(o_2 - e_2)^2}{e_2} = \frac{(115 - 100)^2}{100} + \frac{(85 - 100)^2}{100} = 4.50$$

Since the number of categories, or classes (heads, tails), is $k = 2$, $\nu = k - 1 = 2 - 1 = 1$.

(*a*) The critical value $\chi^2_{.95}$ for 1 degree of freedom is 3.84. Thus, since $4.50 > 3.84$, we reject the hypothesis that the coin is fair at the 0.05 significance level.

(*b*) The critical value $\chi^2_{.99}$ for 1 degree of freedom is 6.63. Thus, since $4.50 < 6.63$, we cannot reject the hypothesis that the coin is fair at the 0.02 significance level.

We conclude that the observed results are *probably significant* and that the coin is *probably not fair.* For a comparison of this method with previous methods used, see Problem 12.3.

Using EXCEL, the *p*-value is given by =CHIDIST(4.5,1), which equals 0.0339. And we see, using the *p*-value approach that the results are significant at 0.05 but not at 0.01. Either of these methods of testing may be used.

12.2 Work Problem 12.1 by using Yates' correction.

SOLUTION

$$\chi^2\,(\text{corrected}) = \frac{(|o_1 - e_1| - 0.5)^2}{e_1} + \frac{(|o_2 - e_2| - 0.5)^2}{e_2} = \frac{(|115 - 100| - 0.5)^2}{100} + \frac{(|85 - 100| - 0.5)^2}{100}$$

$$= \frac{(14.5)^2}{100} + \frac{(14.5)^2}{100} = 4.205$$

Since $4.205 > 3.84$ and $4.205 < 6.63$, the conclusions reached in Problem 12.1 are valid. For a comparison with previous methods, see Problem 12.3.

12.3 Work Problem 12.1 by using the normal approximation to the binomial distribution.

SOLUTION

Under the hypothesis that the coin is fair, the mean and standard deviation of the number of heads expected in 200 tosses of a coin are $\mu = Np = (200)(0.5) = 100$ and $\sigma = \sqrt{Npq} = \sqrt{(200)(0.5)(0.5)} = 7.07$, respectively.

First method

$$115 \text{ heads in standard units} = \frac{115 - 100}{7.07} = 2.12$$

Using the 0.05 significance level and a two-tailed test, we would reject the hypothesis that the coin is fair if the z score were outside the interval -1.96 to 1.96. With the 0.01 level, the corresponding interval would be -2.58 to 2.58. It follows (as in Problem 12.1) that we can reject the hypothesis at the 0.05 level but cannot reject it at the 0.01 level.

Note that the square of the above standard score, $(2.12)^2 = 4.50$, is the same as the value of χ^2 obtained in Problem 12.1. This is always the case for a chi-square test involving two categories (see Problem 12.10).

Second method

Using the correction for continuity, 115 or more heads is equivalent to 114.5 or more heads. Then 114.5 in standard units $= (114.5 - 100)/7.07 = 2.05$. This leads to the same conclusions as in the first method.

Note that the square of this standard score is $(2.05)^2 = 4.20$, agreeing with the value of χ^2 corrected for continuity by using Yates' correction of Problem 12.2. This is always the case for a chi-square test involving two categories in which Yates' correction is applied.

12.4 Table 12.8 shows the observed and expected frequencies in tossing a die 120 times.

(*a*) Test the hypothesis that the die is fair using a 0.05 significance level by calculating χ^2 and giving the 0.05 critical value and comparing the computed test statistic with the critical value.

(*b*) Compute the p-value and compare it with 0.05 to test the hypothesis.

Table 12.8

Die face	1	2	3	4	5	6
Observed frequency	25	17	15	23	24	16
Expected frequency	20	20	20	20	20	20

SOLUTION

$$\chi^2 = \frac{(o_1 - e_1)^2}{e_1} + \frac{(o_2 - e_2)^2}{e_2} + \frac{(o_3 - e_3)^2}{e_3} + \frac{(o_4 - e_4)^2}{e_4} + \frac{(o_5 - e_5)^2}{e_5} + \frac{(o_6 - e_6)^2}{e_6}$$

$$= \frac{(25 - 20)^2}{20} + \frac{(17 - 20)^2}{20} + \frac{(15 - 20)^2}{20} + \frac{(23 - 20)^2}{20} + \frac{(24 - 20)^2}{20} + \frac{(16 - 20)^2}{20} = 5.00$$

(*a*) The 0.05 critical value is given by the EXCEL expression =CHIINV(0.05, 5) or 11.0705. The computed value of the test statistic is 5.00. Since the computed test statistic is not in the 0.05 critical region, do not reject the null that the die is fair.

(*b*) The p-value is given by the EXCEL expression =CHIDIST(5.00, 5) or 0.4159. Since the p-value is not less than 0.05, do not reject the null that the die is fair.

12.5 Table 12.9 shows the distribution of the digits 0, 1, 2, ... , 9 in a random-number table of 250 digits. (*a*) Find the value of the test statistic χ^2, (*b*) find the 0.01 critical value and give your conclusion for $\alpha = 0.01$, and (*c*) find the *p*-value for the value you found in (*a*) and give your conclusion for $\alpha = 0.01$.

Table 12.9

Digit	0	1	2	3	4	5	6	7	8	9
Observed frequency	17	31	29	18	14	20	35	30	20	36
Expected frequency	25	25	25	25	25	25	25	25	25	25

SOLUTION

(*a*) $$\chi^2 = \frac{(17-25)^2}{25} + \frac{(31-25)^2}{25} + \frac{(29-25)^2}{25} + \frac{(18-25)^2}{25} + \cdots + \frac{(36-25)^2}{25} = 23.3$$

(*b*) The 0.01 critical value is given by =CHIINV(0.01, 9) which equals 21.6660. Since the computed value of χ^2 exceeds this value, we reject the hypothesis that the numbers are random.

(*c*) The *p*-value is given by the EXCEL expression =CHIDIST(23.3, 9) which equals 0.0056, which is less than 0.01. By the *p*-value technique, we reject the null.

12.6 In his experiments with peas, Gregor Mendel observed that 315 were round and yellow, 108 were round and green, 101 were wrinkled and yellow, and 32 were wrinkled and green. According to his theory of heredity, the numbers should be in the proportion 9 : 3 : 3 : 1. Is there any evidence to doubt his theory at the (*a*) 0.01 and (*b*) 0.05 significance levels?

SOLUTION

The total number of peas is $315 + 108 + 101 + 32 = 556$. Since the expected numbers are in the proportion 9 : 3 : 3 : 1 (and $9 + 3 + 3 + 1 = 16$), we would expect

$\frac{9}{16}(556) = 312.75$ round and yellow $\qquad \frac{3}{16}(556) = 104.25$ wrinkled and yellow

$\frac{3}{16}(556) = 104.25$ round and green $\qquad \frac{1}{16}(556) = 34.75$ wrinkled and green

Thus $$\chi^2 = \frac{(315-312.75)^2}{312.75} + \frac{(108-104.25)^2}{104.25} + \frac{(101-104.25)^2}{104.25} + \frac{(32-34.75)^2}{34.75} = 0.470$$

Since there are four categories, $k = 4$ and the number of degrees of freedom is $\nu = 4 - 1 = 3$.

(*a*) For $\nu = 3$, $\chi^2_{.99} = 11.3$, and thus we cannot reject the theory at the 0.01 level.

(*b*) For $\nu = 3$, $\chi^2_{.95} = 7.81$, and thus we cannot reject the theory at the 0.05 level.

We conclude that the theory and experiment are in agreement.

Note that for 3 degrees of freedom, $\chi^2_{.05} = 0.352$ and $\chi^2 = 0.470 > 0.352$. Thus, although the agreement is good, the results obtained are subject to a reasonable amount of sampling error.

12.7 An urn contains a very large number of marbles of four different colors: red, orange, yellow, and green. A sample of 12 marbles drawn at random from the urn revealed 2 red, 5 orange, 4 yellow, and 1 green marble. Test the hypothesis that the urn contains equal proportions of the differently colored marbles.

SOLUTION

Under the hypothesis that the urn contains equal proportions of the differently colored marbles, we would expect 3 of each kind in a sample of 12 marbles. Since these expected numbers are less than 5, the chi-square approximation will be in error. To avoid this, we combine categories so that the expected number in each category is at least 5.

If we wish to reject the hypothesis, we should combine categories in such a way that the evidence against the hypothesis shows up best. This is achieved in our case by considering the categories "red or green" and "orange or yellow," for which the sample revealed 3 and 9 marbles, respectively. Since the expected number in each category under the hypothesis of equal proportions is 6, we have

$$\chi^2 = \frac{(3-6)^2}{6} + \frac{(9-6)^2}{6} = 3$$

For $\nu = 2 - 1 = 1$, $\chi^2_{.95} = 3.84$. Thus we cannot reject the hypothesis at the 0.05 significance level (although we can at the 0.10 level). Conceivably the observed results could arise on the basis of chance even when equal proportions of the colors are present.

Another method

Using Yates' correction, we find

$$\chi^2 = \frac{(|3-6| - 0.5)^2}{6} + \frac{(|9-6| - 0.5)^2}{6} = \frac{(2.5)^2}{6} + \frac{(2.5)^2}{6} = 2.1$$

which leads to the same conclusion given above. This is to be expected, of course, since Yates' correction always *reduces* the value of χ^2.

It should be noted that if the χ^2 approximation is used despite the fact that the frequencies are too small, we would obtain

$$\chi^2 = \frac{(2-3)^2}{3} + \frac{(5-3)^2}{3} + \frac{(4-3)^2}{3} + \frac{(1-3)^2}{3} = 3.33$$

Since for $\nu = 4 - 1 = 3$, $\chi^2_{.95} = 7.81$, we would arrive at the same conclusions as above. Unfortunately, the χ^2 approximation for small frequencies is poor; hence, when it is not advisable to combine frequencies, we must resort to the exact probability methods of Chapter 6.

12.8 In 360 tosses of a pair of dice, 74 sevens and 24 elevens are observed. Using the 0.05 significance level, test the hypothesis that the dice are fair.

SOLUTION

A pair of dice can fall 36 ways. A seven can occur in 6 ways, an eleven in 2 ways. Then $\Pr\{\text{seven}\} = \frac{6}{36} = \frac{1}{6}$ and $\Pr\{\text{eleven}\} = \frac{2}{36} = \frac{1}{18}$. Thus in 360 tosses we would expect $\frac{1}{6}(360) = 60$ sevens and $\frac{1}{18}(360) = 20$ elevens, so that

$$\chi^2 = \frac{(74-60)^2}{60} + \frac{(24-20)^2}{20} = 4.07$$

For $\nu = 2 - 1 = 1$, $\chi^2_{.95} = 3.84$. Thus, since $4.07 > 3.84$, we would be inclined to reject the hypothesis that the dice are fair. Using Yates' correction, however, we find

$$\chi^2\,(\text{corrected}) = \frac{(|74-60| - 0.5)^2}{60} + \frac{(|24-20| - 0.5)^2}{20} = \frac{(13.5)^2}{60} + \frac{(3.5)^2}{20} = 3.65$$

Thus on the basis of the corrected χ^2 we could not reject the hypothesis at the 0.05 level.

In general, for large samples such as we have here, results using Yates' correction prove to be more reliable than uncorrected results. However, since even the corrected value of χ^2 lies so close to the critical value, we are hesitant about making decisions one way or the other. In such cases it is perhaps best to increase the sample size by taking more observations if we are interested especially in the 0.05 level for some reason; otherwise, we could reject the hypothesis at some other level (such as 0.10) if this is satisfactory.

12.9 A survey of 320 families with 5 children revealed the distribution shown in Table 12.10. Is the result consistent with the hypothesis that male and female births are equally probable?

Table 12.10

Number of boys and girls	5 boys 0 girls	4 boys 1 girl	3 boys 2 girls	2 boys 3 girls	1 boy 4 girls	0 boys 5 girls	Total
Number of families	18	56	110	88	40	8	320

SOLUTION

Let p = probability of a male birth, and let $q = 1 - p$ = probability of a female birth. Then the probabilities of (5 boys), (4 boys and 1 girl), ..., (5 girls) are given by the terms in the binomial expansion

$$(p+q)^5 = p^5 + 5p^4q + 10p^3q^2 + 10p^2q^3 + 5pq^4 + q^5$$

If $p = q = \frac{1}{2}$, we have

$\Pr\{5 \text{ boys and } 0 \text{ girls}\} = (\frac{1}{2})^5 = \frac{1}{32}$ $\qquad \Pr\{2 \text{ boys and } 3 \text{ girls}\} = 10(\frac{1}{2})^2(\frac{1}{2})^3 = \frac{10}{32}$

$\Pr\{4 \text{ boys and } 1 \text{ girl}\} = 5(\frac{1}{2})^4(\frac{1}{2}) = \frac{5}{32}$ $\qquad \Pr\{1 \text{ boy and } 4 \text{ girls}\} = 5(\frac{1}{2})(\frac{1}{2})^4 = \frac{5}{32}$

$\Pr\{3 \text{ boys and } 2 \text{ girls}\} = 10(\frac{1}{2})^3(\frac{1}{2})^2 = \frac{10}{32}$ $\qquad \Pr\{0 \text{ boys and } 5 \text{ girls}\} = (\frac{1}{2})^5 = \frac{1}{32}$

Then the expected number of families with 5, 4, 3, 2, 1, and 0 boys are obtained by multiplying the above probabilities by 320, and the results are 10, 50, 100, 100, 50, and 10, respectively. Hence

$$\chi^2 = \frac{(18-10)^2}{10} + \frac{(56-50)^2}{50} + \frac{(110-100)^2}{100} + \frac{(88-100)^2}{100} + \frac{(40-50)^2}{50} + \frac{(8-10)^2}{10} = 12.0$$

Since $\chi^2_{.95} = 11.1$ and $\chi^2_{.99} = 15.1$ for $\nu = 6 - 1 = 5$ degrees of freedom, we can reject the hypothesis at the 0.05 but not at the 0.01 significance level. Thus we conclude that the results are probably significant and male and female births are not equally probable.

12.10 In a survey of 500 individuals, it was found that 155 of the 500 rented at least one video from a video rental store during the past week. Test the hypothesis that 25% of the population rented at least one video during the past week using a two-tailed alternative and $\alpha = 0.05$. Perform the test using both the standard normal distribution and the chi-square distribution. Show that the chi-square test involving only two categories is equivalent to the significance test for proportions given in Chapter 10.

SOLUTION

If the null hypothesis is true, then $\mu = Np = 500(0.25) = 125$ and $\sigma = \sqrt{Npq} = \sqrt{500(0.25)(0.75)} = 9.68$. The computed test statistic is $Z = (155 - 125)/9.68 = 3.10$. The critical values are ± 1.96, and the null hypothesis is rejected.

The solution using the chi-square distribution is found by using the results as displayed in Table 12.11.

Table 12.11

Frequency	Rented Video	Did Not Rent Video	Total
Observed	155	345	500
Expected	125	375	500

The computed chi-square statistic is determined as follows:

$$\chi^2 = \frac{(155-125)^2}{125} + \frac{(345-375)^2}{375} = 9.6$$

The critical value for one degree of freedom is 3.84, and the null hypothesis is rejected. Note that $(3.10)^2 = 9.6$ and $(\pm 1.96)^2 = 3.84$ or $Z^2 = \chi^2$. The two procedures are equivalent.

12.11 (*a*) Prove that formula (*1*) of this chapter can be written

$$\chi^2 = \sum \frac{o_j^2}{e_j} - N$$

(*b*) Use the result of part (*a*) to verify the value of χ^2 computed in Problem 12.6.

SOLUTION

(*a*) By definition,

$$\chi^2 = \sum \frac{(o_j - e_j)^2}{e_j} = \sum \left(\frac{o_j^2 - 2o_j e_j + e_j^2}{e_j} \right)$$

$$= \sum \frac{o_j^2}{e_j} - 2\sum o_j + \sum e_j = \sum \frac{o_j^2}{e_j} - 2N + N = \sum \frac{o_j^2}{e_j} - N$$

where formula (*2*) of this chapter has been used.

(*b*)
$$\chi^2 = \sum \frac{o_j^2}{e_j} - N = \frac{(315)^2}{312.75} + \frac{(108)^2}{104.25} + \frac{(101)^2}{104.25} + \frac{(32)^2}{34.75} - 556 = 0.470$$

GOODNESS OF FIT

12.12 A racquetball player plays 3 game sets for exercise several times over the years and keeps records of how he does in the three game sets. For 250 days, his records show that he wins 0 games on 25 days, 1 game on 75 days, 2 games on 125 days, and wins all 3 games on 25 days. Test that X = the number of wins in a 3 game series is binomial distributed at $\alpha = 0.05$.

SOLUTION

The mean number of wins in 3 game sets is $(0 \times 25 + 1 \times 75 + 2 \times 125 + 3 \times 25)/250 = 1.6$. If X is binomial, the mean is $np = 3p$ which is set equal to the statistic 1.6 and solving for p, we find that $p = 0.53$. We wish to test that X is binomial with $n = 3$ and $p = 0.53$. If X is binomial with $p = 0.53$, the distribution of X and the expected number of wins is shown in the following EXCEL output. Note that the binomial probabilities, $p(x)$ are found by entering =BINOMDIST(A2, 3, 0.53, 0) and performing a click-and-drag from B2 to B5. This gives the values shown under $p(x)$.

x	p(x)	expected wins	observed wins
0	0.103823	25.95575	25
1	0.351231	87.80775	75
2	0.396069	99.01725	125
3	0.148877	37.21925	25

The expected wins are found by multiplying the $p(x)$ values by 250.

$$\chi^2 = \frac{(25 - 30.0)^2}{30.0} + \frac{(75 - 87.8)^2}{87.8} + \frac{(125 - 99.0)^2}{99.0} + \frac{(25 - 37.2)^2}{37.2} = 12.73.$$

Since the number of parameters used in estimating the expected frequencies is $m = 1$ (namely, the parameter p of the binomial distribution), $v = k - 1 - m = 4 - 1 - 1 = 2$. The p-value is given by the EXCEL expression =CHIDIST(12.73, 2) = 0.0017 and the hypothesis that the variable X is binomial distributed is rejected.

12.13 The number of hours per week that 200 college students spend on the Internet is grouped into the classes 0 to 3, 4 to 7, 8 to 11, 12 to 15, 16 to 19, 20 to 23, and 24 to 27 with the observed frequencies 12, 25, 36, 45, 34, 31, and 17. The grouped mean and the grouped standard deviation are found from the data. The null hypothesis is that the data are normally distributed. Using the mean and the standard deviation that are found from the grouped data and assuming a normal distribution, the expected frequencies are found, after rounding off, to be the following: 10, 30, 40, 50, 36, 28, and 6.t

(*a*) Find χ^2.
(*b*) How many degrees of freedom does χ^2 have?
(*c*) Use EXCEL to find the 5% critical value and give your conclusion at 5%.
(*d*) Use EXCEL to find the *p*-value for your result.

SOLUTION

(*a*) A portion of the EXCEL worksheet is shown in Fig. 12-1. =(A2-B2)^2/B2 is entered into C2 and a click-and-drag is executed from C2 to C8. =SUM(C2:C8) is entered into C9. We see that $\chi^2 = 22.7325$.

	A	B	C	D
1	observed	expected	(O - E)^2/E	
2	12	10	0.4	
3	25	30	0.8333333	
4	36	40	0.4	
5	45	50	0.5	
6	34	36	0.1111111	
7	31	28	0.3214286	
8	17	6	20.166667	
9			22.73254	
10				

Fig. 12-1 Portion of EXCEL worksheet for Problem 12.13.

(*b*) Since the number of parameters used in estimating the expected frequencies is $m = 2$ (namely, the mean μ and the standard deviation σ of the normal distribution), $v = k - 1 - m = 7 - 1 - 2 = 4$. Note that no classes needed to be combined, since the expected frequencies all exceeded 5.

(*c*) The 5% critical value is given by =CHIINV(0.05, 4) or 9.4877. Reject the null hypothesis that the data came from a normal distribution since 22.73 exceeds the critical value.

(*d*) The *p*-value is given by =CHIDIST(22.7325, 4) or we have *p*-value = 0.000143.

CONTINGENCY TABLES

12.14 Work Problem 10.20 by using the chi-square test. Also work using MINITAB and compare the two solutions.

SOLUTION

The conditions of the problem are presented in Table 12.12(*a*). Under the null hypothesis H_0 that the serum has no effect, we would expect 70 people in each of the groups to recover and 30 in each group not to recover, as shown in Table 12.12(*b*). Note that H_0 is equivalent to the statement that recovery is *independent* of the use of the serum (i.e., the classifications are independent).

Table 12.12(*a*) Frequencies Observed

	Recover	Do Not Recover	Total
Group A (using serum)	75	25	100
Group B (not using serum)	65	35	100
Total	140	60	200

Table 12.12(b) Frequencies Expected under H_0

	Recover	Do Not Recover	Total
Group A (using serum)	70	30	100
Group B (not using serum)	70	30	100
Total	140	60	200

$$\chi^2 = \frac{(75-70)^2}{70} + \frac{(65-70)^2}{70} + \frac{(25-30)^2}{30} + \frac{(35-30)^2}{30} = 2.38$$

To determine the number of degrees of freedom, consider Table 12.13, which is the same as Table 12.12 except that only the totals are shown. It is clear that we have the freedom of placing only one number in any of the four empty cells, since once this is done the numbers in the remaining cells are uniquely determined from the indicated totals. Thus there is 1 degree of freedom.

Table 12.13

	Recover	Do Not Recover	Total
Group A			100
Group B			100
Total	140	60	200

Another method

By formula (see Problem 12.18), $\nu = (h-1)(k-1) = (2-1)(2-1) = 1$. Since $\chi^2_{.95} = 3.84$ for 1 degree of freedom and since $\chi^2 = 2.38 < 3.84$, we conclude that the results are *not significant* at the 0.05 level. We are thus unable to reject H_0 at this level, and we either conclude that the serum is not effective or withhold decision, pending further tests.

Note that $\chi^2 = 2.38$ is the square of the z score, $z = 1.54$, obtained in Problem 10.20. In general the chi-square test involving sample proportions in a 2×2 contingency table is equivalent to a test of significance of differences in proportions using the normal approximation.

Note also that a one-tailed test using χ^2 is equivalent to a two-tailed test using χ since, for example, $\chi^2 > \chi^2_{.95}$ corresponds to $\chi > \chi_{.95}$ or $\chi < -\chi_{.95}$. Since for 2×2 tables χ^2 is the square of the z score, it follows that χ is the same as z for this case. Thus a rejection of a hypothesis at the 0.05 level using χ^2 is equivalent to a rejection in a two-tailed test at the 0.10 level using z.

Chi-Square Test: Recover, Not-recover

```
Expected counts are printed below observed counts
Chi-Square contributions are printed below expected counts

        Recover    Not-recover     Total
1            75             25       100
          70.00          30.00
          0.357          0.833

2            65             35       100
          70.00          30.00
          0.357          0.833

Total       140             60       200
Chi-Sq=2.381, DF=1, P-Value=0.123
```

12.15 Work Problem 12.14 by using Yates' correction.

SOLUTION

$$\chi^2\,(\text{corrected}) = \frac{(|75-70|-0.5)^2}{70} + \frac{(|65-70|-0.5)^2}{70} + \frac{(|25-30|-0.5)^2}{30} + \frac{(|35-30|-0.5)^2}{30} = 1.93$$

Thus the conclusions reached in Problem 12.14 are valid. This could have been realized at once by noting that Yates' correction always decreases the value of χ^2.

12.16 A cellular phone company conducts a survey to determine the ownership of cellular phones in different age groups. The results for 1000 households are shown in Table 12.14. Test the hypothesis that the proportions owning cellular phones are the same for the different age groups.

Table 12.14

Cellular phone	18–24	25–54	55–64	≥ 65	Total
Yes	50	80	70	50	250
No	200	170	180	200	750
Total	250	250	250	250	1000

SOLUTION

Under the hypothesis H_0 that the proportions owning cellular phones are the same for the different age groups, $250/1000 = 25\%$ is an estimate of the percentage owning a cellular phone in each age group, and 75% is an estimate of the percent not owning a cellular phone in each age group. The frequencies expected under H_0 are shown in Table 12.15.

The computed value of the chi-square statistic can be found as illustrated in Table 12.16.

The degrees of freedom for the chi-square distribution is $\nu = (h-1)(k-1) = (2-1)(4-1) = 3$. Since $\chi^2_{.95} = 7.81$, and 14.3 exceeds 7.81, we reject the null hypothesis and conclude that the percentages are not the same for the four age groups.

Table 12.15

Cellular phone	18–24	25–54	55–64	≥ 65	Total
Yes	25% of 250 = 62.5	25% of 250 = 62.5	25% of 250 = 62.5	25% of 250 = 62.5	250
No	75% of 250 = 187.5	75% of 250 = 187.5	75% of 250 = 187.5	75% of 250 = 187.5	750
Total	250	250	250	250	1000

Table 12.16

Row, column	o	e	$(o-e)$	$(o-e)^2$	$(o-e)^2/e$
1, 1	50	62.5	−12.5	156.25	2.5
1, 2	80	62.5	17.5	306.25	4.9
1, 3	70	62.5	7.5	56.25	0.9
1, 4	50	62.5	−12.5	156.25	2.5
2, 1	200	187.5	12.5	156.25	0.8
2, 2	170	187.5	−17.5	306.25	1.6
2, 3	180	187.5	−7.5	56.25	0.3
2, 4	200	187.5	12.5	156.25	0.8
Sum	1000	1000	0		14.3

12.17 Use MINITAB to solve Problem 12.16.

SOLUTION

The MINITAB solution to Problem 12.16 is shown below. The observed and the expected counts are shown along with the computation of the test statistic. Note that the null hypothesis would be rejected for any level of significance exceeding 0.002.

Data Display

```
Row    18-24    25-54    55-64     65 or more

 1       50       80       70         50
 2      200      170      180        200
MTB > chisquare c1-c4
```

Chi-Square Test

```
Expected counts are printed below observed counts

             18-24     25-54     55-64    65 or mo    Total
      1         50        80        70          50      250
             62.50     62.50     62.50       62.50

      2        200       170       180         200      750
            187.50    187.50    187.50      187.50

  Total        250       250       250         250     1000
Chi-Sq =     2.500  +  4.900  +  0.900  +    2.500 +
             0.833  +  1.633  +  0.300  +    0.833 = 14.400
DF = 3, P-Value = 0.002
```

12.18 Show that for an $h \times k$ contingency table the number of degrees of freedom is $(h-1) \times (k-1)$, where $h > 1$ and $k > 1$.

SOLUTION

In a table with h rows and k columns, we can leave out a single number in each row and column, since such numbers can easily be restored from a knowledge of the totals of each column and row. It follows that we have the freedom of placing only $(h-1)(k-1)$ numbers into the table, the others being then automatically determined uniquely. Thus the number of degrees of freedom is $(h-1)(k-1)$. Note that this result holds if the population parameters needed in obtaining the expected frequencies are known.

12.19 (*a*) Prove that for the 2×2 contingency table shown in Table 12.17(*a*),

$$\chi^2 = \frac{N(a_1 b_2 - a_2 b_1)^2}{N_1 N_2 N_A N_B}$$

(*b*) Illustrate the result in part (*a*) with reference to the data of Problem 12.14.

Table 12.17(*a*) Results Observed

	I	II	Total
A	a_1	a_2	N_A
B	b_1	b_2	N_B
Total	N_1	N_2	N

Table 12.17(*b*) Results Expected

	I	II	Total
A	$N_1 N_A/N$	$N_2 N_A/N$	N_A
B	$N_1 N_B/N$	$N_2 N_B/N$	N_B
Total	N_1	N_2	N

SOLUTION

(*a*) As in Problem 12.14, the results expected under a null hypothesis are shown in Table 12.17(*b*). Then

$$\chi^2 = \frac{(a_1 - N_1N_A/N)^2}{N_1N_A/N} + \frac{(a_2 - N_2N_A/N)^2}{N_2N_A/N} + \frac{(b_1 - N_1N_B/N)^2}{N_1N_B/N} + \frac{(b_2 - N_2N_B/N)^2}{N_2N_B/N}$$

But
$$a_1 - \frac{N_1N_A}{N} = a_1 - \frac{(a_1+b_1)(a_1+a_2)}{a_1+b_1+a_2+b_2} = \frac{a_1b_2 - a_2b_1}{N}$$

Similarly
$$a_2 - \frac{N_2N_A}{N} \quad \text{and} \quad b_1 - \frac{N_1N_B}{N} \quad \text{and} \quad b_2 - \frac{N_2N_B}{N}$$

are also equal to
$$\frac{a_1b_2 - a_2b_1}{N}$$

Thus we can write

$$\chi^2 = \frac{N}{N_1N_A}\left(\frac{a_1b_2 - a_2b_1}{N}\right)^2 + \frac{N}{N_2N_A}\left(\frac{a_1b_2 - a_2b_1}{N}\right)^2 + \frac{N}{N_1N_B}\left(\frac{a_1b_2 - a_2b_1}{N}\right)^2 + \frac{N}{N_2N_B}\left(\frac{a_1b_2 - a_2b_1}{N}\right)^2$$

which simplifies to
$$\chi^2 = \frac{N(a_1b_2 - a_2b_1)^2}{N_1N_2N_AN_B}$$

(*b*) In Problem 12.14, $a_1 = 75$, $a_2 = 25$, $b_1 = 65$, $b_2 = 35$, $N_1 = 140$, $N_2 = 60$, $N_A = 100$, $N_B = 100$, and $N = 200$; then, as obtained before,

$$\chi^2 = \frac{200[(75)(35) - (25)(65)]^2}{(140)(60)(100)(100)} = 2.38$$

Using Yates' correction, the result is the same as in Problem 12.15:

$$\chi^2\ (\text{corrected}) = \frac{N(|a_1b_2 - a_2b_1| - \frac{1}{2}N)^2}{N_1N_2N_AN_B} = \frac{200[|(75)(35) - (25)(65)| - 100]^2}{(140)(60)(100)(100)} = 1.93$$

12.20 Nine hundred males and 900 females were asked whether they would prefer more federal programs to assist with childcare. Forty percent of the females and 36 percent of the males responded yes. Test the null hypothesis of equal percentages versus the alternative hypothesis of unequal percentages at $\alpha = 0.05$. Show that a chi-square test involving two sample proportions is equivalent to a significance test of differences using the normal approximation of Chapter 10.

SOLUTION

Under hypothesis H_0,

$$\mu_{P_1-P_2} = 0 \text{ and } \sigma_{P_1-P_2} = \sqrt{pq\left(\frac{1}{N_1} + \frac{1}{N_2}\right)} = \sqrt{(0.38)(0.62)\left(\frac{1}{900} + \frac{1}{900}\right)} = 0.0229$$

where p is estimated by pooling the proportions in the two samples. That is

$$p = \frac{360 + 324}{900 + 900} = 0.38 \quad \text{and} \quad q = 1 - 0.38 = 0.62$$

The normal approximation test statistic is as follows:

$$Z = \frac{P_1 - P_2}{\sigma_{P_1-P_2}} = \frac{0.40 - 0.36}{0.0229} = 1.7467$$

The MINITAB solution for the chi-square analysis is as follows.

Chi-Square Test

```
Expected counts are printed below observed counts

         males     females   Total
1        324       360       684
         342.00    342.00

2        576       549       1116
         558.00    558.00

Total    900       900       1800

Chi-Sq =  0.947 +  0.947 +
         0.581 + 0.581 = 3.056
DF = 1, P-Value = 0.080
```

The square of the normal test statistic is $(1.7467)^2 = 3.056$, the value for the chi-square statistic. The two tests are equivalent. The *p*-values are always the same for the two tests.

COEFFICIENT OF CONTINGENCY

12.21 Find the coefficient of contingency for the data in the contingency table of Problem 12.14.

SOLUTION

$$C = \sqrt{\frac{\chi^2}{\chi^2 + N}} = \sqrt{\frac{2.38}{2.38 + 200}} = \sqrt{0.01176} = 0.1084$$

12.22 Find the maximum value of C for the 2×2 table of Problem 12.14.

SOLUTION

The maximum value of C occurs when the two classifications are perfectly dependent or associated. In such case all those who take the serum will recover, and all those who do not take the serum will not recover. The contingency table then appears as in Table 12.18.

Table 12.18

	Recover	Do Not Recover	Total
Group A (using serum)	100	0	100
Group B (not using serum)	0	100	100
Total	100	100	200

Since the expected cell frequencies, assuming complete independence, are all equal to 50,

$$\chi^2 = \frac{(100 - 50)^2}{50} + \frac{(0 - 50)^2}{50} + \frac{(0 - 50)^2}{50} + \frac{(100 - 50)^2}{50} = 200$$

Thus the maximum value of C is $\sqrt{\chi^2/(\chi^2 + N)} = \sqrt{200/(200 + 200)} = 0.7071$.

In general, for perfect dependence in a contingency table where the number of rows and columns are both equal to k, the only nonzero cell frequencies occur in the diagonal from upper left to lower right of the contingency table. For such cases, $C_{\max} = \sqrt{(k-1)/k}$. (See Problems 12.52 and 12.53.)

CORRELATION OF ATTRIBUTES

12.23 For Table 12.12 of Problem 12.14, find the correlation coefficient (*a*) without and (*b*) with Yates' correction.

SOLUTION

(*a*) Since $\chi^2 = 2.38$, $N = 200$, and $k = 2$, we have

$$r = \sqrt{\frac{\chi^2}{N(k-1)}} = \sqrt{\frac{2.38}{200}} = 0.1091$$

indicating very little correlation between recovery and the use of the serum.

(*b*) From Problem 12.15, r (corrected) $= \sqrt{1.93/200} = 0.0982$.

12.24 Prove that the correlation coefficient for contingency tables, as defined by equation (*12*) of this chapter, lies between 0 and 1.

SOLUTION

By Problem 12.53, the maximum value of $\sqrt{\chi^2/(\chi^2+N)}$ is $\sqrt{(k-1)/k}$. Thus

$$\frac{\chi^2}{\chi^2+N} \le \frac{k-1}{k} \qquad k\chi^2 \le (k-1)(\chi^2+N) \quad k\chi^2 \le k\chi^2 - \chi^2 + kN - N$$

$$\chi^2 \le (k-1)N \qquad \frac{\chi^2}{N(k-1)} \le 1 \quad \text{and} \quad r = \sqrt{\frac{\chi^2}{N(k-1)}} \le 1$$

Since $\chi^2 \ge 0$, $r \ge 0$. Thus $0 \le r \le 1$, as required.

ADDITIVE PROPERTY OF χ^2

12.25 To test a hypothesis H_0, an experiment is performed three times. The resulting values of χ^2 are 2.37, 2.86, and 3.54, each of which corresponds to 1 degree of freedom. Show that while H_0 cannot be rejected at the 0.05 level on the basis of any individual experiment, it can be rejected when the three experiments are combined.

SOLUTION

The value of χ^2 obtained by combining the results of the three experiments is, according to the *additive property*, $\chi^2 = 2.37 + 2.86 + 3.54 = 8.77$ with $1 + 1 + 1 = 3$ degrees of freedom. Since $\chi^2_{.95}$ for 3 degrees of freedom is 7.81, we can reject H_0 at the 0.05 significance level. But since $\chi^2_{.95} = 3.84$ for 1 degree of freedom, we cannot reject H_0 on the basis of any one experiment.

In combining experiments in which values of χ^2 corresponding to 1 degree of freedom are obtained, Yates' correction is omitted since it has a tendency to overcorrect.

Supplementary Problems

THE CHI-SQUARE TEST

12.26 In 60 tosses of a coin, 37 heads and 23 tails were observed. Using significance levels of (*a*) 0.05 and (*b*) 0.01, test the hypothesis that the coin is fair.

12.27 Work Problem 12.26 by using Yates' correction.

12.28 Over a long period of time the grades given by a group of instructors in a particular course have averaged 12% A's, 18% B's, 40% C's, 18% D's, and 12% F's. A new instructor gives 22 A's, 34 B's, 66 C's, 16 D's, and 12 F's during two semesters. Determine at the 0.05 significance level whether the new instructor is following the grade pattern set by the others.

12.29 Three coins were tossed a total of 240 times, and each time the number of heads turning up was observed. The results are shown in Table 12.19, together with the results expected under the hypothesis that the coins are fair. Test this hypothesis at a significance level of 0.05.

Table 12.19

	0 Heads	1 Head	2 Heads	3 Heads
Observed frequency	24	108	95	23
Expected frequency	30	90	90	30

12.30 The number of books borrowed from a public library during a particular week is given in Table 12.20. Test the hypothesis that the number of books borrowed does not depend on the day of the week, using significance levels of (*a*) 0.05 and (*b*) 0.01.

Table 12.20

	Mon	Tue	Wed	Thu	Fri
Number of books borrowed	135	108	120	114	146

12.31 An urn contains 6 red marbles and 3 white ones. Two marbles are selected at random from the urn, their colors are noted, and then the marbles are replaced in the urn. This process is performed a total of 120 times, and the results obtained are shown in Table 12.21.

(*a*) Determine the expected frequencies.

(*b*) Determine at a significance level of 0.05 whether the results obtained are consistent with those expected.

Table 12.21

	0 Red 2 White	1 Red 1 White	2 Red 0 White
Number of drawings	6	53	61

12.32 Two hundred bolts were selected at random from the production of each of four machines. The numbers of defective bolts found were 2, 9, 10, and 3. Determine whether there is a significant difference between the machines, using a significance level of 0.05.

GOODNESS OF FIT

12.33 (*a*) Use the chi-square test to determine the goodness of fit of the data in Table 7.9 of Problem 7.75. (*b*) Is the fit "too good"? Use the 0.05 significance level.

12.34 Use the chi-square test to determine the goodness of fit of the data in (*a*) Table 3.8 of Problem 3.59 and (*b*) Table 3.10 of Problem 3.61. Use a significance level of 0.05, and in each case determine whether the fit is "too good."

12.35 Use the chi-square test to determine the goodness of fit of the data in (*a*) Table 7.9 of Problem 7.75 and (*b*) Table 7.12 of Problem 7.80. Is your result in part (*a*) consistent with that of Problem 12.33?

CONTINGENCY TABLES

12.36 Table 12.22 shows the result of an experiment to investigate the effect of vaccination of laboratory animals against a particular disease. Using the (*a*) 0.01 and (*b*) 0.05 significance levels, test the hypothesis that there is no difference between the vaccinated and unvaccinated groups (i.e., that vaccination and this disease are independent).

12.37 Work Problem 12.36 using Yates' correction.

12.38 Table 12.23 shows the numbers of students in each of two classes, A and B, who passed and failed an examination given to both groups. Using the (*a*) 0.05 and (*b*) 0.01 significance levels, test the hypothesis that there is no difference between the two classes. Work the problem with and without Yates' correction.

Table 12.22

	Got Disease	Did Not Get Disease
Vaccinated	9	42
Not vaccinated	17	28

Table 12.23

	Passed	Failed
Class A	72	17
Class B	64	23

12.39 Of a group of patients who complained that they did not sleep well, some were given sleeping pills while others were given sugar pills (although they all *thought* they were getting sleeping pills). They were later asked whether the pills helped them or not. The results of their responses are shown in Table 12.24. Assuming that all patients told the truth, test the hypothesis that there is no difference between sleeping pills and sugar pills at a significance level of 0.05.

Table 12.24

	Slept Well	Did Not Sleep Well
Took sleeping pills	44	10
Took sugar pills	81	35

12.40 On a particular proposal of national importance, Democrats and Republicans cast their votes as shown in Table 12.25. At significance levels of (*a*) 0.01 and (*b*) 0.05, test the hypothesis that there is no difference between the two parties insofar as this proposal is concerned.

Table 12.25

	In Favor	Opposed	Undecided
Democrats	85	78	37
Republicans	118	61	25

12.41 Table 12.26 shows the relation between the performances of students in mathematics and physics. Test the hypothesis that performance in physics is independent of performance in mathematics, using the (*a*) 0.05 and (*b*) 0.01 significance levels.

Table 12.26

		Mathematics		
		High Grades	Medium Grades	Low Grades
Physics	High grades	56	71	12
	Medium grades	47	163	38
	Low grades	14	42	85

12.42 The results of a survey made to determine whether the age of a driver 21 years of age and older has any effect on the number of automobile accidents in which he is involved (including all minor accidents) are shown in Table 12.27. At significance levels of (*a*) 0.05 and (*b*) 0.01, test the hypothesis that the number of accidents is independent of the age of the driver. What possible sources of difficulty in sampling techniques, as well as other considerations, could affect your conclusions?

Table 12.27

		Age of Driver				
		21–30	31–40	41–50	51–60	61–70
Number of accidents	0	748	821	786	720	672
	1	74	60	51	66	50
	2	31	25	22	16	15
	>2	9	10	6	5	7

12.43 (*a*) Prove that $\chi^2 = \sum (o_j^2/e_j) - N$ for all contingency tables, where N is the total frequency of all cells.

(*b*) Using the result in part (*a*), work Problem 12.41.

12.44 If N_i and N_j denote, respectively, the sum of the frequencies in the ith row and jth columns of a contingency table (the *marginal frequencies*), show that the expected frequency for the cell belonging to the ith row and jth column is $N_i N_j / N$, where N is the total frequency of all cells.

12.45 Prove formula (*9*) of this chapter. (*Hint:* Use Problems 12.43 and 12.44.)

12.46 Extend the result of formula (*9*) of this chapter to $2 \times k$ contingency tables, where $k > 3$.

12.47 Prove formula (*8*) of this chapter.

12.48 By analogy with the ideas developed for $h \times k$ contingency tables, discuss $h \times k \times l$ contingency tables, pointing out their possible applications.

COEFFICIENT OF CONTINGENCY

12.49 Table 12.28 shows the relationship between hair and eye color of a sample of 200 students.

(*a*) Find the coefficient of contingency without and with Yates' correction.

(*b*) Compare the result of part (*a*) with the maximum coefficient of contingency.

Table 12.28

		Hair Color	
		Blonde	Not blonde
Eye color	Blue	49	25
	Not blue	30	96

12.50 Find the coefficient of contingency for the data of (*a*) Problem 12.36 and (*b*) Problem 12.38 without and with Yates' correction.

12.51 Find the coefficient of contingency for the data of Problem 12.41.

12.52 Prove that the maximum coefficient of contingency for a 3×3 table is $\sqrt{\frac{2}{3}} = 0.8165$ approximately.

12.53 Prove that the maximum coefficient of contingency for a $k \times k$ table is $\sqrt{(k-1)/k}$.

CORRELATION OF ATTRIBUTES

12.54 Find the correlation coefficient for the data in Table 12.28.

12.55 Find the correlation coefficient for the data in (*a*) Table 12.22 and (*b*) Table 12.23 without and with Yates' correction.

12.56 Find the correlation coefficient between the mathematics and physics grades in Table 12.26.

12.57 If C is the coefficient of contingency for a $k \times k$ table and r is the corresponding coefficient of correlation, prove that $r = C/\sqrt{(1 - C^2)(k-1)}$.

ADDITIVE PROPERTY OF χ^2

12.58 To test a hypothesis H_0, an experiment is performed five times. The resulting values of χ^2, each corresponding to 4 degrees of freedom, are 8.3, 9.1, 8.9, 7.8, and 8.6, respectively. Show that while H_0 cannot be rejected at the 0.05 level on the basis of each experiment separately, it can be rejected at the 0.005 level on the basis of the combined experiments.

CHAPTER 13

Curve Fitting and the Method of Least Squares

RELATIONSHIP BETWEEN VARIABLES

Very often in practice a relationship is found to exist between two (or more) variables. For example, weights of adult males depend to some degree on their heights, the circumferences of circles depend on their radii, and the pressure of a given mass of gas depends on its temperature and volume.

It is frequently desirable to express this relationship in mathematical form by determining an equation that connects the variables.

CURVE FITTING

To determine an equation that connects variables, a first step is to collect data that show corresponding values of the variables under consideration. For example, suppose that X and Y denote, respectively, the height and weight of adult males; then a sample of N individuals would reveal the heights $X_1, X_2, \ldots, X_N$ and the corresponding weights $Y_1, Y_2, \ldots, Y_N$.

A next step is to plot the points $(X_1, Y_1), (X_2, Y_2), \ldots, (X_N, Y_N)$ on a rectangular coordinate system. The resulting set of points is sometimes called a *scatter diagram*.

From the scatter diagram it is often possible to visualize a smooth curve that approximates the data. Such a curve is called an *approximating curve*. In Fig. 13-1, for example, the data appear to be approximated well by a straight line, and so we say that a *linear relationship* exists between the variables. In Fig. 13-2, however, although a relationship exists between the variables, it is not a linear relationship, and so we call it a *nonlinear relationship*.

The general problem of finding equations of approximating curves that fit given sets of data is called *curve fitting*.

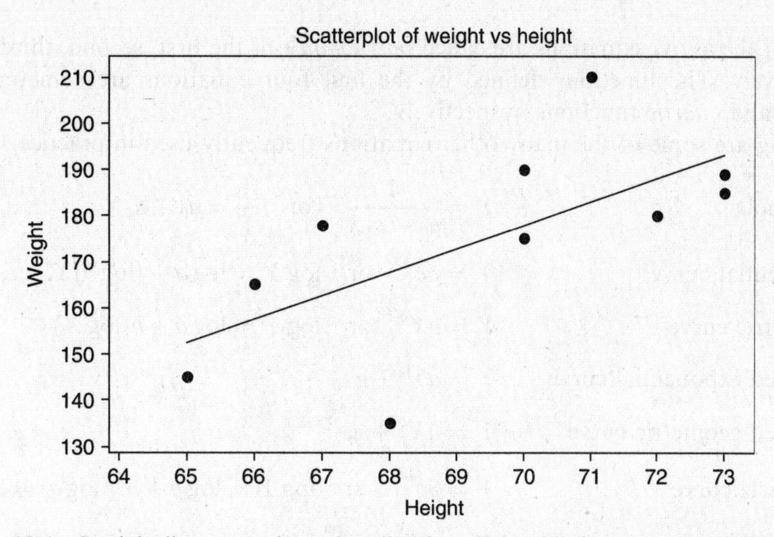

Fig. 13-1 Straight lines sometimes describe the relationship between two variables.

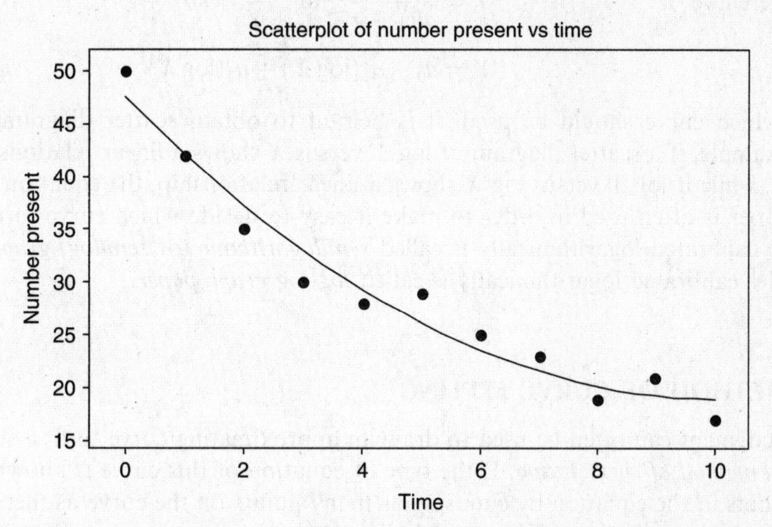

Fig. 13-2 Nonlinear relationships sometimes describe the relationship between two variables.

EQUATIONS OF APPROXIMATING CURVES

Several common types of approximating curves and their equations are listed below for reference purposes. All letters other than X and Y represent constants. The variables X and Y are often referred to as *independent* and *dependent variables*, respectively, although these roles can be interchanged.

Straight line	$Y = a_0 + a_1 X$	*(1)*
Parabola, or quadratic curve	$Y = a_0 + a_1 X + a_2 X^2$	*(2)*
Cubic curve	$Y = a_0 + a_1 X + a_2 X^2 + a_3 X^3$	*(3)*
Quartic curve	$Y = a_0 + a_1 X + a_2 X^2 + a_3 X^3 + a_4 X^4$	*(4)*
nth-Degree curve	$Y = a_0 + a_1 X + a_2 X^2 + \cdots + a_n X^n$	*(5)*

The right sides of the above equations are called *polynomials* of the first, second, third, fourth, and nth degrees, respectively. The functions defined by the first four equations are sometimes called *linear*, *quadratic*, *cubic*, and *quartic* functions, respectively.

The following are some of the many other equations frequently used in practice:

Hyperbola $$Y = \frac{1}{a_0 + a_1 X} \quad \text{or} \quad \frac{1}{Y} = a_0 + a_1 X \tag{6}$$

Exponential curve $$Y = ab^X \quad \text{or} \quad \log Y = \log a + (\log b)X = a_0 + a_1 X \tag{7}$$

Geometric curve $$Y = aX^b \quad \text{or} \quad \log Y = \log a + b(\log X) \tag{8}$$

Modified exponential curve $$Y = ab^X + g \tag{9}$$

Modified geometric curve $$Y = aX^b + g \tag{10}$$

Gompertz curve $$Y = pq^{b^X} \quad \text{or} \quad \log Y = \log p + b^X(\log q) = ab^X + g \tag{11}$$

Modified Gompertz curve $$Y = pq^{b^X} + h \tag{12}$$

Logistic curve $$Y = \frac{1}{ab^X + g} \quad \text{or} \quad \frac{1}{Y} = ab^X + g \tag{13}$$

$$Y = a_0 + a_1(\log X) + a_2(\log X)^2 \tag{14}$$

To decide which curve should be used, it is helpful to obtain scatter diagrams of transformed variables. For example, if a scatter diagram of log Y versus X shows a linear relationship, the equation has the form (*7*), while if log Y versus log X shows a linear relationship, the equation has the form (*8*). Special graph paper is often used in order to make it easy to decide which curve to use. Graph paper having one scale calibrated logarithmically is called *semilogarithmic* (or *semilog*) *graph paper*, and that having both scales calibrated logarithmically is called *log-log graph paper*.

FREEHAND METHOD OF CURVE FITTING

Individual judgment can often be used to draw an approximating curve to fit a set of data. This is called a *freehand method of curve fitting*. If the type of equation of this curve is known, it is possible to obtain the constants in the equation by choosing as many points on the curve as there are constants in the equation. For example, if the curve is a straight line, two points are necessary; if it is a parabola, three points are necessary. The method has the disadvantage that different observers will obtain different curves and equations.

THE STRAIGHT LINE

The simplest type of approximating curve is a straight line, whose equation can be written

$$Y = a_0 + a_1 X \tag{15}$$

Given any two points (X_1, Y_1) and (X_2, Y_2) on the line, the constants a_0 and a_1 can be determined. The resulting equation of the line can be written

$$Y - Y_1 = \left(\frac{Y_2 - Y_1}{X_2 - X_1}\right)(X - X_1) \quad \text{or} \quad Y - Y_1 = m(X - X_1) \tag{16}$$

where $$m = \frac{Y_2 - Y_1}{X_2 - X_1}$$

is called the *slope* of the line and represents the change in Y divided by the corresponding change in X.

When the equation is written in the form (*15*), the constant a_1 is the slope m. The constant a_0, which is the value of Y when $X = 0$, is called the *Y intercept*.

THE METHOD OF LEAST SQUARES

To avoid individual judgment in constructing lines, parabolas, or other approximating curves to fit sets of data, it is necessary to agree on a definition of a "best-fitting line," "best-fitting parabola," etc.

By way of forming a definition, consider Fig. 13-3, in which the data points are given by (X_1, Y_1), $(X_2, Y_2), \ldots, (X_N, Y_N)$. For a given value of X, say X_1, there will be a difference between the value Y_1 and the corresponding value as determined from the curve C. As shown in the figure, we denote this difference by D_1, which is sometimes referred to as a *deviation*, *error*, or *residual* and may be positive, negative, or zero. Similarly, corresponding to the values $X_2, \ldots, X_N$ we obtain the deviations $D_2, \ldots, D_N$.

A measure of the "goodness of fit" of the curve C to the given data is provided by the quantity $D_1^2 + D_2^2 + \cdots + D_N^2$. If this is small, the fit is good; if it is large, the fit is bad. We therefore make the following

> **Definition:** Of all curves approximating a given set of data points, the curve having the property that $D_1^2 + D_2^2 + \cdots + D_N^2$ is a minimum is called a *best-fitting curve.*

A curve having this property is said to fit the data in the *least-squares sense* and is called a *least-squares curve*. Thus a line having this property is called a *least-squares line*, a parabola with this property is called a *least-squares parabola*, etc.

It is customary to employ the above definition when X is the independent variable and Y is the dependent variable. If X is the dependent variable, the definition is modified by considering horizontal instead of vertical deviations, which amounts to an interchange of the X and Y axes. These two definitions generally lead to different least-squares curves. Unless otherwise specified, we shall consider Y the dependent variable and X the independent variable.

It is possible to define another least-squares curve by considering perpendicular distances from each of the data points to the curve instead of either vertical or horizontal distances. However, this is not used very often.

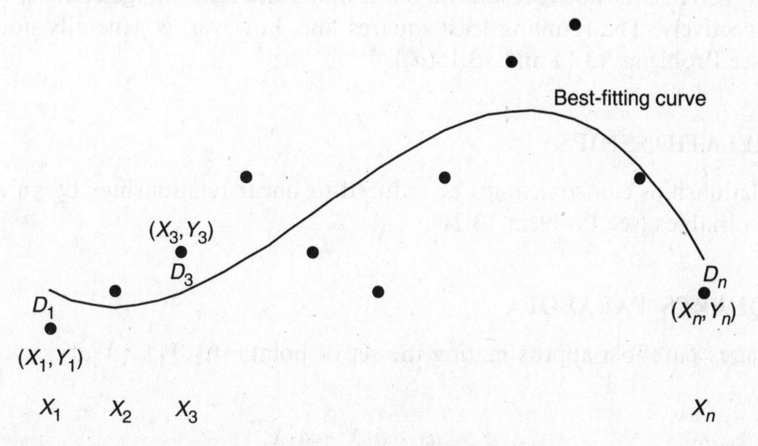

Fig. 13-3 D_1 is the distance from data point (X_1, Y_1) to the best-fitting curve, ..., D_n is the distance from data point (X_n, Y_n) to the best-fitting curve.

THE LEAST-SQUARES LINE

The least-squares line approximating the set of points $(X_1, Y_1), (X_2, Y_2), \ldots, (X_N, Y_N)$ has the equation

$$Y = a_0 + a_1 X \tag{17}$$

where the constants a_0 and a_1 are determined by solving simultaneously the equations

$$\begin{aligned}\sum Y &= a_0 N \qquad + a_1 \sum X \\ \sum XY &= a_0 \sum X + a_1 \sum X^2\end{aligned} \tag{18}$$

which are called the *normal equations for the least-squares line* (*17*). The constants a_0 and a_1 of equations (*18*) can, if desired, be found from the formulas

$$a_0 = \frac{(\sum Y)(\sum X^2) - (\sum X)(\sum XY)}{N \sum X^2 - (\sum X)^2} \qquad a_1 = \frac{N \sum XY - (\sum X)(\sum Y)}{N \sum X^2 - (\sum X)^2} \tag{19}$$

The normal equations (*18*) are easily remembered by observing that the first equation can be obtained formally by summing on both sides of (*17*) [i.e., $\sum Y = \sum (a_0 + a_1 X) = a_0 N + a_1 \sum X$], while the second equation is obtained formally by first multiplying both sides of (*17*) by X and then summing [i.e., $\sum XY = \sum X(a_0 + a_1 X) = a_0 \sum X + a_1 \sum X^2$]. Note that this is not a derivation of the normal equations, but simply a means for remembering them. Note also that in equations (*18*) and (*19*) we have used the short notation $\sum X$, $\sum XY$, etc., in place of $\sum_{j=1}^N X_j$, $\sum_{j=1}^N X_j Y_j$, etc.

The labor involved in finding a least-squares line can sometimes be shortened by transforming the data so that $x = X - \bar{X}$ and $y = Y - \bar{Y}$. The equation of the least-squares line can then be written (see Problem 13.15)

$$y = \left(\frac{\sum xy}{\sum x^2}\right)x \qquad \text{or} \qquad y = \left(\frac{\sum xY}{\sum x^2}\right)x \tag{20}$$

In particular, if X is such that $\sum X = 0$ (i.e., $\bar{X} = 0$), this becomes

$$Y = \bar{Y} + \left(\frac{\sum XY}{\sum X^2}\right)X \tag{21}$$

Equation (*20*) implies that $y = 0$ when $x = 0$; thus the least-squares line passes through the point $(\bar{X}, \bar{Y})$, called the *centroid*, or *center of gravity*, of the data.

If the variable X is taken to be the dependent instead of the independent variable, we write equation (*17*) as $X = b_0 + b_1 Y$. Then the above results hold if X and Y are interchanged and a_0 and a_1 are replaced by b_0 and b_1, respectively. The resulting least-squares line, however, is generally not the same as that obtained above [see Problems 13.11 and 13.15(*d*)].

NONLINEAR RELATIONSHIPS

Nonlinear relationships can sometimes be reduced to linear relationships by an appropriate transformation of the variables (see Problem 13.21).

THE LEAST-SQUARES PARABOLA

The least-squares parabola approximating the set of points (X_1, Y_1), $(X_2, Y_2), \ldots, (X_N, Y_N)$ has the equation

$$Y = a_0 + a_1 X + a_2 X^2 \tag{22}$$

where the constants a_0, a_1, and a_2 are determined by solving simultaneously the equations

$$\begin{aligned}\sum Y &= a_0 N \qquad + a_1 \sum X + a_2 \sum X^2 \\ \sum XY &= a_0 \sum X + a_1 \sum X^2 + a_2 \sum X^3 \\ \sum X^2 Y &= a_0 \sum X^2 + a_1 \sum X^3 + a_2 \sum X^4\end{aligned} \tag{23}$$

called the *normal equations for the least-squares parabola* (*22*).

Equations (*23*) are easily remembered by observing that they can be obtained formally by multiplying equation (*22*) by 1, X, and X^2, respectively, and summing on both sides of the resulting equations. This technique can be extended to obtain normal equations for least-squares cubic curves, least-squares quartic curves, and in general any of the least-squares curves corresponding to equation (*5*).

As in the case of the least-squares line, simplifications of equations (*23*) occur if X is chosen so that $\sum X = 0$. Simplification also occurs by choosing the new variables $x = X - \bar{X}$ and $y = Y - \bar{Y}$.

REGRESSION

Often, on the basis of sample data, we wish to estimate the value of a variable Y corresponding to a given value of a variable X. This can be accomplished by estimating the value of Y from a least-squares curve that fits the sample data. The resulting curve is called a *regression curve of Y on X*, since Y is estimated from X.

If we wanted to estimate the value of X from a given value of Y, we would use a *regression curve of X on Y*, which amounts to interchanging the variables in the scatter diagram so that X is the dependent variable and Y is the independent variable. This is equivalent to replacing the vertical deviations in the definition of the least-squares curve on page 284 with horizontal deviations.

In general, the regression line or curve of Y on X is not the same as the regression line or curve of X on Y.

APPLICATIONS TO TIME SERIES

If the independent variable X is time, the data show the values of Y at various times. Data arranged according to time are called *time series*. The regression line or curve of Y on X in this case is often called a *trend line* or *trend curve* and is often used for purposes of *estimation*, *prediction*, or *forecasting*.

PROBLEMS INVOLVING MORE THAN TWO VARIABLES

Problems involving more than two variables can be treated in a manner analogous to that for two variables. For example, there may be a relationship between the three variables X, Y, and Z that can be described by the equation

$$Z = a_0 + a_1 X + a_2 Y \tag{24}$$

which is called a *linear equation in the variables X, Y, and Z*.

In a three-dimensional rectangular coordinate system this equation represents a plane, and the actual sample points (X_1, Y_1, Z_1), $(X_2, Y_2, Z_2), \ldots, (X_N, Y_N, Z_N)$ may "scatter" not too far from this plane, which we call an *approximating plane*.

By extension of the method of least squares, we can speak of a *least-squares plane* approximating the data. If we are estimating Z from given values of X and Y, this would be called a *regression plane of Z on X and Y*. The normal equations corresponding to the least-squares plane (*24*) are given by

$$\begin{aligned}
\sum Z &= a_0 N + a_1 \sum X + a_2 \sum Y \\
\sum XZ &= a_0 \sum X + a_1 \sum X^2 + a_2 \sum XY \\
\sum YZ &= a_0 \sum Y + a_1 \sum XY + a_2 \sum Y^2
\end{aligned} \tag{25}$$

and can be remembered as being obtained from equation (*24*) by multiplying by 1, X, and Y successively and then summing.

More complicated equations than (*24*) can also be considered. These represent *regression surfaces*. If the number of variables exceeds three, geometric intuition is lost since we then require four-, five-,... n-dimensional spaces.

Problems involving the estimation of a variable from two or more variables are called problems of *multiple regression* and will be considered in more detail in Chapter 15.

Solved Problems

STRAIGHT LINES

13.1 Thirty high school students were surveyed in a study involving time spent on the Internet and their grade point average (GPA). The results are shown in Table 13.1. X is the amount of time spent on the Internet weekly and Y is the GPA of the student.

Table 13.1

Hours	GPA	Hours	GPA	Hours	GPA
11	2.84	9	2.85	25	1.85
5	3.20	5	3.35	6	3.14
22	2.18	14	2.60	9	2.96
23	2.12	18	2.35	20	2.30
20	2.55	6	3.14	14	2.66
20	2.24	9	3.05	19	2.36
10	2.90	24	2.06	21	2.24
19	2.36	25	2.00	7	3.08
15	2.60	12	2.78	11	2.84
18	2.42	6	2.90	20	2.45

Use MINITAB to do the following:

(*a*) Make a scatter plot of the data.

(*b*) Fit a straight line to the data and give the values of a_0 and a_1.

SOLUTION

(*a*) The data are entered into columns `C1` and `C2` of the MINITAB worksheet. `C1` is named `Internet-hours` and `C2` is named `GPA`. The pull-down menu **Stat → Regression → Regression** is given and the results shown in Fig. 13-4 are formed.

(*b*) The value for a_0 is 3.49 and the value for a_1 is -0.0594.

13.2 Solve Problem 13.1 using EXCEL.

SOLUTION

The data is entered into columns A and B of the EXCEL worksheet. The pull-down menu **Tools → Data Analysis → Regression** produces the dialog box in Fig. 13-5 which is filled in as shown. The part of the output which is currently of interest is

Intercept	3.488753
Internet-hours	−0.05935

Scatterplot of GPA vs Internet-hours
3.50
3.25
3.00
2.75
2.50
2.25
2.00
GPA
5
10
15
20
25
Internet-hours

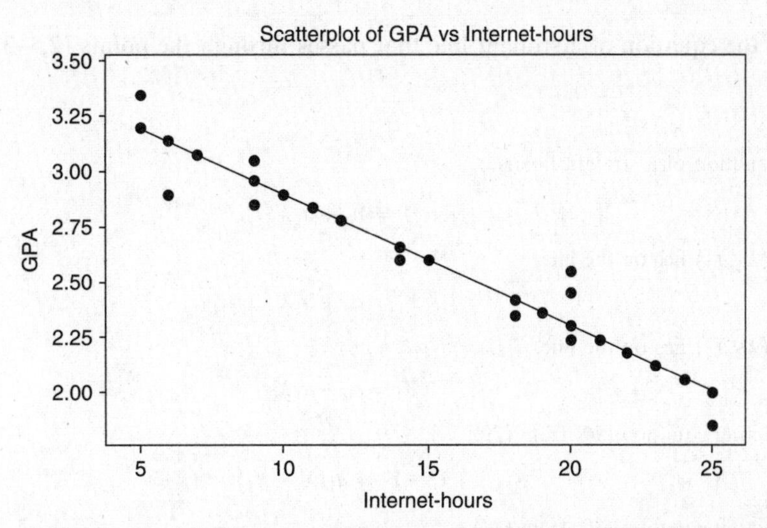

Fig. 13-4 The sum of the squares of the distances of the points from the best-fitting line is a minimum for the line GPA = 3.49 – 0.0594 Internet-hours.

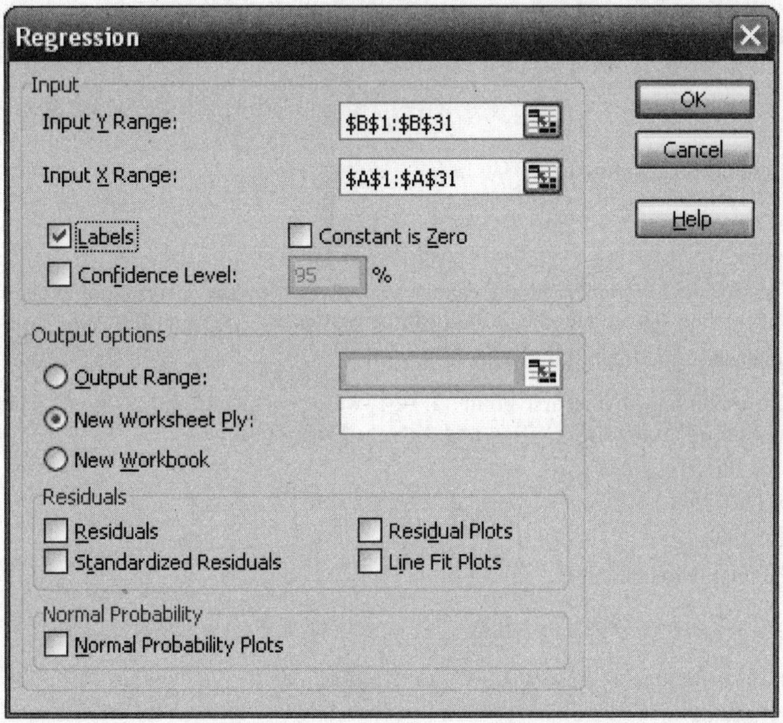

Regression
Input
Input Y Range:
$B$1:$B$31
Input X Range:
$A$1:$A$31
Labels
Constant is Zero
Confidence Level:
95
%
OK
Cancel
Help
Output options
Output Range:
New Worksheet Ply:
New Workbook
Residuals
Residuals
Standardized Residuals
Residual Plots
Line Fit Plots
Normal Probability
Normal Probability Plots

Fig. 13-5 EXCEL dialog box for Problem 13.2.

The constant a_0 is called the *intercept* and a_1 is the *slope*. The same values as given by MINITAB are obtained.

13.3 (*a*) Show that the equation of a straight line that passes through the points (X_1, Y_1) and (X_2, Y_2) is given by

$$Y - Y_1 = \frac{Y_2 - Y_1}{X_2 - X_1}(X - X_1)$$

(*b*) Find the equation of a straight line that passes through the points $(2, -3)$ and $(4, 5)$.

SOLUTION

(*a*) The equation of a straight line is

$$Y = a_0 + a_1 X \tag{29}$$

Since (X_1, Y_1) lies on the line,

$$Y_1 = a_0 + a_1 X_1 \tag{30}$$

Since (X_2, Y_2) lies on the line,

$$Y_2 = a_0 + a_1 X_2 \tag{31}$$

Subtracting equation (*30*) from (*29*),

$$Y - Y_1 = a_1(X - X_1) \tag{32}$$

Subtracting equation (*30*) from (*31*),

$$Y_2 - Y_1 = a_1(X_2 - X_1) \quad \text{or} \quad a_1 = \frac{Y_2 - Y_1}{X_2 - X_1}$$

Substituting this value of a_1 into equation (*32*), we obtain

$$Y - Y_1 = \frac{Y_2 - Y_1}{X_2 - X_1}(X - X_1)$$

as required. The quantity

$$\frac{Y_2 - Y_1}{X_2 - X_1}$$

often abbreviated m, represents the change in Y divided by the corresponding change in X and is the *slope* of the line. The required equation can be written $Y - Y_1 = m(X - X_1)$.

(*b*) **First method** [using the result of part (*a*)]

Corresponding to the first point $(2, -3)$, we have $X_1 = 2$ and $Y_1 = -3$; corresponding to the second point $(4, 5)$, we have $X_2 = 4$ and $Y_2 = 5$. Thus the slope is

$$m = \frac{Y_2 - Y_1}{X_2 - X_1} = \frac{5 - (-3)}{4 - 2} = \frac{8}{2} = 4$$

and the required equation is

$$Y - Y_1 = m(X - X_1) \quad \text{or} \quad Y - (-3) = 4(X - 2)$$

which can be written $Y + 3 = 4(X - 2)$, or $Y = 4X - 11$.

Second method

The equation of a straight line is $Y = a_0 + a_1 X$. Since the point $(2, -3)$ is on the line $-3 = a_0 + 2a_1$, and since the point $(4, 5)$ is on the line, $5 = a_0 + 4a_1$; solving these two equations simultaneously, we obtain $a_1 = 4$ and $a_0 = -11$. Thus the required equation is

$$Y = -11 + 4X \quad \text{or} \quad Y = 4X - 11$$

13.4 Wheat is grown on 9 equal-sized plots. The amount of fertilizer put on each plot is given in Table 13.2 along with the yield of wheat.

Use MINITAB to fit the parabolic curve $Y = a_0 + a_1 X + a_2 X^2$ to the data.

Table 13.2

Amount of Wheat (y)	Fertilizer (x)
2.4	1.2
3.4	2.3
4.4	3.3
5.1	4.1
5.5	4.8
5.2	5.0
4.9	5.5
4.4	6.1
3.9	6.9

SOLUTION

The wheat yield is entered into C1 and the fertilizer is entered into C2. The pull-down **Stat → Regression → Fitted Line Plot** gives the dialog box in Fig. 13-6.

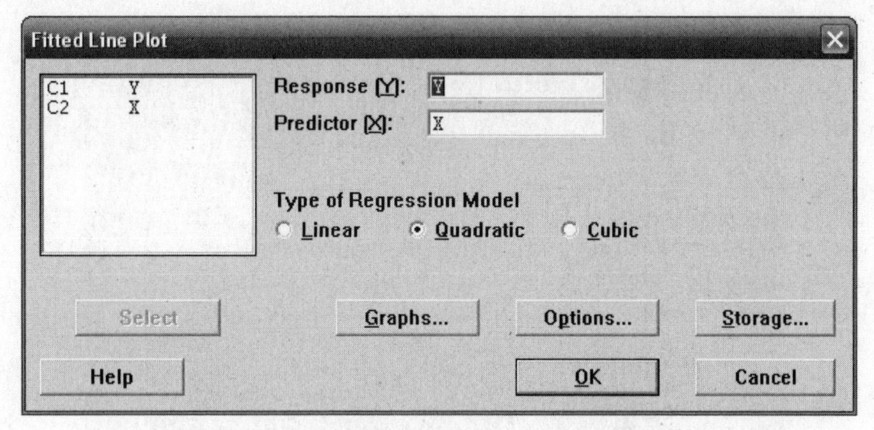

Fig. 13-6 MINITAB dialog box for Problem 13.4.

This dialog box gives the output in Fig. 13-7.

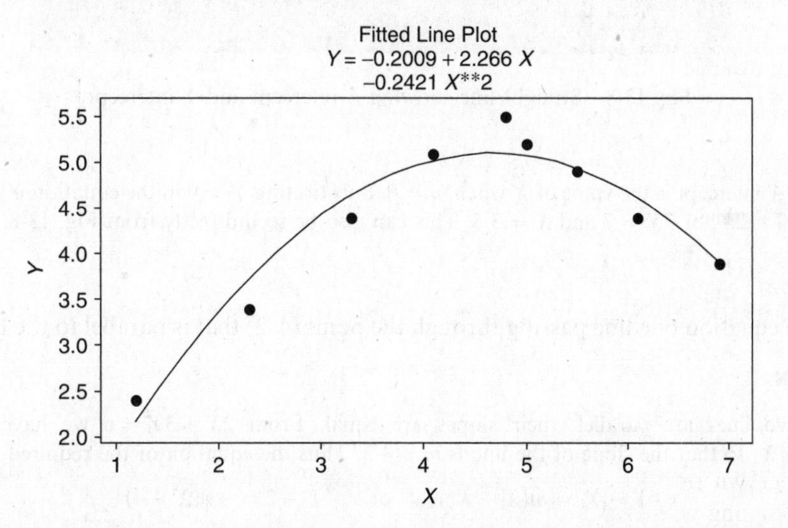

Fig. 13-7 Fitting the least-squares parabolic curve to a set of data using MINITAB.

13.5 Find (*a*) the slope, (*b*) the equation, (*c*) the Y intercept, and (*d*) the X intercept of the line that passes through the points $(1, 5)$ and $(4, -1)$.

SOLUTION

(*a*) $(X_1 = 1,\ Y_1 = 5)$ and $(X_2 = 4,\ Y_2 = -1)$. Thus

$$m = \text{slope} = \frac{Y_2 - Y_1}{X_2 - X_1} = \frac{-1 - 5}{4 - 1} = \frac{-6}{3} = -2$$

The negative sign of the slope indicates that as X increases, Y decreases, as shown in Fig. 13-8.

(*b*) The equation of the line is

$$Y - Y_1 = m(X - X_1) \quad \text{or} \quad Y - 5 = -2(X - 1)$$

That is,
$$Y - 5 = -2X + 2 \quad \text{or} \quad Y = 7 - 2X$$

This can also be obtained by the second method of Problem 13.3(*b*).

(*c*) The Y intercept, which is the value of Y when $X = 0$, is given by $Y = 7 - 2(0) = 7$. This can also be seen directly from Fig. 13-8.

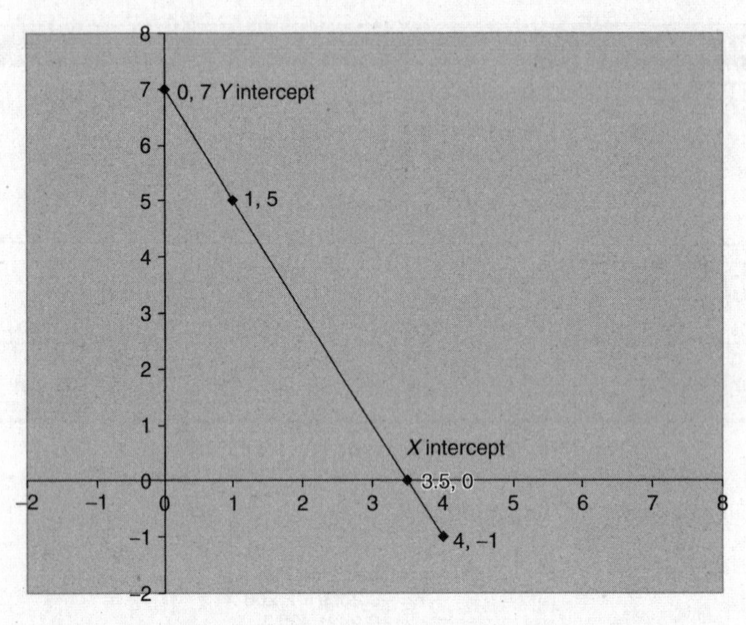

Fig. 13-8 Straight line showing X intercept and Y intercept.

(*d*) The X intercept is the value of X when $Y = 0$. Substituting $Y = 0$ in the equation $Y = 7 - 2X$, we have $0 = 7 - 2X$, or $2X = 7$ and $X = 3.5$. This can also be seen directly from Fig. 13-8.

13.6 Find the equation of a line passing through the point $(4, 2)$ that is parallel to the line $2X + 3Y = 6$.

SOLUTION

If two lines are parallel, their slopes are equal. From $2X + 3Y = 6$ we have $3Y = 6 - 2X$, or $Y = 2 - \frac{2}{3}X$, so that the slope of the line is $m = -\frac{2}{3}$. Thus the equation of the required line is

$$Y - Y_1 = m(X - X_1) \quad \text{or} \quad Y - 2 = -\tfrac{2}{3}(X - 4)$$

which can also be written $2X + 3Y = 14$.

Another method

Any line parallel to $2X + 3Y = 6$ has the equation $2X + 3Y = c$. To find c, let $X = 4$ and $Y = 2$. Then $2(4) + 3(2) = c$, or $c = 14$, and the required equation is $2X + 3Y = 14$.

13.7 Find the equation of a line whose slope is -4 and whose Y intercept is 16.

SOLUTION

In the equation $Y = a_0 + a_1X$, $a_0 = 16$ is the Y intercept and $a_1 = -4$ is the slope. Thus the required equation is $Y = 16 - 4X$.

13.8 (*a*) Construct a straight line that approximates the data of Table 13.3.

(*b*) Find an equation for this line.

Table 13.3

X	1	3	4	6	8	9	11	14
Y	1	2	4	4	5	7	8	9

SOLUTION

(*a*) Plot the points (1, 1), (3, 2), (4, 4), (6, 4), (8, 5), (9, 7), (11, 8), and (14, 9) on a rectangular coordinate system, as shown in Fig. 13-9. A straight line approximating the data is drawn *freehand* in the figure. For a method eliminating the need for individual judgment, see Problem 13.11, which uses the method of least squares.

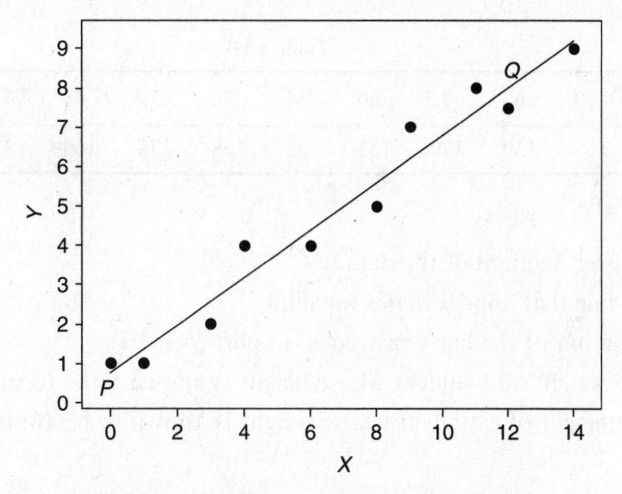

Fig. 13-9 Freehand method of curve fitting.

(*b*) To obtain the equation of the line constructed in part (*a*), choose any two points on the line, such as P and Q; the coordinates of points P and Q, as read from the graph, are approximately (0, 1) and (12, 7.5). The equation of the line is $Y = a_0 + a_1X$. Thus for point (0, 1) we have $1 = a_0 + a_1(0)$, and for point (12, 7.5) we have $7.5 = a_0 + 12a_1$; since the first of these equations gives us $a_0 = 1$, the second gives us $a_1 = 6.5/12 = 0.542$. Thus the required equation is $Y = 1 + 0.542X$.

Another method

$$Y - Y_1 = \frac{Y_2 - Y_1}{X_2 - X_1}(X - X_1) \quad \text{and} \quad Y - 1 = \frac{7.4 - 1}{12 - 0}(X - 0) = 0.542X$$

Thus $Y = 1 + 0.542X$.

13.9 (*a*) Compare the values of Y obtained from the approximating line with those given in Table 13.2.
(*b*) Estimate the value of Y when $X = 10$.

SOLUTION

(*a*) For $X = 1$, $Y = 1 + 0.542(1) = 1.542$, or 1.5. For $X = 3$, $Y = 1 + 0.542(3) = 2.626$ or 2.6. The values of Y corresponding to other values of X can be obtained similarly. The values of Y estimated from the equation $Y = 1 + 0.542X$ are denoted by Y_{est}. These estimated values, together with the actual data from Table 13.3, are shown in Table 13.4.

(*b*) The estimated value of Y when $X = 10$ is $Y = 1 + 0.542(10) = 6.42$, or 6.4.

Table 13.4

X	1	3	4	6	8	9	11	14
Y	1	2	4	4	5	7	8	9
Y_{est}	1.5	2.6	3.2	4.3	5.3	5.9	7.0	8.6

13.10 Table 13.5 shows the heights to the nearest inch (in) and the weights to the nearest pound (lb) of a sample of 12 male students drawn at random from the first-year students at State College.

Table 13.5

Height X (in)	70	63	72	60	66	70	74	65	62	67	65	68
Weight Y (lb)	155	150	180	135	156	168	178	160	132	145	139	152

(*a*) Obtain a scatter diagram of the data.
(*b*) Construct a line that approximates the data.
(*c*) Find the equation of the line constructed in part (*b*).
(*d*) Estimate the weight of a student whose height is known to be 63 in.
(*e*) Estimate the height of a student whose weight is known to be 168 lb.

SOLUTION

(*a*) The scatter diagram, shown in Fig. 13-10, is obtained by plotting the points (70, 155), (63, 150), ..., (68, 152).

(*b*) A straight line that approximates the data is shown dashed in Fig. 13-10. This is but one of the many possible lines that could have been constructed.

(*c*) Choose any two points on the line constructed in part (*b*), such as P and Q, for example. The coordinates of these points as read from the graph are approximately (60, 130) and (72, 170). Thus

$$Y - Y_1 = \frac{Y_2 - Y_1}{X_2 - X_1}(X - X_1) \qquad Y - 130 = \frac{170 - 130}{72 - 60}(X - 60) \qquad Y = \frac{10}{3}X - 70$$

(*d*) If $X = 63$, then $Y = \frac{10}{3}(63) - 70 = 140$ lb.

(*e*) If $Y = 168$, then $168 = \frac{10}{3}X - 70$, $\frac{10}{3}X = 238$, and $X = 71.4$, or 71 in.

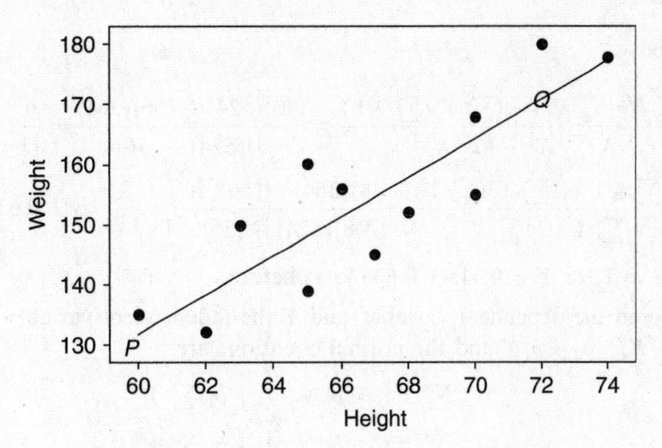

Fig. 13-10 Freehand method of curve fitting.

THE LEAST-SQUARES LINE

13.11 Fit a least-squares line to the data of Problem 13.8 by using (*a*) X as the independent variable and (*b*) X as the dependent variable.

SOLUTION

(*a*) The equation of the line is $Y = a_0 + a_1 X$. The normal equations are

$$\sum Y = a_0 N + a_1 \sum X$$
$$\sum XY = a_0 \sum X + a_1 \sum X^2$$

The work involved in computing the sums can be arranged as in Table 13.6. Although the right-hand column is not needed for this part of the problem, it has been added to the table for use in part (*b*).

Since there are eight pairs of values of X and Y, $N = 8$ and the normal equations become

$$8a_0 + 56a_1 = 40$$
$$56a_0 + 524a_1 = 364$$

Solving simultaneously, $a_0 = \frac{6}{11}$, or 0.545; $a_1 = \frac{7}{11}$, or 0.636; and the required least-squares line is $Y = \frac{6}{11} + \frac{7}{11}X$, or $Y = 0.545 + 0.636X$.

Table 13.6

X	Y	X^2	XY	Y^2
1	1	1	1	1
3	2	9	6	4
4	4	16	16	16
6	4	36	24	16
8	5	64	40	25
9	7	81	63	49
11	8	121	88	64
14	9	196	126	81
$\sum X = 56$	$\sum Y = 40$	$\sum X^2 = 524$	$\sum XY = 364$	$\sum Y^2 = 256$

Another method

$$a_0 = \frac{(\sum Y)(\sum X^2) - (\sum X)(\sum XY)}{N \sum X^2 - (\sum X)^2} = \frac{(40)(524) - (56)(364)}{(8)(524) - (56)^2} = \frac{6}{11} \quad \text{or} \quad 0.545$$

$$a_1 = \frac{N \sum XY - (\sum X)(\sum Y)}{N \sum X^2 - (\sum X)^2} = \frac{(8)(364) - (56)(40)}{(8)(524) - (56)^2} = \frac{7}{11} \quad \text{or} \quad 0.636$$

Thus $Y = a_0 + a_1 X$, or $Y = 0.545 + 0.636X$, as before.

(*b*) If X is considered the dependent variable, and Y the independent variable, the equation of the least-squares line is $X = b_0 + b_1 Y$ and the normal equations are

$$\sum X = b_0 N + b_1 \sum Y$$

$$\sum XY = b_0 \sum Y + b_1 \sum Y^2$$

Then from Table 13.6 the normal equations become

$$8b_0 + 40b_1 = 56$$

$$40b_0 + 256b_1 = 364$$

from which $b_0 = -\frac{1}{2}$, or -0.50, and $b_1 = \frac{3}{2}$, or 1.50. These values can also be obtained from

$$b_0 = \frac{(\sum X)(\sum Y^2) - (\sum Y)(\sum XY)}{N \sum Y^2 - (\sum Y)^2} = \frac{(56)(256) - (40)(364)}{(8)(256) - (40)^2} = -0.50$$

$$b_1 = \frac{N \sum XY - (\sum X)(\sum Y)}{N \sum Y^2 - (\sum Y)^2} = \frac{(8)(364) - (56)(40)}{(8)(256) - (40)^2} = 1.50$$

Thus the required equation of the least-squares line is $X = b_0 + b_1 Y$, or $X = -0.50 + 1.50Y$.

Note that by solving this equation for Y we obtain $Y = \frac{1}{3} + \frac{2}{3}X$, or $Y = 0.333 + 0.667X$, which is not the same as the line obtained in part (*a*).

13.12 For the height/weight data in Problem 13.10 use the statistical package SAS to plot the observed data points and the least-squares line on the same graph.

SOLUTION

In Fig. 13-11, the observed data values are shown as open circles, and the least-squares line is shown as a dashed line.

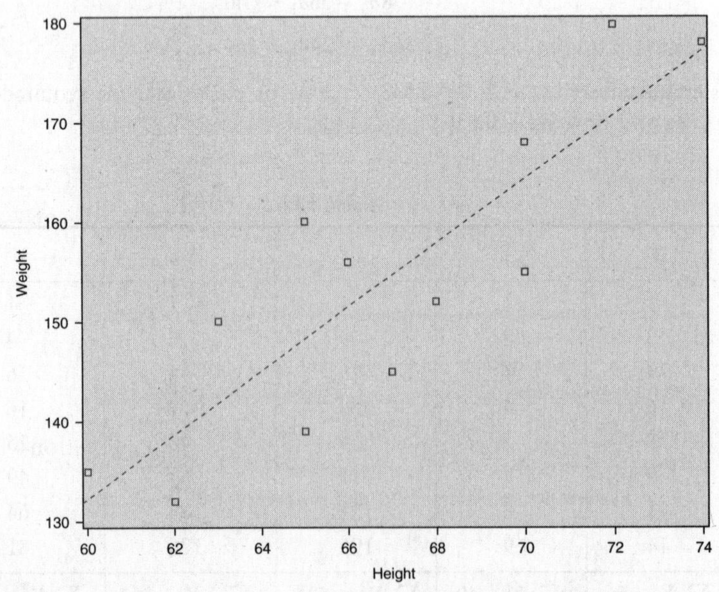

Fig. 13-11 SAS plot of the data points from Table 13.5 and the least-squares line.

13.13 (*a*) Show that the two least-squares lines obtained in Problem 13.11 intersect at point $(\bar{X}, \bar{Y})$.
(*b*) Estimate the value of Y when $X = 12$.
(*c*) Estimate the value of X when $Y = 3$.

SOLUTION

$$\bar{X} = \frac{\sum X}{N} = \frac{56}{8} = 7 \qquad \bar{Y} = \frac{\sum Y}{N} = \frac{40}{8} = 5$$

Thus point $(\bar{X}, \bar{Y})$, called the *centroid*, is (7, 5).

(*a*) Point (7, 5) lies on line $Y = 0.545 + 0.636X$; or, more exactly, $Y = \frac{6}{11} + \frac{7}{11}X$, since $5 = \frac{6}{11} + \frac{7}{11}(7)$. Point (7, 5) lies on line $X = -\frac{1}{2} + \frac{3}{2}Y$, since $7 = -\frac{1}{2} + \frac{3}{2}(5)$.

Another method

The equations of the two lines are $Y = \frac{6}{11} + \frac{7}{11}X$ and $X = -\frac{1}{2} + \frac{3}{2}Y$. Solving simultaneously, we find that $X = 7$ and $Y = 5$. Thus the lines intersect at point (7,5).

(*b*) Putting $X = 12$ into the regression line of Y (Problem 13.11), $Y = 0.545 + 0.636(12) = 8.2$.

(*c*) Putting $Y = 3$ into the regression line of X (Problem 13.11), $X = -0.50 + 1.50(3) = 4.0$.

13.14 Prove that a least-squares line always passes through the point $(\bar{X}, \bar{Y})$.

SOLUTION

Case 1 (X is the independent variable)

The equation of the least-squares line is

$$Y = a_0 + a_1 X \tag{34}$$

A normal equation for the least-squares line is

$$\sum Y = a_0 N + a_1 \sum X \tag{35}$$

Dividing both sides of equation (*35*) by N gives

$$\bar{Y} = a_0 + a_1 \bar{X} \tag{36}$$

Subtracting equation (*36*) from equation (*34*), the least-squares line can be written

$$Y - \bar{Y} = a_1(X - \bar{X}) \tag{37}$$

which shows that the line passes through the point $(\bar{X}, \bar{Y})$.

Case 2 (Y is the independent variable)

Proceeding as in Case 1, but interchanging X and Y and replacing the constants a_0 and a_1 with b_0 and b_1, respectively, we find that the least-squares line can be written

$$X - \bar{X} = b_1(Y - \bar{Y}) \tag{38}$$

which indicates that the line passes through the point $(\bar{X}, \bar{Y})$.

Note that lines (*37*) and (*38*) are not coincident, but intersect in $(\bar{X}, \bar{Y})$.

13.15 (*a*) Considering X to be the independent variable, show that the equation of the least-squares line can be written

$$y = \left(\frac{\sum xy}{\sum x^2}\right)x \qquad \text{or} \qquad y = \left(\frac{\sum xY}{\sum x^2}\right)x$$

where $x = X - \bar{X}$ and $y = Y - \bar{Y}$.

(*b*) If $\bar{X} = 0$, show that the least-squares line in part (*a*) can be written

$$Y = \bar{Y} + \left(\frac{\sum XY}{\sum X^2}\right)X$$

(*c*) Write the equation of the least-squares line corresponding to that in part (*a*) if Y is the independent variable.

(*d*) Verify that the lines in parts (*a*) and (*c*) are not necessarily the same.

SOLUTION

(*a*) Equation (*37*) can be written $y = a_1 x$, where $x = X - \bar{X}$ and $y = Y - \bar{Y}$. Also, from the simultaneous solution of the normal equations (*18*) we have

$$a_1 = \frac{N\sum XY - (\sum X)(\sum Y)}{N\sum X^2 - (\sum X)^2} = \frac{N\sum(x+\bar{X})(y+\bar{Y}) - [\sum(x+\bar{X})][\sum(y+\bar{Y})]}{N\sum(x+\bar{X})^2 - [\sum(x+\bar{X})]^2}$$

$$= \frac{N\sum(xy + x\bar{Y} + \bar{X}y + \bar{X}\bar{Y}) - (\sum x + N\bar{X})(\sum y + N\bar{Y})}{N\sum(x^2 + 2x\bar{X} + \bar{X}^2) - (\sum x + N\bar{X})^2}$$

$$= \frac{N\sum xy + N\bar{Y}\sum x + N\bar{X}\sum y + N^2\bar{X}\bar{Y} - (\sum x + N\bar{X})(\sum y + N\bar{Y})}{N\sum x^2 + 2N\bar{X}\sum x + N^2\bar{X}^2 - (\sum x + N\bar{X})^2}$$

But $\sum x = \sum(X - \bar{X}) = 0$ and $\sum y = \sum(Y - \bar{Y}) = 0$; hence the above simplifies to

$$a_1 = \frac{N\sum xy + N^2\bar{X}\bar{Y} - N^2\bar{X}\bar{Y}}{N\sum x^2 + N^2\bar{X}^2 - N^2\bar{X}^2} = \frac{\sum xy}{\sum x^2}$$

This can also be written

$$a_1 = \frac{\sum xy}{\sum x^2} = \frac{\sum x(Y - \bar{Y})}{\sum x^2} = \frac{\sum xY - \bar{Y}\sum x}{\sum x^2} = \frac{\sum xY}{\sum x^2}$$

Thus the least-squares line is $y = a_1 x$; that is,

$$y = \left(\frac{\sum xy}{\sum x^2}\right)x \quad \text{or} \quad y = \left(\frac{\sum xY}{\sum x^2}\right)x$$

(*b*) If $\bar{X} = 0$, $x = X - \bar{X} = X$. Then from

$$y = \left(\frac{\sum xY}{\sum x^2}\right)$$

we have

$$y = \left(\frac{\sum XY}{\sum X^2}\right)X \quad \text{or} \quad Y = \bar{Y} + \left(\frac{\sum XY}{\sum X^2}\right)X$$

Another method

The normal equations of the least-squares line $Y = a_0 + a_1 X$ are

$$\sum Y = a_0 N + a_1 \sum X \quad \text{and} \quad \sum XY = a_0 \sum X + a_1 \sum X^2$$

If $\bar{X} = (\sum X)/N = 0$, then $\sum X = 0$ and the normal equations become

$$\sum Y = a_0 N \quad \text{and} \quad \sum XY = a_1 \sum X^2$$

from which

$$a_0 = \frac{\sum Y}{N} = \bar{Y} \quad \text{and} \quad a_1 = \frac{\sum XY}{\sum X^2}$$

Thus the required equation of the least-squares line is

$$Y = a_0 + a_1 X \quad \text{or} \quad Y = \bar{Y} + \left(\frac{\sum XY}{\sum X^2}\right)X$$

(c) By interchanging X and Y or x and y, we can show as in part (a) that

$$x = \left(\frac{\sum xy}{\sum y^2}\right) y$$

(d) From part (a), the least-squares line is

$$y = \left(\frac{\sum xy}{\sum x^2}\right) x \tag{39}$$

From part (c), the least-squares line is

$$x = \left(\frac{\sum xy}{\sum y^2}\right) y$$

or

$$y = \left(\frac{\sum y^2}{\sum xy}\right) x \tag{40}$$

Since in general

$$\frac{\sum xy}{\sum x^2} \neq \frac{\sum y^2}{\sum xy}$$

the least-squares lines (39) and (40) are different in general. Note, however, that they intersect at $x = 0$ and $y = 0$ [i.e., at the point $(\bar{X}, \bar{Y})$].

13.16 If $X' = X + A$ and $Y' = Y + B$, where A and B are any constants, prove that

$$a_1 = \frac{N \sum XY - (\sum X)(\sum Y)}{N \sum X^2 - (\sum X)^2} = \frac{N \sum X'Y' - (\sum X')(\sum Y')}{N \sum X'^2 - (\sum X')^2} = a_1'$$

SOLUTION

$$x' = X' - \bar{X}' = (X + A) - (\bar{X} + A) = X - \bar{X} = x$$
$$y' = Y' - \bar{Y}' = (Y + B) - (\bar{Y} + B) = Y - \bar{Y} = y$$

Then

$$\frac{\sum xy}{\sum x^2} = \frac{\sum x'y'}{\sum x'^2}$$

and the result follows from Problem 13.15. A similar result holds for b_1.

This result is useful, since it enables us to simplify calculations in obtaining the regression line by subtracting suitable constants from the variables X and Y (see the second method of Problem 13.17).

Note: The result does not hold if $X' = c_1X + A$ and $Y' = c_2Y + B$ unless $c_1 = c_2$.

13.17 Fit a least-squares line to the data of Problem 13.10 by using (a) X as the independent variable and (b) X as the dependent variable.

SOLUTION

First method

(a) From Problem 13.15(a) the required line is

$$y = \left(\frac{\sum xy}{\sum x^2}\right) x$$

where $x = X - \bar{X}$ and $y = Y - \bar{Y}$. The work involved in computing the sums can be arranged as in Table 13.7. From the first two columns we find $\bar{X} = 802/12 = 66.8$ and $\bar{Y} = 1850/12 = 154.2$. The last column has been added for use in part (b).

Table 13.7

Height X	Weight Y	$x = X - \bar{X}$	$y = Y - \bar{Y}$	xy	x^2	y^2
70	155	3.2	0.8	2.56	10.24	0.64
63	150	−3.8	−4.2	15.96	14.44	17.64
72	180	5.2	25.8	134.16	27.04	665.64
60	135	−6.8	−19.2	130.56	46.24	368.64
66	156	−0.8	1.8	−1.44	0.64	3.24
70	168	3.2	13.8	44.16	10.24	190.44
74	178	7.2	23.8	171.36	51.84	566.44
65	160	−1.8	5.8	−10.44	3.24	33.64
62	132	−4.8	−22.2	106.56	23.04	492.84
67	145	0.2	−9.2	−1.84	0.04	84.64
65	139	−1.8	−15.2	27.36	3.24	231.04
68	152	1.2	−2.2	−2.64	1.44	4.84
$\sum X = 802$ $\bar{X} = 66.8$	$\sum Y = 1850$ $\bar{Y} = 154.2$			$\sum xy = 616.32$	$\sum x^2 = 191.68$	$\sum y^2 = 2659.68$

The required least-squares line is

$$y = \left(\frac{\sum xy}{\sum x^2}\right)x = \frac{616.32}{191.68}x = 3.22x$$

or $Y - 154.2 = 3.22(X - 66.8)$, which can be written $Y = 3.22X - 60.9$. This equation is called the *regression line of Y on X* and is used for estimating Y from given values of X.

(*b*) If X is the dependent variable, the required line is

$$x = \left(\frac{\sum xy}{\sum y^2}\right)y = \frac{616.32}{2659.68}y = 0.232y$$

which can be written $X - 66.8 = 0.232(Y - 154.2)$, or $X = 31.0 + 0.232Y$. This equation is called the *regression line of X on Y* and is used for estimating X from given values of Y.

Note that the method of Problem 13.11 can also be used if desired.

Second method

Using the result of Problem 13.16, we may subtract suitable constants from X and Y. We choose to subtract 65 from X and 150 from Y. Then the results can be arranged as in Table 13.7.

$$a_1 = \frac{N\sum X'Y' - (\sum X')(\sum Y')}{N\sum X'^2 - (\sum X')^2} = \frac{(12)(708) - (22)(50)}{(12)(232) - (22)^2} = 3.22$$

$$b_1 = \frac{N\sum X'Y' - (\sum Y')(\sum X')}{N\sum Y'^2 - (\sum Y')^2} = \frac{(12)(708) - (50)(22)}{(12)(2868) - (50)^2} = 0.232$$

Since $\bar{X} = 65 + 22/12 = 66.8$ and $\bar{Y} = 150 + 50/12 = 154.2$, the regression equations are $Y - 154.2 = 3.22(X - 66.8)$ and $X - 66.8 = 0.232(Y - 154.2)$; that is $Y = 3.22X - 60.9$ and $X = 0.232Y + 31.0$, in agreement with the first method.

13.18 Work Problem 13.17 using MINITAB. Plot the regression line of weight on height and the regression line of height on weight on the same set of axes. Show that the point $(\bar{X}, \bar{Y})$ satisfies both equations. The lines therefore intersect at $(\bar{X}, \bar{Y})$.

SOLUTION

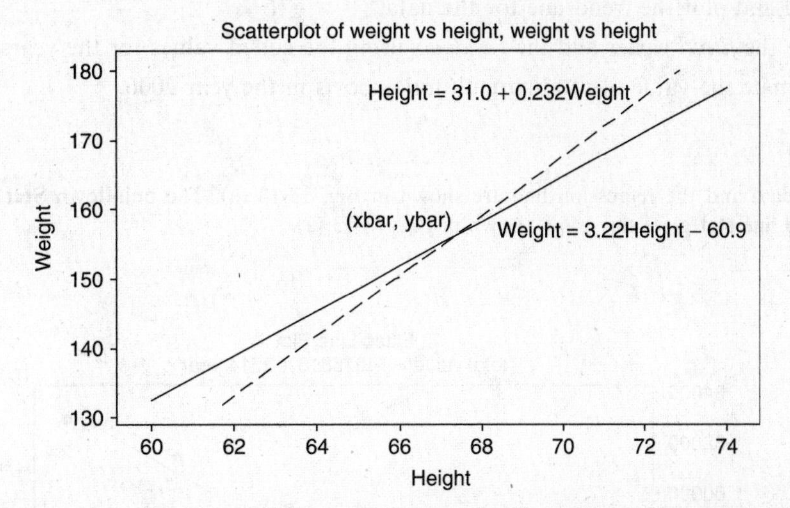

Fig. 13-12 The regression line of weight on height and the regression line of height on weight both pass through the point (*x*bar, *y*bar).

$(\bar{X}, \bar{Y})$ is the same as (`xbar, ybar`) and equals (66.83, 154.17). Note that weight = 3.22(**66.83**)− 60.9 = **154.17** and height = 31.0 + 0.232(**154.17**) = **66.83**. Therefore, both lines pass through (`xbar, ybar`).

Table 13.8

X'	Y'	X'^2	$X'Y'$	Y'^2
5	5	25	25	25
−2	0	4	0	0
7	30	49	210	900
−5	−15	25	75	225
1	6	1	6	36
5	18	25	90	324
9	28	81	252	784
0	10	0	0	100
−3	−18	9	54	324
2	−5	4	−10	25
0	−11	0	0	121
3	2	9	6	4
$\sum X' = 22$	$\sum Y' = 50$	$\sum X'^2 = 232$	$\sum X'Y' = 708$	$\sum Y'^2 = 2868$

APPLICATIONS TO TIME SERIES

13.19 The total agricultural exports in millions of dollars are given in Table 13.9. Use MINITAB to do the following.

Table 13.9

Year	2000	2001	2002	2003	2004	2005
Total value	51246	53659	53115	59364	61383	62958
Coded year	1	2	3	4	5	6

Source: The 2007 Statistical Abstract.

(*a*) Graph the data and show the least-squares regression line.
(*b*) Find and plot the trend line for the data.
(*c*) Give the *fitted values* and the *residuals* using the coded values for the years.
(*d*) Estimate the value of total agricultural exports in the year 2006.

SOLUTION

(*a*) The data and the regression line are shown in Fig. 13-13 (a). The pull-down **Stat → Regression → Fitted line plot** gives the graph shown in Fig. 13-13 (a).

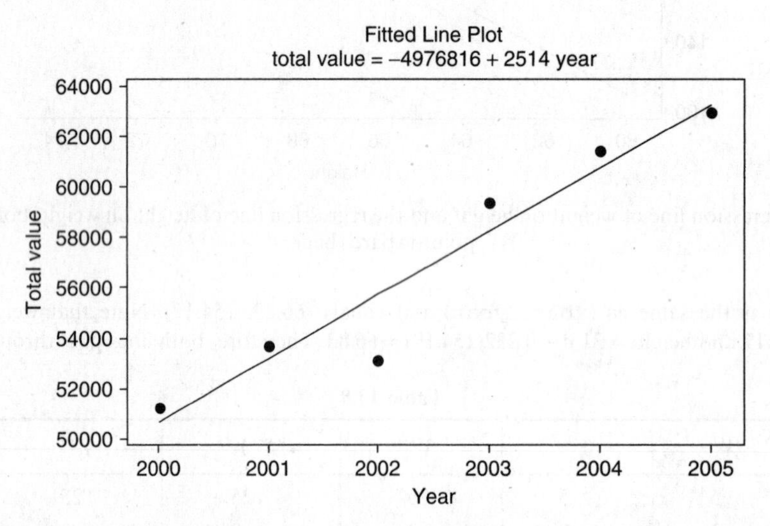

Fig. 13-13 (a) The regression line for total agricultural exports in millions of dollars.

(*b*) The pull-down **Stat → Time series → Trend Analysis** gives the plot shown in Fig. 13-13 (b). It is a different way of looking at the same data. It may be a little easier to work with index numbers rather than years.

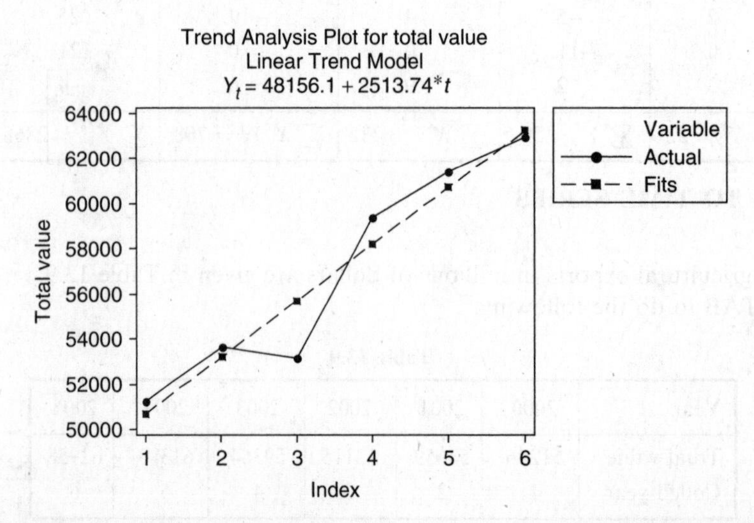

Fig. 13-13 (b) The trend line for total agricultural exports in millions of dollars.

(*c*) Table 13.10 gives the fitted values and the residuals for the data in Table 13.9 using coded values for the years.

Table 13.10

Year coded	Total value	Fitted value	Residual
1	51246	50669.8	576.19
2	53659	53183.6	475.45
3	53115	55697.3	−2582.30
4	59364	58211.0	1152.96
5	61383	60724.8	658.22
6	62958	63238.5	−280.52

(*d*) Using the coded value, the estimated value is $Y_t = 48156.1 + 2513.74(7) = 65752.3$.

13.20 Table 13.11 gives the purchasing power of the dollar as measured by consumer prices according to the U.S. Bureau of Labor Statistics, Survey of Current Business.

Table 13.11

Year	2000	2001	2002	2003	2004	2005
Consumer prices	0.581	0.565	0.556	0.544	0.530	0.512

Source: U.S. Bureau of Labor Statistics, Survey of Current Business.

(*a*) Graph the data and obtain the trend line using MINITAB.
(*b*) Find the equation of the trend line by hand.
(*c*) Estimate the consumer price in 2008 assuming the trend continues for 3 more years.

SOLUTION

(*a*) The solid line in Fig. 13-14 shows a plot of the data in Table 13.11 and the dashed line shows the graph of the least-squares line.

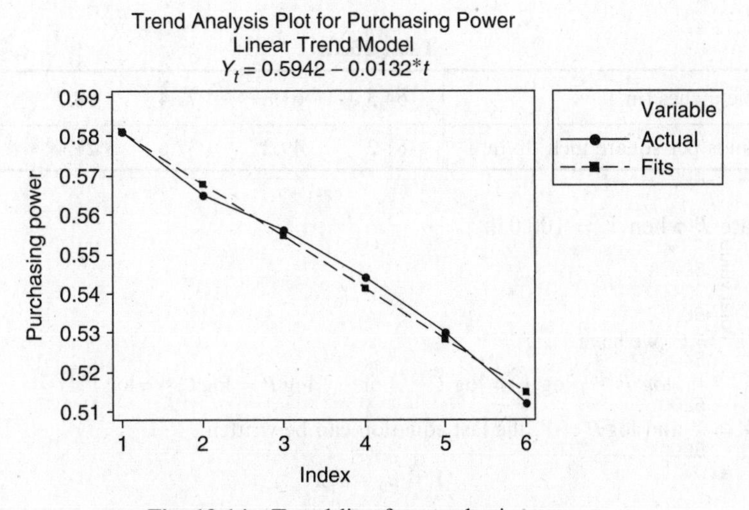

Fig. 13-14 Trend line for purchasing power.

(*b*) The computations for finding the trend line by hand are shown in Table 13.12. The equation is

$$y = \frac{\sum xy}{\sum x^2} x$$

where $x = X - \bar{X}$ and $y = Y - \bar{Y}$, which can be written as $Y - 0.548 = -0.0132(X - 3.5)$ or $Y = -0.0132X + 0.5942$. The computational work saved by statistical software is tremendous as illustrated by this problem.

Table 13.12

Year	X	Y	$x = X - \bar{X}$	$y = Y - \bar{Y}$	x^2	xy
2000	1	0.581	−2.5	0.033	6.25	−0.0825
2001	2	0.565	−1.5	0.017	2.25	−0.0255
2002	3	0.556	−0.5	0.008	0.25	−0.004
2003	4	0.544	0.5	−0.004	0.25	−0.002
2004	5	0.530	1.5	−0.018	2.25	−0.027
2005	6	0.512	2.5	−0.036	6.25	−0.09
	$\Sigma X = 21$	$\Sigma Y = 3.288$			Σx^2	Σxy
	$\bar{X} = 3.5$	$\bar{Y} = 0.548$			17.5	−0.231

(*c*) The estimation for 2008 is obtained by substituting $t = 9$ in the trend line equation. Estimated consumer price is $0.5942 - 0.0132(9) = 0.475$.

NONLINEAR EQUATIONS REDUCIBLE TO LINEAR FORM

13.21 Table 13.13 gives experimental values of the pressure P of a given mass of gas corresponding to various values of the volume V. According to thermodynamic principles, a relationship having the form $PV^{\gamma} = C$, where γ and C are constants, should exist between the variables.

(*a*) Find the values of γ and C.

(*b*) Write the equation connecting P and V.

Table 13.13

Volume V in cubic inches (in^3)	54.3	61.8	72.4	88.7	118.6	194.0
Pressure P in pounds per square inch (lb/in^2)	61.2	49.2	37.6	28.4	19.2	10.1

(*c*) Estimate P when $V = 100.0\,\text{in}^3$.

SOLUTION

Since $PV^{\gamma} = C$, we have

$$\log P + \gamma \log V = \log C \qquad \text{or} \qquad \log P = \log C - \gamma \log V$$

Calling $\log V = X$ and $\log P = Y$, the last equation can be written

$$Y = a_0 + a_1 X \tag{41}$$

where $a_0 = \log C$ and $a_1 = -\gamma$.

Table 13.14 gives $X = \log V$ and $Y = \log P$, corresponding to the values of V and P in Table 13.13, and also indicates the calculations involved in computing the least-squares line (*41*). The normal equations corresponding to the least-squares line (*41*) are

$$\sum Y = a_0 N + a_1 \sum X \quad \text{and} \quad \sum XY = a_0 \sum X + a_1 \sum X^2$$

from which

$$a_0 = \frac{(\sum Y)(\sum X^2) - (\sum X)(\sum XY)}{N \sum X^2 - (\sum X)^2} = 4.20 \qquad a_1 = \frac{N \sum XY - (\sum X)(\sum Y)}{N \sum X^2 - (\sum X)^2} = -1.40$$

Thus $Y = 4.20 - 1.40X$.

(*a*) Since $a_0 = 4.20 = \log C$ and $a_1 = -1.40 = -\gamma$, $C = 1.60 \times 10^4$ and $\gamma = 1.40$.

(*b*) The required equation in terms of P and V can be written $PV^{1.40} = 16{,}000$.

(*c*) When $V = 100$, $X = \log V = 2$ and $Y = \log P = 4.20 - 1.40(2) = 1.40$. Then $P = \text{antilog}\, 1.40 = 25.1\ \text{lb/in}^2$.

Table 13.14

$X = \log V$	$Y = \log P$	X^2	XY
1.7348	1.7868	3.0095	3.0997
1.7910	1.6946	3.2077	3.0350
1.8597	1.5752	3.4585	2.9294
1.9479	1.4533	3.7943	2.8309
2.0741	1.2833	4.3019	2.6617
2.2878	1.0043	5.2340	2.2976
$\sum X = 11.6953$	$\sum Y = 8.7975$	$\sum X^2 = 23.0059$	$\sum XY = 16.8543$

13.22 Use MINITAB to assist in the solution of Problem 13.21.

SOLUTION

The transformations $X = \log_t(V)$ and $Y = \log_t(P)$ converts the problem to a linear fit problem. The calculator in MINITAB is used to take common logarithms of volume and pressure. The following are in columns C1 through C4 of the MINITAB worksheet.

V	*P*	Log10*V*	Log10*P*
54.3	61.2	1.73480	1.78675
61.8	49.2	1.79099	1.69197
72.4	37.6	1.85974	1.57519
88.7	28.4	1.94792	1.45332
118.6	19.2	2.07408	1.28330
194.0	10.1	2.28780	1.00432

The least-squares fit gives the following: $\log_{10}(P) = 4.199 - 1.402 \log_{10}(V)$. See Fig. 13-15. $a_0 = \log C$ and $a_1 = -\gamma$. Taking antilogs gives $C = 10^{a_0}$ and $\gamma = -a_1$ or $C = 15812$ and $\gamma = 1.402$. The nonlinear equation is $PV^{1.402} = 15812$.

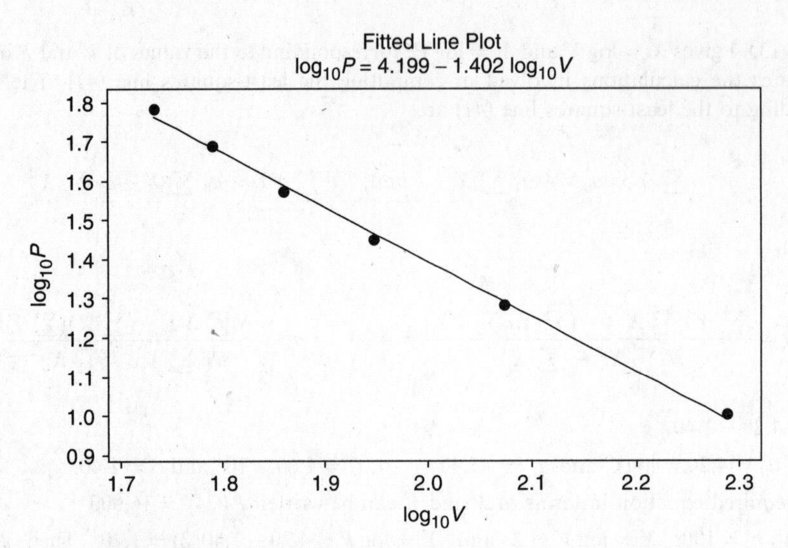

Fig. 13-15 Reducing nonlinear equations to linear form.

13.23 Table 13.15 gives the population of the United States at 5-year intervals from 1960 to 2005 in millions. Fit a straight line as well as a parabola to the data and comment on the two fits. Use both models to predict the United States population in 2010.

Table 13.15

Year	1960	1965	1970	1975	1980	1985	1990	1995	2000	2005
Population	181	194	205	216	228	238	250	267	282	297

Source: U.S. Bureau of Census.

SOLUTION

A partial printout of the MINITAB solution for the least-squares line and least-squares parabola is given below.

```
Year  Population   x     xsquare

1960      181      1       1
1965      194      2       4
1970      205      3       9
1975      216      4      16
1980      228      5      25
1985      238      6      36
1990      250      7      49
1995      267      8      64
2000      282      9      81
2005      297     10     100
```

The straight line model is as follows:
The regression equation is

Population = 166 + 12.6 x

The quadratic model is as follows:
The regression equation is
Population = 174 + 9.3 x - 0.326 x²

Table 13.16 gives the fitted values and residuals for the straight line fit to the data.

Table 13.16

Year	Population	Fitted value	Residual
1960	181	179.018	1.98182
1965	194	191.636	2.36364
1970	205	204.255	0.74545
1975	216	216.873	−0.87273
1980	228	229.491	−1.49091
1985	238	242.109	−4.10909
1990	250	254.727	−4.72727
1995	267	267.345	−0.34545
2000	282	279.964	2.03636
2005	297	292.582	4.41818

Table 13.17 gives the fitted values and residuals for the parabolic fit to the data. The sum of squares of the residuals for the straight line is 76.073 and the sum of squares of the residuals for the parabola is 20.042. It appears that, overall, the parabola fits the data better than the straight line.

Table 13.17

Year	Population	Fitted value	Residual
1960	181	182.927	−1.92727
1965	194	192.939	1.06061
1970	205	203.603	1.39697
1975	216	214.918	1.08182
1980	228	226.885	1.11515
1985	238	239.503	−1.50303
1990	250	252.773	−2.77273
1995	267	266.694	0.30606
2000	282	281.267	0.73333
2005	297	296.491	0.50909

To predict the population in the year 2010, note that the coded value for 2010 is 11. The straight line predicted value is population $= 166 + 12.6x = 166 + 138.6 = 304.6$ million and the parabola model predicts population $= 174 + 9.03x + 0.326x^2 = 174 + 99.33 + 39.446 = 312.776$.

Supplementary Problems

STRAIGHT LINES

13.24 If $3X + 2Y = 18$, find (*a*) X when $Y = 3$, (*b*) Y when $X = 2$, (*c*) X when $Y = -5$, (*d*) Y when $X = -1$, (*e*) the X intercept, and (*f*) the Y intercept.

13.25 Construct a graph of the equations (*a*) $Y = 3X - 5$ and (*b*) $X + 2Y = 4$ on the same set of axes. In what point do the graphs intersect?

13.26 (*a*) Find an equation for the straight line passing through the points $(3, -2)$ and $(-1, 6)$.
(*b*) Determine the X and Y intercepts of the line in part (*a*).
(*c*) Find the value of Y corresponding to $X = 3$ and to $X = 5$.
(*d*) Verify your answers to parts (*a*), (*b*), and (*c*) directly from a graph.

13.27 Find an equation for the straight line whose slope is $\frac{2}{3}$ and whose Y intercept is -3.

13.28 (*a*) Find the slope and Y intercept of the line whose equation is $3X - 5Y = 20$.
(*b*) What is the equation of a line which is parallel to the line in part (*a*) and which passes through the point $(2, -1)$?

13.29 Find (*a*) the slope, (*b*) the Y intercept, and (*c*) the equation of the line passing through the points $(5, 4)$ and $(2, 8)$.

13.30 Find the equation of a straight line whose X and Y intercepts are 3 and -5, respectively.

13.31 A temperature of 100 degrees Celsius (°C) corresponds to 212 degrees Fahrenheit (°F), while a temperature of 0°C corresponds to 32°F. Assuming that a linear relationship exists between Celsius and Fahrenheit temperatures, find (*a*) the equation connecting Celsius and Fahrenheit temperatures, (*b*) the Fahrenheit temperature corresponding to 80°C, and (*c*) the Celsius temperature corresponding to 68°F.

THE LEAST-SQUARES LINE

13.32 Fit a least-squares line to the data in Table 13.18, using (*a*) X as the independent variable and (*b*) X as the dependent variable. Graph the data and the least-squares lines, using the same set of coordinate axes.

Table 13.18

X	3	5	6	8	9	11
Y	2	3	4	6	5	8

13.33 For the data of Problem 13.32, find (*a*) the values of Y when $X = 5$ and $X = 12$ and (*b*) the value of X when $Y = 7$.

13.34 (*a*) Use the freehand method to obtain an equation for a line fitting the data of Problem 13.32.
(*b*) Using the result of part (*a*), answer Problem 13.33.

13.35 Table 13.19 shows the final grades in algebra and physics obtained by 10 students selected at random from a large group of students.

(*a*) Graph the data.
(*b*) Find a least-squares line fitting the data, using X as the independent variable.

(c) Find a least-squares line fitting the data, using Y as the independent variable.

(d) If a student receives a grade of 75 in algebra, what is her expected grade in physics?

(e) If a student receives a grade of 95 in physics, what is her expected grade in algebra?

Table 13.19

Algebra (X)	75	80	93	65	87	71	98	68	84	77
Physics (Y)	82	78	86	72	91	80	95	72	89	74

13.36 Table 13.20 shows the birth rate per 1000 population during the years 1998 through 2004.

(a) Graph the data.

(b) Find the least-squares line fitting the data. Code the years 1998 through 2004 as the whole numbers 1 through 7.

(c) Compute the trend values (fitted values) and the residuals.

(d) Predict the birth rate in 2010, assuming the present trend continues.

Table 13.20

Year	1998	1999	2000	2001	2002	2003	2004
Birth rate per 1000	14.3	14.2	14.4	14.1	13.9	14.1	14.0

Source: U.S. National Center for Health Statistics, Vital Statistics of the United States, annual; National Vital Statistics Reports and unpublished data.

13.37 Table 13.21 shows the number in thousands of the United States population 85 years and over for the years 1999 through 2005.

(a) Graph the data.

(b) Find the least-squares line fitting the data. Code the years 1999 through 2005 as the whole numbers 1 through 7.

(c) Compute the trend values (fitted values) and the residuals.

(d) Predict the number of individuals 85 years and older in 2010, assuming the present trend continues.

Table 13.21

Year	1999	2000	2001	2002	2003	2004	2005
85 and over	4154	4240	4418	4547	4716	4867	5096

Source: U.S. Bureau of Census.

LEAST-SQUARES CURVES

13.38 Fit a least-squares parabola, $Y = a_0 + a_1X + a_2X^2$, to the data in Table 13.22.

Table 13.22

X	0	1	2	3	4	5	6
Y	2.4	2.1	3.2	5.6	9.3	14.6	21.9

13.39 The total time required to bring an automobile to a stop after one perceives danger is the reaction time (the time between recognizing danger and applying the brakes) plus the braking time (the time for stopping after applying the brakes). Table 13.23 gives the stopping distance D (in feet, of ft) of an automobile traveling at speeds V (in miles per hour, or mi/h) from the instant that danger is perceived.

(*a*) Graph D against V.

(*b*) Fit a least-squares parabola of the form $D = a_0 + a_1 V + a_2 V^2$ to the data.

(*c*) Estimate D when $V = 45$ mi/h and 80 mi/h.

Table 13.23

Speed V (mi/h)	20	30	40	50	60	70
Stopping distance D (ft)	54	90	138	206	292	396

13.40 Table 13.24 shows the male and female populations of the United States during the years 1940 through 2005 in millions. It also shows the years coded and the differences which equals male minus female.

(*a*) Graph the data points and the linear least-squares best fit.

(*b*) Graph the data points and the quadratic least-squares best fit.

(*c*) Graph the data points and the cubic least-squares best fit.

(*d*) Give the fitted values and the residuals using each of the three models. Give the sum of squares for residuals for all three models.

(*e*) Use each of the three models to predict the difference in 2010.

Table 13.24

Year	1940	1950	1960	1970	1980	1990	2000	2005
Coded	0	1	2	3	4	5	6	6.5
Male	66.1	75.2	88.3	98.9	110.1	121.2	138.1	146.0
Female	65.6	76.1	91.0	104.3	116.5	127.5	143.4	150.4
Difference	0.5	−0.9	−2.7	−5.4	−6.4	−6.3	−5.3	−4.4

Source: U.S. Bureau of Census.

13.41 Work Problem 13.40 using the ratio of females to males instead of differences.

13.42 Work Problem 13.40 by fitting a least squares parabola to the differences.

13.43 The number Y of bacteria per unit volume present in a culture after X hours is given in Table 13.25.

Table 13.25

Number of hours (X)	0	1	2	3	4	5	6
Number of bacteria per unit volume (Y)	32	47	65	92	132	190	275

(*a*) Graph the data on semilog graph paper, using the logarithmic scale for Y and the arithmetic scale for X.

(*b*) Fit a least-squares curve of the form $Y = ab^X$ to the data and explain why this particular equation should yield good results.

(*c*) Compare the values of Y obtained from this equation with the actual values.

(*d*) Estimate the value of Y when $X = 7$.

13.44 In Problem 13.43, show how a graph on semilog graph paper can be used to obtain the required equation without employing the method of least squares.

Correlation Theory

CORRELATION AND REGRESSION

In Chapter 13 we considered the problem of *regression*, or *estimation*, of one variable (the dependent variable) from one or more related variables (the independent variables). In this chapter we consider the closely related problem of *correlation*, or the degree of relationship between variables, which seeks to determine *how well* a linear or other equation describes or explains the relationship between variables.

If all values of the variables satisfy an equation exactly, we say that the variables are *perfectly correlated* or that there is *perfect correlation* between them. Thus the circumferences C and radii r of all circles are perfectly correlated since $C = 2\pi r$. If two dice are tossed simultaneously 100 times, there is no relationship between corresponding points on each die (unless the dice are loaded); that is, they are *uncorrelated*. Such variables as the height and weight of individuals would show *some* correlation.

When only two variables are involved, we speak of *simple correlation* and *simple regression*. When more than two variables are involved, we speak of *multiple correlation* and *multiple regression*. This chapter considers only simple correlation. Multiple correlation and regression are considered in Chapter 15.

LINEAR CORRELATION

If X and Y denote the two variables under consideration, a *scatter diagram* shows the location of points (X, Y) on a rectangular coordinate system. If all points in this scatter diagram seem to lie near a line, as in Figs. 14-1(a) and 14-1(b), the correlation is called *linear*. In such cases, as we have seen in Chapter 13, a linear equation is appropriate for purposes of regression (or estimation).

If Y tends to increase as X increases, as in Fig. 14-1(a), the correlation is called *positive*, or *direct, correlation*. If Y tends to decrease as X increases, as in Fig. 14-1(b), the correlation is called *negative*, or *inverse, correlation*.

If all points seem to lie near some curve, the correlation is called *nonlinear*, and a nonlinear equation is appropriate for regression, as we have seen in Chapter 13. It is clear that nonlinear correlation can be sometimes positive and sometimes negative.

If there is no relationship indicated between the variables, as in Fig. 14-1(c), we say that there is *no correlation* between them (i.e., they are *uncorrelated*).

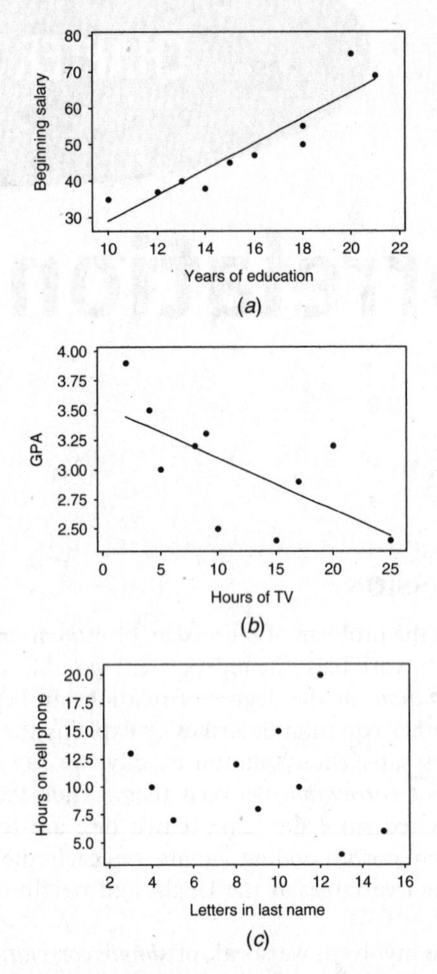

Fig. 14-1 Examples of positive correlation, negative correlation and no correlation. (*a*) Beginning salary and years of formal education are positively correlated; (*b*) Grade point average (GPA) and hours spent watching TV are negatively correlated; (*c*) There is no correlation between hours on a cell phone and letters in last name.

MEASURES OF CORRELATION

We can determine in a *qualitative* manner how well a given line or curve describes the relationship between variables by direct observation of the scatter diagram itself. For example, it is seen that a straight line is far more helpful in describing the relation between X and Y for the data of Fig. 14-1(*a*) than for the data of Fig. 14-1(*b*) because of the fact that there is less scattering about the line of Fig. 14-1(*a*).

If we are to deal with the problem of scattering of sample data about lines or curves in a *quantitative* manner, it will be necessary for us to devise *measures of correlation*

THE LEAST-SQUARES REGRESSION LINES

We first consider the problem of how well a straight line explains the relationship between two variables. To do this, we shall need the equations for the least-squares regression lines obtained in Chapter 13. As we have seen, the least-squares regression line of Y on X is

$$Y = a_0 + a_1 X \tag{1}$$

where a_0 and a_1 are obtained from the normal equations

$$\begin{aligned} \sum Y &= a_0 N + a_1 \sum X \\ \sum XY &= a_0 \sum X + a_1 \sum X^2 \end{aligned} \tag{2}$$

which yield

$$\begin{aligned} a_0 &= \frac{(\sum Y)(\sum X^2) - (\sum X)(\sum XY)}{N \sum X^2 - (\sum X)^2} \\ a_1 &= \frac{N \sum XY - (\sum X)(\sum Y)}{N \sum X^2 - (\sum X)^2} \end{aligned} \tag{3}$$

Similarly, the regression line of X on Y is given by

$$X = b_0 + b_1 Y \tag{4}$$

where b_0 and b_1 are obtained from the normal equations

$$\begin{aligned} \sum X &= b_0 N + b_1 \sum Y \\ \sum XY &= b_0 \sum X + b_1 \sum Y^2 \end{aligned} \tag{5}$$

which yield

$$\begin{aligned} b_0 &= \frac{(\sum X)(\sum Y^2) - (\sum Y)(\sum XY)}{N \sum Y^2 - (\sum Y)^2} \\ b_1 &= \frac{N \sum XY - (\sum X)(\sum Y)}{N \sum Y^2 - (\sum Y)^2} \end{aligned} \tag{6}$$

Equations (*1*) and (*4*) can also be written, respectively, as

$$y = \left(\frac{\sum xy}{\sum x^2}\right)x \quad \text{and} \quad x = \left(\frac{\sum xy}{\sum y^2}\right)y \tag{7}$$

where $x = X - \bar{X}$ and $y = Y - \bar{Y}$.

The regression equations are identical if and only if all points of the scatter diagram lie on a line. In such case there is *perfect linear correlation* between X and Y.

STANDARD ERROR OF ESTIMATE

If we let Y_{est} represent the value of Y for given values of X as estimated from equation (*1*), a measure of the scatter about the regression line of Y on X is supplied by the quantity

$$s_{Y.X} = \sqrt{\frac{\sum (Y - Y_{est})^2}{N}} \tag{8}$$

which is called the *standard error of estimate of Y on X*.

If the regression line (4) is used, an analogous standard error of estimate of X on Y is defined by

$$s_{X.Y} = \sqrt{\frac{\sum (X - X_{\text{est}})^2}{N}} \tag{9}$$

In general, $s_{Y.X} \neq s_{X.Y}$.

Equation (8) can be written

$$s_{Y.X}^2 = \frac{\sum Y^2 - a_0 \sum Y - a_1 \sum XY}{N} \tag{10}$$

which may be more suitable for computation (see Problem 14.3). A similar expression exists for equation (9).

The standard error of estimate has properties analogous to those of the standard deviation. For example, if we construct lines parallel to the regression line of Y on X at respective vertical distances $s_{Y.X}$, $2s_{Y.X}$, and $3s_{Y.X}$ from it, we should find, if N is large enough, that there would be included between these lines about 68%, 95%, and 99.7% of the sample points.

Just as a modified standard deviation given by

$$\hat{s} = \sqrt{\frac{N}{N-1}}\, s$$

was found useful for small samples, so a modified standard error of estimate given by

$$\hat{s}_{Y.X} = \sqrt{\frac{N}{N-2}}\, s_{Y.X}$$

is useful. For this reason, some statisticians prefer to define equation (8) or (9) with $N - 2$ replacing N in the denominator.

EXPLAINED AND UNEXPLAINED VARIATION

The *total variation* of Y is defined as $\sum (Y - \bar{Y})^2$: that is, the sum of the squares of the deviations of the values of Y from the mean $\bar{Y}$. As shown in Problem 14.7, this can be written

$$\sum (Y - \bar{Y})^2 = \sum (Y - Y_{\text{est}})^2 + \sum (Y_{\text{est}} - \bar{Y})^2 \tag{11}$$

The first term on the right of equation (11) is called the *unexplained variation*, while the second term is called the *explained variation*—so called because the deviations $Y_{\text{est}} - \bar{Y}$ have a definite pattern, while the deviations $Y - Y_{\text{est}}$ behave in a random or unpredictable manner. Similar results hold for the variable X.

COEFFICIENT OF CORRELATION

The ratio of the explained variation to the total variation is called the *coefficient of determination*. If there is zero explained variation (i.e., the total variation is all unexplained), this ratio is 0. If there is zero unexplained variation (i.e., the total variation is all explained), the ratio is 1. In other cases the ratio lies between 0 and 1. Since the ratio is always nonnegative, we denote it by r^2. The quantity r, called the *coefficient of correlation* (or briefly *correlation coefficient*), is given by

$$r = \pm\sqrt{\frac{\text{explained variation}}{\text{total variation}}} = \pm\sqrt{\frac{\sum (Y_{\text{est}} - \bar{Y})^2}{\sum (Y - \bar{Y})^2}} \tag{12}$$

and varies between −1 and +1. The + and − signs are used for positive linear correlation and negative linear correlation, respectively. Note that r is a dimensionless quantity; that is, it does not depend on the units employed.

By using equations (*8*) and (*11*) and the fact that the standard deviation of Y is

$$s_Y = \sqrt{\frac{\sum (Y - \bar{Y})^2}{N}} \tag{13}$$

we find that equation (*12*) can be written, disregarding the sign, as

$$r = \sqrt{1 - \frac{s_{Y.X}^2}{s_Y^2}} \quad \text{or} \quad s_{Y.X} = s_Y \sqrt{1 - r^2} \tag{14}$$

Similar equations exist when X and Y are interchanged.

For the case of linear correlation, the quantity r is the same regardless of whether X or Y is considered the independent variable. Thus r is a very good measure of the linear correlation between two variables.

REMARKS CONCERNING THE CORRELATION COEFFICIENT

The definitions of the correlation coefficient in equations (*12*) and (*14*) are quite general and can be used for nonlinear relationships as well as for linear ones, the only differences being that Y_{est} is computed from a nonlinear regression equation in place of a linear equation and that the + and − signs are omitted. In such case equation (*8*), defining the standard error of estimate, is perfectly general. Equation (*10*), however, which applies to linear regression only, must be modified. If, for example, the estimating equation is

$$Y = a_0 + a_1 X + a_2 X^2 + \cdots + a_{n-1} X^{n-1} \tag{15}$$

then equation (*10*) is replaced by

$$s_{Y.X}^2 = \frac{\sum Y^2 - a_0 \sum Y - a_1 \sum XY - \cdots - a_{n-1} \sum X^{n-1} Y}{N} \tag{16}$$

In such case the *modified standard error of estimate* (discussed earlier in this chapter) is

$$\hat{s}_{Y.X} = \sqrt{\frac{N}{N-n}}\, s_{Y.X}$$

where the quantity $N - n$ is called the number of *degrees of freedom*.

It must be emphasized that in every case the computed value of r measures the degree of the relationship relative to the type of equation that is actually assumed. Thus if a linear equation is assumed and equation (*12*) or (*14*) yields a value of r near zero, it means that there is almost no *linear correlation* between the variables. However, it does not mean that there is no correlation at all, since there may actually be a high *nonlinear correlation* between the variables. In other words, the correlation coefficient measures the goodness of fit between (1) the equation actually assumed and (2) the data. Unless otherwise specified, the term *correlation coefficient* is used to mean *linear correlation coefficient*.

It should also be pointed out that a high correlation coefficient (i.e., near 1 or −1) does not necessarily indicate a direct dependence of the variables. Thus there may be a high correlation between the number of books published each year and the number of thunderstorms each year. Such examples are sometimes referred to as *nonsense*, or *spurious, correlations*.

PRODUCT-MOMENT FORMULA FOR THE LINEAR CORRELATION COEFFICIENT

If a linear relationship between two variables is assumed, equation (*12*) becomes

$$r = \frac{\sum xy}{\sqrt{(\sum x^2)(\sum y^2)}} \tag{17}$$

where $x = X - \bar{X}$ and $y = Y - \bar{Y}$ (see Problem 14.10). This formula, which automatically gives the proper sign of r, is called the *product-moment formula* and clearly shows the symmetry between X and Y.

If we write

$$s_{XY} = \frac{\sum xy}{N} \qquad s_X = \sqrt{\frac{\sum x^2}{N}} \qquad s_Y = \sqrt{\frac{\sum y^2}{N}} \tag{18}$$

then s_X and s_Y will be recognized as the standard deviations of the variables X and Y, respectively, while s_X^2 and s_Y^2 are their variances. The new quantity s_{XY} is called the *covariance* of X and Y. In terms of the symbols of formulas (*18*), formula (*17*) can be written

$$r = \frac{s_{XY}}{s_X s_Y} \tag{19}$$

Note that r is not only independent of the choice of units of X and Y, but is also independent of the choice of origin.

SHORT COMPUTATIONAL FORMULAS

Formula (*17*) can be written in the equivalent form

$$r = \frac{N \sum XY - (\sum X)(\sum Y)}{\sqrt{[n \sum X^2 - (\sum X)^2][N \sum Y^2 - (\sum Y)^2]}} \tag{20}$$

which is often used in computing r.

For data grouped as in a *bivariate frequency table*, or *bivariate frequency distribution* (see Problem 14.17), it is convenient to use a *coding method* as in previous chapters. In such case, formula (*20*) can be written

$$r = \frac{N \sum f u_X u_Y - (\sum f_X u_X)(\sum f_Y u_Y)}{\sqrt{[N \sum f_X u_X^2 - (\sum f_X u_X)^2][N \sum f_Y u_Y^2 - (\sum f_Y u_Y)^2]}} \tag{21}$$

(see Problem 14.18). For convenience in calculations using this formula, a *correlation table* is used (see Problem 14.19).

For grouped data, formulas (*18*) can be written

$$s_{XY} = c_X c_Y \left[\frac{\sum f u_X u_Y}{N} - \left(\frac{\sum f_X u_X}{N}\right)\left(\frac{\sum f_Y u_Y}{N}\right)\right] \tag{22}$$

$$s_X = c_X \sqrt{\frac{\sum f_X u_X^2}{N} - \left(\frac{\sum f_X u_X}{N}\right)^2} \tag{23}$$

$$s_Y = c_Y \sqrt{\frac{\sum f_Y u_Y^2}{N} - \left(\frac{\sum f_Y u_Y}{N}\right)^2} \tag{24}$$

where c_X and c_Y are the class-interval widths (assumed constant) corresponding to the variables X and Y, respectively. Note that (*23*) and (*24*) are equivalent to formula (*11*) of Chapter 4.

Formula (*19*) is seen to be equivalent to (*21*) if results (*22*) to (*24*) are used.

REGRESSION LINES AND THE LINEAR CORRELATION COEFFICIENT

The equation of the least-squares line $Y = a_0 + a_1X$, the regression line of Y on X, can be written

$$Y - \bar{Y} = \frac{rs_Y}{s_X}(X - \bar{X}) \quad \text{or} \quad y = \frac{rs_Y}{s_X}x \tag{25}$$

Similarly, the regression line of X on Y, $X = b_0 + b_1Y$, can be written

$$X - \bar{X} = \frac{rs_X}{s_Y}(Y - \bar{Y}) \quad \text{or} \quad x = \frac{rs_X}{s_Y}y \tag{26}$$

The slopes of the lines in equations (*25*) and (*26*) are equal if and only if $r = \pm 1$. In such case the two lines are identical and there is perfect linear correlation between the variables X and Y. If $r = 0$, the lines are at right angles and there is no linear correlation between X and Y. Thus the linear correlation coefficient measures the departure of the two regression lines.

Note that if equations (*25*) and (*26*) are written $Y = a_0 + a_1X$ and $X = b_0 + b_1Y$, respectively, then $a_1b_1 = r^2$ (see Problem 14.22).

CORRELATION OF TIME SERIES

If each of the variables X and Y depends on time, it is possible that a relationship may exist between X and Y even though such relationship is not necessarily one of direct dependence and may produce "nonsense correlation." The correlation coefficient is obtained simply by considering the pairs of values (X, Y) corresponding to the various times and proceeding as usual, making use of the above formulas (see Problem 14.28).

It is possible to attempt to correlate values of a variable X at certain times with corresponding values of X at earlier times. Such correlation is often called *autocorrelation.*

CORRELATION OF ATTRIBUTES

The methods described in this chapter do not enable us to consider the correlation of variables that are nonnumerical by nature, such as the *attributes* of individuals (e.g., hair color, eye color, etc.). For a discussion of the correlation of attributes, see Chapter 12.

SAMPLING THEORY OF CORRELATION

The N pairs of values (X, Y) of two variables can be thought of as samples from a population of all such pairs that are possible. Since two variables are involved, this is called a *bivariate population,* which we assume to be a *bivariate normal distribution.*

We can think of a theoretical population coefficient of correlation, denoted by ρ, which is estimated by the sample correlation coefficient r. Tests of significance or hypotheses concerning various values of ρ require knowledge of the sampling distribution of r. For $\rho = 0$ this distribution is symmetrical, and a statistic involving Student's distribution can be used. For $\rho \neq 0$, the distribution is skewed; in such case a transformation developed by Fisher produces a statistic that is approximately normally distributed. The following tests summarize the procedures involved:

1. **Test of Hypothesis** $\rho = 0$. Here we use the fact that the statistic

$$t = \frac{r\sqrt{N-2}}{\sqrt{1-r^2}} \tag{27}$$

has Student's distribution with $\nu = N - 2$ degrees of freedom (see Problems 14.31 and 14.32).

2. **Test of Hypothesis** $\rho = \rho_0 \neq 0$. Here we use the fact that the statistic

$$Z = \frac{1}{2}\log_e\left(\frac{1+r}{1-r}\right) = 1.1513\log_{10}\left(\frac{1+r}{1-r}\right) \tag{28}$$

where $e = 2.71828\ldots$, is approximately normally distributed with mean and standard deviation given by

$$\mu_Z = \frac{1}{2}\log_e\left(\frac{1+\rho_0}{1-\rho_0}\right) = 1.1513\log_{10}\left(\frac{1+\rho_0}{1-\rho_0}\right) \qquad \sigma_Z = \frac{1}{\sqrt{N-3}} \tag{29}$$

Equations (*28*) and (*29*) can also be used to find confidence limits for correlation coefficients (see Problems 14.33 and 14.34). Equation (*28*) is called *Fisher's Z transformation.*

3. **Significance of a Difference between Correlation Coefficients.** To determine whether two correlation coefficients, r_1 and r_2, drawn from samples of sizes N_1 and N_2, respectively, differ significantly from each other, we compute Z_1 and Z_2 corresponding to r_1 and r_2 by using equation (*28*). We then use the fact that the test statistic

$$z = \frac{Z_1 - Z_2 - \mu_{Z_1-Z_2}}{\sigma_{Z_1-Z_2}} \tag{30}$$

where

$$\mu_{Z_1-Z_2} = \mu_{Z_1} - \mu_{Z_2}$$

and

$$\sigma_{Z_1-Z_2} = \sqrt{\sigma^2_{Z_1} + \sigma^2_{Z_2}} = \sqrt{\frac{1}{N_1-3} + \frac{1}{N_2-3}}$$

is normally distributed (see Problem 14.35).

SAMPLING THEORY OF REGRESSION

The regression equation $Y = a_0 + a_1X$ is obtained on the basis of sample data. We are often interested in the corresponding regression equation for the population from which the sample was drawn. The following are three tests concerning such a population:

1. **Test of Hypothesis** $a_1 = A_1$. To test the hypothesis that the regression coefficient a_1 is equal to some specified value A_1, we use the fact that the statistic

$$t = \frac{a_1 - A_1}{s_{Y.X}/s_X}\sqrt{N-2} \tag{31}$$

has Student's distribution with $N-2$ degrees of freedom. This can also be used to find confidence intervals for population regression coefficients from sample values (see Problems 14.36 and 14.37).

2. **Test of Hypothesis for Predicted Values.** Let Y_0 denote the predicted value of Y corresponding to $X = X_0$ as estimated from the sample regression equation (i.e., $Y_0 = a_0 + a_1X_0$). Let Y_p denote the predicted value of Y corresponding to $X = X_0$ for the population. Then the statistic

$$t = \frac{Y_0 - Y_p}{s_{Y.X}\sqrt{N+1+(X_0-\bar{X})^2/s_X^2}}\sqrt{N-2} = \frac{Y_0 - Y_p}{\hat{s}_{X.Y}\sqrt{1+1/N+(X_0-\bar{X})^2/(Ns_X^2)}} \tag{32}$$

has Student's distribution with $N-2$ degrees of freedom. From this, confidence limits for predicted population values can be found (see Problem 14.38).

3. **Test of Hypothesis for Predicted Mean Values.** Let Y_0 denote the predicted value of Y corresponding to $X = X_0$ as estimated from the sample regression equation (i.e., $Y_0 = a_0 + a_1 X_0$). Let $\bar{Y}_p$ denote the predicted *mean value* of Y corresponding to $X = X_0$ for the population. Then the statistic

$$t = \frac{Y_0 - \bar{Y}_p}{s_{Y.X}\sqrt{1 + (X_0 - \bar{X})^2/s_X^2}}\sqrt{N-2} = \frac{Y_0 - \bar{Y}_p}{\hat{s}_{Y.X}\sqrt{1/N + (X_0 - \bar{X})^2/(Ns_X^2)}} \tag{33}$$

has Student's distribution with $N - 2$ degrees of freedom. From this, confidence limits for predicted mean population values can be found (see Problem 14.39).

Solved Problems

SCATTER DIAGRAMS AND REGRESSION LINES

14.1 Table 14.1 shows [in inches (in)] the respective heights X and Y of a sample of 12 fathers and their oldest son.

(*a*) Construct a scatter diagram of the data.

(*b*) Find the least-squares regression line of the height of the father on the height of the son by solving the normal equations and by using SPSS.

(*c*) Find the least-squares regression line of the height of the son on the height of the father by solving the normal equations and by using STATISTIX.

Table 14.1

Height X of father (in)	65	63	67	64	68	62	70	66	68	67	69	71
Height Y of son (in)	68	66	68	65	69	66	68	65	71	67	68	70

SOLUTION

(*a*) The scatter diagram is obtained by plotting the points (X, Y) on a rectangular coordinate system, as shown in Fig. 14-2.

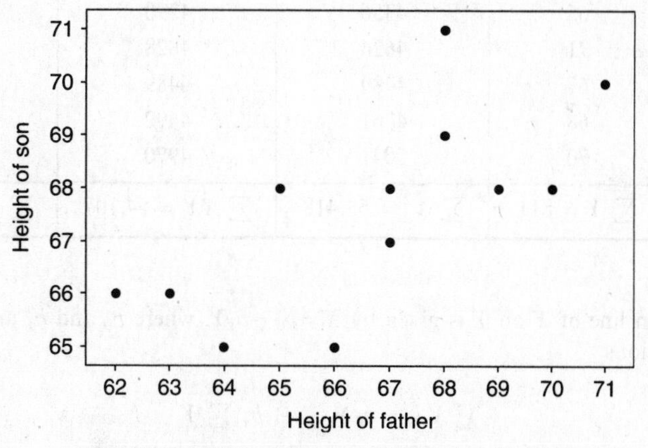

Fig. 14-2 Scatter diagram of the data in Table 14.1.

(*b*) The regression line of Y on X is given by $Y = a_0 + a_1 X$, where a_0 and a_1 are obtained by solving the normal equations.

$$\sum Y = a_0 N + a_1 \sum X$$
$$\sum XY = a_0 \sum X + a_1 \sum X^2$$

The sums are shown in Table 14.2, from which the normal equations become

$$12a_0 + 800a_1 = 811$$
$$800a_0 + 53418a_1 = 54107$$

from which we find that $a_0 = 35.82$ and $a_1 = 0.476$, and thus $Y = 35.82 + 0.476X$.

The SPSS pull-down **Analyze → Regression → Linear** gives the following partial output:

Coefficients[a]

Model	Unstandardized Coefficients		Standardized Coefficients	t	Sig.
	B	Std. Error	Beta		
1 (Constant)	35.825	10.178		3.520	.006
Htfather	.476	.153	.703	3.123	.011

[a]Dependent Variable: Htson.

Opposite the word (`Constant`) is the value of a_0 and opposite `Htfather` is the value of a_1.

Table 14.2

X	Y	X^2	XY	Y^2
65	68	4225	4420	4624
63	66	3969	4158	4356
67	68	4489	4556	4624
64	65	4096	4160	4225
68	69	4624	4692	4761
62	66	3844	4092	4356
70	68	4900	4760	4624
66	65	4356	4290	4225
68	71	4624	4828	5041
67	67	4489	4489	4489
69	68	4761	4692	4624
71	70	5041	4970	4900
$\sum X = 800$	$\sum Y = 811$	$\sum X^2 = 53{,}418$	$\sum XY = 54{,}107$	$\sum Y^2 = 54{,}849$

(*c*) The regression line of X on Y is given by $X = b_0 + b_1 Y$, where b_0 and b_1 are obtained by solving the normal equations

$$\sum X = b_0 N + b_1 \sum Y$$
$$\sum XY = b_0 \sum Y + b_1 \sum Y^2$$

Using the sums in Table 14.2, these become

$$12b_0 + 811b_1 = 800$$
$$811b_0 + 54849b_1 = 54107$$

from which we find that $b_0 = -3.38$ and $b_1 = 1.036$, and thus $X = -3.38 + 1.036Y$

The STATISTIX pull down **Statistics → Linear models → Linear regression** gives the following partial output:

```
Statistix 8.0
Unweighted Least Squares Linear Regression of Htfather

Predictor
Variable      Coefficient    Std Error         T           P
Constant       -3.37687       22.4377        -0.15       0.8834
Htson           1.03640        0.33188        3.12       0.0108
```

Opposite the word `constant` is the value of $b_0 = -3.37687$ and opposite `Htson` is the value of $b_1 = 1.0364$.

14.2 Work Problem 14.1 using MINITAB. Construct tables giving the fitted values, Y_{est}, and the residuals. Find the sum of squares for the residuals for both regression lines.

SOLUTION

The least-squares regression line of Y on X will be found first. A part of the MINITAB output is shown below. Table 14.3 gives the fitted values, the residuals, and the squares of the residuals for the regression line of Y on X.

Table 14.3

X	Y	Fitted value Y_{est}	Residual $Y - Y_{\text{est}}$	Residual squared
65	68	66.79	1.21	1.47
63	66	65.84	0.16	0.03
67	68	67.74	0.26	0.07
64	65	66.31	−1.31	1.72
68	69	68.22	0.78	0.61
62	66	65.36	0.64	0.41
70	68	69.17	−1.17	1.37
66	65	67.27	−2.27	5.13
68	71	68.22	2.78	7.74
67	67	67.74	−0.74	0.55
69	68	68.69	−0.69	0.48
71	70	69.65	0.35	0.12
			Sum = 0	Sum = 19.70

```
MTB > Regress 'Y' on 1 predictor 'X'
```

Regression Analysis

```
The regression equation is Y = 35.8 + 0.476 X
```

The Minitab output for finding the least-squares regression line of X on Y is as follows:

```
MTB > Regress 'X' on 1 predictor 'Y'
```

Regression Analysis

```
The regression equation is X = -3.4 + 1.04 Y
```

Table 14.4 gives the fitted values, the residuals, and the squares of the residuals for the regression line of X on Y.

Table 14.4

X	Y	Fitted Value X_{est}	Residual $X - X_{est}$	Residual Squared
65	68	67.10	−2.10	4.40
63	66	65.03	−2.03	4.10
67	68	67.10	−0.10	0.01
64	65	63.99	0.01	0.00
68	69	68.13	−0.13	0.02
62	66	65.03	−3.03	9.15
70	68	67.10	2.90	8.42
66	65	63.99	2.01	4.04
68	71	70.21	−2.21	4.87
67	67	66.06	0.94	0.88
69	68	67.10	1.90	3.62
71	70	69.17	1.83	3.34
			Sum = 0	Sum = 42.85

The comparison of the sums of squares of residuals indicates that the fit for the least-squares regression line of Y on X is much better than the fit for the least-squares regression line of X on Y. Recall that the smaller the sums of squares of residuals, the better the regression model fits the data. The height of the father is a better predictor of the height of the son than the height of the son is of the height of the father.

STANDARD ERROR OF ESTIMATE

14.3 If the regression line of Y on X is given by $Y = a_0 + a_1 X$, prove that the standard error of estimate $s_{Y.X}$ is given by

$$s^2_{Y.X} = \frac{\sum Y^2 - a_0 \sum Y - a_1 \sum XY}{N}$$

SOLUTION

The values of Y as estimated from the regression line are given by $Y_{est} = a_0 + a_1 X$. Thus

$$s^2_{Y.X} = \frac{\sum (Y - Y_{est})^2}{N} = \frac{\sum (Y - a_0 - a_1 X)^2}{N}$$

$$= \frac{\sum Y(Y - a_0 - a_1 X) - a_0 \sum (Y - a_0 - a_1 X) - a_1 \sum X(Y - a_0 - a_1 X)}{N}$$

But $$\sum (Y - a_0 - a_1 X) = \sum Y - a_0 N - a_1 \sum X = 0$$

and $$\sum X(Y - a_0 - a_1 X) = \sum XY - a_0 \sum X - a_1 \sum X^2 = 0$$

since from the normal equations

$$\sum Y = a_0 N + a_1 \sum X$$

$$\sum XY = a_0 \sum X + a_1 \sum X^2$$

Thus $$s^2_{Y.X} = \frac{\sum Y(Y - a_0 - a_1 X)}{N} = \frac{\sum Y^2 - a_0 \sum Y - a_1 \sum XY}{N}$$

This result can be extended to nonlinear regression equations.

14.4 If $x = X - \bar{X}$ and $y = Y - \bar{Y}$, show that the result of Problem 14.3 can be written

$$s^2_{Y.X} = \frac{\sum y^2 - a_1 \sum xy}{N}$$

SOLUTION

From Problem 14.3, with $X = x + \bar{X}$ and $Y = y + \bar{Y}$, we have

$$\begin{aligned} Ns^2_{Y.X} &= \sum Y^2 - a_0 \sum Y - a_1 \sum XY = \sum (y + \bar{Y})^2 - a_0 \sum (y + \bar{Y}) - a_1 \sum (x + \bar{X})(y + \bar{Y}) \\ &= \sum (y^2 + 2y\bar{Y} + \bar{Y}^2) - a_0(\sum y + N\bar{Y}) - a_1 \sum (xy + \bar{X}y + x\bar{Y} + \bar{X}\bar{Y}) \\ &= \sum y^2 + 2\bar{Y} \sum y + N\bar{Y}^2 - a_0 N\bar{Y} - a_1 \sum xy - a_1\bar{X} \sum y - a_1 \bar{Y} \sum x - a_1 N\bar{X}\bar{Y} \\ &= \sum y^2 + N\bar{Y}^2 - a_0 N\bar{Y} - a_1 \sum xy - a_1 N\bar{X}\bar{Y} \\ &= \sum y^2 - a_1 \sum xy + N\bar{Y}(\bar{Y} - a_0 - a_1\bar{X}) \\ &= \sum y^2 - a_1 \sum xy \end{aligned}$$

where we have used the results $\sum x = 0$, $\sum y = 0$, and $\bar{Y} = a_0 + a_1\bar{X}$ (which follows on dividing both sides of the normal equation $\sum Y = a_0 N + a_1 \sum X$ by N).

14.5 Compute the standard error of estimate, $s_{Y.X}$, for the data of Problem 14.1 by using (*a*) the definition and (*b*) the result of Problem 14.4.

SOLUTION

(*a*) From Problem 14.1(*b*) the regression line of Y on X is $Y = 35.82 + 0.476X$. Table 14.5 lists the actual values of Y (from Table 14.1) and the estimated values of Y, denoted by Y_{est}, as obtained from the regression line; for example, corresponding to $X = 65$ we have $Y_{est} = 35.82 + 0.476(65) = 66.76$. Also listed are the values $Y - Y_{est}$, which are needed in computing $s_{Y.X}$:

$$s^2_{Y.X} = \frac{\sum (Y - Y_{est})}{N} = \frac{(1.24)^2 + (0.19)^2 + \cdots + (0.38)^2}{12} = 1.642$$

and $s_{Y.X} = \sqrt{1.1642} = 1.28$ in.

(*b*) From Problems 14.1, 14.2, and 14.4

$$s^2_{Y.X} = \frac{\sum y^2 - a_1 \sum xy}{N} = \frac{38.92 - 0.476(40.34)}{12} = 1.643$$

and $s_{Y.X} = \sqrt{1.643} = 1.28$ in.

Table 14.5

X	65	63	67	64	68	62	70	66	68	67	69	71
Y	68	66	68	65	69	66	68	65	71	67	68	70
Y_{est}	66.76	65.81	67.71	66.28	68.19	65.33	69.14	67.24	68.19	67.71	68.66	69.62
$Y - Y_{est}$	1.24	0.19	0.29	−1.28	0.81	0.67	−1.14	−2.24	2.81	−0.71	−0.66	0.38

14.6 (*a*) Construct two lines which are parallel to the regression line of Problem 14.1 and which are at a vertical distance $S_{Y.X}$ from it.

(*b*) Determine the percentage of data points falling between these two lines.

SOLUTION

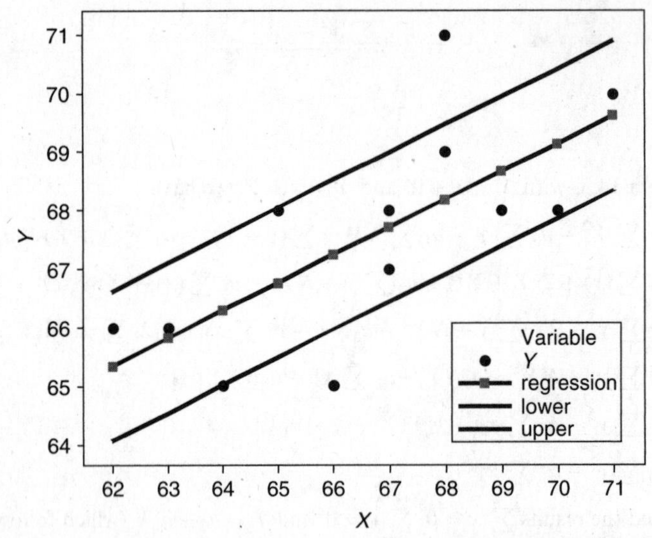

Fig. 14-3 Sixty-six percent of the data points are within $S_{Y.X}$ of the regression line.

(*a*) The regression line $Y = 35.82 + 0.476X$, as obtained in Problem 14.1, is shown as the line with the squares on it. It is the middle of the three lines in Fig. 14-3. There are two other lines shown in Fig. 14-3. They are at a vertical distance $S_{Y.X} = 1.28$ from the regression line. They are called the lower and upper lines.

(*b*) The twelve data points are shown as black circles in Fig. 14-3. Eight of twelve or 66.7% of the data points are between the upper and lower lines. Two data points are outside the lines and two are on the lines.

EXPLAINED AND UNEXPLAINED VARIATION

14.7 Prove that $\sum (Y - \bar{Y})^2 = \sum (Y - Y_{\text{est}})^2 + \sum (Y_{\text{est}} - \bar{Y})^2$.

SOLUTION

Squaring both sides of $Y - \bar{Y} = (Y - Y_{\text{est}}) + (Y_{\text{est}} - \bar{Y})$ and then summing, we have

$$\sum (Y - \bar{Y})^2 = \sum (Y - Y_{\text{est}})^2 + \sum (Y_{\text{est}} - \bar{Y})^2 + 2 \sum (Y - Y_{\text{est}})(Y_{\text{est}} - \bar{Y})$$

The required result follows at once if we can show that the last sum is zero; in the case of linear regression, this is so since

$$\begin{aligned}\sum (Y - Y_{\text{est}})(Y_{\text{est}} - \bar{Y}) &= \sum (Y - a_0 - a_1 X)(a_0 + a_1 X - \bar{Y}) \\ &= a_0 \sum (Y - a_0 - a_1 X) + a_1 \sum X(Y - a_0 - a_1 X) - \bar{Y} \sum (Y - a_0 - a_1 X) = 0\end{aligned}$$

because of the normal equations $\sum (Y - a_0 - a_1 X) = 0$ and $\sum X(Y - a_0 - a_1 X) = 0$.

The result can similarly be shown valid for nonlinear regression by using a least-squares curve given by $Y_{\text{est}} = a_0 + a_1 X + a_2 X^2 + \cdots + a_n X^n$.

14.8 Compute (*a*) the total variation, (*b*) the unexplained variation, and (*c*) the explained variation for the data in Problem 14.1.

SOLUTION

The least-squares regression line is $Y_{\text{est}} = 35.8 + 0.476X$. From Table 14.6, we see that the total variation $= \sum (Y - \bar{Y})^2 = 38.917$, the unexplained variation $= \sum (Y - Y_{\text{est}})^2 = 19.703$, and the explained variation $= \sum (Y_{\text{est}} - \bar{Y})^2 = 19.214$.

Table 14.6

Y	Y_{est}	$(Y-\bar{Y})^2$	$(Y-Y_{est})^2$	$(Y_{est}-\bar{Y})^2$
68	66.7894	0.1739	1.46562	0.62985
66	65.8366	2.5059	0.02669	3.04986
68	67.7421	0.1739	0.06650	0.02532
65	66.3130	6.6719	1.72395	1.61292
69	68.2185	2.0079	0.61074	0.40387
66	65.3602	2.5059	0.40930	4.94068
68	69.1713	0.1739	1.37185	2.52257
65	67.2657	6.6719	5.13361	0.10065
71	68.2185	11.6759	7.73672	0.40387
67	67.7421	0.3399	0.55075	0.02532
68	68.6949	0.1739	0.48286	1.23628
70	69.6476	5.8419	0.12416	4.26273
$\bar{Y}=67.5833$		Sum = 38.917	Sum = 19.703	Sum = 19.214

The following output from MINITAB gives these same sums of squares. They are shown in bold. Note the tremendous amount of computation that the software saves the user.

```
MTB > Regress 'Y' 1 'X';
SUBC> Constant;
SUBC> Brief 1.
```

Regression Analysis

```
The regression equation is
Y = 35.8 + 0.476 X

Analysis of Variance

Source            DF        SS        MS       F        P
Regression         1    19.214    19.214    9.75    0.011
Residual Error    10    19.703     1.970
Total             11    38.917
```

COEFFICIENT OF CORRELATION

14.9 Use the results of Problem 14.8 to find (*a*) the coefficient of determination and (*b*) the coefficient of correlation.

SOLUTION

(*a*) Coefficient of determination $= r^2 = \dfrac{\text{explained variation}}{\text{total variation}} = \dfrac{19.214}{38.917} = 0.4937$

(*b*) Coefficient of correlation $= r = \pm\sqrt{0.4937} = \pm 0.7027$

Since X and Y are directly related, we choose the plus sign and have two decimal places $r = 0.70$.

14.10 Prove that for linear regression the coefficient of correlation between the variables X and Y can be written

$$r = \frac{\sum xy}{\sqrt{(\sum x^2)(\sum y^2)}}$$

where $x = X - \bar{X}$ and $y = Y - \bar{Y}$.

SOLUTION

The least-squares regression line of Y on X can be written $Y_{\text{est}} = a_0 + a_1 X$ or $y_{\text{est}} = a_1 x$, where [see Problem 13.15(*a*)]

$$a_1 = \frac{\sum xy}{\sum x^2} \quad \text{and} \quad y_{\text{est}} = Y_{\text{est}} - \bar{Y}$$

Then

$$r^2 = \frac{\text{explained variation}}{\text{total variation}} = \frac{\sum (Y_{\text{est}} - \bar{Y})^2}{\sum (Y - \bar{Y})^2} = \frac{\sum y_{\text{est}}^2}{\sum y^2}$$

$$= \frac{\sum a_1^2 x^2}{\sum y^2} = \frac{a_1^2 \sum x^2}{\sum y^2} = \left(\frac{\sum xy}{\sum x^2}\right)^2 \frac{\sum x^2}{\sum y^2} = \frac{(\sum xy)^2}{(\sum x^2)(\sum y^2)}$$

and

$$r = \pm \frac{\sum xy}{\sqrt{(\sum x^2)(\sum y^2)}}$$

However, since the quantity

$$\frac{\sum xy}{\sqrt{(\sum x^2)(\sum y^2)}}$$

is positive when y_{est} increases as x increases (i.e., positive linear correlation) and negative when y_{est} decreases as x increases (i.e., negative linear correlation), it *automatically* has the correct sign associated with it. Hence we define the coefficient of linear correlation to be

$$r = \frac{\sum xy}{\sqrt{(\sum x^2)(\sum y^2)}}$$

This is often called the *product-moment formula* for the linear correlation coefficient.

PRODUCT-MOMENT FORMULA FOR THE LINEAR CORRELATION COEFFICIENT

14.11 Find the coefficient of linear correlation between the variables X and Y presented in Table 14.7.

Table 14.7

X	1	3	4	6	8	9	11	14
Y	1	2	4	4	5	7	8	9

SOLUTION

The work involved in the computation can be organized as in Table 14.8.

$$r = \frac{\sum xy}{\sqrt{(\sum x^2)(\sum y^2)}} = \frac{84}{\sqrt{(132)(56)}} = 0.977$$

This shows that there is a very high linear correlation between the variables, as we have already observed in Problems 13.8 and 13.12.

Table 14.8

X	Y	$x = X - \bar{X}$	$y = Y - \bar{Y}$	x^2	xy	y^2
1	1	−6	−4	36	24	16
3	2	−4	−3	16	12	9
4	4	−3	−1	9	3	1
6	4	−1	−1	1	1	1
8	5	1	0	1	0	0
9	7	2	2	4	4	4
11	8	4	3	16	12	9
14	9	7	4	49	28	16
$\sum X = 56$ $\bar{X} = 56/8 = 7$	$\sum Y = 40$ $\bar{Y} = 40/8 = 5$			$\sum x^2 = 132$	$\sum xy = 84$	$\sum y^2 = 56$

14.12 In order to investigate the connection between grade point average (GPA) and hours of TV watched per week, the pairs of data in Table 14.9 and Fig. 14-4 were collected and EXCEL was used to make a scatter plot of the data. The data was collected on 10 high school students and X is the number of hours the subject spends watching TV per week (TV hours) and Y is their GPA:

Table 14.9

TV hours	GPA
20	2.35
5	3.8
8	3.5
10	2.75
13	3.25
7	3.4
13	2.9
5	3.5
25	2.25
14	2.75

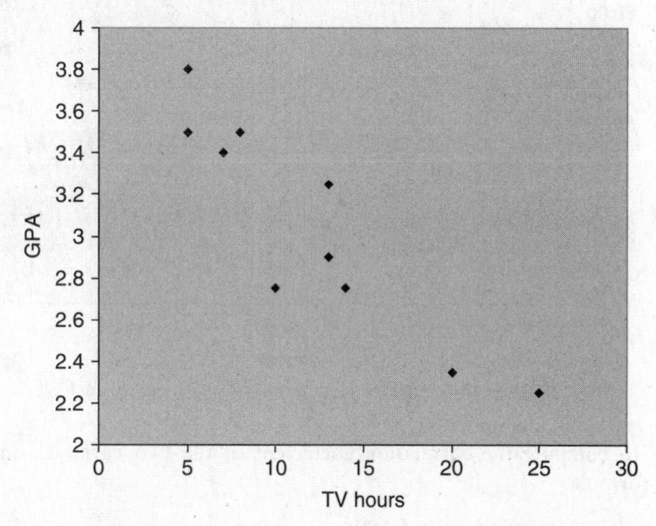

Fig. 14-4 EXCEL scatter plot of data in Problem 14.12.

Use EXCEL to compute the correlation coefficient of the two variables and verify it by using the product-moment formula.

SOLUTION

The EXCEL function `=CORREL(E2:E11,F2:F11)` is used to find the correlation coefficient, if the TV hours are in `E2:E11` and the GPA values are in `F2:F11`. The correlation coefficient is –0.9097. The negative value means that the two variables are negatively correlated. That is, as more hours of TV are watched, the lower is the GPA.

14.13 A study recorded the starting salary (in thousands), Y, and years of education, X, for 10 workers. The data and a SPSS scatter plot are given in Table 14.10 and Fig. 14-5.

Table 14.10

Starting salary	Years of Education
35	12
46	16
48	16
50	15
40	13
65	19
28	10
37	12
49	17
55	14

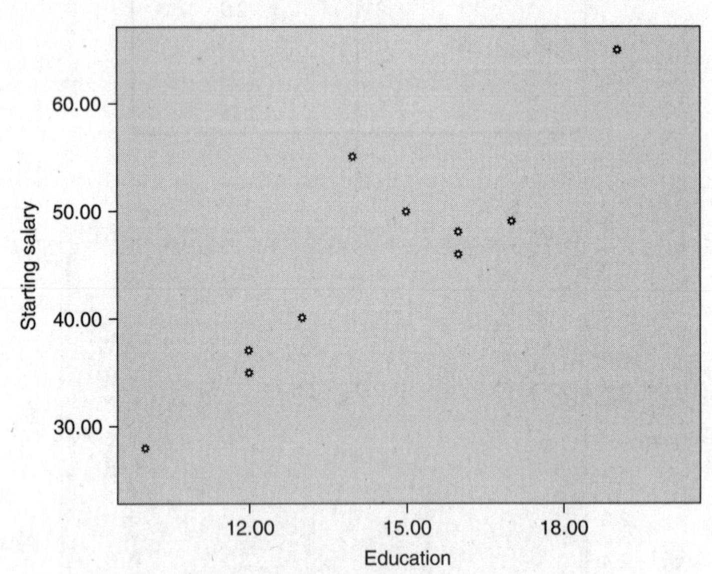

Fig. 14-5 SPSS scatter plot for Problem 14.13.

Use SPSS to compute the correlation coefficient of the two variables and verify it by using the product-moment formula.

SOLUTION

Correlations

		startsal	education
startsal	Pearson Correlation	1	.891**
	Sig. (2-tailed)		.001
	N	10	10
education	Pearson Correlation	.891**	1
	Sig. (2-tailed)	.001	
	N	10	10

**Correlation is significant at the 0.001 level (2-tailed).

The SPSS pull-down **Analyze** → **Correlate** → **Bivariate** gives the correlation by the product-moment formula. It is also called the Pearson correlation.

The output above gives the coefficient of correlation, $r = 0.891$.

14.14 A study recorded the hours per week on a cell phone, Y, and letters in the last name, X, for 10 students. The data and a STATISTIX scatter plot are given in Table 14.11 and Fig. 14-6.

Table 14.11

Hours on Cell Phone	Letters in Last Name
6	13
6	11
3	12
17	7
19	14
14	4
15	4
3	13
13	4
7	9

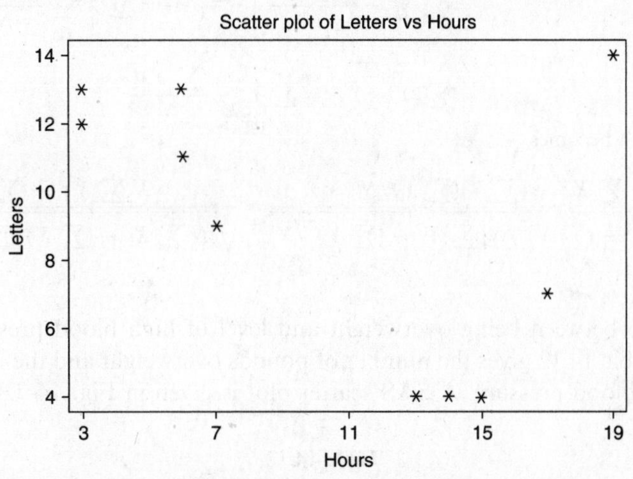

Fig. 14-6 STATISTIX scatter plot of data in Table 14.11.

Use STATISTIX to compute the correlation coefficient of the two variables and verify it by using the product-moment formula.

SOLUTION

The pull-down "**Statistics** → **Linear models** → **correlations(Pearson)**" gives the following output:

```
Statistix 8.0
Correlations (Pearson)
                    Hours

Letters           -0.4701
P-VALUE            0.1704
```

The coefficient of correlation is $r = -0.4701$. There is no significant correlation between these two variables.

14.15 Show that the linear correlation coefficient is given by

$$r = \frac{N \sum XY - (\sum X)(\sum Y)}{\sqrt{[N \sum X^2 - (\sum X)^2][N \sum Y^2 - (\sum Y)^2]}}$$

SOLUTION

Writing $x = X - \bar{X}$ and $y = Y - \bar{Y}$ in the result of Problem 14.10, we have

$$r = \frac{\sum xy}{\sqrt{(\sum x^2)(\sum y^2)}} = \frac{\sum (X - \bar{X})(Y - \bar{Y})}{\sqrt{[\sum (X - \bar{X})^2][\sum (Y - \bar{Y})^2]}} \tag{34}$$

But

$$\begin{aligned} \sum (X - \bar{X})(Y - \bar{Y}) &= \sum (XY - \bar{X}Y - X\bar{Y} + \bar{X}\bar{Y}) = \sum XY - \bar{X} \sum Y - \bar{Y} \sum X + N\bar{X}\bar{Y} \\ &= \sum XY - N\bar{X}\bar{Y} - N\bar{Y}\bar{X} + N\bar{X}\bar{Y} = \sum XY - N\bar{X}\bar{Y} \\ &= \sum XY - \frac{(\sum X)(\sum Y)}{N} \end{aligned}$$

since $\bar{X} = (\sum X)/N$ and $\bar{Y} = (\sum Y)/N$. Similarly,

$$\begin{aligned} \sum (X - \bar{X})^2 &= \sum (X^2 - 2X\bar{X} + \bar{X}^2) = \sum X^2 - 2\bar{X} \sum X + N\bar{X}^2 \\ &= \sum X^2 - \frac{2(\sum X)^2}{N} + \frac{(\sum X)^2}{N} = \sum X^2 - \frac{(\sum X)^2}{N} \end{aligned}$$

and

$$\sum (Y - \bar{Y})^2 = \sum Y^2 - \frac{(\sum Y)^2}{N}$$

Thus equation (*34*) becomes

$$r = \frac{\sum XY - (\sum X)(\sum Y)/N}{\sqrt{[\sum X^2 - (\sum X)^2/N][\sum Y^2 - (\sum Y)^2/N]}} = \frac{N \sum XY - (\sum X)(\sum Y)}{\sqrt{[N \sum X^2 - (\sum X)^2][N \sum Y^2 - (\sum Y)^2]}}$$

14.16 The relationship between being overweight and level of high blood pressure was researched in obese adults. Table 14.12 gives the number of pounds overweight and the number of units over 80 of the diastolic blood pressure. A SAS scatter plot is given in Fig. 14-7.

Table 14.12

Pounds over Weight	Units over 80
75	15
86	13
88	10
125	27
75	20
30	5
47	8
150	31
114	28
68	22

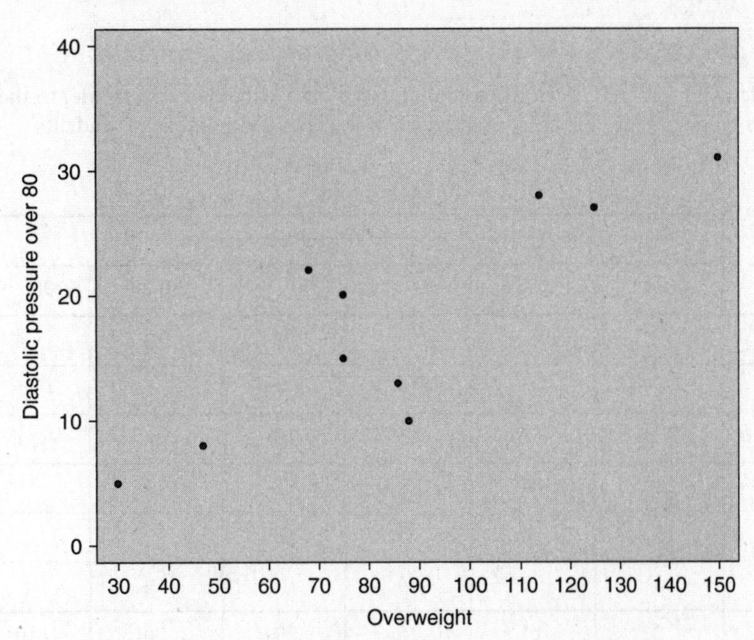

Fig. 14-7 SAS scatter plot for Problem 14.16.

Use SAS to compute the correlation coefficient of the two variables and verify it by using the product-moment formula.

SOLUTION

The SAS pull-down **Statistics** → **Descriptive** → **Correlations** gives the correlation procedure, part of which is shown.

```
                    The CORR Procedure
                      Overwt                  Over80

Overwt               1.00000                 0.85536
Overwt                                       0.0016
Over80               0.85536                 1.00000
Over80               0.0016
```

The output gives the coefficient of correlation as 0.85536. There is a significant correlation between how much overweight the individual is and how far over 80 their diastolic blood pressure is.

CORRELATION COEFFICIENT FOR GROUPED DATA

14.17 Table 14.13 shows the frequency distributions of the final grades of 100 students in mathematics and physics. Referring to this table, determine:

(*a*) The number of students who received grades of 70–79 in mathematics and 80–89 in physics.

(*b*) The percentage of students with mathematics grades below 70.

(*c*) The number of students who received a grade of 70 or more in physics and of less than 80 in mathematics.

(*d*) The percentage of students who passed at least one of the subjects; assume that the minimum passing grade is 60.

SOLUTION

(*a*) In Table 14.13, proceed down the column headed 70–79 (mathematics grade) to the row marked 80–89 (physics grade), where the entry is 4, which is the required number of students.

Table 14.13

		Mathematics Grades						
		40–49	50–59	60–69	70–79	80–89	90–99	Total
Physics Grades	90–99				2	4	4	10
	80–89			1	4	6	5	16
	70–79			5	10	8	1	24
	60–69	1	4	9	5	2		21
	50–59	3	6	6	2			17
	40–49	3	5	4				12
	Total	7	15	25	23	20	10	100

(*b*) The total number of students with mathematics grades below 70 is the number with grades 40–49 + the number with grades 50–59 + the number with grades 60–69 = 7 + 15 + 25 = 47. Thus the required percentage of students is 47/100 = 47%.

(*c*) The required number of students is the total of the entries in Table 14.14 (which represents part of Table 14.13). Thus the required number of students is 1 + 5 + 2 + 4 + 10 = 22.

(*d*) Table 14.15 (taken from Table 14.13) shows that the number of students with grades below 60 in both mathematics and physics is 3 + 3 + 6 + 5 = 17. Thus the number of students with grades 60 or over in either physics or mathematics or in both is 100 − 17 = 83, and the required percentage is 83/100 = 83%.

Table 14.14

		Mathematics Grades	
		60–69	70–79
Physics Grades	90–99		2
	80–89	1	4
	70–79	5	10

Table 14.15

		Mathematics Grades	
		40–49	50–59
Physics Grades	50–59	3	6
	40–49	3	5

Table 14.13 is sometimes called a *bivariate frequency table*, or *bivariate frequency distribution*. Each square in the table is called a *cell* and corresponds to a pair of classes or class intervals. The number indicated in the cell is called the *cell frequency*. For example, in part (*a*) the number 4 is the frequency of the cell corresponding to the pair of class intervals 70–79 in mathematics and 80–89 in physics.

The totals indicated in the last row and last column are called *marginal totals*, or *marginal frequencies*. They correspond, respectively, to the class frequencies of the separate frequency distributions of the mathematics and physics grades.

14.18 Show how to modify the formula of Problem 14.15 for the case of data grouped as in the bivariate frequency table (Table 14.13).

SOLUTION

For grouped data, we can consider the various values of the variables X and Y as coinciding with the class marks, while f_X and f_Y are the corresponding class frequencies, or marginal frequencies, shown in the last row and column of the bivariate frequency table. If we let f represent the various cell frequencies corresponding to the pairs of class marks (X, Y), then we can replace the formula of Problem 14.15 with

$$r = \frac{N \sum fXY - (\sum f_X X)(\sum f_Y Y)}{\sqrt{[N \sum f_X X^2 - (\sum f_X X)^2][N \sum f_Y Y^2 - (\sum f_Y Y)^2]}} \tag{35}$$

If we let $X = A + c_X u_X$ and $Y = B + c_Y u_Y$, where c_X and c_Y are the class-interval widths (assumed constant) and A and B are arbitrary class marks corresponding to the variables, formula (*35*) becomes formula (*21*) of this chapter:

$$r = \frac{N \sum f u_X u_Y - (\sum f_X u_X)(\sum f_Y u_Y)}{\sqrt{[N \sum f_X u_X^2 - (\sum f_X u_X)^2][N \sum f_Y u_Y^2 - (\sum f_Y u_Y)^2]}} \tag{21}$$

This is the *coding method* used in previous chapters as a short method for computing means, standard deviations, and higher moments.

14.19 Find the coefficient of linear correlation of the mathematics and physics grades of Problem 14.17.

SOLUTION

We use formula (*21*). The work can be arranged as in Table 14.16, which is called a *correlation table*. The sums $\sum f_X$, $\sum f_X u_X$, $\sum f_X u_X^2$, $\sum f_Y$, $\sum f_Y u_Y$, and $\sum f_Y u_Y^2$ are obtained by using the coding method, as in earlier chapters.

The number in the corner of each cell in Table 14.16 represents the product $f u_X u_Y$, where f is the cell frequency. The sum of these corner numbers in each row is indicated in the corresponding row of the last column. The sum of these corner numbers in each column is indicated in the corresponding column of the last row. The final totals of the last row and last column are equal and represent $\sum f u_X u_Y$.

From Table 14.16 we have

$$r = \frac{N \sum f u_X u_Y - (\sum f_X u_X)(\sum f_Y u_Y)}{\sqrt{[N \sum f_X u_X^2 - (\sum f_X u_X)^2][N \sum f_Y u_Y^2 - (\sum f_Y u_Y)^2]}}$$

$$= \frac{(100)(125) - (64)(-55)}{\sqrt{[(100(236) - (64)^2][(100)(253) - (-55)^2]}} = \frac{16{,}020}{\sqrt{(19{,}504)(22{,}275)}} = 0.7686$$

14.20 Use Table 14.16 to compute (*a*) s_X, (*b*) s_Y, and (*c*) s_{XY} and thus to verify the formula $r = s_{XY}/(s_X s_Y)$.

SOLUTION

(*a*) $$s_X = c_X \sqrt{\frac{\sum f_X u_X^2}{N} - \left(\frac{\sum f_X u_X}{N}\right)^2} = 10\sqrt{\frac{236}{100} - \left(\frac{64}{100}\right)^2} = 13.966$$

(*b*) $$s_Y = c_Y \sqrt{\frac{\sum f_Y u_Y^2}{N} - \left(\frac{\sum f_Y u_Y}{N}\right)^2} = 10\sqrt{\frac{253}{100} - \left(\frac{-55}{100}\right)^2} = 14.925$$

(*c*) $$s_{XY} = c_X c_Y \left[\frac{\sum f u_X u_Y}{N} - \left(\frac{\sum f_X u_X}{N}\right)\left(\frac{\sum f_Y u_Y}{N}\right)\right] = (10)(10)\left[\frac{125}{100} - \left(\frac{64}{100}\right)\left(\frac{-55}{100}\right)\right] = 160.20$$

Thus the standard deviations of the mathematics and physics grades are 14.0 and 14.9, respectively, while their covariance is 160.2. The correlation coefficient r is therefore

Table 14.16

		Mathematics Grades X									
	X	44.5	54.5	64.5	74.5	84.5	94.5				
Y	u_Y \ u_X	−2	−1	0	1	2	3	f_Y	$f_Y u_Y$	$f_Y u_Y^2$	Sum of corner numbers in each row
Physics Grades Y: 94.5	2				2 [4]	4 [16]	4 [24]	10	20	40	44
84.5	1			1 [0]	4 [4]	6 [12]	5 [15]	16	16	16	31
74.5	0			5 [0]	10 [0]	8 [0]	1 [0]	24	0	0	0
64.5	−1	1 [2]	4 [4]	9 [0]	5 [−5]	2 [−4]		21	−21	21	−3
54.5	−2	3 [12]	6 [12]	6 [0]	2 [−4]			17	−34	68	20
44.5	−3	3 [18]	5 [15]	4 [0]				12	−36	108	33
f_X		7	15	25	23	20	10	$\sum f_X = \sum f_Y = N = 100$	$\sum f_Y u_Y = -55$	$\sum f_Y u_Y^2 = 253$	$\sum f u_X u_Y = 125$
$f_X u_X$		−14	−15	0	23	40	30	$\sum f_X u_X = 64$			
$f_X u_X^2$		28	15	0	23	80	90	$\sum f_X u_X^2 = 236$			
Sum of corner numbers in each column		32	31	0	−1	24	39	$\sum f u_X u_Y = 125$	Check		

$$r = \frac{s_{XY}}{s_X s_Y} = \frac{160.20}{(13.966)(14.925)} = 0.7686$$

agreeing with Problem 14.19.

REGRESSION LINES AND THE CORRELATION COEFFICIENT

14.21 Prove that the regression lines of Y on X and of X on Y have equations given, respectively, by (*a*) $Y - \bar{Y} = (rs_Y/s_X)(X - \bar{Y})$ and (*b*) $X - \bar{X} = (rs_X/s_Y)(Y - \bar{Y})$.

SOLUTION

(*a*) From Problem 13.15(*a*), the regression line of Y on X has the equation

$$y = \left(\frac{\sum xy}{\sum x^2}\right)x \qquad \text{or} \qquad Y - \bar{Y} = \left(\frac{\sum xy}{\sum x^2}\right)(X - \bar{X})$$

Then, since $$r = \frac{\sum xy}{\sqrt{(\sum x^2)(\sum y^2)}} \qquad \text{(see Problem 14.10)}$$

we have $$\frac{\sum xy}{\sum x^2} = \frac{r\sqrt{(\sum x^2)(\sum y^2)}}{\sum x^2} = \frac{r\sqrt{\sum y^2}}{\sqrt{\sum x^2}} = \frac{rs_Y}{s_X}$$

and the required result follows.

(*b*) This follows by interchanging X and Y in part (*a*).

14.22 If, the regression lines of Y on X and of X on Y are given, respectively, by $Y = a_0 + a_1 X$ and $X = b_0 + b_1 Y$, prove that $a_1 b_1 = r^2$.

SOLUTION

From Problem 14.21, parts (*a*) and (*b*),

$$a_1 = \frac{rs_Y}{s_X} \quad \text{and} \quad b_1 = \frac{rs_X}{s_Y}$$

Thus $$a_1 b_1 = \left(\frac{rs_Y}{s_X}\right)\left(\frac{rs_X}{s_Y}\right) = r^2$$

This result can be taken as the starting point for a definition of the linear correlation coefficient.

14.23 Use the result of Problem 14.22 to find the linear correlation coefficient for the data of Problem 14.1.

SOLUTION

From Problem 14.1 [parts (*b*) and (*c*), respectively] $a_1 = 484/1016 = 0.476$ and $b_1 = 484/467 = 1.036$. Thus $Y^2 = a_1 b_1 = (384/1016)(484/467)$ and $r = 0.7027$.

14.24 For the data of Problem 14.19, write the equations of the regression lines of (*a*) Y on X and (*b*) X on Y.

SOLUTION

From the correlation table (Table 14.16) of Problem 14.19 we have

$$\bar{X} = A + c_X \frac{\sum f_X u_X}{N} = 64.5 + \frac{(10)(64)}{100} = 70.9$$

$$\bar{Y} = B + c_Y \frac{\sum f_Y u_Y}{N} = 74.5 + \frac{(10)(-55)}{100} = 69.0$$

From the results of Problem 14.20, $s_X = 13.966$, $s_Y = 14.925$, and $r = 0.7686$. We now use Problem 14.21, parts (*a*) and (*b*), to obtain the equations of the regression lines.

(*a*) $$Y - \bar{Y} = \frac{rs_Y}{s_X}(X - \bar{X}) \qquad Y - 69.0 = \frac{(0.7686)(14.925)}{13.966}(X - 70.9) = 0.821(X - 70.9)$$

(*b*) $$X - \bar{X} = \frac{rs_X}{s_Y}(Y - \bar{Y}) \qquad X - 70.9 = \frac{(0.7686)(13.966)}{14.925}(Y - 69.0) = 0.719(Y - 69.0)$$

14.25 For the data of Problem 14.19, compute the standard errors of estimate (*a*) $s_{Y.X}$ and (*b*) $s_{X.Y}$. Use the results of Problem 14.20.

SOLUTION

(*a*) $s_{Y.X} = s_Y\sqrt{1 - r^2} = 14.925\sqrt{1 - (0.7686)^2} = 9.548$

(*b*) $s_{X.Y} = s_X\sqrt{1 - r^2} = 13.966\sqrt{1 - (0.7686)^2} = 8.934$

14.26 Table 14.17 shows the United States consumer price indexes for food and medical-care costs during the years 2000 through 2006 compared with prices in the base years, 1982 to 1984 (mean taken as 100). Compute the correlation coefficient between the two indexes and give the MINITAB computation of the coefficient.

Table 14.17

Year	2000	2001	2002	2003	2004	2005	2006
Food	167.8	173.1	176.2	180.0	186.2	190.7	195.2
Medical	260.8	272.8	285.6	297.1	310.1	323.2	336.2

Source: Bureau of Labor Statistics.

SOLUTION

Denoting the index numbers for food and medical care as X and Y, respectively, the calculation of the correlation coefficient can be organized as in Table 14.18. (Note that the year is used only to specify the corresponding values of X and Y.)

Table 14.18

X	Y	$x = X - \bar{X}$	$y = Y - \bar{Y}$	x^2	xy	y^2
167.8	260.8	−13.5	−37.2	182.25	502.20	1383.84
173.1	272.8	− 8.2	−25.2	67.24	206.64	635.04
176.2	285.6	− 5.1	−12.4	26.01	63.24	153.76
180.0	297.1	− 1.3	− 0.9	1.69	1.17	0.81
186.2	310.1	4.9	12.1	24.01	59.29	46.41
190.7	323.2	9.4	25.2	88.36	236.88	635.04
195.2	336.2	13.9	38.2	193.21	530.98	1459.24
$\bar{X} = 181.3$	$\bar{Y} = 298.0$			Sum = 582.77	Sum = 1600.4	Sum = 4414.14

Then by the product-moment formula

$$r = \frac{\sum xy}{\sqrt{(\sum x^2)(\sum y^2)}} = \frac{1600.4}{\sqrt{(582.77)(4414.14)}} = 0.998$$

After putting the X values in C1 and the Y values in C2, the MINITAB command `correlation C1 C2` produces the correlation coefficient which is the same that we computed.

```
Correlations: X, Y

Pearson correlation of X and Y = 0.998
P-Value = 0.000
```

NONLINEAR CORRELATION

14.27 Fit a least-squares parabola of the form $Y = a_0 + a_1X + a_2X^2$ to the set of data in Table 14.19. Also give the MINITAB solution.

SOLUTION

The normal equations (*23*) of Chapter 13 are

$$\begin{aligned} \sum Y &= a_0 N + a_1 \sum X + a_2 \sum X^2 \\ \sum XY &= a_0 \sum X + a_1 \sum X^2 + a_2 \sum X^3 \\ \sum X^2Y &= a_0 \sum X^2 + a_1 \sum X^3 + a_2 \sum X^4 \end{aligned} \tag{36}$$

Table 14.19

X	1.2	1.8	3.1	4.9	5.7	7.1	8.6	9.8
Y	4.5	5.9	7.0	7.8	7.2	6.8	4.5	2.7

The work involved in computing the sums can be arranged as in Table 14.20. Then, since $N = 8$, the normal equations (*36*) become

$$\begin{aligned} 8a_0 + 42.2a_1 + 291.20a_2 &= 46.4 \\ 42.2a_0 + 291.20a_1 + 2275.35a_2 &= 230.42 \\ 291.20a_0 + 2275.35a_1 + 18971.92a_2 &= 1449.00 \end{aligned} \tag{37}$$

Solving, $a_0 = 2.588$, $a_1 = 2.065$, and $a_2 = -0.2110$; hence the required least-squares parabola has the equation

$$Y = 2.588 + 2.065X - 0.2110X^2$$

Table 14.20

X	Y	X^2	X^3	X^4	XY	X^2Y
1.2	4.5	1.44	1.73	2.08	5.40	6.48
1.8	5.9	3.24	5.83	10.49	10.62	19.12
3.1	7.0	9.61	29.79	92.35	21.70	67.27
4.9	7.8	24.01	117.65	576.48	38.22	187.28
5.7	7.2	32.49	185.19	1055.58	41.04	233.93
7.1	6.8	50.41	357.91	2541.16	48.28	342.79
8.6	4.5	73.96	636.06	5470.12	38.70	332.82
9.8	2.7	96.04	941.19	9223.66	26.46	259.31
$\sum X = 42.2$	$\sum Y = 46.4$	$\sum X^2 = 291.20$	$\sum X^3 = 2275.35$	$\sum X^4 = 18{,}971.92$	$\sum XY = 230.42$	$\sum X^2Y = 1449.00$

The values for Y are entered into `C1`, the Values for X are entered into `C2`, and the values for X^2 are entered into `C3`. The MINITAB pull-down **Stat → Regression → Regression** is given. The least-squares parabola is given as part of the output as follows.

```
The regression equation is Y=2.59+2.06 X-0.211 Xsquared
```

This is the same solution as given by solving the normal equations.

14.28 Use the least-squares parabola of Problem 14.27 to estimate the values of Y from the given values of X.

SOLUTION

For $X = 1.2$, $Y_{est} = 2.588 + 2.065(1.2) - 0.2110(1.2)^2 = 4.762$. Other estimated values are obtained similarly. The results are shown in Table 14.21 together with the actual values of Y.

Table 14.21

Y_{est}	4.762	5.621	6.962	7.640	7.503	6.613	4.741	2.561
Y	4.5	5.9	7.0	7.8	7.2	6.8	4.5	2.7

14.29 (*a*) Find the linear correlation coefficient between the variables X and Y of Problem 14.27.

(*b*) Find the nonlinear correlation coefficient between these variables, assuming the parabolic relationship obtained in Problem 14.27.

(*c*) Explain the difference between the correlation coefficients obtained in parts (*a*) and (*b*).

(*d*) What percentage of the total variation remains unexplained by assuming a parabolic relationship between X and Y?

SOLUTION

(*a*) Using the calculations already obtained in Table 14.20 and the added fact that $\sum Y^2 = 290.52$, we find that

$$r = \frac{N\sum XY - (\sum X)(\sum Y)}{\sqrt{[N\sum X^2 - (\sum X)^2][N\sum Y^2 - (\sum Y)^2]}}$$

$$= \frac{(8)(230.42) - (42.2)(46.4)}{\sqrt{[(8)(291.20) - (42.2)^2][(8)(290.52) - (46.4)^2]}} = -0.3743$$

(*b*) From Table 14.20, $\bar{Y} = (\sum Y)/N = 46.4/8 = 5.80$; thus the total variation is $\sum(Y - \bar{Y})^2 = 21.40$. From Table 14.21, the explained variation is $\sum(Y_{est} - \bar{Y})^2 = 21.02$. Thus

$$r^2 = \frac{\text{explained variation}}{\text{total variation}} = \frac{21.02}{21.40} = 0.9822 \quad \text{and} \quad r = 0.9911 \quad \text{or} \quad 0.99$$

(*c*) The fact that part (*a*) shows a linear correlation coefficient of only -0.3743 indicates that there is practically no *linear relationship* between X and Y. However, there is a very good *nonlinear relationship* supplied by the parabola of Problem 14.27, as indicated by the fact that the correlation coefficient in part (*b*) is 0.99.

(*d*)

$$\frac{\text{Unexplained variation}}{\text{Total variation}} = 1 - r^2 = 1 - 0.9822 = 0.0178$$

Thus 1.78% of the total variation remains unexplained. This could be due to random fluctuations or to an additional variable that has not been considered.

14.30 Find (*a*) s_Y and (*b*) $s_{Y.X}$ for the data of Problem 14.27.

SOLUTION

(*a*) From Problem 14.29(*a*), $\sum(Y - \bar{Y})^2 = 21.40$. Thus the standard deviation of Y is

$$s_Y = \sqrt{\frac{\sum(Y - \bar{Y})^2}{N}} = \sqrt{\frac{21.40}{8}} = 1.636 \quad \text{or} \quad 1.64$$

(*b*) **First method**

Using part (*a*) and Problem 14.29(*b*), the standard error of estimate of Y on X is

$$s_{Y.X} = s_Y\sqrt{1 - r^2} = 1.636\sqrt{1 - (0.9911)^2} = 0.218 \quad \text{or} \quad 0.22$$

Second method

Using Problem 14.29,

$$s_{Y.X} = \sqrt{\frac{\sum (Y - Y_{\text{est}})^2}{N}} = \sqrt{\frac{\text{unexplained variation}}{N}} = \sqrt{\frac{21.40 - 21.02}{8}} = 0.218 \quad \text{or} \quad 0.22$$

Third method

Using Problem 14.27 and the additional calculation $\sum Y^2 = 290.52$, we have

$$s_{Y.X} = \sqrt{\frac{\sum Y^2 - a_0 \sum Y - a_1 \sum XY - a_2 \sum X^2 Y}{N}} = 0.218 \quad \text{or} \quad 0.22$$

SAMPLING THEORY OF CORRELATION

14.31 A correlation coefficient based on a sample of size 18 was computed to be 0.32. Can we conclude at significance levels of (*a*) 0.05 and (*b*) 0.01 that the corresponding population correlation coefficient differs from zero?

SOLUTION

We wish to decide between the hypotheses $H_0 : \rho = 0$ and $H_1 : \rho > 0$.

$$t = \frac{r\sqrt{N-2}}{\sqrt{1-r^2}} = \frac{0.32\sqrt{18-2}}{\sqrt{1-(0.32)^2}} = 1.35$$

(*a*) Using a one-tailed test of Student's distribution at the 0.05 level, we would reject H_0 if $t > t_{.95} = 1.75$ for $(18 - 2) = 16$ degrees of freedom. Thus we cannot reject H_0 at the 0.05 level.

(*b*) Since we cannot reject H_0 at the 0.05 level, we certainly cannot reject it at the 0.01 level.

14.32 What is the minimum sample size necessary in order that we may conclude that a correlation coefficient of 0.32 differs significantly from zero at the 0.05 level?

SOLUTION

Using a one-tailed test of Student's distribution at the 0.05 level, the minimum value of N must be such that

$$\frac{0.32\sqrt{N-2}}{\sqrt{1-(0.32)^2}} = t_{.95}$$

for $N - 2$ degrees of freedom. For an infinite number of degrees of freedom, $t_{.95} = 1.64$ and hence $N = 25.6$.

For $N = 26$: $\quad \nu = 24 \quad t_{.95} = 1.71 \quad t = 0.32\sqrt{24}/\sqrt{1-(0.32)^2} = 1.65$

For $N = 27$: $\quad \nu = 25 \quad t_{.95} = 1.71 \quad t = 0.32\sqrt{25}/\sqrt{1-(0.32)^2} = 1.69$

For $N = 28$: $\quad \nu = 26 \quad t_{.95} = 1.71 \quad t = 0.32\sqrt{26}/\sqrt{1-(0.32)^2} = 1.72$

Thus the minimum sample size is $N = 28$.

14.33 A correlation coefficient on a sample of size 24 was computed to be $r = 0.75$. At the 0.05 significance level, can we reject the hypothesis that the population correlation coefficient is as small as (*a*) $\rho = 0.60$ and (*b*) $\rho = 0.50$?

SOLUTION

(*a*)
$$Z = 1.1513 \log\left(\frac{1+0.75}{1-0.75}\right) = 0.9730 \quad \mu_Z = 1.1513 \log\left(\frac{1+0.60}{1-0.60}\right) = 0.6932$$

and
$$\sigma_Z = \frac{1}{\sqrt{N-3}} = \frac{1}{\sqrt{21}} = 0.2182$$

Thus
$$z = \frac{Z - \mu_Z}{\sigma_Z} = \frac{0.9730 - 0.6932}{0.2182} = 1.28$$

Using a one-tailed test of the normal distribution at the 0.05 level, we would reject the hypothesis only if z were greater than 1.64. Thus we cannot reject the hypothesis that the population correlation coefficient is as small as 0.60.

(*b*) If $\rho = 0.50$, then $\mu_Z = 1.1513 \log 3 = 0.5493$ and $z = (0.9730 - 0.5493)/0.2182 = 1.94$. Thus we can reject the hypothesis that the population correlation coefficient is as small as $\rho = 0.50$ at the 0.05 level.

14.34 The correlation coefficient between the final grades in physics and mathematics for a group of 21 students was computed to be 0.80. Find the 95% confidence limits for this coefficient.

SOLUTION

Since $r = 0.80$ and $N = 21$, the 95% confidence limits for μ_Z are given by

$$Z \pm 1.96\sigma_Z = 1.1513 \log\left(\frac{1+r}{1-r}\right) \pm 1.96\left(\frac{1}{\sqrt{N-3}}\right) = 1.0986 \pm 0.4620$$

Thus μ_Z has the 95% confidence interval 0.5366 to 1.5606. Now if

$$\mu_Z = 1.1513 \log\left(\frac{1+\rho}{1-\rho}\right) = 0.5366 \quad \text{then} \quad \rho = 0.4904$$

and if
$$\mu_Z = 1.1513 \log\left(\frac{1+\rho}{1-\rho}\right) = 1.5606 \quad \text{then} \quad \rho = 0.9155$$

Thus the 95% confidence limits for ρ are 0.49 and 0.92.

14.35 Two correlation coefficients obtained from samples of size $N_1 = 28$ and $N_3 = 35$ were computed to be $r_1 = 0.50$ and $r_2 = 0.30$, respectively. Is there a significant difference between the two coefficients at the 0.05 level?

SOLUTION

$$Z_1 = 1.1513 \log\left(\frac{1+r_1}{1-r_1}\right) = 0.5493 \qquad Z_2 = 1.1513 \log\left(\frac{1+r_2}{1-r_2}\right) = 0.3095$$

and
$$\sigma_{Z_1 - Z_2} = \sqrt{\frac{1}{N_1 - 3} + \frac{1}{N_2 - 3}} = 0.2669$$

We wish to decide between the hypotheses $H_0: \mu_{Z_1} = \mu_{Z_2}$ and $H_1: \mu_{Z_1} \neq \mu_{Z_2}$. Under hypothesis H_0,

$$z = \frac{Z_1 - Z_2 - (\mu_{Z_1} - \mu_{Z_2})}{\sigma_{Z_1 - Z_2}} = \frac{0.5493 - 0.3095 - 0}{0.2669} = 0.8985$$

Using a two-tailed test of the normal distribution, we would reject H_0 only if $z > 1.96$ or $z < -1.96$. Thus we cannot reject H_0, and we conclude that the results are not significantly different at the 0.05 level.

SAMPLING THEORY OF REGRESSION

14.36 In Problem 14.1 we found the regression equation of Y on X to be $Y = 35.82 + 0.476X$. Test the null hypothesis at the 0.05 significance level that the regression coefficient of the population regression equation is 0.180 versus the alternative hypothesis that the regression coefficient exceeds 0.180. Perform the test without the aid of computer software as well as with the aid of MINITAB computer software.

SOLUTION

$$t = \frac{a_1 - A_1}{S_{Y.X}/S_X}\sqrt{N-2} = \frac{0.476 - 0.180}{1.28/2.66}\sqrt{12-2} = 1.95$$

since $S_{Y.X} = 1.28$ (computed in Problem 14.5) and $S_X = \sqrt{(\sum x^2)/N} = \sqrt{84.68/12} = 2.66$. Using a one-tailed test of Student's distribution at the 0.05 level, we would reject the hypothesis that the regression coefficient is 0.180 if $t > t_{.95} = 1.81$ for $(12 - 2) = 10$ degrees of freedom. Thus, we reject the null hypothesis.

The MINITAB output for this problem is as follows.

```
MTB > Regress 'Y' 1 'X';
SUBC> Constant;
SUBC> Predict c7.
```

Regression Analysis

```
The regression equation is
Y = 35.8 + 0.476 X

Predictor        Coef        StDev        T         P
Constant        35.82        10.18     3.52     0.006
X              0.4764       0.1525     3.12     0.011

S = 1.404    R-Sq = 49.4%    R-Sq(adj) = 44.3%

Analysis of Variance
Source          DF        SS        MS        F        P
Regression       1    19.214    19.214     9.75    0.011
Residual Error  10    19.703     1.970
Total           11    38.917

Predicted Values
Fit      StDev Fit      95.0% CI              95.0% PI
66.789   0.478       (65.724,  67.855)   (63.485,  70.094)
69.171   0.650       (67.723,  70.620)   (65.724,  72.618)
```

The following portion of the output gives the information needed to perform the test of hypothesis.

Predictor	Coef	StDev	T	P
Constant	35.82	10.18	3.52	0.006
X	**0.4764**	**0.1525**	**3.12**	0.011

The computed test statistic is found as follows:

$$t = \frac{0.4764 - 0.180}{0.1525} = 1.94$$

The computed t value shown in the output, **3.12**, is used for testing the null hypothesis that the regression coefficient is 0. To test any other value for the regression coefficient requires a computation like the one shown. To test that the regression coefficient is 0.25, for example, the computed value of the test statistic would equal

$$t = \frac{0.4764 - 0.25}{0.1525} = 1.48$$

The null hypothesis that the regression coefficient equals 0.25 would not be rejected.

14.37 Find the 95% confidence limits for the regression coefficient of Problem 14.36. Set the confidence interval without the aid of any computer software as well as with the aid of MINITAB computer software.

SOLUTION

The confidence interval may be expressed as

$$a_1 \pm \frac{t}{\sqrt{N-2}}\left(\frac{S_{Y.X}}{S_X}\right)$$

Thus the 95% confidence limits for A_1 (obtained by setting $t = \pm t_{.975} = \pm 2.23$ for $12 - 2 = 10$ degrees of freedom) are given by

$$a_1 \pm \frac{2.23}{\sqrt{12-2}}\left(\frac{S_{Y.X}}{S_X}\right) = 0.476 \pm \frac{2.23}{\sqrt{10}}\left(\frac{1.28}{2.66}\right) = 0.476 \pm 0.340$$

That is, we are 95% confident that A_1 lies between 0.136 and 0.816.

The following portion of the MINITAB output from Problem 14.36 gives the information needed to set the 95% confidence interval.

```
Predictor      Coef        StDev        T        P
Constant      35.82        10.18      3.52    0.006
X            0.4764       0.1525      3.12    0.011
```

The term

$$\frac{1}{\sqrt{N-2}}\left(\frac{S_{Y.X}}{S_X}\right)$$

is sometimes called the standard error associated with the estimated regression coefficient. The value for this standard error is shown in the output as `0.1525`. To find the 95% confidence interval, we multiply this standard error by $t_{.975}$ and then add and subtract this term from $a_1 = 0.476$ to obtain the following confidence interval for A_1:

$$0.476 \pm 2.23(0.1525) = 0.476 \pm 0.340$$

14.38 In Problem 14.1, find the 95% confidence limits for the heights of sons whose fathers' heights are (*a*) 65.0 and (*b*) 70.0 in. Set the confidence interval without the aid of any computer software as well as with the aid of MINITAB computer software.

SOLUTION

Since $t_{.975} = 2.23$ for $(12 - 2) = 10$ degrees of freedom, the 95% confidence limits for Y_P are given by

$$Y_0 \pm \frac{2.23}{\sqrt{N-2}} S_{Y.X}\sqrt{N + 1 + \frac{(X_0 - \bar{X})^2}{S_X^2}}$$

where $Y_0 = 35.82 + 0.476X_0$, $S_{Y.X} = 1.28$, $S_X = 2.66$, and $N = 12$.

(*a*) If $X_0 = 65.0$, then $Y_0 = 66.76$ in. Also, $(X_0 - \bar{X})^2 = (65.0 - 66.67)^2 = 2.78$. Thus the 95% confidence limits are

$$66.76 \pm \frac{2.23}{\sqrt{10}}(1.28)\sqrt{12 + 1 + \frac{2.78}{2.66^2}} = 66.76 \pm 3.30 \text{ in}$$

That is, we can be 95% confident that the sons' heights are between 63.46 and 70.06 in.

(*b*) If $X_0 = 70.0$, then $Y_0 = 69.14$ in. Also, $(X_0 - \bar{X})^2 = (70.0 - 66.67)^2 = 11.09$. Thus the 95% confidence limits are computed to be 69.14 ± 3.45 in; that is, we can be 95% confident that the sons' heights are between 65.69 and 72.59 in.

The following portion of the MINITAB output found in Problem 14.36 gives the confidence limits for the sons' heights.

```
Predicted Values
Fit      StDev Fit       95.0% CI              95.0% PI
66.789     0.478    (65.724,    67.855)   (63.485,    70.094)
69.171     0.650    (67.723,    70.620)   (65.724,    72.618)
```

The confidence interval for individuals are sometimes referred to as prediction intervals. The 95% prediction intervals are shown in bold. These intervals agree with those computed above except for rounding errors.

14.39 In Problem 14.1, find the 95% confidence limits for the mean heights of sons whose fathers' heights are (*a*) 65.0 in and (*b*) 70.0 in. Set the confidence interval without the aid of any computer software as well as with the aid of MINITAB computer software.

SOLUTION

Since $t_{.975} = 2.23$ for 10 degrees of freedom, the 95% confidence limits for $\bar{Y}_P$ are given by

$$Y_0 \pm \frac{2.23}{\sqrt{10}} S_{Y.X} \sqrt{1 + \frac{(X_0 - \bar{X})^2}{S_X^2}}$$

where $Y_0 = 35.82 + 0.476X_0$, $S_{Y.X} = 1.28$, and $S_X = 2.66$.

(*a*) For $X_0 = 65.0$, we find the confidence limits to be 66.76 ± 1.07 or 65.7 and 67.8.

(*b*) For $X_0 = 70.0$, we find the confidence limits to be 69.14 ± 1.45 or 67.7 and 70.6.

The following portion of the MINITAB output found in Problem 14.36 gives the confidence limits for the mean heights.

```
Predicted Values
Fit      StDev Fit       95.0% CI              95.0% PI
66.789     0.478    (65.724,    67.855)   (63.485,    70.094)
69.171     0.650    (67.723,    70.620)   (65.724,    72.618)
```

Supplementary Problems

LINEAR REGRESSION AND CORRELATION

14.40 Table 14.22 shows the first two grades (denoted by X and Y, respectively) of 10 students on two short quizzes in biology.

(*a*) Construct a scatter diagram.

(*b*) Find the least-squares regression line of Y on X.

(*c*) Find the least-squares regression line of X on Y.

(*d*) Graph the two regression lines of parts (*b*) and (*c*) on the scatter diagram of part (*a*).

14.41 Find (*a*) $s_{Y.X}$ and (*b*) $s_{X.Y}$ for the data in Table 14.22.

Table 14.22

Grade on first quiz (X)	6	5	8	8	7	6	10	4	9	7
Grade on second quiz (Y)	8	7	7	10	5	8	10	6	8	6

14.42 Compute (*a*) the total variation in Y, (*b*) the unexplained variation in Y, and (*c*) the explained variation in Y for the data of Problem 14.40.

14.43 Use the results of Problem 14.42 to find the correlation coefficient between the two sets of quiz grades of Problem 14.40.

14.44 Find the correlation coefficient between the two sets of quiz grades in Problem 14.40 by using the product-moment formula, and compare this finding with the correlation coefficient given by SPSS, SAS, STATISTIX, MINITAB, and EXCEL.

14.45 Find the covariance for the data of Problem 14.40(*a*) directly and (*b*) by using the formula $s_{XY} = rs_X s_Y$ and the result of Problem 14.43 or Problem 14.44.

14.46 Table 14.23 shows the ages X and the systolic blood pressures Y of 12 women.

(*a*) Find the correlation coefficient between X and Y using the product-moment formula, EXCEL, MINITAB, SAS, SPSS, and STATISTIX.

(*b*) Determine the least-squares regression equation of Y on X by solving the normal equations, and by using EXCEL, MINITAB, SAS, SPSS, and STATISTIX.

(*c*) Estimate the blood pressure of a woman whose age is 45 years.

Table 14.23

Age (X)	56	42	72	36	63	47	55	49	38	42	68	60
Blood pressure (Y)	147	125	160	118	149	128	150	145	115	140	152	155

14.47 Find the correlation coefficients for the data of (*a*) Problem 13.32 and (*b*) Problem 13.35.

14.48 The correlation coefficient between two variables X and Y is $r = 0.60$. If $s_X = 1.50$, $s_Y = 2.00$, $\bar{X} = 10$, and $\bar{Y} = 20$, find the equations of the regression lines of (*a*) Y on X and (*b*) X on Y.

14.49 Compute (*a*) $s_{Y.X}$ and (*b*) $s_{X.Y}$ for the data of Problem 14.48.

14.50 If $s_{Y.X} = 3$ and $s_Y = 5$, find r.

14.51 If the correlation coefficient between X and Y is 0.50, what percentage of the total variation remains unexplained by the regression equation?

14.52 (*a*) Prove that the equation of the regression line of Y on X can be written

$$Y - \bar{Y} = \frac{s_{XY}}{s_X^2}(X - \bar{X})$$

(*b*) Write the analogous equation for the regression line of X on Y.

14.53 (*a*) Compute the correlation coefficient between the corresponding values of X and Y given in Table 14.24.

Table 14.24

X	2	4	5	6	8	11
Y	18	12	10	8	7	5

(*b*) Multiply each X value in the table by 2 and add 6. Multiply each Y value in the table by 3 and subtract 15. Find the correlation coefficient between the two new sets of values, explaining why you do or do not obtain the same result as in part (*a*).

14.54 (*a*) Find the regression equations of Y on X for the data considered in Problem 14.53, parts (*a*) and (*b*).

(*b*) Discuss the relationship between these regression equations.

14.55 (*a*) Prove that the correlation coefficient between X and Y can be written

$$r = \frac{\overline{XY} - \bar{X}\bar{Y}}{\sqrt{[\overline{X^2} - \bar{X}^2][\overline{Y^2} - \bar{Y}^2]}}$$

(*b*) Using this method, work Problem 14.1.

14.56 Prove that a correlation coefficient is independent of the choice of origin of the variables or the units in which they are expressed. (*Hint:* Assume that $X' = c_1X + A$ and $Y' = c_2Y + B$, where c_1, c_2, A, and B are any constants, and prove that the correlation coefficient between X' and Y' is the same as that between X and Y.)

14.57 (*a*) Prove that, for linear regression,

$$\frac{s_{Y.X}^2}{s_Y^2} = \frac{s_{X.Y}^2}{s_X^2}$$

(*b*) Does the result hold for nonlinear regression?

CORRELATION COEFFICIENT FOR GROUPED DATA

14.58 Find the correlation coefficient between the heights and weights of the 300 U.S. adult males given in Table 14.25, a frequency table.

Table 14.25

Weights Y (lb)	Heights X (in) 59–62	63–66	67–70	71–74	75–78
90–109	2	1			
110–129	7	8	4	2	
130–149	5	15	22	7	1
150–169	2	12	63	19	5
170–189		7	28	32	12
190–209		2	10	20	7
210–229			1	4	2

14.59 (*a*) Find the least-squares regression equation of Y on X for the data of Problem 14.58.

(*b*) Estimate the weights of two men whose heights are 64 and 72 in, respectively.

14.60 Find (*a*) $s_{Y.X}$ and (*b*) $s_{X.Y}$ for the data of Problem 14.58.

14.61 Establish formula (*21*) of this chapter for the correlation coefficient of grouped data.

CORRELATION OF TIME SERIES

14.62 Table 14.26 shows the average annual expenditures per consumer unit for health care and the per capita income for the years 1999 through 2004. Find the correlation coefficient.

Table 14.26

Year	1999	2000	2001	2002	2003	2004
Health care cost	1959	2066	2182	2350	2416	2574
Per capita income	27939	29845	30574	30810	31484	33050

Source: Bureau of Labor Statistics and U.S. Bureau of Economic Analysis.

14.63 Table 14.27 shows the average temperature and precipitation in a city for the month of July during the years 2000 through 2006. Find the correlation coefficient.

Table 14.27

Year	2000	2001	2002	2003	2004	2005	2006
Temperature (°F)	78.1	71.8	75.6	72.7	75.3	73.6	75.1
Precipitation (in)	6.23	3.64	3.42	2.84	1.83	2.82	4.04

SAMPLING THEORY OF CORRELATION

14.64 A correlation coefficient based on a sample of size 27 was computed to be 0.40. Can we conclude at significance levels of (*a*) 0.05 and (*b*) 0.01, that the corresponding population correlation coefficient differs from zero?

14.65 A correlation coefficient based on a sample of size 35 was computed to be 0.50. At the 0.05 significance level, can we reject the hypothesis that the population correlation coefficient is (*a*) as small as $\rho = 0.30$ and (*b*) as large as $\rho = 0.70$?

14.66 Find the (*a*) 95% and (*b*) 99% confidence limits for a correlation coefficient that is computed to be 0.60 from a sample of size 28.

14.67 Work Problem 14.66 if the sample size is 52.

14.68 Find the 95% confidence limits for the correlation coefficients computed in (*a*) Problem 14.46 and (*b*) Problem 14.58.

14.69 Two correlation coefficients obtained from samples of size 23 and 28 were computed to be 0.80 and 0.95, respectively. Can we conclude at levels of (*a*) 0.05 and (*b*) 0.01 that there is a significant difference between the two coefficients?

SAMPLING THEORY OF REGRESSION

14.70 On the basis of a sample of size 27, a regression equation of Y on X was found to be $Y = 25.0 + 2.00X$. If $s_{Y.X} = 1.50$, $s_X = 3.00$, and $\bar{X} = 7.50$, find the (*a*) 95% and (*b*) 99% confidence limits for the regression coefficient.

14.71 In Problem 14.70, test the hypothesis that the population regression coefficient at the 0.01 significance level is (*a*) as low as 1.70 and (*b*) as high as 2.20.

14.72 In Problem 14.70, find the (*a*) 95% and (*b*) 99% confidence limits for Y when $X = 6.00$.

14.73 In Problem 14.70, find the (*a*) 95% and (*b*) 99% confidence limits for the mean of all values of Y corresponding to $X = 6.00$.

14.74 Referring to Problem 14.46, find the 95% confidence limits for (*a*) the regression coefficient of Y on X, (*b*) the blood pressures of all women who are 45 years old, and (*c*) the mean of the blood pressures of all women who are 45 years old.

Multiple and Partial Correlation

MULTIPLE CORRELATION

The degree of relationship existing between three or more variables is called *multiple correlation*. The fundamental principles involved in problems of multiple correlation are analogous to those of simple correlation, as treated in Chapter 14.

SUBSCRIPT NOTATION

To allow for generalizations to large numbers of variables, it is convenient to adopt a notation involving subscripts.

We shall let $X_1, X_2, X_3, \ldots$ denote the variables under consideration. Then we can let $X_{11}, X_{12}, X_{13}, \ldots$ denote the values assumed by the variable X_1, and $X_{21}, X_{22}, X_{23}, \ldots$ denote the values assumed by the variable X_2, and so on. With this notation, a sum such as $X_{21} + X_{22} + X_{23} + \cdots + X_{2N}$ could be written $\sum_{j=1}^{N} X_{2j}$, $\sum_j X_{2j}$, or simply $\sum X_2$. When no ambiguity can result, we use the last notation. In such case the mean of X_2 is written $\bar{X}_2 = \sum X_2/N$.

REGRESSION EQUATIONS AND REGRESSION PLANES

A *regression equation* is an equation for estimating a dependent variable, say X_1, from the independent variables $X_2, X_3, \ldots$ and is called a *regression equation of* X_1 *on* $X_2, X_3, \ldots$. In functional notation this is sometimes written briefly as $X_1 = F(X_2, X_3, \ldots)$ (read "X_1 is a function of X_2, X_3, and so on").

For the case of three variables, the simplest regression equation of X_1 on X_2 and X_3 has the form

$$X_1 = b_{1.23} + b_{12.3}X_2 + b_{13.2}X_3 \qquad (1)$$

where $b_{1.23}, b_{12.3}$, and $b_{13.2}$ are constants. If we keep X_3 constant in equation (*1*), the graph of X_1 versus X_2 is a straight line with slope $b_{12.3}$. If we keep X_2 constant, the graph of X_1 versus X_3 is a straight line with slope $b_{13.2}$. It is clear that the subscripts after the dot indicate the variables held constant in each case.

Due to the fact that X_1 varies partially because of variation in X_2 and partially because of variation in X_3, we call $b_{12.3}$ and $b_{13.2}$ the *partial regression coefficients* of X_1 on X_2 keeping X_3 constant and of X_1 on X_3 keeping X_2 constant, respectively.

Equation (*1*) is called a *linear regression equation* of X_1 on X_2 and X_3. In a three-dimensional rectangular coordinate system it represents a plane called a *regression plane* and is a generalization of the regression line for two variables, as considered in Chapter 13.

NORMAL EQUATIONS FOR THE LEAST-SQUARES REGRESSION PLANE

Just as there exist least-squares regression lines approximating a set of N data points (X, Y) in a two-dimensional scatter diagram, so also there exist *least-squares regression planes* fitting a set of N data points (X_1, X_2, X_3) in a three-dimensional scatter diagram.

The least-squares regression plane of X_1 on X_2 and X_3 has the equation (*1*) where $b_{1.23}$, $b_{12.3}$, and $b_{13.2}$ are determined by solving simultaneously the *normal equations*

$$\begin{aligned} \sum X_1 &= b_{1.23}N + b_{12.3}\sum X_2 + b_{13.2}\sum X_3 \\ \sum X_1X_2 &= b_{1.23}\sum X_2 + b_{12.3}\sum X_2^2 + b_{13.2}\sum X_2X_3 \\ \sum X_1X_3 &= b_{1.23}\sum X_3 + b_{12.3}\sum X_2X_3 + b_{13.2}\sum X_3^2 \end{aligned} \tag{2}$$

These can be obtained formally by multiplying both sides of equation (*1*) by 1, X_2, and X_3 successively and summing on both sides.

Unless otherwise specified, whenever we refer to a regression equation it will be assumed that the least-squares regression equation is meant.

If $x_1 = X_1 - \bar{X}_1$, $x_2 = X_2 - \bar{X}_2$, and $x_3 = X_3 - \bar{X}_3$, the regression equation of X_1 on X_2 and X_3 can be written more simply as

$$x_1 = b_{12.3}x_2 + b_{13.2}x_3 \tag{3}$$

where $b_{12.3}$ and $b_{13.2}$ are obtained by solving simultaneously the equations

$$\begin{aligned} \sum x_1x_2 &= b_{12.3}\sum x_2^2 + b_{13.2}\sum x_2x_3 \\ \sum x_1x_3 &= b_{12.3}\sum x_2x_3 + b_{13.2}\sum x_3^2 \end{aligned} \tag{4}$$

These equations which are equivalent to the normal equations (*2*) can be obtained formally by multiplying both sides of equation (*3*) by x_2 and x_3 successively and summing on both sides (see Problem 15.8).

REGRESSION PLANES AND CORRELATION COEFFICIENTS

If the linear correlation coefficients between variables X_1 and X_2, X_1 and X_3, and X_2 and X_3, as computed in Chapter 14, are denoted respectively by r_{12}, r_{13}, and r_{23} (sometimes called *zero-order correlation coefficients*), then the least-squares regression plane has the equation

$$\frac{x_1}{s_1} = \left(\frac{r_{12} - r_{13}r_{23}}{1 - r_{23}^2}\right)\frac{x_2}{s_2} + \left(\frac{r_{13} - r_{12}r_{23}}{1 - r_{23}^2}\right)\frac{x_3}{s_3} \tag{5}$$

where $x_1 = X - \bar{X}_1$, $x_2 = X_2 - \bar{X}_2$, and $x_3 = X_3 - \bar{X}_3$ and where s_1, s_2, and s_3 are the standard deviations of X_1, X_2, and X_3, respectively (see Problem 15.9).

Note that if the variable X_3 is nonexistent and if $X_1 = Y$ and $X_2 = X$, then equation (*5*) reduces to equation (*25*) of Chapter 14.

STANDARD ERROR OF ESTIMATE

By an obvious generalization of equation (8) of Chapter 14, we can define the *standard error of estimate* of X_1 on X_2 and X_3 by

$$s_{1.23} = \sqrt{\frac{\sum(X_1 - X_{1,\text{est}})^2}{N}} \tag{6}$$

where $X_{1,\text{est}}$ indicates the estimated values of X_1 as calculated from the regression equations (1) or (5).

In terms of the correlation coefficients r_{12}, r_{13}, and r_{23}, the standard error of estimate can also be computed from the result

$$s_{1.23} = s_1\sqrt{\frac{1 - r_{12}^2 - r_{13}^2 - r_{23}^2 + 2r_{12}r_{13}r_{23}}{1 - r_{23}^2}} \tag{7}$$

The sampling interpretation of the standard error of estimate for two variables as given on page 313 for the case when N is large can be extended to three dimensions by replacing the lines parallel to the regression line with planes parallel to the regression plane. A better estimate of the population standard error of estimate is given by $\hat{s}_{1.23} = \sqrt{N/(N-3)}s_{1.23}$.

COEFFICIENT OF MULTIPLE CORRELATION

The coefficient of multiple correlation is defined by an extension of equation (12) or (14) of Chapter 14. In the case of two independent variables, for example, the *coefficient of multiple correlation* is given by

$$R_{1.23} = \sqrt{1 - \frac{s_{1.23}^2}{s_1^2}} \tag{8}$$

where s_1 is the standard deviation of the variable X_1 and $s_{1.23}$ is given by equation (6) or (7). The quantity $R_{1.23}^2$ is called the *coefficient of multiple determination.*

When a linear regression equation is used, the coefficient of multiple correlation is called the *coefficient of linear multiple correlation.* Unless otherwise specified, whenever we refer to multiple correlation, we shall imply linear multiple correlation.

In terms of r_{12}, r_{13}, and r_{23}, equation (8) can also be written

$$R_{1.23} = \sqrt{\frac{r_{12}^2 + r_{13}^2 - 2r_{12}r_{13}r_{23}}{1 - r_{23}^2}} \tag{9}$$

A coefficient of multiple correlation, such as $R_{1.23}$, lies between 0 and 1. The closer it is to 1, the better is the linear relationship between the variables. The closer it is to 0, the worse is the linear relationship. If the coefficient of multiple correlation is 1, the correlation is called *perfect.* Although a correlation coefficient of 0 indicates no linear relationship between the variables, it is possible that a *nonlinear relationship* may exist.

CHANGE OF DEPENDENT VARIABLE

The above results hold when X_1 is considered the dependent variable. However, if we want to consider X_3 (for example) to be the dependent variable instead of X_1, we would only have to replace the subscripts 1 with 3, and 3 with 1, in the formulas already obtained. For example, the regression equation of X_3 on X_1 and X_2 would be

$$\frac{x_3}{s_3} = \left(\frac{r_{23} - r_{13}r_{12}}{1 - r_{12}^2}\right)\frac{x_2}{s_2} + \left(\frac{r_{13} - r_{23}r_{12}}{1 - r_{12}^2}\right)\frac{x_1}{s_1} \tag{10}$$

as obtained from equation (*5*), using the results $r_{32} = r_{23}$, $r_{31} = r_{13}$, and $r_{21} = r_{12}$.

GENERALIZATIONS TO MORE THAN THREE VARIABLES

These are obtained by analogy with the above results. For example, the linear regression equations of X_1 on X_2, X_3, and X_4 can be written

$$X_1 = b_{1.234} + b_{12.34}X_2 + b_{13.24}X_3 + b_{14.23}X_4 \tag{11}$$

and represents a *hyperplane in four-dimensional space*. By formally multiplying both sides of equation (*11*) by 1, X_2, X_3, and X_4 successively and then summing on both sides, we obtain the normal equations for determining $b_{1.234}$, $b_{12.34}$, $b_{13.24}$, and $b_{14.23}$; substituting these in equation (*11*) then gives us the *least-squares regression equation of X_1 on X_2, X_3, and X_4*. This least-squares regression equation can be written in a form similar to that of equation (*5*). (See Problem 15.41.)

PARTIAL CORRELATION

It is often important to measure the correlation between a dependent variable and one particular independent variable when all other variables involved are kept constant; that is, when the effects of all other variables are removed (often indicated by the phrase "other things being equal"). This can be obtained by defining a *coefficient of partial correlation*, as in equation (*12*) of Chapter 14, except that we must consider the explained and unexplained variations that arise both with and without the particular independent variable.

If we denote by $r_{12.3}$ the coefficient of partial correlation between X_1 and X_2 keeping X_3 constant, we find that

$$r_{12.3} = \frac{r_{12} - r_{13}r_{23}}{\sqrt{(1 - r_{13}^2)(1 - r_{23}^2)}} \tag{12}$$

Similarly, if $r_{12.34}$ is the coefficient of partial correlation between X_1 and X_2 keeping X_3 and X_4 constant, then

$$r_{12.34} = \frac{r_{12.4} - r_{13.4}r_{23.4}}{\sqrt{(1 - r_{13.4}^2)(1 - r_{23.4}^2)}} = \frac{r_{12.3} - r_{14.3}r_{24.3}}{\sqrt{(1 - r_{14.3}^2)(1 - r_{24.3}^2)}} \tag{13}$$

These results are useful since by means of them any partial correlation coefficient can ultimately be made to depend on the correlation coefficients r_{12}, r_{23}, etc. (i.e., the *zero-order correlation coefficients*).

In the case of two variables, X and Y, if the two regression lines have equations $Y = a_0 + a_1X$ and $X = b_0 + b_1Y$, we have seen that $r^2 = a_1b_1$ (see Problem 14.22). This result can be generalized. For example, if

$$X_1 = b_{1.234} + b_{12.34}X_2 + b_{13.24}X_3 + b_{14.23}X_4 \tag{14}$$

and

$$X_4 = b_{4.123} + b_{41.23}X_1 + b_{42.13}X_2 + b_{43.12}X_3 \tag{15}$$

are linear regression equations of X_1 on X_2, X_3, and X_4 and of X_4 on X_1, X_2, and X_3, respectively, then

$$r_{14.23}^2 = b_{14.23}b_{41.23} \tag{16}$$

(see Problem 15.18). This can be taken as the starting point for a definition of linear partial correlation coefficients.

RELATIONSHIPS BETWEEN MULTIPLE AND PARTIAL CORRELATION COEFFICIENTS

Interesting results connecting the multiple correlation coefficients can be found. For example, we find that

$$1 - R_{1.23}^2 = (1 - r_{12}^2)(1 - r_{13.2}^2) \tag{17}$$

$$1 - R_{1.234}^2 = (1 - r_{12}^2)(1 - r_{13.2}^2)(1 - r_{14.23}^2) \tag{18}$$

Generalizations of these results are easily made.

NONLINEAR MULTIPLE REGRESSION

The above results for linear multiple regression can be extended to nonlinear multiple regression. Coefficients of multiple and partial correlation can then be defined by methods similar to those given above.

Solved Problems

REGRESSION EQUATIONS INVOLVING THREE VARIABLES

15.1 Using an appropriate subscript notation, write the regression equations of (*a*), X_2 on X_1 and X_3; (*b*) X_3 on X_1, X_2, and X_4; and (*c*) X_5 on X_1, X_2, X_3, and X_4.

SOLUTION

(*a*) $X_2 = b_{2.13} + b_{21.3}X_1 + b_{23.1}X_3$

(*b*) $X_3 = b_{3.124} + b_{31.24}X_1 + b_{32.14}X_2 + b_{34.12}X_4$

(*c*) $X_5 = b_{5.1234} + b_{51.234}X_1 + b_{52.134}X_2 + b_{53.124}X_3 + b_{54.123}X_4$

15.2 Write the normal equations corresponding to the regression equations (*a*) $X_3 = b_{3.12} + b_{31.2}X_1 + b_{32.1}X_2$ and (*b*) $X_1 = b_{1.234} + b_{12.34}X_2 + b_{13.24}X_3 + b_{14.23}X_4$.

SOLUTION

(*a*) Multiply the equation successively by 1, X_1, and X_2, and sum on both sides. The normal equations are

$$\begin{aligned}
\sum X_3 &= b_{3.12}N + b_{31.2}X_1 + b_{32.1}\sum X_2 \\
\sum X_1X_3 &= b_{3.12}\sum X_1 + b_{31.2}\sum X_1^2 + b_{32.1}\sum X_1X_2 \\
\sum X_2X_3 &= b_{3.12}\sum X_2 + b_{31.2}\sum X_1X_2 + b_{32.1}\sum X_2^2
\end{aligned}$$

(*b*) Multiply the equation successively by 1, X_2, X_3, and X_4, and sum on both sides. The normal equations are

$$\begin{aligned}
\sum X_1 &= b_{1.234}N + b_{12.34}\sum X_2 + b_{13.24}\sum X_3 + b_{14.23}\sum X_4 \\
\sum X_1X_2 &= b_{1.234}\sum X_2 + b_{12.34}\sum X_2^2 + b_{13.24}\sum X_2X_3 + b_{14.23}\sum X_2X_4 \\
\sum X_1X_3 &= b_{1.234}\sum X_3 + b_{12.34}\sum X_2X_3 + b_{13.24}\sum X_3^2 + b_{14.23}\sum X_3X_4 \\
\sum X_1X_4 &= b_{1.234}\sum X_4 + b_{12.34}\sum X_2X_4 + b_{13.24}\sum X_3X_4 + b_{14.23}\sum X_4^2
\end{aligned}$$

Note that these are not derivations of the normal equations, but only formal means for remembering them.

The number of normal equations is equal to the number of unknown constants.

15.3 Table 15.1 shows the weights X_1 to the nearest pound (lb), the heights X_2 to the nearest inch (in), and the ages X_3 to the nearest year of 12 boys.

(*a*) Find the least-squares regression equation of X_1 on X_2 and X_3.

(*b*) Determine the estimated values of X_1 from the given values of X_2 and X_3.

(*c*) Estimate the weight of a boy who is 9 years old and 54 in tall.

(*d*) Find the least-squares regression equation using EXCEL, MINITAB, SPSS and STATISTIX.

Table 15.1

Weight (X_1)	64	71	53	67	55	58	77	57	56	51	76	68
Height (X_2)	57	59	49	62	51	50	55	48	52	42	61	57
Age (X_3)	8	10	6	11	8	7	10	9	10	6	12	9

SOLUTION

(*a*) The linear regression equation of X_1 on X_2 and X_3 can be written

$$X_1 = b_{1.23} + b_{12.3}X_2 + b_{13.2}X_3$$

The normal equations of the least-squares regression equation are

$$\begin{aligned}
\sum X_1 &= b_{1.23}N + b_{12.3}\sum X_2 + b_{13.2}\sum X_3 \\
\sum X_1X_2 &= b_{1.23}\sum X_2 + b_{12.3}\sum X_2^2 + b_{13.2}\sum X_2X_3 \\
\sum X_1X_3 &= b_{1.23}\sum X_3 + b_{12.3}\sum X_2X_3 + b_{13.2}\sum X_3^2
\end{aligned} \tag{19}$$

The work involved in computing the sums can be arranged as in Table 15.2. (Although the column headed X_1^2 is not needed at present, it has been added for future reference.) Using Table 15.2, the normal equations (*19*) become

$$\begin{aligned}
12b_{1.23} + 643b_{12.3} + 106b_{13.2} &= 753 \\
643b_{1.23} + 34{,}843b_{12.3} + 5{,}779b_{13.2} &= 40{,}830 \\
106b_{1.23} + 5{,}779b_{12.3} + 976b_{13.2} &= 6{,}796
\end{aligned} \tag{20}$$

Solving, $b_{1.23} = 3.6512$, $b_{12.3} = 0.8546$, and $b_{13.2} = 1.5063$, and the required regression equation is

$$X_1 = 3.6512 + 0.8546X_2 + 1.5063X_3 \quad \text{or} \quad X_1 = 3.65 + 0.855X_2 + 1.506X_3 \tag{21}$$

Table 15.2

X_1	X_2	X_3	X_1^2	X_2^2	X_3^2	X_1X_2	X_1X_3	X_2X_3
64	57	8	4096	3249	64	3648	512	456
71	59	10	5041	3481	100	4189	710	590
53	49	6	2809	2401	36	2597	318	294
67	62	11	4489	3844	121	4154	737	682
55	51	8	3025	2601	64	2805	440	408
58	50	7	3364	2500	49	2900	406	350
77	55	10	5929	3025	100	4235	770	550
57	48	9	3249	2304	81	2736	513	432
56	52	10	3136	2704	100	2912	560	520
51	42	6	2601	1764	36	2142	306	252
76	61	12	5776	3721	144	4636	912	732
68	57	9	4624	3249	81	3876	612	513
$\sum X_1 = 753$	$\sum X_2 = 643$	$\sum X_3 = 106$	$\sum X_1^2 = 48{,}139$	$\sum X_2^2 = 34{,}843$	$\sum X_3^2 = 976$	$\sum X_1X_2 = 40{,}830$	$\sum X_1X_3 = 6796$	$\sum X_2X_3 = 5779$

For another method, which avoids solving simultaneous equations, see Problem 15.6.

(*b*) Using the regression equation (*21*), we obtain the estimated values of X_1, denoted by $X_{1,\text{est}}$, by substituting the corresponding values of X_2 and X_3. For example, substituting $X_2 = 57$ and $X_3 = 8$ in (*21*), we find $X_{1,\text{est}} = 64.414$.

The other estimated values of X_1 are obtained similarly. They are given in Table 15.3 together with the sample values of X_1.

Table 15.3

$X_{1,\text{est}}$	64.414	69.136	54.564	73.206	59.286	56.925	65.717	58.229	63.153	48.582	73.857	65.920
X_1	64	71	53	67	55	58	77	57	56	51	76	68

(*c*) Putting $X_2 = 54$ and $X_3 = 9$ in equation (*21*), the estimated weight is $X_{1,\text{est}} = 63.356$, or about 63 lb.

(*d*) Part of the EXCEL output is shown in Fig. 15-1. The pull-down **Tools → Data analysis → Regression** is used to give the output. The coefficients $b_{1.23} = 3.6512$, $b_{12.3} = 0.8546$, and $b_{13.2} = 1.5063$ are shown in bold in the output.

Part of the MINITAB output is `the regression equation is` $X1 = 3.7 + 0.855X2 + 1.51X3$. The pull-down **Stat → Regression → Regression** is used after the data are entered into `C1 - C3`.

A portion of the SPSS output is shown in Fig. 15-2. The pull-down **analyse → Regression → Linear** gives the output. The coefficients $b_{1.23} = 3.651$, $b_{12.3} = 0.855$, and $b_{13.2} = 1.506$ are shown in the output under the column titled `Unstandardized Coefficients`.

Part of the STATISTIX output is shown in Fig. 15-3. The pull-down **Statistics → Linear models → Linear Regression** gives the output.

X1	X2	X3
64	57	8
71	59	10
53	49	6
67	62	11
55	51	8
58	50	7
77	55	10
57	48	9
56	52	10
51	42	6
76	61	12
68	57	9

SUMMARY OUTPUT

Regression Statistics	
Multiple R	0.841757
R Square	0.708554
Adjusted R Square	0.643789
Standard Error	5.363215
Observations	12

ANOVA

	df
Regression	2
Residual	9
Total	11

	Coefficients
Intercept	**3.651216**
X2	**0.85461**
X3	**1.506332**

Fig. 15-1 EXCEL output for Problem 15.3(*d*).

Coefficients[a]

Model		Unstandardized Coefficients B	Unstandardized Coefficients Std. Error	Standardized Coefficients Beta	t	Sig.
1	(Constant)	3.651	16.168		.226	.826
	X2	.855	.452	.565	1.892	.091
	X3	1.506	1.414	.318	1.065	.315

[a] Dependent Variable: X1

Fig. 15-2 SPSS output for Problem 15.3(*d*).

Statistix 8.0

Unweighted Least Squares Linear Regression of X1

Predictor Variables	**Coefficient**	**Std Error**	**T**	**P**	**VIF**
Constant	**3.65122**	16.1678	0.23	0.8264	
X2	**0.85461**	0.45166	1.89	0.0910	2.8
X3	**1.50633**	1.41427	1.07	0.3146	2.8

Fig. 15-3 STATISTIX output for Problem 15.3(*d*).

The software solutions are the same as the normal equations solution.

15.4 Calculate the standard deviations (*a*) s_1, (*b*) s_2, and (*c*) s_3 for the data of Problem 15.3.

SOLUTION

(*a*) The quantity s_1 is the standard deviation of the variable X_1. Then, using Table 15.2 of Problem 15.3 and the methods of Chapter 4, we find

$$s_1 = \sqrt{\frac{\sum X_1^2}{N} - \left(\frac{\sum X_1}{N}\right)^2} = \sqrt{\frac{48{,}139}{12} - \left(\frac{753}{12}\right)^2} = 8.6035 \quad \text{or} \quad 8.6\,\text{lb}$$

(*b*)
$$s_2 = \sqrt{\frac{\sum X_2^2}{N} - \left(\frac{\sum X_2}{N}\right)^2} = \sqrt{\frac{34{,}843}{12} - \left(\frac{643}{12}\right)^2} = 5.6930 \quad \text{or} \quad 5.7\,\text{in}$$

(*c*)
$$s_3 = \sqrt{\frac{\sum X_3^2}{N} - \left(\frac{\sum X_3}{N}\right)^2} = \sqrt{\frac{976}{12} - \left(\frac{106}{12}\right)^2} = 1.8181 \quad \text{or} \quad 1.8\,\text{years}$$

15.5 Compute (*a*) r_{12}, (*b*) r_{13}, and (*c*) r_{23} for the data of Problem 15.3. Compute the three correlations by using EXCEL, MINITAB, SPSS, and STATISTIX.

SOLUTION

(*a*) The quantity r_{12} is the linear correlation coefficient between the variables X_1 and X_2, ignoring the variable X_3. Then, using the methods of Chapter 14, we have

$$r_{12} = \frac{N\sum X_1X_2 - (\sum X_1)(\sum X_2)}{\sqrt{[N\sum X_1^2 - (\sum X_1)^2][N\sum X_2^2 - (\sum X_2)^2]}}$$

$$= \frac{(12)(40{,}830) - (753)(643)}{\sqrt{[(12)(48{,}139) - (753)^2][(12)(34{,}843) - (643)^2]}} = 0.8196 \quad \text{or} \quad 0.82$$

(*b*) and (*c*) Using corresponding formulas, we obtain $r_{12} = 0.7698$, or 0.77, and $r_{23} = 0.7984$, or 0.80.

(*d*) Using EXCEL, we have:

A	B	C	D	E
X1	X2	X3	0.819645	=CORREL(A2:A13,B2:B13)
64	57	8	0.769817	=CORREL(A2:A13,C2:C13)
71	59	10	0.798407	=CORREL(B2:B13,C2:C13)
53	49	6		
67	62	11		
55	51	8		
58	50	7		
77	55	10		
57	48	9		
56	52	10		
51	42	6		
76	61	12		
68	57	9		

We find that r_{12} is in D1, r_{13} is in D2, and r_{23} is in D3. The functions we used to produce the correlations in EXCEL are shown in E1, E2, and E3.

Using the MINITAB pull-down **Stat → Basic Statistics → Correlation** gives the following output.

```
Correlations: X1, X2, X3
        X1       X2
X2    0.820
      0.001

X3    0.770    0.798
      0.003    0.002

Cell Contents: Pearson correlation
                P-Value
```

The correlation r_{12} is at the intersection of X1 and X2 and is seen to be 0.820. The value below it is 0.001. It is the *p*-value for testing no population correlation between X1 and X2. Because this *p*-value is less than 0.05, we would reject the null hypothesis of no population correlation between height (X2) and weight (X1). All the other correlations and their *p*-values are read similarly.

Using the SPSS pull-down **Analyze → Correlate → Bivariate** gives the following output which is read similar to MINITAB.

Correlations

		X1	X2	X3
X1	Pearson Correlation	1	.820**	.770**
	Sig. (2-tailed)		.001	.003
	N	12	12	12
X2	Pearson Correlation	.820**	1	.798**
	Sig. (2-tailed)	.001		.002
	N	12	12	12
X3	Pearson Correlation	.770**	.798**	1
	Sig. (2-tailed)	.003	.002	
	N	12	12	12

**Correlation is significant at the 0.01 level (2-tailed).

Using the STATISTIX pull-down **Statistics → Linear models → Correlation** gives the following output which is similar to the other software outputs.

```
Statistix 8.0
Correlations (Pearson)
                X1          X2
X2          0.8196
P-VALUE     0.0011

X3          0.7698      0.7984
            0.0034      0.0018
```

Once again, we see that the software saves us a great deal of time doing the computations for us.

15.6 Work Problem 15.3(*a*) by using equation (*5*) of this chapter and the results of Problems 15.4 and 15.5.

SOLUTION

The regression equation of X_1 on X_2 and X_3 is, on multiplying both sides of equation (*5*) by s_1,

$$x_1 = \left(\frac{r_{12} - r_{13}r_{23}}{1 - r_{23}^2}\right)\left(\frac{s_1}{s_2}\right)x_2 + \left(\frac{r_{13} - r_{12}r_{23}}{1 - r_{23}^2}\right)\left(\frac{s_1}{s_3}\right)x_3 \tag{22}$$

where $x_1 = X_1 - \bar{X}_1$, $x_2 = X_2 - \bar{X}_2$, and $x_3 = X_3 - \bar{X}_3$. Using the results of Problems 15.4 and 15.5, equation (*22*) becomes

$$x_1 = 0.8546x_2 + 1.5063x_3$$

Since $\bar{X}_1 = \frac{\sum X_1}{N} = \frac{753}{12} = 62.750 \qquad \bar{X}_2 = \frac{\sum X_2}{N} = 53.583 \qquad \text{and} \qquad \bar{X}_3 = 8.833$

(from Table 15.2 of Problem 15.3), the required equation can be written

$$X_1 - 62.750 = 0.8546(X_2 - 53.583) + 1.506(X_3 - 8.833)$$

agreeing with the result of Problem 15.3(*a*).

15.7 For the data of Problem 15.3, determine (*a*) the average increase in weight per inch of increase in height for boys of the same age and (*b*) the average increase in weight per year for boys having the same height.

SOLUTION

From the regression equation obtained in Problem 15.3(*a*) or 15.6 we see that the answer to (*a*) is 0.8546, or about 0.9 lb, and that the answer to (*b*) is 1.5063, or about 1.5 lb.

15.8 Show that equations (*3*) and (*4*) of this chapter follow from equations (*1*) and (*2*).

SOLUTION

From the first of equations (*2*), on dividing both sides by N, we have

$$\bar{X}_1 = b_{1.23} + b_{12.3}\bar{X}_2 + b_{13.2}\bar{X}_3 \tag{23}$$

Subtracting equation (*23*) from equation (*1*) gives

$$X_1 - \bar{X}_1 = b_{12.3}(X_2 - \bar{X}_2) + b_{13.2}(X_3 - \bar{X}_3)$$

or

$$x_1 = b_{12.3}x_2 + b_{13.2}x_3 \tag{24}$$

which is equation (*3*).

Let $X_1 = x_1 + \bar{X}_1$, $X_2 = x_2 + \bar{X}_2$, and $X_3 = x_3 + \bar{X}_3$ in the second and third of equations (*2*). Then after some algebraic simplifications, using the results $\sum x_1 = \sum x_2 = \sum x_3 = 0$, they become

$$\sum x_1x_2 = b_{12.3}\sum x_2^2 + b_{13.2}\sum x_2x_3 + N\bar{X}_2[b_{1.23} + b_{12.3}\bar{X}_2 + b_{13.2}\bar{X}_3 - \bar{X}_1] \tag{25}$$

$$\sum x_1x_3 = b_{12.3}\sum x_2x_3 + b_{13.2}\sum x_3^2 + N\bar{X}_3[b_{1.23} + b_{12.3}\bar{X}_2 + b_{13.2}\bar{X}_3 - \bar{X}_1] \tag{26}$$

which reduce to equations (*4*) since the quantities in brackets on the right-hand sides of equations (*25*) and (*26*) are zero because of equation (*1*).

15.9 Establish equation (*5*), repeated here:

$$\frac{x_1}{s_1} = \left(\frac{r_{12} - r_{13}r_{23}}{1 - r_{23}^2}\right)\frac{x_2}{s_2} + \left(\frac{r_{13} - r_{12}r_{23}}{1 - r_{23}^2}\right)\frac{x_3}{s_3} \tag{5}$$

SOLUTION

From equations (*25*) and (*26*)

$$\begin{aligned} b_{12.3}\sum x_2^2 + b_{13.2}\sum x_2x_3 &= \sum x_1x_2 \\ b_{12.3}\sum x_2x_3 + b_{13.2}\sum x_3^2 &= \sum x_1x_3 \end{aligned} \tag{27}$$

Since

$$s_2^2 = \frac{\sum x_2^2}{N} \quad \text{and} \quad s_3^2 = \frac{\sum x_3^2}{N}$$

$\sum x_2^2 = Ns_2^2$ and $\sum x_3^2 = Ns_3^2$. Since

$$r_{23} = \frac{\sum x_2x_3}{\sqrt{(\sum x_2^2)(\sum x_3^2)}} = \frac{\sum x_2x_3}{Ns_2s_3}$$

$\sum x_2x_3 = Ns_2s_3r_{23}$. Similarly, $\sum x_1x_2 = Ns_1s_2r_{12}$ and $\sum x_1x_3 = Ns_1s_3r_{13}$.

Substituting in *(27)* and simplifying, we find

$$\begin{aligned} b_{12.3}s_2 \quad + b_{13.2}s_3r_{23} &= s_1r_{12} \\ b_{12.3}s_2r_{23} + b_{13.2}s_3 \quad &= s_1r_{13} \end{aligned} \tag{28}$$

Solving equations *(28)* simultaneously, we have

$$b_{12.3} = \left(\frac{r_{12} - r_{13}r_{23}}{1 - r_{23}^2}\right)\left(\frac{s_1}{s_2}\right) \quad \text{and} \quad b_{13.2} = \left(\frac{r_{13} - r_{12}r_{23}}{1 - r_{23}^2}\right)\left(\frac{s_1}{s_3}\right)$$

Substituting these in the equation $x_1 = b_{12.3}x_2 + b_{13.2}x_3$ [equation *(24)*] and dividing by s_1 yields the required result.

STANDARD ERROR OF ESTIMATE

15.10 Compute the standard error of estimate of X_1 on X_2 and X_3 for the data of Problem 15.3.

SOLUTION

From Table 15.3 of Problem 15.3 we have

$$s_{1.23} = \sqrt{\frac{\sum(X_1 - X_{1,\text{est}})^2}{N}}$$

$$= \sqrt{\frac{(64 - 64.414)^2 + (71 - 69.136)^2 + \cdots + (68 - 65.920)^2}{12}} = 4.6447 \quad \text{or} \quad 4.6 \text{ lb}$$

The population standard error of estimate is estimated by $\hat{s}_{1.23} = \sqrt{N/(N-3)}s_{1.23} = 5.3$ lb in this case.

15.11 To obtain the result of Problem 15.10, use

$$s_{12.3} = s_1\sqrt{\frac{1 - r_{12}^2 - r_{13}^2 - r_{23}^2 + 2r_{12}r_{13}r_{23}}{1 - r_{23}^2}}$$

SOLUTION

From Problems 15.4*(a)* and 15.5 we have

$$s_{1.23} = 8.6035\sqrt{\frac{1 - (0.8196)^2 - (0.7698)^2 - (0.7984)^2 + 2(0.8196)(0.7698)(0.7984)}{1 - (0.7984)^2}} = 4.6 \text{ lb}$$

Note that by the method of this problem the standard error of estimate can be found without using the regression equation.

COEFFICIENT OF MULTIPLE CORRELATION

15.12 Compute the coefficient of linear multiple correlation of X_1 on X_2 and X_3 for the data of Problem 15.3. Refer to the MINITAB output in the solution of Problem 15.3 to determine the coefficient of linear multiple correlation.

SOLUTION

First method

From the results of Problems 15.4(*a*) and 15.10 we have

$$R_{1.23} = \sqrt{1 - \frac{s_{1.23}^2}{s_1^2}} = \sqrt{1 - \frac{(4.6447)^2}{(8.6035)^2}} = 0.8418$$

Second method

From the results of Problem 15.5 we have

$$R_{1.23} = \sqrt{\frac{r_{12}^2 + r_{13}^2 - 2r_{12}r_{13}r_{23}}{1 - r_{23}^2}} = \sqrt{\frac{(0.8196)^2 + (0.7698)^2 - 2(0.8196)(0.7698)(0.7984)}{1 - (0.7984)^2}} = 0.8418$$

Note that the coefficient of multiple correlation, $R_{1.23}$, is larger than either of the coefficients r_{12} or r_{13} (see Problem 15.5). This is always true and is in fact to be expected, since by taking into account additional relevant independent variables we should arrive at a better relationship between the variables.

The following part of the MINITAB output given in the solution of Problem 15.3, `R-Sq = 70.9%`, gives the square of the coefficient of linear multiple correlation. The coefficient of linear multiple correlation is the square rooot of this value. That is $R_{1.23} = \sqrt{0.709} = 0.842$.

15.13 Compute the coefficient of multiple determination of X_1 on X_2 and X_3 for the data of Problem 15.3. Refer to the MINITAB output in the solution of Problem 15.3 to determine the coefficient of multiple determination.

SOLUTION

The coefficient of multiple determination of X_1 on X_2 and X_3 is

$$R_{1.23}^2 = (0.8418)^2 = 0.7086$$

using Problem 15.12. Thus about 71% of the total variation in X_1 is explained by using the regression equation.

The coefficient of multiple determination is read directly from the MINITAB output given in the solution of Problem 15.3 as `R-Sq = 70.9%`.

15.14 For the data of Problem 15.3, calculate (*a*) $R_{2.13}$ and (*b*) $R_{3.12}$ and compare their values with the value of $R_{1.23}$.

SOLUTION

(*a*) $$R_{2.13} = \sqrt{\frac{r_{12}^2 + r_{23}^2 - 2r_{12}r_{13}r_{23}}{1 - r_{13}^2}} = \sqrt{\frac{(0.8196)^2 + (0.7984)^2 - 2(0.8196)(0.7698)(0.7984)}{1 - (0.7698)^3}} = 0.8606$$

(*b*) $$R_{3.12} = \sqrt{\frac{r_{13}^2 + r_{23}^2 - 2r_{12}r_{13}r_{23}}{1 - r_{12}^2}} = \sqrt{\frac{(0.7698)^2 + (0.7984)^2 - 2(0.8196)(0.7698)(0.7984)}{1 - (0.8196)^2}} = 0.8234$$

This problem illustrates the fact that, in general, $R_{2.13}$, $R_{3.12}$, and $R_{1.23}$ are not necessarily equal, as seen by comparison with Problem 15.12.

15.15 If $R_{1.23} = 1$, prove that (*a*) $R_{2.13} = 1$ and (*b*) $R_{3.12} = 1$.

SOLUTION

$$R_{1.23} = \sqrt{\frac{r_{12}^2 + r_{13}^2 - 2r_{12}r_{13}r_{23}}{1 - r_{23}^2}} \tag{29}$$

and

$$R_{2.13} = \sqrt{\frac{r_{12}^2 + r_{23}^2 - 2r_{12}r_{13}r_{23}}{1 - r_{13}^2}} \tag{30}$$

(*a*) In equation (*29*), setting $R_{1.23} = 1$ and squaring both sides, $r_{12}^2 + r_{13}^2 - 2r_{12}r_{13}r_{23} = 1 - r_{23}^2$. Then

$$r_{12}^2 + r_{23}^2 - 2r_{12}r_{13}r_{23} = 1 - r_{13}^2 \qquad \text{or} \qquad \frac{r_{12}^2 + r_{23}^2 - 2r_{12}r_{13}r_{23}}{1 - r_{13}^2} = 1$$

That is, $R_{2.13}^2 = 1$ or $R_{2.13} = 1$, since the coefficient of multiple correlation is considered nonnegative.

(*b*) $R_{3.12} = 1$ follows from part (*a*) by interchanging subscripts 2 and 3 in the result $R_{2.13} = 1$.

15.16 If $R_{1.23} = 0$, does it necessarily follow that $R_{2.13} = 0$?

SOLUTION

From equation (*29*), $R_{1.23} = 0$ if and only if

$$r_{12}^2 + r_{13}^2 - 2r_{12}r_{13}r_{23} = 0 \qquad \text{or} \qquad 2r_{12}r_{13}r_{23} = r_{12}^2 + r_{13}^2$$

Then from equation (*30*) we have

$$R_{2.13} = \sqrt{\frac{r_{12}^2 + r_{23}^2 - (r_{12}^2 + r_{13}^2)}{1 - r_{13}^2}} = \sqrt{\frac{r_{23}^2 - r_{13}^2}{1 - r_{13}^2}}$$

which is not necessarily zero.

PARTIAL CORRELATION

15.17 For the data of Problem 15.3, compute the coefficients of linear partial correlation $r_{12.3}$, $r_{13.2}$, and $r_{23.1}$. Also use STATISTIX to do these same computations.

SOLUTION

$$r_{12.3} = \frac{r_{12} - r_{13}r_{23}}{\sqrt{(1 - r_{13}^2)(1 - r_{23}^2)}} \qquad r_{13.2} = \frac{r_{13} - r_{12}r_{23}}{\sqrt{(1 - r_{12}^2)(1 - r_{23}^2)}} \qquad r_{23.1} = \frac{r_{23} - r_{12}r_{13}}{\sqrt{(1 - r_{12}^2)(1 - r_{13}^2)}}$$

Using the results of Problem 15.5, we find that $r_{12.3} = 0.5334$, $r_{13.2} = 0.3346$, and $r_{23.1} = 0.4580$. It follows that for boys of the same age, the correlation coefficient between weight and height is 0.53; for boys of the same height, the correlation coefficient between weight and age is only 0.33. Since these results are based on a small sample of only 12 boys, they are of course not as reliable as those which would be obtained from a larger sample.

Use the pull-down **Statistics → Linear models → Partial Correlations** to obtain the dialog box in Fig. 15-4. Fill it in as shown. It is requesting $r_{12.3}$. The following output is obtained.

```
Statistix 8.

Partial Correlations with X1
Controlled for X3

X2                  0.5335
```

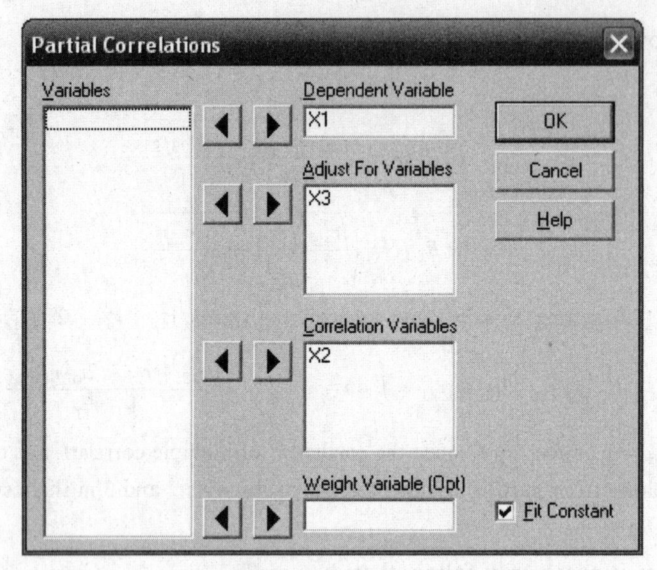

Fig. 15-4 STATISTIX dialog box for Problem 15.17.

Similarly, STATISTIX may be used to find the other two partial correlations.

15.18 If $X_1 = b_{1.23} + b_{12.3}X_2 + b_{13.2}X_3$ and $X_3 = b_{3.12} + b_{32.1}X_2 + b_{31.2}X_1$ are the regression equations of X_1 on X_2 and X_3 and of X_3 on X_2 and X_1, respectively, prove that $r_{13.2}^2 = b_{13.2}b_{31.2}$.

SOLUTION

The regression equation of X_1 on X_2 and X_3 can be written [see equation (*5*) of this chapter]

$$X_1 = \bar{X}_1 = \left(\frac{r_{12} - r_{13}r_{23}}{1 - r_{23}^2}\right)\left(\frac{s_1}{s_2}\right)(X_2 - \bar{X}_2) + \left(\frac{r_{13} - r_{12}r_{23}}{1 - r_{23}^2}\right)\left(\frac{s_1}{s_3}\right)(X_3 - \bar{X}_3) \tag{31}$$

The regression equation of X_3 on X_2 and X_1 can be written [see equation (*10*)]

$$X_3 - \bar{X}_3 = \left(\frac{r_{23} - r_{13}r_{12}}{1 - r_{12}^2}\right)\left(\frac{s_3}{s_2}\right)(X_2 - \bar{X}_2) + \left(\frac{r_{13} - r_{23}r_{12}}{1 - r_{12}^2}\right)\left(\frac{s_3}{s_1}\right)(X_1 - \bar{X}_1) \tag{32}$$

From equations (*31*) and (*32*) the coefficients of X_3 and X_1 are, respectively,

$$b_{13.2} = \left(\frac{r_{13} - r_{12}r_{23}}{1 - r_{23}^2}\right)\left(\frac{s_1}{s_3}\right) \quad \text{and} \quad b_{31.2} = \left(\frac{r_{13} - r_{23}r_{12}}{1 - r_{12}^2}\right)\left(\frac{s_3}{s_1}\right)$$

Thus

$$b_{13.2}b_{31.2} = \frac{(r_{13} - r_{12}r_{23})^2}{(1 - r_{23}^2)(1 - r_{12}^2)} = r_{13.2}^2$$

15.19 If $r_{12.3} = 0$, prove that

$$(a) \quad r_{13.2} = r_{13}\sqrt{\frac{1 - r_{23}^2}{1 - r_{12}^2}} \qquad (b) \quad r_{23.1} = r_{23}\sqrt{\frac{1 - r_{13}^2}{1 - r_{12}^2}}$$

SOLUTION

If

$$r_{12.3} = \frac{r_{12} - r_{13}r_{23}}{\sqrt{(1 - r_{13}^2)(1 - r_{23}^2)}} = 0$$

we have $r_{12} = r_{13}r_{23}$.

(a)
$$r_{13.2} = \frac{r_{13} - r_{12}r_{23}}{\sqrt{(1 - r_{12}^2)(1 - r_{23}^2)}} = \frac{r_{13} - (r_{13}r_{23})r_{23}}{\sqrt{(1 - r_{12}^2)(1 - r_{23}^2)}} = \frac{r_{13}(1 - r_{23}^2)}{\sqrt{(1 - r_{12}^2)(1 - r_{23}^2)}} = r_{13}\sqrt{\frac{1 - r_{23}^2}{1 - r_{12}^2}}$$

(b) Interchange the subscripts 1 and 2 in the result of part (a).

MULTIPLE AND PARTIAL CORRELATION INVOLVING FOUR OR MORE VARIABLES

15.20 A college entrance examination consisted of three tests: in mathematics, English, and general knowledge. To test the ability of the examination to predict performance in a statistics course, data concerning a sample of 200 students were gathered and analyzed. Letting

X_1 = grade in statistics course X_3 = score on English test

X_2 = score on mathematics test X_4 = score on general knowledge test

the following calculations were obtained:

$$\bar{X}_1 = 75 \quad s_1 = 10 \quad \bar{X}_2 = 24 \quad s_2 = 5$$
$$\bar{X}_3 = 15 \quad s_3 = 3 \quad \bar{X}_4 = 36 \quad s_4 = 6$$
$$r_{12} = 0.90 \quad r_{13} = 0.75 \quad r_{14} = 0.80 \quad r_{23} = 0.70 \quad r_{24} = 0.70 \quad r_{34} = 0.85$$

Find the least-squares regression equation of X_1 on X_2, X_3, and X_4.

SOLUTION

Generalizing the result of Problem 15.8, we can write the least-squares regression equation of X_1 on X_2, X_3, and X_4 in the form

$$x_1 = b_{12.34}x_2 + b_{13.24}x_3 + b_{14.23}x_4 \tag{33}$$

where $b_{12.34}$, $b_{13.24}$, and $b_{14.23}$ can be obtained from the normal equations

$$\begin{aligned} \sum x_1x_2 &= b_{12.34}\sum x_2^2 + b_{13.24}\sum x_2x_3 + b_{14.23}\sum x_2x_4 \\ \sum x_1x_3 &= b_{12.34}\sum x_2x_3 + b_{13.24}\sum x_3^2 + b_{14.23}\sum x_3x_4 \\ \sum x_1x_4 &= b_{12.34}\sum x_2x_4 + b_{13.24}\sum x_3x_4 + b_{14.23}\sum x_4^2 \end{aligned} \tag{34}$$

and where $x_1 = X_1 - \bar{X}_1$, $x_2 = X_2 - \bar{X}_2$, $x_3 = X_3 - \bar{X}_3$, and $x_4 = X_4 - \bar{X}_4$.

From the given data, we find

$$\sum x_2^2 - Ns_2^2 = 5000 \qquad \sum x_1x_2 = Ns_1s_2r_{12} = 9000 \qquad \sum x_2x_3 = Ns_1s_3r_{23} = 2100$$
$$\sum x_3^2 - Ns_3^2 = 1800 \qquad \sum x_1x_3 = Ns_1s_3r_{13} = 4500 \qquad \sum x_2x_4 = Ns_2s_4r_{24} = 4200$$
$$\sum x_4^2 - Ns_4^2 = 7200 \qquad \sum x_1x_4 = Ns_1s_4r_{14} = 9600 \qquad \sum x_3x_4 = Ns_3s_4r_{34} = 3060$$

Putting these results into equations (34) and solving, we obtain

$$b_{12.34} = 1.3333 \qquad b_{13.24} = 0.0000 \qquad b_{14.23} = 0.5556 \tag{35}$$

which, when substituted in equation (33), yield the required regression equation

$$x_1 = 1.3333x_2 + 0.0000x_3 + 0.5556x_4$$

or
$$X_1 - 75 = 1.3333(X_2 - 24) + 0.5556(X_4 - 27) \tag{36}$$

or
$$X_1 = 22.9999 + 1.3333X_2 + 0.5556X_4$$

An exact solution of equations (34) yields $b_{12.34} = \frac{4}{3}$, $b_{13.24} = 0$, and $b_{14.23} = \frac{5}{9}$, so that the regression equation can also be written

$$X_1 = 23 + \tfrac{4}{3}X_2 + \tfrac{5}{9}X_4 \tag{37}$$

It is interesting to note that the regression equation does not involve the score in English, namely, X_3. This does not mean that one's knowledge of English has no bearing on proficiency in statistics. Instead, it means that the need for English, insofar as prediction of the statistics grade is concerned, is amply evidenced by the scores achieved on the other tests.

15.21 Two students taking the college entrance examination of Problem 15.20 receive respective scores of (*a*) 30 in mathematics, 18 in English, and 32 in general knowledge; and (*b*) 18 in mathematics, 20 in English, and 36 in general knowledge. What would be their predicted grades in statistics?

SOLUTION

(*a*) Substituting $X_2 = 30$, $X_3 = 18$, and $X_4 = 32$ in equation (37), the predicted grade in statistics is $X_1 = 81$.

(*b*) Proceeding as in part (*a*) with $X_2 = 18$, $X_3 = 20$, and $X_4 = 36$, we find $X_1 = 67$.

15.22 For the data of Problem 15.20, find the partial correlation coefficients (*a*) $r_{12.34}$, (*b*) $r_{13.24}$, and (*c*) $r_{14.23}$.

SOLUTION

(*a*) and (*b*)

$$r_{12.4} = \frac{r_{12} - r_{14}r_{24}}{\sqrt{(1 - r_{14}^2)(1 - r_{24}^2)}} \qquad r_{13.4} = \frac{r_{13} - r_{14}r_{34}}{\sqrt{(1 - r_{14}^2)(1 - r_{34}^2)}} \qquad r_{23.4} = \frac{r_{23} - r_{24}r_{34}}{\sqrt{(1 - r_{24}^2)(1 - r_{34}^2)}}$$

Substituting the values from Problem 15.20, we obtain $r_{12.4} = 0.7935$, $r_{13.4} = 0.2215$, and $r_{23.4} = 0.2791$. Thus

$$r_{12.34} = \frac{r_{12.4} - r_{13.4}r_{23.4}}{\sqrt{(1 - r_{13.4}^2)(1 - r_{23.4}^2)}} = 0.7814 \qquad \text{and} \qquad r_{13.24} = \frac{r_{13.4} - r_{12.4}r_{23.4}}{\sqrt{(1 - r_{12.4}^2)(1 - r_{23.4}^2)}} = 0.0000$$

(*c*)

$$r_{14.3} = \frac{r_{14} - r_{13}r_{34}}{\sqrt{(1 - r_{13}^2)(1 - r_{34}^2)}} \qquad r_{12.3} = \frac{r_{12} - r_{13}r_{23}}{\sqrt{(1 - r_{13}^2)(1 - r_{23}^2)}} \qquad r_{24.3} = \frac{r_{24} - r_{23}r_{34}}{\sqrt{(1 - r_{23}^2)(1 - r_{34}^2)}}$$

Substituting the values from Problem 15.20, we obtain $r_{14.3} = 0.4664$, $r_{12.3} = 0.7939$, and $r_{24.3} = 0.2791$. Thus

$$r_{14.23} = \frac{r_{14.3} - r_{12.3}r_{24.3}}{\sqrt{(1 - r_{12.3}^2)(1 - r_{24.3}^2)}} = 0.4193$$

15.23 Interpret the partial correlation coefficients (*a*) $r_{12.4}$, (*b*) $r_{13.4}$, (*c*) $r_{12.34}$, (*d*) $r_{14.3}$, and (*e*) $r_{14.23}$ obtained in Problem 15.22.

SOLUTION

(*a*) $r_{12.4} = 0.7935$ represents the (linear) correlation coefficient between statistics grades and mathematics scores for students having the same general knowledge scores. In obtaining this coefficient, scores in English (as well as other factors that have not been taken into account) are not considered, as is evidenced by the fact that the subscript 3 is omitted.

(*b*) $r_{13.4} = 0.2215$ represents the correlation coefficient between statistics grades and English scores for students having the same general knowledge scores. Here, scores in mathematics have not been considered.

(*c*) $r_{12.34} = 0.7814$ represents the correlation coefficient between statistics grades and mathematics scores for students having both the same English scores and general knowledge scores.

(*d*) $r_{14.3} = 0.4664$ represents the correlation coefficient between statistics grades and general knowledge scores for students having the same English scores.

(*e*) $r_{14.23} = 0.4193$ represents the correlation coefficient between statistics grades and general knowledge scores for students having both the same mathematics scores and English scores.

15.24 (*a*) For the data of Problem 15.20, show that

$$\frac{r_{12.4} - r_{13.4}r_{23.4}}{\sqrt{(1 - r_{13.4}^2)(1 - r_{23.4}^2)}} = \frac{r_{12.3} - r_{14.3}r_{24.3}}{\sqrt{(1 - r_{14.3}^2)(1 - r_{24.3}^2)}} \tag{38}$$

(*b*) Explain the significance of the equality in part (*a*).

SOLUTION

(*a*) The left-hand side of equation (*38*) is evaluated in Problem 15.22(*a*) yielding the result 0.7814. To evaluate the right-hand side of equation (*38*), use the results of Problem 15.22(*c*); again, the result is 0.7814. Thus the equality holds in this special case. It can be shown by direct algebraic processes that the equality holds in general.

(*b*) The left-hand side of equation (*38*) is $r_{12.34}$, and the right-hand side is $r_{12.43}$. Since $r_{12.34}$ is the correlation between variables X_1 and X_2 keeping X_3 and X_4 constant, while $r_{12.43}$ is the correlation between X_1 and X_2 keeping X_4 and X_3 constant, it is at once evident why the equality should hold.

15.25 For the data of Problem 15.20, find (*a*) the multiple correlation coefficient $R_{1.234}$ and (*b*) the standard error of estimate $s_{1.234}$.

SOLUTION

(*a*)

$$1 - R_{1.234}^2 = (1 - r_{12}^2)(1 - r_{13.2}^2)(1 - r_{14.23}^2) \qquad \text{or} \qquad R_{1.234} = 0.9310$$

since $r_{12} = 0.90$ from Problem 15.20, $r_{14.23} = 0.4193$ from Problem 15.22(*c*), and

$$r_{13.2} = \frac{r_{13} - r_{12}r_{23}}{\sqrt{(1 - r_{12}^2)(1 - r_{23}^2)}} = \frac{0.75 - (0.90)(0.70)}{\sqrt{[1 - (0.90)^2][1 - (0.70)^2)]}} = 0.3855$$

Another method

Interchanging subscripts 2 and 4 in the first equation yields

$$1 - R_{1.234}^2 = (1 - r_{14}^2)(1 - r_{13.4}^2)(1 - r_{12.34}^2) \qquad \text{or} \qquad R_{1.234} = 0.9310$$

where the results of Problem 15.22(*a*) are used directly.

(*b*)

$$R_{1.234} = \sqrt{\frac{1 - s_{1.234}^2}{s_1^2}} \qquad \text{or} \qquad s_{1.234} = s_1\sqrt{1 - R_{1.234}^2} = 10\sqrt{1 - (0.9310)^2} = 3.650$$

Compare with equation (*8*) of this chapter.

Supplementary Problems

REGRESSION EQUATIONS INVOLVING THREE VARIABLES

15.26 Using an appropriate subscript notation, write the regression equations of (*a*) X_3 on X_1 and X_2 and (*b*) X_4 on X_1, X_2, X_3, and X_5.

15.27 Write the normal equations corresponding to the regression equations of (*a*) X_2 on X_1 and X_3 and (*b*) X_5 on X_1, X_2, X_3, and X_4.

15.28 Table 15.4 shows the corresponding values of three variables: X_1, X_2, and X_3.

(*a*) Find the least-squares regression equation of X_3 on X_1 and X_2.

(*b*) Estimate X_3 when $X_1 = 10$ and $X_2 = 6$.

Table 15.4

X_1	3	5	6	8	12	14
X_2	16	10	7	4	3	2
X_3	90	72	54	42	30	12

15.29 An instructor of mathematics wished to determine the relationship of grades on a final examination to grades on two quizzes given during the semester. Calling X_1, X_2, and X_3 the grades of a student on the first quiz, second quiz, and final examination, respectively, he made the following computations for a total of 120 students:

$$\bar{X}_1 = 6.8 \qquad \bar{X}_2 = 7.0 \qquad \bar{X}_3 = 74$$

$$s_1 = 1.0 \qquad s_2 = 0.80 \qquad s_2 = 9.0$$

$$r_{12} = 0.60 \qquad r_{13} = 0.70 \qquad r_{23} = 0.65$$

(*a*) Find the least-squares regression equation of X_3 on X_1 and X_2.

(*b*) Estimate the final grades of two students whose respective scores on the two quizzes were (1) 9 and 7 and (2) 4 and 8.

15.30 The data in Table 15.5 give the price in thousands (X_1), the number of bedrooms (X_2), and the number of baths (X_3) for 10 houses. Use the normal equations to find the least-squares regression equation of X_1 on X_2 and X_3. Use EXCEL, MINITAB, SAS, SPSS, and STATISTIX to find the least-squares regression equation of X_1 on X_2 and X_3. Use the least-squares regression equation of X_1 on X_2 and X_3 to estimate the price of a home having 5 bedrooms and 4 baths.

Table 15.5

Price	Bedrooms	Baths
165	3	2
200	3	3
225	4	3
180	2	3
202	4	2
250	4	4
275	3	4
300	5	3
155	2	2
230	4	4

STANDARD ERROR OF ESTIMATE

15.31 For the data of Problem 15.28, find the standard error of estimate of X_3 on X_1 and X_2.

15.32 For the data of Problem 15.29, find the standard error of estimate of (*a*) X_3 on X_1 and X_2 and (*b*) X_1 on X_2 and X_3.

COEFFICIENT OF MULTIPLE CORRELATION

15.33 For the data of Problem 15.28, compute the coefficient of linear multiple correlation of X_3 on X_1 and X_2.

15.34 For the data of Problem 15.29, compute (*a*) $R_{3.12}$, (*b*) $R_{1.23}$, and (*c*) $R_{2.13}$.

15.35 (*a*) If $r_{12} = r_{13} = r_{23} = r \neq 1$, show that

$$R_{1.23} = R_{2.31} = R_{3.12} = \frac{r\sqrt{2}}{\sqrt{1+r}}$$

(*b*) Discuss the case $r = 1$.

15.36 If $R_{1.23} = 0$, prove that $|r_{23}| \geq |r_{12}|$ and $|r_{23}| \geq |r_{13}|$ and interpret.

PARTIAL CORRELATION

15.37 For the data of Problem 15.28, compute the coefficients of linear partial correlation $r_{12.3}$, $r_{13.2}$, and $r_{23.1}$. Also use STATISTIX to do these same computations.

15.38 Work Problem 15.37 for the data of Problem 15.29.

15.39 If $r_{12} = r_{13} = r_{23} = r \neq 1$, show that $r_{12.3} = r_{13.2} = r_{23.1} = r/(1+r)$. Discuss the case $r = 1$.

15.40 If $r_{12.3} = 1$, show that (*a*) $|r_{13.2}| = 1$, (*b*) $|r_{23.1}| = 1$, (*c*) $R_{1.23} = 1$, and (*d*) $s_{1.23} = 0$.

MULTIPLE AND PARTIAL CORRELATION INVOLVING FOUR OR MORE VARIABLES

15.41 Show that the regression equation of X_4 on X_1, X_2, and X_3 can be written

$$\frac{x_4}{s_4} = a_1\left(\frac{x_1}{s_1}\right) + a_2\left(\frac{x_2}{s_2}\right) + a_3\left(\frac{x_3}{s_3}\right)$$

where a_1, a_2, and a_3 are determined by solving simultaneously the equations

$$a_1 r_{11} + a_2 r_{12} + a_3 r_{13} = r_{14}$$
$$a_1 r_{21} + a_2 r_{22} + a_3 r_{23} = r_{24}$$
$$a_1 r_{31} + a_2 r_{32} + a_3 r_{33} = r_{34}$$

and where $x_j = X_j - \bar{X}_j$, $r_{jj} = 1$, and $j = 1, 2, 3$, and 4. Generalize to the case of more than four variables.

15.42 Given $\bar{X}_1 = 20$, $\bar{X}_2 = 36$, $\bar{X}_3 = 12$, $\bar{X}_4 = 80$, $s_1 = 1.0$, $s_2 = 2.0$, $s_3 = 1.5$, $s_4 = 6.0$, $r_{12} = -0.20$, $r_{13} = 0.40$, $r_{23} = 0.50$, $r_{14} = 0.40$, $r_{24} = 0.30$, and $r_{34} = -0.10$, (*a*) find the regression equation of X_4 on X_1, X_2, and X_3, and (*b*) estimate X_4 when $X_1 = 15$, $X_2 = 40$, and $X_3 = 14$.

15.43 Find (*a*) $r_{41.23}$, (*b*) $r_{42.13}$, and (*c*) $r_{43.12}$ for the data of Problem 15.42 and interpret your results.

15.44 For the data of Problem 15.42, find (*a*) $R_{4.123}$ and (*b*) $s_{4.123}$.

15.45 The amount spent on medical expenses per year is correlated with other health factors for 15 adult males. A study collected the medical expenses per year, Y, as well as information on the following independent variables,

$$X_1 = \begin{cases} 0, & \text{if a non-smoker} \\ 1, & \text{if a smoker} \end{cases} \qquad X_2 = \text{money spent on alcohol per week,}$$

$$X_3 = \text{hours spent exercising per week,}$$

$$X_4 = \begin{cases} 0, & \text{dietary knowledge is low} \\ 1, & \text{dietary knowledge is average} \\ 2, & \text{dietary knowledge is high} \end{cases}$$

$$X_5 = \text{weight} \quad X_6 = \text{age}$$

The notation given in this problem will be encountered in many statistics books. Y will be used for the dependent variable and X with subscripts for the independent variables. Using the data in Table 15.6, find the regression equation of Y on X_1 through X_6 by solving the normal equations and contrast this with the EXCEL, MINITAB, SAS, SPSS, and STATISTIX solutions.

Table 15.6

Medcost	Smoker	Alcohol	Exercise	Dietary	Weight	Age
2100	0	20	5	1	185	50
2378	1	25	0	1	200	42
1657	0	10	10	2	175	37
2584	1	20	5	2	225	54
2658	1	25	0	1	220	32
1842	0	0	10	1	165	34
2786	1	25	5	0	225	30
2178	0	10	10	1	180	41
3198	1	30	0	1	225	31
1782	0	5	10	0	180	45
2399	0	25	12	2	225	45
2423	0	15	15	0	220	33
3700	1	25	0	1	275	43
2892	1	30	5	1	230	42
2350	1	30	10	1	245	40

Analysis of Variance

THE PURPOSE OF ANALYSIS OF VARIANCE

In Chapter 8 we used sampling theory to test the significance of differences between two sampling means. We assumed that the two populations from which the samples were drawn had the same variance. In many situations there is a need to test the significance of differences between three or more sampling means or, equivalently, to test the null hypothesis that the sample means are all equal.

EXAMPLE 1. Suppose that in an agricultural experiment four different chemical treatments of soil produced mean wheat yields of 28, 22, 18, and 24 bushels per acre, respectively. Is there a significant difference in these means, or is the observed spread due simply to chance?

Problems such as this can be solved by using an important technique known as *analysis of variance*, developed by Fisher. It makes use of the F distribution already considered in Chapter 11.

ONE-WAY CLASSIFICATION, OR ONE-FACTOR EXPERIMENTS

In a *one-factor experiment*, measurements (or observations) are obtained for a independent groups of samples, where the number of measurements in each group is b. We speak of *a treatments*, each of which has *b repetitions*, or *b replications*. In Example 1, $a = 4$.

The results of a one-factor experiment can be presented in a table having a rows and b columns, as shown in Table 16.1. Here X_{jk} denotes the measurement in the jth row and kth column, where $j = 1, 2, \ldots, a$ and where $k = 1, 2, \ldots, b$. For example, X_{35} refers to the fifth measurement for the third treatment.

Table 16.1

Treatment 1	$X_{11}, X_{12}, \ldots, X_{1b}$	$\bar{X}_{1.}$
Treatment 2	$X_{21}, X_{22}, \ldots, X_{2b}$	$\bar{X}_{2.}$
⋮	⋮	⋮
Treatment a	$X_{a1}, X_{a2}, \ldots, X_{ab}$	$\bar{X}_{a.}$

We shall denote by $\bar{X}_{j.}$ the mean of the measurements in the jth row. We have

$$\bar{X}_{j.} = \frac{1}{b}\sum_{k=1}^{b} X_{jk} \qquad j = 1, 2, \ldots, a \tag{1}$$

The dot in $\bar{X}_{j.}$ is used to show that the index k has been summed out. The values $\bar{X}_{j.}$ are called *group means, treatment means*, or *row means*. The *grand mean*, or *overall mean*, is the mean of all the measurements in all the groups and is denoted by $\bar{X}$:

$$\bar{X} = \frac{1}{ab}\sum_{j=1}^{a}\sum_{k=1}^{b} X_{jk} \tag{2}$$

TOTAL VARIATION, VARIATION WITHIN TREATMENTS, AND VARIATION BETWEEN TREATMENTS

We define the *total variation*, denoted by V, as the sum of the squares of the deviations of each measurement from the grand mean $\bar{X}$

$$\text{Total variation} = V = \sum_{j,k}(X_{jk} - \bar{X})^2 \tag{3}$$

By writing the identity

$$X_{jk} - \bar{X} = (X_{jk} - \bar{X}_{j.}) + (\bar{X}_{j.} - \bar{X}) \tag{4}$$

and then squaring and summing over j and k, we have (see Problem 16.1)

$$\sum_{j,k}(X_{jk} - \bar{X})^2 = \sum_{j,k}(X_{jk} - \bar{X}_{j.})^2 + \sum_{j,k}(\bar{X}_{j.} - \bar{X})^2 \tag{5}$$

or

$$\sum_{j,k}(X_{jk} - \bar{X})^2 = \sum_{j,k}(X_{jk} - \bar{X}_{j.})^2 + b\sum_{j}(\bar{X}_{j.} - \bar{X})^2 \tag{6}$$

We call the first summation on the right-hand side of equations (5) and (6) the *variation within treatments* (since it involves the squares of the deviations of X_{jk} from the treatment means $\bar{X}_{j.}$) and denote it by V_W. Thus

$$V_W = \sum_{j,k}(X_{jk} - \bar{X}_{j.})^2 \tag{7}$$

The second summation on the right-hand side of equations (5) and (6) is called the *variation between treatments* (since it involves the squares of the deviations of the various treatment means $\bar{X}_{j.}$ from the grand mean $\bar{X}$) and is denoted by V_B. Thus

$$V_B = \sum_{j,k}(\bar{X}_{j.} - \bar{X})^2 = b\sum_{j}(\bar{X}_{j} - \bar{X})^2 \tag{8}$$

Equations (5) and (6) can thus be written

$$V = V_W + V_B \tag{9}$$

SHORTCUT METHODS FOR OBTAINING VARIATIONS

To minimize the labor of computing the above variations, the following forms are convenient:

$$V = \sum_{j,k} X_{jk}^2 - \frac{T^2}{ab} \tag{10}$$

$$V_B = \frac{1}{b}\sum_{j} T_{j.}^2 - \frac{T^2}{ab} \tag{11}$$

$$V_W = V - V_B \tag{12}$$

where T is the total of all values X_{jk} and where $T_{j.}$ is the total of all values in the jth treatment:

$$T = \sum_{j,k} X_{jk} \qquad T_{j.} = \sum_{k} X_{jk} \tag{13}$$

In practice, it is convenient to subtract some fixed value from all the data in the table in order to simplify the calculation; this has no effect on the final results.

MATHEMATICAL MODEL FOR ANALYSIS OF VARIANCE

We can consider each row of Table 16.1 to be a random sample of size b from the population for that particular treatment. The X_{jk} will differ from the population mean μ_j for the jth treatment by a *chance error*, or *random error*, which we denote by ε_{jk}; thus

$$X_{jk} = \mu_j + \varepsilon_{jk} \tag{14}$$

These errors are assumed to be normally distributed with mean 0 and variance σ^2. If μ is the mean of the population for all treatments and if we let $\alpha_j = \mu_j - \mu$, so that $\mu_j = \mu + \alpha_j$, then equation (*14*) becomes

$$X_{jk} = \mu + \alpha_j + \varepsilon_{jk} \tag{15}$$

where $\sum_j \alpha_j = 0$ (see Problem 16.9). From equation (*15*) and the assumption that the ε_{jk} are normally distributed with mean 0 and variance σ^2, we conclude that the X_{jk} can be considered random variables that are normally distributed with mean μ and variance σ^2.

The null hypothesis that all treatment means are equal is given by ($H_0: \alpha_j = 0$; $j = 1, 2, \ldots, a$) or, equivalently, by ($H_0: \mu_j = \mu$; $j = 1, 2, \ldots, a$). If H_0 is true, the treatment populations will all have the same normal distribution (i.e., with the same mean and variance). In such cases there is just one treatment population (i.e., all treatments are statistically identical); in other words, there is no significant difference between the treatments.

EXPECTED VALUES OF THE VARIATIONS

It can be shown (see Problem 16.10) that the expected values of V_W, V_B, and V are given by

$$E(V_W) = a(b-1)\sigma^2 \tag{16}$$

$$E(V_B) = (a-1)\sigma^2 + b\sum_j \alpha_j^2 \tag{17}$$

$$E(V) = (ab-1)\sigma^2 + b\sum_j \alpha_j^2 \tag{18}$$

From equation (*16*) it follows that

$$E\left[\frac{V_W}{a(b-1)}\right] = \sigma^2 \tag{19}$$

so that

$$\hat{S}_W^2 = \frac{V_W}{a(b-1)} \tag{20}$$

is always a best (unbiased) estimate of σ^2 regardless of whether H_0 is true. On the other hand, we see from equations (*16*) and (*18*) that only if H_0 is true (i.e., $\alpha_j = 0$) will we have

$$E\left(\frac{V_B}{a-1}\right) = \sigma^2 \quad \text{and} \quad E\left(\frac{V}{ab-1}\right) = \sigma^2 \tag{21}$$

so that only in such case will

$$\hat{S}_B^2 = \frac{V_B}{a-1} \qquad \text{and} \qquad \hat{S}^2 = \frac{V}{ab-1} \tag{22}$$

provide unbiased estimates of σ^2. If H_0 is not true, however, then from equation (*16*) we have

$$E(\hat{S}_B^2) = \sigma^2 + \frac{b}{a-1}\sum_j \alpha_j^2 \tag{23}$$

DISTRIBUTIONS OF THE VARIATIONS

Using the additive property of chi-square, we can prove the following fundamental theorems concerning the distributions of the variations V_W, V_B, and V:

> **Theorem 1:** V_W/σ^2 is chi-square distributed with $a(b-1)$ degrees of freedom.
>
> **Theorem 2:** Under the null hypothesis H_0, V_B/σ^2 and V/σ^2 are chi-square distributed with $a-1$ and $ab-1$ degrees of freedom, respectively.

It is important to emphasize that Theorem 1 is valid whether or not H_0 is assumed, whereas Theorem 2 is valid only if H_0 is assumed.

THE *F* TEST FOR THE NULL HYPOTHESIS OF EQUAL MEANS

If the null hypothesis H_0 is not true (i.e., if the treatment means are not equal), we see from equation (*23*) that we can expect $\hat{S}_B^2$ to be greater than σ^2, with the effect becoming more pronounced as the discrepancy between the means increases. On the other hand, from equations (*19*) and (*20*) we can expect $\hat{S}_W^2$ to be equal to σ^2 regardless of whether the means are equal. It follows that a good statistic for testing hypothesis H_0 is provided by $\hat{S}_B^2/\hat{S}_W^2$. If this statistic is significantly large, we can conclude that there is a significant difference between the treatment means and can thus reject H_0; otherwise, we can either accept H_0 or reserve judgment, pending further analysis.

In order to use the $\hat{S}_B^2/\hat{S}_W^2$ statistic, we must know its sampling distribution. This is provided by Theorem 3.

> **Theorem 3:** The statistic $F = \hat{S}_B^2/\hat{S}_W^2$ has the F distribution with $a-1$ and $a(b-1)$ degrees of freedom.

Theorem 3 enables us to test the null hypothesis at some specified significance level by using a one-tailed test of the F distribution (discussed in Chapter 11).

ANALYSIS-OF-VARIANCE TABLES

The calculations required for the above test are summarized in Table 16.2, which is called an *analysis-of-variance table*. In practice, we would compute V and V_B by using either the long method [equations (*3*) and (*8*)] or the short method [equations (*10*) and (*11*)] and then by computing $V_W = V - V_B$. It should be noted that the degrees of freedom for the total variation (i.e., $ab-1$) are equal to the sum of the degrees of freedom for the between-treatments and within-treatments variations.

Table 16.2

Variation	Degrees of Freedom	Mean Square	F
Between treatments, $V_B = b\sum_j (\bar{X}_{j.} - \bar{X})^2$	$a - 1$	$\hat{S}_B^2 = \frac{V_B}{a-1}$	$\frac{\hat{S}_B^2}{\hat{S}_W^2}$
Within treatments, $V_W = V - V_B$	$a(b-1)$	$\hat{S}_W^2 = \frac{V_W}{a(b-1)}$	with $a - 1$ and $a(b-1)$ degrees of freedom
Total, $V = V_B + V_W = \sum_{j,k} (X_{jk} - \bar{X})^2$	$ab - 1$		

MODIFICATIONS FOR UNEQUAL NUMBERS OF OBSERVATIONS

In case the treatments $1, \ldots, a$ have different numbers of observations—equal to $N_1, \ldots, N_a$, respectively—the above results are easily modified, Thus we obtain

$$V = \sum_{j,k}(X_{jk} - \bar{X})^2 = \sum_{j,k} X_{jk}^2 - \frac{T^2}{N} \tag{24}$$

$$V_B = \sum_{j,k}(\bar{X}_{j.} - \bar{X})^2 = \sum_j N_j(\bar{X}_{j.} - \bar{X})^2 = \sum_j \frac{T_{j.}^2}{N_j} - \frac{T^2}{N} \tag{25}$$

$$V_W = V - V_B \tag{26}$$

where $\sum_{j,k}$ denotes the summation over k from 1 to N_j and then the summation over j from 1 to a. Table 16.3 is the analysis-of-variance table for this case.

Table 16.3

Variation	Degrees of Freedom	Mean Square	F
Between treatments $V_B = \sum_j N_j(\bar{X}_{j.} - \bar{X})^2$	$a - 1$	$\hat{S}_B^2 = \frac{V_B}{a-1}$	$\frac{\hat{S}_B^2}{\hat{S}_W^2}$
Within treatments, $V_W = V - V_B$	$N - a$	$\hat{S}_W^2 = \frac{V_W}{N-a}$	with $a - 1$ and $N - a$ degrees of freedom
Total, $V = V_B + V_W = \sum_{j,k} (X_{jk} - \bar{X})^2$	$N - 1$		

TWO-WAY CLASSIFICATION, OR TWO-FACTOR EXPERIMENTS

The ideas of analysis of variance for one-way classification, or one-factor experiments, can be generalized. Example 2 illustrates the procedure for *two-way classification*, or *two-factor experiments*.

EXAMPLE 2. Suppose that an agricultural experiment consists of examining the yields per acre of 4 different varieties of wheat, where each variety is grown on 5 different plots of land. Thus a total of $(4)(5) = 20$ plots are needed. It is convenient in such case to combine the plots into *blocks*, say 4 plots to a block, with a different variety of wheat grown on each plot within a block. Thus 5 blocks would be required here.

In this case there are two classifications, or factors, since there may be differences in yield per acre due to (1) the particular type of wheat grown or (2) the particular block used (which may involve different soil fertility, etc.).

By analogy with the agricultural experiment of Example 2, we often refer to the two factors in an experiment as *treatments* and *blocks*, but of course we could simply refer to them as factor 1 and factor 2.

NOTATION FOR TWO-FACTOR EXPERIMENTS

Assuming that we have a treatments and b blocks, we construct Table 16.4, where it is supposed that there is one experimental value (such as yield per acre) corresponding to each treatment and block. For treatment j and block k, we denote this value by X_{jk}. The mean of the entries in the jth row is denoted by $\bar{X}_{j.}$, where $j = 1, \ldots, a$, while the mean of the entries in the kth column is denoted by $\bar{X}_{.k}$, where $k = 1, \ldots, b$. The overall, or grand, mean is denoted by $\bar{X}$. In symbols,

$$\bar{X}_{j.} = \frac{1}{b}\sum_{k=1}^{b} X_{jk} \qquad \bar{X}_{.k} = \frac{1}{a}\sum_{j=1}^{a} X_{jk} \qquad \bar{X} = \frac{1}{ab}\sum_{j,k} X_{jk} \tag{27}$$

Table 16.4

	Block				
	1	2	$\cdots$	b	
Treatment 1	X_{11}	X_{12}	$\cdots$	X_{1b}	$\bar{X}_{1.}$
Treatment 2	X_{21}	X_{22}	$\cdots$	X_{2b}	$\bar{X}_{2.}$
$\vdots$	$\vdots$	$\vdots$	$\vdots$	$\vdots$	$\vdots$
Treatment a	X_{a1}	X_{a2}	$\cdots$	X_{ab}	$\bar{X}_{a.}$
	$\bar{X}_{.1}$	$\bar{X}_{.2}$		$\bar{X}_{.b}$	

VARIATIONS FOR TWO-FACTOR EXPERIMENTS

As in the case of one-factor experiments, we can define variations for two-factor experiments. We first define the *total variation*, as in equation (*3*), to be

$$V = \sum_{j,k}(X_{jk} - \bar{X})^2 \tag{28}$$

By writing the identity

$$X_{jk} - \bar{X} = (X_{jk} - \bar{X}_{j.} - \bar{X}_{.k} + \bar{X}) + (\bar{X}_{j.} - \bar{X}) + (\bar{X}_{.k} - \bar{X}) \tag{29}$$

and then squaring and summing over j and k, we can show that

$$V = V_E + V_R + V_C \tag{30}$$

where $V_E = \text{variation due to error or chance} = \sum_{j,k}(X_{jk} - \bar{X}_{j.} - \bar{X}_{.k} + \bar{X})^2$

$$V_R = \text{variation between rows (treatments)} = b\sum_{j=1}^{a}(\bar{X}_{j.} - \bar{X})^2$$

$$V_C = \text{variation between columns (blocks)} = a\sum_{k=1}^{b}(\bar{X}_{.k} - \bar{X})^2$$

The variation due to error or chance is also known as the *residual variation* or *random variation*.

The following, analogous to equations (*10*), (*11*), and (*12*), are shortcut formulas for computation:

$$V = \sum_{jk} X_{jk}^2 - \frac{T^2}{ab} \tag{31}$$

$$V_R = \frac{1}{b}\sum_{j=1}^{a} T_{j.}^2 - \frac{T^2}{ab} \tag{32}$$

$$V_C = \frac{1}{a}\sum_{k=1}^{b} T_{.k}^2 - \frac{T^2}{ab} \tag{33}$$

$$V_E = V - V_R - V_C \tag{34}$$

where $T_{j.}$ is the total of entries in the jth row, $T_{.k}$ is the total of entries in the kth column, and T is the total of all entries.

ANALYSIS OF VARIANCE FOR TWO-FACTOR EXPERIMENTS

The generalization of the mathematical model for one-factor experiments given by equation (*15*) leads us to assume for two-factor experiments that

$$X_{jk} = \mu + \alpha_j + \beta_k + \varepsilon_{jk} \tag{35}$$

where $\sum \alpha_j = 0$ and $\sum \beta_k = 0$. Here μ is the population grand mean, α_j is that part of X_{jk} due to the different treatments (sometimes called the *treatment effects*), β_k is that part of X_{jk} due to the different blocks (sometimes called the *block effects*), and ε_{jk} is that part of X_{jk} due to chance or error. As before, we assume that the ε_{jk} are normally distributed with mean 0 and variance σ^2, so that the X_{jk} are also normally distributed with mean μ, and variance σ^2.

Corresponding to results (*16*), (*17*), and (*18*), we can prove that the expectations of the variations are given by

$$E(V_E) = (a-1)(b-1)\sigma^2 \tag{36}$$

$$E(V_R) = (a-1)\sigma^2 + b\sum_j \alpha_j^2 \tag{37}$$

$$E(V_C) = (b-1)\sigma^2 + a\sum_k \beta_k^2 \tag{38}$$

$$E(V) = (ab-1)\sigma^2 + b\sum_j \alpha_j^2 + a\sum_k \beta_k^2 \tag{39}$$

There are two null hypotheses that we would want to test:

$H_0^{(1)}$: All treatment (row) means are equal; that is, $\alpha_j = 0$, and $j = 1, \ldots, a$.

$H_0^{(2)}$: All block (column) means are equal; that is, $\beta_k = 0$, and $k = 1, \ldots, b$.

We see from equation (*38*) that, without regard to $H_0^{(1)}$ or $H_0^{(2)}$, a best (unbiased) estimate of σ^2 is provided by

$$\hat{S}_E^2 = \frac{V_E}{(a-1)(b-1)} \qquad \text{that is,} \qquad E(\hat{S}_E^2) = \sigma^2 \tag{40}$$

Also, if hypotheses $H_0^{(1)}$ and $H_0^{(2)}$ are true, then

$$\hat{S}_R^2 = \frac{V_R}{a-1} \qquad \hat{S}_C^2 = \frac{V_C}{b-1} \qquad \hat{S}^2 = \frac{V}{ab-1} \tag{41}$$

will be unbiased estimates of σ^2. If $H_0^{(1)}$ and $H_0^{(2)}$ are not true, however, then from equations (*36*) and (*37*), respectively, we have

$$E(\hat{S}_R^2) = \sigma^2 + \frac{b}{a-1}\sum_j \alpha_j^2 \tag{42}$$

$$E(\hat{S}_C^2) = \sigma^2 + \frac{a}{b-1}\sum_k \beta_k^2 \tag{43}$$

The following theorems are similar to Theorems 1 and 2:

Theorem 4: V_E/σ^2 is chi-square distributed with $(a-1)(b-1)$ degrees of freedom, without regard to $H_0^{(1)}$ or $H_0^{(2)}$.

Theorem 5: Under hypothesis $H_0^{(1)}$, V_R/σ^2 is chi-square distributed with $a-1$ degrees of freedom. Under hypothesis $H_0^{(2)}$, V_C/σ^2 is chi-square distributed with $b-1$ degrees of freedom. Under both hypotheses, $H_0^{(1)}$ and $H_0^{(2)}$, V/σ^2 is chi-square distributed with $ab-1$ degrees of freedom.

To test hypothesis $H_0^{(1)}$, it is natural to consider the statistic $\hat{S}_R^2/\hat{S}_E^2$ since we can see from equation (*42*) that $\hat{S}_R^2$ is expected to differ significantly from σ^2 if the row (treatment) means are significantly different. Similarly, to test hypothesis $H_0^{(2)}$, we consider the statistic $\hat{S}_C^2/\hat{S}_E^2$. The distributions of $\hat{S}_R^2/\hat{S}_E^2$ and $\hat{S}_C^2/\hat{S}_E^2$ are given in Theorem 6, which is analogous to Theorem 3.

Theorem 6: Under hypothesis $H_0^{(1)}$, the statistic $\hat{S}_R^2/\hat{S}_E^2$ has the F distribution with $a-1$ and $(a-1)(b-1)$ degrees of freedom. Under hypothesis $H_0^{(2)}$, the statistic $\hat{S}_C^2/\hat{S}_E^2$ has the F distribution with $b-1$ and $(a-1)(b-1)$ degrees of freedom.

Theorem 6 enables us to accept or reject $H_0^{(1)}$ or $H_0^{(2)}$ at specified significance levels. For convenience, as in the one-factor case, an analysis-of-variance table can be constructed as shown in Table 16.5.

TWO-FACTOR EXPERIMENTS WITH REPLICATION

In Table 16.4 there is only one entry corresponding to a given treatment and a given block. More information regarding the factors can often be obtained by repeating the experiment, a process called *replication*. In such case there will be more than one entry corresponding to a given treatment and a given block. We shall suppose that there are c entries for every position; appropriate changes can be made when the replication numbers are not all equal.

Because of replication, an appropriate model must be used to replace that given by equation (*35*). We use

$$X_{jkl} = \mu + \alpha_j + \beta_k + \gamma_{jk} + \varepsilon_{jkl} \tag{44}$$

where the subscripts j, k, and l of X_{jkl} correspond to the jth row (or treatment), the kth column (or block), and the lth repetition (or replication), respectively. In equation (*44*) the μ, α_j, and β_k are defined

Table 16.5

Variation	Degrees of Freedom	Mean Square	F
Between treatments, $V_R = b\sum_j (\bar{X}_{j.} - \bar{X})^2$	$a-1$	$\hat{S}_R^2 = \frac{V_R}{a-1}$	$\hat{S}_R^2/\hat{S}_E^2$ with $a-1$ and $(a-1)(b-1)$ degrees of freedom
Between blocks, $V_C = a\sum_k (\bar{X}_{.k} - \bar{X})^2$	$b-1$	$\hat{S}_C^2 = \frac{V_C}{b-1}$	$\hat{S}_C^2/\hat{S}_E^2$ with $b-1$ and $(a-1)(b-1)$ degrees of freedom
Residual or random, $V_E = V - V_R - V_C$	$(a-1)(b-1)$	$\hat{S}_E^2 = \frac{V_E}{(a-1)(b-1)}$	
Total, $V = V_R + V_C + V_E$ $= \sum_{j,k} (X_{jk} - \bar{X})^2$	$ab-1$		

as before; ε_{jkl} is a chance or error term, while the γ_{jk} denote the row-column (or treatment-block) *interaction effects*, often simply called *interactions*. We have the restrictions

$$\sum_j \alpha_j = 0 \qquad \sum_k \beta_k = 0 \qquad \sum_j \gamma_{jk} = 0 \qquad \sum_k \gamma_{jk} = 0 \tag{45}$$

and the X_{jkl} are assumed to be normally distributed with mean μ and variance σ^2.

As before, the total variation V of all the data can be broken up into variations due to rows V_R, columns V_C, interaction V_I, and random or residual error V_E:

$$V = V_R + V_C + V_I + V_E \tag{46}$$

where

$$V = \sum_{j,k,l} (X_{jkl} - \bar{X})^2 \tag{47}$$

$$V_R = bc\sum_{j=1}^{a} (\bar{X}_{j..} - \bar{X})^2 \tag{48}$$

$$V_C = ac\sum_{k=1}^{b} (\bar{X}_{.k.} - \bar{X})^2 \tag{49}$$

$$V_I = c\sum_{j,k} (\bar{X}_{jk.} - \bar{X}_{j..} - \bar{X}_{.k.} + \bar{X})^2 \tag{50}$$

$$V_E = \sum_{j,k,l} (X_{jkl} - \bar{X}_{jk.})^2 \tag{51}$$

In these results the dots in the subscripts have meanings analogous to those given before; thus, for example,

$$\bar{X}_{j..} = \frac{1}{bc}\sum_{k,l} X_{jkl} = \frac{1}{b}\sum_k \bar{X}_{jk.} \tag{52}$$

The expected values of the variations can be found as before. Using the appropriate number of degrees of freedom for each source of variation, we can set up the analysis-of-variance table as shown in

Table 16.6. The F ratios in the last column of Table 16.6 can be used to test the null hypotheses:

$H_0^{(1)}$: All treatment (row) means are equal; that is, $\alpha_j = 0$.

$H_0^{(2)}$: All block (column) means are equal; that is, $\beta_k = 0$.

$H_0^{(3)}$: There are no interactions between treatments and blocks; that is, $\gamma_{jk} = 0$.

Table 16.6

Variation	Degrees of Freedom	Mean Square	F
Between treatments, V_R	$a-1$	$\hat{S}_R^2 = \dfrac{V_R}{a-1}$	$\hat{S}_R^2/\hat{S}_E^2$ with $a-1$ and $ab(c-1)$ degrees of freedom
Between blocks, V_C	$b-1$	$\hat{S}_C^2 = \dfrac{V_C}{b-1}$	$\hat{S}_C^2/\hat{S}_E^2$ with $b-1$ and $ab(c-1)$ degrees of freedom
Interaction, V_I	$(a-1)(b-1)$	$\hat{S}_I^2 = \dfrac{V_I}{(a-1)(b-1)}$	$\hat{S}_I^2/\hat{S}_E^2$ with $(a-1)(b-1)$ and $ab(c-1)$ degrees of freedom
Residual or random, V_E	$ab(c-1)$	$\hat{S}_E^2 = \dfrac{V_E}{ab(c-1)}$	
Total, V	$abc-1$		

From a practical point of view we should first decide whether or not $H_0^{(3)}$ can be rejected at an appropriate level of significance by using the F ratio $\hat{S}_I^2/\hat{S}_E^2$ of Table 16.6. Two possible cases then arise:

1. **$H_0^{(3)}$ Cannot Be Rejected.** In this case we can conclude that the interactions are not too large. We can then test $H_0^{(1)}$ and $H_0^{(2)}$ by using the F ratios $\hat{S}_R^2/\hat{S}_E^2$ and $\hat{S}_C^2/\hat{S}_E^2$, respectively, as shown in Table 16.6. Some statisticians recommend pooling the variations in this case by taking the total of $V_I + V_E$ and dividing it by the total corresponding degrees of freedom $(a-1)(b-1) + ab(c-1)$ and using this value to replace the denominator $\hat{S}_E^2$ in the F test.
2. **$H_0^{(3)}$ Can Be Rejected.** In this case we can conclude that the interactions are significantly large. Differences in factors would then be of importance only if they were large compared with such interactions. For this reason, many statisticians recommend that $H_0^{(1)}$ and $H_0^{(2)}$ be tested by using the F ratios $\hat{S}_R^2/\hat{S}_I^2$ and $\hat{S}_C^2/\hat{S}_I^2$ rather than those given in Table 16.6. We, too, shall use this alternative procedure.

The analysis of variance with replication is most easily performed by first totaling the replication values that correspond to particular treatments (rows) and blocks (columns). This produces a two-factor table with single entries, which can be analyzed as in Table 16.5. This procedure is illustrated in Problem 16.16.

EXPERIMENTAL DESIGN

The techniques of analysis of variance discussed above are employed after the results of an experiment have been obtained. However, in order to gain as much information as possible, the design of an experiment must be planned carefully in advance; this is often referred to as the *design of the experiment*. The following are some important examples of experimental design:

1. **Complete Randomization.** Suppose that we have an agricultural experiment as in Example 1. To design such an experiment, we could divide the land into $4 \times 4 = 16$ plots (indicated in Fig. 16-1 by squares, although physically any shape can be used) and assign each treatment (indicated by A, B, C, and D) to four blocks chosen completely at random. The purpose of the randomization is to eliminate various sources of error, such as soil fertility.

D	A	C	C
B	D	B	A
D	C	B	D
A	B	C	A

Fig. 16-1 Complete randomization.

Blocks	Treatments			
I	C	B	A	D
II	A	B	D	C
III	B	C	D	A
IV	A	D	C	B

Fig. 16-2 Randomized blocks.

	Factor 1			
Factor 2	D	B	C	A
	B	D	A	C
	C	A	D	B
	A	C	B	D

Fig. 16-3 Latin square.

B_γ	A_β	D_δ	C_α
A_δ	B_α	C_γ	D_β
D_α	C_δ	B_β	A_γ
C_β	D_γ	A_α	B_δ

Fig. 16-4 Graeco-Latin square.

2. **Randomized Blocks.** When, as in Example 2, it is necessary to have a complete set of treatments for each block, the treatments A, B, C, and D are introduced in random order within each block: I, II, III, and IV (i.e., the rows in Fig. 16-2), and for this reason the blocks are referred to as *randomized blocks*. This type of design is used when it is desired to control *one source of error or variability*: namely, the difference in blocks.
3. **Latin Squares.** For some purposes it is necessary to control *two sources of error or variability* at the same time, such as the difference in rows and the difference in columns. In the experiment of Example 1, for instance, errors in different rows and columns could be due to changes in soil fertility in different parts of the land. In such case it is desirable that each treatment occur once in each row and once in each column, as in Fig. 16-3. The arrangement is called a *Latin square* from the fact that the Latin letters A, B, C, and D are used.
4. **Graeco-Latin Squares.** If it is necessary to control *three sources of error or variability*, a *Graeco-Latin square* is used, as shown in Fig. 16-4. Such a square is essentially two Latin squares superimposed on each other, with the Latin letters A, B, C, and D used for one square and the Greek letters α, β, γ, and δ used for the other square. The additional requirement that must be met is that each Latin letter must be used once and only once with each Greek letter; when this requirement is met, the square is said to be *orthogonal*.

Solved Problems

ONE-WAY CLASSIFICATION, OR ONE-FACTOR EXPERIMENTS

16.1 Prove that $V = V_W + V_B$; that is,

$$\sum_{j,k}(X_{jk} - \bar{X})^2 = \sum_{j,k}(X_{jk} - \bar{X}_{j.})^2 + \sum_{j,k}(\bar{X}_{j.} - \bar{X})^2$$

SOLUTION

We have
$$X_{jk} - \bar{X} = (X_{jk} - \bar{X}_{j.}) + (\bar{X}_{j.} - \bar{X})$$

Then, squaring and summing over j and k, we obtain

$$\sum_{j,k}(X_{jk} - \bar{X})^2 = \sum_{j,k}(X_{jk} - \bar{X}_{j.})^2 + \sum_{j,k}(\bar{X}_{j.} - \bar{X})^2 + 2\sum_{j,k}(X_{jk} - \bar{X}_{j.})(\bar{X}_{j.} - \bar{X})$$

To prove the required result, we must show that the last summation is zero. In order to do this, we proceed as follows:

$$\sum_{j,k}(X_{jk} - \bar{X}_{j.})(\bar{X}_{j.} - \bar{X}) = \sum_{j=1}^{a}(\bar{X}_{j.} - \bar{X})\left[\sum_{k=1}^{b}(X_{jk} - \bar{X}_{j.})\right]$$
$$= \sum_{j=1}^{a}(\bar{X}_{j.} - \bar{X})\left[\left(\sum_{k=1}^{b}X_{jk}\right) - b\bar{X}_{j.}\right] = 0$$

since
$$\bar{X}_{j.} = \frac{1}{b}\sum_{k=1}^{b}X_{jk}$$

16.2 Verify that (*a*) $T = ab\bar{X}$, (*b*) $T_{j.} = b\bar{X}_{j.}$, and (*c*) $\sum_j T_{j.} = ab\bar{X}$, using the notation on page 362.

SOLUTION

(*a*)
$$T = \sum_{j,k}X_{jk} = ab\left(\frac{1}{ab}\sum_{j,k}X_{jk}\right) = ab\bar{X}$$

(*b*)
$$T_{j.} = \sum_{k}X_{jk} = b\left(\frac{1}{b}\sum_{k}X_{jk}\right) = b\bar{X}_{j.}$$

(*c*) Since $T_{j.} = \sum_k X_{jk}$, by part (*a*) we have

$$\sum_{j}T_{j.} = \sum_{j}\sum_{k}X_{jk} = T = ab\bar{X}$$

16.3 Verify the shortcut formulas (*10*), (*11*), and (*12*) of this chapter.

SOLUTION

We have
$$V = \sum_{j,k}(X_{jk} - \bar{X})^2 = \sum_{j,k}(X_{jk}^2 - 2\bar{X}X_{jk} + \bar{X}^2)$$
$$= \sum_{j,k}X_{jk}^2 - 2\bar{X}\sum_{j,k}X_{jk} + ab\bar{X}^2$$
$$= \sum_{j,k}X_{jk}^2 - 2\bar{X}(ab\bar{X}) + ab\bar{X}^2$$
$$= \sum_{j,k}X_{jk}^2 - ab\bar{X}^2$$
$$= \sum_{j,k}X_{jk}^2 - \frac{T^2}{ab}$$

using Problem 16.2(*a*) in the third and last lines above. Similarly,

$$\begin{aligned} V_B &= \sum_{j,k} (\bar{X}_{j.} - \bar{X})^2 = \sum_{j,k} (\bar{X}_{j.}^2 - 2\bar{X}\bar{X}_{j.} + \bar{X}^2) \\ &= \sum_{j,k} \bar{X}_{j.}^2 - 2\bar{X} \sum_{j,k} \bar{X}_{j.} + ab\bar{X}^2 \\ &= \sum_{j,k} \left(\frac{T_{j.}}{b}\right)^2 - 2\bar{X} \sum_{j,k} \frac{T_{j.}}{b} + ab\bar{X}^2 \\ &= \frac{1}{b^2} \sum_{j=1}^{a} \sum_{k=1}^{b} T_{j.}^2 - 2\bar{X}(ab\bar{X}) + ab\bar{X}^2 \\ &= \frac{1}{b} \sum_{j=1}^{a} T_{j.}^2 - ab\bar{X}^2 \\ &= \frac{1}{b} \sum_{j=1}^{a} T_{j.}^2 - \frac{T^2}{ab} \end{aligned}$$

using Problem 16.2(*b*) in the third line and Problem 16.2(*a*) in the last line. Finally, equation (*12*) follows from that fact that $V = V_W + V_B$, or $V_W = V - V_B$.

16.4 Table 16.7 shows the yields in bushels per acre of a certain variety of wheat grown in a particular type of soil treated with chemicals *A*, *B*, or *C*. Find (*a*) the mean yields for the different treatments, (*b*) the grand mean for all treatments, (*c*) the total variation, (*d*) the variation between treatments, and (*e*) the variation within treatments. Use the long method. (*f*) Give the EXCEL analysis for the data shown in Table 16.7.

Table 16.7

A	48	49	50	49
B	47	49	48	48
C	49	51	50	50

Table 16.8

3	4	5	4
2	4	3	3
4	6	5	5

SOLUTION

To simplify the arithmetic, we may subtract some suitable number, say 45, from all the data without affecting the values of the variations. We then obtain the data of Table 16.8.

(*a*) The treatment (row) means for Table 16.8 are given, respectively, by

$$\bar{X}_{1.} = \tfrac{1}{4}(3+4+5+4) = 4 \qquad \bar{X}_{2.} = \tfrac{1}{4}(2+4+3+3) = 3 \qquad \bar{X}_{3.} = \tfrac{1}{4}(4+6+5+5) = 5$$

Thus the mean yields, obtained by adding 45 to these, are 49, 48, and 50 bushels per acre for *A*, *B*, and *C*, respectively.

(*b*) The grand mean for all treatments is

$$\bar{X} = \tfrac{1}{12}(3+4+5+4+2+4+3+3+4+6+5+5) = 4$$

Thus the grand mean for the original set of data is $45 + 4 = 49$ bushels per acre.

(*c*) The total variation is

$$V = \sum_{j,k} (X_{jk} - X)^2 = (3-4)^2 + (4-4)^2 + (5-4)^2 + (4-4)^2 + (2-4)^2 + (4-4)^2$$
$$+ (3-4)^2 + (3-4)^2 + (4-4)^2 + (6-4)^2 + (5-4)^2 + (5-4)^2 = 14$$

(*d*) The variation between treatments is

$$V_B = b\sum_j (\bar{X}_{j.} - \bar{X})^2 = 4[(4-4)^2 + (3-4)^2 + (5-4)^2] = 8$$

(*e*) The variation within treatments is

$$V_W = V - V_B = 14 - 8 = 6$$

Another method

$$V_W = \sum_{j,k}(X_{jk} - X_{j.})^2 = (3-4)^2 + (4-4)^2 + (5-4)^2 + (4-4)^2 + (2-3)^2 + (4-3)^2$$
$$+ (3-3)^2 + (3-3)^2 + (4-5)^2 + (6-5)^2 + (5-5)^2 + (5-5)^2 = 6$$

Note: Table 16.9 is the analysis-of-variance table for Problems 16.4, 16.5, and 16.6.

Table 16.9

Variation	Degree of Freedom	Mean Square	F
Between treatments, $V_B = 8$	$a - 1 = 2$	$\hat{S}_B^2 = \frac{8}{2} = 4$	$\frac{\hat{S}_B^2}{\hat{S}_W^2} = \frac{4}{2/3} = 6$
Within treatments, $V_W = V - V_B$ $= 14 - 8 = 6$	$a(b-1) = (3)(3) = 9$	$\hat{S}_W^2 = \frac{6}{9} = \frac{2}{3}$	with 2 and 9 degrees of freedom
Total, $V = 14$	$ab - 1 = (3)(4) - 1$ $= 11$		

(*f*) The EXCEL pull-down **Tools → Data analysis → Anova single factor** gives the below analysis. The p-value tells us that the means for the three varieties are different at $\alpha = 0.05$.

A	B	C
48	47	49
49	49	51
50	48	50
49	48	50

Anova: Single Factor

SUMMARY

Groups	*Count*	*Sum*	*Average*	*Variance*
A	4	196	49	0.666667
B	4	192	48	0.666667
C	4	200	50	0.666667

ANOVA

Source of Variation	*SS*	*df*	*MS*	*F*	*P-value*
Between Groups	8	2	4	6	0.022085
Within Groups	6	9	0.666667		
Total	14	11			

Figure 16-5 shows a MINITAB dot plot for the yields of the three varieties of wheat. Figure 16-6 shows a MINITAB box plot for the yields of the three varieties of wheat. The EXCEL analysis and the MINITAB plots tell us that variety C significantly out yields variety B.

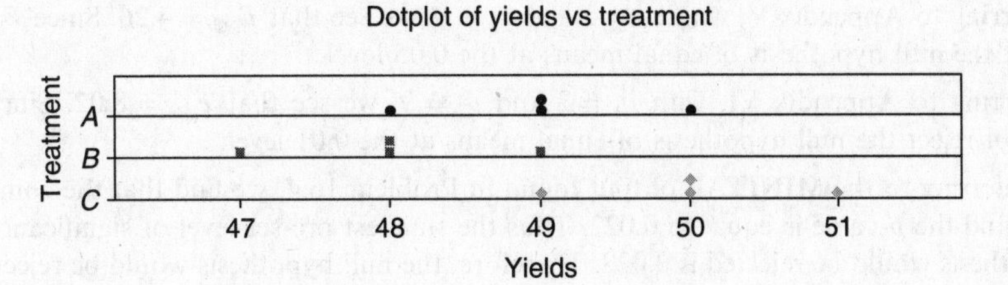

Fig. 16-5 A MINITAB dot plot for the yields of the three varieties of wheat.

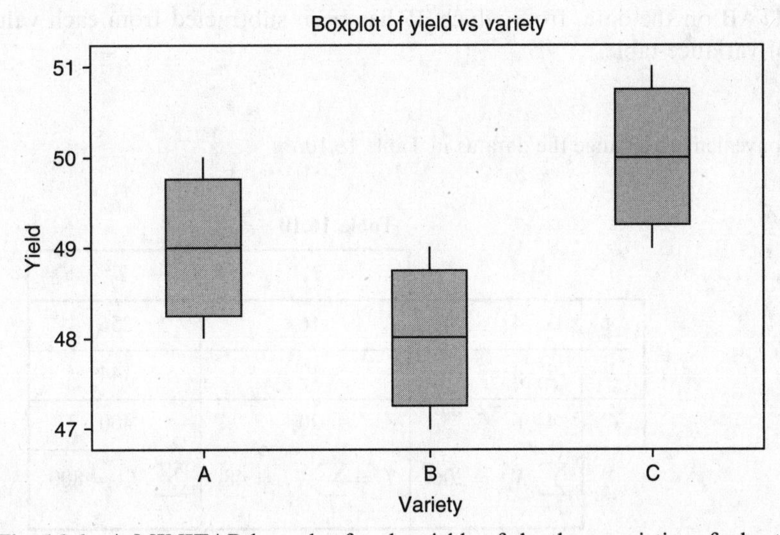

Fig. 16-6 A MINITAB box plot for the yields of the three varieties of wheat.

16.5 Referring to Problem 16.4, find an unbiased estimate of the population variance σ^2 from (*a*) the variation between treatments under the null hypothesis of equal treatment means and (*b*) the variation within treatments. (*c*) Refer to the MINITAB output given in the solution to Problem 16.4 and locate the variance estimates computed in parts (*a*) and (*b*).

SOLUTION

(*a*)
$$\hat{S}_B^2 = \frac{V_B}{a-1} = \frac{8}{3-1} = 4$$

(*b*)
$$\hat{S}_W^2 = \frac{V_W}{a(b-1)} = \frac{6}{3(4-1)} = \frac{2}{3}$$

(*c*) The variance estimate $\hat{S}_B^2$ is the same as the Factor mean square in the MINITAB output. That is Factor MS = 4.000 is the same as $\hat{S}_B^2$.

The variance estimate $\hat{S}_W^2$ is the same as the Error mean square in the MINITAB output. That is Error MS = 4.000 is the same as $\hat{S}_W^2$.

16.6 Referring to Problem 16.4, can we reject the null hypothesis of equal means at significance levels of (*a*) 0.05 and (*b*) 0.01? (*c*) Refer to the MINITAB output given in the solution to Problem 16.4 to test the null hypothesis of equal means.

SOLUTION

We have
$$F = \frac{\hat{S}_B^2}{\hat{S}_W^2} = \frac{4}{2/3} = 6$$
with $a - 1 = 3 - 1 = 2$ degrees of freedom and $a(b - 1) = 3(4 - 1) = 9$ degrees of freedom.

(*a*) Referring to Appendix V, with $\nu_1 = 2$ and $\nu_2 = 9$, we see that $F_{.95} = 4.26$. Since $F = 6 > F_{.95}$, we can reject the null hypothesis of equal means at the 0.05 level.

(*b*) Referring to Appendix VI, with $\nu_1 = 2$ and $\nu_2 = 9$, we see that $F_{.99} = 8.02$. Since $F = 6 < F_{.99}$, we cannot reject the null hypothesis of equal means at the 0.01 level.

(*c*) By referring to the MINITAB output found in Problem 16.4, we find that the computed value of F is 6.00 and the p-value is equal to 0.022. Thus the smallest pre-set level of significance at which the null hypothesis would be rejected is 0.022. Therefore, the null hypothesis would be rejected at 0.05, but not at 0.01.

16.7 Use the shortcut formulas (*10*), (*11*), and (*12*) to obtain the results of Problem 16.4. In addition, use MINITAB on the data, from which 45 has been subtracted from each value, to obtain the analysis-of-variance-table.

SOLUTION

It is convenient to arrange the data as in Table 16.10.

Table 16.10

		$T_{j.}$	$T_{j.}^2$
A	3 4 5 4	16	256
B	2 4 3 3	12	144
C	4 6 5 5	20	400
	$\sum_{j,k} X_{jk}^2 = 206$	$T = \sum_j T_{j.} = 48$	$\sum_j T_{j.}^2 = 800$

(*a*) Using formula (*10*), we have
$$\sum_{j,k} X_{jk}^2 = 9 + 16 + 25 + 16 + 4 + 16 + 9 + 9 + 16 + 36 + 25 + 25 = 206$$
and
$$T = 3 + 4 + 5 + 4 + 2 + 4 + 3 + 3 + 4 + 6 + 5 + 5 = 48$$
Thus
$$V = \sum_{j,k} X_{jk}^2 - \frac{T^2}{ab} = 206 - \frac{(48)^2}{(3)(4)} = 206 - 192 = 14$$

(*b*) The totals of the rows are
$$T_{1.} = 3 + 4 + 5 + 4 = 16 \qquad T_{2.} = 2 + 4 + 3 + 3 = 12 \qquad T_{3.} = 4 + 6 + 5 + 5 = 20$$
and
$$T = 16 + 12 + 20 = 48$$
Thus, using formula (*11*), we have
$$V_B = \frac{1}{b}\sum_j T_{j.}^2 - \frac{T^2}{ab} = \frac{1}{4}(16^2 + 12^2 + 20^2) - \frac{(48)^2}{(3)(4)} = 200 - 192 = 8$$

(*c*) Using formula (*12*), we have
$$V_W = V - V_B = 14 - 8 = 6$$

The results agree with those obtained in Problem 16.4, and from this point on the analysis proceeds as before.

The MINITAB pull-down **Stat → Anova → Oneway** gives the following output. Note the differences in terminology used. The variation within treatments is called `Within Groups` in the EXCEL output and `Error` in the MINITAB output. The variation between treatments is called `Between Groups` in the EXCEL output and `Factor` in the MINITAB output. You have to get used to the different terminology used by the various software packages.

One-way ANOVA: A, B, C

```
Source   DF      SS      MS      F      P
Factor    2   8.000   4.000   6.00  0.022
Error     9    6.00   0.667
Total    11  14.000

S=0.8165      R-Sq=57.14%       R-Sq(adj)=47.62%
```

16.8 A company wishes to purchase one of five different machines: A, B, C, D, or E. In an experiment designed to test whether there is a difference in the machines' performance, each of five experienced operators works on each of the machines for equal times. Table 16.11 shows the numbers of units produced per machine. Test the hypothesis that there is no difference between the machines at significance levels of (*a*) 0.05 and (*b*) 0.01. (*c*) Give the STATISTIX solution to the problem and test the hypothesis that there is no difference between the machines using the *p*-value approach. Use $\alpha = 0.05$.

Table 16.11

A	68	72	77	42	53
B	72	53	63	53	48
C	60	82	64	75	72
D	48	61	57	64	50
E	64	65	70	68	53

Table 16.12

						$T_{j\cdot}$	$T_{j\cdot}^2$
A	8	12	17	−18	−7	12	144
B	12	−7	3	−7	−12	−11	121
C	0	22	4	15	12	53	2809
D	−12	1	−3	4	−10	−20	400
E	4	5	10	8	−7	20	400
			$\sum X_{jk}^2 = 2658$			54	3874

SOLUTION

Subtract a suitable number, say 60, from all the data to obtain Table 16.12. Then

$$V = 2658 - \frac{(54)^2}{(5)(5)} = 2658 - 116.64 = 2541.36$$

and

$$V_B = \frac{3874}{5} - \frac{(54)^2}{(5)(4)} = 774.8 - 116.64 = 658.16$$

We now form Table 16.13. For 4 and 20 degrees of freedom, we have $F_{.95} = 2.87$. Thus we cannot reject the null hypothesis at the 0.05 level and therefore certainly cannot reject it at the 0.01 level.

The STATISTIX pull-down **Statistics → One, two, multi-sample tests → One-way Anova** gives the following output.

```
Statistix 8.
One-Way AOV for: A B C D E

Source      DF         SS          MS            F         P
Between      4     658.16     164.540         1.75    0.1792
Within      20    1883.20      94.160
Total       24    2541.36

Grand Mean 62.160       CV 15.61

Variable       Mean
A            62.400
B            57.800
C            70.600
D            56.000
E            64.000
```

The *p*-value is 0.1792. There are no significant differences in the population means.

Table 16.13

Variation	Degrees of Freedom	Mean square	F
Between treatments $V_B = 658.2$	$a - 1 = 4$	$\hat{S}_B^2 = \dfrac{658.2}{4} = 164.5$	$F = \dfrac{164.55}{94.16} = 1.75$
Within treatments $V_W = 1883.2$	$a(b-1) = (5)(4) = 20$	$\hat{S}_W^2 = \dfrac{1883.2}{20} = 94.16$	
Total $V = 2514.4$	$ab - 1 = 24$		

MODIFICATIONS FOR UNEQUAL NUMBERS OF OBSERVATIONS

16.9 Table 16.14 shows the lifetimes in hours of samples from three different types of television tubes manufactured by a company. Using the long method, determine whether there is a difference between the three types at significance levels of (*a*) 0.05 and (*b*) 0.01.

Table 16.14

Sample 1	407	411	409		
Sample 2	404	406	408	405	402
Sample 3	410	408	406	408	

SOLUTION

It is convenient to subtract a suitable number from the data, say 400, obtaining Table 16.15. This table shows the row totals, the sample (or group) means, and the grand mean. Thus we have

$$V = \sum_{j,k} (X_{jk} - \bar{X})^2 = (7-7)^2 + (11-7)^2 + \cdots + (8-7)^2 = 72$$

$$V_B = \sum_{j,k} (\bar{X}j. - \bar{X})^2 = \sum_j N_j(\bar{X}_{j.} - \bar{X})^2 = 3(9-7)^2 + 5(7-5)^2 + 4(8-7)^2 = 36$$

$$V_W = V - V_B = 72 - 36 = 36$$

We can also obtain V_W directly by observing that it is equal to

$$(7-9)^2 + (11-9)^2 + (9-9)^2 + (4-5)^2 + (6-5)^2 + (8-5)^2 + (5-5)^2$$
$$+ (2-5)^2 + (10-8)^2 + (8-8)^2 + (6-8)^2 + (8-8)^2$$

Table 16.15

						Total	Mean
Sample 1	7	11	9			27	9
Sample 2	4	6	8	5	2	25	5
Sample 3	10	8	6	8		32	8
	$\bar{X}$ = grand mean = $\frac{84}{12} = 7$						

The data can be summarized as in Table 16.16, the analysis-of-variance table. Now for 2 and 9 degrees of freedom, we find from Appendix V that $F_{.95} = 4.26$ and from Appendix VI that $F_{.99} = 8.02$. Thus we can reject the hypothesis of equal means (i.e., no difference between the three types of tubes) at the 0.05 level but not at the 0.01 level.

Table 16.16

Variation	Degree of Freedom	Mean Square	F
$V_B = 36$	$a - 1 = 2$	$\hat{S}_B^2 = \frac{36}{2} = 18$	$\frac{\hat{S}_B^2}{\hat{S}_W^2} = \frac{18}{4}$
$V_W = 36$	$N - a = 9$	$\hat{S}_W^2 = \frac{36}{9} = 4$	$= 4.5$

16.10 Work Problem 16.9 by using the shortcut formulas included in equations (*24*), (*25*), and (*26*). In addition, give the SAS solution to the problem.

SOLUTION

From Table 16.15 we have $N_1 = 3$, $N_2 = 5$, $N_3 = 4$, $N = 12$, $T_{1.} = 27$, $T_{2.} = 25$, $T_{3.} = 32$, and $T = 84$. Thus we have

$$V = \sum_{j,k} X_{Jk}^2 - \frac{T^2}{N} = 7^2 + 11^2 + \cdots + 6^2 + 8^2 - \frac{(84)^2}{12} = 72$$

$$V_B = \sum_j \frac{T_{j.}^2}{N_j} - \frac{T^2}{N} = \frac{(27)^2}{3} + \frac{(25)^2}{5} + \frac{(32)^2}{4} - \frac{(84)^2}{12} = 36$$

$$V_W = V - V_B = 36$$

Using these, the analysis of variance then proceeds as in Problem 16.9.

The SAS pull-down **Statistics** → **ANOVA** → **Oneway ANOVA** gives the following output.

```
The ANOVA Procedure
                         Class Level Information
                         Class      Levels    Values
                         Sample_       3      1 2 3

                    Number of Observations Read    12
                    Number of Observations Used    12

                           The ANOVA Procedure
Dependent Variable: lifetime

Source            DF    Sum of Squares   Mean Square    F value        Pr > F
Model              2       36.00000000   18.00000000       4.50        0.0442
Error              9       36.00000000    4.00000000
Corrected Total   11       72.00000000

                          R-Square        coeff Var     Root MSE  lifetime Mean
                          0.500000         0.491400     2.000000       407.0000

Source            DF          Anova SS   Mean Square    F Value        Pr > F
Sample_            2       36.00000000   18.00000000       4.50        0.0442
```

Note that SAS calls the variation between treatments `model` and variation within treatments it calls `error`. The test statistic is called *F* value and equals 4.50. The *p*-value is represented $Pr > F$ and this equals 0.0442. At $\alpha = 0.05$ the lifetimes would be declared unequal.

TWO-WAY CLASSIFICATION, OR TWO-FACTOR EXPERIMENTS

16.11 Table 16.17 shows the yields per acre of four different plant crops grown on lots treated with three different types of fertilizer. Using the long method, determine at the 0.01 significance level whether there is a difference in yield per acre (*a*) due to the fertilizers and (*b*) due to the crops. (*c*) Give the MINITAB solution to this two-factor experiment.

SOLUTION

Compute the row totals, the row means, the column totals, the column means, the grand total, and the grand mean, as shown in Table 16.18. From this table we obtain:

The variation of row means from the grand mean is

$$V_R = 4[(6.2 - 6.8)^2 + (8.3 - 6.8)^2 + (5.9 - 6.8)^2] = 13.68$$

The variation of column means from the grand mean is

$$V_C = 3[(6.4 - 6.8)^2 + (7.0 - 6.8)^2 + (7.5 - 6.8)^2 + (6.3 - 6.8)^2] = 2.82$$

Table 16.17

	Crop I	Crop II	Crop III	Crop IV
Fertilizer *A*	4.5	6.4	7.2	6.7
Fertilizer *B*	8.8	7.8	9.6	7.0
Fertilizer *C*	5.9	6.8	5.7	5.2

Table 16.18

	Crop I	Crop II	Crop III	Crop IV	Row Total	Row Mean
Fertilizer *A*	4.5	6.4	7.2	6.7	24.8	6.2
Fertilizer *B*	8.8	7.8	9.6	7.0	33.2	8.3
Fertilizer *C*	5.9	6.8	5.7	5.2	23.6	5.9
Column total	19.2	21.0	22.5	18.9	Grand total = 81.6	
Column mean	6.4	7.0	7.5	6.3	Grand mean = 6.8	

The total variation is

$$V = (4.5 - 6.8)^2 + (6.4 - 6.8)^2 + (7.2 - 6.8)^2 + (6.7 - 6.8)^2$$
$$+ (8.8 - 6.8)^2 + (7.8 - 6.8)^2 + (9.6 - 6.8)^2 + (7.0 - 6.8)^2$$
$$+ (5.9 - 6.8)^2 + (6.8 - 6.8)^2 + (5.7 - 6.8)^2 + (5.2 - 6.8)^2 = 23.08$$

The random variation is

$$V_E = V - V_R - V_C = 6.58$$

This leads to the analysis of variance in Table 16.19.

Table 16.19

Variation	Degree of Freedom	Mean Square	F
$V_R = 13.68$	2	$\hat{S}_R^2 = 6.84$	$\hat{S}_R^2/\hat{S}_E^2 = 6.24$ with 2 and 6 degrees of freedom
$V_C = 2.82$	3	$\hat{S}_C^2 = 0.94$	$\hat{S}_C^2/\hat{S}_E^2 = 0.86$ with 3 and 6 degrees of freedom
$V_E = 6.58$	6	$\hat{S}_E^2 = 1.097$	
$V = 23.08$	11		

At the 0.05 significance level with 2 and 6 degrees of freedom, $F_{.95} = 5.14$. Then, since $6.24 > 5.14$, we can reject the hypothesis that the row means are equal and conclude that at the 0.05 level there is a significant difference in yield due to the fertilizers.

Since the F value corresponding to the differences in column means is less than 1, we can conclude that there is no significant difference in yield due to the crops.

(*c*) The data structure for the MINITAB worksheet is given first, followed by the MINITAB analysis of the two-factor experiment.

```
Row   Crop   Fertilizer   Yield
  1     1        1         4.5
  2     1        2         8.8
  3     1        3         5.9
  4     2        1         6.4
  5     2        2         7.8
  6     2        3         6.8
  7     3        1         7.2
  8     3        2         9.6
  9     3        3         5.7
 10     4        1         6.7
 11     4        2         7.0
 12     4        3         5.2

MTB > Twoway 'Yield' 'Crop' 'Fertilizer';
SUBC > Means 'Crop' 'Fertilizer'.
```

Two-way Analysis of Variance

```
Analysis of Variance for Yield
Source       DF        SS        MS        F        P
Crop          3      2.82      0.94     0.86    0.512
Fertiliz      2     13.68      6.84     6.24    0.034
Error         6      6.58      1.10
Total        11     23.08

                      Individual 95% CI
Crop        Mean    --+---------+---------+---------+---------
1           6.40     (--------------*--------------)
2           7.00           (--------------*--------------)
3           7.50              (--------------*--------------)
4           6.30    (--------------*--------------)
                    --+---------+---------+---------+---------
                    5.00      6.00      7.00      8.00

                      Individual 95% CI
Fertiliz    Mean    --+---------+---------+---------+---------
1           6.20      (----------*----------)
2           8.30                     (----------*----------)
3           5.90    (----------*----------)
                    --+---------+---------+---------+---------
                     4.80      6.00      7.20      8.40
```

The data structure in the worksheet must correspond exactly to the data as given in Table 16.17. The first row, `1 1 4.5`, corresponds to Crop 1, Fertilizer 1, and Yield 4.5, the second row, `1 2 8.8`, corresponds to Crop 1, Fertilizer 2, and Yield 8.8, etc. A mistake that is often made in using statistical software is to set up the data structure in the worksheet incorrectly. Make certain that the data given in a table like Table 16.17 and the data structure in the worksheet correspond in a one-to-one manner. Note that the two-way analysis of variance table given in the MINITAB output contains the same information given in Table 16.19. The *p*-values given in the MINITAB output allow the researcher to test the hypothesis of interest without consulting tables of the *F* distribution to find critical values. The *p*-value for crops is 0.512. This is the minimum level of significance for which we could reject a difference in mean yield for crops. The mean yields for the four crops are not statistically significant at 0.05 or 0.01. The *p*-value for fertilizers is 0.034. This tells us that the mean yields for the three fertilizers are statistically different at 0.05 but not statistically different at 0.01.

The confidence intervals for the means of the four crops shown in the MINITAB output reinforce our conclusion of no difference in mean yields for the four different crops. The confidence intervals for the three fertilizers indicates that it is likely that Fertilizer *B* produces higher mean yields than either Fertilizer *A* or *C*.

16.12 Use the short computational formulas to obtain the results of Problem 16.11. In addition, give the SPSS solution to Problem 16.11.

SOLUTION

From Table 16.18 we have

$$\sum_{j,k} X_{jk}^2 = (4.5)^2 + (6.4)^2 + \cdots + (5.2)^2 = 577.96$$

$$T = 24.8 + 33.2 + 23.6 = 81.6$$

$$\sum T_{j.}^2 = (24.8)^2 + (33.2)^2 + (23.6)^2 = 2274.24$$

$$\sum T_{.k}^2 = (19.2)^2 + (21.0)^2 + (22.5)^2 + (18.9)^2 = 1673.10$$

Then

$$V = \sum_{j,k} X_{jk}^2 - \frac{T^2}{ab} = 577.96 - 554.88 = 23.08$$

$$V_R = \frac{1}{b}\sum T_{j.}^2 - \frac{T^2}{ab} = \frac{1}{4}(2274.24) - 554.88 = 13.68$$

$$V_C = \frac{1}{a}\sum T_{.k}^2 - \frac{T^2}{ab} = \frac{1}{3}(1673.10) - 554.88 = 2.82$$

$$V_E = V - V_R - V_C = 23.08 - 13.68 - 2.82 = 6.58$$

in agreement with Problem 16.11.

The SPSS pull-down **Analyze → General Linear Model → Univariate** gives the following output.

Tests of Between-Subjects Effects

Dependent Variable: Yield

Source	Type 1 Sum of Squares	df	Mean Square	F	Sig.
Corrected Model	16.500[a]	5	3.300	3.009	.106
Intercept	554.880	1	554.880	505.970	.000
Crop	2.820	3	.940	.857	.512
Fert	13.680	2	6.840	6.237	.034
Error	6.580	6	1.097		
Total	577.960	12			
Corrected Total	23.080	11			

[a] R Squared = .715 (Adjusted R Squared = .477)

Note that the test statistic is given by F and for crops the F value is 0.857 with the corresponding p-value of 0.512. The F value for fertilizer is 6.237 with the corresponding p-value of 0.034. These values correspond with the values in Table 16.19 as well as the MINITAB output in Problem 16.11.

TWO-FACTOR EXPERIMENTS WITH REPLICATION

16.13 A manufacturer wishes to determine the effectiveness of four types of machines (A, B, C, and D) in the production of bolts. To accomplish this, the numbers of defective bolts produced by each machine in the days of a given week are obtained for each of two shifts; the results are shown in Table 16.20. Perform an analysis of variance to determine at the 0.05 significance level whether this is a difference (*a*) between the machines and (*b*) between the shifts. (*c*) Use MINITAB to perform the analysis of variance and test for differences between machines and differences between shifts using the *p*-value approach.

Table 16.20

	First Shift					Second Shift				
Machine	Mon.	Tue.	Wed.	Thu.	Fri.	Mon.	Tue.	Wed.	Thu.	Fri.
A	6	4	5	5	4	5	7	4	6	8
B	10	8	7	7	9	7	9	12	8	8
C	7	5	6	5	9	9	7	5	4	6
D	8	4	6	5	5	5	7	9	7	10

SOLUTION

The data can be equivalently organized as in Table 16.21. In this table the two main factors are indicated: the machine and the shift. Note that two shifts have been indicated for each machine. The days of the week can be considered to be replicates (or repetitions) of performance for each machine for the two shifts. The total variation for all the data of Table 16.21 is

$$V = 6^2 + 4^2 + 5^2 + \cdots + 7^2 + 10^2 - \frac{(268)^2}{40} = 1946 - 1795.6 = 150.4$$

Table 16.21

Factor I: Machine	Factor II: Shift	Replicates Mon.	Tue.	Wed.	Thu.	Fri.	Total
A	1	6	4	5	5	4	24
	2	5	7	4	6	8	30
B	1	10	8	7	7	9	41
	2	7	9	12	8	8	44
C	1	7	5	6	5	9	32
	2	9	7	5	4	6	31
D	1	8	4	6	5	5	28
	2	5	7	9	7	10	38
	Total	57	51	54	47	59	268

In order to consider the two main factors (the machine and the shift), we limit our attention to the total of replication values corresponding to each combination of factors. These are arranged in Table 16.22, which is thus a two-factor table with single entries. The total variation for Table 16.22, which we shall call the *subtotal variation* V_S, is given by

$$V_S = \frac{(24)^2}{5} + \frac{(41)^2}{5} + \frac{(32)^2}{5} + \frac{(28)^2}{5} + \frac{(30)^2}{5} + \frac{(44)^2}{5} + \frac{(31)^2}{5} + \frac{(38)^2}{5} - \frac{(268)^2}{40}$$
$$= 1861.2 - 1795.6 = 65.6$$

The variation between rows is given by

$$V_R = \frac{(54)^2}{10} + \frac{(85)^2}{10} + \frac{(63)^2}{10} + \frac{(66)^2}{10} - \frac{(268)^2}{40} = 1846.6 - 1795.6 = 51.0$$

Table 16.22

Machine	First Shift	Second Shift	Total
A	24	30	54
B	41	44	85
C	32	31	63
D	28	38	66
Total	125	143	268

The variation between columns is given by

$$V_C = \frac{(125)^2}{20} + \frac{(143)^2}{20} - \frac{(268)^2}{40} = 1803.7 - 1795.6 = 8.1$$

If we now subtract from the subtotal variation V_S the sum of the variations between the rows and columns $(V_R + V_C)$, we obtain the variation due to the *interaction* between the rows and columns. This is given by

$$V_I = V_S - V_R - V_C = 65.6 - 51.0 - 8.1 = 6.5$$

Finally, the residual variation, which we can think of as the random or error variation V_E (provided that we believe that the various days of the week do not provide any important differences), is found by subtracting the subtotal variation (i.e., the sum of the row, column, and interaction variations) from the total variation V. This yields

$$V_E = V - (V_R + V_C + V_I) = V - V_S = 150.4 - 65.6 = 84.8$$

These variations are shown in Table 16.23, the analysis of variance. The table also gives the number of degrees of freedom corresponding to each type of variation. Thus, since there are four rows in Table 16.22,

Table 16.23

Variation	Degrees of Freedom	Mean Square	F
Rows (machines), $V_R = 51.0$	3	$\hat{S}_R^2 = 17.0$	$\frac{17.0}{2.65} = 6.42$
Columns (shifts), $V_C = 8.1$	1	$\hat{S}_C^2 = 8.1$	$\frac{8.1}{2.65} = 3.06$
Interaction, $V_I = 6.5$	3	$\hat{S}_I^2 = 2.167$	$\frac{2.167}{2.65} = 0.817$
Subtotal, $V_S = 65.6$	7		
Random or residual, $V_E = 84.8$	32	$\hat{S}_E^2 = 2.65$	
Total, $V = 150.4$	39		

the variation due to rows has $4 - 1 = 3$ degrees of freedom, while the variation due to the two columns has $2 - 1 = 1$ degree of freedom. To find the degrees of freedom due to interaction, we note that there are eight entries in Table 16.22; thus the total degrees of freedom are $8 - 1 = 7$. Since 3 of these 7 degrees of freedom are due to rows and 1 is due to columns, the remainder $[7 - (3 + 1) = 3]$ are due to interaction. Since there are 40 entries in the original Table 16.21, the total degrees of freedom are $40 - 1 = 39$. Thus the degrees of freedom due to random or residual variation are $39 - 7 = 32$.

First, we must determine whether there is any significant interaction. The interpolated critical value for the F distribution with 3 and 32 degrees of freedom is 2.90. The computed F value for interaction is 0.817 and is not significant. There is a significant difference between machines, since the computed F value for machines is 6.42 and the critical value is 2.90. The critical value for shifts is 4.15. The computed value of F for shifts is 3.06. There is no difference in defects due to shifts.

The data structure for MINITAB is shown below. Compare the data structure with the data given in Table 16.21 to see how the two sets of data compare.

Row	Machine	Shift	Defects
1	1	1	6
2	1	1	4
3	1	1	5
4	1	1	5
5	1	1	4
6	1	2	5
7	1	2	7
8	1	2	4
9	1	2	6
10	1	2	8
11	2	1	10
12	2	1	8
13	2	1	7
14	2	1	7
15	2	1	9
16	2	2	7
17	2	2	9
18	2	2	12
19	2	2	8
20	2	2	8
21	3	1	7
22	3	1	5
23	3	1	6
24	3	1	5
25	3	1	9
26	3	2	9
27	3	2	7
28	3	2	5
29	3	2	4
30	3	2	6
31	4	1	8
32	4	1	4
33	4	1	6
34	4	1	5
35	4	1	5
36	4	2	5
37	4	2	7
38	4	2	9
39	4	2	7
40	4	2	10

The command `MTB > Twoway 'Defects' 'Machine' 'Shifts'` is used to produce the two-way analysis of variance. The p-value for interaction is 0.494. This is the minimum level of significance for which the null hypothesis would be rejected. Clearly, there is not a significant amount of interaction between shifts and machines. The p-value for shifts is 0.090. Since this exceeds 0.050, the mean number of defects for the two shifts is not significantly different. The *p*-value for machines is 0.002. The mean numbers of defects for the four machines are significantly different at the 0.050 level of significance.

```
MTB > Twoway 'Defects' 'Machine' 'Shift'
```

Two-way Analysis of Variance

Analysis of Variance for Defects

Source	DF	SS	MS	F	P
Machine	3	51.00	17.00	6.42	0.002
Shift	1	8.10	8.10	3.06	0.090
Interaction	3	6.50	2.17	0.82	0.494
Error	32	84.80	2.65		
Total	39	150.40			

Figure 16-7 gives the interaction plot for shifts and machines. The plot indicates possible interaction between shifts and machines. However, the *p*-value for interaction in the analysis of variance table tells us that there is not a significant amount of interaction. When interaction is not present, the broken line graphs for shift 1 and shift 2 are parallel. The main effects plot shown in Fig. 16-8 indicates that machine 1 produced the fewest defects on the average in this experiment and that machine 2 produced the most. There were more defects produced on shift 2 than shift 1. However, the analysis of variance tells us that this difference is not statistically significant.

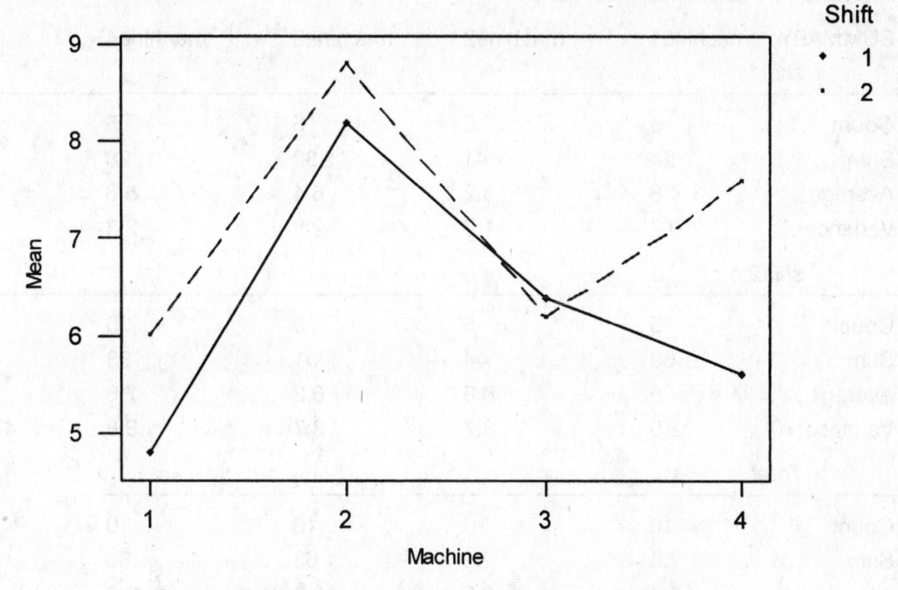

Fig. 16-7 Interaction plot—data means for defects.

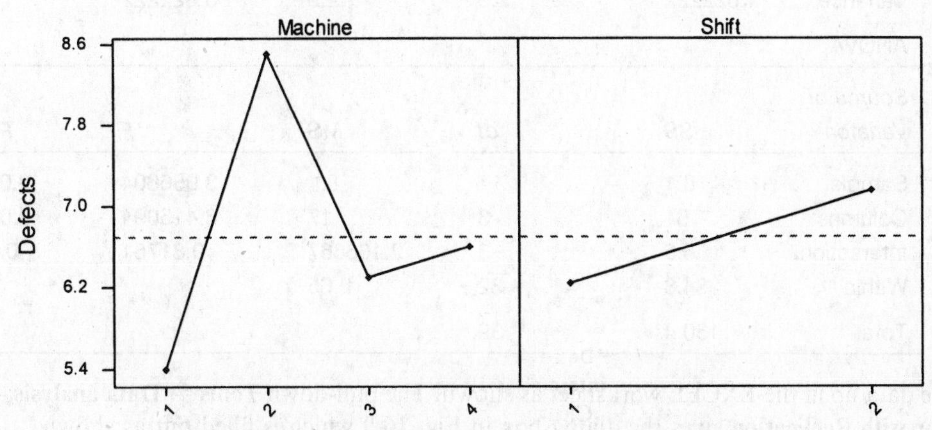

Fig. 16-8 Main effects plot—data means for defects.

16.14 Use EXCEL to do Problem 16.13.

SOLUTION

A	B	C	D	E
	machine1	machine2	machine3	machine4
shift1	6	10	7	8
shift1	4	8	5	4
shift1	5	7	6	6
shift1	5	7	5	5
shift1	4	9	9	5
shift2	5	7	9	5
shift2	7	9	7	7
shift2	4	12	5	9
shift2	6	8	4	7
shift2	8	8	6	10

Anova: Two-Factor With Replication

SUMMARY	machine1	machine2	machine3	machine4	Total
shift1					
Count	5	5	5	5	20
Sum	24	41	32	28	125
Average	4.8	8.2	6.4	5.6	6.25
Variance	0.7	1.7	2.8	2.3	3.25
shift2					
Count	5	5	5	5	20
Sum	30	44	31	38	143
Average	6	8.8	6.2	7.6	7.15
Variance	2.5	3.7	3.7	3.8	4.239474
Total					
Count	10	10	10	10	
Sum	54	85	63	66	
Average	5.4	8.5	6.3	6.6	
Variance	1.822222	2.5	2.9	3.822222	

ANOVA

Source of Variaton	*SS*	*df*	*MS*	*F*	*P-value*
Sample	8.1	1	8.1	3.056604	0.089999
Columns	51	3	17	6.415094	0.001584
Interaction	6.5	3	2.166667	0.81761	0.49371
Within	84.8	32	2.65		
Total	150.4	39			

Set the data up in the EXCEL worksheet as shown. The pull-down **Tools → Data analysis → Anova: Two-Factor with Replication** gives the dialog box in Fig. 16-9 which is filled out as shown.

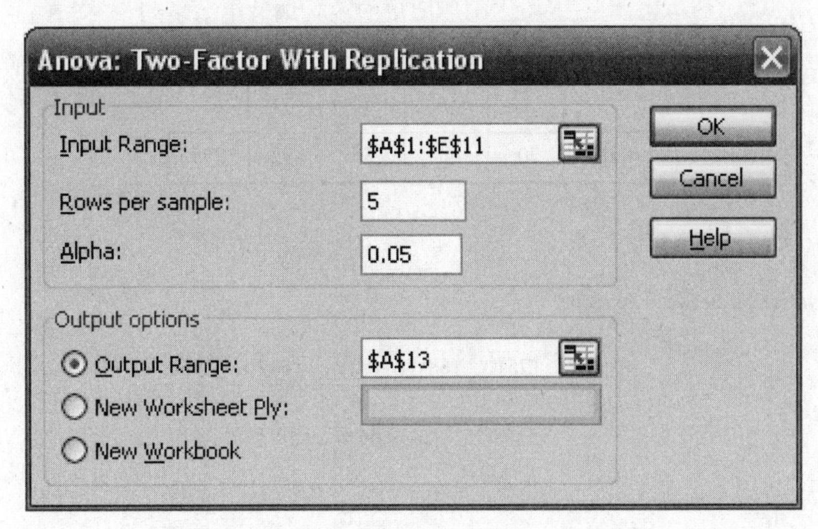

Fig. 16-9 EXCEL dialog box for Problem 16.14.

In the ANOVA, identify sample with shift and machine with column. Compare this output with the MINITAB output of Problem 16.13.

LATIN SQUARES

16.15 A farmer wishes to test the effects of four different fertilizers (*A*, *B*, *C*, and *D*) on the yield of wheat. In order to eliminate sources of error due to variability in soil fertility, he uses the fertilizers in a Latin-square arrangement, as shown in Table 16.24, where the numbers indicate yields in bushels per unit area. Perform an analysis of variance to determine whether there is a difference between the fertilizers at significance levels of (*a*) 0.05 and (*b*) 0.01. (*c*) Give the MINITAB solution to this Latin-square design. (*d*) Give the STATISTIX solution to this Latin-square design.

SOLUTION

We first obtain totals for the rows and columns, as shown in Table 16.25. We also obtain total yields for each of the fertilizers, as shown in Table 16.26. The total variation and the variations for the rows, columns, and treatments are then obtained as usual. We find:

The total variation is

$$V = (18)^2 + (21)^2 + (25)^5 + \cdots + (10)^2 + (17) - \frac{(295)^2}{16}$$

$$= 5769 - 5439.06 = 329.94$$

Table 16.24

A 18	*C* 21	*D* 25	*B* 11
D 22	*B* 12	*A* 15	*C* 19
B 15	*A* 20	*C* 23	*D* 24
C 22	*D* 21	*B* 10	*A* 17

Table 16.25

					Total
	A 18	*C* 21	*D* 25	*B* 11	75
	D 22	*B* 12	*A* 15	*C* 19	68
	B 15	*A* 20	*C* 23	*D* 24	82
	C 22	*D* 21	*B* 10	*A* 17	70
Total	77	74	73	71	295

Table 16.26

	A	B	C	D	
Total	70	48	85	92	295

The variation between rows is

$$V_R = \frac{(75)^2}{4} + \frac{(68)^2}{4} + \frac{(82)^2}{4} + \frac{(70)^2}{4} - \frac{(295)^2}{16}$$
$$= 5468.25 - 5439.06 = 29.19$$

The variation between columns is

$$V_C = \frac{(77)^2}{4} + \frac{(74)^2}{4} + \frac{(73)^2}{4} + \frac{(71)^2}{4} - \frac{(295)^2}{16}$$
$$= 5443.75 - 5439.06 = 4.69$$

The variation between treatments is

$$V_B = \frac{(70)^2}{4} + \frac{(48)^2}{4} + \frac{(85)^2}{4} + \frac{(92)^2}{4} - \frac{(295)^2}{16} = 5723.25 - 5439.06 = 284.19$$

Table 16.27 shows the analysis of variance.

Table 16.27

Variation	Degrees of Freedom	Mean Square	F
Rows, 29.19	3	9.73	4.92
Columns, 4.69	3	1.563	0.79
Treatments, 284.19	3	94.73	47.9
Residuals, 11.87	6	1.978	
Total, 329.94	15		

(*a*) Since $F_{.95,3,6} = 4.76$, we can reject at the 0.05 level the hypothesis that there are equal row means. It follows that at the 0.05 level there is a difference in the fertility of the soil from one row to another.

Since the F value for columns is less than 1, we conclude that there is no difference in soil fertility in the columns.

Since the F value for treatments is $47.9 > 4.76$, we can conclude that there is a difference between fertilizers.

(*b*) Since $F_{.99,3,6} = 9.78$, we can accept the hypothesis that there is no difference in soil fertility in the rows (or the columns) at the 0.01 level. However, we must still conclude that there is a difference between fertilizers at the 0.01 level.

(*c*) The structure of the data file for the MINITAB worksheet is given first.

Row	Rows	Columns	Treatment	Yield
1	1	1	1	18
2	1	2	3	21
3	1	3	4	25
4	1	4	2	11
5	2	1	4	22
6	2	2	2	12
7	2	3	1	15
8	2	4	3	19
9	3	1	2	15
10	3	2	1	20
11	3	3	3	23
12	3	4	4	24
13	4	1	3	22
14	4	2	4	21
15	4	3	2	10
16	4	4	1	17

Note that rows and columns in the farm layout are numbered 1 through 4. Fertilizers *A* through *D* in Table 16.24 are coded as 1 through 4, respectively, in the worksheet. The MINITAB pull-down **Stat → ANOVA → General Linear Model** gives the following output.

General Linear Model: Yield versus Rows, Columns, Treatment

Factor	Type	Levels	Values
Rows	fixed	4	1, 2, 3, 4
Columns	fixed	4	1, 2, 3, 4
Treatment	fixed	4	1, 2, 3, 4

Analysis of Variance for Yield, using Adjusted SS for Tests

Source	DF	Seq SS	Adj SS	Adj MS	F	P
Rows	3	29.188	29.188	9.729	4.92	0.047
Columns	3	4.688	4.687	1.562	0.79	0.542
Treatment	3	284.188	284.188	94.729	47.86	0.000
Error	6	11.875	11.875	1.979		
Total	15	329.938				

S=1.40683 R-Sq=96.40% R-sq(adj)-91.00%

The MINITAB output is the same as that computed by hand above. It indicates a difference in fertility from row to row at the 0.05 level but not at the 0.01 level. There is no difference in fertility from column to column. There is a difference in the four fertilizers at the 0.01 level.

(*d*) The STATISTIX pull-down **Statistics → Linear Models → Analysis of Variance → Latin Square Design** gives the following output.

Statistix 8.0

Latin Square AOV Table for Yield

Source	DF	SS	MS	F	P
Rows	3	29.188	9.7292		
Columns	3	4.688	1.5625		
Treatment	3	284.188	94.7292	47.86	0.0001
Error	6	11.875	1.9792		
Total	15	329.938			

GRAECO-LATIN SQUARES

16.16 It is of interest to determine whether there is any significant difference in mileage per gallon between gasolines *A*, *B*, *C*, and *D*. Design an experiment that uses four different drivers, four different cars, and four different roads.

SOLUTION

Since the same number of gasolines, drivers, cars, and roads is involved (four), we can use a Graeco-Latin square. Suppose that the different cars are represented by the rows and that the different drivers are represented by the columns, as shown in Table 16.28. We now assign the different gasolines (*A*, *B*, *C*, and *D*) to the rows and columns at random, subject only to the requirement that each letter appears just once in each row and just once in each column. Thus each driver will have an opportunity to drive each car and to use each type of gasoline, and no car will be driven twice with the same gasoline.

We now assign at random the four roads to be used, denoted by α, β, γ, and δ, subjecting them to the same requirement imposed on the Latin squares. Thus each driver will also have the opportunity to drive along each of the roads. Table 16.28 shows one possible arrangement.

Table 16.28

	Driver			
	1	2	3	4
Car 1	B_γ	A_β	D_δ	C_α
Car 2	A_δ	B_α	C_γ	D_β
Car 3	D_α	C_δ	B_β	A_γ
Car 4	C_β	D_γ	A_α	B_δ

16.17 Suppose that in carrying out the experiment of Problem 16.16 the numbers of miles per gallon are as given in Table 16.29. Use analysis of variance to determine whether there are any differences at the 0.05 significance level. Use MINITAB to obtain the analysis of variance and use the *p*-values provided by MINITAB to test for any differences at the 0.05 significance level.

Table 16.29

	Driver			
	1	2	3	4
Car 1	B_γ 19	A_β 16	D_δ 16	C_α 14
Car 2	A_δ 15	B_α 18	C_γ 11	D_β 15
Car 3	D_α 14	C_δ 11	B_β 21	A_γ 16
Car 4	C_β 16	D_γ 16	A_α 15	B_δ 23

SOLUTION

We first obtain the row and column totals, as shown in Table 16.30. We then obtain the totals for each Latin letter and for each Greek letter, as follows:

$$A \text{ total: } 15+16+15+16=62$$

$$B \text{ total: } 19+18+21+23=81$$

$$C \text{ total: } 16+11+11+14=52$$

$$D \text{ total: } 14+16+16+15=61$$

$$\alpha \text{ total: } 14+18+15+14=61$$

$$\beta \text{ total: } 16+16+21+15=68$$

$$\gamma \text{ total: } 19+16+11+16=62$$

$$\delta \text{ total: } 15+11+16+23=65$$

We now compute the variations corresponding to all of these, using the shortcut method:

$$\text{Rows: } \frac{(65)^2}{4}+\frac{(59)^2}{4}+\frac{(62)^2}{4}+\frac{(70)^2}{4}-\frac{(256)^2}{16}=4112.50-4096=16.50$$

$$\text{Columns: } \frac{(64)^2}{4}+\frac{(61)^2}{4}+\frac{(63)^2}{4}+\frac{(68)^2}{4}-\frac{(256)^2}{16}=4102.50-4096=6.50$$

$$\text{Gasolines } (A,B,C,D)\text{: } \frac{(62)^2}{4}+\frac{(81)^2}{4}+\frac{(52)^2}{4}+\frac{(61)^2}{4}-\frac{(256)^2}{16}=4207.50-4096=111.50$$

$$\text{Roads } (\alpha,\beta,\gamma,\delta)\text{: } \frac{(61)^2}{4}+\frac{(68)^2}{4}+\frac{(62)^2}{4}+\frac{(65)^2}{4}-\frac{(256)^2}{16}=4103.50-4096=7.50$$

The total variation is

$$(19)^2+(16)^2+(16)^2+\cdots+(15)^2+(23)^2-\frac{(256)^2}{16}=4244-4096=148.00$$

so that the variation due to error is

$$148.00-16.50-6.50-111.50-7.50=6.00$$

The results are shown in Table 16.31, the analysis of variance. The total number of degrees of freedom is N^2-1 for an $N \times N$ square. Each of the rows, columns, Latin letters, and Greek letters has $N-1$ degrees of freedom. Thus the degrees of freedom for error are $N^2-1-4(N-1)=(N-1)(N-3)$. In our case, $N=4$.

Table 16.30

					Total
	B_γ 19	A_β 16	D_δ 16	C_α 14	65
	A_δ 15	B_α 18	C_γ 11	D_β 15	59
	D_α 14	C_δ 11	B_β 21	A_γ 16	62
	C_β 16	D_γ 16	A_α 15	B_δ 23	70
Total	64	61	63	68	256

Table 16.31

Variation	Degrees of Freedom	Mean Square	F
Rows (cars), 16.50	3	5.500	$\frac{5.500}{2.000} = 2.75$
Columns (drivers), 6.50	3	2.167	$\frac{2.167}{2.000} = 1.08$
Gasolines (A, B, C, D), 111.50	3	37.167	$\frac{37.167}{2.000} = 18.6$
Roads ($\alpha, \beta, \gamma, \delta$), 7.50	3	2.500	$\frac{2.500}{2.000} = 1.25$
Error, 6.00	3	2.000	
Total, 148.00	15		

We have $F_{.95,3,3} = 9.28$ and $F_{.99,3,3} = 29.5$. Thus we can reject the hypothesis that the gasolines are the same at the 0.05 level but not at the 0.01 level.

The structure of the data file for the MINITAB worksheet is given first.

```
Row    Car    Driver    Gasoline    Road    MPG
  1      1       1          2         3      19
  2      1       2          1         2      16
  3      1       3          4         4      16
  4      1       4          3         1      14
  5      2       1          1         4      15
  6      2       2          2         1      18
  7      2       3          3         3      11
  8      2       4          4         2      15
  9      3       1          4         1      14
 10      3       2          3         4      11
 11      3       3          2         2      21
 12      3       4          1         3      16
 13      4       1          3         2      16
 14      4       2          4         3      16
 15      4       3          1         1      15
 16      4       4          2         4      23
```

Note that cars and drivers are numbered the same in the MINITAB worksheet as in Table 16.29. Gasoline brands A through D in Table 16.29 are coded as 1 through 4, respectively, in the worksheet. Roads α, β, γ, and δ, in Table 16.29 are coded as 1, 2, 3, and 4 in the worksheet. The MINITAB pull-down **Stat → ANOVA → General Linear Model** gives the following output.

General Linear Model: Mpg versus Car, Driver, Gasoline, Road

```
Factor     Type    Levels    Values
Car        fixed        4    1, 2, 3, 4
Driver     fixed        4    1, 2, 3, 4
Gasoline   fixed        4    1, 2, 3, 4
Road       fixed        4    1, 2, 3, 4
```

```
Analysis of Variance for Mpg, using Adjusted SS for Tests

Source      DF      Seq SS     Adj SS     Adj MS         F        P
Car          3      16.500     16.500      5.500      2.75    0.214
Driver       3       6.500      6.500      2.167      1.08    0.475
Gasoline     3     111.500    111.500     37.167     18.58    0.019
Road         3       7.500      7.500      2.500      1.25    0.429
Error        3       6.000      6.000      2.000
Total       15     148.000
```

The column labeled `Seq MS` in the MINITAB output is the same as the `Mean Square` column in Table 16.31. `The computed` F values in the MINITAB output are the same as those given in Table 16.31. The p-values for cars, drivers, gasoline brands, and roads are 0.214, 0.475, 0.019, and 0.429, respectively. Recall that a p-value is the minimum value for a pre-set level of significance at which the hypothesis of equal means for a given factor can be rejected. The p-values indicate no differences for cars, drivers, or roads at the 0.01 or 0.05 levels. The means for gasoline brands are statistically different at the 0.05 level but not at the 0.01 level. Further investigation of the means for the brands would indicate how the means differ.

MISCELLANEOUS PROBLEMS

16.18 Prove [as in equation (*15*) of this chapter] that $\sum_j \alpha_j = 0$.

SOLUTION

The treatment population means μ_j and the total population mean μ are related by

$$\mu = \frac{1}{a}\sum_j \mu_j \tag{53}$$

Then, since $\alpha_j = \mu_j - \mu$, we have, using equation (*53*),

$$\sum_j \alpha_j = \sum_j (\mu_j - \mu) = \sum_j \mu_j - a\mu = 0 \tag{54}$$

16.19 Derive (*a*) equation (*16*) and (*b*) equation (*17*) of this chapter.

SOLUTION

(*a*) By definition, we have

$$V_W = \sum_{j,k}(X_{jk} - \bar{X}_{j.})^2 = b\sum_{j=1}^{a}\left[\frac{1}{b}\sum_{k=1}^{b}(X_{jk} - \bar{X}_{j.})^2\right] = b\sum_{j=1}^{a} S_j^2$$

where S_j^2 is the sample variance for the jth treatment. Then, since the sample size is b,

$$E(V_W) = b\sum_{j=1}^{a} E(S_j^2) = b\sum_{j=1}^{a}\left(\frac{b-1}{b}\sigma^2\right) = a(b-1)\sigma^2$$

(*b*) By definition

$$V_B = b\sum_{j=1}^{a}(\bar{X}_{j.} - \bar{X})^2 = b\sum_{j=1}^{a}\bar{X}_{j.}^2 - 2b\bar{X}\sum_{j=1}^{a}\bar{X}_{j.} + ab\bar{X}^2 = b\sum_{j=1}^{a}\bar{X}_{j.}^2 - ab\bar{X}^2$$

since $\bar{X} = (\sum_j \bar{X}_{j.})/a$. Then, omitting the summation index, we have

$$E(V_B) = b\sum E(\bar{X}_{j.}^2) - abE(\bar{X}^2) \tag{55}$$

Now for any random variable $U, E(U^2) = \text{var}(U) + [E(U)]^2$, where var ($U$) denotes the variance of U. Thus

$$E(\bar{X}_{j.}^2) = \text{var}(\bar{X}_{j.}) + [E(\bar{X}_{j.})]^2 \tag{56}$$

$$E(\bar{X}^2) = \text{var}(\bar{X}) + [E(\bar{X})]^2 \tag{57}$$

But since the treatment populations are normal with mean $\mu_j = \mu + \alpha_j$, we have

$$\text{var}(\bar{X}_{j.}) = \frac{\sigma^2}{b} \tag{58}$$

$$\text{var}(\bar{X}) = \frac{\sigma^2}{ab} \tag{59}$$

$$E(\bar{X}_{j.}) = \mu_j = \mu + \alpha_j \tag{60}$$

$$E(\bar{X}) = \mu \tag{61}$$

Using results (*56*) to (*61*) together with result (*53*), we have

$$\begin{aligned} E(V_B) &= b\sum\left[\frac{\sigma^2}{b} + (\mu + \alpha_j)^2\right] - ab\left[\frac{\sigma^2}{ab} + \mu^2\right] \\ &= a\sigma^2 + b\sum(\mu + \alpha_j)^2 - \sigma^2 - ab\mu^2 \\ &= (a-1)\sigma^2 + ab\mu^2 + 2b\mu\sum\alpha_j + b\sum\alpha_j^2 + ab\mu^2 \\ &= (a-1)\sigma^2 + b\sum\alpha_j^2 \end{aligned}$$

16.20 Prove Theorem 1 in this chapter.

SOLUTION

As shown in Problem 16.19,

$$V_W = b\sum_{j=1}^{a} S_j^2 \qquad \text{or} \qquad \frac{V_W}{\sigma^2} = \sum_{j=1}^{a}\frac{bS_j^2}{\sigma^2}$$

where S_j^2 is the sample variance for samples of size b drawn from the population of treatment j. We see that bS_j^2/σ^2 has a chi-square distribution with $b-1$ degrees of freedom. Thus, since the variances S_j^2 are independent, we conclude from page 264 that V_W/σ^2 is chi-square-distributed with $a(b-1)$ degrees of freedom.

Supplementary Problems

ONE-WAY CLASSIFICATION, OR ONE-FACTOR EXPERIMENTS

In all of the supplementary problems, the reader is encouraged to work the problems "by hand" using the equations given in the chapter before using the suggested software to work the problem. This will increase your understanding of the ANOVA technique as well as help you appreciate the power of the software.

16.21 An experiment is performed to determine the yields of five different varieties of wheat: *A*, *B*, *C*, *D*, and *E*. Four plots of land are assigned to each variety, and the yields (in bushels per acre) are as shown in Table 16.32. Assuming that the plots are of similar fertility and that the varieties are assigned to the plots at random, determine whether there is a difference between the yields at significance levels of (*a*) 0.05 and (*b*) 0.01. (*c*) Give the MINITAB analysis for this one-way classification or one-factor experiment.

Table 16.32

A	20	12	15	19
B	17	14	12	15
C	23	16	18	14
D	15	17	20	12
E	21	14	17	18

16.22 A company wishes to test four different types of tires: *A*, *B*, *C*, and *D*. The tires' lifetimes, as determined from their treads, are given (in thousands of miles) in Table 16.33, where each type has been tried on six similar automobiles assigned at random to the tires. Determine whether there is a significant difference between the tires at the (*a*) 0.05 and (*b*) 0.01 levels. (*c*) Give the STATISTIX analysis for this one-way classification or one-factor experiment.

Table 16.33

A	33	38	36	40	31	35
B	32	40	42	38	30	34
C	31	37	35	33	34	30
D	29	34	32	30	33	31

16.23 A teacher wishes to test three different teaching methods: I, II, and III. To do this, three groups of five students each are chosen at random, and each group is taught by a different method. The same examination is then given to all the students, and the grades in Table 16.34 are obtained. Determine whether there is a difference between the teaching methods at significance levels of (*a*) 0.05 and (*b*) 0.01. (*c*) Give the EXCEL analysis for this one-way classification or one-factor experiment.

Table 16.34

Method I	75	62	71	58	73
Method II	81	85	68	92	90
Method III	73	79	60	75	81

MODIFICATIONS FOR UNEQUAL NUMBERS OF OBSERVATIONS

16.24 Table 16.35 gives the numbers of miles to the gallon obtained by similar automobiles using five different brands of gasoline. Determine whether there is a difference between the brands at significance levels of (*a*) 0.05 and (*b*) 0.01. (*c*) Give the SPSS analysis for this one-way classification or one-factor experiment.

Table 16.35

Brand *A*	12	15	14	11	15
Brand *B*	14	12	15		
Brand *C*	11	12	10	14	
Brand *D*	15	18	16	17	14
Brand *E*	10	12	14	12	

Table 16.36

Mathematics	72	80	83	75	
Science	81	74	77		
English	88	82	90	87	80
Economics	74	71	77	70	

16.25 During one semester a student received grades in various subjects, as shown in Table 16.36. Determine whether there is a significant difference between the student's grades at the (*a*) 0.05 and (*b*) 0.01 levels. (*c*) Give the SAS analysis for this one-way classification or one-factor experiment.

TWO-WAY CLASSIFICATION, OR TWO-FACTOR EXPERIMENTS

16.26 Articles manufactured by a company are produced by three operators using three different machines. The manufacturer wishes to determine whether there is a difference (*a*) between the operators and (*b*) between the machines. An experiment is performed to determine the number of articles per day produced by each operator using each machine; the results are shown in Table 16.37. Provide the desired information using MINITAB and a significance level of 0.05.

Table 16.37

	Operator		
	1	2	3
Machine *A*	23	27	24
Machine *B*	34	30	28
Machine *C*	28	25	27

Table 16.38

	Type of Corn			
	I	II	III	IV
Block *A*	12	15	10	14
Block *B*	15	19	12	11
Block *C*	14	18	15	12
Block *D*	11	16	12	16
Block *E*	16	17	11	14

16.27 Use EXCEL to work Problem 16.26 at the 0.01 significance level.

16.28 Seeds of four different types of corn are planted in five blocks. Each block is divided into four plots, which are then randomly assigned to the four types. Determine at the 0.05 significance level whether the yields in bushels per acre, as shown in Table 16.38, vary significantly with differences in (*a*) the soil (i.e., the five blocks) and (*b*) the type of corn. Use SPSS to build your ANOVA.

16.29 Work Problem 16.28 at the 0.01 significance level. Use STATISTIX to build your ANOVA.

16.30 Suppose that in Problem 16.22 the first observation for each type of tire is made using one particular kind of automobile, the second observation is made using a second particular kind, and so on. Determine at the 0.05

significance level whether there is a difference (*a*) between the types of tires and (*b*) between the kinds of automobiles. Use SAS to build your ANOVA.

16.31 Use MINITAB to work Problem 16.30 at the 0.01 level of significance.

16.32 Suppose that in Problem 16.23 the first entry for each teaching method corresponds to a student at one particular school, the second method corresponds to a student at another school, and so on. Test the hypothesis at the 0.05 significance level that there is a difference (*a*) between the teaching methods and (*b*) between the schools. Use STATISTIX to build your ANOVA.

16.33 An experiment is performed to test whether the hair color and heights of adult female students in the United States have any bearing on scholastic achievement. The results are given in Table 16.39, where the numbers indicate individuals in the top 10% of those graduating. Analyze the experiment at a significance level of 0.05. Use EXCEL to build your ANOVA.

Table 16.39

	Redhead	Blonde	Brunette
Tall	75	78	80
Medium	81	76	79
Short	73	75	77

Table 16.40

A	16	18	20	23
B	15	17	16	19
C	21	19	18	21
D	18	22	21	23
E	17	18	24	20

16.34 Work Problem 16.33 at the 0.01 significance level. Use SPSS to build your ANOVA and compare your output with the EXCEL output of Problem 16.33.

TWO-FACTOR EXPERIMENTS WITH REPLICATION

16.35 Suppose that the experiment of Problem 16.21 was carried out in the southern part of the United States and that the columns of Table 16.32 now indicate four different types of fertilizer, while a similar experiment performed in the western part gives the results shown in Table 16.40. Determine at the 0.05 significance level whether there is a difference in yields due to (*a*) the fertilizers and (*b*) the locations. Use MINITAB to build your ANOVA.

16.36 Work Problem 16.35 at the 0.01 significance level. Use STATISTIX to build your ANOVA and compare your output with the MINITAB output of Problem 16.35.

16.37 Table 16.41 gives the number of articles produced by four different operators working on two different types of machines, I and II, on different days of the week. Determine at the 0.05 level whether there are significant

Table 16.41

	Machine I					Machine II				
	Mon.	Tue.	Wed.	Thu.	Fri.	Mon.	Tue.	Wed.	Thu.	Fri.
Operator *A*	15	18	17	20	12	14	16	18	17	15
Operator *B*	12	16	14	18	11	11	15	12	16	12
Operator *C*	14	17	18	16	13	12	14	16	14	11
Operator *D*	19	16	21	23	18	17	15	18	20	17

differences (*a*) between the operators and (*b*) between the machines. Use SAS and MINITAB to build your ANOVA.

LATIN SQUARES

16.38 An experiment is performed to test the effect on corn yield of four different fertilizer treatments (*A*, *B*, *C*, and *D*) and of soil variations in two perpendicular directions. The Latin square of Table 16.42 is obtained, where the numbers show the corn yield per unit area. Test at the 0.01 significance level the hypothesis that there is no difference between (*a*) the fertilizers and (*b*) the soil variations. Use STATISTIX to build your ANOVA.

Table 16.42

C 8	*A* 10	*D* 12	*B* 11
A 14	*C* 12	*B* 11	*D* 15
D 10	*B* 14	*C* 16	*A* 10
B 7	*D* 16	*A* 14	*C* 12

Table 16.43

E 75	*W* 78	*M* 80
M 81	*E* 76	*W* 79
W 73	*M* 75	*E* 77

16.39 Work Problem 16.38 at the 0.05 significance level. Use MINITAB to build your ANOVA and compare your output with the STATISTIX output of Problem 16.38.

16.40 Referring to Problem 16.33, suppose that we introduce an additional factor—giving the section *E*, *M*, or *W* of the United States in which a student was born, as shown in Table 16.43. Determine at the 0.05 level whether there is a significant difference in the scholastic achievements of female students due to differences in (*a*) height, (*b*) hair color, and (*c*) birthplace. Use SPSS to build your ANOVA.

GRAECO-LATIN SQUARES

16.41 In order to produce a superior type of chicken feed, four different quantities of each of two chemicals are added to the basic ingredients. The different quantities of the first chemical are indicated by *A*, *B*, *C*, and *D*, while those of the second chemical are indicated by α, β, γ, and δ. The feed is given to baby chicks arranged in groups according to four different initial weights (W_1, W_2, W_3, and W_4) and four different species (S_1, S_2, S_3, and S_4). The increases in weight per unit time are given in the Graeco-Latin square of Table 16.44. Perform an analysis of variance of the experiment at the 0.05 significance level, stating any conclusions that can be drawn. Use MINITAB to build your ANOVA.

Table 16.44

	W_1	W_2	W_3	W_4
S_1	C_γ 8	B_β 6	A_α 5	D_δ 6
S_2	A_δ 4	D_α 3	C_β 7	B_γ 3
S_3	D_β 5	A_γ 6	B_δ 5	C_α 6
S_4	B_α 6	C_δ 10	D_γ 10	A_β 8

16.42 Four different types of cable (T_1, T_2, T_3, and T_4) are manufactured by each of four companies (C_1, C_2, C_3, and C_4). Four operators (*A*, *B*, *C*, and *D*) using four different machines (α, β, γ, and δ) measure the cable strengths. The average strengths obtained are given in the Graeco-Latin square of Table 16.45. Perform an analysis of variance at the 0.05 significance level, stating any conclusions that can be drawn. Use SPSS to build your ANOVA.

MISCELLANEOUS PROBLEMS

16.43 Table 16.46 gives data on the accumulated rust on iron that has been treated with chemical A, B, or C, respectively. Determine at the (*a*) 0.05 and (*b*) 0.01 levels whether there is a significant difference in the treatments. Use EXCEL to build your ANOVA.

Table 16.45

	C_1	C_2	C_3	C_4
T_1	A_β 164	B_γ 181	C_α 193	D_δ 160
T_2	C_δ 171	D_α 162	A_γ 183	B_β 145
T_3	D_γ 198	C_β 212	B_δ 207	A_α 188
T_4	B_α 157	A_δ 172	D_β 166	C_γ 136

Table 16.46

A	3	5	4	4
B	4	2	3	3
C	6	4	5	5

16.44 An experiment measures the intelligence quotients (IQs) of adult male students of tall, short, and medium stature. The results are given in Table 16.47. Determine at the (*a*) 0.05 and (*b*) 0.01 significance levels whether there is any difference in the IQ scores relative to the height differences. Use MINITAB to build your ANOVA.

Table 16.47

Tall	110	105	118	112	90	
Short	95	103	115	107		
Medium	108	112	93	104	96	102

16.45 Prove results (*10*), (*11*), and (*12*) of this chapter.

16.46 An examination is given to determine whether veterans or nonveterans of different IQs performed better. The scores obtained are shown in Table 16.48. Determine at the 0.05 significance level whether there is a difference in scores due to differences in (*a*) veteran status and (*b*) IQ. Use SPSS to build your ANOVA.

Table 16.48

	Test Score		
	High IQ	Medium IQ	Low IQ
Veteran	90	81	74
Nonveteran	85	78	70

16.47 Use STATISTIX to work Problem 16.46 at the 0.01 significance level.

16.48 Table 16.49 shows test scores for a sample of college students who are from different parts of the country and who have different IQs. Analyze the table at the 0.05 significance level and state your conclusions. Use MINITAB to build your ANOVA.

16.49 Use SAS to work Problem 16.48 at the 0.01 significance level.

16.50 In Problem 16.37, can you determine whether there is a significant difference in the number of articles produced on different days of the week? Explain.

Table 16.49

	Test Score		
	High IQ	Medium IQ	Low IQ
East	88	80	72
West	84	78	75
South	86	82	70
North and central	80	75	79

16.51 In analysis-of-variance calculations it is known that a suitable constant can be added or subtracted from each entry without affecting the conclusions. Is this also true if each entry is multiplied or divided by a suitable constant? Justify your answer.

16.52 Derive results *(24)*, *(25)*, and *(26)* for unequal numbers of observations.

16.53 Suppose that the results in Table 16.46 of Problem 16.43 hold for the northeastern part of the United States, while the corresponding results for the western part are those given in Table 16.50. Determine at the 0.05 significance level whether there are differences due to (*a*) chemicals and (*b*) location. Use MINITAB to build your ANOVA.

Table 16.50

A	5	4	6	3
B	3	4	2	3
C	5	7	4	6

Table 16.51

A	17	14	18	12
B	20	10	20	15
C	18	15	16	17
D	12	11	14	11
E	15	12	19	14

16.54 Referring to Problems 16.21 and 16.35, suppose that an additional experiment performed in the northeastern part of the United States produced the results given in Table 16.51. Determine at the 0.05 significance level whether there is a difference in yields due to (*a*) the fertilizers and (*b*) the three locations. Use STATISTIX to build your ANOVA.

16.55 Work Problem 16.54 at the 0.01 significance level. Use MINITAB to build your ANOVA.

16.56 Perform an analysis of variance on the Latin square of Table 16.52 at the 0.05 significance level and state your conclusions. Use SPSS to build your ANOVA.

Table 16.52

	Factor 1		
Factor 2	*B* 16	*C* 21	*A* 15
	A 18	*B* 23	*C* 14
	C 15	*A* 18	*B* 12

16.57 Make up an experiment leading to the Latin square of Table 16.52.

16.58 Perform an analysis of variance on the Graeco-Latin square of Table 16.53 at the 0.05 significance level and state your conclusions. Use SPSS to build your ANOVA.

Table 16.53

	Factor 1			
Factor 2	A_γ 6	B_β 12	C_δ 4	D_α 18
	B_δ 3	A_α 8	D_γ 15	C_β 14
	D_β 15	C_γ 20	B_α 9	A_δ 5
	C_α 16	D_δ 6	A_β 17	B_γ 7

16.59 Make up an experiment leading to the Graeco-Latin square of Table 16.53.

16.60 Describe how to use analysis-of-variance techniques for three-factor experiments with replications.

16.61 Make up and solve a problem that illustrates the procedure in Problem 16.60.

16.62 Prove (*a*) equation (*30*) and (*b*) results (*31*) to (*34*) of this chapter.

16.63 In practice, would you expect to find (*a*) a 2×2 Latin square and (*b*) a 3×3 Graeco-Latin square? Explain.

Nonparametric Tests

INTRODUCTION

Most tests of hypotheses and significance (or decision rules) considered in previous chapters require various assumptions about the distribution of the population from which the samples are drawn. For example, the one-way classification of Chapter 16 requires that the populations be normally distributed and have equal standard deviations.

Situations arise in practice in which such assumptions may not be justified or in which there is doubt that they apply, as in the case where a population may be highly skewed. Because of this, statisticians have devised various tests and methods that are independent of population distributions and associated parameters. These are called *nonparametric tests*.

Nonparametric tests can be used as shortcut replacements for more complicated tests. They are especially valuable in dealing with nonnumerical data, such as arise when consumers rank cereals or other products in order of preference.

THE SIGN TEST

Consider Table 17.1, which shows the numbers of defective bolts produced by two different types of machines (I and II) on 12 consecutive days and which assumes that the machines have the same total output per day. We wish to test the hypothesis H_0 that there is no difference between the machines: that the observed differences between the machines in terms of the numbers of defective bolts they produce are merely the result of chance, which is to say that the samples come from the same population.

Table 17.1

Day	1	2	3	4	5	6	7	8	9	10	11	12
Machine I	47	56	54	49	36	48	51	38	61	49	56	52
Machine II	71	63	45	64	50	55	42	46	53	57	75	60

A simple nonparametric test in the case of such paired samples is provided by the *sign test*. This test consists of taking the difference between the numbers of defective bolts for each day and writing only the *sign* of the difference; for instance, for day 1 we have 47–71, which is negative. In this way we obtain from Table 17.1 the sequence of signs

$$- \quad - \quad + \quad - \quad - \quad - \quad + \quad - \quad + \quad - \quad - \quad - \qquad (1)$$

(i.e., 3 pluses and 9 minuses). Now if it is just as likely to get a + as a −, we would expect to get 6 of each. The test of H_0 is thus equivalent to that of whether a coin is fair if 12 tosses result in 3 heads (+) and 9 tails (−). This involves the binomial distribution of Chapter 7. Problem 17.1 shows that by using a two-tailed test of this distribution at the 0.05 significance level, we cannot reject H_0; that is, there is no difference between the machines at this level.

Remark 1: If on some day the machines produced the same number of defective bolts, a difference of *zero* would appear in sequence (*1*). In such case we can omit these sample values and use 11 instead of 12 observations.

Remark 2: A normal approximation to the binomial distribution, using a correction for continuity, can also be used (see Problem 17.2).

Although the sign test is particularly useful for paired samples, as in Table 17.1, it can also be used for problems involving single samples (see Problems 17.3 and 17.4).

THE MANN–WHITNEY *U* TEST

Consider Table 17.2, which shows the strengths of cables made from two different alloys, I and II. In this table we have two samples: 8 cables of alloy I and 10 cables of alloy II. We would like to decide whether or not there is a difference between the samples or, equivalently, whether or not they come from the same population. Although this problem can be worked by using the t test of Chapter 11, a non-parametric test called the *Mann–Whitney U test*, or briefly the *U test*, is useful. This test consists of the following steps:

Table 17.2

Alloy I				Alloy II				
18.3	16.4	22.7	17.8	12.6	14.1	20.5	10.7	15.9
18.9	25.3	16.1	24.2	19.6	12.9	15.2	11.8	14.7

Step 1. Combine all sample values in an array from the smallest to the largest, and assign ranks (in this case from 1 to 18) to all these values. If two or more sample values are identical (i.e., there are *tie scores*, or briefly *ties*), the sample values are each assigned a rank equal to the *mean* of the ranks that would otherwise be assigned. If the entry 18.9 in Table 17.2 were 18.3, two identical values 18.3 would occupy ranks 12 and 13 in the array so that the rank assigned to each would be $\frac{1}{2}(12 + 13) = 12.5$.

Step 2. Find the sum of the ranks for each of the samples. Denote these sums by R_1, and R_2, where N_1 and N_2 are the respective sample sizes. For convenience, choose N_1 as the smaller size if they are unequal, so that $N_1 \leq N_2$. A significant difference between the rank sums R_1 and R_2 implies a significant difference between the samples.

Step 3. To test the difference between the rank sums, use the statistic

$$U = N_1 N_2 + \frac{N_1(N_1 + 1)}{2} - R_1 \tag{2}$$

corresponding to sample 1. The sampling distribution of U is symmetrical and has a mean and variance given, respectively, by the formulas

$$\mu_U = \frac{N_1 N_2}{2} \qquad \sigma_U^2 = \frac{N_1 N_2 (N_1 + N_2 + 1)}{12} \tag{3}$$

If N_1 and N_2 are both at least equal to 8, it turns out that the distribution of U is nearly normal, so that

$$z = \frac{U - \mu_U}{\sigma_U} \tag{4}$$

is normally distributed with mean 0 and variance 1. Using Appendix II, we can then decide whether the samples are significantly different. Problem 17.5 shows that there is a significant difference between the cables at the 0.05 level.

Remark 3: A value corresponding to sample 2 is given by the statistic

$$U = N_1 N_2 + \frac{N_2(N_2+1)}{2} - R_2 \tag{5}$$

and has the same sampling distribution as statistic (2), with the mean and variance of formulas (3). Statistic (5) is related to statistic (2), for if U_1 and U_2 are the values corresponding to statistics (2) and (5), respectively, then we have the result

$$U_1 + U_2 = N_1 N_2 \tag{6}$$

We also have

$$R_1 + R_2 = \frac{N(N+1)}{2} \tag{7}$$

where $N = N_1 + N_2$. Result (7) can provide a check for calculations.

Remark 4: The statistic U in equation (2) is the total number of times that sample 1 values precede sample 2 values when all sample values are arranged in increasing order of magnitude. This provides an alternative *counting method* for finding U.

THE KRUSKAL–WALLIS *H* TEST

The U test is a nonparametric test for deciding whether or not two samples come from the same population. A generalization of this for k samples is provided by the *Kruskal–Wallis H test*, or briefly the *H test*.

This test may be described thus: Suppose that we have k samples of sizes $N_1, N_2, \ldots, N_k$, with the total size of all samples taken together being given by $N = N_1 + N_2 + \cdots + N_k$. Suppose further that the data from all the samples taken together are ranked and that the sums of the ranks for the k samples are $R_1, R_2, \ldots, R_k$, respectively. If we define the statistic

$$H = \frac{12}{N(N+1)} \sum_{j=1}^{k} \frac{R_j^2}{N_j} - 3(N+1) \tag{8}$$

then it can be shown that the sampling distribution of H is very nearly a *chi-square distribution* with $k-1$ degrees of freedom, provided that $N_1, N_2, \ldots, N_k$ are all at least 5.

The H test provides a nonparametric method in the *analysis of variance* for one-way classification, or one-factor experiments, and generalizations can be made.

THE *H* TEST CORRECTED FOR TIES

In case there are too many ties among the observations in the sample data, the value of H given by statistic (8) is smaller than it should be. The corrected value of H, denoted by H_c, is obtained by dividing the value given in statistic (*8*) by the correction factor

$$1 - \frac{\sum (T^3 - T)}{N^3 - N} \tag{9}$$

where T is the number of ties corresponding to each observation and where the sum is taken over all the observations. If there are no ties, then $T = 0$ and factor (*9*) reduces to 1, so that no correction is needed. In practice, the correction is usually negligible (i.e., it is not enough to warrant a change in the decision).

THE RUNS TEST FOR RANDOMNESS

Although the word "random" has been used many times in this book (such as in "random sampling" and "tossing a coin at random"), no previous chapter has given any test for randomness. A nonparametric test for randomness is provided by the *theory of runs.*

To understand what a run is, consider a sequence made up of two symbols, a and b, such as

$$a\,a \mid b\,b\,b \mid a \mid b\,b \mid a\,a\,a\,a\,a \mid b\,b\,b \mid a\,a\,a\,a \mid \tag{10}$$

In tossing a coin, for example, a could represent "heads" and b could represent "tails." Or in sampling the bolts produced by a machine, a could represent "defective" and b could represent "nondefective."

A *run* is defined as a set of identical (or related) symbols contained between two different symbols or no symbol (such as at the beginning or end of the sequence). Proceeding from left to right in sequence (*10*), the first run, indicated by a vertical bar, consists of two a's; similarly, the second run consists of three b's, the third run consists of one a, etc. There are seven runs in all.

It seems clear that some relationship exists between randomness and the number of runs. Thus for the sequence

$$a \mid b \mid a \mid b \mid a \mid b \mid a \mid b \mid a \mid b \mid a \mid b \mid \tag{11}$$

there is a *cyclic pattern*, in which we go from a to b, back to a again, etc., which we could hardly believe to be random. In such case we have *too many* runs (in fact, we have the maximum number possible for the given number of a's and b's).

On the other hand, for the sequence

$$a\,a\,a\,a\,a\,a \mid b\,b\,b\,b \mid a\,a\,a\,a\,a \mid b\,b\,b \mid \tag{12}$$

there seems to be a *trend pattern*, in which the a's and b's are grouped (or clustered) together. In such case there are *too few* runs, and we would not consider the sequence to be random.

Thus a sequence would be considered nonrandom if there are either too many or too few runs, and random otherwise. To quantify this idea, suppose that we form all possible sequences consisting of N_1 a's and N_2 b's, for a total of N symbols in all ($N_1 + N_2 = N$). The collection of all these sequences provides us with a sampling distribution: Each sequence has an associated number of runs, denoted by V. In this way we are led to the sampling distribution of the statistic V. It can be shown that this sampling distribution has a mean and variance given, respectively, by the formulas

$$\mu_V = \frac{2N_1N_2}{N_1 + N_2} + 1 \qquad \sigma_V^2 = \frac{2N_1N_2(2N_1N_2 - N_1 - N_2)}{(N_1 + N_2)^2(N_1 + N_2 - 1)} \tag{13}$$

By using formulas (*13*), we can test the hypothesis of randomness at appropriate levels of significance. It turns out that if both N_1 and N_2 are at least equal to 8, then the sampling distribution of V is very nearly a normal distribution. Thus

$$z = \frac{V - \mu_V}{\sigma_V} \tag{14}$$

is normally distributed with mean 0 and variance 1, and thus Appendix II can be used.

FURTHER APPLICATIONS OF THE RUNS TEST

The following are other applications of the runs test to statistical problems:

1. **Above- and Below-Median Test for Randomness of Numerical Data.** To determine whether numerical data (such as collected in a sample) are random, first place the data in the *same order* in which they were collected. Then find the median of the data and replace each entry with the letter a or b according to whether its value is *above* or *below* the median. If a value is the same as the median, omit it from the sample. The sample is random or not according to whether the sequence of a's and b's is random or not. (See Problem 17.20.)
2. **Differences in Populations from Which Samples Are Drawn.** Suppose that two samples of sizes m and n are denoted by $a_1, a_2, \ldots, a_m$ and $b_1, b_2, \ldots, b_n$, respectively. To decide whether the samples do or do not come from the same population, first arrange all $m + n$ sample values in a sequence of increasing values. If some values are the same, they should be ordered by a random process (such as by using random numbers). If the resulting sequence is random, we can conclude that the samples are not really different and thus come from the same population; if the sequence is not random, no such conclusion can be drawn. This test can provide an alternative to the Mann–Whitney U test. (See Problem 17.21.)

SPEARMAN'S RANK CORRELATION

Nonparametric methods can also be used to measure the correlation of two variables, X and Y. Instead of using precise values of the variables, or when such precision is unavailable, the data may be ranked from 1 to N in order of size, importance, etc. If X and Y are ranked in such a manner, the *coefficient of rank correlation*, or *Spearman's formula for rank correlation* (as it is often called), is given by

$$r_S = 1 - \frac{6 \sum D^2}{N(N^2 - 1)} \tag{15}$$

where D denotes the differences between the ranks of corresponding values of X and Y, and where N is the number of pairs of values (X, Y) in the data.

Solved Problems

THE SIGN TEST

17.1 Referring to Table 17.1, test the hypothesis H_0 that there is no difference between machines I and II against the alternative hypothesis H_1 that there is a difference at the 0.05 significance level.

SOLUTION

Figure 17-1 shows the binomial distribution of the probabilities of X heads in 12 tosses of a coin as areas under rectangles at $X = 0, 1, \ldots, 12$. Superimposed on the binomial distribution is the normal distribution, shown as a dashed curve. The mean of the binomial distribution is $\mu = Np = 12(0.5) = 6$. The standard deviation is $\sigma = \sqrt{Npq} = \sqrt{12(0.5)(0.5)} = \sqrt{3} = 1.73$. The normal curve also has mean $= 6$ and standard deviation $= 1.73$. From chapter 7 the binomial probability of X heads is

$$\Pr\{X\} = \binom{12}{X}\left(\frac{1}{2}\right)^X\left(\frac{1}{2}\right)^{12-X} = \binom{12}{X}\left(\frac{1}{2}\right)^{12}$$

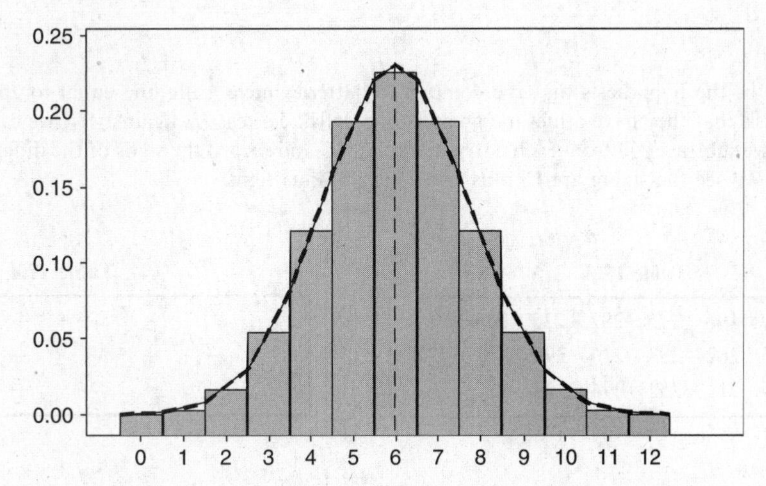

Fig. 17-1 Binomial distribution (areas under rectangles) and normal approximation (dashed curve) to the binomial distribution.

The probabilities may be found by using EXCEL. The p-value corresponding to the outcome $X=3$ is $2P\{X \leq 3\}$ which using EXCEL is `=2*BINOMDIST(3,12,0.5,1)` or 0.146. (The p-value is twice the area in the left tail of the binomial distribution.) Since this area exceeds 0.05, we are unable to reject the null at level of significance 0.05. We thus conclude that there is no difference in the two machines at this level.

17.2 Work Problem 17.1 by using a normal approximation to the binomial distribution.

SOLUTION

For a normal approximation to the binomial distribution, we use the fact that the z score corresponding to the number of heads is

$$z = \frac{X - \mu}{\sigma} = \frac{X - Np}{\sqrt{Npq}}$$

Because the variable X for the binomial distribution is discrete while that for a normal distribution is continuous, we make a *correction for continuity* (for example, 3 heads are really a value between 2.5 and 3.5 heads). This amounts to decreasing X by 0.5 if $X > Np$ and to increasing X by 0.5 if $X < Np$. Now $N = 12$, $\mu = Np = (12)(0.5) = 6$, and $\sigma = \sqrt{Npq} = \sqrt{(12)(0.5)(0.5)} = 1.73$, so that

$$z = \frac{(3 + 0.5) - 6}{1.73} = -1.45$$

The p-value is twice the area to the left of -1.45. Using EXCEL, we evaluate `=2*NORMSDIST(-1.45)` and get 0.147. Looking at Figure 17-1, the approximate p-value is the area under the standard normal curve from -1.45 to the left and it is doubled because the hypothesis is two-tailed. Notice how close the two p-values are to one another. The binomial area under the rectangles is 0.146 and the area under the approximating standard normal curve is 0.147.

17.3 The PQR Company claims that the lifetime of a type of battery that it manufactures is more than 250 hours (h). A consumer advocate wishing to determine whether the claim is justified measures the lifetimes of 24 of the company's batteries; the results are listed in Table 17.3. Assuming the sample to be random, determine whether the company's claim is justified at the 0.05 significance level. Work the problem first by hand, supplying all the details for the sign test. Follow this with the MINITAB solution to the problem.

SOLUTION

Let H_0 be the hypothesis that the company's batteries have a lifetime equal to 250 h, and let H_1 be the hypothesis that they have a lifetime greater than 250 h. To test H_0 against H_1, we can use the sign test. To do this, we subtract 250 from each entry in Table 17.3 and record the signs of the differences, as shown in Table 17.4. We see that there are 15 plus signs and 9 minus signs.

Table 17.3

271	230	198	275	282	225	284	219
253	216	262	288	236	291	253	224
264	295	211	252	294	243	272	268

Table 17.4

+	−	−	+	+	−	+	−
+	−	+	+	−	+	+	−
+	+	−	+	+	−	+	+

Using the normal approximation to the binomial distribution, the z score is

$$z = \frac{(15 - 0.5) - 24(0.5)}{\sqrt{24(0.5)(0.5)}} = 1.02.$$

Note the correction for continuity $(15 - 0.5) = 14.5$ when 0.5 is subtracted form 15. The p-value is the area to the right of 1.02 on the z-curve, also called the standard normal curve. (See Fig. 17-2.)

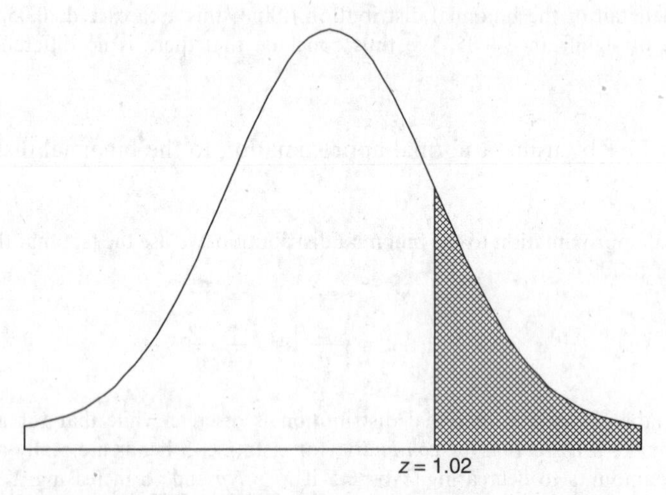

Fig. 17-2 The p-value is the area to the right of $z = 1.02$.

The p-value is the area to the right of $z = 1.02$ or, using EXCEL, the area is given by `=1-NORMSDIST(1.02)` or 0.1537. Since the p-value > 0.05, the companies claim cannot be justified.

The solution using MINITAB is as follows. The data in Table 17.3 are entered into column 1 of the MINITAB worksheet and the column is named Lifetime. The pull-down **Stat → Non-parametrics → 1-sample sign** gives the following output.

Sign Test for Median: Lifetime

```
Sign test of median = 250.0 versus > 250.0

            N   Below   Equal   Above        P   Median
Lifetime   24       9       0      15   0.1537    257.5
```

Note that the information given here is the same that we arrived at earlier in the solution of this problem.

17.4 A sample of 40 grades from a statewide examination is shown in Table 17.5. Test the hypothesis at the 0.05 significance level that the median grade for all participants is (*a*) 66 and (*b*) 75. Work the problem first by hand, supplying all the details for the sign test. Follow this with the MINITAB solution to the problem.

Table 17.5

71	67	55	64	82	66	74	58	79	61
78	46	84	93	72	54	78	86	48	52
67	95	70	43	70	73	57	64	60	83
73	40	78	70	64	86	76	62	95	66

SOLUTION

(*a*) Subtracting 66 from all the entries of Table 17.5 and retaining only the associated signs gives us Table 17.6, in which we see that there are 23 pluses, 15 minuses, and 2 zeros. Discarding the 2 zeros, our sample consists of 38 signs: 23 pluses and 15 minuses. Using a two-tailed test of the normal distribution with probabilities $\frac{1}{2}(0.05) = 0.025$ in each tail (Fig. 17-3), we adopt the following decision rule:

Accept the hypothesis if $-1.96 \leq z \leq 1.96$.

Reject the hypothesis otherwise.

Since $$z = \frac{X - Np}{\sqrt{Npq}} = \frac{(23 - 0.5) - (38)(0.5)}{\sqrt{(38)(0.5)(0.5)}} = 1.14$$

we accept the hypothesis that the median is 66 at the 0.05 level.

Table 17.6

+	+	−	−	+	0	+	−	+	−
+	−	+	+	+	−	+	+	−	−
+	+	+	−	+	+	−	−	−	+
+	−	+	+	−	+	+	−	+	0

Note that we could also have used 15, the number of minus signs. In this case

$$z = \frac{(15 + 0.5) - (38)(0.5)}{\sqrt{(38)(0.5)(0.5)}} = -1.14$$

with the same conclusion.

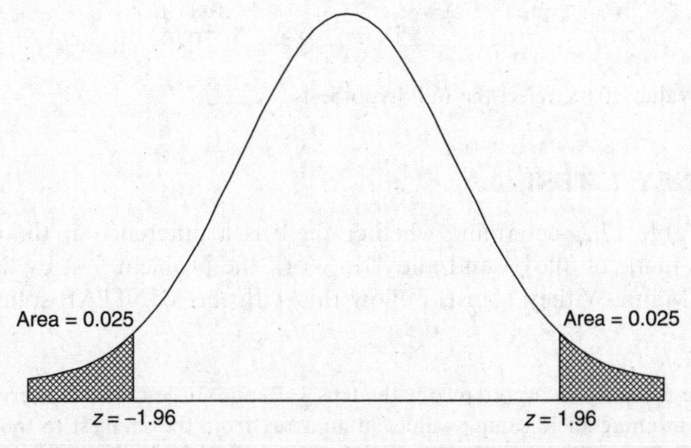

Fig. 17-3 Two-tailed test showing the critical region with $\alpha = 0.05$.

(*b*) Subtracting 75 from all the entries in Table 17.5 gives us Table 17.7, in which there are 13 pluses and 27 minuses. Since

$$z = \frac{(13 + 0.5) - (40)(0.5)}{\sqrt{(40)(0.5)(0.5)}} = -2.06$$

we reject the hypothesis that the median is 75 at the 0.05 level.

Table 17.7

−	−	−	−	+	−	−	−	+	−
+	−	+	+	−	−	+	+	−	−
−	+	−	−	−	−	−	−	−	+
−	−	+	−	−	+	+	−	+	−

Using this method, we can arrive at a 95% confidence interval for the median grade on the examination. (See Problem 17.30.)

Note that the above solution uses the classical method of testing hypothesis. The classical method uses $\alpha = 0.05$ to find the rejection region for the test. [$z < -1.96$ or $z > 1.96$]. Next, the test statistic is computed [$z = -1.14$ in part (a) and $z = -2.06$ in part (b)]. If the test statistic falls in the rejection region, then reject the null. If the test statistic does not fall in the rejection region, do not reject the null.

The MINITAB solution uses the *p*-value approach. Calculate the *p*-value and, if it is <0.05, reject the null hypothesis. If the *p*-value >0.05, do not reject the null hypothesis. The same decision will always be reached using either the classical method or the *p*-value method.

The solution using MINITAB proceeds as follows. To test that the median equals 66, the output is

Sign Test for Median: Grade

```
Sign test of median = 66.00 versus not = 66.00

         N   Below   Equal   Above         P   Median
Grade   40      15       2      23    0.2559    70.00
```

Since the *p*-value >0.05, do not reject the null hypothesis.
To test that the median equals 75, the output is

Sign Test for Median: Grade

```
Sign test of median = 75.00 versus not = 75.00

         N   Below   Equal   Above         P   Median
Grade   40      27       0      13    0.0385    70.00
```

Since the *p*-value <0.05, reject the null hypothesis.

THE MANN–WHITNEY *U* TEST

17.5 Referring to Table 17.2, determine whether there is a difference at the 0.05 significance level between cables made of alloy I and alloy II. Work the problem first by hand, supplying all the details for the Mann–Witney *U* test. Follow this with the MINITAB solution to the problem.

SOLUTION

We organize the work in accordance with steps 1, 2, and 3 (described earlier in this chapter):

Step 1. Combining all 18 sample values in an array from the smallest to the largest gives us the first line of Table 17.8. These values are numbered 1 to 18 in the second line, which gives us the ranks.

Table 17.8

10.7	11.8	12.6	12.9	14.1	14.7	15.2	15.9	16.1	16.4	17.8	18.3	18.9	19.6	20.5	22.7	24.2	25.3
1	2	3	4	5	6	7	8	9	10	11	12	13	14	15	16	17	18

Step 2. To find the sum of the ranks for each sample, rewrite Table 17.2 by using the associated ranks from Table 17.8; this gives us Table 17.9. The sum of the ranks is 106 for alloy I and 65 for alloy II.

Table 17.9

Alloy I		Alloy II	
Cable Strength	Rank	Cable Strength	Rank
18.3	12	12.6	3
16.4	10	14.1	5
22.7	16	20.5	15
17.8	11	10.7	1
18.9	13	15.9	8
25.3	18	19.6	14
16.1	9	12.9	4
24.2	17	15.2	7
	Sum 106	11.8	2
		14.7	6
			Sum 65

Step 3. Since the alloy I sample has the smaller size, $N_1 = 8$ and $N_2 = 10$. The corresponding sums of the ranks are $R_1 = 106$ and $R_2 = 65$. Then

$$U = N_1N_2 + \frac{N_1(N_1+1)}{2} - R_1 = (8)(10) + \frac{(8)(9)}{2} - 106 = 10$$

$$\mu_U = \frac{N_1N_2}{2} = \frac{(8)(10)}{2} = 40 \qquad \sigma_U^2 = \frac{N_1N_2(N_1+N_2+1)}{12} = \frac{(8)(10)(19)}{12} = 126.67$$

Thus $\sigma_U = 11.25$ and

$$z = \frac{U - \mu_U}{\sigma_U} = \frac{10-40}{11.25} = -2.67$$

Since the hypothesis H_0 that we are testing is whether there is *no* difference between the alloys, a two-tailed test is required. For the 0.05 significance level, we have the decision rule:

Accept H_0 if $-1.96 \le z \le 1.96$.
Reject H_0 otherwise.

Because $z = -2.67$, we reject H_0 and conclude that there is a difference between the alloys at the 0.05 level.

The MINITAB solution to the problem is as follows: First, the data values for alloy I and alloy II are entered into columns C1 and C2, respectively, and the columns are named `AlloyI` and `AlloyII`. The pull-down **Stat → Nonparametrics → Mann-Whitney** gives the following output.

```
Mann-Whitney Test and CI: AlloyI, AlloyII

          N   Median
AlloyI    8   18.600
AlloyII  10   14.400
```

```
Point estimate for ETA1-ETA2 is      4.800
95.4 Percent CI for ETA1-ETA2 is (2.000, 9.401)
W = 106.0
Test of ETA1 = ETA2  vs  ETA1 not = ETA2 is significant at 0.0088
```

The MINITAB output gives the median strength of the cables for each sample, a point estimate of the difference in population medians (ETA1-ETA2), a confidence interval for the difference in population medians, the sum of the ranks for the first named variable ($W = 106$), and the two-tailed p-value $= 0.0088$. Since the p-value <0.05, the null hypothesis would be rejected. We would conclude that alloy 1 results in stronger cables.

17.6 Verify results (*6*) and (*7*) of this chapter for the data of Problem 17.5.

SOLUTION

(*a*) Since samples 1 and 2 yield values for U given by

$$U_1 = N_1N_2 + \frac{N_1(N_1+1)}{2} - R_1 = (8)(10) + \frac{(8)(9)}{2} - 106 = 10$$

$$U_2 = N_1N_2 + \frac{N_2(N_2+1)}{2} - R_2 = (8)(10) + \frac{(10)(11)}{2} - 65 = 70$$

we have $U_1 + U_2 = 10 + 70 = 80$, and $N_1N_2 = (8)(10) = 80$.

(*b*) Since $R_1 = 106$ and $R_2 = 65$, we have $R_1 + R_2 = 106 + 65 = 171$ and

$$\frac{N(N+1)}{2} = \frac{(N_1+N_2)(N_1+N_2+1)}{2} = \frac{(18)(19)}{2} = 171$$

17.7 Work Problem 17.5 by using the statistic U for the alloy II sample.

SOLUTION

For the alloy II sample,

$$U = N_1N_2 + \frac{N_2(N_2+1)}{2} - R_2 = (8)(10) + \frac{(10)(11)}{2} - 65 = 70$$

so that

$$z = \frac{U - \mu_U}{\sigma_U} = \frac{70 - 40}{11.25} = 2.67$$

This value of z is the *negative* of the z in Problem 17.5, and the right-hand tail of the normal distribution is used instead of the left-hand tail. Since this value of z also lies outside $-1.96 \le z \le 1.96$, the conclusion is the same as that for Problem 17.5.

17.8 A professor has two classes in psychology: a morning class of 9 students, and an afternoon class of 12 students. On a final examination scheduled at the same time for all students, the classes received the grades shown in Table 17.10. Can one conclude at the 0.05 significance level that the morning class performed worse than the afternoon class? Work the problem first by hand, supplying all the details for the Mann–Whitney U test. Follow this with the MINITAB solution to the problem.

Table 17.10

Morning class	73	87	79	75	82	66	95	75	70			
Afternoon class	86	81	84	88	90	85	84	92	83	91	53	84

SOLUTION

Step 1. Table 17.11 shows the array of grades and ranks. Note that the rank for the two grades of 75 is $\frac{1}{2}(5+6) = 5.5$, while the rank for the three grades of 84 is $\frac{1}{3}(11+12+13) = 12$.

Table 17.11

53	66	70	73	75 75	79	81	82	83	84 84 84	85	86	87	88	90	91	92	95
1	2	3	4	5.5	7	8	9	10	12	14	15	16	17	18	19	20	21

Step 2. Rewriting Table 17.10 in terms of ranks gives us Table 17.12.
Check: $R_1 = 73$, $R_2 = 158$, and $N = N_1 + N_2 = 9 + 12 = 21$; thus $R_1 + R_2 = 73 + 158 = 231$ and

$$\frac{N(N+1)}{2} = \frac{(21)(22)}{2} = 231 = R_1 + R_2$$

Table 17.12

													Sum of Ranks
Morning class	4	16	7	5.5	9	2	21	5.5	3				73
Afternoon class	15	8	12	17	18	14	12	20	10	19	1	12	158

Step 3.

$$U = N_1 N_2 + \frac{N_1(N_1+1)}{2} - R_1 = (9)(12) + \frac{(9)(10)}{2} - 73 = 80$$

$$\mu_U = \frac{N_1 N_2}{2} = \frac{(9)(12)}{2} = 54 \qquad \sigma_U^2 = \frac{N_1 N_2 (N_1 + N_2 + 1)}{12} = \frac{(9)(12)(22)}{12} = 198$$

Thus

$$z = \frac{U - \mu_U}{\sigma_U} = \frac{80 - 54}{14.07} = 1.85$$

The one-tailed p-value is given by the EXCEL expression `=1-NORMSDIST(1.85)` which yields 0.0322. Since the p-value <0.05, we can conclude that the morning class performs worse than the afternoon class.

The MINITAB solution to the problem is as follows. First the data values for the morning and afternoon classes are entered into columns `C1` and `C2` and the columns are named morning and afternoon. The pull-down **Stat → Nonparametrics → Mann-Whitney** gives the following output.

Mann-Whitney Test and CI: Morning, Afternoon

```
             N   Median
Morning      9    75.00
Afternoon   12    84.50

Point estimate for ETA1-ETA2 is -9.00
95.7 Percent CI for ETA1-ETA2 is (-15.00, 2.00)
W = 73.0
Test of ETA1 = ETA2 vs ETA1 < ETA2 is significant at 0.0350
The test is significant at 0.0348 (adjusted for ties)
```

The MINITAB output gives the median grade for each sample, a point estimate of the difference in population medians, a confidence interval for the difference in population medians, the sum of ranks for the first named variable (in this case, the morning class), and the one-tail p-value = 0.0350. Since the p-value is less than 0.05, the null hypothesis would be rejected. We conclude that the morning class performs worse than the afternoon class.

17.9 Find U for the data of Table 17.13 by using (*a*) formula (*2*) of this chapter and (*b*) the counting method (as described in Remark 4 of this chapter).

SOLUTION

(*a*) Arranging the data from both samples in an array in increasing order of magnitude and assigning ranks from 1 to 5 gives us Table 17.14. Replacing the data of Table 17.13 with the corresponding ranks gives us Table 17.15, from which the sums of the ranks are $R_1 = 5$ and $R_2 = 10$. Since $N_1 = 2$ and $N_2 = 3$, the value of U for sample 1 is

$$U = N_1 N_2 + \frac{N_1(N_1 + 1)}{2} - R_1 = (2)(3) + \frac{(2)(3)}{2} - 5 = 4$$

The value of U for sample 2 can be found similarly to be $U = 2$.

Table 17.13

Sample 1	22	10	
Sample 2	17	25	14

Table 17.14

Data	10	14	17	22	25
Rank	1	2	3	4	5

Table 17.15

				Sum of Ranks
Sample 1	4	1		5
Sample 2	3	5	2	10

(*b*) Let us replace the sample values in Table 17.14 with I or II, depending on whether the value belongs to sample 1 or 2. Then the first line of Table 17.14 becomes

Data	I	II	II	I	II

From this we see that

Number of sample 1 values preceding first sample 2 value = 1
Number of sample 1 values preceding second sample 2 value = 1
Number of sample 1 values preceding third sample 2 value = 2
Total = 4

Thus the value of U corresponding to the first sample is 4.

Similarly, we have

Number of sample 2 values preceding first sample 1 value = 0
Number of sample 2 values preceding second sample 1 value = 2
Total = 2

Thus the value of U corresponding to the second sample is 2.

Note that since $N_1 = 2$ and $N_2 = 3$, these values satisfy $U_1 + U_2 = N_1 N_2$; that is, $4 + 2 = (2)(3) = 6$.

17.10 A population consists of the values 7, 12, and 15. Two samples are drawn without replacement from this population: sample 1, consisting of one value, and sample 2, consisting of two values. (Between them, the two samples exhaust the population.)

(*a*) Find the sampling distribution of U and its graph.

(*b*) Find the mean and variance of the distribution in part (*a*).

(*c*) Verify the results found in part (*b*) by using formulas (*3*) of this chapter.

SOLUTION

(*a*) We choose sampling without replacement to avoid ties—which would occur if, for example, the value 12 were to appear in both samples.

There are $3 \cdot 2 = 6$ possibilities for choosing the samples, as shown in Table 17.16. It should be noted that we could just as easily use ranks 1, 2, and 3 instead of 7, 12, and 15. The value U in Table 17.16 is that found for sample 1, but if U for sample 2 were used, the distribution would be the same.

Table 17.16

Sample 1	Sample 2	U
7	12 15	2
7	15 12	2
12	7 15	1
12	15 7	1
15	7 12	0
15	12 7	0

A graph of this distribution is shown in Fig. 17-4, where f is the frequency. The probability distribution of U can also be graphed; in this case $\Pr\{U=0\} = \Pr\{U=1\} = \Pr\{U=2\} = \frac{1}{3}$. The required graph is the same as that shown in Fig. 17-4, but with ordinates 1 and 2 replaced by $\frac{1}{6}$ and $\frac{1}{3}$, respectively.

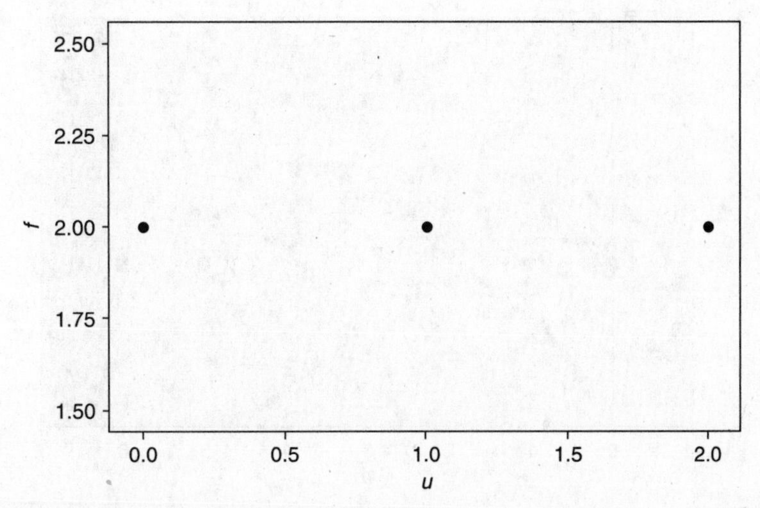

Fig. 17-4 MINITAB plot of the sampling distribution for U and $N_1 = 1$ and $N_2 = 2$.

(*b*) The mean and variance found from Table 17.16 are given by

$$\mu_U = \frac{2+2+1+1+0+0}{6} = 1$$

$$\sigma_U^2 = \frac{(2-1)^2 + (2-1)^2 + (1-1)^2 + (1-1)^2 + (0-1)^2 + (0-1)^2}{6} = \frac{2}{3}$$

(*c*) By formulas (*3*),

$$\mu_U = \frac{N_1 N_2}{2} = \frac{(1)(2)}{2} = 1$$

$$\sigma_U^2 = \frac{N_1 N_2 (N_1 + N_2 + 1)}{12} = \frac{(1)(2)(1+2+1)}{12} = \frac{2}{3}$$

showing agreement with part (*a*).

17.11 (*a*) Find the sampling distribution of U in Problem 17.9 and graph it.

(*b*) Graph the corresponding probability distribution of U.

(*c*) Obtain the mean and variance of U directly from the results of part (*a*).

(*d*) Verify part (*c*) by using formulas (*3*) of this chapter.

SOLUTION

(*a*) In this case there are $5 \cdot 4 \cdot 3 \cdot 2 = 120$ possibilities for choosing values for the two samples and the method of Problem 17.9 is too laborious. To simplify the procedure, let us concentrate on the smaller sample (of size $N_1 = 2$) and the possible sums of the ranks, R_1. The sum of the ranks for sample 1 is the *smallest* when the sample consists of the two lowest-ranking numbers (1, 2); then $R_1 = 1 + 2 = 3$. Similarly, the sum of the ranks for sample 1 is the *largest* when the sample consists of the two highest-ranking numbers (4, 5); then $R_1 = 4 + 5 = 9$. Thus R_1, varies from 3 to 9.

Column 1 of Table 17.17 lists these values of R_1 (from 3 to 9), and column 2 shows the corresponding sample 1 values, whose sum is R_1. Column 3 gives the frequency (or number) of samples with sum R_1; for example, there are $f = 2$ samples with $R_1 = 5$. Since $N_1 = 2$ and $N_2 = 3$, we have

$$U = N_1 N_2 + \frac{N_1(N_1+1)}{2} - R_1 = (2)(3) + \frac{(2)(3)}{2} - R_1 = 9 - R_1$$

From this we find the corresponding values of U in column 4 of the table; note that as R_1 varies from 3 to 9, U varies from 6 to 0. The sampling distribution is provided by columns 3 and 4, and the graph is shown in Fig. 17-5.

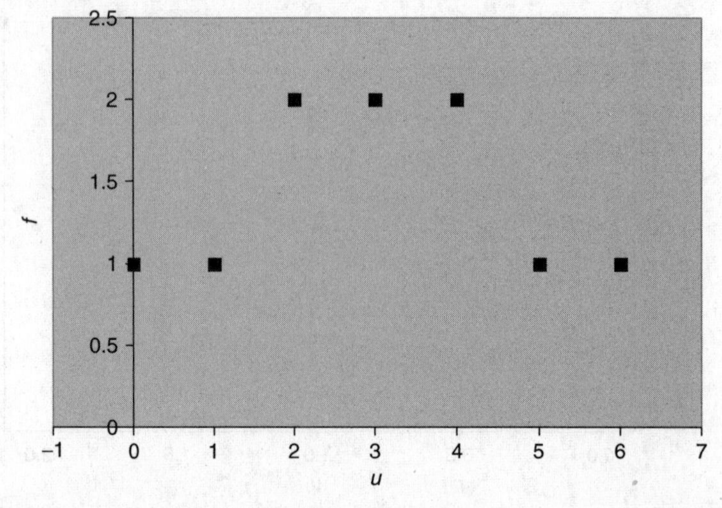

Fig. 17-5 EXCEL plot of the sampling distribution for U and $N_1 = 2$ and $N_2 = 3$.

(*b*) The probability that $U = R_1$ (i.e., $\Pr\{U = R_1\}$) is shown in column 5 of Table 17.17 and is obtained by finding the relative frequency. The relative frequency is found by dividing each frequency f by the sum of all the frequencies, or 10; for example, $\Pr\{U = 5\} = \frac{2}{10} = 0.2$. The graph of the probability distribution is shown in Fig. 17-6.

Table 17.17

R_1	Sample 1 Values	f	U	$\Pr\{U = R_1\}$
3	(1, 2)	1	6	0.1
4	(1, 3)	1	5	0.1
5	(1, 4), (2, 3)	2	4	0.2
6	(1, 5), (2, 4)	2	3	0.2
7	(2, 5), (3, 4)	2	2	0.2
8	(3, 5)	1	1	0.1
9	(4, 5)	1	0	0.1

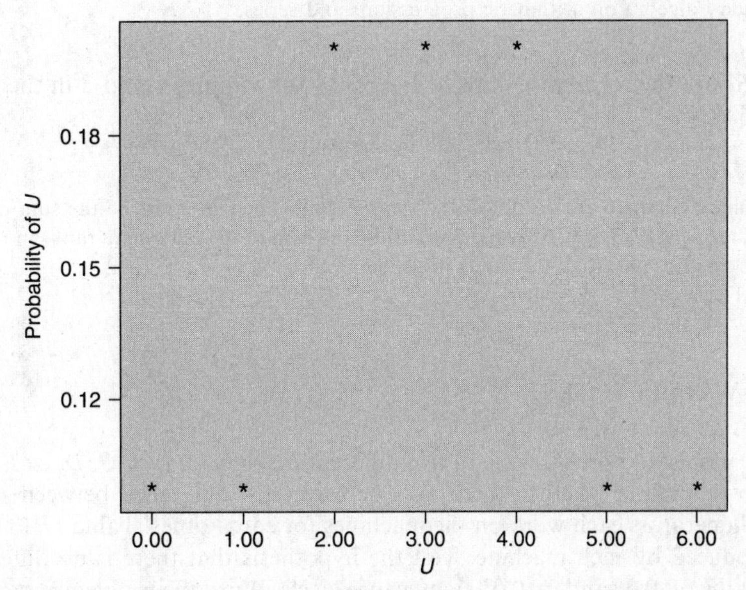

Fig. 17-6 SPSS plot of the probability distribution for U with $N_1 = 2$ and $N_2 = 3$.

(*c*) From columns 3 and 4 of Table 17.17 we have

$$\mu_U = \bar{U} = \frac{\sum fU}{\sum f} = \frac{(1)(6) + (1)(5) + (2)(4) + (2)(3) + (2)(2) + (1)(1) + (1)(0)}{1 + 1 + 2 + 2 + 2 + 1 + 1} = 3$$

$$\sigma_U^2 = \frac{\sum f(U - \bar{U})^2}{\sum f}$$

$$= \frac{(1)(6-3)^2 + (1)(5-3)^2 + (2)(4-3)^2 + (2)(3-3)^2 + (2)(2-3)^2 + (1)(1-3)^2 + (1)(0-3)^2}{10} = 3$$

Another method

$$\sigma_U^2 = \overline{U^2} - \bar{U}^2 = \frac{(1)(6)^2 + (1)(5)^2 + (2)(4)^2 + (2)(3)^2 + (2)(2)^2 + (1)(1)^2 + (1)(0)^2}{10} - (3)^2 = 3$$

(*d*) By formulas (*3*), using $N_1 = 2$ and $N_2 = 3$, we have

$$\mu_U = \frac{N_1 N_2}{2} = \frac{(2)(3)}{2} = 3 \qquad \sigma_U^2 = \frac{N_1 N_2(N_1 + N_2 + 1)}{12} = \frac{(2)(3)(6)}{12} = 3$$

17.12 If N numbers in a set are ranked from 1 to N, prove that the sum of the ranks is $[N(N+1)]/2$.

SOLUTION

Let R be the sum of the ranks. Then we have

$$R = 1 + 2 + 3 + \cdots + (N-1) + N \tag{16}$$

$$R = N + (N-1) + (N-2) + \cdots + 2 + 1 \tag{17}$$

where the sum in equation (*17*) is obtained by writing the sum in (*16*) backward. Adding equations (*16*) and (*17*) gives

$$2R = (N+1) + (N+1) + (N+1) + \cdots + (N+1) + (N+1) = N(N+1)$$

since $(N+1)$ occurs N times in the sum; thus $R = [N(N+1)]/2$. This can also be obtained by using a result from elementary algebra on arithmetic progressions and series.

17.13 If R_1 and R_2 are the respective sums of the ranks for samples 1 and 2 in the U test, prove that $R_1 + R_2 = [N(N+1)]/2$.

SOLUTION

We assume that there are no ties in the sample data. Then R_1 must be the sum of some of the ranks (numbers) in the set $1, 2, 3, \ldots, N$, while R_2 must be the sum of the remaining ranks in the set. Thus the sum $R_1 + R_2$ must be the sum of all the ranks in the set; that is, $R_1 + R_2 = 1 + 2 + 3 + \cdots + N = [N(N+1)]/2$ by Problem 17.12.

THE KRUSKAL–WALLIS *H* TEST

17.14 A company wishes to purchase one of five different machines: *A*, *B*, *C*, *D*, or *E*. In an experiment designed to determine whether there is a performance difference between the machines, five experienced operators each work on the machines for equal times. Table 17.18 shows the number of units produced by each machine. Test the hypothesis that there is no difference between the machines at the (*a*) 0.05 and (*b*) 0.01 significance levels. Work the problem first by hand, supplying all the details for the Kruskal–Wallis *H* test. Follow this with the MINITAB solution to the problem.

Table 17.18

A	68	72	77	42	53
B	72	53	63	53	48
C	60	82	64	75	72
D	48	61	57	64	50
E	64	65	70	68	53

Table 17.19

						Sum of Ranks
A	17.5	21	24	1	6.5	70
B	21	6.5	12	6.5	2.5	48.5
C	10	25	14	23	21	93
D	2.5	11	9	14	4	40.5
E	14	16	19	17.5	6.5	73

SOLUTION

Since there are five samples (A, B, C, D, and E), $k = 5$. And since each sample consists of five values, we have $N_1 = N_2 = N_3 = N_4 = N_5 = 5$, and $N = N_1 + N_2 + N_3 + N_4 + N_5 = 25$. By arranging all the values in increasing order of magnitude and assigning appropriate ranks to the ties, we replace Table 17.18 with Table 17.19, the right-hand column of which shows the sum of the ranks. We see from Table 17.19 that $R_1 = 70$, $R_2 = 48.5$, $R_3 = 93$, $R_4 = 40.5$, and $R_5 = 73$. Thus

$$H = \frac{12}{N(N+1)} \sum_{j=1}^{k} \frac{R_j^2}{N_j} - 3(N+1)$$

$$= \frac{12}{(25)(26)} \left[\frac{(70)^2}{5} + \frac{(48.5)^2}{5} + \frac{(93)^2}{5} + \frac{(40)^2}{5} + \frac{(73)^2}{5} \right] - 3(26) = 6.44$$

For $k - 1 = 4$ degrees of freedom at the 0.05 significance level, from Appendix IV we have $\chi^2_{.95} = 9.49$. Since $6.44 < 9.49$, we cannot reject the hypothesis of no difference between the machines at the 0.05 level and therefore certainly cannot reject it at the 0.01 level. In other words, we can accept the hypothesis (or reserve judgment) that there is no difference between the machines at both levels.

Note that we have already worked this problem by using analysis of variance (see Problem 16.8) and have arrived at the same conclusion.

The solution to the problem by using MINITAB proceeds as follows. First the data need to be entered in the worksheet in stacked form. The data structure is as follows:

Row	Machine	Units
1	1	68
2	1	72
3	1	77
4	1	42
5	1	53
6	2	72
7	2	53
8	2	63
9	2	53
10	2	48
11	3	60
12	3	82
13	3	64
14	3	75
15	3	72
16	4	48
17	4	61
18	4	57
19	4	64
20	4	50
21	5	64
22	5	65
23	5	70
24	5	68
25	5	53

The pull-down **Stat** → **Nonparametrics** → **Kruskal-Wallis** gives the following output:

Kruskal-Wallis Test: Units versus Machine

Machine	N	Median	Ave Rank	Z
1	5	68.00	14.0	0.34
2	5	53.00	9.7	-1.12
3	5	72.00	18.6	1.90
4	5	57.00	8.1	-1.66
5	5	65.00	14.6	0.54
Overall	25		13.0	

```
H = 6.44   DF = 4   P = 0.168
H = 6.49   DF = 4   P = 0.165 (adjusted for ties)
```

Note that two *p*-values are provided. One is adjusted for ties that occur in the ranking procedure and one is not. It is clear that the machines are not statistically different with respect to the number of units produced at either the 0.05 or the 0.01 levels since either *p*-value far exceeds both levels.

17.15 Work Problem 17.14 using the software package STATISTIX and the software package SPSS.

SOLUTION

The data file is structured as in problem 17.14. The STATISTIX pull-down **Statistics → One, Two, Multi-sample tests → Kruskal-Wallis One-way AOV Test** gives the following output. Compare this with the output in Problem 17.14.

Statistix 8.0

Kruskal-Wallis One-Way Nonparametric AOV for Units by Machine

Machine	Mean Rank	Sample Size
1	14.0	5
2	9.7	5
3	18.6	5
4	8.1	5
5	14.6	5
Total	13.0	25

Kruskal-Wallis Statistic	6.4949
p-Value, Using Chi-Squared Approximation	0.1651

The STATISTIX output compares with the MINITAB output adjusted for ties.
The SPSS output also compares with the MINITAB output adjusted for ties.

Kruskal-Wallis Test

Ranks

	Machine	N	Mean Rank
Units	1.00	5	14.00
	2.00	5	9.70
	3.00	5	18.60
	4.00	5	8.10
	5.00	5	14.60
	Total	25	

Test Statistics[a,b]

	Units
Chi-Square	6.495
df	4
Asymp. Sig.	.165

[a]Kruskal Wallis Test

[b]Grouping Variable: Machine

17.16 Table 17.20 gives the number of DVD's rented during the past year for random samples of teachers, lawyers, and physicians. Use the Kruskal–Wallis H test procedure in SAS to test the null hypothesis that the distributions of rentals are the same for the three professions. Test at the 0.01 level.

Table 17.20

Teachers	Lawyers	Physicians
18	2	14
4	16	30
5	21	11
9	24	1
20	5	7
26	2	5
7	50	14
17	10	7
43	7	16
20	49	14
24	35	27
7	1	19
34	45	15
30	6	22
45	9	20
2	24	10
45	36	
9	50	
	44	
	3	

SOLUTION

The data are entered in two columns. One column contains 1, 2, or 3 for teacher, lawyer, and physician. The other column contains the number of DVD's rented. The SAS pull-down **Statistics → ANOVA → Nonparametric Oneway ANOVA** gives the following output.

```
                    The NPAR1WAY Procedure

           Wilcoxon Scores (Rank Sums) for Variable number
                 Classified by Variable Profession

                      Sum of      Expected      Std Dev         Mean
Profession      N     Scores      Under H0      Under H0        Score
ffffffffffffffffffffffffffffffffffffffffffffffffffffffffffffffffffff
1              18     520.50       495.0      54.442630      29.916667
2              20     574.50       550.0      55.770695      28.725000
3              16     390.00       440.0      52.735539      24.375000

                  Average scores were used for ties.

                       Kruskal-Wallis Test

                   Chi-Square         0.9004
                   DF                      2
                   Pr > Chi-Sqaure    0.6375
```

The p-value is indicated as `Pr > Chi-Square 0.6375`. Since the p-value is much greater than 0.05, do not reject the null hypothesis.

THE RUNS TEST FOR RANDOMNESS

17.17 In 30 tosses of a coin the following sequence of heads (H) and tails (T) is obtained:

H T T H T H H H T H H T T H T

H T H H T H T T H T H H T H T

(*a*) Determine the number of runs, V.

(*b*) Test at the 0.05 significance level whether the sequence is random.

Work the problem first by hand, supplying all the details of the runs test for randomness. Follow this with the MINITAB solution to the problem.

SOLUTION

(*a*) Using a vertical bar to indicate a run, we see from

H | T T | H | T | H H H | T | H H | T T | H | T |

H | T | H H | T | H | T T | H | T | H H | T | H | T |

that the number of runs is $V = 22$.

(*b*) There are $N_1 = 16$ heads and $N_2 = 14$ tails in the given sample of tosses, and from part (*a*) the number of runs is $V = 22$. Thus from formulas (*13*) of this chapter we have

$$\mu_V = \frac{2(16)(14)}{16+14} + 1 = 15.93 \qquad \sigma_V^2 = \frac{2(16)(14)[2(16)(14) - 16 - 14]}{(16+14)^2(16+14-1)} = 7.175$$

or $\sigma_V = 2.679$. The z score corresponding to $V = 22$ runs is therefore

$$z = \frac{V - \mu_V}{\sigma_V} = \frac{22 - 15.93}{2.679} = 2.27$$

Now for a two-tailed test at the 0.05 significance level, we would accept the hypothesis H_0 of randomness if $-1.96 \le z \le 1.96$ and would reject it otherwise (see Fig. 17-7). Since the calculated value of z is $2.27 > 1.96$, we conclude that the tosses are not random at the 0.05 level. The test shows that there are *too many* runs, indicating a *cyclic pattern* in the tosses.

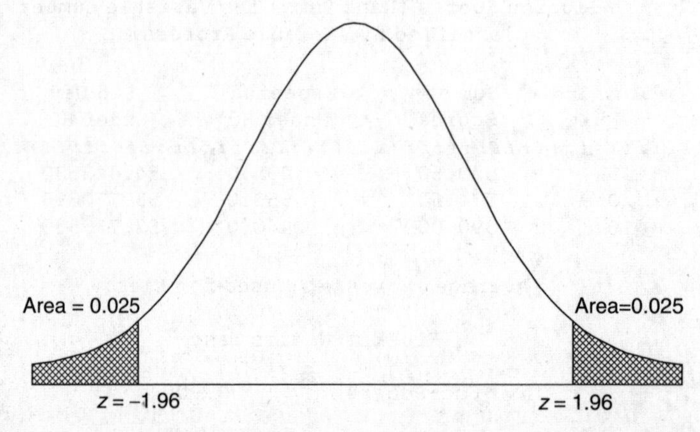

Fig. 17-7 Rejection region under the standard normal curve for level of significance equal to 0.05.

If the correction for continuity is used, the above z score is replaced by

$$z = \frac{(22 - 0.5) - 15.93}{2.679} = 2.08$$

and the same conclusion is reached.

The solution to the problem by using MINITAB proceeds as follows. The data are entered into column C1 as follows. Each head is represented by the number 1 and each tail is represented by the number 0. Column 1 is named Coin. The MINITAB pull-down **Stat → Nonparametrics → Runs Test** gives the following output.

Runs Test: Coin

```
Runs test for Coin
Runs above and below K = 0.533333

The observed number of runs = 22
The expected number of runs = 15.9333
16 Observations above K    14 below
p-value = 0.024
```

The value shown for K is the mean of the zeros and ones in column 1. The number of observations above and below K will be the number of heads and tails in the 30 tosses of the coin. The p-value is equal to 0.0235. Since this p-value is less than 0.05, we reject the null hypothesis. The number of runs is not random.

17.18 A sample of 48 tools produced by a machine shows the following sequence of good (G) and defective (D) tools:

G G G G G G D D G G G G G G G G

G G D D D D G G G G G G D G G G

G G G G G G D D G G G G G D G G

Test the randomness of the sequence at the 0.05 significance level. Also use SPSS to test the randomness of the sequence.

SOLUTION

The numbers of D's and G's are $N_1 = 10$ and $N_2 = 38$, respectively, and the number of runs is $V = 11$. Thus the mean and variance are given by

$$\mu_V = \frac{2(10)(38)}{10+38} + 1 = 16.83 \qquad \sigma_V^2 = \frac{2(10)(38)[2(10)(38) - 10 - 38]}{(10+38)^2(10+38-1)} = 4.997$$

so that $\sigma_V = 2.235$.

For a two-tailed test at the 0.05 level, we would accept the hypothesis H_0 of randomness if $-1.96 \le z \le 1.96$ (see Fig. 17-7) and would reject it otherwise. Since the z score corresponding to $V = 11$ is

$$z = \frac{V - \mu_V}{\sigma_V} = \frac{11 - 16.83}{2.235} = -2.61$$

and $-2.61 < -1.96$, we can reject H_0 at the 0.05 level.

The test shows that there are *too few* runs, indicating a clustering (or bunching) of defective tools. In other words, there seems to be a *trend pattern* in the production of defective tools. Further examination of the production process is warranted.

The pull-down **Analyze → Nonparametric Tests → Runs** gives the following SPSS output. If the G's are replaced by 1's and the D's are replaced by 0's, the test value, 0.7917, is the mean of these values. The other values are $N_1 = 10$, $N_2 = 38$, Sum = 48, and $V = 11$. The computed value of z has the correction for continuity in it. This causes the z value to differ from the z value computed above. The term `Asymp. Sig. (2-tailed)` is the two-tailed p-value corresponding to $z = -2.386$. We see that the SPSS output has the same information as we computed by hand. We simply have to know how to interpret it.

Runs Test

	Quality
Test Value[a]	.7917
Cases < Test Value	10
Cases > = Test Value	38
Total Cases	48
Number of Runs	11
Z	−2.386
Asymp. Sig. (2-tailed)	.017

[a]Mean

17.19 (*a*) Form all possible sequences consisting of three *a*'s and two *b*'s and give the numbers of runs, V, corresponding to each sequence.

(*b*) Obtain the sampling distribution of V and its graph.

(*c*) Obtain the probability distribution of V and its graph.

SOLUTION

(*a*) The number of possible sequences consisting of three *a*'s and two *b*'s is

$$\binom{5}{2} = \frac{5!}{2!3!} = 10$$

These sequences are shown in Table 17.21, along with the number of runs corresponding to each sequence.

(*b*) The sampling distribution of V is given in Table 17.22 (obtained from Table 17.21), where V denotes the number of runs and f denotes the frequency. For example, Table 17.22 shows that there is one 5, four 4's, etc. The corresponding graph is shown in Fig. 17-8.

Table 17.21

Sequence	Runs (V)
a a a b b	2
a a b a b	4
a a b b a	3
a b a b a	5
a b b a a	3
a b a a b	4
b b a a a	2
b a b a a	4
b a a a b	3
b a a b a	4

Table 17.22

V	f
2	2
3	3
4	4
5	1

(c) The probability distribution of V, graphed in Fig. 17-9, is obtained from Table 17.22 by dividing each frequency by the total frequency $2+3+4+1=10$. For example, $\Pr\{V=5\}=\frac{1}{10}=0.1$.

17.20 Find (a) the mean and (b) the variance of the number of runs in Problem 17.19 directly from the results obtained there.

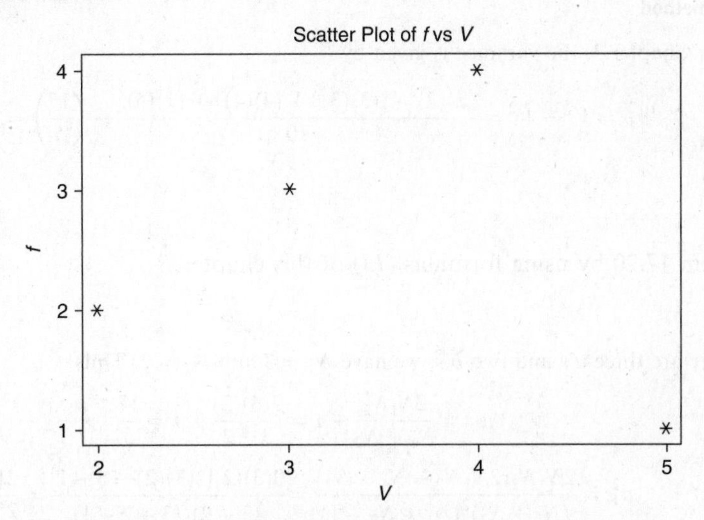

Fig. 17-8 STATISTIX graph of the sampling distribution of V.

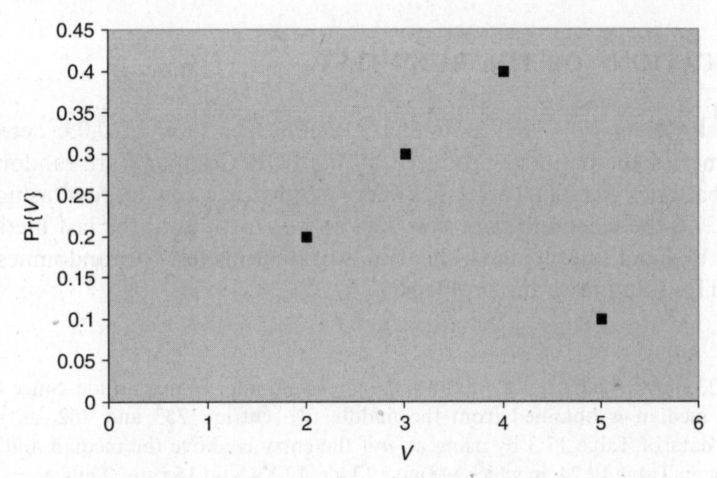

Fig. 17-9 EXCEL graph of the probability distribution of V.

SOLUTION

(a) From Table 17.21 we have

$$\mu_V=\frac{2+4+3+5+3+4+2+4+3+4}{10}=\frac{17}{5}$$

Another method

From Table 17.21 the grouped-data method gives

$$\mu_V=\frac{\sum fV}{\sum f}=\frac{(2)(2)+(3)(3)+(4)(4)+(1)(5)}{2+3+4+1}=\frac{17}{5}$$

(*b*) Using the grouped-data method for computing the variance, from Table 17.22 we have

$$\sigma_V^2 = \frac{\sum f(V-\bar{V})^2}{\sum f} = \frac{1}{10}\left[(2)\left(2-\frac{17}{5}\right)^2 + (3)\left(3-\frac{17}{5}\right)^2 + (4)\left(4-\frac{17}{5}\right)^2 + (1)\left(5-\frac{17}{5}\right)^2\right] = \frac{21}{25}$$

Another method

As in Chapter 3, the variance is given by

$$\sigma_V^2 = \overline{V^2} - \bar{V}^2 = \frac{(2)(2)^2 + (3)(3)^2 + (4)(4)^2 + (1)(5)^2}{10} - \left(\frac{17}{5}\right)^2 = \frac{21}{25}$$

17.21 Work Problem 17.20 by using formulas (*13*) of this chapter.

SOLUTION

Since there are three *a*'s and two *b*'s, we have $N_1 = 3$ and $N_2 = 2$. Thus

(*a*)
$$\mu_V = \frac{2N_1N_2}{N_1+N_2} + 1 = \frac{2(3)(2)}{3+2} + 1 = \frac{17}{5}$$

(*b*)
$$\sigma_V^2 = \frac{2N_1N_2(2N_1N_2 - N_1 - N_2)}{(N_1+N_2)^2(N_1+N_2-1)} = \frac{2(3)(2)[2(3)(2)-3-2]}{(3+2)^2(3+2-1)} = \frac{21}{25}$$

FURTHER APPLICATIONS OF THE RUNS TEST

17.22 Referring to Problem 17.3, and assuming a significance level of 0.05, determine whether the sample lifetimes of the batteries produced by the PQR Company are random. Assume the lifetimes of the batteries given in Table 17.3 were recorded in a row by row fashion. That is, the first lifetime was 217, the second lifetime was 230, and so forth until the last lifetime, 268. Work the problem first by hand, supplying all the details of the runs test for randomness. Follow this with the STATISTIX solution to the problem.

SOLUTION

Table 17.23 shows the batteries' lifetimes in increasing order of magnitude. Since there are 24 entries in the table, the median is obtained from the middle two entries, 253 and 262, as $\frac{1}{2}(253+262) = 257.5$. Rewriting the data of Table 17.3 by using an *a* if the entry is above the median and a *b* if it is below the median, we obtain Table 17.24, in which we have 12 *a*'s, 12 *b*'s, and 15 runs. Thus $N_1 = 12$, $N_2 = 12$, $N = 24$, $V = 15$, and we have

$$\mu_V = \frac{2N_1N_2}{N_1+N_2} + 1 = \frac{2(12)(12)}{12+12} + 1 = 13 \qquad \sigma_V^2 = \frac{2(12)(12)(264)}{(24)^2(23)} = 5.739$$

so that
$$z = \frac{V - \mu_V}{\sigma_V} = \frac{15-13}{2.396} = 0.835$$

Table 17.23

198	211	216	219	224	225	230	236
243	252	253	253	262	264	268	271
272	275	282	284	288	291	294	295

Table 17.24

a	*b*	*b*	*a*	*a*	*b*	*a*	*b*
b	*b*	*a*	*a*	*b*	*a*	*b*	*b*
a	*a*	*b*	*b*	*a*	*b*	*a*	*a*

Using a two-tailed test at the 0.05 significance level, we would accept the hypothesis of randomness if $-1.96 \leq z \leq 1.96$. Since 0.835 falls within this range, we conclude that the sample is random.

The STATISTIX analysis proceeds as follows. The lifetimes are entered into column 1 in the order in which they were collected. The column is named `Lifetime`. The pull-down **Statistics → Randomness/ Normality Tests → Runs Test** gives the following output.

```
Statistix 8.0

Runs Test for Lifetimes

Median                          257.50
Values Above the Median             12
Values below the Median             12
Values Tied with the Median          0
Runs Above the Median                8
Runs Below the Median                7
Total Number of Runs                15
Expected Number of Runs           13.0

p-Value, Two-Tailed Test                 0.5264
Probability of getting 15 or fewer runs  0.8496
Probability of getting 14 or more runs   0.2632
```

The large *p*-value indicates that the number of runs may be regarded as being random.

17.23 Work Problem 17.5 by using the runs test for randomness.

SOLUTION

The arrangement of all values from both samples already appears in line 1 of Table 17.8. Using the symbols *a* and *b* for the data from samples I and II, respectively, the arrangement becomes

$$b \quad b \quad b \quad b \quad b \quad b \quad b \quad b \quad a \quad a \quad a \quad a \quad a \quad b \quad b \quad a \quad a \quad a$$

Since there are four runs, we have $V = 4$, $N_1 = 8$, and $N_2 = 10$. Then

$$\mu_V = \frac{2N_1N_2}{N_1 + N_2} + 1 = \frac{2(8)(10)}{18} + 1 = 9.889$$

$$\sigma_V^2 = \frac{2N_1N_2(2N_1N_2 - N_1 - N_2)}{(N_1 + N_2)^2(N_1 + N_2 - 1)} + \frac{2(8)(10)(142)}{(18)^2(17)} = 4.125$$

so that
$$z = \frac{V - \mu_V}{\sigma_V} = \frac{4 - 9.889}{2.031} = -2.90$$

If H_0 is the hypothesis that there is no difference between the alloys, it is also the hypothesis that the above sequence is random. We would accept this hypothesis if $-1.96 \leq z \leq 1.96$ and would reject it otherwise. Since $z = -2.90$ lies outside this interval, we reject H_0 and reach the same conclusion as for Problem 17.5.

Note that if a correction is made for continuity,

$$z = \frac{V - \mu_V}{\sigma_V} = \frac{(4 + 0.5) - 9.889}{2.031} = -2.65$$

and we reach the same conclusion.

RANK CORRELATION

17.24 Table 17.25 shows how 10 students, arranged in alphabetical order, were ranked according to their achievements in both the laboratory and lecture sections of a biology course. Find the coefficient of rank correlation. Use SPSS to compute Spearman's rank correlation.

Table 17.25

Laboratory	8	3	9	2	7	10	4	6	1	5
Lecture	9	5	10	1	8	7	3	4	2	6

SOLUTION

The difference in ranks, D, in the laboratory and lecture sections for each student is given in Table 17.26, which also gives D^2 and $\sum D^2$. Thus

$$r_S = 1 - \frac{6 \sum D^2}{N(N^2 - 1)} = 1 - \frac{6(24)}{10(10^2 - 1)} = 0.8545$$

indicating that there is a marked relationship between the achievements in the course's laboratory and lecture sections.

Table 17.26

Difference of ranks (D)	−1	−2	−1	1	−1	3	1	2	−1	−1	
D^2	1	4	1	1	1	9	1	4	1	1	$\sum D^2 = 24$

Enter the data from Table 17.26 in columns named `Lab` and `Lecture`. The pull-down **Analyze → Correlate → Bivariate** results in the following output.

Correlations

			Lab	Lecture
Spearman's rho	Lab	Correlation Coefficient	1.000	.855
		Sig. (2-tailed)	.	.002
		N	10	10
	Lecture	Correlation Coefficient	.855	1.000
		Sig. (2-tailed)	.002	
		N	10	10

The output indicates a significant correlation between performance in `Lecture` and `Lab`.

17.25 Table 17.27 shows the heights of a sample of 12 fathers and their oldest adult sons. Find the coefficient of rank correlation. Work the problem first by hand, supplying all the details in finding the coefficient of rank correlation. Follow this with the SAS solution to the problem.

Table 17.27

Height of father (inches)	65	63	67	64	68	62	70	66	68	67	69	71
Height of son (inches)	68	66	68	65	69	66	68	65	71	67	68	70

SOLUTION

Arranged in ascending order of magnitude, the fathers' heights are

$$62\quad 63\quad 64\quad 65\quad 66\quad 67\quad 67\quad 68\quad 68\quad 69\quad 71 \tag{18}$$

Since the sixth and seventh places in this array represent the same height [67 inches (in)], we assign a mean rank $\frac{1}{2}(6+7) = 6.5$ to these places. Similarly, the eighth and ninth places are assigned the rank $\frac{1}{2}(8+9) = 8.5$. Thus the fathers' heights are assigned the ranks

$$1\quad 2\quad 3\quad 4\quad 5\quad 6.5\quad 6.5\quad 8.5\quad 8.5\quad 10\quad 11\quad 12 \tag{19}$$

Similarly, arranged in ascending order of magnitude, the sons' heights are

$$65\quad 65\quad 66\quad 66\quad 67\quad 68\quad 68\quad 68\quad 68\quad 69\quad 70\quad 71 \tag{20}$$

and since the sixth, seventh, eighth, and ninth places represent the same height (68 in), we assign the mean rank $\frac{1}{4}(6+7+8+9) = 7.5$ to these places. Thus the sons' heights are assigned the ranks

$$1.5\quad 1.5\quad 3.5\quad 3.5\quad 5\quad 7.5\quad 7.5\quad 7.5\quad 7.5\quad 10\quad 11\quad 12 \tag{21}$$

Using the correspondences *(18)* and *(19)*, and *(20)* and *(21)*, we can replace Table 17.27 with Table 17.28. Table 17.29 shows the difference in ranks, D, and the computations of D^2 and $\sum D^2$, whereby

$$r_S = 1 - \frac{6\sum D^2}{N(N^2-1)} = 1 - \frac{6(72.50)}{12(12^2-1)} = 0.7465$$

Table 17.28

Rank of father	4	2	6.5	3	8.5	1	11	5	8.5	6.5	10	12
Rank of son	7.5	3.5	7.5	1.5	10	3.5	7.5	1.5	12	5	7.5	11

Table 17.29

D	−3.5	−1.5	−1.0	1.5	−1.5	−2.5	3.5	3.5	−3.5	1.5	2.5	1.0	
D^2	12.25	2.25	1.00	2.25	2.25	6.25	12.25	12.25	12.25	2.25	6.25	1.00	$\sum D^2 = 72.50$

This result agrees well with the correlation coefficient obtained by other methods.

Enter the fathers' heights in the first column and the sons' heights in the second column and give the SAS pull-down command **Statistics → Descriptive → Correlations**. The following output is obtained.

```
                        The CORR Procedure
                   2 Variables:    Fatherht Sonht

                        Simple Statistics

Variable        N        Mean      Std Dev       Median      Minimum      Maximum
  Fatherht     12    66.66667     2.77434      67.00000     62.00000     71.00000
  Sonht        12    67.58333     1.88092      68.00000     65.00000     71.00000
```

```
Spearman Correlation Coefficient, N = 12
        prob > |r| under H0: Rho = 0
                    Fatherht     Sonht
      Fatherht      1.00000      0.74026
                                 0.0059
      Sonht         0.74026      1.00000
                                 0.0059
```

In addition to the simple statistics, SAS gives the Spearman correlation coefficient as 0.74. The p-value, 0.0059, can be used to test the null hypothesis that the population coefficient of rank correlation is equal to 0 versus the alternative that the population coefficient of rank correlation is different from 0. We conclude that there is a relationship between the heights of fathers and sons in the population.

Supplementary Problems

THE SIGN TEST

17.26 A company claims that if its product is added to an automobile's gasoline tank, the mileage per gallon will improve. To test the claim, 15 different automobiles are chosen and the mileage per gallon with and without the additive is measured; the results are shown in Table 17.30. Assuming that the driving conditions are the same, determine whether there is a difference due to the additive at significance levels of (*a*) 0.05 and (*b*) 0.01.

Table 17.30

With additive	34.7	28.3	19.6	25.1	15.7	24.5	28.7	23.5	27.7	32.1	29.6	22.4	25.7	28.1	24.3
Without additive	31.4	27.2	20.4	24.6	14.9	22.3	26.8	24.1	26.2	31.4	28.8	23.1	24.0	27.3	22.9

17.27 Can one conclude at the 0.05 significance level that the mileage per gallon achieved in Problem 17.26 is *better* with the additive than without it?

17.28 A weight-loss club advertises that a special program that it has designed will produce a weight loss of at least 6% in 1 month if followed precisely. To test the club's claim, 36 adults undertake the program. Of these, 25 realize the desired loss, 6 gain weight, and the rest remain essentially unchanged. Determine at the 0.05 significance level whether the program is effective.

17.29 A training manager claims that by giving a special course to company sales personnel, the company's annual sales will increase. To test this claim, the course is given to 24 people. Of these 24, the sales of 16 increase, those of 6 decrease, and those of 2 remain unchanged. Test at the 0.05 significance level the hypothesis that the course increased the company's sales.

17.30 The MW Soda Company sets up "taste tests" in 27 locations around the country in order to determine the public's relative preference for two brands of cola, A and B. In eight locations brand A is preferred over brand B, in 17 locations brand B is preferred over brand A, and in the remaining locations there is indifference. Can one conclude at the 0.05 significance level that brand B is preferred over brand A?

17.31 The breaking strengths of a random sample of 25 ropes made by a manufacturer are given in Table 17.31. On the basis of this sample, test at the 0.05 significance level the manufacturer's claim that the breaking strength of a rope is (*a*) 25, (*b*) 30, (*c*) 35, and (*d*) 40.

Table 17.31

41	28	35	38	23
37	32	24	46	30
25	36	22	41	37
43	27	34	27	36
42	33	28	31	24

17.32 Show how to obtain 95% confidence limits for the data in Problem 17.4.

17.33 Make up and solve a problem involving the sign test.

THE MANN–WHITNEY *U* TEST

17.34 Instructors *A* and *B* both teach a first course in chemistry at XYZ University. On a common final examination, their students received the grades shown in Table 17.32. Test at the 0.05 significance level the hypothesis that there is no difference between the two instructors' grades.

Table 17.32

A	88	75	92	71	63	84	55	64	82	96				
B	72	65	84	53	76	80	51	60	57	85	94	87	73	61

17.35 Referring to Problem 17.34, can one conclude at the 0.01 significance level that the students' grades in the morning class are worse than those in the afternoon class?

17.36 A farmer wishes to determine whether there is a difference in yields between two different varieties of wheat, I and II. Table 17.33 shows the production of wheat per unit area using the two varieties. Can the farmer conclude at significance levels of (*a*) 0.05 and (*b*) 0.01 that a difference exists?

Table 17.33

Wheat I	15.9	15.3	16.4	14.9	15.3	16.0	14.6	15.3	14.5	16.6	16.0
Wheat II	16.4	16.8	17.1	16.9	18.0	15.6	18.1	17.2	15.4		

17.37 Can the farmer of Problem 17.36 conclude at the 0.05 level that wheat II produces a larger yield than wheat I?

17.38 A company wishes to determine whether there is a difference between two brands of gasoline, *A* and *B*. Table 17.34 shows the distances traveled per gallon for each brand. Can we conclude at the 0.05 significance level (*a*) that there is a difference between the brands and (*b*) that brand *B* is better than brand *A*?

Table 17.34

A	30.4	28.7	29.2	32.5	31.7	29.5	30.8	31.1	30.7	31.8
B	33.5	29.8	30.1	31.4	33.8	30.9	31.3	29.6	32.8	33.0

17.39 Can the *U* test be used to determine whether there is a difference between machines I and II of Table 17.1? Explain.

17.40 Make up and solve a problem using the U test.

17.41 Find U for the data of Table 17.35, using (*a*) the formula method and (*b*) the counting method.

17.42 Work Problem 17.41 for the data of Table 17.36.

Table 17.35

Sample 1	15	25
Sample 2	20	32

Table 17.36

Sample 1	40	27	30	56
Sample 2	10	35		

17.43 A population consists of the values 2, 5, 9, and 12. Two samples are drawn from this population, the first consisting of one of these values and the second consisting of the other three values.

(*a*) Obtain the sampling distribution of U and its graph.

(*b*) Obtain the mean and variance of this distribution, both directly and by formula.

17.44 Prove that $U_1 + U_2 = N_1 N_2$.

17.45 Prove that $R_1 + R_2 = [N(N+1)]/2$ for the case where the number of ties is (*a*) 1, (*b*) 2, and (*c*) any number.

17.46 If $N_1 = 14$, $N_2 = 12$, and $R_1 = 105$, find (*a*) R_2, (*b*) U_1, and (*c*) U_2.

17.47 If $N_1 = 10$, $N_2 = 16$, and $U_2 = 60$, find (*a*) R_1, (*b*) R_2, and (*c*) U_1.

17.48 What is the largest number of the values N_1, N_2, R_1, R_2, U_1, and U_2 that can be determined from the remaining ones? Prove your answer.

THE KRUSKAL–WALLIS H TEST

17.49 An experiment is performed to determine the yields of five different varieties of wheat: A, B, C, D, and E. Four plots of land are assigned to each variety. The yields (in bushels per acre) are shown in Table 17.37. Assuming that the plots have similar fertility and that the varieties are assigned to the plots at random, determine whether there is a significant difference between the yields at the (*a*) 0.05 and (*b*) 0.01 levels.

Table 17.37

A	20	12	15	19
B	17	14	12	15
C	23	16	18	14
D	15	17	20	12
E	21	14	17	18

Table 17.38

A	33	38	36	40	31	35
B	32	40	42	38	30	34
C	31	37	35	33	34	30
D	27	33	32	29	31	28

17.50 A company wishes to test four different types of tires: A, B, C, and D. The lifetimes of the tires, as determined from their treads, are given (in thousands of miles) in Table 17.38; each type has been tried on six similar automobiles assigned to the tires at random. Determine whether there is a significant difference between the tires at the (*a*) 0.05 and (*b*) 0.01 levels.

17.51 A teacher wishes to test three different teaching methods: I, II, and III. To do this, the teacher chooses at random three groups of five students each and teaches each group by a different method. The same examination is then given to all the students, and the grades in Table 17.39 are obtained. Determine at the (*a*) 0.05 and (*b*) 0.01 significance levels whether there is a difference between the teaching methods.

Table 17.39

Method I	78	62	71	58	73
Method II	76	85	77	90	87
Method III	74	79	60	75	80

17.52 During one semester a student received in various subjects the grades shown in Table 17.40. Test at the (*a*) 0.05 and (*b*) 0.01 significance levels whether there is a difference between the grades in these subjects.

Table 17.40

Mathematics	72	80	83	75	
Science	81	74	77		
English	88	82	90	87	80
Economics	74	71	77	70	

17.53 Using the H test, work (*a*) Problem 16.9, (*b*) Problem 16.21, and (*c*) Problem 16.22.

17.54 Using the H test, work (*a*) Problem 16.23, (*b*) Problem 16.24, and (*c*) Problem 16.25.

THE RUNS TEST FOR RANDOMNESS

17.55 Determine the number of runs, V, for each of these sequences:

(*a*) A B A B B A A A B B A B

(*b*) H H T H H H T T T T H H T H H T H T

17.56 Twenty-five individuals were sampled as to whether they liked or did not like a product (indicated by Y and N, respectively). The resulting sample is shown by the following sequence:

Y Y N N N N Y Y Y N Y N N Y N N N N N Y Y Y Y N N

(*a*) Determine the number of runs, V.

(*b*) Test at the 0.05 significance level whether the responses are random.

17.57 Use the runs test on sequences (*10*) and (*11*) in this chapter, and state any conclusions about randomness.

17.58 (*a*) Form all possible sequences consisting of two *a*'s and one *b*, and give the number of runs, V, corresponding to each sequence.

(*b*) Obtain the sampling distribution of V and its graph.

(*c*) Obtain the probability distribution of V and its graph.

17.59 In Problem 17.58, find the mean and variance of V (*a*) directly from the sampling distribution and (*b*) by formula.

17.60 Work Problems 17.58 and 17.59 for the cases in which there are (*a*) two *a*'s and two *b*'s, (*b*) one *a* and three *b*'s, and (*c*) one *a* and four *b*'s.

17.61 Work Problems 17.58 and 17.59 for the case in which there are two *a*'s and four *b*'s.

FURTHER APPLICATIONS OF THE RUNS TEST

17.62 Assuming a significance level of 0.05, determine whether the sample of 40 grades in Table 17.5 is random.

17.63 The closing prices of a stock on 25 successive days are given in Table 17.41. Determine at the 0.05 significance level whether the prices are random.

Table 17.41

10.375	11.125	10.875	10.625	11.500
11.625	11.250	11.375	10.750	11.000
10.875	10.750	11.500	11.250	12.125
11.875	11.375	11.875	11.125	11.750
11.375	12.125	11.750	11.500	12.250

17.64 The first digits of $\sqrt{2}$ are 1.41421 35623 73095 0488···. What conclusions can you draw concerning the randomness of the digits?

17.65 What conclusions can you draw concerning the randomness of the following digits?

(*a*) $\sqrt{3} = 1.73205\ 08075\ 68877\ 2935\cdots$

(*b*) $\pi = 3.14159\ 26535\ 89793\ 2643\cdots$

17.66 In Problem 17.62, show that the *p*-value using the normal approximation is 0.105.

17.67 In Problem 17.63, show that the *p*-value using the normal approximation is 0.168.

17.68 In Problem 17.64, show that the *p*-value using the normal approximation is 0.485.

RANK CORRELATION

17.69 In a contest, two judges were asked to rank eight candidates (numbered 1 through 8) in order of preference. The judges submitted the choices shown in Table 17.42.

(*a*) Find the coefficient of rank correlation.

(*b*) Decide how well the judges agreed in their choices.

Table 17.42

First judge	5	2	8	1	4	6	3	7
Second judge	4	5	7	3	2	8	1	6

17.70 Table 14.17 is reproduced below and gives the United States consumer price indexes for food and medical-care costs during the years 2000 through 2006 compared with prices in the base years, 1982 through 1984 (mean taken as 100).

Year	2000	2001	2002	2003	2004	2005	2006
Food	167.8	173.1	176.2	180.0	186.2	190.7	195.2
Medical	260.8	272.8	285.6	297.1	310.1	323.2	336.2

Source: Bureau of Labor Statistics.

Find the Spearman rank correlation for the data and the Pearson correlation coefficient.

17.71 The rank correlation coefficient is derived by using the ranked data in the product-moment formula of Chapter 14. Illustrate this by using both methods to work a problem.

17.72 Can the rank correlation coefficient be found for grouped data? Explain this, and illustrate your answer with an example.

Statistical Process Control and Process Capability

GENERAL DISCUSSION OF CONTROL CHARTS

Variation in any process is due to *common causes* or *special causes*. The natural variation that exists in materials, machinery, and people gives rise to common causes of variation. In industrial settings, special causes, also known as *assignable causes*, are due to excessive tool wear, a new operator, a change of materials, a new supplier, etc. One of the purposes of *control charts* is to locate and, if possible, eliminate special causes of variation. The general structure of a control chart consists of *control limits* and a *centerline* as shown in Fig. 18-1. There are two control limits, called an *upper control limit* or UCL and a *lower control limit* or LCL.

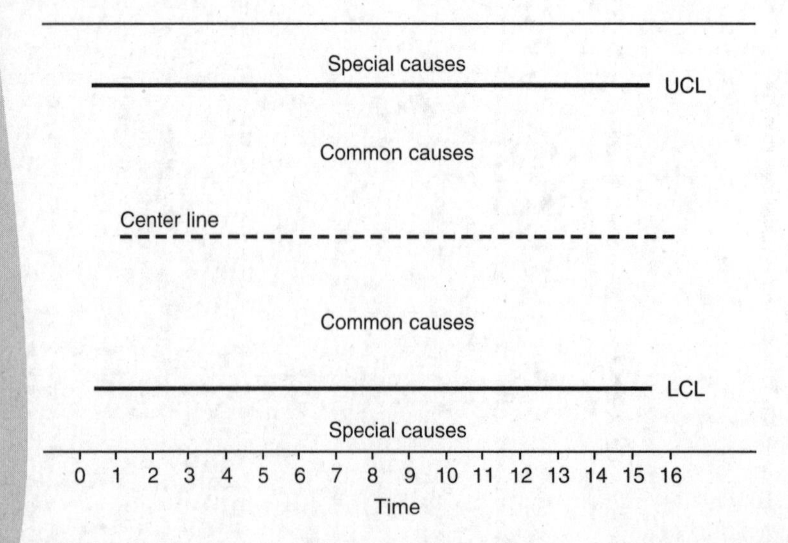

Fig. 18-1 Control charts are of two types (variable and attribute control charts).

When a point on the control chart falls outside the control limits, the process is said to be out of statistical control. There are other anomalous patterns besides a point outside the control limits that also indicate a process that is out of control. These will be discussed later. It is desirable for a process to be in control so that its' behavior is predictable.

VARIABLES AND ATTRIBUTES CONTROL CHARTS

Control charts may be divided into either *variables control charts* or *attributes control charts.* The terms "variables" and "attributes" are associated with the type of data being collected on the process. When measuring characteristics such as time, weight, volume, length, pressure drop, concentration, etc., we consider such data to be continuous and refer to it as *variables data.* When counting the number of defective items in a sample or the number of defects associated with a particular type of item, the resulting data are called *attributes data.* Variables data are considered to be of a higher level than attributes data. Table 18.1 gives the names of many of the various variables and attributes control charts and the statistics plotted on the chart.

Table 18.1

Chart Type	Statistics Plotted
X-bar and *R* chart	Averages and ranges of subgroups of variables data
X-bar and Sigma chart	Averages and standard deviations of subgroups of variables data
Median chart	Median of subgroups of variables data
Individuals chart	Individual measurements
Cusum chart	Cumulative sum of each $\bar{X}$ minus the nominal
Zone chart	Zone weights
EWMA chart	Exponentially weighted moving average
P-chart	Ratio of defective items to total number inspected
NP-chart	Actual number of defective items
C-chart	Number of defects per item for a constant sample size
U-chart	Number of defects per item for varying sample size

The charts above the dashed line in Table 18.1 are variables control charts and the charts below the dashed line are attributes control charts. We shall discuss some of the more basic charts. MINITAB will be used to construct the charts. Today, charting is almost always accomplished by the use of statistical software such as MINITAB.

X-BAR AND *R* CHARTS

The general idea of an *X*-bar chart can be understood by considering a process having mean μ and standard deviation σ. Suppose the process is monitored by taking periodic samples, called *subgroups*, of size n and computing the sample mean, $\bar{X}$, for each sample. The central limit theorem assures us that the mean of the sample mean is μ and the standard deviation of the sample mean is $\sigma/\sqrt{n}$. The centerline for the sample means is taken to be μ and the upper and lower control limits are taken to be $3(\sigma/\sqrt{n})$ above and below the centerline. The lower control limit is given by equation (*1*):

$$\text{LCL} = \mu - 3(\sigma/\sqrt{n}) \tag{1}$$

The upper control limit is given by equation (2):

$$\text{UCL} = \mu + 3(\sigma/\sqrt{n}) \qquad (2)$$

For a normally distributed process, a subgroup mean will fall between the limits, given in (1) and (2), 99.7% of the time. In practice, the process mean and the process standard deviation are unknown and need to be estimated. The process mean is estimated by using the mean of the periodic sample means. This is given by equation (3), where m is the number of periodic samples of size n selected.

$$\bar{\bar{X}} = \frac{\sum \bar{X}}{m} \qquad (3)$$

The mean, $\bar{\bar{X}}$, can also be found by summing all the data and then dividing by mn. The process standard deviation is estimated by pooling the subgroup variances, averaging the subgroup standard deviations or ranges, or by sometimes using a historical value of σ.

EXAMPLE 1. Data are obtained on the width of a product. Five observations per time period are sampled for 20 periods. The data are shown below in Table 18.2. The number of periodic samples is $m = 20$, the sample size or subgroup size is $n = 5$, the sum of all the data is 199.84, and centerline is $\bar{\bar{X}} = 1.998$. The MINITAB pull-down menu "**Stat** ⇒ **Control charts** ⇒ ***Xbar***" was used to produce the control chart shown in Fig. 18-2. The data in Table 18.2 are stacked into a single column before applying the above pull-down menu sequence.

Table 18.2

1	2	3	4	5	6	7	8	9	10
2.000	2.007	1.987	1.989	1.997	1.983	1.966	2.004	2.009	1.991
1.988	1.988	1.983	1.989	2.018	1.972	1.982	1.998	1.994	1.989
1.975	2.002	2.006	1.997	1.999	2.002	1.995	2.011	2.020	2.000
1.994	1.978	2.019	1.976	1.990	1.991	2.020	1.991	2.000	2.016
1.991	2.012	2.021	2.007	2.003	1.997	2.008	1.972	2.006	2.037

11	12	13	14	15	16	17	18	19	20
2.004	1.988	1.996	1.999	2.018	1.986	2.002	1.988	2.011	1.998
1.980	1.991	2.005	1.984	2.009	2.010	1.969	2.031	1.976	2.003
1.998	2.003	1.996	1.988	2.023	2.012	2.018	1.978	1.998	2.016
1.994	1.997	2.008	2.011	2.010	2.013	1.984	1.987	2.023	1.996
2.006	1.985	2.007	2.005	1.993	1.988	1.990	1.990	1.998	2.009

The standard deviation for the process may be estimated in four different ways: By using the average of the 20 subgroup ranges, by using the average of the 20 subgroup standard deviations, by pooling the 20 subgroup variances, or by using a historical value for σ, if one is known. MINITAB allows for all four options. The 20 means for the samples shown in Table 18.2 are plotted in Fig. 18-2. The chart indicates a process that is in control. The individual means randomly vary about the centerline and none fall outside the control limits.

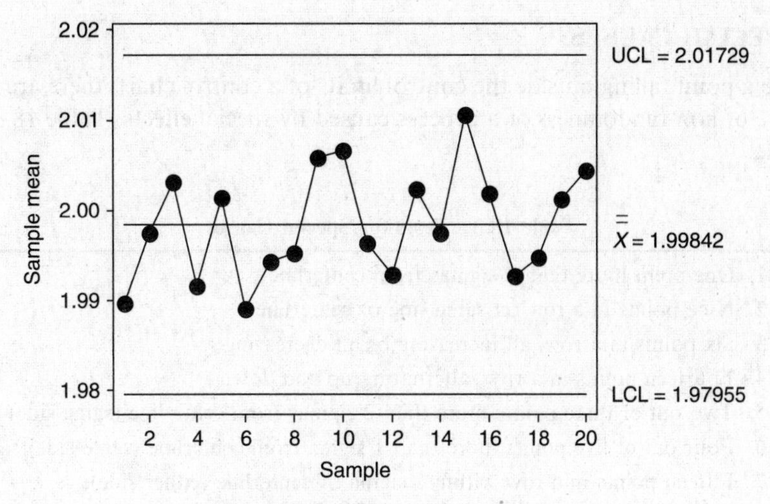

Fig. 18-2 *X*-bar chart for width.

The *R chart* is used to track process variation. The range, R, is computed for each of the m subgroups. The centerline for the R chart is given by equation (4).

$$\bar{R} = \frac{\sum R}{m} \tag{4}$$

As with the *X*-bar chart, several different methods are used to estimate the standard deviation of the process.

EXAMPLE 2. For the data in Table 18.2, the range of the first subgroup is $R_1 = 2.000 - 1.975 = 0.025$ and the range for the second subgroup is $R_2 = 2.012 - 1.978 = 0.034$. The 20 ranges are: 0.025, 0.034, 0.038, 0.031, 0.028, 0.030, 0.054, 0.039, 0.026, 0.048, 0.026, 0.018, 0.012, 0.027, 0.030, 0.027, 0.049, 0.053, 0.047 and 0.020. The mean of these 20 ranges is 0.0327. A MINITAB plot of these ranges is shown in Fig. 18-3. The R chart does not indicate any unusual patterns with respect to variability. The MINITAB pull-down menu "**Stat → Control charts → *R* chart**" is used to produce the control chart shown in Fig. 18-3. The data in Table 18.2 are stacked into a single column before applying the above pull-down menu sequence.

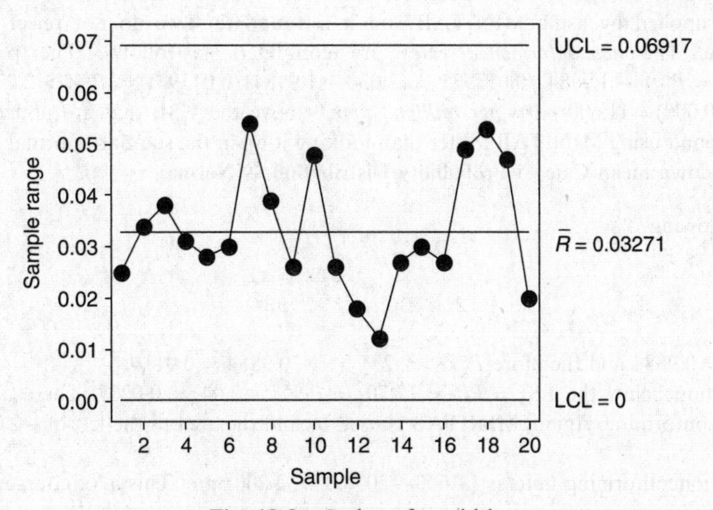

Fig. 18-3 *R* chart for width.

TESTS FOR SPECIAL CAUSES

In addition to a point falling outside the control limits of a control chart, there are other indications that are suggestive of non-randomness of a process caused by special effects. Table 18.3 gives eight tests for special causes.

Table 18.3 Tests for Special Causes

1. One point more than 3 sigmas from centerline
2. Nine points in a row on same side of centerline
3. Six points in a row, all increasing or all decreasing
4. Fourteen points in a row, alternating up and down
5. Two out of three points more than 2 sigmas from centerline (same side)
6. Four out of five points more than 1 sigma from centerline (same side)
7. Fifteen points in a row within 1 sigma of centerline (either side)
8. Eight points in a row more than 1 sigma from centerline (either side)

PROCESS CAPABILITY

To perform a capability analysis on a process, the process needs to be in statistical control. It is usually assumed that the process characteristic being measured is normally distributed. This may be checked out using tests for normality such as the Kolmogorov–Smirnov test, the Ryan–Joiner test, or the Anderson–Darling test. Process capability compares process performance with process requirements. Process requirements determine *specification limits.* LSL and USL represent the *lower specification limit* and the *upper specification limit*.

The data used to determine whether a process is in statistical control may be used to do the capability analysis. The 3-sigma distance on either side of the mean is called the *process spread.* The mean and standard deviation for the process characteristic may be estimated from the data gathered for the statistical process control study.

EXAMPLE 3. As we saw in Example 2, the data in Table 18.2 come from a process that is in statistical control. We found the estimate of the process mean to be 1.9984. The standard deviation of the 100 observations is found to equal 0.013931. Suppose the specification limits are LSL = 1.970 and USL = 2.030. The Kolmogorov–Smirnov test for normality is applied by using MINITAB and it is found that we do not reject the normality of the process characteristic. The *nonconformance rates* are computed as follows. The proportion above the USL = $P(X > 2.030) = P[(X - 1.9984)/0.013931 > (2.030 - 1.9984)/0.013931] = P(Z > 2.27) = 0.0116$. That is, there are 0.0116(1,000,000) = 11,600 *parts per million (ppm)* above the USL that are nonconforming. Note that $P(Z > 2.27)$ may be found using MINITAB rather than looking it up in the standard normal tables. This is done as follows. Use the pull-down mean **Calc → Probability Distribution → Normal**.

$X = 2.27$ gives the following

```
     x      P(X <= x)
2.2700       0.9884
```

We have $P(Z < 2.27) = 0.9884$ and therefore $P(Z > 2.27) = 1 - 0.9884 = 0.0116$.
Similarly, the proportion below the LSL = $P(X < 1.970) = P(Z < -2.04) = 0.0207$. There are 20,700 ppm below the LSL that are nonconforming. Again, MINITAB is used to find the area to the left of -2.04 under the standard normal curve.
The total number of nonconforming units is 11,600 + 20,700 = 32,300 ppm. This is of course an unacceptably high number of nonconforming units.

Suppose $\hat{\mu}$ represents the estimated mean for the process characteristic and $\hat{\sigma}$ represents the estimated standard deviation for the process characteristic, then the nonconformance rates are estimated as follows: The proportion above the USL equals

$$P(X > \text{USL}) = P\left(Z > \frac{\text{USL} - \hat{\mu}}{\hat{\sigma}}\right)$$

and the proportion below the LSL equals

$$P(X < \text{LSL}) = P\left(Z < \frac{\text{LSL} - \hat{\mu}}{\hat{\sigma}}\right)$$

The *process capability index* measures the process's potential for meeting specifications, and is defined as follows:

$$C_P = \frac{\text{allowable spread}}{\text{measured spread}} = \frac{\text{USL} - \text{LSL}}{6\hat{\sigma}} \tag{5}$$

EXAMPLE 4. For the process data in Table 18.2, $\text{USL} - \text{LSL} = 2.030 - 1.970 = 0.060$, $6\hat{\sigma} = 6(0.013931) = 0.083586$, and $C_P = 0.060/0.083586 = 0.72$.

The C_{PK} *index* measures the process performance, and is defined as follows:

$$C_{PK} = \text{minimum}\left\{\frac{\text{USL} - \hat{\mu}}{3\hat{\sigma}}, \ \frac{\hat{\mu} - \text{LSL}}{3\hat{\sigma}}\right\} \tag{6}$$

EXAMPLE 5. For the process data in Example 1,

$$C_{PK} = \text{minimum}\left\{\frac{2.030 - 1.9984}{3(0.013931)}, \ \frac{1.9984 - 1.970}{3(0.013931)}\right\} = \text{minimum}\,\{0.76, 0.68\} = 0.68$$

For processes with only a lower specification limit, the *lower capability index* C_{PL} is defined as follows:

$$C_{PL} = \frac{\hat{\mu} - \text{LSL}}{3\hat{\sigma}} \tag{7}$$

For processes with only an upper specification limit, the *upper capability index* C_{PU} is defined as follows:

$$C_{PU} = \frac{\text{USL} - \hat{\mu}}{3\hat{\sigma}} \tag{8}$$

Then C_{PK} may be defined in terms of C_{PL} and C_{PU} as follows:

$$C_{PK} = \min\{C_{PL}, C_{PU}\} \tag{9}$$

The relationship between nonconformance rates and C_{PL} and C_{PU} are obtained as follows:

$$P(X < \text{LSL}) = P\left(Z < \frac{\text{LSL} - \hat{\mu}}{\hat{\sigma}}\right) = P(Z < -3C_{PL}), \text{ since } -3C_{PL} = \frac{\text{LSL} - \hat{\mu}}{\hat{\sigma}}$$

$$P(X > \text{USL}) = P\left(Z > \frac{\text{USL} - \hat{\mu}}{\hat{\sigma}}\right) = P(Z > 3C_{PU}), \text{ since } 3C_{PU} = \frac{\text{USL} - \hat{\mu}}{\hat{\sigma}}$$

EXAMPLE 6. Suppose that $C_{PL} = 1.1$, then the proportion nonconforming is $P(Z < -3(1.1)) = P(Z < -3.3)$. This may be found using MINITAB as follows. Give the pull-down "**Calc⇒ Probability Distribution ⇒ Normal.**" The cumulative area to the left of -3.3 is given as:

```
Cumulative Distribution Function

Normal with mean = 0 and standard deviation = 1
       x      P(X <= x)
    -3.3      0.00048348
```

There would be $1{,}000{,}000 \times 0.00048348 = 483$ ppm nonconforming. Using this technique, a table relating nonconformance rate to capability index can be constructed. This is given in Table 18.4.

Table 18.4

C_{PL} or C_{PU}	Proportion Nonconforming	ppm
0.1	0.38208867	382089
0.2	0.27425308	274253
0.3	0.18406010	184060
0.4	0.11506974	115070
0.5	0.06680723	66807
0.6	0.03593027	35930
0.7	0.01786436	17864
0.8	0.00819753	8198
0.9	0.00346702	3467
1.0	0.00134997	1350
1.1	0.00048348	483
1.2	0.00015915	159
1.3	0.00004812	48
1.4	0.00001335	13
1.5	0.00000340	3
1.6	0.00000079	1
1.7	0.00000017	0
1.8	0.00000003	0
1.9	0.00000001	0
2.0	0.00000000	0

EXAMPLE 7. A capability analysis using MINITAB and the data in Table 18.2 may be obtained by using the following pull-down menus in MINITAB **Stat ⇒ Quality tools ⇒ Capability Analysis (Normal)**. The MINITAB output is shown in Fig. 18-4. The output gives nonconformance rates, capability indexes, and several other measures. The quantities found in Examples 3, 4, and 5 are very close to the corresponding measures shown in the figure. The differences are due to round-off error as well as different methods of estimating certain parameters. The graph is very instructive. It shows the distribution of sample measurements as a histogram. The population distribution of process measurements is shown as the normal curve. The tail areas under the normal curve to the right of the USL and to the left of the LSL represent the percentage of nonconforming products. By multiplying the sum of these percentages times one million, we get the ppm non-conformance rate for the process.

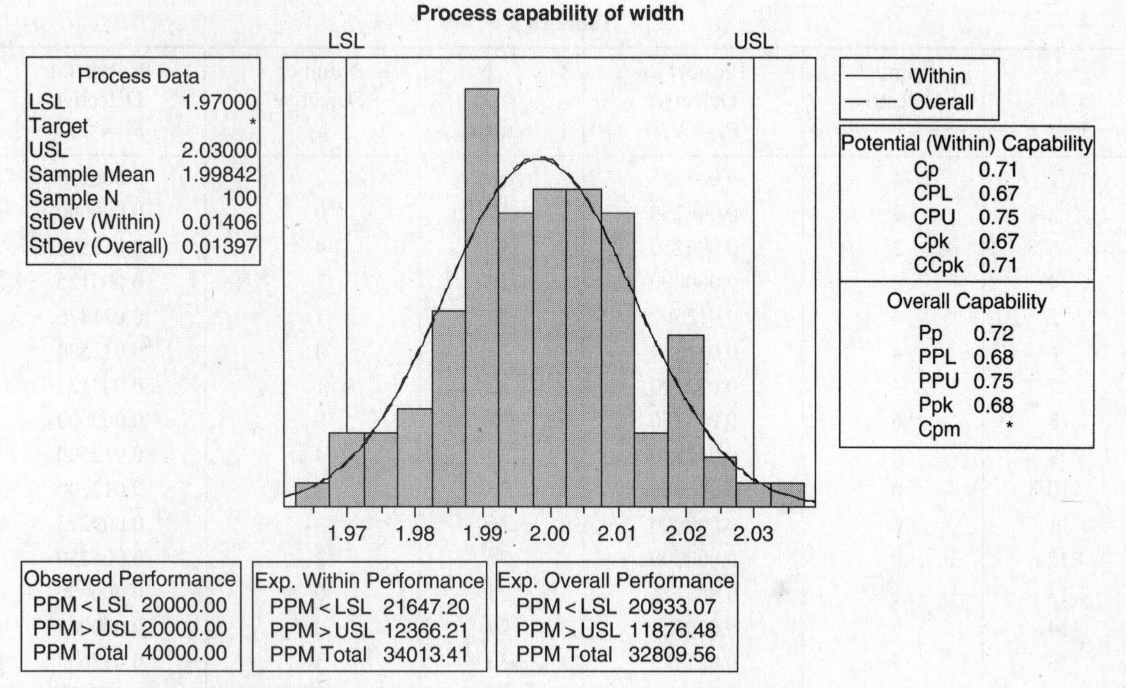

Fig. 18-4 Several capability measures are given for the process.

P- AND *NP*-CHARTS

When mass-produced products are categorized or classified, the resulting data are called *attributes data*. After establishing standards that a product must satisfy, specifications are determined. An item not meeting specifications is called a *non-conforming item*. A nonconforming item that is not usable is called a *defective item*. A defective item is considered to be more serious than a nonconforming item. An item might be nonconforming because of a scratch or a discoloration, but not be a defective item. The failure of a performance test would likely cause the product to be classified as defective as well as nonconforming. Flaws found on a single item are called *nonconformities*. Nonrepairable flaws are called *defects*.

Four different control charts are used when dealing with attributes data. The four charts are the *P*-, *NP*-, *C*-, and *U*-chart. The *P*- and the *NP*-charts are based on the binomial distribution and the *C*- and *U*-charts are based on the Poisson distribution. The *P*-chart is used to monitor the proportion of nonconforming items being produced by a process. The *P*-chart and the notation used to describe it are illustrated in Example 8.

EXAMPLE 8. Suppose 20 respirator masks are examined every thirty minutes and the number of defective units are recorded per 8-hour shift. The total number examined on a given shift is equal to $n = 20(16) = 320$. Table 18.5 gives the results for 30 such shifts. The center line for the *p*-chart is equal to the proportion of defectives for the 30 shifts, and is given by total number of defectives divided by the total number examined for the 30 shifts, or

$$\bar{p} = 72/9600 = 0.0075.$$

The standard deviation associated with the binomial distribution, which underlies this chart, is

$$\sqrt{\frac{\bar{p}(1-\bar{p})}{n}} = \sqrt{\frac{0.0075 \times 0.9925}{320}} = 0.004823.$$

The 3-sigma control limits for this process are

Table 18.5

Shift #	Number Defective X_i	Proportion Defective $P_i = X/n$	Shift #	Number Defective X_i	Proportion Defective $P_i = X/n$
1	1	0.003125	16	2	0.006250
2	2	0.006250	17	0	0.000000
3	2	0.006250	18	4	0.012500
4	0	0.000000	19	1	0.003125
5	4	0.012500	20	7	0.021875
6	4	0.012500	21	4	0.012500
7	4	0.012500	22	1	0.003125
8	6	0.018750	23	0	0.000000
9	4	0.012500	24	4	0.012500
10	0	0.000000	25	4	0.012500
11	0	0.000000	26	3	0.009375
12	0	0.000000	27	2	0.006250
13	1	0.003125	28	0	0.000000
14	0	0.000000	29	0	0.000000
15	7	0.021875	30	5	0.015625

$$\bar{p} \pm 3\sqrt{\frac{\bar{p}(1-\bar{p})}{n}} \tag{10}$$

The lower control limit is LCL = 0.0075 − 3(0.004823) = −0.006969. When the LCL is negative, it is taken to be zero since the proportion defective in a sample can never be negative. The upper control limit is UCL = 0.0075 + 3 (0.004823) = 0.021969.

The MINITAB solution to obtaining the *P*-chart for this process is given by using the pull-down menus **Stat → Control charts → P**. The *P*-chart is shown in Fig. 18-5. Even though it appears that samples 15 and

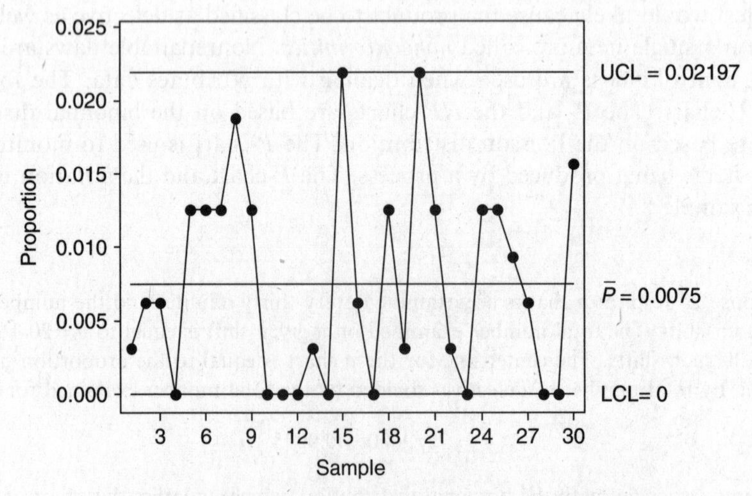

Fig. 18-5 The *P*-chart monitors the percent defective.

20 indicate the presence of a special cause, when the proportion defective for samples 15 and 20 (both equal to 0.021875) are compared with the UCL = 0.021969, it is seen that the points are not beyond the UCL.

The *NP-chart* monitors the number of defectives rather than the proportion of defectives. The *NP*-chart is considered by many to be preferable to the *P*-chart because the number defective is easier for quality technicians and operators to understand than is the proportion defective. The centerline for the *NP*-chart is given by $n\bar{p}$ and the 3-sigma control limits are

$$n\bar{p} \pm 3\sqrt{n\bar{p}(1-\bar{p})} \tag{11}$$

EXAMPLE 9. For the data in Table 18.5, the centerline is given by $n\bar{p} = 320(.0075) = 2.4$ and the control limits are $\text{LCL} = 2.4 - 4.63 = -2.23$, which we take as 0, and $\text{UCL} = 2.4 + 4.63 = 7.03$. If 8 or more defectives are found on a given shift, the process is out of control. The MINITAB solution is found by using the pull-down sequence

Stat → Control charts → NP

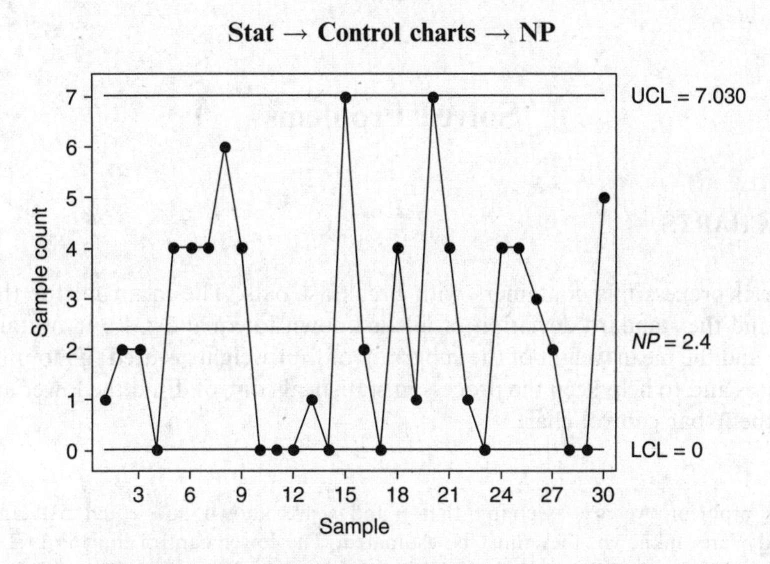

Fig. 18-6 The NP-chart monitors the number defective.

The number defective per sample needs to be entered in some column of the worksheet prior to executing the pull-down sequence. The MINITAB *NP*-chart is shown in Fig. 18-6.

OTHER CONTROL CHARTS

This chapter serves as only an introduction to the use of control charts to assist in statistical process control. Table 18.1 gives a listing of many of the various control charts in use in industrial settings today. To expedite calculations on the shop floor, the *median chart* is sometimes used. The medians of the samples are plotted rather than the means of the samples. If the sample size is odd, then the median is simply the middle value in the ordered sample values.

For low-volume production runs, *individuals charts* are often used. In this case, the subgroup or sample consists of a single observation. Individuals charts are sometimes referred to as *X* charts.

A *zone chart* is divided into four zones. Zone 1 is defined as values within 1 standard deviation of the mean, zone 2 is defined as values between 1 and 2 standard deviations of the mean, zone 3 is defined as values between 2 and 3 standard deviations of the mean, and zone 4 as values 3 or more standard deviations from the mean. Weights are assigned to the four zones. Weights for points on the same side of the centerline are added. When a cumulative sum is equal to or greater than the weight assigned to zone 4, this is taken as a signal that the process is out of control. The cumulative sum is set equal to 0 after signaling a process out of control, or when the next plotted point crosses the centerline.

The exponentially weighted moving average (*EWMA chart*) is an alternative to the individuals or *X*-bar chart that provides a quicker response to a shift in the process average. The EWMA chart incorporates information from all previous subgroups, not only the current subgroup.

Cumulative sums of deviations from a process target value are utilized by a *Cusum chart*. Both the EWMA chart and the Cusum chart allow for quick detection of process shifts.

When we are concerned with the number of nonconformities or defects in a product rather than simply determining whether the product is defective or non defective, we use a *C-chart* or a *U-chart*. When using these charts, it is important to define an *inspection unit*. The inspection unit is defined as the fixed unit of output to be sampled and examined for nonconformities. When there is only one inspection unit per sample, the *C*-chart is used, and when the number of inspection units per sample vary, the *U*-chart is used.

Solved Problems

X-BAR AND *R* CHARTS

18.1 An industrial process fills containers with breakfast oats. The mean fill for the process is 510 grams (g) and the standard deviation of fills is known to equal 5 g. Four containers are selected every hour and the mean weight of the subgroup of four weights is used to monitor the process for special causes and to help keep the process in statistical control. Find the lower and upper control limits for the *X*-bar control chart.

SOLUTION

In this problem, we are assuming that μ and σ are known and equal 510 and 5, respectively. When μ and σ are unknown, they must be estimated. The lower control limit is $\text{LCL} = \mu - 3(\sigma/\sqrt{n}) = 510 - 3(2.5) = 502.5$ and the upper control limit is $\text{UCL} = \mu + 3(\sigma/\sqrt{n}) = 510 + 3(2.5) = 517.5$

Table 18.6

Period 1	Period 2	Period 3	Period 4	Period 5	Period 6	Period 7	Period 8	Period 9	Period 10
2.000	2.007	1.987	1.989	1.997	1.983	1.966	2.004	2.009	1.991
1.988	1.988	1.983	1.989	2.018	1.972	1.982	1.998	1.994	1.989
1.975	2.002	2.006	1.997	1.999	2.002	1.995	2.011	2.020	2.000
1.994	1.978	2.019	1.976	1.990	1.991	2.020	1.991	2.000	2.016
1.991	2.012	2.021	2.007	2.003	1.997	2.008	1.972	2.006	2.037

Period 11	Period 12	Period 13	Period 14	Period 15	Period 16	Period 17	Period 18	Period 19	Period 20
2.004	1.988	1.996	1.999	2.018	2.025	2.002	1.988	2.011	1.998
1.980	1.991	2.005	1.984	2.009	2.022	1.969	2.031	1.976	2.003
1.998	2.003	1.996	1.988	2.023	2.035	2.018	1.978	1.998	2.016
1.994	1.997	2.008	2.011	2.010	2.013	1.984	1.987	2.023	1.996
2.006	1.985	2.007	2.005	1.993	2.020	1.990	1.990	1.998	2.009

18.2 Table 18.6 contains the widths of a product taken at 20 time periods. The control limits for an X-bar chart are LCL = 1.981 and UCL = 2.018. Are there any of the subgroup means outside the control limits?

SOLUTION

The means for the 20 subgroups are 1.9896, 1.9974, 2.0032, 1.9916, 2.0014, 1.9890, 1.9942, 1.9952, 2.0058, 2.0066, 1.9964, 1.9928, 2.0024, 1.9974, 2.0106, **2.0230**, 1.9926, 1.9948, 2.0012, and 2.0044, respectively. The sixteenth mean, 2.0230, is outside the upper control limit. All others are within the control limits.

18.3 Refer to Problem 18.2. It was determined that a spill occurred on the shop floor just before the sixteenth subgroup was selected. This subgroup was eliminated and the control limits were re-computed and found to be LCL = 1.979 and UCL = 2.017. Are there any of the means other than the mean for subgroup 16 outside the new limits.

SOLUTION

None of the means given in Problem 18.2 other than the sixteenth one fall outside the new limits. Assuming that the new chart does not fail any of the other tests for special causes given in table 18.3, the control limits given in this problem could be used to monitor the process.

18.4 Verify the control limits given in Problem 18.2. Estimate the standard deviation of the process by pooling the 20 sample variances.

SOLUTION

The mean of the 100 sample observations is 1.999. One way to find the pooled variance for the 20 samples is to treat the 20 samples, each consisting of 5 observations, as a one-way classification. The within or error mean square is equal to the pooled variance of the 20 samples. The MINITAB analysis as a one-way design gave the following analysis of variance table.

```
Analysis of Variance
Source  DF      SS          MS        F      P
Factor  19   0.006342   0.000334   1.75   0.044
Error   80   0.015245   0.000191
Total   99   0.021587
```

The estimate of the standard deviation is $\sqrt{0.000191} = 0.01382$. The lower control limit is LCL = $1.999 - 3(0.01382/\sqrt{5}) = 1.981$ and the upper control limit is UCL = $1.999 + 3(0.01382/\sqrt{5}) = 2.018$.

TESTS FOR SPECIAL CAUSES

18.5 Table 18.7 contains data from 20 subgroups, each of size 5. The X-bar chart is given in Fig. 18-7. What effect did a change to a new supplier at time period 10 have on the process? Which test for special causes in Table 18.3 did the process fail?

SOLUTION

The control chart in Fig. 18-7 shows that the change to the new supplier caused an increase in the width. This shift after time period 10 is apparent. The 6 shown on the graph in Fig. 18-7 indicates test 6 given in Table 18.3 was failed. Four out of five points were more than 1 sigma from the centerline (same side). The five points correspond to subgroups 4 through 8.

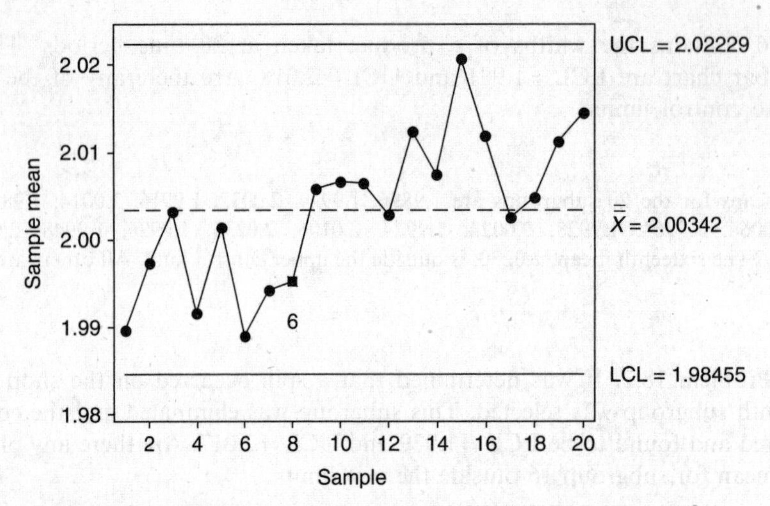

Fig. 18-7 MINITAB points where test—6 in Table 18.3 failed.

Table 18.7

Period 1	Period 2	Period 3	Period 4	Period 5	Period 6	Period 7	Period 8	Period 9	Period 10
2.000	2.007	1.987	1.989	1.997	1.983	1.966	2.004	2.009	1.991
1.988	1.988	1.983	1.989	2.018	1.972	1.982	1.998	1.994	1.989
1.975	2.002	2.006	1.997	1.999	2.002	1.995	2.011	2.020	2.000
1.994	1.978	2.019	1.976	1.990	1.991	2.020	1.991	2.000	2.016
1.991	2.012	2.021	2.007	2.003	1.997	2.008	1.972	2.006	2.037

Period 11	Period 12	Period 13	Period 14	Period 15	Period 16	Period 17	Period 18	Period 19	Period 20
2.014	1.998	2.006	2.009	2.028	1.996	2.012	1.998	2.021	2.008
1.990	2.001	2.015	1.994	2.019	2.020	1.979	2.041	1.986	2.013
2.008	2.013	2.006	1.998	2.033	2.022	2.028	1.988	2.008	2.026
2.004	2.007	2.018	2.021	2.020	2.023	1.994	1.997	2.033	2.006
2.016	1.995	2.017	2.015	2.003	1.998	2.000	2.000	2.008	2.019

PROCESS CAPABILITY

18.6 Refer to Problem 18.2. After determining that a special cause was associated with subgroup 16, we eliminate this subgroup. The mean width is estimated by finding the mean of the data from the other 19 subgroups and the standard deviation is estimated by finding the standard deviation of the same data. If the specification limits are LSL = 1.960 and USL = 2.040, find the lower capability index, the upper capability index, and the C_{PK} index.

SOLUTION

Using the 95 measurements after excluding subgroup 16, we find that $\hat{\mu} = 1.9982$ and $\hat{\sigma} = 0.01400$. The lower capability index is

$$C_{PL} = \frac{\hat{\mu} - \text{LSL}}{3\hat{\sigma}} = \frac{1.9982 - 1.960}{0.0420} = 0.910$$

the upper capability index is

$$C_{PU} = \frac{\text{USL} - \hat{\mu}}{3\hat{\sigma}} = \frac{2.040 - 1.9982}{0.042} = 0.995$$

and $C_{PK} = \min\{C_{PL}, C_{PU}\} = 0.91$.

18.7 Refer to Problem 18.1. (*a*) Find the percentage nonconforming if LSL = 495 and USL = 525. (*b*) Find the percentage nonconforming if LSL = 490 and USL = 530.

SOLUTION

(*a*) Assuming the fills are normally distributed, the area under the normal curve below the LSL is found by using the EXCEL command `=NORMDIST(495,510,5,1)` which gives 0.001350. By symmetry, the area under the normal curve above the USL is also 0.001350. The total area outside the specification limits is 0.002700. The ppm nonconforming is 0.002700(1,000,000) = 2700.

(*b*) The area under the normal curve for LSL = 490 and USL = 530 is found similarly to be 0.000032 + 0.000032 = 0.000064. The ppm is found to be 0.000064(1,000,000) = 64.

P- AND *NP*-CHARTS

18.8 Printed circuit boards are inspected for defective soldering. Five hundred circuit boards per day are tested for a 30-day period. The number defective per day are shown in Table 18.8. Construct a *P*-chart and locate any special causes.

Table 18.8

Day	1	2	3	4	5	6	7	8	9	10
# Defective	2	0	2	5	2	4	5	1	2	3
Day	11	12	13	14	15	16	17	18	19	20
# Defective	3	2	0	4	3	8	10	4	4	5
Day	21	22	23	24	25	26	27	28	29	30
# Defective	2	4	3	2	3	3	2	1	1	2

SOLUTION

The confidence limits are

$$\bar{p} = 3\sqrt{\frac{\bar{p}(1-\bar{p})}{n}}$$

The centerline is $\bar{p} = 92/15{,}000 = 0.00613$ and the standard deviation is

$$\sqrt{\frac{\bar{p}(1-\bar{p})}{n}} = \sqrt{\frac{(0.00613)(0.99387)}{500}} = 0.00349$$

The lower control limit is $0.00613 - 0.01047 = -0.00434$, and is taken to equal 0, since proportions cannot be negative. The upper control limit is $0.00613 + 0.01047 = 0.0166$. The proportion of defectives on day 17 is equal to $P_{17} = 10/500 = 0.02$ and is the only daily proportion to exceed the upper limit.

18.9 Give the control limits for an *NP*-chart for the data in Problem 18.8.

SOLUTION

The control limits for the number defective are $n\bar{p} = 3\sqrt{n\bar{p}(1-\bar{p})}$. The centerline is $n\bar{p} = 3.067$. The lower limit is 0 and the upper limit is 8.304.

18.10 Suppose respirator masks are packaged in either boxes of 25 or 50 per box. At each 30 minute interval during a shift a box is randomly chosen and the number of defectives in the box determined. The box may either contain 25 or 50 masks. The number checked per shift will vary between 400 and 800. The data are shown in Table 18.9. Use MINITAB to find the control chart for the proportion defective.

Table 18.9

Shift #	Sample Size n_i	Number Defective X_i	Proportion Defective $P_i = X_i/n_i$
1	400	3	0.0075
2	575	7	0.0122
3	400	1	0.0025
4	800	7	0.0088
5	475	2	0.0042
6	575	0	0.0000
7	400	8	0.0200
8	625	1	0.0016
9	775	10	0.0129
10	425	8	0.0188
11	400	7	0.0175
12	400	3	0.0075
13	625	6	0.0096
14	800	5	0.0063
15	800	4	0.0050
16	800	7	0.0088
17	475	9	0.0189
18	800	9	0.0113
19	750	9	0.0120
20	475	2	0.0042

SOLUTION

When the sample sizes vary in monitoring a process for defectives, the centerline remains the same, that is, it is the proportion of defectives over all samples. The standard deviation, however, changes from sample to sample and gives control limits consisting of stair-stepped control limits. The control limits are

$$\bar{p} = 3\sqrt{\frac{\bar{p}(1-\bar{p})}{n_i}}$$

The centerline is $\bar{p} = 108/11{,}775 = 0.009172$. For the first subgroup, we have $n_i = 400$

$$\sqrt{\frac{\bar{p}(1-\bar{p})}{n_i}} = \sqrt{\frac{(0.009172)(0.990828)}{400}} = 0.004767$$

and $3(0.004767) = 0.014301$. The lower limit for subgroup 1 is 0 and the upper limit is $0.009172 + 0.014301 = 0.023473$. The limits for the remaining shifts are determined similarly. These changing limits give rise to the stair-stepped upper control limits shown in Fig. 18-8.

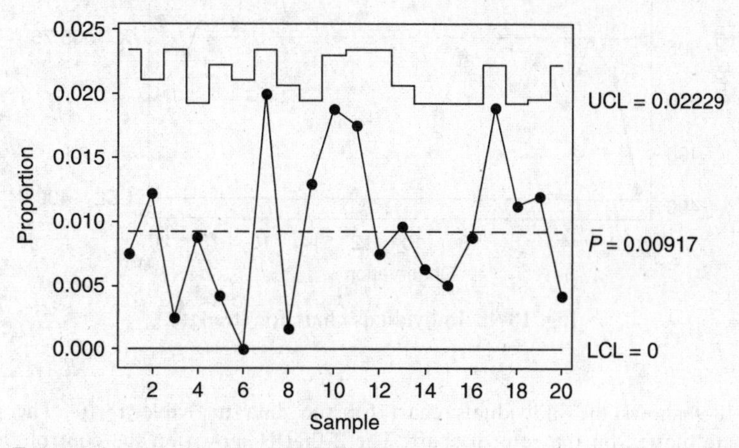

Fig. 18-8 *P*-chart with unequal sample sizes.

OTHER CONTROL CHARTS

18.11 When measurements are expensive, data are available at a slow rate, or when output at any point is fairly homogeneous, an *individuals chart with moving range* may be indicated. The data consist of single measurements taken at different points in time. The centerline is the mean of all the individual measurements, and variation is estimated by using *moving ranges*. Traditionally, moving ranges have been calculated by subtracting adjacent data values and taking the absolute value of the result. Table 18.10 gives the coded breaking strength measurements of an expensive cable used in aircraft. One cable per day is selected from the production process and tested. Give the MINITAB-generated individuals chart and interpret the output.

Table 18.10

Day	1	2	3	4	5	6	7	8	9	10
Strength	491.5	502.0	505.5	499.6	504.1	501.3	503.5	504.3	498.5	508.8
Day	11	12	13	14	15	16	17	18	19	20
Strength	515.4	508.0	506.0	510.9	507.6	519.1	506.9	510.9	503.9	507.4

SOLUTION

The following pull-down menus are used: **Stats → Control charts → individuals.**

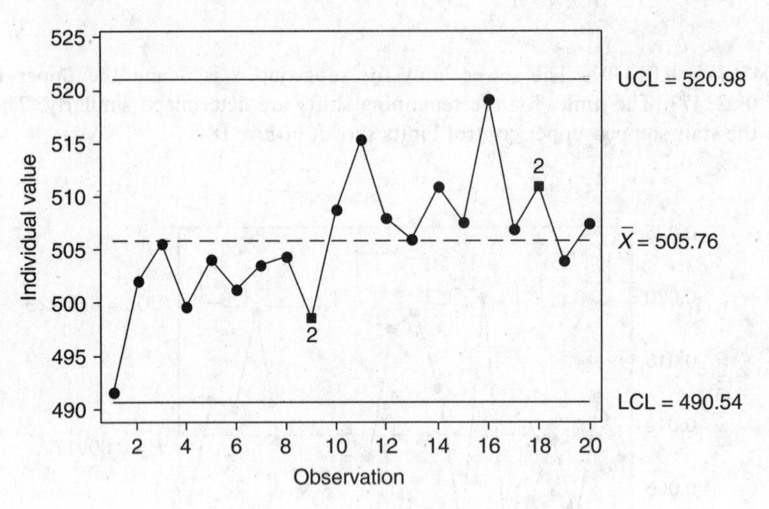

Fig. 18-9 Individuals chart for strength.

Figure 18-9 shows the individuals chart for the data in Table 18.10. The individual values in Table 18.10 are plotted on the control chart. The 2 that is shown on the control chart for weeks 9 and 18 corresponds to the second test for special causes given in Table 18.3. This indication of a special cause corresponds to nine points in a row on the same side of the centerline. An increase in the process temperature at time period 10 resulted in an increase in breaking strength. This change in breaking strength resulted in points below the centerline prior to period 10 and mostly above the centerline after period 10.

18.12 The *exponentially weighted moving average, EWMA chart* is used to detect small shifts from a target value, *t*. The points on the EWMA chart are given by the following equation:

$$\hat{x}_i = w\bar{x}_i + (1 - w)\hat{x}_{i-1}$$

To illustrate the use of this equation, suppose the data in Table 18.7 were selected from a process that has target value equal to 2.000. The stating value $\hat{x}_0$ is chosen to equal the target value, 2.000. The weight w is usually chosen to be between 0.10 and 0.30. MINITAB uses the value 0.20 as a default. The first point on the EWMA chart would be $\hat{x}_1 = w\bar{x}_1 + (1 - w)\hat{x}_0 = 0.20(1.9896) + 0.80(2.000) = 1.9979$. The second point on the chart would be $\hat{x}_2 = w\bar{x}_2 + (1 - w)\hat{x}_1 = 0.20(1.9974) + 0.80(1.9979) = 1.9978$, and so forth. The MINITAB analysis is obtained by using the following pull-down menu **Stat → Control charts → EWMA**. The target value is supplied to MINITAB. The output is shown in Fig. 18-10. By referring to Fig. 18-10, determine for which subgroups the process shifted from the target value.

SOLUTION

The graph of the $\hat{x}_i$ values crosses the upper control limit at the time point 15. This is the point at which we would conclude that the process had shifted away from the target value. Note that the EWMA chart has stair-stepped control limits.

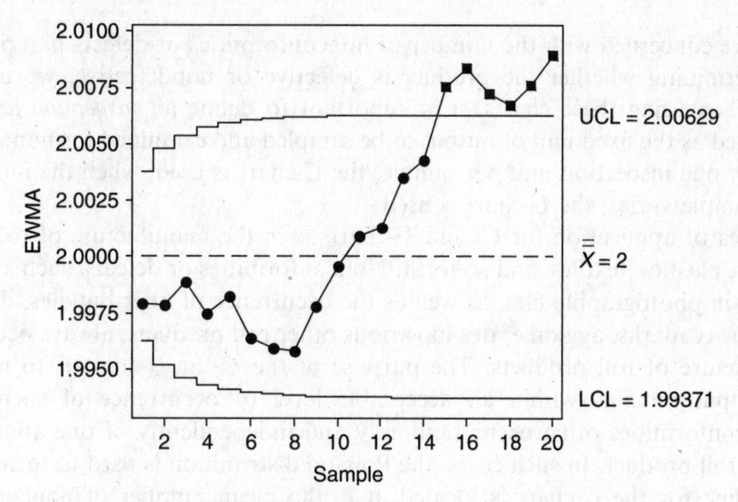

Fig. 18-10 Exponentially Weighted Moving Average chart.

18.13 A *zone chart* is divided into four zones. Zone 1 is defined as values within 1 standard deviation of the mean, zone 2 is defined as values between 1 and 2 standard deviations of the mean, zone 3 is defined as values between 2 and 3 standard deviations of the mean, and zone 4 as values 3 or more standard deviations from the mean. The default weights assigned to the zones by MINITAB are 0, 2, 4, and 8 for zones 1 through 4. Weights for points on the same side of the centerline are added. When a cumulative sum is equal to or greater than the weight assigned to zone 4, this is taken as a signal that the process is out of control. The cumulative sum is set equal to 0 after signaling a process out of control, or when the next plotted point crosses the centerline. Figure 18-11 shows the MINITAB analysis using a zone chart for the data in Table 18.6. The pull-down menus needed to produce this chart are **Stat → Control charts → Zone.** What out of control points does the zone chart find?

SOLUTION

Subgroup 16 corresponds to an out of control point. The zone score corresponding to subgroup 16 is 10, and since this exceeds the score assigned to zone 4, this locates an out of control time period in the process.

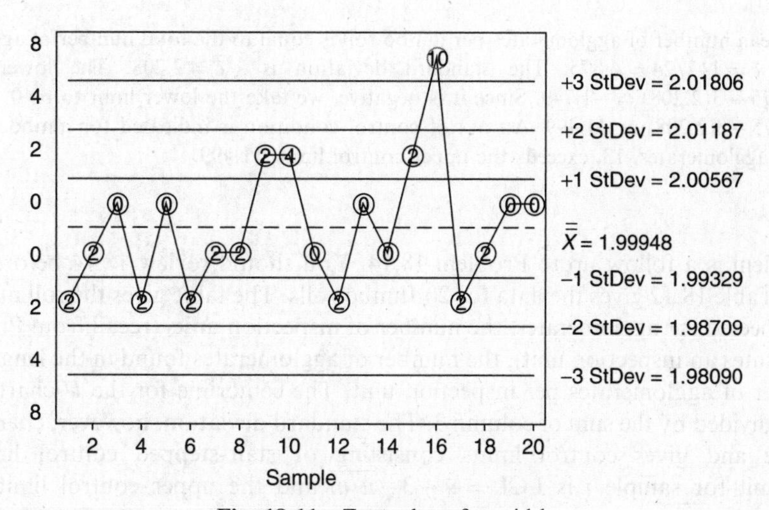

Fig. 18-11 Zone chart for width.

18.14 When we are concerned with the number of nonconformities or defects in a product rather than simply determining whether the product is defective or nondefective, we use a *C-chart* or a *U-chart*. When using these charts, it is important to define an *inspection unit*. The inspection unit is defined as the fixed unit of output to be sampled and examined for nonconformities. When there is only one inspection unit per sample, the C-chart is used; when the number of inspection units per sample varies, the U-chart is used.

One area of application for C- and U-charts is in the manufacture of roll products such as paper, films, plastics, textiles, and so forth. Nonconformities or defects, such as the occurrence of black spots in photographic film, as well as the occurrence of fiber bundles, dirt spots, pinholes, static electricity marks, agglomerates in various other roll products, always occur at some level in the manufacture of roll products. The purpose of the C- or U-chart is to make sure that the process output remains within an acceptable level of occurrence of such nonconformities. These nonconformities often occur randomly and independently of one another over the total area of the roll product. In such cases, the Poisson distribution is used to form the control chart. The centerline for the C-chart is located at $\bar{c}$, the mean number of nonconformities over all subgroups. The standard deviation of the Poisson distribution is $\sqrt{\bar{c}}$ and therefore the 3 sigma control limits are $\bar{c} \pm 3\sqrt{\bar{c}}$. That is, the lower limit is $\text{LCL} = \bar{c} - 3\sqrt{\bar{c}}$ and the upper limit is $\text{UCL} = \bar{c} + 3\sqrt{\bar{c}}$.

When a coating is applied to a material, small nonconformities called agglomerates sometimes occur. The number of agglomerates in a length of 5 feet (ft) are recorded for a jumbo roll of product. The results for 24 such rolls are given in Table 18.11. Are there any points outside the 3-sigma control limits?

Table 18.11

Jumbo roll #	1	2	3	4	5	6	7	8	9	10	11	12
Agglomerates	3	3	6	0	7	5	3	6	3	5	2	2
Jumbo roll #	13	14	15	16	17	18	19	20	21	22	23	24
Agglomerates	2	7	6	4	7	8	5	13	7	3	3	7

SOLUTION

The mean number of agglomerates per jumbo roll is equal to the total number of agglomerates divided by 24 or $\bar{c} = 117/24 = 4.875$. The standard deviation is $\sqrt{\bar{c}} = 2.208$. The lower control limit is $\text{LCL} = 4.875 - 3(2.208) = -1.749$. Since it is negative, we take the lower limit to be 0. The upper limit is $\text{UCL} = 4.875 + 3(2.208)$ or 11.499. An out of control condition is indicated for jumbo roll # 20 since the number of agglomerates, 13, exceeds the upper control limit, 11.499.

18.15 This problem is a follow up to Problem 18.14. You should review 18.14 before attempting this problem. Table 18.12 gives the data for 20 Jumbo rolls. The table gives the roll number, the length of roll inspected for agglomerates, the number of inspection units (recall from Problem 18.14 that 5 ft constitutes an inspection unit), the number of agglomerates found in the length inspected, and the number of agglomerates per inspection unit. The centerline for the U-chart is $\bar{u}$, the sum of column 4 divided by the sum of column 3. The standard deviation, however, changes from sample to sample and gives control limits consisting of stair-stepped control limits. The lower control limit for sample i is $\text{LCL} = \bar{u} - 3\sqrt{\bar{u}/n_i}$ and the upper control limit for sample i is $\text{UCL} = \bar{u} + 3\sqrt{\bar{u}/n_i}$.

Table 18.12

Jumbo roll #	Length Inspected	# of Inspection Units, n_i	# of Agglomerates	$u_i =$ Col. 4/Col. 3
1	5.0	1.0	6	6.00
2	5.0	1.0	4	4.00
3	5.0	1.0	6	6.00
4	5.0	1.0	2	2.00
5	5.0	1.0	3	3.00
6	10.0	2.0	8	4.00
7	7.5	1.5	6	4.00
8	15.0	3.0	6	2.00
9	10.0	2.0	10	5.00
10	7.5	1.5	6	4.00
11	5.0	1.0	4	4.00
12	5.0	1.0	7	7.00
13	5.0	1.0	5	5.00
14	15.0	3.0	8	2.67
15	5.0	1.0	3	3.00
16	5.0	1.0	5	5.00
17	15.0	3.0	10	3.33
18	5.0	1.0	1	1.00
19	15.0	3.0	8	2.67
20	15.0	3.0	15	5.00

Use MINITAB to construct the control chart for this problem and determine if the process is in control.

SOLUTION

The centerline for the U-chart is $\bar{u}$, the sum of column 4 divided by the sum of column 3. The standard deviation, however, changes from sample to sample and gives control limits consisting of stair-stepped control limits. The lower control limit for sample i is $\text{LCL} = \bar{u} - 3\sqrt{\bar{u}/n_i}$ and the upper control limit

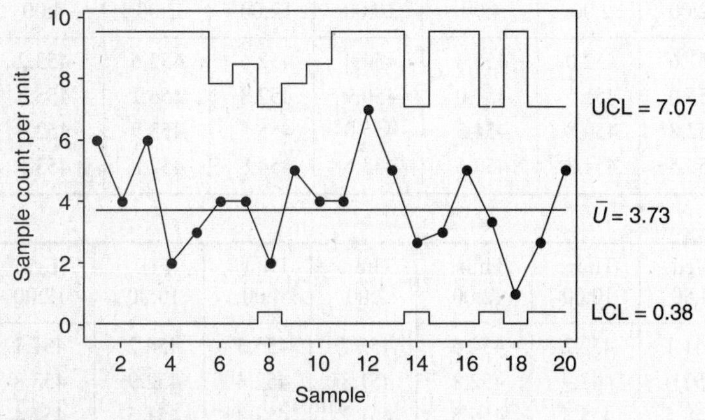

Fig. 18-12 U-chart for agglomerates.

for sample i is UCL $= \bar{u} + 3\sqrt{\bar{u}/n_i}$. The centerline for the above data is $\bar{u} = 123/33 = 3.73$. The MINITAB solution is obtained by the pull-down sequence **Stat → Control Charts →U.**

The information, required by MINITAB to create the *U*-chart, is that given in columns 3 and 4 of Table 18.12. The *U*-chart for the data in Table 18.12 is shown in Fig. 18-12. The control chart does not indicate any out of control points.

Supplementary Problems

X-BAR AND *R* CHARTS

18.16 The data from ten subgroups, each of size 4, is shown in Table 18.13. Compute $\bar{X}$ and R for each subgroup as well as, $\bar{\bar{X}}$, and $\bar{R}$. Plot the $\bar{X}$values on a graph along with the centerline corresponding to $\bar{\bar{X}}$. On another graph, plot the R-values along with the centerline corresponding to $\bar{R}$.

Table 18.13

Subgroup	Subgroup Observations			
1	13	11	13	16
2	11	12	20	15
3	16	18	20	15
4	13	15	18	12
5	12	19	11	12
6	14	10	19	16
7	12	13	20	10
8	17	17	12	14
9	15	12	16	17
10	20	13	18	17

18.17 A frozen food company packages 1 pound (lb) packages (454 g) of green beans. Every two hours, 4 of the packages are selected and the weight is determined to the nearest tenth of a gram. Table 18.14 gives the data for a one-week period.

Table 18.14

Mon. 10:00	Mon. 12:00	Mon. 2:00	Mon. 4:00	Tue. 10:00	Tue. 12:00	Tue. 2:00	Tue. 4:00	Wed. 10:00	Wed. 12:00
453.0	451.6	452.0	455.4	454.8	452.6	453.6	453.2	453.0	451.6
454.5	455.0	451.5	453.0	450.9	452.8	456.1	455.8	451.4	456.0
452.6	452.8	450.8	454.3	455.0	455.5	453.9	452.0	452.5	455.0
451.8	453.5	454.8	450.6	453.6	454.8	454.8	453.5	452.1	453.0

Wed. 2:00	Wed. 4:00	Thur. 10:00	Thur. 12:00	Thur. 2:00	Thur. 4:00	Fri. 10:00	Fri. 12:00	Fri. 2:00	Fri. 4:00
454.7	451.1	452.2	454.0	455.7	455.3	454.2	451.1	455.7	450.7
451.4	452.6	448.9	452.8	451.8	452.4	452.9	453.8	455.3	452.5
450.9	448.5	455.3	455.5	451.2	452.3	451.5	452.4	455.4	454.1
455.8	454.4	453.9	453.8	452.8	452.3	455.8	454.3	453.7	454.2

Use the method discussed in Problem 18.4 to estimate the standard deviation by pooling the variances of the 20 samples. Use this estimate to find the control limits for an X-bar chart. Are any of the 20 subgroup means outside the control limits?

18.18 The control limits for the R chart for the data in Table 18.14 are LCL = 0 and UCL = 8.205. Are any of the subgroup ranges outside the 3 sigma limits?

18.19 The process that fills the 1 lb packages of green beans discussed in Problem 18.17 is modified in hopes of reducing the variability in the weights of the packages. After the modification was implemented and in use for a short time, a new set of weekly data was collected and the ranges of the new subgroups were plotted using the control limits given in Problem 18.18. The new data are given in Table 18.15. Does it appear that the variability has been reduced? If the variability has been reduced, find new control limits for the X-bar chart using the data in Table 18.15.

Table 18.15

Mon. 10:00	Mon. 12:00	Mon. 2:00	Mon. 4:00	Tue. 10:00	Tue. 12:00	Tue. 2:00	Tue. 4:00	Wed. 10:00	Wed. 12:00
454.9	454.2	454.4	454.7	454.3	454.2	454.6	453.6	454.4	454.6
452.7	453.6	453.6	453.9	454.2	452.8	454.5	453.2	455.0	454.1
457.0	454.4	453.6	454.6	454.2	453.3	454.3	453.6	454.6	453.3
454.2	453.9	454.3	453.9	453.4	453.3	454.9	453.1	454.1	454.3

Wed. 2:00	Wed. 4:00	Thur. 10:00	Thur. 12:00	Thur. 2:00	Thur. 4:00	Fri. 10:00	Fri. 12:00	Fri. 2:00	Fri. 4:00
453.0	453.9	453.8	455.1	454.2	454.4	455.1	455.7	452.2	455.4
454.0	454.2	453.6	453.3	453.0	452.6	454.6	452.8	453.7	452.8
452.9	454.3	454.1	454.7	453.8	454.9	454.1	453.8	454.4	454.7
454.2	454.7	454.7	453.9	453.9	454.2	454.6	454.9	454.5	455.1

TESTS FOR SPECIAL CAUSES

18.20 Operators making adjustments to machinery too frequently is a problem in industrial processes. Table 18.16 contains a set of data (20 samples each of size 5) in which this is the case. Find the control limits for an X-bar chart and then form the X-bar chart and check the 8 tests for special causes given in Table 18.3.

Table 18.16

1	2	3	4	5	6	7	8	9	10
2.006	2.001	1.993	1.983	2.003	1.977	1.972	1.998	2.015	1.985
1.994	1.982	1.989	1.983	2.024	1.966	1.988	1.992	2.000	1.983
1.981	1.996	2.012	1.991	2.005	1.996	2.001	2.005	2.026	1.994
2.000	1.972	2.025	1.970	1.996	1.985	2.026	1.985	2.006	2.010
1.997	2.006	2.027	2.001	2.009	1.991	2.014	1.966	2.012	2.031

11	12	13	14	15	16	17	18	19	20
2.010	1.982	2.002	1.993	2.024	1.980	2.008	1.982	2.017	1.992
1.986	1.985	2.011	1.978	2.015	2.004	1.975	2.025	1.982	1.997
2.004	1.997	2.002	1.982	2.029	2.006	2.024	1.972	2.004	2.010
2.000	1.991	2.014	2.005	2.016	2.007	1.990	1.981	2.029	1.990
2.012	1.979	2.013	1.999	1.999	1.982	1.996	1.984	2.004	2.003

PROCESS CAPABILITY

18.21 Suppose the specification limits for the frozen food packages in Problem 18.17 are LSL = 450 g and USL = 458 g. Use the estimates of μ and σ obtained in Problem 18.17 to find C_{PK}. Also estimate the ppm not meeting the specifications.

18.22 In Problem 18.21, compute C_{PK} and estimate the ppm nonconforming after the modifications made in Problem 18.19 have been made.

P- AND *NP*-CHARTS

18.23 A Company produces fuses for automobile electrical systems. Five hundred of the fuses are tested per day for 30 days. Table 18.17 gives the number of defective fuses found per day for the 30 days. Determine the centerline and the upper and lower control limits for a *P*-chart. Does the process appear to be in statistical control? If the process is in statistical control, give a point estimate for the ppm defective rate.

Table 18.17

Day	1	2	3	4	5	6	7	8	9	10
# Defective	3	3	3	3	1	1	1	1	6	1
Day	11	12	13	14	15	16	17	18	19	20
# Defective	1	1	5	4	6	3	6	2	7	3
Day	21	22	23	24	25	26	27	28	29	30
# Defective	2	3	6	1	2	3	1	4	4	5

18.24 Suppose in Problem 18.23, the fuse manufacturer decided to use an *NP*-chart rather than a *P*-chart. Find the centerline and the upper and lower control limits for the chart.

18.25 Scottie Long, the manager of the meat department of a large grocery chain store, is interested in the percentage of packages of hamburger meat that have a slight discoloration. Varying numbers of packages are inspected on a daily basis and the number with a slight discoloration is recorded. The data are shown in Table 18.18. Give the stair-stepped upper control limits for the 20 subgroups.

Table 18.18

Day	Subgroup Size	Number Discolored	Percent Discolored
1	100	1	1.00
2	150	1	0.67
3	100	0	0.00
4	200	1	0.50
5	200	1	0.50
6	150	0	0.00
7	100	0	0.00
8	100	0	0.00
9	150	0	0.00
10	200	2	1.00
11	100	1	1.00
12	200	1	0.50
13	150	3	2.00
14	200	2	1.00
15	150	1	0.67
16	200	1	0.50
17	150	4	2.67
18	150	0	0.00
19	150	0	0.00
20	150	2	1.33

OTHER CONTROL CHARTS

18.26 Review Problem 18.11 prior to working this problem. Hourly readings of the temperature of an oven, used for bread making, are obtained for 24 hours. The baking temperature is critical to the process and the oven is operated constantly during each shift. The data are shown in Table 18.19. An individual chart is used to help monitor the temperature of the process. Find the centerline and the moving ranges corresponding to using adjacent pairs of measurements. How are the control limits found?

Table 18.19

Hour	1	2	3	4	5	6	7	8	9	10	11	12
Temperature	350.0	350.0	349.8	350.4	349.6	350.0	349.7	349.8	349.4	349.8	350.7	350.9
Hour	13	14	15	16	17	18	19	20	21	22	23	24
Temperature	349.8	350.3	348.8	351.6	350.0	349.7	349.8	348.6	350.5	350.3	349.1	350.0

18.27 Review Problem 18.12 prior to working this problem. Use MINITAB to construct a EWMA chart for the data in Table 18.14. Using a target value of 454 g, what does the chart indicate concerning the process?

18.28 Review the discussion of a zone chart in Problem 18.13 before working this problem. Construct a zone chart for the data in Table 18.16. Does the zone chart indicate any out of control conditions? What shortcoming of the zone chart does this problem show?

18.29 Work Problem 18.15 prior to working this problem. Construct the stair-stepped control limits for the u chart in Problem 18.15.

18.30 A *Pareto chart* is often used in quality control. A Pareto chart is a bar graph that lists the defects that are observed in descending order. The most frequently occurring defects are listed first, followed by those that occur less frequently. By the use of such charts, areas of concern can be identified and efforts made to correct those defects that account for the largest percent of defects. The following defects are noted for respirator masks inspected during a given time period: Discoloration, loose strap, dents, tears, and pinholes. The results are shown in Table 18.20.

Table 18.20

discoloration	discoloration	discoloration
strap	strap	strap
discoloration	dent	strap
discoloration	strap	discoloration
strap	discoloration	discoloration
discoloration	discoloration	dent
discoloration	dent	tear
tear	pinhole	discoloration
dent	discoloration	pinhole
discoloration	tear	tear

Figure 18-13 is a Pareto chart generated by MINITAB. The data given in Table 18.20 are entered into column 1 of the worksheet. The pull-down menus needed to construct this chart are as follows: "**Stat → Quality tools → Pareto charts**." By referring to the Pareto chart, what type defect should receive the most attention? What two types of defects should receive the most attention?

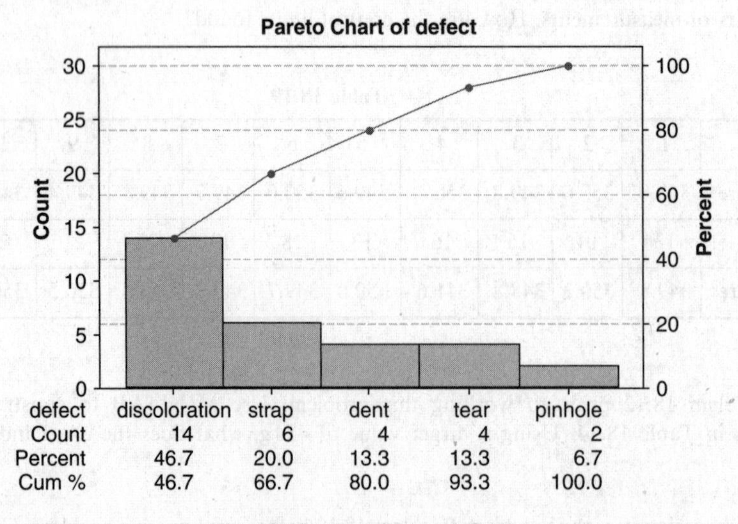

Fig. 18-13 Respirator defects.

Answers to Supplementary Problems

CHAPTER 1

1.46 (*a*) Continuous; (*b*) continuous; (*c*) discrete; (*d*) discrete; (*e*) discrete.

1.47 (*a*) Zero upward; continuous. (*b*) 2, 3, . . . ; discrete.
(*c*) Single, married, divorced, separated, widowed; discrete. (*d*) Zero upward; continuous.
(*e*) 0, 1, 2, . . . ; discrete.

1.48 (*a*) 3300; (*b*) 5.8; (*c*) 0.004; (*d*) 46.74; (*e*) 126.00; (*f*) 4,000,000; (*g*) 148; (*h*) 0.000099; (*i*) 2180; (*j*) 43.88.

1.49 (*a*) 1,325,000; (*b*) 0.0041872; (*c*) 0.0000280; (*d*) 7,300,000,000; (*e*) 0.0003487; (*f*) 18.50.

1.50 (*a*) 3; (*b*) 4; (*c*) 7; (*d*) 3; (*e*) 8; (*f*) unlimited; (*g*) 3; (*h*) 3; (*i*) 4; (*j*) 5.

1.51 (*a*) 0.005 million bu, or 5000 bu; three. (*b*) 0.000000005 cm, or 5×10^{-9} cm; four. (*c*) 0.5 ft; four.
(*d*) 0.05×10^8 m, or 5×10^6 m; two. (*e*) 0.5 mi/sec; six. (*f*) 0.5 thousand mi/sec, or 500 mi/sec; three.

1.52 (*a*) 3.17×10^{-4}; (*b*) 4.280×10^8; (*c*) 2.160000×10^4; (*d*) 9.810×10^{-6}; (*e*) 7.32×10^5; (*f*) 1.80×10^{-3}.

1.53 (*a*) 374; (*b*) 14.0.

1.54 (*a*) 280 (two significant figures), 2.8 hundred, or 2.8×10^2; (*b*) 178.9;
(*c*) 250,000 (three significant figures), 250 thousand, or 2.50×10^5; (*d*) 53.0; (*e*) 5.461; (*f*) 9.05;
(*g*) 11.54; (*h*) 5,745,000 (four significant figures), 5745 thousand, 5.745 million, or 5.745×10^6; (*i*) 1.2;
(*j*) 4157.

1.55 (*a*) -11; (*b*) 2; (*c*) $\frac{35}{8}$, or 4.375; (*d*) 21; (*e*) 3; (*f*) -16; (*g*) $\sqrt{98}$, or 9.89961 approximately; (*h*) $-7/\sqrt{34}$, or -1.20049 approximately; (*i*) 32; (*j*) $10/\sqrt{17}$, or 2.42536 approximately.

1.56 (*a*) 22, 18, 14, 10, 6, 2, -2, -6, and -10; (*b*) 19.6, 16.4, 13.2, 2.8, -0.8, -4, and -8.4; (*c*) -1.2, 30, $10-4\sqrt{2}=4.34$ approximately, and $10+4\pi=22.57$ approximately; (*d*) 3, 1, 5, 2.1, -1.5, 2.5, and 0; (*e*) $X=\frac{1}{4}(10-Y)$.

1.57 (*a*) -5; (*b*) -24; (*c*) 8.

1.58 (*a*) -8; (*b*) 4; (*c*) -16.

1.76 (*a*) -4; (*b*) 2; (*c*) 5; (*d*) $\frac{3}{4}$; (*e*) 1; (*f*) -7.

1.77 (*a*) $a=3$, $b=4$; (*b*) $a=-2$, $b=6$; (*c*) $X=-0.2$, $Y=-1.2$; (*d*) $A=\frac{184}{7}=26.28571$ approximately, $B=\frac{110}{7}=15.71429$ approximately; (*e*) $a=2$, $b=3$, $c=5$; (*f*) $X=-1$, $Y=3$, $Z=-2$; (*g*) $U=0.4$, $V=-0.8$, $W=0.3$.

1.78 (*b*) $(2,-3)$; i.e., $X=2$, $Y=-3$.

1.79 (*a*) 2, -2.5; (*b*) 2.1 and -0.8 approximately.

1.80 (*a*) $\dfrac{4\pm\sqrt{76}}{6}$, or 2.12 and -0.79 approximately.

(*b*) 2 and -2.5.

(*c*) 0.549 and -2.549 approximately.

(*d*) $\dfrac{-8\pm\sqrt{-36}}{2}=\dfrac{-8\pm\sqrt{36}\sqrt{-1}}{2}=\dfrac{-8\pm 6i}{2}=-4\pm 3i$, where $i=\sqrt{-1}$.

These roots are *complex numbers* and will not show up when a graphic procedure is employed.

1.81 (*a*) $-6.15<-4.3<-1.5<1.52<2.37$; (*b*) $2.37>1.52>-1.5>-4.3>-6.15$.

1.82 (*a*) $30\le N\le 50$; (*b*) $S\ge 7$; (*c*) $-4\le X<3$; (*d*) $P\le 5$; (*e*) $X-Y>2$.

1.83 (*a*) $X\ge 4$; (*b*) $X>3$; (*c*) $N<5$; (*d*) $Y\le 1$; (*e*) $-8\le X\le 7$; (*f*) $-1.8\le N<3$; (*g*) $2\le a<22$.

1.84 (*a*) 1; (*b*) 2; (*c*) 3; (*d*) -1; (*e*) -2.

1.85 (*a*) 1.0000; (*b*) 2.3026; (*c*) 4.6052; (*d*) 6.9076; (*e*) -2.3026.

1.86 (*a*) 1; (*b*) 2; (*c*) 3; (*d*) 4; (*e*) 5.

1.87 The EXCEL command is shown below the answer.

```
1.160964    1.974636     2.9974102     1.068622     1.056642
=LOG(5,4)   =LOG(24,5)   =LOG(215,6)   =LOG(8,7)    =LOG(9,8)
```

1.88

```
> evalf(log[4](5));     1.160964047
> evalf(log[5](24));    1.974635869
> evalf(log[6](215));   2.997410155
> evalf(log[7](8));     1.068621561
> evalf(log[8](9));     1.056641667
```

1.89 $\ln\left(\dfrac{a^3b^4}{c^5}\right)=3\ln(a)+4\ln(b)-5\ln(c)$

1.90 $\log\left(\dfrac{xyz}{w^3}\right)=\log(x)+\log(y)+\log(z)-3\log(w)$

1.91 $5\ln(a) - 4\ln(b) + \ln(c) + \ln(d) = \ln\left(\frac{a^5cd}{b^4}\right)$.

1.92 $\log(u) + \log(v) + \log(w) - 2\log(x) - 3\log(y) - 4\log(z) = \log\left(\frac{uvw}{x^2y^3z^4}\right)$.

1.93 104/3.

1.94 2, −5/3.

1.95 $-\frac{5}{2} - \frac{\sqrt{7}}{2}i$ and $-\frac{5}{2} + \frac{\sqrt{7}}{2}i$.

1.96 165.13.

1.97 471.71.

1.98 402.14.

1.99 2.363.

1.100 0.617.

CHAPTER 2

2.19 (*b*) 62.

2.20 (*a*) 799; (*b*) 1000; (*c*) 949.5; (*d*) 1099.5 and 1199.5; (*e*) 100 (hours); (*f*) 76; (*g*) $\frac{62}{400} = 0.155$, or 15.5%; (*h*) 29.5%; (*i*) 19.0%; (*j*) 78.0%.

2.25 (*a*) 24%; (*b*) 11%; (*c*) 46%.

2.26 (*a*) 0.003 in; (*b*) 0.3195, 0.3225, 0.3255, . . . , 0.3375 in.
(*c*) 0.320–0.322, 0.323–0.325, 0.326–0.328, . . . , 0.335–0.337 in.

2.31 (*a*) Each is 5 years; (*b*) four (although strictly speaking the last class has no specified size); (*c*) one; (*d*) (85–94); (*e*) 7 years and 17 years; (*f*) 14.5 years and 19.5 years; (*g*) 49.3% and 87.3%; (*h*) 45.1%; (*i*) cannot be determined.

2.33 19.3, 19.3, 19.1, 18.6, 17.5, 19.1, 21.5, 22.5, 20.7, 18.3, 14.0, 11.4, 10.1, 18.6, 11.4, and 3.7. (These will not add to 265 million because of the rounding errors in the percentages.)

2.34 (*b*) 0.295; (*c*) 0.19; (*d*) 0.

CHAPTER 3

3.47 (*a*) $X_1 + X_2 + X_3 + X_4 + 8$
(*b*) $f_1X_1^2 + f_2X_2^2 + f_3X_3^2 + f_4X_4^2 + f_5X_5^2$
(*c*) $U_1(U_1 + 6) + U_2(U_2 + 6) + U_3(U_3 + 6)$
(*d*) $Y_1^2 + Y_2^2 + \cdots + Y_N^2 - 4N$
(*e*) $4X_1Y_1 + 4Y_2Y_2 + 4X_3Y_3 + 4X_4Y_4$.

3.48 (*a*) $\sum_{j=1}^{3}(X_j + 3)^3$; (*b*) $\sum_{j=1}^{15} f_j(Y_j - a)^2$; (*c*) $\sum_{j=1}^{N}(2X_j - 3Y_j)$;

$(d)\ \sum_{j=1}^{8}\left(\frac{X_j}{Y_j}-1\right)^2; \quad (e)\ \frac{\sum_{j=1}^{12} f_j a_j^2}{\sum_{j=1}^{12} f_j}.$

3.51 (*a*) 20; (*b*) −37; (*c*) 53; (*d*) 6; (*e*) 226; (*f*) −62; (*g*) $\frac{25}{12}$.

3.52 (*a*) −1; (*b*) 23.

3.53 86.

3.54 0.50 second.

3.55 8.25.

3.56 (*a*) 82; (*b*) 79.

3.57 78.

3.58 66.7% males and 33.3% females.

3.59 11.09 tons.

3.60 501.0.

3.61 0.72642 cm.

3.62 26.2.

3.63 715 minutes.

3.64 (*b*) 1.7349 cm.

3.65 (*a*) Mean = 5.4, median = 5; (*b*) mean = 19.91, median = 19.85.

3.66 85.

3.67 0.51 second.

3.68 8.

3.69 11.07 tons.

3.70 490.6.

3.71 0.72638 cm.

3.72 25.4.

3.73 Approximately 78.3 years.

3.74 35.7 years.

3.75 708.3 minutes.

3.76 (*a*) Mean = 8.9, median = 9, mode = 7.
(*b*) Mean = 6.4, median = 6. Since each of the numbers 4, 5, 6, 8, and 10 occurs twice, we can consider these to be the five modes; however, it is more reasonable to conclude in this case that no mode exists.

3.77 It does not exist.

3.78 0.53 second.

3.79 10.

3.80 11.06 tons.

3.81 462.

3.82 0.72632 cm.

3.83 23.5.

3.84 668.7 minutes.

3.85 (*a*) 35–39; (*b*) 75 to 84.

3.86 (*a*) Using formula (9), mode = 11.1 Using formula (10), mode = 11.03
(*b*) Using formula (9), mode = 0.7264 Using formula (10), mode = 0.7263
(*c*) Using formula (9), mode = 23.5 Using formula (10), mode = 23.8
(*d*) Using formula (9), mode = 668.7 Using formula (10), mode = 694.9.

3.88 (*a*) 8.4; (*b*) 4.23.

3.89 (*a*) $G = 8$; (*b*) $\bar{X} = 12.4$.

3.90 (*a*) 4.14; (*b*) 45.8.

3.91 (*a*) 11.07 tons; (*b*) 499.5.

3.92 18.9%.

3.93 (*a*) 1.01%; (*b*) 238.2 million; (*c*) 276.9 million.

3.94 $1586.87.

3.95 $1608.44.

3.96 3.6 and 14.4.

3.97 (*a*) 3.0; (*b*) 4.48.

3.98 (*a*) 3; (*b*) 0; (*c*) 0.

3.100 (*a*) 11.04; (*b*) 498.2.

3.101 38.3 mi/h.

3.102 (*b*) 420 mi/h.

3.104 (*a*) 25; (*b*) 3.55.

3.107 (*a*) Lower quartile = $Q_1 = 67$, middle quartile = Q_2 = median = 75, and upper quartile = $Q_3 = 83$.
(*b*) 25% scored 67 or lower (or 75% scored 67 or higher), 50% scored 75 or lower (or 50% scored 75 or higher), and 75% scored 83 or lower (or 25% scored 83 or higher).

3.108 (*a*) $Q_1 = 10.55$ tons, $Q_2 = 11.07$ tons, and $Q_3 = 11.57$ tons; (*b*) $Q_1 = 469.3$, $Q_2 = 490.6$, and $Q_3 = 523.3$.

3.109 Arithmetic mean, Median, Mode, Q_2, P_{50}, and D_5.

3.110 (*a*) 10.15 tons; (*b*) 11.78 tons; (*c*) 10.55 tons; (*d*) 11.57 tons.

3.112 (*a*) 83; (*b*) 64.

CHAPTER 4

4.33 (*a*) 9; (*b*) 4.273.

4.34 4.0 tons.

4.35 0.0036 cm.

4.36 7.88 kg.

4.37 20 weeks.

4.38 (*a*) 18.2; (*b*) 3.58; (*c*) 6.21; (*d*) 0; (*e*) $\sqrt{2} = 1.414$ approximately; (*f*) 1.88.

4.39 (*a*) 2; (*b*) 0.85.

4.40 (*a*) 2.2; (*b*) 1.317.

4.41 0.576 ton.

4.42 (*a*) 0.00437 cm; (*b*) 60.0%, 85.2%, and 96.4%.

4.43 (*a*) 3.0; (*b*) 2.8.

4.44 (*a*) 31.2; (*b*) 30.6.

4.45 (*a*) 6.0; (*b*) 6.0.

4.46 4.21 weeks.

4.48 (*a*) 0.51 ton; (*b*) 27.0; (*c*) 12.

4.49 3.5 weeks.

4.52 (*a*) 1.63 tons; (*b*) 33.6 or 34.

4.53 The 10–90 percentile range equals $189,500 and 80% of the selling prices are in the range $130,250 ± $94,750.

4.56 (*a*) 2.16; (*b*) 0.90; (*c*) 0.484.

4.58 45.

4.59 (*a*) 0.733 ton; (*b*) 38.60; (*c*) 12.1.

4.61 (*a*) $\bar{X} = 2.47$; (*b*) $s = 1.11$.

4.62 $s = 5.2$ and Range/4 = 5.

4.63 (*a*) 0.00576 cm; (*b*) 72.1%, 93.3%, and 99.76%.

4.64 (*a*) 0.719 ton; (*b*) 38.24; (*c*) 11.8.

4.65 (*a*) 0.000569 cm; (*b*) 71.6%, 93.0%, and 99.68%.

4.66 (*a*) 146.8 lb and 12.9 lb.

4.67 (*a*) 1.7349 cm and 0.00495 cm.

4.74 (*a*) 15; (*b*) 12.

4.75 (*a*) Statistics; (*b*) algebra.

4.76 (*a*) 6.6%; (*b*) 19.0%.

4.77 0.15.

4.78 0.20.

4.79 Algebra.

4.80 0.19, −1.75, 1.17, 0.68, −0.29.

CHAPTER 5

5.15 (*a*) 6; (*b*) 40; (*c*) 288; (*d*) 2188.

5.16 (*a*) 0; (*b*) 4; (*c*) 0; (*d*) 25.86.

5.17 (*a*) −1; (*b*) 5; (*c*) −91; (*d*) 53.

5.19 0, 26.25, 0, 1193.1.

5.21 7.

5.22 (*a*) 0, 6, 19, 42; (*b*) −4, 22, −117, 560; (*c*) 1, 7, 38, 155.

5.23 0, 0.2344, −0.0586, 0.0696.

5.25 (*a*) $m_1 = 0$; (*b*) $m_2 = pq$; (*c*) $m_3 = pq(q - p)$; (*d*) $m_4 = pq(p^2 - pq + q^2)$.

5.27 $m_1 = 0$, $m_2 = 5.97$, $m_3 = -0.397$, $m_4 = 89.22$.

5.29 m_1 (corrected) = 0, m_2 (corrected) = 5.440, m_3 (corrected) = −0.5920, m_4 (corrected) = 76.2332.

5.30 (*a*) $m_1 = 0$, $m_2 = 0.53743$, $m_3 = 0.36206$, $m_4 = 0.84914$;
(*b*) m_2 (corrected) $= 0.51660$, m_4 (corrected) $= 0.78378$.

5.31 (*a*) 0; (*b*) 52.95; (*c*) 92.35; (*d*) 7158.20; (*e*) 26.2; (*f*) 7.28; (*g*) 739.58; (*h*) 22,247; (*i*) 706,428; (*j*) 24,545.

5.32 (*a*) −0.2464; (*b*) −0.2464.

5.33 0.9190.

5.34 First distribution.

5.35 (*a*) 0.040; (*b*) 0.074.

5.36 (*a*) −0.02; (*b*) −0.13.

5.37

	Distribution		
Pearson's coefficient of skewness	1	2	3
First coefficient	0.770	0	−0.770
Second coefficient	1.094	0	−1.094

5.38 (*a*) 2.62; (*b*) 2.58.

5.39 (*a*) 2.94; (*b*) 2.94.

5.40 (*a*) Second; (*b*) first.

5.41 (*a*) Second; (*b*) neither; (*c*) first.

5.42 (*a*) Greater than 1875; (*b*) equal to 1875; (*c*) less than 1875.

5.43 (*a*) 0.313.

CHAPTER 6

6.40 (*a*) $\frac{5}{26}$; (*b*) $\frac{5}{36}$; (*c*) 0.98; (*d*) $\frac{2}{9}$; (*e*) $\frac{7}{8}$.

6.41 (*a*) Probability of a king on the first draw and no king on the second draw.
(*b*) Probability of either a king on the first draw or a king on the second draw, or both.
(*c*) No king on the first draw or no king on the second draw, or both (i.e., no king on the first and second draws).
(*d*) Probability of a king on the third draw, given that a king was drawn on the first draw but not on the second draw.
(*e*) No king on the first, second, and third draws.
(*f*) Probability either of a king on the first draw and a king on the second draw or of no king on the second draw and a king on the third draw.

6.42 (*a*) $\frac{1}{3}$; (*b*) $\frac{3}{5}$; (*c*) $\frac{11}{15}$; (*d*) $\frac{2}{5}$; (*e*) $\frac{4}{5}$.

6.43 (*a*) $\frac{4}{25}$; (*b*) $\frac{4}{75}$; (*c*) $\frac{16}{25}$; (*d*) $\frac{64}{225}$; (*e*) $\frac{11}{15}$; (*f*) $\frac{1}{5}$; (*g*) $\frac{104}{225}$; (*h*) $\frac{221}{225}$; (*i*) $\frac{6}{25}$; (*j*) $\frac{52}{225}$.

6.44 (*a*) $\frac{29}{185}$; (*b*) $\frac{2}{37}$; (*c*) $\frac{118}{185}$; (*d*) $\frac{52}{185}$; (*e*) $\frac{11}{15}$; (*f*) $\frac{1}{5}$; (*g*) $\frac{86}{185}$; (*h*) $\frac{182}{185}$; (*i*) $\frac{9}{37}$; (*j*) $\frac{26}{111}$.

6.45 (*a*) $\frac{5}{18}$; (*b*) $\frac{11}{36}$; (*c*) $\frac{1}{36}$.

6.46 (*a*) $\frac{47}{52}$; (*b*) $\frac{16}{221}$; (*c*) $\frac{15}{34}$; (*d*) $\frac{13}{17}$; (*e*) $\frac{210}{221}$; (*f*) $\frac{10}{13}$; (*g*) $\frac{40}{51}$; (*h*) $\frac{77}{442}$.

6.47 $\frac{5}{18}$.

6.48 (*a*) 81:44; (*b*) 21:4.

6.49 $\frac{19}{42}$.

6.50 (*a*) $\frac{2}{5}$; (*b*) $\frac{1}{5}$; (*c*) $\frac{4}{15}$; (*d*) $\frac{13}{15}$.

6.51 (*a*) 37.5%; (*b*) 93.75%; (*c*) 6.25%; (*d*) 68.75%.

6.52 (*a*)

X	0	1	2	3	4
$p(X)$	$\frac{1}{16}$	$\frac{4}{16}$	$\frac{6}{16}$	$\frac{4}{16}$	$\frac{1}{16}$

6.53 (*a*) $\frac{1}{48}$; (*b*) $\frac{7}{24}$; (*c*) $\frac{3}{4}$; (*d*) $\frac{1}{6}$.

6.54 (*a*)

X	0	1	2	3
$p(X)$	$\frac{1}{6}$	$\frac{1}{2}$	$\frac{3}{10}$	$\frac{1}{30}$

6.55 (*a*)

X	3	4	5	6	7	8	9	10	11	12	13	14	15	16	17	18
$p(X)$*	1	3	6	10	15	21	25	27	27	25	21	15	10	6	3	1

*All the $p(x)$ values have a divisor of 216.

(*b*) 0.532407.

6.56 \$9.

6.57 \$4.80 per day.

6.58 *A* contributes \$12.50; *B* contributes \$7.50.c

6.59 (*a*) 7; (*b*) 590; (*c*) 541; (*d*) 10,900.

6.60 (*a*) 1.2; (*b*) 0.56; (*c*) $\sqrt{0.56} = 0.75$ approximately.

6.63 10.5.

6.64 (*a*) 12; (*b*) 2520; (*c*) 720;
(*d*) =PERMUT(4,2), =PERMUT(7,5), =PERMUT(10,3).

6.65 n=5.

6.66 60.

6.67 (*a*) 5040; (*b*) 720; (*c*) 240.

6.68 (*a*) 8400; (*b*) 2520.

6.69 (*a*) 32,805; (*b*) 11,664.

6.70 26.

6.71 (*a*) 120; (*b*) 72; (*c*) 12.

6.72 (*a*) 35; (*b*) 70; (*c*) 45;
(*d*) `=COMBIN(7,3)`, `=COMBIN(8,4)`, `=COMBIN(10,8)`.

6.73 $n = 6$.

6.74 210.

6.75 840.

6.76 (*a*) 42,000; (*b*) 7000.

6.77 (*a*) 120; (*b*) 12,600.

6.78 (*a*) 150; (*b*) 45; (*c*) 100.

6.79 (*a*) 17; (*b*) 163.

6.81 2.95×10^{25}.

6.83 (*a*) $\frac{6}{5525}$; (*b*) $\frac{22}{425}$; (*c*) $\frac{169}{425}$; (*d*) $\frac{73}{5525}$.

6.84 $\frac{171}{1296}$.

6.85 (*a*) 0.59049; (*b*) 0.32805; (*c*) 0.08866.

6.86 (*b*) $\frac{3}{4}$; (*c*) $\frac{7}{8}$.

6.87 (*a*) 8; (*b*) 78; (*c*) 86; (*d*) 102; (*e*) 20; (*f*) 142.

6.90 $\frac{1}{3}$.

6.91 1/3,838,380 (i.e., the odds against winning are 3,838,379 to 1).

6.92 (*a*) 658,007 to 1; (*b*) 91,389 to 1; (*c*) 9879 to 1.

6.93 (*a*) 649,739 to 1; (*b*) 71,192 to 1; (*c*) 4164 to 1; (*d*) 693 to 1.

6.94 $\frac{11}{36}$.

6.95 $\frac{1}{4}$.

6.96

X	3	4	5	6	7	8	9	10	11	12
$p(X)$*	1	3	6	10	12	12	10	5	3	1

*All the $p(x)$ values have a divisor of 64.

6.97 7.5.

6.98 70%.

6.99 $(0.5)(0.01) + (0.3)(0.02) + (0.2)(0.03) = 0.017.$

6.100 $\dfrac{0.2(0.03)}{0.017} = 0.35.$

CHAPTER 7

7.35 (*a*) 5040; (*b*) 210; (*c*) 126; (*d*) 165; (*e*) 6.

7.36 (*a*) $q^7 + 7q^6p + 21q^5p^2 + 35q^4p^3 + 35q^3p^4 + 21q^2p^5 + 7qp^6 + p^7$
(*b*) $q^{10} + 10q^9p + 45q^8p^2 + 120q^7p^3 + 210q^6p^4 + 252q^5p^5 + 210q^4p^6 + 120q^3p^7 + 45q^2p^8 + 10qp^9 + p^{10}$

7.37 (*a*) $\frac{1}{64}$; (*b*) $\frac{3}{32}$; (*c*) $\frac{15}{64}$; (*d*) $\frac{5}{16}$; (*e*) $\frac{15}{64}$; (*f*) $\frac{3}{32}$; (*g*) $\frac{1}{64}$;

(*h*) **Probability Density Function**

```
Binomial with n=6 and p=0.5
          X          P(X = x )
          0          0.015625
          1          0.093750
          2          0.234375
          3          0.312500
          4          0.234375
          5          0.093750
          6          0.015625
```

7.38 (*a*) $\frac{57}{64}$; (*b*) $\frac{21}{32}$;
(*c*) `1-BINOMDIST(1,6,0.5,1)` or `0.890625, =BINOMDIST(3,6,0.5,1) = 0.65625.`

7.39 (*a*) $\frac{1}{4}$; (*b*) $\frac{5}{16}$; (*c*) $\frac{11}{16}$; (*d*) $\frac{5}{8}$.

7.40 (*a*) 250; (*b*) 25; (*c*) 500.

7.41 (*a*) $\frac{17}{162}$; (*b*) $\frac{1}{324}$.

7.42 $\frac{64}{243}$.

7.43 $\frac{193}{512}$.

7.44 (*a*) $\frac{32}{243}$; (*b*) $\frac{192}{243}$; (*c*) $\frac{40}{243}$; (*d*) $\frac{242}{243}$;
(*e*)

```
a     0.131691     =BINOMDIST(5,5,0.66667,0)
b     0.790128     =1-BINOMDIST(2,5,0.66667,1)
c     0.164606     =BINOMDIST(2,5,0.66667,0)
d     0.995885     =1-BINOMDIST(0,5,0.66667,0).
```

7.45 (*a*) 42; (*b*) 3.550; (*c*) −0.1127; (*d*) 2.927.

7.47 (*a*) $Npq(q - p)$; (*b*) $Npq(1 - 6pq) + 3N^2p^2q^2$.

7.49 (*a*) 1.5 and −1.6; (*b*) 72 and 90.

7.50 (*a*) 75.4; (*b*) 9.

7.51 (*a*) 0.8767; (*b*) 0.0786; (*c*) 0.2991;
(*d*)

```
a    0.8767328    =NORMSDIST(2.4)-NORMSDIST(-1.2)
b    0.0786066    =NORMSDIST(1.87)-NORMSDIST(1.23)
c    0.2991508    =NORMSDIST(-0.5)-NORMSDIST(-2.35).
```

7.52 (*a*) 0.0375; (*b*) 0.7123; (*c*) 0.9265; (*d*) 0.0154; (*e*) 0.7251; (*f*) 0.0395;
(*g*)

```
a    0.037538     =NORMSDIST(-1.78)
b    0.7122603    =NORMSDIST(0.56)
c    0.9264707    =1-NORMSDIST(-1.45)
d    0.0153863    =1-NORMSDIST(2.16)
e    0.7251362    =NORMSDIST(1.53)-NORMSDIST(-0.8)
f    0.0394927    =NORMSDIST(-2.52)+(1-NORMSDIST(1.83)).
```

7.53 (*a*) 0.9495; (*b*) 0.9500; (*c*) 0.6826.

7.54 (*a*) 0.75; (*b*) −1.86; (*c*) 2.08; (*d*) 1.625 or 0.849; (*e*) ±1.645.

7.55 −0.995.

7.56 (*a*) 0.0317; (*b*) 0.3790; (*c*) 0.1989;
(*d*)

```
a    0.03174 =NORMDIST(2.25,0,1,0)
b    0.37903 =NORMDIST(-0.32,0,1,0)
c    0.19886 =NORMDIST(-1.18,0,1,0).
```

7.57 (*a*) 4.78%; (*b*) 25.25%; (*c*) 58.89%.

7.58 (*a*) 2.28%; (*b*) 68.27%; (*c*) 0.14%.

7.59 84.

7.60 (*a*) 61.7%; (*b*) 54.7%.

7.61 (*a*) 95.4%; (*b*) 23.0%; (*c*) 93.3%.

7.62 (*a*) 1.15; (*b*) 0.77.

7.63 (*a*) 0.9962; (*b*) 0.0687; (*c*) 0.0286; (*d*) 0.0558.

7.64 (*a*) 0.2511; (*b*) 0.1342.

7.65 (*a*) 0.0567; (*b*) 0.9198; (*c*) 0.6404; (*d*) 0.0079.

7.66 0.0089.

7.67 (*a*) 0.04979; (*b*) 0.1494; (*c*) 0.2241; (*d*) 0.2241; (*e*) 0.1680; (*f*) 0.1008.

7.68 (*a*) 0.0838; (*b*) 0.5976; (*c*) 0.4232.

7.69 (*a*) 0.05610; (*b*) 0.06131.

7.70 (*a*) 0.00248; (*b*) 0.04462; (*c*) 0.1607; (*d*) 0.1033; (*e*) 0.6964; (*f*) 0.0620.

7.71 (*a*) 0.08208; (*b*) 0.2052; (*c*) 0.2565; (*d*) 0.2138; (*e*) 0.8911; (*f*) 0.0142.

7.72 (*a*) $\frac{5}{3888}$; (*b*) $\frac{5}{324}$.

7.73 (*a*) 0.0348; (*b*) 0.000295.

7.74 $\frac{1}{16}$.

7.75 $p(X) = \binom{4}{X}(0.32)^X(0.68)^{4-X}$. The expected frequencies are 32, 60, 43, 13, and 2, respectively.

7.76

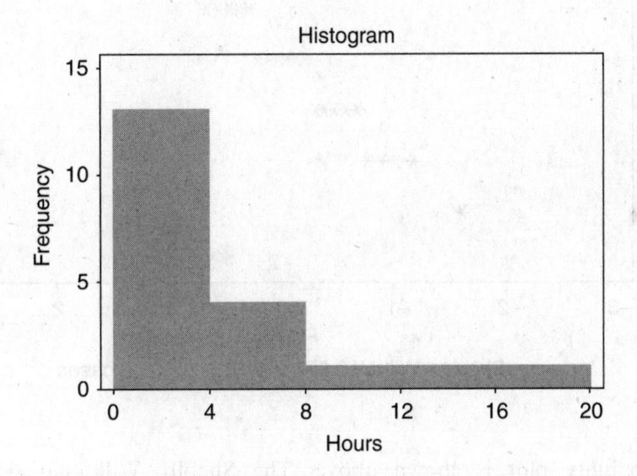

The histogram shows a skew to the data, indicating non-normality.

The Shapiro–Wilt test of STATISTIX indicates non-normality.

7.77 The STATISTIX histogram strongly indicates normality.

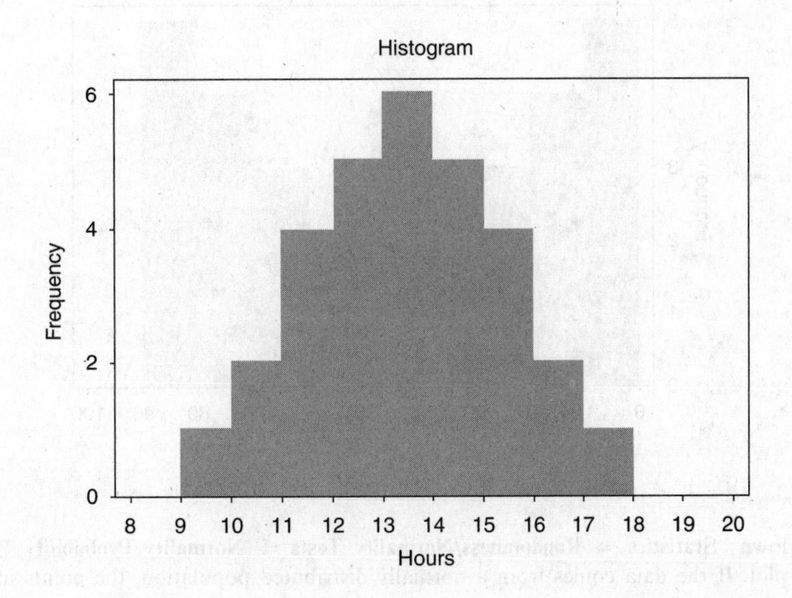

The pull-down "**Statistics ⇒ Randomness/Normality Tests ⇒ Normality Probability Plot**" leads to the following plot. If the data are from a normally distributed population, the points in the plot tend to fall along a straight line and $P(W)$ tends to be larger than 0.05. If $P(W) < 0.05$, generally normality is rejected.

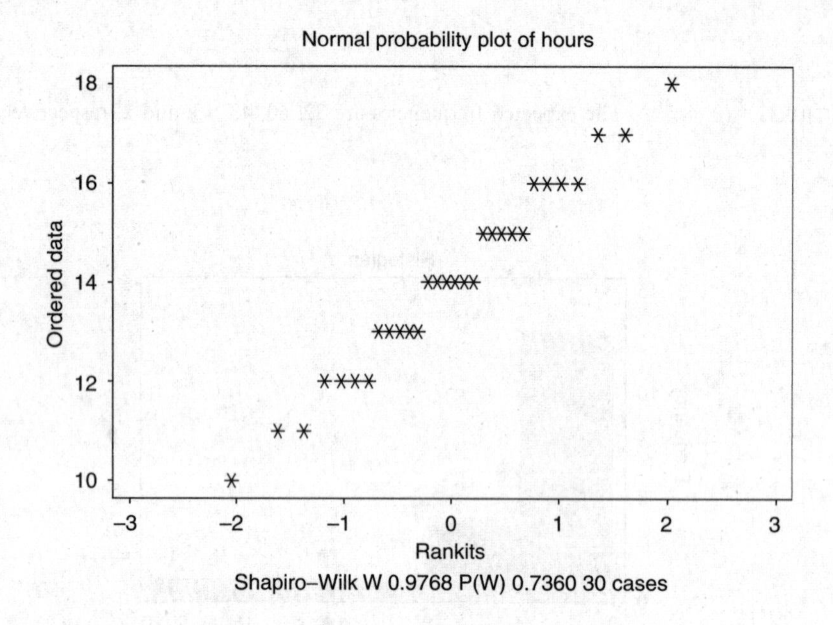

The normal probability plot is shown above. The Shapiro–Wilk statistic along with the p-value is shown. $P(W) = 0.7360$. Since the p-value is considerably larger than 0.05, normality of the data is not rejected.

7.78 The following histogram of the test scores in Table 7.11, produced by STATISTIX, has a U-shape. Hence, it is called a U-shape distribution.

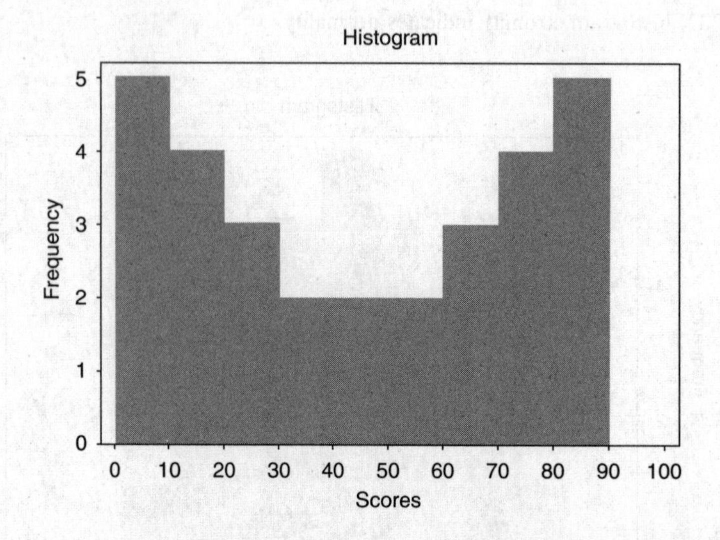

The pull-down "**Statistics ⇒ Randomness/Normality Tests ⇒ Normality Probability Plot**" leads to the following plot. If the data comes from a normally distributed population, the points in the plot tend to fall along a straight line and $P(W)$ tends to be larger than 0.05. If $P(W) < 0.05$, generally normality is rejected.

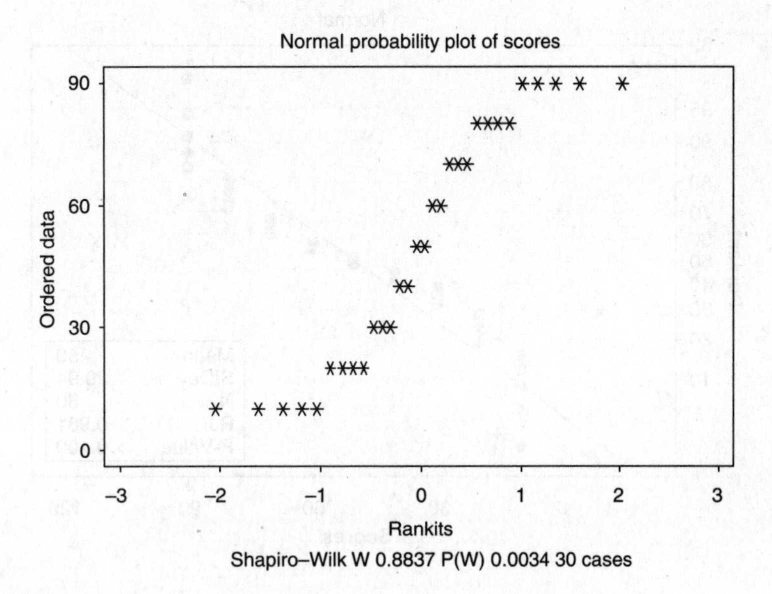

The normal probability plot is shown above. The Shapiro–Wilk statistic along with the p-value is shown. $P(W) = 0.0034$. Since the p-value is less than 0.05, normality of the data is rejected.

7.79 In addition to the Kolmogorov–Smirnov test in MINITAB and the Shapiro–Wilk test in STATISTIX, there are two other tests of normality that we shall discuss that are available for testing for normality. They are the Ryan–Joiner and the Anderson–Darling tests. Basically all four calculate a test statistic and each test statistic has a p-value associated with it. Generally, the following rule is used. If the p-value is <0.05, normality is rejected. The following graphic is given when doing the Anderson–Darling test. In this case, the p-value is 0.006 and I would reject the null hypothesis that the data came from a normal distribution.

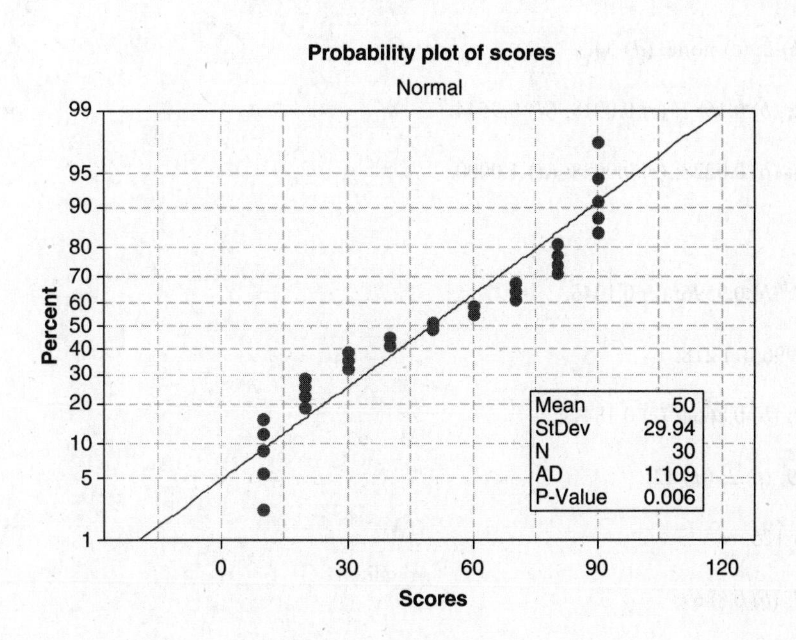

Note that if the Ryan–Joiner test is used you do not reject normality.

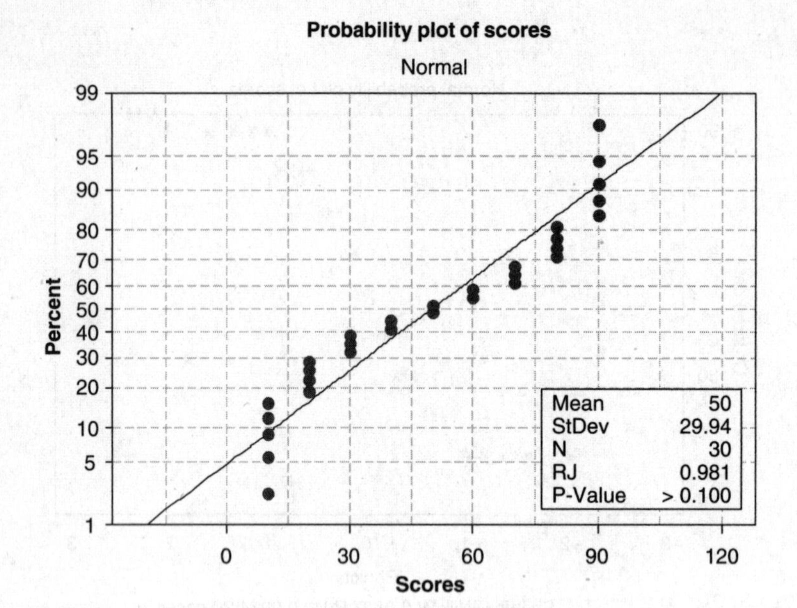

7.80 $p(X) = \dfrac{(0.61)^X e^{-0.61}}{X!}$. The expected frequencies are 108.7, 66.3, 20.2, 4.1, and 0.7, respectively.

CHAPTER 8

8.21 (*a*) 9.0; (*b*) 4.47; (*c*) 9.0; (*d*) 3.16.

8.22 (*a*) 9.0; (*b*) 4.47; (*c*) 9.0; (*d*) 2.58.

8.23 (*a*) $\mu_{\bar{X}} = 22.40$ g, $\sigma_{\bar{X}} = 0.008$ g; (*b*) $\mu_{\bar{X}} = 22.40$ g, $\sigma_{\bar{X}} =$ slightly less than 0.008 g.

8.24 (*a*) $\mu_{\bar{X}} = 22.40$ g, $\sigma_{\bar{X}} = 0.008$ g; (*b*) $\mu_{\bar{X}} = 22.40$ g, $\sigma_{\bar{X}} = 0.0057$ g.

8.25 (*a*) 237; (*b*) 2; (*c*) none; (*d*) 34.

8.26 (*a*) 0.4972; (*b*) 0.1587; (*c*) 0.0918; (*d*) 0.9544.

8.27 (*a*) 0.8164; (*b*) 0.0228; (*c*) 0.0038; (*d*) 1.0000.

8.28 0.0026.

8.34 (*a*) 0.0029; (*b*) 0.9596; (*c*) 0.1446.

8.35 (*a*) 2; (*b*) 996; (*c*) 218.

8.36 (*a*) 0.0179; (*b*) 0.8664; (*c*) 0.1841.

8.37 (*a*) 6; (*b*) 9; (*c*) 2; (*d*) 12.

8.39 (*a*) 19; (*b*) 125.

8.40 (*a*) 0.0077; (*b*) 0.8869.

8.41 (*a*) 0.0028; (*b*) 0.9172.

8.42 (*a*) 0.2150; (*b*) 0.0064, 0.4504.

8.43 0.0482.

8.44 0.0188.

8.45 0.0410.

8.47 (*a*) 118.79 g; (*b*) 0.74 g.

8.48 0.0228.

8.49 $\mu = 12$ and $\sigma^2 = 10.8$.

A	B	C	D
first	second	mean	probability
6	6	6	0.01
6	9	7.5	0.02
6	12	9	0.04
6	15	10.5	0.02
6	18	12	0.01
9	6	7.5	0.02
9	9	9	0.04
9	12	10.5	0.08
9	15	12	0.04
9	18	13.5	0.02
12	6	9	0.04
12	9	10.5	0.08
12	12	12	0.16
12	15	13.5	0.08
12	18	15	0.04
15	6	10.5	0.02
15	9	12	0.04
15	12	13.5	0.08
15	15	15	0.04
15	18	16.5	0.02
18	6	12	0.01
18	9	13.5	0.02
18	12	15	0.04
18	15	16.5	0.02
18	18	18	0.01
			1

8.50 **Probability distribution for x-bar with $n = 2$.**

D	E	F	G	H
probability	xbar	p(xbar)		
0.01		6	0.01	D2
0.02		7.5	0.04	D3+D7

0.04	9	0.12	D4+D8+D12
0.02	10.5	0.2	D5+D9+D13+D17
0.01	12	0.26	D6+D10+D14+D18+D22
0.02	13.5	0.2	D11+D15+D19+D23
0.04	15	0.12	D16+D20+D24
0.08	16.5	0.04	D21+D25
0.04	18	0.01	D26
0.02		1	SUM(G2:G10)
0.04			
0.08			
0.16			
0.08			
0.04			
0.02			
0.04			
0.08			
0.04			
0.02			
0.01			
0.02			
0.04			
0.02			
0.01			

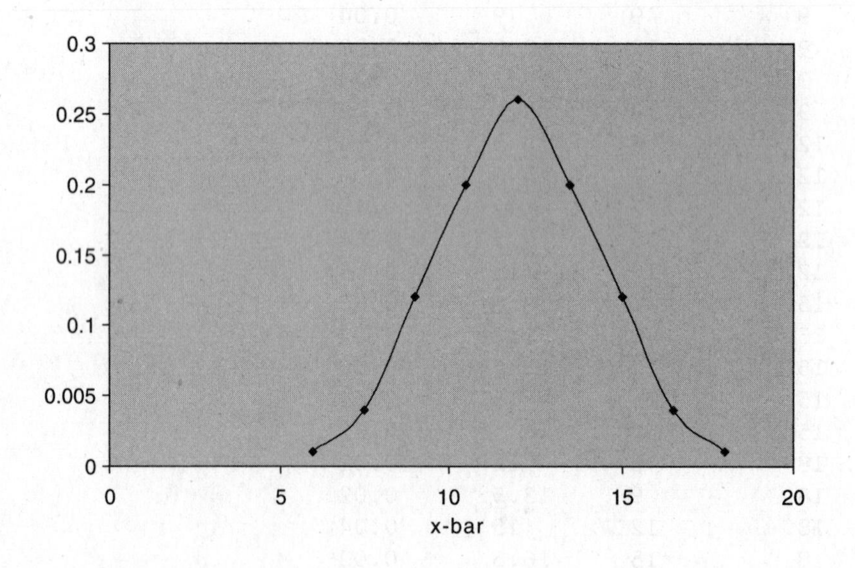

8.51 Mean(x-bar) = 12 Var (x-bar) = 5.4.

8.52

xbar	P(xbar)
6	0.001
7	0.006
8	0.024
9	0.062

```
10    0.123
11     0.18
12    0.208
13     0.18
14    0.123
15    0.062
16    0.024
17    0.006
18    0.001
```

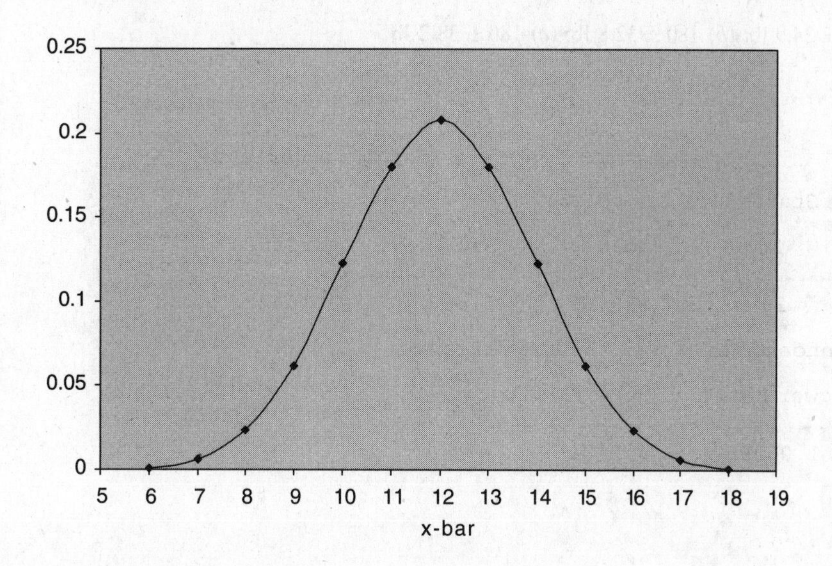

CHAPTER 9

9.21 (*a*) 9.5 kg; (*b*) 0.74 kg^2; (*c*) 0.78 kg and 0.86 kg, respectively.

9.22 (*a*) 1200 h; (*b*) 105.4 h.

9.23 (*a*) Estimates of population standard deviations for sample sizes of 30, 50, and 100 tubes are 101.7 h, 101.0 h, and 100.5 h, respectively; estimates of population means are 1200 h in all cases.

9.24 (*a*) 11.09 ± 0.18 tons; (*b*) 11.09 ± 0.24 tons.

9.25 (*a*) 0.72642 ± 0.000095 in; (*b*) 0.72642 ± 0.000085 in; (*c*) 0.72642 ± 0.000072 in; (*d*) 0.72642 ± 0.000060 in.

9.26 (*a*) 0.72642 ± 0.000025 in; (*b*) 0.000025 in.

9.27 (*a*) At least 97; (*b*) at least 68; (*c*) at least 167; (*d*) at least 225.

9.28 80% CI for mean: (286.064, 322.856).

9.29 (*a*) 2400 ± 45 lb, 2400 ± 59 lb; (*b*) 87.6%.

9.30 (*a*) 0.70 ± 0.12, 0.69 ± 0.11; (*b*) 0.70 ± 0.15, 0.68 ± 0.15; (*c*) 0.70 ± 0.18, 0.67 ± 0.17.

9.31 97.5% CI for difference: (0.352477, 0.421523).

9.32 (*a*) 16,400; (*b*) 27,100; (*c*) 38,420; (*d*) 66,000.

9.33 (*a*) 1.07 ± 0.09 h; (*b*) 1.07 ± 0.12 h.

9.34 85% CI for difference: (−7.99550, −0.20948).

9.35 95% CI for difference: (−0.0918959, −0.00610414).

9.36 (*a*) 180 ± 24.9 lb; (*b*) 180 ± 32.8 lb; (*c*) 180 ± 38.2 lb.

9.37

```
                One Sample Chi-square Test for a Variance
Sample Statistics for volume

          N          Mean        Std. Dev.   Variance
      ---------------------------------------------------
         20         180.65       1.4677      2.1542

99% Confidence Interval for the variance

     Lower Limit    Upper Limit
     --------       ----------
       1.06085       5.98045
```

9.38

```
Two Sample Test for variances of units within line
Sample Statistics
      line
     Group        N         Mean           Std. Dev.        Varinace
     ------------------------------------------------------------------
       1          13        104.9231       12.189           148.5769
       2          15        101.2667       5.5737           31.06667
    95% Confidence Interval of the Ratio of Two Variances

             Lower Limit         Upper Limit
               ---------         ----------
               1.568             15.334
```

CHAPTER 10

10.29 (*a*) 0.2604.

(*b*)

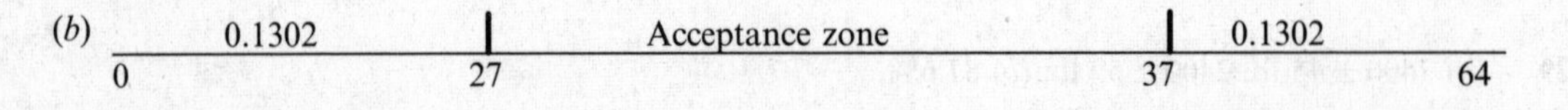

$\alpha = 0.1302 + 0.1302$.

10.30 (*a*) Reject the null hypothesis if $X \leq 21$ or $X \geq 43$, where X = number of red marbles drawn; (*b*) 0.99186; (*c*) *Reject if* $X \leq 23$ *or* $X \geq 41$.

10.31 (*a*) H_o: $p = 0.5$ H_a: $p > 0.5$; (*b*) One-tailed test; (*c*) Reject the null hypothesis if $X \geq 40$; (*d*) Reject the null hypothesis if $X \geq 41$.

10.32 (*a*) Two-tailed *p*-value = `2*(1-BINOMDIST(22,100,0.16666,1)` = 0.126 > 0.05. Do not reject the null at the 0.05 level.
(*b*) One-tailed *p*-value = `1-BINOMDIST(22,100,0.16666,1)` = 0.063 > 0.05. Do not reject the null at the 0.05 level.

10.33 Using either a two-tailed or a one-tailed test, we cannot reject the hypothesis at the 0.01 level.

10.34 H_o: $p \geq 0.95$ H_a: $p < 0.95$
p-value = P{X ≤ 182 from 200 pieces when $p = 0.95$} = `BINOMDIST(182,200,0.95,1)` = 0.012.
Fail to reject at 0.01 but, reject at 0.05.

10.35 Test statistic = 2.63, critical values ±1.7805, reject the null hypothesis.

10.36 Test statistic = −3.39, 0.10 critical value = −1.28155, 0.025 critical value = −1.96. The result is significant at $\alpha = 0.10$ and $\alpha = 0.025$.

10.37 Test statistic = 6.46, critical value = 1.8808, conclude that $\mu > 25.5$.

10.38 `=NORMSINV(0.9)` = 1.2815 for $\alpha = 0.1$, `=NORMSINV(0.99)` = 2.3263 for $\alpha = 0.01$ and `=NORMSINV(0.999)` = 3.0902 for $\alpha = 0.001$.

10.39 *p*-value = $P\{X \leq 3\} + P\{X \geq 12\} = 0.0352$.

10.40 *p*-value = $P\{Z < -2.63\} + P\{Z > 2.63\} = 0.0085$.

10.41 *p*-value = $P\{Z < -3.39\} = 0.00035$.

10.42 *p*-value = $P\{Z > 6.46\}$ = 5.23515E-11.

10.43 (*a*) 8.64 ± 0.96 oz; (*b*) 8.64 ± 0.83 oz; (*c*) 8.64 ± 0.63 oz.

10.44 The upper control limits are (*a*) 12 and (*b*) 10.

10.45 Test statistic = −5.59, *p*-value = 0.000. Reject the null since the *p*-value < α.

10.46 Test statistic = −1.58, *p*-value = 0.059. Do not reject for α = 0.05, do reject for α = 0.10.

10.47 Test statistic = −1.73, *p*-value = 0.042. Reject for α = 0.05, do not reject for α = 0.01.

10.48 A one-tailed test shows that the new fertilizer is superior at both levels of significance.

10.49 (*a*) Test statistic = 1.35, *p*-value = 0.176, unable to reject the null at α = 0.05.
(*b*) Test statistic = 1.35, *p*-value = 0.088, unable to reject the null at α = 0.05.

10.50 (*a*) Test statistic = 1.81, *p*-value = 0.07, unable to reject the null at α = 0.05.
(*b*) Test statistic = 1.81, *p*-value = 0.0035, reject the null at α = 0.05.

10.51 `=1-BINOMDIST(10,15,0.5,1)` or 0.059235.

10.52 =BINOMDIST(2,20,0.5,1) + 1-BINOMDIST(17,20,0.5,1) or 0.000402.

10.53 =BINOMDIST(10,15,0.6,1) or 0.7827.

10.54 =BINOMDIST(17,20,0.9,1)-BINOMDIST(2,20,0.9,1) or 0.3231.

10.55 *p*-value is given by =1-BINOMDIST(9,15,0.5,1) which equals 0.1509. Do not reject the null because the value of $\alpha = 0.0592$ and the *p*-value is not less then α.

10.56 α=BINOMDIST(4,20,0.5,1) +1-BINOMDIST(15,20,0.5,1) or 0.0118
p-value =BINOMDIST(3,20,0.5,1) +1-BINOMDIST(16,20,0.5,1) or 0.0026.
Reject the null because the *p*-value$<\alpha$.

10.57 =1-BINOMDIST(3,30,0.03,1) or 0.0119.

10.58 =BINOMDIST(3,30,0.04,1) or 0.9694.

10.59 α=1-BINOMDIST(5,20,0.16667,1) or 0.1018
p-value =1-BINOMDIST(6,20,0.16667,1) or 0.0371.

CHAPTER 11

11.20 (*a*) 2.60; (*b*) 1.75; (*c*) 1.34; (*d*) 2.95; (*e*) 2.13.

11.21 (*a*) 3.75; (*b*) 2.68; (*c*) 2.48; (*d*) 2.39; (*e*) 2.33.
(*a*) =TINV(0.02,4) or 3.7469; (*b*) 2.6810; (*c*) 2.4851; (*d*) 2.3901; (*e*) 3.3515.

11.22 (*a*) 1.71; (*b*) 2.09; (*c*) 4.03; (*d*) −0.128.

11.23 (*a*) 1.81; (*b*) 2.76; (*c*) −0.879; (*d*) −1.37.

11.24 (*a*) ±4.60; (*b*) ±3.06; (*c*) ±2.79; (*d*) ±2.75; (*e*) ±2.70.

11.25 (*a*) 7.38 ± 0.79; (*b*) 7.38 ± 1.11;
(*c*) (6.59214, 8.16786) (6.26825, 8.49175).

11.26 (*a*) 7.38 ± 0.70; (*b*) 7.38 ± 0.92.

11.27 (*a*) 0.298 ± 0.030 second; (*b*) 0.298 ± 0.049 second.

11.28 A two-tailed test shows that there is no evidence at either the 0.05 or 0.01 level to indicate that the mean lifetime has changed.

11.29 A one-tailed test shows no decrease in the mean at either the 0.05 or 0.01 level.

11.30 A two-tailed test at both levels shows that the product does not meet the required specifications.

11.31 A one-tailed test at both levels shows that the mean copper content is higher than the specifications require.

11.32 A one-tailed test shows that the process should not be introduced if the significance level adopted is 0.01 but it should be introduced if the significance level adopted is 0.05.

11.33 A one-tailed test shows that brand *A* is better than brand *B* at the 0.05 significance level.

11.34 Using a two-tailed test at the 0.05 significance level, we would not conclude on the basis of the samples that there is a difference in acidity between the two types.

11.35 Using a one-tailed test at the 0.05 significance level, we would conclude that the first group is not superior to the second.

11.36 (*a*) 21.0; (*b*) 26.2; (*c*) 23.3; (*d*)`=CHIINV(0.05,12)` or 21.0261 `=CHIINV(0.01,12)` or 26.2170 `=CHIINV(0.025,12)` or 23.3367.

11.37 (*a*) 15.5; (*b*) 30.1; (*c*) 41.3; (*d*) 55.8.

11.38 (*a*) 20.1; (*b*) 36.2; (*c*) 48.3; (*d*) 63.7.

11.39 (*a*) $\chi_1^2 = 9.59$, and $\chi_2^2 = 34.2$.

11.40 (*a*) 16.0; (*b*) 6.35; (*c*) assuming equal areas in the two tails, $\chi_1^2 = 2.17$ and $\chi_2^2 = 14.1$.

11.41 (*a*) 87.0 to 230.9 h; (*b*) 78.1 to 288.5 h.

11.42 (*a*) 95.6 to 170.4 h; (*b*) 88.9 to 190.8 h.

11.43 (*a*) 122.5; (*b*) 179.2; (*c*) `=CHIINV(0.95,150)` or 122.6918; (*d*) `=CHIINV(0.05,150)` or 179.5806.

11.44 (*a*) 207.7; (*b*) 295.2; (*c*) `=CHIINV(0.975,250)` or 208.0978; (*d*) `=CHIINV(0.025,250)` or 295.6886.

11.46 (*a*) 106.1 to 140.5 h; (*b*) 102.1 to 148.1 h.

11.47 105.5 to 139.6 h.

11.48 On the basis of the given sample, the apparent increase in variability is not significant at either level.

11.49 The apparent decrease in variability is significant at the 0.05 level, but not at the 0.01 level.

11.50 (*a*) 3.07; (*b*) 4.02; (*c*) 2.11; (*d*) 2.83.

11.51 (*a*) `=FINV(0.05,8,10)` or 3.0717; (*b*) `=FINV(0.01,24,11)` or 4.0209; (*c*) `=FINV(0.05,15,24)` or 2.1077; (*d*) `=FINV(0.01,20,22)` or 2.8274.

11.52 The sample 1 variance is significantly greater at the 0.05 level, but not at the 0.01 level.

11.53 (*a*) Yes; (*b*) no.

CHAPTER 12

12.26 The hypothesis cannot be rejected at either level.

12.27 The conclusion is the same as before.

12.28 The new instructor is not following the grade pattern of the others. (The fact that the grades happen to be better than average *may* be due to better teaching ability or lower standards, or both.)

12.29 There is no reason to reject the hypothesis that the coins are fair.

12.30 There is no reason to reject the hypothesis at either level.

12.31 (*a*) 10, 60, and 50, respectively;
(*b*) the hypothesis that the results are the same as those expected cannot be rejected at the 0.05 level.

12.32 The difference is significant at the 0.05 level.

12.33 (*a*) The fit is good; (*b*) no.

12.34 (*a*) The fit is "too good"; (*b*) the fit is poor at the 0.05 level.

12.35 (*a*) The fit is very poor at the 0.05 level; since the binomial distribution gives a good fit of the data, this is consistent with Problem 12.33.
(*b*) The fit is good, but not "too good."

12.36 The hypothesis can be rejected at the 0.05 level, but not at the 0.01 level.

12.37 The conclusion is the same as before.

12.38 The hypothesis cannot be rejected at either level.

12.39 The hypothesis cannot be rejected at the 0.05 level.

12.40 The hypothesis can be rejected at both levels.

12.41 The hypothesis can be rejected at both levels.

12.42 The hypothesis cannot be rejected at either level.

12.49 (*a*) 0.3863 (unconnected), and 0.3779 (with Yates' correction).

12.50 (*a*) 0.2205, 0.1985 (corrected); (*b*) 0.0872, 0.0738 (corrected).

12.51 0.4651.

12.54 (*a*) 0.4188, 0.4082 (corrected).

12.55 (*a*) 0.2261, 0.2026 (corrected); (*b*) 0.0875, 0.0740 (corrected).

12.56 0.3715.

CHAPTER 13

13.24 (*a*) 4; (*b*) 6; (*c*) $\frac{28}{3}$; (*d*) 10.5; (*e*) 6; (*f*) 9.

13.25 (2, 1).

13.26 (*a*) $2X + Y = 4$; (*b*) X intercept $= 2$, Y intercept $= 4$; (*c*) -2, -6.

13.27 $Y = \frac{2}{3}X - 3$, or $2X - 3Y = 9$.

13.28 (*a*) Slope $= \frac{3}{5}$, Y intercept $= -4$; (*b*) $3X - 5Y = 11$.

13.29 (*a*) $-\frac{4}{3}$; (*b*) $\frac{32}{3}$; (*c*) $4X + 3Y = 32$.

13.30 $X/3 + Y/(-5) = 1$, or $5X - 3Y = 15$.

13.31 (*a*) $°\text{F} = \frac{9}{5}°\text{C} + 32$; (*b*) 176°F; (*c*) 20°C.

13.32 (*a*) $Y = -\frac{1}{3} + \frac{5}{7}X$, or $Y = -0.333 + 0.714X$; (*b*) $X = 1 + \frac{9}{7}Y$, or $X = 1.00 + 1.29Y$.

13.33 (*a*) 3.24; 8.24; (*b*) 10.00.

13.35 (*b*) $Y = 29.13 + 0.661X$; (*c*) $X = -14.39 + 1.15Y$; (*d*) 79; (*e*) 95.

13.36 (*a*) and (*b*)

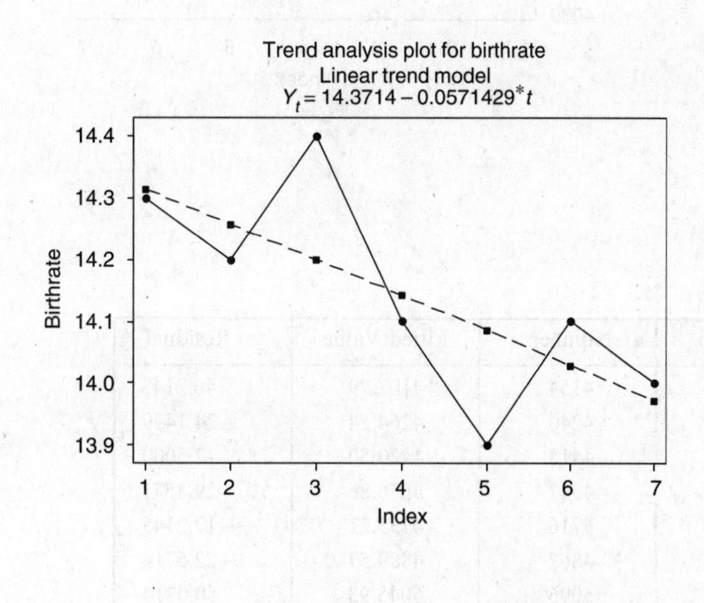

(*c*)

Year	Birth Rate	Fitted Value	Residual
1998	14.3	14.3143	−0.014286
1999	14.2	14.2571	−0.057143
2000	14.4	14.2000	0.200000
2001	14.1	14.1429	−0.042857
2002	13.9	14.0857	−0.185714
2003	14.1	14.0286	0.071429
2004	14.0	13.9714	0.028571

(*d*) 14.3714−0.0571429(13) = 13.6.

13.37 (*a*) and (*b*)

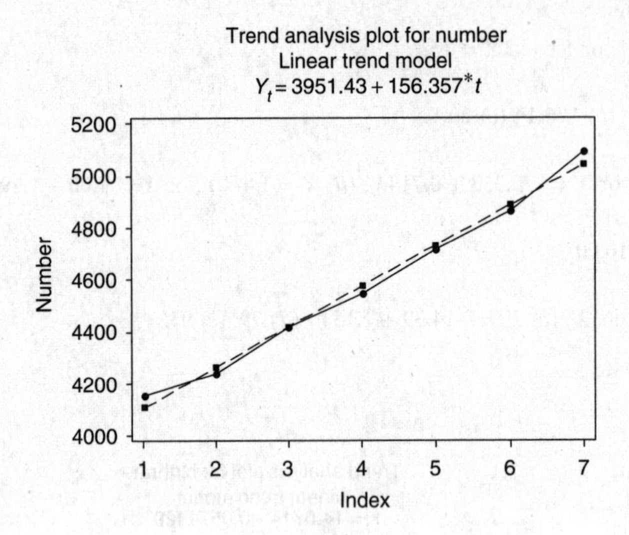

(*c*)

Year	Number	Fitted Value	Residual
1999	4154	4107.79	46.2143
2000	4240	4264.14	−24.1429
2001	4418	4420.50	−2.5000
2002	4547	4576.86	−29.8571
2003	4716	4733.21	−17.2143
2004	4867	4889.57	−22.5714
2005	5096	5045.93	50.0714

(*d*) $3951.43 + 156.357(12) = 5827.7$.

13.38 $Y = 5.51 + 3.20(X - 3) + 0.733(X - 3)^2$, or $Y = 2.51 - 1.20X + 0.733X^2$.

13.39 (*b*) $D = 41.77 - 1.096V + 0.08786V^2$; (*c*) 170 ft, 516 ft.

13.40

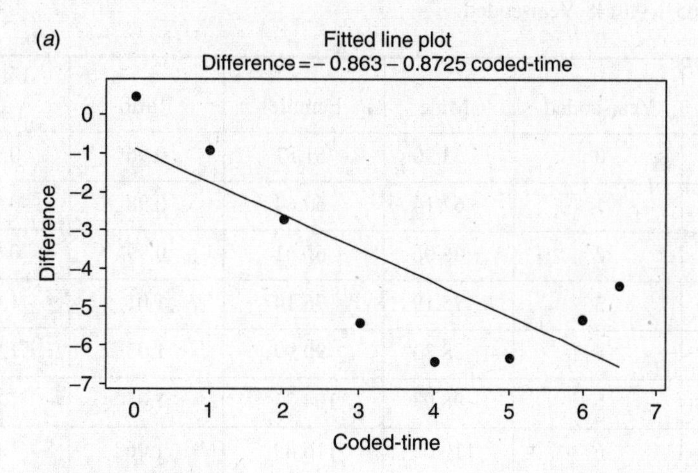

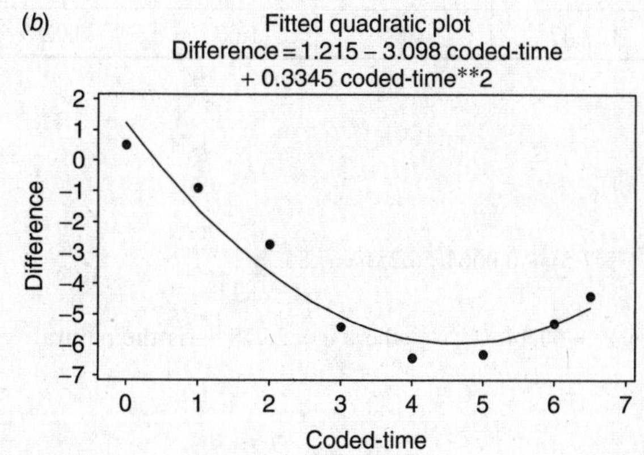

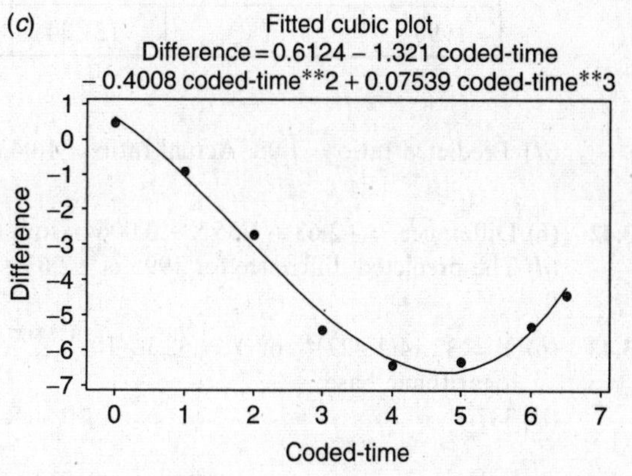

(*d*)

	Linear Model		Quadratic Model		Cubic Model	
Year	Residual	Fitted	Residual	Fitted	Residual	Fitted
1940	1.363	0.863	−0.715	1.215	−0.112	0.612
1950	0.836	1.736	0.649	−1.549	0.134	−1.034
1960	−0.092	0.092	0.943	−3.643	0.330	−3.030
1970	−1.919	1.919	−0.332	−5.068	−0.476	−4.924
1980	−2.047	2.047	−0.575	−5.825	−0.139	−6.261
1990	−1.074	1.074	−0.388	−5.912	0.292	−6.592
2000	0.798	6.098	0.030	−5.330	0.162	−5.462
2005	2.134	6.534	0.388	−4.788	−0.191	−4.209
	SSQ = 16.782		SSQ = 2.565		SSQ = 0.533	

(*e*)

Linear: $-0.863 - 0.8725(7) = -6.97$

Quadratic: $1.215 - 3.098(7) + 0.3345(7^2) = -4.08$

Cubic: $0.6124 - 1.321(7) - 0.4008(49) + 0.0754(343) = -2.41.$

13.41 (*b*) Ratio $= 0.965 + 0.0148$ Year-coded.

(*c*)

Year	Year-coded	Male	Female	Ratio	Fitted Value	Residual
1920	0	53.90	51.81	0.96	0.97	−0.00
1930	1	62.14	60.64	0.98	0.98	−0.00
1940	2	66.06	65.61	0.99	0.99	−0.00
1950	3	75.19	76.14	1.01	1.01	0.00
1960	4	88.33	90.99	1.03	1.02	0.01
1970	5	98.93	104.31	1.05	1.04	0.01
1980	6	110.05	116.49	1.06	1.05	0.00
1990	7	121.24	127.47	1.05	1.07	−0.02

(*d*) Predicted ratio $= 1.08$. Actual ratio $= 1.04$.

13.42 (*b*) Difference $= -2.63 + 1.35$ x $+ 0.0064$ xsquare.
(*d*) The predicted difference for 1995 is $-2.63 + 1.35(7.5) + 0.0064(56.25) = 7.86$.

13.43 (*b*) $Y = 32.14(1.427)^X$, or $Y = 32.14(10)^{0.1544X}$, or $Y = 32.14e^{0.3556X}$, where $e = 2.718\cdots$ is the natural logarithmic base.
(*d*) 387.

CHAPTER 14

14.40 (*b*) $Y = 4.000 + 0.500X$; (*c*) $X = 2.408 + 0.612Y$.

14.41 (*a*) 1.304; (*b*) 1.443.

14.42 (*a*) 24.50; (*b*) 17.00; (*c*) 7.50.

14.43 0.5533.

14.44 The EXCEL solution is =`CORREL(A2:A11,B2:B11)` or 0.553.

14.45 1.5.

14.46 (*a*) 0.8961; (*b*) $Y = 80.78 + 1.138X$; (*c*) 132.

14.47 (*a*) 0.958; (*b*) 0.872.

14.48 (*a*) $Y = 0.8X + 12$; (*b*) $X = 0.45Y + 1$.

14.49 (*a*) 1.60; (*b*) 1.20.

14.50 ±0.80.

14.51 75%.

14.53 You get the same answer for both parts, namely −0.9203.

14.54 (*a*) $Y = 18.04 - 1.34X$, $Y = 51.18 - 2.01X$.

14.58 0.5440.

14.59 (*a*) $Y = 4.44X - 142.22$; (*b*) 141.9 lb and 177.5 lb, respectively.

14.60 (*a*) 16.92 lb; (*b*) 2.07 in.

14.62 Pearson correlation of C1 and C2 = 0.957.

14.63 Pearson correlation of C1 and C2 = 0.582.

14.64 (*a*) Yes; (*b*) no.

14.65 (*a*) No ; (*b*) yes.

14.66 (*a*) 0.2923 and 0.7951; (*b*) 0.1763 and 0.8361.

14.67 (*a*) 0.3912 and 0.7500; (*b*) 0.3146 and 0.7861.

14.68 (*a*) 0.7096 and 0.9653; (*b*) 0.4961 and 0.7235.

14.69 (*a*) Yes; (*b*) no.

14.70 (*a*) 2.00 ± 0.21; (*b*) 2.00 ± 0.28.

14.71 (*a*) Using a one-tailed test, we can reject the hypothesis.
(*b*) Using a one-tailed test, we cannot reject the hypothesis.

14.72 (*a*) 37.0 ± 3.28; (*b*) 37.0 ± 4.45.

14.73 (*a*) 37.0 ± 0.69; (*b*) 37.0 ± 0.94.

14.74 (*a*) 1.138 ± 0.398; (*b*) 132.0 ± 16.6; (*c*) 132.0 ± 5.4.

CHAPTER 15

15.26 (*a*) $X_3 = b_{3.12} + b_{31.2}X_1 + b_{32.1}X_2$; (*b*) $X_4 = b_{4.1235} + b_{41.235}X_1 + b_{42.135}X_2 + b_{43.125}X_3$.

15.28 (*a*) $X_3 = 61.40 - 3.65X_1 + 2.54X_2$; (*b*) 40.

15.29 (*a*) $X_3 - 74 = 4.36(X_1 - 6.8) + 4.04(X_2 - 7.0)$, or $X_3 = 16.07 + 4.36X_1 + 4.04X_2$; (*b*) 84 and 66.

15.30 The output is abridged in all cases.

EXCEL

Price	Bedrooms	Baths
165	3	2
200	3	3
225	4	3
180	2	3
202	4	2
250	4	4
275	3	4
300	5	3
155	2	2
230	4	4

SUMMARY OUTPUT

Regression Statistics	
Multiple R	0.877519262
R Square	0.770040055
Adjusted R Square	0.704337213
Standard Error	25.62718211
Observations	10

	Coefficients
Intercept	**32.94827586**
Bedrooms	**28.64655172**
Baths	**29.28448276**

MINITAB

Regression Analysis: Price versus Bedrooms, Baths

The regression equation is

Price = 32.9 + 28.6 Bedrooms + 29.3 Baths

R-Sq = 77.0% R-Sq(adj) = 70.4%

SAS

Root MSE	25.62718	R-Square	0.7700
Dependent Mean	218.20000	Adj R-Sq	0.7043
Coeff Var	11.74481		

Parameter Estimates

Variable	Label	DF	Parameter Estimate	Standard Error	t Value	Pr > |t|
Intercept	Intercept	1	32.94828	39.24247	0.84	0.4289
Bedrooms	Bedrooms	1	28.64655	9.21547	3.11	0.0171
Baths	Baths	1	29.28448	10.90389	2.69	0.0313

SPSS

Coefficients[a]

Model		Unstandardized Coefficients		Standardized Coefficients		
		B	Std. Error	Beta	t	Sig.
1	(Constant)	32.948	39.242		.840	.429
	Bedrooms	28.647	9.215	.587	3.109	.017
	Baths	29.284	10.904	.507	2.686	.031

[a]Dependent Variable: Price

STATISTIX

Statistix 8.0

Unweighted Least Squares Linear Regression of Price

Predictor Variables	**Coefficients**	**Std Error**	**T**	**P**	**VIF**
Constant	**32.9483**	39.2425	0.84	0.4289	
Bedrooms	**28.6466**	9.21547	3.11	0.0171	1.1
Baths	**29.2845**	10.9039	2.69	0.0313	1.1
R-Squared	0.7700				

Estimated Price = 32.9 + 28.6(5) + 29.3(4) = 293.1 thousand.

15.31 3.12.

15.32 (*a*) 5.883; (*b*) 0.6882.

15.33 0.9927.

15.34 (*a*) 0.7567; (*b*) 0.7255; (*c*) 0.6810.

15.37 The STATISTIX pull down "**Statistics ⇒ Linear models ⇒ Partial Correlations**" is used. The following dialog box is filled as shown.

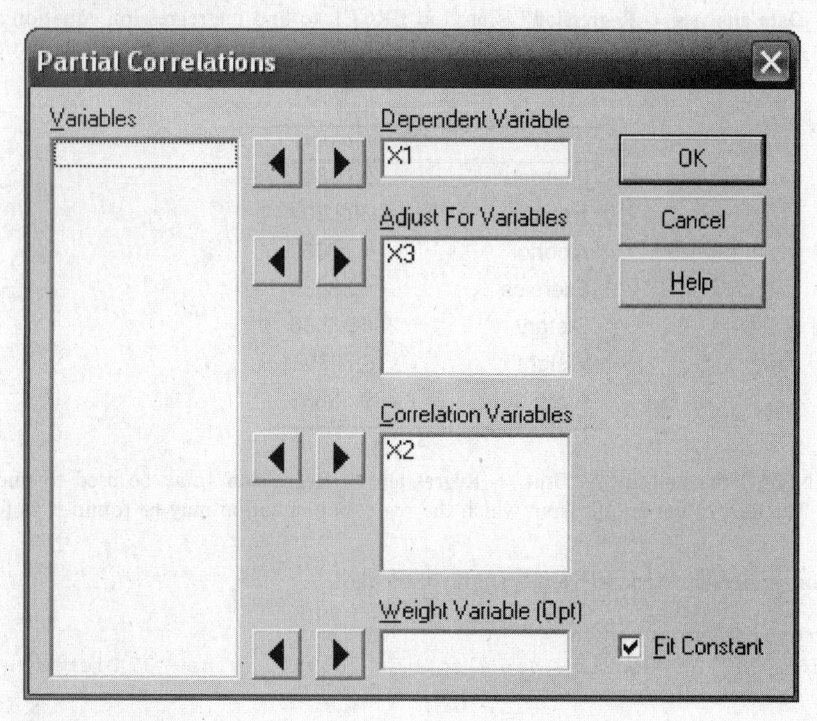

The following output results.

Statistix 8.0

Partial Correlations with X1

Controlled for X3

X2 0.5950

Similarly, it is found that:

```
Statistix 8.0
Partial Correlations with X1
Controlled for X2

X3          -0.8995
```

and

```
Statistix 8.0
Partial Correlations with X2
Controlled for X1

X3            0.8727
```

15.38 (*a*) 0.2672; (*b*) 0.5099; (*c*) 0.4026.

15.42 (*a*) $X_4 = 6X_1 + 3X_2 - 4X_3 - 100$; (*b*) 54.

15.43 (*a*) 0.8710; (*b*) 0.8587; (*c*) −0.8426.

15.44 (*a*) 0.8947; (*b*) 2.680.

15.45 Any of the following solutions will give the same coefficients as solving the normal equations. The pull-down "**Tools ⇒ Data analysis ⇒ Regression**" is used in EXCEL to find the regression equation as well as other regression measurements. The part of the output from which the regression equation may be found is as follows.

	Coefficients
Intercept	−25.3355
smoker	−302.904
Alcohol	−4.57069
Exercise	−60.8839
Dietary	−36.8586
Weight	16.76998
Age	−9.52833

In MINITAB the pull-down "**Stat ⇒ Regression ⇒ Regression**" may be used to find the regression equation. The part of the output from which the regression equation may be found is as follows.

Regression Analysis: Medcost versus smoker, alcohol, . . .

```
The regression equation is
Medcost = - 25 - 303 smoker - 4.6 alcohol - 60.9 Exercise - 37 Dietary + 16.8 weight
  - 9.53 Age

    Predictor       Coef      SE Coef        T       P
    Constant       -25.3       644.8      -0.04    0.970
    smoker        -302.9       256.1      -1.18    0.271
    alcohol        -4.57       11.89      -0.38    0.711
    Exercise      -60.88       19.75      -3.08    0.015
```

```
Dietary        -36.9        104.0       -0.35    0.732
weight         16.770       3.561        4.71    0.002
Age            -9.528       9.571       -1.00    0.349
```

In SAS, the pull-down "**Statistics ⇒ Regression ⇒ Linear**" may be used to find the regression equation. The part of the output from which the regression equation may be found is as follows.

```
                          The REG Procedure
                           Model: MODEL1
                  Dependent Variable: Medcost Medcost

           Number of Observations Read                      16
           Number of Observations Used                      15
           Number of Observations with Missing Values        1

     Root MSE              224.41971    R-Square     0.9029
           Dependent Mean       2461.80000   Adj R-Sq   0.8301
           Coeff Var               9.11608

                         Parameter Estimates

Variable     Label       DF    Parameter     Standard    t value     Pr > |t|
                               Estimate      Error
Intercept    Intercept   1      -25.33552    644.84408    -0.04       0.9696
smoker       smoker      1     -302.90395    256.11003    -1.18       0.2709
alcohol      alcohol     1       -4.57069     11.88579    -0.38       0.7106
Exercise     Exercise    1      -60.88386     19.75371    -3.08       0.0151
Dietary      Dietary     1      -36.85858    104.04736    -0.35       0.7323
weight       weight      1       16.76998      3.56074     4.71       0.0015
Age          Age         1       -9.52833      9.57104    -1.00       0.3486
```

In SPSS, the pull-down "**Analyze ⇒ Regression ⇒ Linear**" may be used to find the regression equation. The part of the output from which the regression equation may be found is as follows. Look under the unstandardized coefficients.

Coefficients[a]

Model		Unstandardized Coefficients		Standardized Coefficients		
		B	Std. Error	Beta	t	Sig.
1	(Constant)	−25.336	644.844		−.039	.970
	smoker	−302.904	256.110	−.287	−1.183	.271
	alcohol	−4.571	11.886	−.080	−.385	.711
	exercise	−60.884	19.754	−.553	−3.082	.015
	dietary	−36.859	104.047	−.044	−.354	.732
	weight	−16.770	3.561	.927	4.710	.002
	age	−9.528	9.571	−.124	−.996	.349

[a]Dependent Variable: medcost

In STATISTIX, the pull-down "**Statistics ⇒ Linear models ⇒ Linear regression**" may be used to find the regression equation. The part of the output from which the regression equation may be found is as follows. Look under the coefficients column.

Statistix 8.0

Unweighted Least Squares Linear Regression of Medcost

Predictor Variables	**Coefficient**	**Std Error**	**T**	**P**	**VIF**
Constant	**−25.3355**	644.844	−0.04	0.9696	
Age	**−9.52833**	9.57104	−1.00	0.3486	1.3
Dietary	**−36.8586**	104.047	−0.35	0.7323	1.3
Exercise	**−60.8839**	19.7537	−3.08	0.0151	2.6
alcohol	**−4.57069**	11.8858	−0.38	0.7106	3.6
smoker	**−302.904**	256.110	−1.18	0.2709	4.9
weight	**16.7700**	3.56074	4.71	0.0015	3.2

CHAPTER 16

16.21 There is no significant difference in the five varieties at the 0.05 or 0.01 levels. The MINITAB analysis is as follows.

One-way ANOVA: A, B, C, D, E

Source	DF	SS	MS	F	P
Factor	4	27.2	6.8	0.65	0.638
Error	15	157.8	10.5		
Total	19	185.0			

S = 3.243   R-Sq = 14.71%   R-Sq(adj) = 0.00%

```
                                  Individual 95% CIs For Mean Based on
                                  Pooled StDev
Level  N    Mean   StDev   ---+---------+---------+---------+-----
A      4  16.500   3.697          (----------*----------)
B      4  14.500   2.082     (----------*----------)
C      4  17.750   3.862             (----------*----------)
D      4  16.000   3.367          (----------*----------)
E      4  17.500   2.887             (----------*----------)
                           ---+---------+---------+---------+-----
                            12.0      15.0      18.0      21.0
```

16.22 There is no difference in the four types of tires at the 0.05 or 0.01 level. The STATISTIX analysis is as follows.

Statistix 8.0

Completely Randomized AOV for Mileage

Source	**DF**	**SS**	**MS**	**F**	**P**
Type	3	77.500	25.8333	2.39	0.0992
Error	20	216.333	10.8167		
Total	23	293.833			

Grand Mean 34.083   CV 9.65

	Chi-Sq	DF	P
Bartlett's Test of Equal Variances	4.13	3	0.2476
Cochran's Q	0.5177		
Largest Var / Smallest Var	6.4000		

```
Component of variance for between groups        2.50278
Effective cell size                                 6.0

Type      Mean
A         35.500
B         36.000
C         33.333
D         31.500
```

16.23 There is a difference in the three teaching methods at the 0.05 level, but not at the 0.01 level. The EXCEL analysis is as follows:

MethodI	MethodII	MethodIII
75	81	73
62	85	79
71	68	60
58	92	75
73	90	81

Anova: Single Factor

SUMMARY

Groups	*Count*	*Sum*	*Average*	*Variance*
MethodI	5	339	67.8	54.7
MethodII	5	416	83.2	90.7
MethodIII	5	368	73.6	67.8

ANOVA

Source of Variation	*SS*	*df*	*MS*	*F*	*P-value*
Between Groups	604.9333	2	302.4667	4.256098	0.040088
Within Groups	852.8	12	71.06667		
Total	1457.733	14			

16.24 There is a difference in the five brands at the 0.05 level but not at the 0.01 level. The SPSS analysis is follows.

ANOVA

mpg

	Sum of Squares	df	Mean Square	F	Sig.
Between Groups	52.621	4	13.155	4.718	.010
Within Groups	44.617	16	2.789		
Total	97.238	20			

16.25 There is a difference in the four courses at both levels. The SAS analysis is follows.

```
                    The ANOVA Procedure
                 Class Level Information
              Class        Levels      Values
              Subject           4     1 2 3 4

        Number of Observations Read            16
        Number of Observations Used            16
```

```
                         The ANOVA Procedure
Dependent Variable: Grade
                                  Sum of
Source              DF           Squares     Mean Square    F Value    Pr > F

Model                3       365.5708333     121.8569444       7.35    0.0047
Error               12       198.8666667      16.5722222
Corrected Total     15       564.4375000
```

16.26 There is no difference in the operators or machines at the 0.05 level. The MINITAB analysis is follows.

```
Two-way ANOVA: Number versus Machine, Operator
Source     DF    SS      MS      F      P
Machine     2    56    28.0   4.31  0.101
Operator    2     6     3.0   0.46  0.660
Error       4    26     6.5
Total       8    88
S = 2.550    R-Sq = 70.45%    R-Sq(adj) = 40.91%
```

16.27 There is no difference in the operators or machines at the 0.01 level. The EXCEL analysis is follows. Compare this EXCEL analysis and the MINITAB analysis in the previous problem.

	Operator1	Operator2	Operator3
machine1	23	27	24
machine2	34	30	28
machine3	28	25	27

Anova: Two-Factor Without Replication

SUMMARY	*Count*	*Sum*	*Average*	*Variance*
Row 1	3	74	24.66667	4.333333
Row 2	3	92	30.66667	9.333333
Row 3	3	80	26.66667	2.333333
Column 1	3	85	28.33333	30.33333
Column 2	3	82	27.33333	6.333333
Column 3	3	79	26.33333	4.333333

ANOVA

Source of Variation	*SS*	*df*	*MS*	*F*	*P-value*
Rows	56	2	28	4.307692	0.100535
Columns	6	2	3	0.461538	0.660156
Error	26	4	6.5		
Total	88	8			

16.28 The *p*-value is called `sig.` in SPSS. The *p*-value for blocks is 0.640 and the *p*-value for the type of corn is 0.011, which is less than 0.05 and is therefore significant. There is no difference in the blocks at the 0.05 level. There are differences in the yield due to the type of corn at the 0.05 level. The SPSS analysis is as follows.

Tests of Between-Subjects Effects

Dependent Variable: yield

Source	Type III Sum of Squares	df	Mean Square	F	Sig.
Corrected Model	77.600[a]	7	11.086	2.867	.053
Intercept	3920.000	1	3920.000	1013.793	.000
block	10.000	4	2.500	.647	.640
type	67.600	3	22.533	5.828	.011
Error	46.400	12	3.867		
Total	4044.000	20			
Corrected Total	124.000	19			

[a]R Squared = .626 (Adjusted R Squared = .408)

16.29 At the 0.01 significance level, there is no difference in yield due to blocks or type of corn. Compare this STATISTIX output with the SPSS output is Problem 16.28.

```
Statistix 8.0
Randomized Complete Block AOV Table for yield

Source    DF        SS         MS        F         P
block      4    10.000     2.5000     5.83    0.0108
type       3    67.600    22.5333
Error     12    46.400     3.8667
Total     19   124.000

Grand Mean 14.000           CV 14.05

Means of yield for type
type      Mean
1        13.600
2        17.000
3        12.000
4        13.400
```

16.30 SAS represents the p-value as `Pr > F`. Referring to the last two lines of output, we see that both autos and tire brands are significant at the 0.05 level.

```
                         The GLM Procedure
                      Class Level Information
                Class        Levels        Values
                Auto            6          1 2 3 4 5 6
                Brand           4          1 2 3 4
                         The GLM Procedure
  Dependent Variable: lifetime
                                    Sum of
Source                  DF         Squares      Mean Square     F Value     Pr > F
Model                    8     201.3333333       25.1666667        4.08     0.0092
Error                   15      92.5000000        6.1666667
Corrected Total         23     293.8333333
```

Source	DF	Type III SS	Mean Square	F Value	Pr > F
Auto	5	123.8333333	24.7666667	4.02	0.0164
Brand	3	77.5000000	25.8333333	4.19	0.0243

16.31 Compare the MINITAB analysis in this problem with the SAS analysis in Problem 16.30. At the 0.01 significance level, there is no difference in either auto or brand since the *p*-values are 0.016 and 0.024 and both of these exceed 0.01.

Two-way ANOVA: lifetime versus Auto, Brand

Source	DF	SS	MS	F	P
Auto	5	123.833	24.7667	4.02	0.016
Brand	3	77.500	25.8333	4.19	0.024
Error	15	92.500	6.1667		
Total	23	293.833			

S = 2.483 R-Sq = 68.52% R-Sq(adj) = 51.73%

16.32 The STATISTIX output is shown. The *p*-value of 0.3171 indicates no difference in schools at 0.05 level of significance.

Statistix 8.0

Randomized Complete Block AOV Table for Grade

Source	**DF**	**SS**	**MS**	**F**	**P**
Method	2	604.93	302.467		
School	4	351.07	87.767	1.40	0.3171
Error	8	501.73	62.717		
Total	14	1457.73			

The *F* value for teaching methods is 4.82 and the *p*-value for teaching methods is 0.0423. Therefore, there is a difference between teaching methods at the 0.05 significance level.

16.33 The EXCEL output indicates that neither hair color nor heights of adult females have any bearing on scholastic achievement at significance level 0.05. The *p*-values for hair color is 0.4534 and the *p*-value for height is 0.2602.

	Redhead	Blonde	Brunette
Tall	75	78	80
Medium	81	76	79
Short	73	75	77

Anova: Two-Factor Without Replication

SUMMARY	*Count*	*Sum*	*Average*	*Variance*
Row 1	3	233	77.66667	6.333333
Row 2	3	236	78.66667	6.333333
Row 3	3	225	75	4
Column 1	3	229	76.33333	17.33333
Column 2	3	229	76.33333	2.333333
Column 3	3	236	78.66667	2.333333

ANOVA

Source of Variation	*SS*	*df*	*MS*	*F*	*P-value*
Rows	21.55556	2	10.77778	1.920792	0.260203
Columns	10.88889	2	5.444444	0.970297	0.453378
Error	22.44444	4	5.611111		
Total	54.88889	8			

16.34 In the following SPSS output, the *p*-value is called `Sig`. The *p*-value for hair is 0.453 and the *p*-value for height is 0.260. These are the same *p*-values obtained in the EXCEL analysis in Problem 16.33. Since neither of these are less than 0.01, they are not different at level of significance 0.01. That is, the scores are not different for different hair colors nor are they different for different heights.

Tests of Between-Subjects Effects

Dependent Variable: Score

Source	Type III Sum of Squares	df	Mean Square	F	Sig.
Corrected Model	32.444[a]	4	8.111	1.446	.365
Intercept	53515.111	1	53515.111	9537.347	.000
Hair	10.889	2	5.444	.970	.453
Height	21.556	2	10.778	1.921	.260
Error	22.444	4	5.611		
Total	53570.000	9			
Corrected Total	54.889	8			

[a] R Squared = .591 (Adjusted R Squared = .182)

16.35 The MINITAB output shows that at the 0.05 level of significance there is a difference due to location but no difference due to fertilizers. There is significant interaction at the 0.05 level.

```
ANOVA: yield versus location, fertilizer
Factor        Type     Levels   Values
location      fixed         2   1, 2
fertilizer    fixed         4   1, 2, 3, 4

Analysis of Variance for yield

Source               DF        SS        MS       F       P
location              1    81.225    81.225   12.26   0.001
fertilizer            3    18.875     6.292    0.95   0.428
location*fertilizer   3    78.275    26.092    3.94   0.017
Error                32   212.000     6.625
Total                39   390.375
```

16.36 The STATISTIX output shows that at the 0.01 level of significance there is a difference due to location but no difference due to fertilizers. There is no significant interaction at the 0.01 level.

Statistix 8.0

Analysis of Variance Table for yield

Source	**DF**	**SS**	**MS**	**F**	**P**
fertilize	3	18.875	6.2917	0.95	0.4283
location	1	81.225	81.2250	12.26	0.0014
fertilize*location	3	78.275	26.0917	3.94	0.0169
Error	32	212.000	6.6250		
Total	39	390.375			

16.37 The following SAS output gives the p-value for machines as 0.0664, the p-value for operators as 0.0004, and the p-value for interaction as 0.8024. At the 0.05 significance level, only operators are significant.

The GLM Procedure

Class Level Information

Class	Levels	Values
Operator	4	1 2 3 4
Machine	2	1 2

Number of Observations Read 40

Number of Observations Used 40

The GLM Procedure

Dependent Variable: Articles

Source	DF	Sum of Squares	Mean Square	F Value	Pr > F
Model	7	154.8000000	22.1142857	4.08	0.0027
Error	32	173.6000000	5.4250000		
Corrected Tot	39	328.4000000			

Source	DF	Type III SS	Mean Square	F Value	Pr>F
Machine	1	19.6000000	19.6000000	3.61	0.0664
Operator	3	129.8000000	43.2666667	7.98	0.0004
Operator*Machine	3	5.4000000	1.8000000	0.33	0.8024

The following MINITAB output is the same as the SAS output.

ANOVA: Articles versus Machine, Operator

Factor	Type	Levels	Values
Machine	fixed	2	1, 2
Operator	fixed	4	1, 2, 3, 4

Analysis of Variance for Articles

Source	DF	SS	MS	F	P
Machine	1	19.600	19.600	3.61	0.066
Operator	3	129.800	43.267	7.98	0.000
Machine*Operator	3	5.400	1.800	0.33	0.802
Error	32	173.600	5.425		
Total	39	328.400			

16.38 The following STATISTIX output gives the p-values for soil variations in two perpendicular directions as 0.5658 and 0.3633 and the p-value for treatments as 0.6802. At the 0.01 significance level, none of the three are significant.

Statistix 8.0

Latin Square AOV Table for Yield

Source	**DF**	**SS**	**MS**	**F**	**P**
Row	3	17.500	5.8333	0.74	0.5658
Column	3	30.500	10.1667	1.28	0.3633
Treatment	3	12.500	4.1667	0.53	0.6802
Error	6	47.500	7.9167		
Total	15	108.000			

16.39 The following MINITAB output is the same as the STATISTIX output in Problem 16.38. None of the factors are significant at 0.05.

General Linear Model: Yield versus Row, Column, Treatment

Factor	Type	Levels	Values
Row	fixed	4	1, 2, 3, 4
Column	fixed	4	1, 2, 3, 4
Treatment	fixed	4	1, 2, 3, 4

Analysis of Variance for yield, using Adjusted SS for Tests

Source	DF	Seq SS	Adj SS	Adj MS	F	P
Row	3	17.500	17.500	5.833	0.74	0.567
Column	3	30.500	30.500	10.167	1.28	0.362
Treatment	3	12.500	12.500	4.167	0.53	0.680
Error	6	47.500	47.500	7.917		
Total	15	108.000				

16.40 The SPSS output shows no difference in scholastic achievements due to hair color, height, or birthplace at the 0.05 significance level.

Tests of Between-Subjects Effects

Dependent Variable: Score

Source	Type III Sum of Squares	df	Mean Square	F	Sig.
Corrected Model	44.000[a]	6	7.333	1.347	.485
Intercept	53515.111	1	53515.111	9829.306	.000
Hair	10.889	2	5.444	1.000	.500
Height	21.556	2	10.778	1.980	.336
Birthpl	11.556	2	5.778	1.061	.485
Error	10.889	2	5.444		
Total	53570.000	9			
Corrected Total	54.889	8			

[a] R Squared = .802 (Adjusted R Squared = .206)

16.41 The MINITAB analysis shows that there are significant differences in terms of the species of chicken and the quantities of the first chemical, but not in terms of the second chemical or of the chick's initial weights. Note that the *p*-value for species is 0.009 and the *p*-value for chemical is 0.032.

```
General Linear Model: Wtgain versus Weight, Species, ...
Factor       Type    Levels      Values
Weight       fixed        4      1, 2, 3, 4
Species      fixed        4      1, 2, 3, 4
Chemical1    fixed        4      1, 2, 3, 4
Chemical2    fixed        4      1, 2, 3, 4

Analysis of Variance for Wtgain, using Adjusted SS for Tests

Source       DF    Seq SS     Adj SS     Adj MS       F       P
Weight        3    2.7500     2.7500     0.9167    2.20   0.267
Species       3   38.2500    38.2500    12.7500   30.60   0.009
Chemical1     3   16.2500    16.2500     5.4167   13.00   0.032
Chemical2     3    7.2500     7.2500     2.4167    5.80   0.091
Error         3    1.2500     1.2500     0.4167
Total        15   65.7500
```

16.42 The *p*-values are called `Sig.` in SPSS. There are significant differences in cable strength due to types of cable, but there are no significant differences due to operators, machines, or companies.

Tests of Between-Subjects Effects

Dependent Variable: Strength

Source	Type III Sum of Squares	df	Mean Square	F	Sig.
Corrected Model	6579.750[a]	12	548.313	5.489	.094
Intercept	488251.563	1	488251.563	4887.607	.000
Type	4326.188	3	1442.063	14.436	.027
Company	2066.188	3	688.729	6.894	.074
Operator	120.688	3	40.229	.403	.763
Machine	66.688	3	22.229	.223	.876
Error	299.688	3	99.896		
Total	495131.000	16			
Corrected Total	6879.438	15			

[a] R Squared = .956 (Adjusted R Squared = .782)

16.43 There is a significant difference in the three treatments at the 0.05 significance level, but not at the 0.01 level. The EXCEL analysis is given below.

A	B	C
3	4	6
5	2	4
4	3	5
4	3	5

Anova: Single Factor

SUMMARY

Groups	*Count*	*Sum*	*Average*	*Variance*
A	4	16	4	0.666667
B	4	12	3	0.666667
C	4	20	5	0.666667

ANOVA

Source of Variation	*SS*	*df*	*MS*	*F*	*P-value*
Between Groups	8	2	4	6	0.022085
Within Groups	6	9	0.666667		
Total	14	11			

16.44 The MINITAB *p*-value is 0.700. There is no difference in the IQ due to height.

```
One-way ANOVA: Tall, Short, Medium

Source   DF      SS     MS     F      P
Factor    2    55.8   27.9  0.37  0.700
Error    12   911.5   76.0
Total    14   967.3

S = 8.715   R-Sq = 5.77%   R-Sq(adj) = 0.00%

                               Individual 95% CIs For Mean Based on
                               Pooled StDev
Level    N     Mean   StDev   ---+---------+---------+---------+-----
Tall     5   107.00   10.58          (-----------*----------)
Short    4   105.00    8.33       (-----------*----------)
Medium   6   102.50    7.15    (-----------*----------)
                              ----+---------+---------+---------+-----
                                96.0     102.0     108.0     114.0
```

16.46 The *p*-value is called `Sig.` in the SPSS output. At the 0.05 level, there is a significant difference in examination scores due both to veteran status and to IQ.

Tests of Between-Subjects Effects

Dependent Variable: Score

Source	Type III Sum of Squares	df	Mean Square	F	Sig.
Corrected Model	264.333[a]	3	88.111	176.222	.006
Intercept	38080.667	1	38080.667	76161.333	.000
Veteran	24.000	1	24.000	48.000	.020
IQ	240.333	2	120.167	240.333	.004
Error	1.000	2	.500		
Total	38346.000	6			
Corrected Total	265.333	5			

[a] R Squared = .996 (Adjusted R Squared = .991)

16.47 The STATISTIX analysis indicates that at the 0.01 level the difference in examination scores due to veteran status is not significant, but the difference due to the IQ is significant.

```
Statistix 8.0

Randomized Complete Block AOV Table for Score

Source     DF        SS        MS       F        P
Veteran     1    24.000    24.000   48.00    0.020
IQ          2   240.333   120.167  240.33   0.0041
```

Error	2	1.000	0.500
Total	5	265.333	

16.48 The MINITAB analysis indicates that at the 0.05 level the differences in examination scores due to location are not significant, but the differences due to the IQ are significant.

Two-way ANOVA: Testscore versus Location, IQ

Source	DF	SS	MS	F	P
Location	3	6.250	2.083	0.12	0.943
IQ	2	221.167	110.583	6.54	0.031
Error	6	101.500	16.917		
Total	11	328.917			

16.49 The SAS analysis indicates that at the 0.01 level the differences in examination scores due to location are not significant, and the differences due to the IQ are not significant. Remember the *p*-value is written as Pr>F.

The GLM Procedure
Class Level Information

Class	Levels	Values
Location	4	1 2 3 4
IQ	3	1 2 3

Number of Observations Read	12
Number of Observations Used	12

The GLM Procedure

Source	DF	Type III SS	Mean Square	F Value	Pr>F
Location	3	6.2500000	2.0833333	0.12	0.9430
IQ	2	221.1666667	110.5833333	6.54	0.0311

16.53 The MINITAB analysis indicates that at the 0.05 level the differences in rust scores due to location are not significant, but the differences due to the chemicals are significant. There is no significant interaction between location and chemicals.

ANOVA: rust versus location, chemical

Factor	Type	Levels	Values
location	fixed	2	1, 2
chemical	fixed	3	1, 2, 3

Analysis of Variance for rust

Source	DF	SS	MS	F	P
location	1	0.667	0.667	0.67	0.425
chemical	2	20.333	10.167	10.17	0.001
location*chemical	2	0.333	0.167	0.17	0.848
Error	18	18.000	1.000		
Total	23	39.333			

16.54 The STATISTIX analysis indicates that at the 0.05 level the differences in yields due to location are significant, but the differences due to varieties are not significant. There is no significant interaction between location and varieties.

Statistix 8.0

Analysis of Variance Table for Yield

Source	***DF***	***SS***	***MS***	***F***	***P***
Variety	4	35.333	8.8333	1.07	0.3822
location	2	191.433	95.7167	11.60	0.0001
Variety*location	8	82.567	10.3208	1.25	0.2928
Error	45	371.250	8.2500		
Total	59	680.583			

16.55 The MINITAB analysis indicates that at the 0.01 level the differences in yields due to location are significant, but the differences due to varieties are not significant. There is no significant interaction between location and varieties.

General Linear Model: Yield versus location, Variety

Factor	Type	Levels	Values
location	fixed	3	1, 2, 3
variety	fixed	5	1, 2, 3, 4, 5

Analysis of Variance for Yield, using Adjusted SS for Tests

Source	DF	Seq SS	Adj SS	Adj MS	F	P
location	2	191.433	191.433	95.717	11.60	0.000
Variety	4	35.333	35.333	8.833	1.07	0.382
location*Variety	8	82.567	82.567	10.321	1.25	0.293
Error	45	371.250	371.250	8.250		
Total	59	680.583				

16.56 Referring to the SPSS ANOVA, and realizing that `Sig.` is the same as the *p*-value in SPSS, we see that `Factor1`, `Factor2`, and `Treatment` do not have significant effects on the response variable at the 0.05 level, since the *p*-values are greater than 0.05 for all three.

Tests of Between-Subjects Effects

Dependent Variable: Response

Source	Type III Sum of Squares	df	Mean Square	F	Sig.
Corrected Model	92.667[a]	6	15.444	7.316	.125
Intercept	2567.111	1	2567.111	1216.000	.001
Factor1	74.889	2	37.444	17.737	.053
Factor2	17.556	2	8.778	4.158	.194
Treatment	.222	2	.111	.053	.950
Error	4.222	2	2.111		
Total	2664.000	9			
Corrected Total	96.889	8			

[a]R Squared = .956 (Adjusted R Squared = .826)

16.58 Referring to the SPSS ANOVA, and realizing that `Sig.` is the same as the *p*-value in SPSS, we see that `Factor1, Factor2, Latin Treatment` and `Greek Treatment` do not have significant effects on the response variable at the 0.05 level, since the *p*-values are greater than 0.05 for all four.

Tests of Between-Subjects Effects

Dependent Variable: Response

Source	Type III Sum of Squares	df	Mean Square	F	Sig.
Corrected Model	362.750[a]	12	30.229	.924	.607
Intercept	1914.063	1	1914.063	58.482	.005
Factor1	5.188	3	1.729	.053	.981
Factor2	15.188	3	5.063	.155	.920
Latin	108.188	3	36.063	1.102	.469
Greek	234.188	3	78.063	2.385	.247
Error	98.188	3	32.729		
Total	2375.000	16			
Corrected Total	460.938	15			

[a] R Squared = .787 (Adjusted R Squared = −.065)

CHAPTER 17

17.26 For a two-tailed alternative, the *p*-value is 0.0352. There is a difference due to the additive at the 0.05 level, but not at the 0.01 level. The *p*-value is obtained using the binomial distribution and not the normal approximation to the binomial.

17.27 For a one-tailed alternative, the *p*-value is 0.0176. Since this *p*-value is less than 0.05, reject the null hypothesis that there is no difference due to the additive.

17.28 The *p*-value using EXCEL, is given by `=1 - BINOMDIST(24,31,0.5,1)` or 0.00044. The program is effective at the 0.05 level of significance.

17.29 The *p*-value using EXCEL, is given by `=1 - BINOMDIST(15,22,0.5,1)` or 0.0262. The program is effective at the 0.05 level.

17.30 The *p*-value using EXCEL, is given by `=1 - BINOMDIST(16,25,0.5,1)` or 0.0539. We cannot conclude that brand *B* is preferred over brand *A* at the 0.05 level.

17.31

	BS = 25	BS = 30	BS = 35	BS = 40
41	16	11	6	1
37	12	7	2	−3
25	0	−5	−10	−15
43	18	13	8	3
42	17	12	7	2
28	3	−2	−7	−12
32	7	2	−3	−8
36	11	6	1	−4
27	2	−3	−8	−13
33	8	3	−2	−7

35	10	5	0	−5
24	−1	−6	−11	−16
22	−3	−8	−13	−18
34	9	4	−1	−6
28	3	−2	−7	−12
38	13	8	3	−2
46	21	16	11	6
41	16	11	6	1
27	2	−3	−8	−13
31	6	1	−4	−9
23	−2	−7	−12	−17
30	5	0	−5	−10
37	12	7	2	−3
36	11	6	1	−4
24	−1	−6	−11	−16

0.00154388	2*BINOMDIST(4,24,0.5,1)	Reject null
0.307456255	2*BINOMDIST(9,24,0.5,1)	Do not reject null
0.541256189	2*BINOMDIST(10,24,0.5,1)	Do not reject null
0.004077315	2*BINOMDIST(5,25,0.5,1)	Reject null

17.34 Sum of ranks of the smaller sample = 141.5 and the sum of ranks for the larger sample = 158.5. Two-tailed *p*-value = 0.3488. Do not reject the null hypothesis of no difference at 0.05 level, since *p*-value > 0.05.

17.35 Cannot reject the one-sided null hypothesis in Problem 17.34 at the 0.01 level.

17.36 Sum of ranks of the smaller sample = 132.5 and the sum of ranks for the larger sample = 77.5. Two-tailed *p*-value = 0.0044. Reject the null hypothesis of no difference at both the 0.01 and the 0.05 level, since *p*-value < 0.05.

17.37 The farmer of Problem 17.36 can conclude that wheat II produces a larger yield than wheat I at the 0.05 level.

17.38 Sum of ranks for brand $A = 86.0$ and the sum of ranks for brand $B = 124.0$. Two-tailed *p*-value = 0.1620. (*a*) Do not reject the null hypothesis of no difference between the two brands versus there is a difference at the 0.05 level, since *p*-value > 0.05. (*b*) Cannot conclude that brand *B* is better than brand *A* at the 0.05 level since the one-tailed *p*-value (0.081) > 0.05.

17.39 Yes, the *U* test as well as the sign test may be used to determine if there is a difference between the two machines.

17.41 3.

17.42 6.

17.46 (*a*) 246; (*b*) 168; (*c*) 0.

17.47 (*a*) 236; (*b*) 115; (*c*) 100.

17.49 $H = 2.59$, $DF = 4$, $P = 0.629$. There is no significant difference in yields of the five varieties at the 0.05 or the 0.01 level since the *p*-value is greater than 0.01 and 0.05.

17.50 $H = 8.42$, $DF = 3$, $P = 0.038$. There is a significant difference in the four brands of tires at the 0.05 level, but not at the 0.01 level since the *p*-value is such that $0.01 < p\text{-value} < 0.05$.

17.51 $H = 6.54$, $DF = 2$, $P = 0.038$. There is a significant difference in the three teaching methods at the 0.05 level, but not at the 0.01 level since the p-value is such that $0.01 < p\text{-value} < 0.05$.

17.52 $H = 9.22$, $DF = 3$, $P = 0.026$. There is a significant difference in the four subjects at the 0.05 level, but not at the 0.01 level since the p-value is such that $0.01 < p\text{-value} < 0.05$.

17.53 (*a*) $H = 7.88$, $DF = 8$, $P = 0.446$. There is no significant differences in the three TV tube lifetimes at the 0.01 or the 0.05 levels since the $p\text{-value} > 0.01$ and 0.05.
(*b*) $H = 2.59$, $DF = 4$, $P = 0.629$. There is no significant differences in the five varieties of wheat at the 0.01 or the 0.05 levels since the $p\text{-value} > 0.01$ and 0.05.
(*c*) $H = 5.70$, $DF = 3$, $P = 0.127$. There is no difference in the four brands of tires at either the 0.01 or the 0.05 levels since the $p\text{-value} > 0.01$ as well as > 0.05.

17.54 (*a*) $H = 5.65$, $DF = 2$, $P = 0.059$. There is no difference in the three methods of teaching at either the 0.01 or the 0.05 levels since the $p\text{-value} > 0.01$ as well as > 0.05.
(*b*) $H = 10.25$, $DF = 4$, $P = 0.036$. There is a difference in the five brands of gasoline at the 0.05 level, but not at the 0.01 level since $0.01 < p\text{-value} < 0.05$.
(*c*) $H = 9.22$, $DF = 3$, $P = 0.026$. There is a difference in the four subjects at the 0.05 level, but not at the 0.01 level since $0.01 < p\text{-value} < 0.05$.

17.55 (*a*) 8; (*b*) 10.

17.56 (*a*) The number of runs $V = 10$.
(*b*) The randomness test is based on the standard normal. The mean is

$$\mu_V = \frac{2N_1N_2}{N_1 + N_2} + 1 = \frac{2(11)(14)}{25} + 1 = 13.32$$

and the standard deviation is

$$\sigma_V = \sqrt{\frac{2(11)(14)\{2(11)(14) - 11 - 14\}}{25^2(24)}} = 2.41$$

The computed Z is

$$Z = \frac{10 - 13.32}{2.41} = -1.38$$

Using EXCEL the p-value is = `2*NORMSDIST(-1.38)` or 0.1676. Due to the large p-value, we do not doubt randomness.

17.57 (*a*) Even though the number of runs is below what we expect, the p-value is not less than 0.05. We do not reject randomness of sequence (*10*).

Runs Test: coin

```
Runs test for coin
Runs above and below K = 0.4
The observed number of runs = 7
The expected number of runs = 10.6
8 observations above K, 12 below
* N is small, so the following approximation may be invalid.
P-value = 0.084
```

(*b*) The number of runs is above what we expect. We reject randomness of the sequence (*11*).

Runs Test: coin

```
Runs test for coin
Runs above and below K = 0.5
```

```
The observed number of runs = 12
The expected number of runs = 7
6 observations above K, 6 below
* N is small, so the following approximation may be invalid.
P-value = 0.002
```

17.58 (*a*)

Sequence			Runs
a	a	b	2
a	b	a	3
b	a	a	2

(*b*)

Sampling distribution

V	f
2	2
3	1

(*c*)

Probability distribution

V	Pr{V}
2	0.667
3	0.333

17.59 Mean = 2.333 Variance = 0.222

17.60 (*a*)

Sequence				Runs
a	a	b	b	2
a	b	a	b	4
a	b	b	a	3
b	b	a	a	2
b	a	b	a	4
b	a	a	b	3

Sampling distribution

V	f
2	2
3	2
4	2

Probability distribution

V	Pr{V}
2	0.333
3	0.333
4	0.333

Mean V 3
Variance V 0.667

(*b*)

Sequence				Runs
a	b	b	b	2
b	a	b	b	3

b	b	a	b	3
b	b	b	a	3

Sampling distribution

V	f
2	1
3	3

Probability distribution

V	Pr{V}
2	0.25
3	0.75

Mean V 2.75
Variance V 0.188

(*c*)

Sequence					Runs
a	b	b	b	b	2
b	a	b	b	b	3
b	b	a	b	b	3
b	b	b	a	b	3
b	b	b	b	a	2

Sampling distribution

V	f
2	2
3	3

Probability distribution

V	Pr{V}
2	0.4
3	0.6

Mean V 2.6
Variance V 0.24

17.61 (*a*)

Sequence						Runs
a	a	b	b	b	b	2
a	b	a	b	b	b	4
a	b	b	a	b	b	4
a	b	b	b	a	b	4
a	b	b	b	b	a	3
b	a	a	b	b	b	3
b	a	b	a	b	b	5
b	a	b	b	a	b	5
b	a	b	b	b	a	4
b	b	a	a	b	b	3
b	b	a	b	a	b	5
b	b	a	b	b	a	4
b	b	b	b	a	a	2
b	b	b	a	b	a	4
b	b	b	a	a	b	3

(*b*)

Sampling distribution

V	f
2	2
3	4
4	6
5	3

(*c*)

Probability distribution

V	Pr{V}
2	0.133
3	0.267
4	0.4
5	0.2

Mean V	3.667
Variance V	0.888

17.62 Assume the rows are read in one at a time. That is row 1 is followed by row 2, is followed by row 3, and finally row 4.

Runs Test: Grade

```
Runs test for Grade
Runs above and below K = 69
The observed number of runs = 26
The expected number of runs = 20.95
21 observations above K, 19 below
P-value = 0.105
```

The grades may be assumed to have been recorded randomly at the 0.05 level.

17.63 Assume the data is recorded row by row.

```
Runs test for price
Runs above and below K = 11.36
The observed number of runs = 10
The expected number of runs = 13.32
14 observations above K, 11 below
P-value = 0.168
```

The prices may be assumed random at the 0.05 level.

17.64 In the digits following the decimal, let 0 represent an even digit and 1 represent and odd digit.

```
Runs test for digit
Runs above and below K = 0.473684
The observed number of runs = 9
The expected number of runs = 10.4737
9 observations above K, 10 below
* N is small, so the following approximation may be invalid.
P-value = 0.485
```

The digits may be assumed to be random at the 0.05 level.

17.65 The digits may be assumed random at the 0.05 level.

17.66 Using the normal approximation, the computed Z value is 1.62. The computed p-value using EXCEL is = `2*NORMSDIST(-1.62)` or 0.105.

17.67 Using the normal approximation, the computed Z value is -1.38. The computed p-value using EXCEL is = `2*NORMSDIST(-1.38)` or 0.168.

17.68 Using the normal approximation, the computed Z value is -1.38. The computed p-value using EXCEL is = `2*NORMSDIST(-0.70)` or 0.484.

17.70 Spearman rank correlation = 1.0 and Pearson correlation coefficient = 0.998.

CHAPTER 18

18.16 Subgroup means: 13.25 14.50 17.25 14.50 13.50 14.75 13.75 15.00 15.00 17.00

Subgroup ranges: 5 9 5 6 8 9 10 5 5 7

$\bar{\bar{X}} = 14.85$, $\bar{R} = 6.9$.

18.17 The pooled estimate of σ is 1.741. LCL = 450.7, UCL = 455.9. None of the subgroup means is outside the control limits.

18.18 No.

18.19 The plot indicates that variability has been reduced. The new control limits are LCL = 452.9 and UCL = 455.2. It also appears that the process is centered closer to the target after the modification.

18.20 The control limits are LCL = 1.980 and UCL = 2.017. Periods 4, 5, and 6 fail test 5. Periods 15 through 20 fail test 4. Each of these periods is the end of 14 points in a row, alternating up and down.

18.21 $C_{PK} = 0.63$. ppm non-conforming = 32, 487.

18.22 $C_{PK} = 1.72$. ppm non-conforming = less than 1.

18.23 Centerline = 0.006133, LCL = 0, UCL = 0.01661. Process is in control. ppm = 6,133.

18.24 Centerline = 3.067, LCL = 0, UCL = 8.304.

18.25 0.032 0.027 0.032 0.024 0.024 0.027 0.032 0.032 0.027 0.024 0.032 0.024 0.027 0.024 0.027 0.024 0.027 0.027 0.027 0.027

18.26 Centerline = $\bar{X} = 349.9$.

Moving ranges: 0.0 0.2 0.6 0.8 0.4 0.3 0.1 0.4 0.4 0.9 0.2 1.1 0.5 1.5 2.8 1.6 0.3 0.1 1.2 1.9 0.2 1.2 0.9

Mean of the moving ranges given above = $\bar{R}_M = 0.765$.

Individuals chart control limits: $\bar{X} \pm 3(\bar{R}_M/d_2)$. d_2 is a control chart constant that is available from tables in many different sources and in this case is equal to 1.128. LCL = 347.9 and UCL = 352.0.

18.27 The EWMA chart indicates that the process means are consistently below the target value. The means for subgroups 12 and 13 drop below the lower control limits. The subgroup means beyond 13 are above the lower control limit; however, the process mean is still consistently below the target value.

18.28 A zone chart does not indicate any out of control conditions. However, as seen in Problem 19.20, there are 14 points in a row alternating up and down. Because of the way the zone chart operates, it will not indicate this condition.

18.29 The 20 lower control limits are: 0.00 0.00 0.00 0.00 0.00 0.00 0.00 0.38 0.00 0.00 0.00 0.00 0.00 0.38 0.00 0.00 038 0.00 0.38 0.38.

The 20 upper control limits are: 9.52 9.52 9.52 9.52 9.52 7.82 8.46 7.07 7.82 8.46 9.52 9.52 9.52 7.07 9.52 9.52 7.07 9.52 7.07 7.07.

18.30 Discoloration; discoloration and loose straps.

APPENDIXES

Appendix I

Ordinates (*Y*) of the Standard Normal Curve at *z*

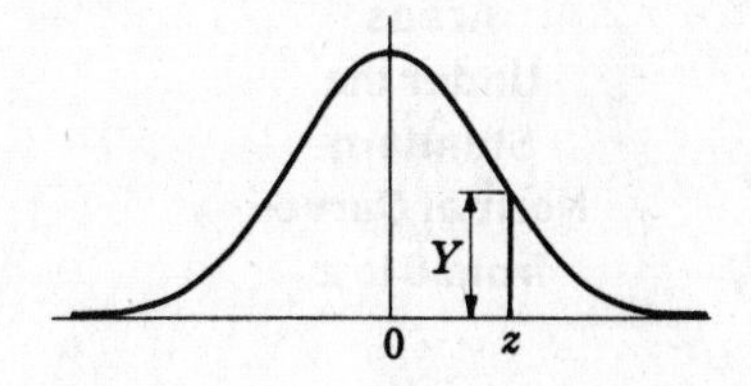

z	0	1	2	3	4	5	6	7	8	9
0.0	.3989	.3989	.3989	.3988	.3986	.3984	.3982	.3980	.3977	.3973
0.1	.3970	.3965	.3961	.3956	.3951	.3945	.3939	.3932	.3925	.3918
0.2	.3910	.3902	.3894	.3885	.3876	.3867	.3857	.3847	.3836	.3825
0.3	.3814	.3802	.3790	.3778	.3765	.3752	.3739	.3725	.3712	.3697
0.4	.3683	.3668	.3653	.3637	.3621	.3605	.3589	.3572	.3555	.3538
0.5	.3521	.3503	.3485	.3467	.3448	.3429	.3410	.3391	.3372	.3352
0.6	.3332	.3312	.3292	.3271	.3251	.3230	.3209	.3187	.3166	.3144
0.7	.3123	.3101	.3079	.3056	.3034	.3011	.2989	.2966	.2943	.2920
0.8	.2897	.2874	.2850	.2827	.2803	.2780	.2756	.2732	.2709	.2685
0.9	.2661	.2637	.2613	.2589	.2565	.2541	.2516	.2492	.2468	.2444
1.0	.2420	.2396	.2371	.2347	.2323	.2299	.2275	.2251	.2227	.2203
1.1	.2179	.2155	.2131	.2107	.2083	.2059	.2036	.2012	.1989	.1965
1.2	.1942	.1919	.1895	.1872	.1849	.1826	.1804	.1781	.1758	.1736
1.3	.1714	.1691	.1669	.1647	.1626	.1604	.1582	.1561	.1539	.1518
1.4	.1497	.1476	.1456	.1435	.1415	.1394	.1374	.1354	.1334	.1315
1.5	.1295	.1276	.1257	.1238	.1219	.1200	.1182	.1163	.1145	.1127
1.6	.1109	.1092	.1074	.1057	.1040	.1023	.1006	.0989	.0973	.0957
1.7	.0940	.0925	.0909	.0893	.0878	.0863	.0848	.0833	.0818	.0804
1.8	.0790	.0775	.0761	.0748	.0734	.0721	.0707	.0694	.0681	.0669
1.9	.0656	.0644	.0632	.0620	.0608	.0596	.0584	.0573	.0562	.0551
2.0	.0540	.0529	.0519	.0508	.0498	.0488	.0478	.0468	.0459	.0449
2.1	.0440	.0431	.0422	.0413	.0404	.0396	.0387	.0379	.0371	.0363
2.2	.0355	.0347	.0339	.0332	.0325	.0317	.0310	.0303	.0297	.0290
2.3	.0283	.0277	.0270	.0264	.0258	.0252	.0246	.0241	.0235	.0229
2.4	.0224	.0219	.0213	.0208	.0203	.0198	.0194	.0189	.0184	.0180
2.5	.0175	.0171	.0167	.0163	.0158	.0154	.0151	.0147	.0143	.0139
2.6	.0136	.0132	.0129	.0126	.0122	.0119	.0116	.0113	.0110	.0107
2.7	.0104	.0101	.0099	.0096	.0093	.0091	.0088	.0086	.0084	.0081
2.8	.0079	.0077	.0075	.0073	.0071	.0069	.0067	.0065	.0063	.0061
2.9	.0060	.0058	.0056	.0055	.0053	.0051	.0050	.0048	.0047	.0046
3.0	.0044	.0043	.0042	.0040	.0039	.0038	.0037	.0036	.0035	.0034
3.1	.0033	.0032	.0031	.0030	.0029	.0028	.0027	.0026	.0025	.0025
3.2	.0024	.0023	.0022	.0022	.0021	.0020	.0020	.0019	.0018	.0018
3.3	.0017	.0017	.0016	.0016	.0015	.0015	.0014	.0014	.0013	.0013
3.4	.0012	.0012	.0012	.0011	.0011	.0010	.0010	.0010	.0009	.0009
3.5	.0009	.0008	.0008	.0008	.0008	.0007	.0007	.0007	.0007	.0006
3.6	.0006	.0006	.0006	.0005	.0005	.0005	.0005	.0005	.0005	.0004
3.7	.0004	.0004	.0004	.0004	.0004	.0004	.0003	.0003	.0003	.0003
3.8	.0003	.0003	.0003	.0003	.0003	.0002	.0002	.0002	.0002	.0002
3.9	.0002	.0002	.0002	.0002	.0002	.0002	.0002	.0002	.0001	.0001

Appendix II

Areas Under the Standard Normal Curve from 0 to *z*

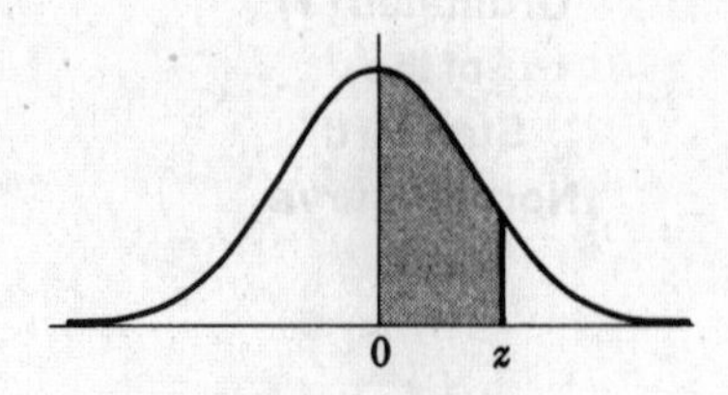

z	0	1	2	3	4	5	6	7	8	9
0.0	.0000	.0040	.0080	.0120	.0160	.0199	.0239	.0279	.0319	.0359
0.1	.0398	.0438	.0478	.0517	.0557	.0596	.0636	.0675	.0714	.0754
0.2	.0793	.0832	.0871	.0910	.0948	.0987	.1026	.1064	.1103	.1141
0.3	.1179	.1217	.1255	.1293	.1331	.1368	.1406	.1443	.1480	.1517
0.4	.1554	.1591	.1628	.1664	.1700	.1736	.1772	.1808	.1844	.1879
0.5	.1915	.1950	.1985	.2019	.2054	.2088	.2123	.2157	.2190	.2224
0.6	.2258	.2291	.2324	.2357	.2389	.2422	.2454	.2486	.2518	.2549
0.7	.2580	.2612	.2642	.2673	.2704	.2734	.2764	.2794	.2823	.2852
0.8	.2881	.2910	.2939	.2967	.2996	.3023	.3051	.3078	.3106	.3133
0.9	.3159	.3186	.3212	.3238	.3264	.3289	.3315	.3340	.3365	.3389
1.0	.3413	.3438	.3461	.3485	.3508	.3531	.3554	.3577	.3599	.3621
1.1	.3643	.3665	.3686	.3708	.3729	.3749	.3770	.3790	.3810	.3830
1.2	.3849	.3869	.3888	.3907	.3925	.3944	.3962	.3980	.3997	.4015
1.3	.4032	.4049	.4066	.4082	.4099	.4115	.4131	.4147	.4162	.4177
1.4	.4192	.4207	.4222	.4236	.4251	.4265	.4279	.4292	.4306	.4319
1.5	.4332	.4345	.4357	.4370	.4382	.4394	.4406	.4418	.4429	.4441
1.6	.4452	.4463	.4474	.4484	.4495	.4505	.4515	.4525	.4535	.4545
1.7	.4554	.4564	.4573	.4582	.4591	.4599	.4608	.4616	.4625	.4633
1.8	.4641	.4649	.4656	.4664	.4671	.4678	.4686	.4693	.4699	.4706
1.9	.4713	.4719	.4726	.4732	.4738	.4744	.4750	.4756	.4761	.4767
2.0	.4772	.4778	.4783	.4788	.4793	.4798	.4803	.4808	.4812	.4817
2.1	.4821	.4826	.4830	.4834	.4838	.4842	.4846	.4850	.4854	.4857
2.2	.4861	.4864	.4868	.4871	.4875	.4878	.4881	.4884	.4887	.4890
2.3	.4893	.4896	.4898	.4901	.4904	.4906	.4909	.4911	.4913	.4916
2.4	.4918	.4920	.4922	.4925	.4927	.4929	.4931	.4932	.4934	.4936
2.5	.4938	.4940	.4941	.4943	.4945	.4946	.4948	.4949	.4951	.4952
2.6	.4953	.4955	.4956	.4957	.4959	.4960	.4961	.4962	.4963	.4964
2.7	.4965	.4966	.4967	.4968	.4969	.4970	.4971	.4972	.4973	.4974
2.8	.4974	.4975	.4976	.4977	.4977	.4978	.4979	.4979	.4980	.4981
2.9	.4981	.4982	.4982	.4983	.4984	.4984	.4985	.4985	.4986	4986
3.0	.4987	.4987	.4987	.4988	.4988	.4989	.4989	.4989	.4990	.4990
3.1	.4990	.4991	.4991	.4991	.4992	.4992	.4992	.4992	.4993	.4993
3.2	.4993	.4993	.4994	.4994	.4994	.4994	.4994	.4995	.4995	.4995
3.3	.4995	.4995	.4995	.4996	.4996	.4996	.4996	.4996	.4996	.4997
3.4	.4997	.4997	.4997	.4997	.4997	.4997	.4997	.4997	.4997	.4998
3.5	.4998	.4998	.4998	.4998	.4998	.4998	.4998	.4998	.4998	.4998
3.6	.4998	.4998	.4999	.4999	.4999	.4999	.4999	.4999	.4999	.4999
3.7	.4999	.4999	.4999	.4999	.4999	.4999	.4999	.4999	.4999	.4999
3.8	.4999	.4999	.4999	.4999	.4999	.4999	.4999	.4999	.4999	.4999
3.9	.5000	.5000	.5000	.5000	.5000	.5000	.5000	.5000	.5000	.5000

Appendix III

Percentile Values (t_p)
for
Student's t Distribution
with ν Degrees of Freedom
(shaded area = p)

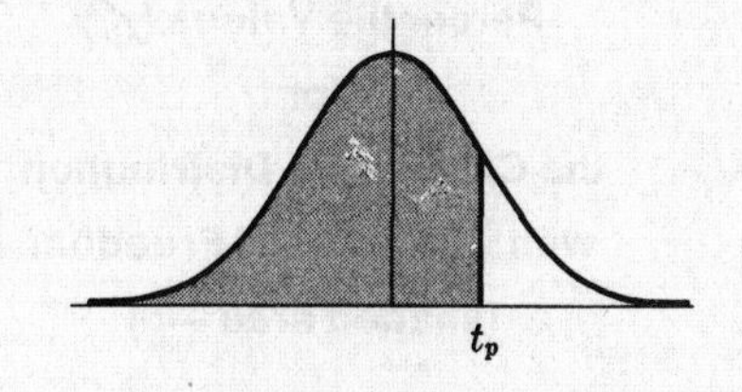

ν	$t_{.995}$	$t_{.99}$	$t_{.975}$	$t_{.95}$	$t_{.90}$	$t_{.80}$	$t_{.75}$	$t_{.70}$	$t_{.60}$	$t_{.55}$
1	63.66	31.82	12.71	6.31	3.08	1.376	1.000	.727	.325	.158
2	9.92	6.96	4.30	2.92	1.89	1.061	.816	.617	.289	.142
3	5.84	4.54	3.18	2.35	1.64	.978	.765	.584	.277	.137
4	4.60	3.75	2.78	2.13	1.53	.941	.741	.569	.271	.134
5	4.03	3.36	2.57	2.02	1.48	.920	.727	.559	.267	.132
6	3.71	3.14	2.45	1.94	1.44	.906	.718	.553	.265	.131
7	3.50	3.00	2.36	1.90	1.42	.896	.711	.549	.263	.130
8	3.36	2.90	2.31	1.86	1.40	.889	.706	.546	.262	.130
9	3.25	2.82	2.26	1.83	1.38	.883	.703	.543	.261	.129
10	3.17	2.76	2.23	1.81	1.37	.879	.700	.542	.260	.129
11	3.11	2.72	2.20	1.80	1.36	.876	.697	.540	.260	.129
12	3.06	2.68	2.18	1.78	1.36	.873	.695	.539	.259	.128
13	3.01	2.65	2.16	1.77	1.35	.870	.694	.538	.259	.128
14	2.98	2.62	2.14	1.76	1.34	.868	.692	.537	.258	.128
15	2.95	2.60	2.13	1.75	1.34	.866	.691	.536	.258	.128
16	2.92	2.58	2.12	1.75	1.34	.865	.690	.535	.258	.128
17	2.90	2.57	2.11	1.74	1.33	.863	.689	.534	.257	.128
18	2.88	2.55	2.10	1.73	1.33	.862	.688	.534	.257	.127
19	2.86	2.54	2.09	1.73	1.33	.861	.688	.533	.257	.127
20	2.84	2.53	2.09	1.72	1.32	.860	.687	.533	.257	.127
21	2.83	2.52	2.08	1.72	1.32	.859	.686	.532	.257	.127
22	2.82	2.51	2.07	1.72	1.32	.858	.686	.532	.256	.127
23	2.81	2.50	2.07	1.71	1.32	.858	.685	.532	.256	.127
24	2.80	2.49	2.06	1.71	1.32	.857	.685	.531	.256	.127
25	2.79	2.48	2.06	1.71	1.32	.856	.684	.531	.256	.127
26	2.78	2.48	2.06	1.71	1.32	.856	.684	.531	.256	.127
27	2.77	2.47	2.05	1.70	1.31	.855	.684	.531	.256	.127
28	2.76	2.47	2.05	1.70	1.31	.855	.683	.530	.256	.127
29	2.76	2.46	2.04	1.70	1.31	.854	.683	.530	.256	.127
30	2.75	2.46	2.04	1.70	1.31	.854	.683	.530	.256	.127
40	2.70	2.42	2.02	1.68	1.30	.851	.681	.529	.255	.126
60	2.66	2.39	2.00	1.67	1.30	.848	.679	.527	.254	.126
120	2.62	2.36	1.98	1.66	1.29	.845	.677	.526	.254	.126
∞	2.58	2.33	1.96	1.645	1.28	.842	.674	.524	.253	.126

Source: R. A. Fisher and F. Yates, *Statistical Tables for Biological, Agricultural and Medical Research* (5th edition), Table III, Oliver and Boyd Ltd., Edinburgh, by permission of the authors and publishers.

Appendix IV

Percentile Values (χ_p^2)
for
the Chi-Square Distribution
with ν Degrees of Freedom
(shaded area = p)

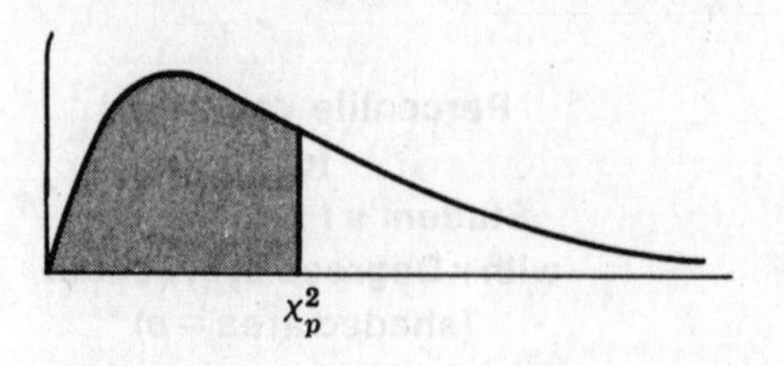

ν	$\chi^2_{.995}$	$\chi^2_{.99}$	$\chi^2_{.975}$	$\chi^2_{.95}$	$\chi^2_{.90}$	$\chi^2_{.75}$	$\chi^2_{.50}$	$\chi^2_{.25}$	$\chi^2_{.10}$	$\chi^2_{.05}$	$\chi^2_{.025}$	$\chi^2_{.01}$	$\chi^2_{.005}$
1	7.88	6.63	5.02	3.84	2.71	1.32	.455	.102	.0158	.0039	.0010	.0002	.0000
2	10.6	9.21	7.38	5.99	4.61	2.77	1.39	.575	.211	.103	.0506	.0201	.0100
3	12.8	11.3	9.35	7.81	6.25	4.11	2.37	1.21	.584	.352	.216	.115	.072
4	14.9	13.3	11.1	9.49	7.78	5.39	3.36	1.92	1.06	.711	.484	.297	.207
5	16.7	15.1	12.8	11.1	9.24	6.63	4.35	2.67	1.61	1.15	.831	.554	.412
6	18.5	16.8	14.4	12.6	10.6	7.84	5.35	3.45	2.20	1.64	1.24	.872	.676
7	20.3	18.5	16.0	14.1	12.0	9.04	6.35	4.25	2.83	2.17	1.69	1.24	.989
8	22.0	20.1	17.5	15.5	13.4	10.2	7.34	5.07	3.49	2.73	2.18	1.65	1.34
9	23.6	21.7	19.0	16.9	14.7	11.4	8.34	5.90	4.17	3.33	2.70	2.09	1.73
10	25.2	23.2	20.5	18.3	16.0	12.5	9.34	6.74	4.87	3.94	3.25	2.56	2.16
11	26.8	24.7	21.9	19.7	17.3	13.7	10.3	7.58	5.58	4.57	3.82	3.05	2.60
12	28.3	26.2	23.3	21.0	18.5	14.8	11.3	8.44	6.30	5.23	4.40	3.57	3.07
13	29.8	27.7	24.7	22.4	19.8	16.0	12.3	9.30	7.04	5.89	5.01	4.11	3.57
14	31.3	29.1	26.1	23.7	21.1	17.1	13.3	10.2	7.79	6.57	5.63	4.66	4.07
15	32.8	30.6	27.5	25.0	22.3	18.2	14.3	11.0	8.55	7.26	6.26	5.23	4.60
16	34.3	32.0	28.8	26.3	23.5	19.4	15.3	11.9	9.31	7.96	6.91	5.81	5.14
17	35.7	33.4	30.2	27.6	24.8	20.5	16.3	12.8	10.1	8.67	7.56	6.41	5.70
18	37.2	34.8	31.5	28.9	26.0	21.6	17.3	13.7	10.9	9.39	8.23	7.01	6.26
19	38.6	36.2	32.9	30.1	27.2	22.7	18.3	14.6	11.7	10.1	8.91	7.63	6.84
20	40.0	37.6	34.2	31.4	28.4	23.8	19.3	15.5	12.4	10.9	9.59	8.26	7.43
21	41.4	38.9	35.5	32.7	29.6	24.9	20.3	16.3	13.2	11.6	10.3	8.90	8.03
22	42.8	40.3	36.8	33.9	30.8	26.0	21.3	17.2	14.0	12.3	11.0	9.54	8.64
23	44.2	41.6	38.1	35.2	32.0	27.1	22.3	18.1	14.8	13.1	11.7	10.2	9.26
24	45.6	43.0	39.4	36.4	33.2	28.2	23.3	19.0	15.7	13.8	12.4	10.9	9.89
25	46.9	44.3	40.6	37.7	34.4	29.3	24.3	19.9	16.5	14.6	13.1	11.5	10.5
26	48.3	45.6	41.9	38.9	35.6	30.4	25.3	20.8	17.3	15.4	13.8	12.2	11.2
27	49.6	47.0	43.2	40.1	36.7	31.5	26.3	21.7	18.1	16.2	14.6	12.9	11.8
28	51.0	48.3	44.5	41.3	37.9	32.6	27.3	22.7	18.9	16.9	15.3	13.6	12.5
29	52.3	49.6	45.7	42.6	39.1	33.7	28.3	23.6	19.8	17.7	16.0	14.3	13.1
30	53.7	50.9	47.0	43.8	40.3	34.8	29.3	24.5	20.6	18.5	16.8	15.0	13.8
40	66.8	63.7	59.3	55.8	51.8	46.6	39.3	33.7	29.1	26.5	24.4	22.2	20.7
50	79.5	76.2	71.4	67.5	63.2	56.3	49.3	42.9	37.7	34.8	32.4	29.7	28.0
60	92.0	88.4	83.3	79.1	74.4	67.0	59.3	52.3	46.5	43.2	40.5	37.5	35.5
70	104.2	100.4	95.0	90.5	85.5	77.6	69.3	61.7	55.3	51.7	48.8	45.4	43.3
80	116.3	112.3	106.6	101.9	96.6	88.1	79.3	71.1	64.3	60.4	57.2	53.5	51.2
90	128.3	124.1	118.1	113.1	107.6	98.6	89.3	80.6	73.3	69.1	65.6	61.8	59.2
100	140.2	135.8	129.6	124.3	118.5	109.1	99.3	90.1	82.4	77.9	74.2	70.1	67.3

Source: Catherine M. Thompson, *Table of percentage points of the χ^2 distribution*, Biometrika, Vol. 32 (1941), by permission of the author and publisher.

Appendix V

95th Percentile Values for the F Distribution
(ν_1 degrees of freedom in numerator)
(ν_2 degrees of freedom in denominator)

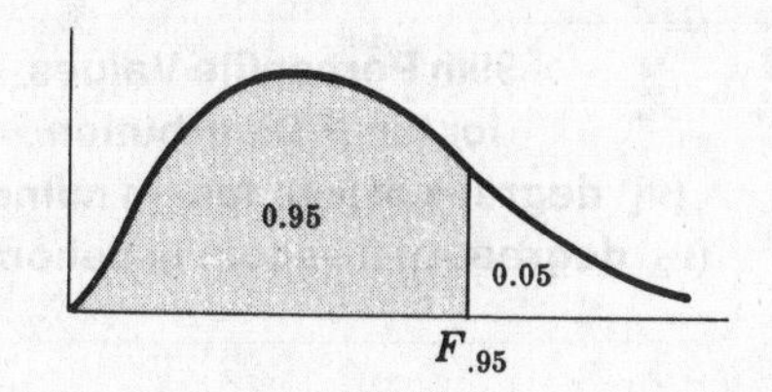

ν_2 \ ν_1	1	2	3	4	5	6	7	8	9	10	12	15	20	24	30	40	60	120	∞
1	161	200	216	225	230	234	237	239	241	242	244	246	248	249	250	251	252	253	254
2	18.5	19.0	19.2	19.2	19.3	19.3	19.4	19.4	19.4	19.4	19.4	19.4	19.4	19.5	19.5	19.5	19.5	19.5	19.5
3	10.1	9.55	9.28	9.12	9.01	8.94	8.89	8.85	8.81	8.79	8.74	8.70	8.66	8.64	8.62	8.59	8.57	8.55	8.53
4	7.71	6.94	6.59	6.39	6.26	6.16	6.09	6.04	6.00	5.96	5.91	5.86	5.80	5.77	5.75	5.72	5.69	5.66	5.63
5	6.61	5.79	5.41	5.19	5.05	4.95	4.88	4.82	4.77	4.74	4.68	4.62	4.56	4.53	4.50	4.46	4.43	4.40	4.37
6	5.99	5.14	4.76	4.53	4.39	4.28	4.21	4.15	4.10	4.06	4.00	3.94	3.87	3.84	3.81	3.77	3.74	3.70	3.67
7	5.59	4.74	4.35	4.12	3.97	3.87	3.79	3.73	3.68	3.64	3.57	3.51	3.44	3.41	3.38	3.34	3.30	3.27	3.23
8	5.32	4.46	4.07	3.84	3.69	3.58	3.50	3.44	3.39	3.35	3.28	3.22	3.15	3.12	3.08	3.04	3.01	2.97	2.93
9	5.12	4.26	3.86	3.63	3.48	3.37	3.29	3.23	3.18	3.14	3.07	3.01	2.94	2.90	2.86	2.83	2.79	2.75	2.71
10	4.96	4.10	3.71	3.48	3.33	3.22	3.14	3.07	3.02	2.98	2.91	2.85	2.77	2.74	2.70	2.66	2.62	2.58	2.54
11	4.84	3.98	3.59	3.36	3.20	3.09	3.01	2.95	2.90	2.85	2.79	2.72	2.65	2.61	2.57	2.53	2.49	2.45	2.40
12	4.75	3.89	3.49	3.26	3.11	3.00	2.91	2.85	2.80	2.75	2.69	2.62	2.54	2.51	2.47	2.43	2.38	2.34	2.30
13	4.67	3.81	3.41	3.18	3.03	2.92	2.83	2.77	2.71	2.67	2.60	2.53	2.46	2.42	2.38	2.34	2.30	2.25	2.21
14	4.60	3.74	3.34	3.11	2.96	2.85	2.76	2.70	2.65	2.60	2.53	2.46	2.39	2.35	2.31	2.27	2.22	2.18	2.13
15	4.54	3.68	3.29	3.06	2.90	2.79	2.71	2.64	2.59	2.54	2.48	2.40	2.33	2.29	2.25	2.20	2.16	2.11	2.07
16	4.49	3.63	3.24	3.01	2.85	2.74	2.66	2.59	2.54	2.49	2.42	2.35	2.28	2.24	2.19	2.15	2.11	2.06	2.01
17	4.45	3.59	3.20	2.96	2.81	2.70	2.61	2.55	2.49	2.45	2.38	2.31	2.23	2.19	2.15	2.10	2.06	2.01	1.96
18	4.41	3.55	3.16	2.93	2.77	2.66	2.58	2.51	2.46	2.41	2.34	2.27	2.19	2.15	2.11	2.06	2.02	1.97	1.92
19	4.38	3.52	3.13	2.90	2.74	2.63	2.54	2.48	2.42	2.38	2.31	2.23	2.16	2.11	2.07	2.03	1.98	1.93	1.88
20	4.35	3.49	3.10	2.87	2.71	2.60	2.51	2.45	2.39	2.35	2.28	2.20	2.12	2.08	2.04	1.99	1.95	1.90	1.84
21	4.32	3.47	3.07	2.84	2.68	2.57	2.49	2.42	2.37	2.32	2.25	2.18	2.10	2.05	2.01	1.96	1.92	1.87	1.81
22	4.30	3.44	3.05	2.82	2.66	2.55	2.46	2.40	2.34	2.30	2.23	2.15	2.07	2.03	1.98	1.94	1.89	1.84	1.78
23	4.28	3.42	3.03	2.80	2.64	2.53	2.44	2.37	2.32	2.27	2.20	2.13	2.05	2.01	1.96	1.91	1.86	1.81	1.76
24	4.26	3.40	3.01	2.78	2.62	2.51	2.42	2.36	2.30	2.25	2.18	2.11	2.03	1.98	1.94	1.89	1.84	1.79	1.73
25	4.24	3.39	2.99	2.76	2.60	2.49	2.40	2.34	2.28	2.24	2.16	2.09	2.01	1.96	1.92	1.87	1.82	1.77	1.71
26	4.23	3.37	2.98	2.74	2.59	2.47	2.39	2.32	2.27	2.22	2.15	2.07	1.99	1.95	1.90	1.85	1.80	1.75	1.69
27	4.21	3.35	2.96	2.73	2.57	2.46	2.37	2.31	2.25	2.20	2.13	2.06	1.97	1.93	1.88	1.84	1.79	1.73	1.67
28	4.20	3.34	2.95	2.71	2.56	2.45	2.36	2.29	2.24	2.19	2.12	2.04	1.96	1.91	1.87	1.82	1.77	1.71	1.65
29	4.18	3.33	2.93	2.70	2.55	2.43	2.35	2.28	2.22	2.18	2.10	2.03	1.94	1.90	1.85	1.81	1.75	1.70	1.64
30	4.17	3.32	2.92	2.69	2.53	2.42	2.33	2.27	2.21	2.16	2.09	2.01	1.93	1.89	1.84	1.79	1.74	1.68	1.62
40	4.08	3.23	2.84	2.61	2.45	2.34	2.25	2.18	2.12	2.08	2.00	1.92	1.84	1.79	1.74	1.69	1.64	1.58	1.51
60	4.00	3.15	2.76	2.53	2.37	2.25	2.17	2.10	2.04	1.99	1.92	1.84	1.75	1.70	1.65	1.59	1.53	1.47	1.39
120	3.92	3.07	2.68	2.45	2.29	2.18	2.09	2.02	1.96	1.91	1.83	1.75	1.66	1.61	1.55	1.50	1.43	1.35	1.25
∞	3.84	3.00	2.60	2.37	2.21	2.10	2.01	1.94	1.88	1.83	1.75	1.67	1.57	1.52	1.46	1.39	1.32	1.22	1.00

Source: E. S. Pearson and H. O. Hartley, *Biometrika Tables for Statisticians*, Vol. 2 (1972), Table 5, page 178, by permission.

Appendix VI

99th Percentile Values for the F Distribution
(ν_1 degrees of freedom in numerator)
(ν_2 degrees of freedom in denominator)

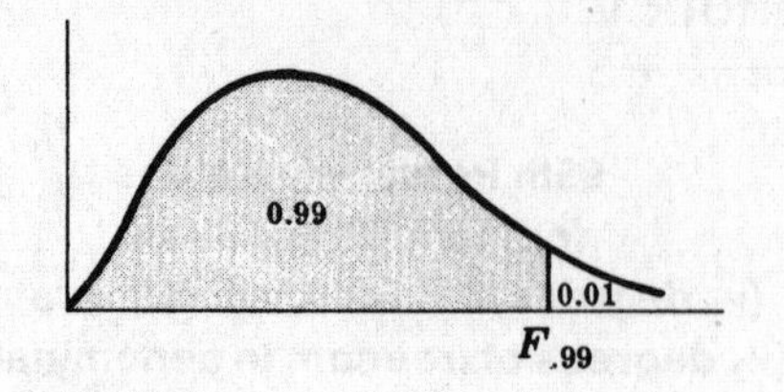

ν_2 \ ν_1	1	2	3	4	5	6	7	8	9	10	12	15	20	24	30	40	60	120	∞
1	4052	5000	5403	5625	5764	5859	5928	5981	6023	6056	6106	6157	6209	6235	6261	6287	6313	6339	6366
2	98.5	99.0	99.2	99.2	99.3	99.3	99.4	99.4	99.4	99.4	99.4	99.4	99.4	99.5	99.5	99.5	99.5	99.5	99.5
3	34.1	30.8	29.5	28.7	28.2	27.9	27.7	27.5	27.3	27.2	27.1	26.9	26.7	26.6	26.5	26.4	26.3	26.2	26.1
4	21.2	18.0	16.7	16.0	15.5	15.2	15.0	14.8	14.7	14.5	14.4	14.2	14.0	13.9	13.8	13.7	13.7	13.6	13.5
5	16.3	13.3	12.1	11.4	11.0	10.7	10.5	10.3	10.2	10.1	9.89	9.72	9.55	9.47	9.38	9.29	9.20	9.11	9.02
6	13.7	10.9	9.78	9.15	8.75	8.47	8.26	8.10	7.98	7.87	7.72	7.56	7.40	7.31	7.23	7.14	7.06	6.97	6.88
7	12.2	9.55	8.45	7.85	7.46	7.19	6.99	6.84	6.72	6.62	6.47	6.31	6.16	6.07	5.99	5.91	5.82	5.74	5.65
8	11.3	8.65	7.59	7.01	6.63	6.37	6.18	6.03	5.91	5.81	5.67	5.52	5.36	5.28	5.20	5.12	5.03	4.95	4.86
9	10.6	8.02	6.99	6.42	6.06	5.80	5.61	5.47	5.35	5.26	5.11	4.96	4.81	4.73	4.65	4.57	4.48	4.40	4.31
10	10.0	7.56	6.55	5.99	5.64	5.39	5.20	5.06	4.94	4.85	4.71	4.56	4.41	4.33	4.25	4.17	4.08	4.00	3.91
11	9.65	7.21	6.22	5.67	5.32	5.07	4.89	4.74	4.63	4.54	4.40	4.25	4.10	4.02	3.94	3.86	3.78	3.69	3.60
12	9.33	6.93	5.95	5.41	5.06	4.82	4.64	4.50	4.39	4.30	4.16	4.01	3.86	3.78	3.70	3.62	3.54	3.45	3.36
13	9.07	6.70	5.74	5.21	4.86	4.62	4.44	4.30	4.19	4.10	3.96	3.82	3.66	3.59	3.51	3.43	3.34	3.25	3.17
14	8.86	6.51	5.56	5.04	4.70	4.46	4.28	4.14	4.03	3.94	3.80	3.66	3.51	3.43	3.35	3.27	3.18	3.09	3.00
15	8.68	6.36	5.42	4.89	4.56	4.32	4.14	4.00	3.89	3.80	3.67	3.52	3.37	3.29	3.21	3.13	3.05	2.96	2.87
16	8.53	6.23	5.29	4.77	4.44	4.20	4.03	3.89	3.78	3.69	3.55	3.41	3.26	3.18	3.10	3.02	2.93	2.84	2.75
17	8.40	6.11	5.19	4.67	4.34	4.10	3.93	3.79	3.68	3.59	3.46	3.31	3.16	3.08	3.00	2.92	2.83	2.75	2.65
18	8.29	6.01	5.09	4.58	4.25	4.01	3.84	3.71	3.60	3.51	3.37	3.23	3.08	3.00	2.92	2.84	2.75	2.66	2.57
19	8.18	5.93	5.01	4.50	4.17	3.94	3.77	3.63	3.52	3.43	3.30	3.15	3.00	2.92	2.84	2.76	2.67	2.58	2.49
20	8.10	5.85	4.94	4.43	4.10	3.87	3.70	3.56	3.46	3.37	3.23	3.09	2.94	2.86	2.78	2.69	2.61	2.52	2.42
21	8.02	5.78	4.87	4.37	4.04	3.81	3.64	3.51	3.40	3.31	3.17	3.03	2.88	2.80	2.72	2.64	2.55	2.46	2.36
22	7.95	5.72	4.82	4.31	3.99	3.76	3.59	3.45	3.35	3.26	3.12	2.98	2.83	2.75	2.67	2.58	2.50	2.40	2.31
23	7.88	5.66	4.76	4.26	3.94	3.71	3.54	3.41	3.30	3.21	3.07	2.93	2.78	2.70	2.62	2.54	2.45	2.35	2.26
24	7.82	5.61	4.72	4.22	3.90	3.67	3.50	3.36	3.26	3.17	3.03	2.89	2.74	2.66	2.58	2.49	2.40	2.31	2.21
25	7.77	5.57	4.68	4.18	3.86	3.63	3.46	3.32	3.22	3.13	2.99	2.85	2.70	2.62	2.54	2.45	2.36	2.27	2.17
26	7.72	5.53	4.64	4.14	3.82	3.59	3.42	3.29	3.18	3.09	2.96	2.82	2.66	2.58	2.50	2.42	2.33	2.23	2.13
27	7.68	5.49	4.60	4.11	3.78	3.56	3.39	3.26	3.15	3.06	2.93	2.78	2.63	2.55	2.47	2.38	2.29	2.20	2.10
28	7.64	5.45	4.57	4.07	3.75	3.53	3.36	3.23	3.12	3.03	2.90	2.75	2.60	2.52	2.44	2.35	2.26	2.17	2.06
29	7.60	5.42	4.54	4.04	3.73	3.50	3.33	3.20	3.09	3.00	2.87	2.73	2.57	2.49	2.41	2.33	2.23	2.14	2.03
30	7.56	5.39	4.51	4.02	3.70	3.47	3.30	3.17	3.07	2.98	2.84	2.70	2.55	2.47	2.39	2.30	2.21	2.11	2.01
40	7.31	5.18	4.31	3.83	3.51	3.29	3.12	2.99	2.89	2.80	2.66	2.52	2.37	2.29	2.20	2.11	2.02	1.92	1.80
60	7.08	4.98	4.13	3.65	3.34	3.12	2.95	2.82	2.72	2.63	2.50	2.35	2.20	2.12	2.03	1.94	1.84	1.73	1.60
120	6.85	4.79	3.95	3.48	3.17	2.96	2.79	2.66	2.56	2.47	2.34	2.19	2.03	1.95	1.86	1.76	1.66	1.53	1.38
∞	6.63	4.61	3.78	3.32	3.02	2.80	2.64	2.51	2.41	2.32	2.18	2.04	1.88	1.79	1.70	1.59	1.47	1.32	1.00

Source: E. S. Pearson and H. O. Hartley, *Biometrika Tables for Statisticians*, Vol. 2 (1972), Table 5, page 180, by permission.

Four-Place Common Logarithms

N	0	1	2	3	4	5	6	7	8	9	Proportional Parts 1	2	3	4	5	6	7	8	9
10	0000	0043	0086	0128	0170	0212	0253	0294	0334	0374	4	8	12	17	21	25	29	33	37
11	0414	0453	0492	0531	0569	0607	0645	0682	0719	0755	4	8	11	15	19	23	26	30	34
12	0792	0828	0864	0899	0934	0969	1004	1038	1072	1106	3	7	10	14	17	21	24	28	31
13	1139	1173	1206	1239	1271	1303	1335	1367	1399	1430	3	6	10	13	16	19	23	26	29
14	1461	1492	1523	1553	1584	1614	1644	1673	1703	1732	3	6	9	12	15	18	21	24	27
15	1761	1790	1818	1847	1875	1903	1931	1959	1987	2014	3	6	8	11	14	17	20	22	25
16	2041	2068	2095	2122	2148	2175	2201	2227	2253	2279	3	5	8	11	13	16	18	21	24
17	2304	2330	2355	2380	2405	2430	2455	2480	2504	2529	2	5	7	10	12	15	17	20	22
18	2553	2577	2601	2625	2648	2672	2695	2718	2742	2765	2	5	7	9	12	14	16	19	21
19	2788	2810	2833	2856	2878	2900	2923	2945	2967	2989	2	4	7	9	11	13	16	18	20
20	3010	3032	3054	3075	3096	3118	3139	3160	3181	3201	2	4	6	8	11	13	15	17	19
21	3222	3243	3263	3284	3304	3324	3345	3365	3385	3404	2	4	6	8	10	12	14	16	18
22	3424	3444	3464	3483	3502	3522	3541	3560	3579	3598	2	4	6	8	10	12	14	15	17
23	3617	3636	3655	3674	3692	3711	3729	3747	3766	3784	2	4	6	7	9	11	13	15	17
24	3802	3820	3838	3856	3874	3892	3909	3927	3945	3962	2	4	5	7	9	11	12	14	16
25	3979	3997	4014	4031	4048	4065	4082	4099	4116	4133	2	3	5	7	9	10	12	14	15
26	4150	4166	4183	4200	4216	4232	4249	4265	4281	4298	2	3	5	7	8	10	11	13	15
27	4314	4330	4346	4362	4378	4393	4409	4425	4440	4456	2	3	5	6	8	9	11	13	14
28	4472	4487	4502	4518	4533	4548	4564	4579	4594	4609	2	3	5	6	8	9	11	12	14
29	4624	4639	4654	4669	4683	4698	4713	4728	4742	4757	1	3	4	6	7	9	10	12	13
30	4771	4786	4800	4814	4829	4843	4857	4871	4886	4900	1	3	4	6	7	9	10	11	13
31	4914	4928	4942	4955	4969	4983	4997	5011	5024	5038	1	3	4	6	7	8	10	11	12
32	5051	5065	5079	5092	5105	5119	5132	5145	5159	5172	1	3	4	5	7	8	9	11	12
33	5185	5198	5211	5224	5237	5250	5263	5276	5289	5302	1	3	4	5	6	8	9	10	12
34	5315	5328	5340	5353	5366	5378	5391	5403	5416	5428	1	3	4	5	6	8	9	10	11
35	5441	5453	5465	5478	5490	5502	5514	5527	5539	5551	1	2	4	5	6	7	9	10	11
36	5563	5575	5587	5599	5611	5623	5635	5647	5658	5670	1	2	4	5	6	7	8	10	11
37	5682	5694	5705	5717	5729	5740	5752	5763	5775	5786	1	2	3	5	6	7	8	9	10
38	5798	5809	5821	5832	5843	5855	5866	5877	5888	5899	1	2	3	5	6	7	8	9	10
39	5911	5922	5933	5944	5955	5966	5977	5988	5999	6010	1	2	3	4	5	7	8	9	10
40	6021	6031	6042	6053	6064	6075	6085	6096	6107	6117	1	2	3	4	5	6	8	9	10
41	6128	6138	6149	6160	6170	6180	6191	6201	6212	6222	1	2	3	4	5	6	7	8	9
42	6232	6243	6253	6263	6274	6284	6294	6304	6314	6325	1	2	3	4	5	6	7	8	9
43	6335	6345	6355	6365	6375	6385	6395	6405	6415	6425	1	2	3	4	5	6	7	8	9
44	6435	6444	6454	6464	6474	6484	6493	6503	6513	6522	1	2	3	4	5	6	7	8	9
45	6532	6542	6551	6561	6571	6580	6590	6599	6609	6618	1	2	3	4	5	6	7	8	9
46	6628	6637	6646	6656	6665	6675	6684	6693	6702	6712	1	2	3	4	5	6	7	7	8
47	6721	6730	6739	6749	6758	6767	6776	6785	6794	6803	1	2	3	4	5	5	6	7	8
48	6812	6821	6830	6839	6848	6857	6866	6875	6884	6893	1	2	3	4	4	5	6	7	8
49	6902	6911	6920	6928	6937	6946	6955	6964	6972	6981	1	2	3	4	4	5	6	7	8
50	6990	6998	7007	7016	7024	7033	7042	7050	7059	7067	1	2	3	3	4	5	6	7	8
51	7076	7084	7093	7101	7110	7118	7126	7135	7143	7152	1	2	3	3	4	5	6	7	8
52	7160	7168	7177	7185	7193	7202	7210	7218	7226	7235	1	2	2	3	4	5	6	7	7
53	7243	7251	7259	7267	7275	7284	7292	7300	7308	7316	1	2	2	3	4	5	6	6	7
54	7324	7332	7340	7348	7356	7364	7372	7380	7388	7396	1	2	2	3	4	5	6	6	7
N	0	1	2	3	4	5	6	7	8	9	1	2	3	4	5	6	7	8	9

Four-Place Common Logarithms (continued)

N	0	1	2	3	4	5	6	7	8	9	Proportional Parts 1	2	3	4	5	6	7	8	9
55	7404	7412	7419	7427	7435	7443	7451	7459	7466	7474	1	2	2	3	4	5	5	6	7
56	7482	7490	7497	7505	7513	7520	7528	7536	7543	7551	1	2	2	3	4	5	5	6	7
57	7559	7566	7574	7582	7589	7597	7604	7612	7619	7627	1	2	2	3	4	5	5	6	7
58	7634	7642	7649	7657	7664	7672	7679	7686	7694	7701	1	1	2	3	4	4	5	6	7
59	7709	7716	7723	7731	7738	7745	7752	7760	7767	7774	1	1	2	3	4	4	5	6	7
60	7782	7789	7796	7803	7810	7818	7825	7832	7839	7846	1	1	2	3	4	4	5	6	6
61	7853	7860	7868	7875	7882	7889	7896	7903	7910	7917	1	1	2	3	4	4	5	6	6
62	7924	7931	7938	7945	7952	7959	7966	7973	7980	7987	1	1	2	3	3	4	5	6	6
63	7993	8000	8007	8014	8021	8028	8035	8041	8048	8055	1	1	2	3	3	4	5	5	6
64	8062	8069	8075	8082	8089	8096	8102	8109	8116	8122	1	1	2	3	3	4	5	5	6
65	8129	8136	8142	8149	8156	8162	8169	8176	8182	8189	1	1	2	3	3	4	5	5	6
66	8195	8202	8209	8215	8222	8228	8235	8241	8248	8254	1	1	2	3	3	4	5	5	6
67	8261	8267	8274	8280	8287	8293	8299	8306	8312	8319	1	1	2	3	3	4	5	5	6
68	8325	8331	8338	8344	8351	8357	8363	8370	8376	8382	1	1	2	3	3	4	4	5	6
69	8388	8395	8401	8407	8414	8420	8426	8432	8439	8445	1	1	2	2	3	4	4	5	6
70	8451	8457	8463	8470	8476	8482	8488	8494	8500	8506	1	1	2	2	3	4	4	5	6
71	8513	8519	8525	8531	8537	8543	8549	8555	8561	8567	1	1	2	2	3	4	4	5	5
72	8573	8579	8585	8591	8597	8603	8609	8615	8621	8627	1	1	2	2	3	4	4	5	5
73	8633	8639	8645	8651	8657	8663	8669	8675	8681	8686	1	1	2	2	3	4	4	5	5
74	8692	8698	8704	8710	8716	8722	8727	8733	8739	8745	1	1	2	2	3	4	4	5	5
75	8751	8756	8762	8768	8774	8779	8785	8791	8797	8802	1	1	2	2	3	3	4	5	5
76	8808	8814	8820	8825	8831	8837	8842	8848	8854	8859	1	1	2	2	3	3	4	5	5
77	8865	8871	8876	8882	8887	8893	8899	8904	8910	8915	1	1	2	2	3	3	4	4	5
78	8921	8927	8932	8938	8943	8949	8954	8960	8965	8971	1	1	2	2	3	3	4	4	5
79	8976	8982	8987	8993	8998	9004	9009	9015	9020	9025	1	1	2	2	3	3	4	4	5
80	9031	9036	9042	9047	9053	9058	9063	9069	9074	9079	1	1	2	2	3	3	4	4	5
81	9085	9090	9096	9101	9106	9112	9117	9122	9128	9133	1	1	2	2	3	3	4	4	5
82	9138	9143	9149	9154	9159	9165	9170	9175	9180	9186	1	1	2	2	3	3	4	4	5
83	9191	9196	9201	9206	9212	9217	9222	9227	9232	9238	1	1	2	2	3	3	4	4	5
84	9243	9248	9253	9258	9263	9269	9274	9279	9284	9289	1	1	2	2	3	3	4	4	5
85	9294	9299	9304	9309	9315	9320	9325	9330	9335	9340	1	1	2	2	3	3	4	4	5
86	9345	9350	9355	9360	9365	9370	9375	9380	9385	9390	1	1	2	2	3	3	4	4	5
87	9395	9400	9405	9410	9415	9420	9425	9430	9435	9440	0	1	1	2	2	3	3	4	4
88	9445	9450	9455	9460	9465	9469	9474	9479	9484	9489	0	1	1	2	2	3	3	4	4
89	9494	9499	9504	9509	9513	9518	9523	9528	9533	9538	0	1	1	2	2	3	3	4	4
90	9542	9547	9552	9557	9562	9566	9571	9576	9581	9586	0	1	1	2	2	3	3	4	4
91	9590	9595	9600	9605	9609	9614	9619	9624	9628	9633	0	1	1	2	2	3	3	4	4
92	9638	9643	9647	9652	9657	9661	9666	9671	9675	9680	0	1	1	2	2	3	3	4	4
93	9685	9689	9694	9699	9703	9708	9713	9717	9722	9727	0	1	1	2	2	3	3	4	4
94	9731	9736	9741	9745	9750	9754	9759	9763	9768	9773	0	1	1	2	2	3	3	4	4
95	9777	9782	9786	9791	9795	9800	9805	9809	9814	9818	0	1	1	2	2	3	3	4	4
96	9823	9827	9832	9836	9841	9845	9850	9854	9859	9863	0	1	1	2	2	3	3	4	4
97	9868	9872	9877	9881	9886	9890	9894	9899	9903	9908	0	1	1	2	2	3	3	4	4
98	9912	9917	9921	9926	9930	9934	9939	9943	9948	9952	0	1	1	2	2	3	3	4	4
99	9956	9961	9965	9969	9974	9978	9983	9987	9991	9996	0	1	1	2	2	3	3	3	4
N	0	1	2	3	4	5	6	7	8	9	1	2	3	4	5	6	7	8	9

Appendix VIII

Values of $e^{-\lambda}$

$(0 < \lambda < 1)$

λ	0	1	2	3	4	5	6	7	8	9
0.0	1.0000	.9900	.9802	.9704	.9608	.9512	.9418	.9324	.9231	.9139
0.1	.9048	.8958	.8869	.8781	.8694	.8607	.8521	.8437	.8353	.8270
0.2	.8187	.8106	.8025	.7945	.7866	.7788	.7711	.7634	.7558	.7483
0.3	.7408	.7334	.7261	.7189	.7118	.7047	.6977	.6907	.6839	.6771
0.4	.6703	.6636	.6570	.6505	.6440	.6376	.6313	.6250	.6188	.6126
0.5	.6065	.6005	.5945	.5886	.5827	.5770	.5712	.5655	.5599	.5543
0.6	.5488	.5434	.5379	.5326	.5273	.5220	.5169	.5117	.5066	.5016
0.7	.4966	.4916	.4868	.4819	.4771	.4724	.4677	.4630	.4584	.4538
0.8	.4493	.4449	.4404	.4360	.4317	.4274	.4232	.4190	.4148	.4107
0.9	.4066	.4025	.3985	.3946	.3906	.3867	.3829	.3791	.3753	.3716

$(\lambda = 1, 2, 3, \ldots, 10)$

λ	1	2	3	4	5	6	7	8	9	10
$e^{-\lambda}$	.36788	.13534	.04979	.01832	.006738	.002479	.000912	.000335	.000123	.000045

Note: To obtain values of $e^{-\lambda}$ for other values of λ, use the laws of exponents.
Example: $e^{-3.48} = (e^{-3.00})(e^{-.48}) = (0.04979)(0.6188) = 0.03081$.

Appendix IX

Random Numbers

51772	74640	42331	29044	46621	62898	93582	04186	19640	87056
24033	23491	83587	06568	21960	21387	76105	10863	97453	90581
45939	60173	52078	25424	11645	55870	56974	37428	93507	94271
30586	02133	75797	45406	31041	86707	12973	17169	88116	42187
03585	79353	81938	82322	96799	85659	36081	50884	14070	74950
64937	03355	95863	20790	65304	55189	00745	65253	11822	15804
15630	64759	51135	98527	62586	41889	25439	88036	24034	67283
09448	56301	57683	30277	94623	85418	68829	06652	41982	49159
21631	91157	77331	60710	52290	16835	48653	71590	16159	14676
91097	17480	29414	06829	87843	28195	27279	47152	35683	47280
50532	25496	95652	42457	73547	76552	50020	24819	52984	76168
07136	40876	79971	54195	25708	51817	36732	72484	94923	75936
27989	64728	10744	08396	56242	90985	28868	99431	50995	20507
85184	73949	36601	46253	00477	25234	09908	36574	72139	70185
54398	21154	97810	36764	32869	11785	55261	59009	38714	38723
65544	34371	09591	07839	58892	92843	72828	91341	84821	63886
08263	65952	85762	64236	39238	18776	84303	99247	46149	03229
39817	67906	48236	16057	81812	15815	63700	85915	19219	45943
62257	04077	79443	95203	02479	30763	92486	54083	23631	05825
53298	90276	62545	21944	16530	03878	07516	95715	02526	33537

Abscissa, 4
Absolute dispersion, 100
Absolute value, 95
Acceptance region, 247 (*see also* Hypotheses)
Alternative hypothesis, 245
Analysis of variance, 362–401
  mathematical model for, 403
  one-factor experiments using, 403
  purpose of, 403
  tables, 406
  two-factor experiments using, 407
  using Graeco-Latin squares, 413
  using Latin squares, 413
Approximating curves, equations of, 317
Areas:
  of chi-square distribution, 277
  of *F* distribution, 279
  of normal distribution, 173
  of *t* distribution, 275
Arithmetic mean, 61
  assumed or guessed, 63
  Charlier's check for, 99
  coding method for computing, 63
  computed from grouped data, 63
  effect of extreme values on, 71
  for population and for sample, 144
  long and short methods for computing, 63
  of arithmetic means, 63
  probability distributions, 142
  properties of, 63
  relation to median and mode, 64
  relation to geometric and harmonic means, 66
  weighted, 62
Arithmetic progression:
  moments of, 136
  variance of, 120
Arrays, 37
Assignable causes, 480
Asymptotically normal, 204
Attributes control chart, 481
Attributes data, 481
Attributes, correlation of, 298
Autocorrelation, 351
Average, 62

Bar charts or graphs, component part, 4
Base, 2
  of common logarithms, 6
  of natural logarithms, 6
Bayes' rule or theorem, 170
Bernoulli distribution (*see* Binomial distribution)
Best estimate, 228
Biased estimates, 227
Bimodal frequency curve, 41
Binomial coefficients, 173
  Pascal's triangle for, 180
Binomial distribution, 172
  fitting of data, 195
  properties of, 172
  relation to normal distribution, 174
  relation to Poisson distribution, 175
  tests of hypotheses using, 250
Binomial expansion or formula, 173
Bivariate:
  frequency distribution or table, 366
  normal distribution, 351
  population, 351
Blocks, randomized, 413

Categories, 37
C-chart, 490
Cell frequencies, 496
Center of gravity, 320
Centerline, 480
Central limit theorem, 204
Centroid, 320
Charlier's check, 99
  for mean and variance, 99
  for moments, 124
Chi-square, 294
  additive property of, 299
  definition of, 294
  for goodness of fit, 295
  formulas for in contingency tables, 297
  test, 295
  Yates' correction for, 297
Chi-square distribution, 297 (*see also* Chi-square)
  confidence intervals using, 278
  tests of hypothesis and significance, 294

Circular graph (*see* Pie graph)
Class, 37 (*see also* Class intervals)
Class boundaries, lower and upper, 38
Class frequency, 37
Class intervals, 38
  median, 64
  modal, 43
  open, 38
  unequal, 51
  width or size of, 38
Class length, size or width, 38
Class limits, 38
  lower and upper, 38
  true, 38
Class mark, 38
Coding methods, 63
  for correlation coefficient, 350
  for mean, 63
  for moment, 124
  for standard deviation, 98
Combinations, 146
Combinatorial analysis, 146
Common causes, 480
Common logarithms, 6
Complete randomization, 413
Compound event, 143
Compound interest formula, 85
Computations, 3
  rules for, 3
  rules for, using logarithms, 7
Conditional probability, 140
Confidence coefficients, 228
Confidence interval:
  for differences and sums, 230
  for means, 229
  for proportions, 229
  for standard deviations, 230
  in correlation and regression, 374
  using chi-square distribution, 278
  using normal distribution, 229–230
  using t distribution, 276
Confidence levels, table of, 229
Confidence limits, 229
Constant, 1
Contingency tables, 296
  correlation coefficient from, 298
  formulas for chi-square in, 298
Contingency, coefficient of, 310
Continuous data, 1
  graphical representation of, 57
Continuous probability distributions, 143
Continuous variable, 1
Control charts, 480
Control limits, 480
Coordinates, rectangular, 4
Correlation, 345
  auto-, 351
Correlation (*Contd.*):
  coefficient of (*see* Correlation coefficient)
  linear, 345
  measures of, 346
  of attributes, 311
  partial, 385
  positive and negative, 345
  rank, 450
  simple, 345
  spurious, 549
Correlation coefficient, 348
  for grouped data, 350
  from contingency tables, 298
  product-moment formula for, 350
  regression lines and, 351
  sampling theory and, 351
Correlation table, 350
Counting method in Mann–Whitney U test, 447
Countings or enumerations, 2
CP index, 485
CPK index, 485
Critical region, 247
Critical values, 248
Cubic curve, 317
Cumulative frequency, 40
Cumulative probability distributions, 143
Cumulative rounding errors, 2
Curve fitting, 316
  freehand method of, 318
  least-squares method of, 319
Cusum chart, 490
Cyclic pattern in runs test, 449

Data:
  grouped, 38
  raw, 37
  spread or variation of, 89
Deciles, 66
  from grouped data, 87
  standard errors for, 206
Decision rules, 246 (*see also* Statistical decisions)
Deductive statistics, 1
Defective item, 487
Defects, 487
Degrees of freedom, 276
Density function, 144
Dependent events, 140
Dependent variable, 4
  change of in regression equation, 384
Descriptive statistics, 1
Design of experiments, 412
Determination:
  coefficient of, 348
  coefficient of multiple, 384
Deviation from arithmetic mean, 63
Diagrams (*see* Graphs)
Dimensionless moments, 124

Discrete data, 2
  graphical representation of, 54
Discrete probability distributions, 142
Discrete variable, 1
Dispersion, 87 (*see also* Variation)
  absolute, 95
  coefficient of, 100
  measures of, 95
  relative, 100
Distribution function, 142
Domain of variable, 1

Efficient estimates and estimators, 228
Empirical probability, 139
Empirical relation between mean, median, and mode, 64
Empirical relation between measures of dispersion, 100
Enumerations, 2
Equations, 5
  equivalent, 5
  left and right hand members of, 5
  of approximating curves, 317
  quadratic, 35
  regression, 382
  simultaneous, 25, 34
  solution of, 5
  transposition in, 26
Errors:
  grouping, 39
  probable, 230
  rounding, 2
Estimates (*see also* Estimation)
  biased and unbiased, 227
  efficient and inefficient, 228
  point and interval, 228
Estimation, 227
Euler diagram, 146
Events, 140
  compound, 140
  dependent, 140
  independent, 140
  mutually exclusive, 141
EWMA chart, 490
Exact or small sampling theory, 181
Expectation, mathematical, 144
Expected or theoretical frequencies, 294
Experimental design, 412
Explained variation, 348
Exponent, 2
Exponential curve, 318

F distribution, 279 (*see also* Analysis of variance)
Factorial, 143
  Stirling's formula for, 146

Failure, 139
Fitting of data, 19 (*see also* Curve fitting)
  by binomial distribution, 195
  by normal distribution, 196
  by Poisson distribution, 198
  using probability graph paper, 177
Four-dimensional space, 385
Freehand method of curve fitting, 318
Frequency (*see also* Class frequency)
  cumulative, 40
  modal, 41
  relative, 40
Frequency curves, 41
  relative, 41
  types of, 41
Frequency distributions, 37
  cumulative, 40
  percentage or relative, 39
  rule for forming, 38
Frequency function, 142
Frequency polygons, 39
  percentage or relative, 39
  smoothed, 41
Frequency table (*see also* Frequency distributions)
  cumulative, 40
  relative, 40
Function, 4
  distribution, 142
  frequency, 142
  linear, 318
  multiple-valued, 4
  probability, 139
  single-valued, 4

Geometric curve, 318
Geometric mean, 65
  from grouped data, 83
  relation to arithmetic and harmonic mean, 66
  suitability for averaging ratios, 84
  weighted, 84
Gompertz curve, 318
Goodness of fit test, 177 (*see also* Fitting of data)
Gossett, 276
Graeco-Latin squares, 413
Grand mean, 404
Graph, 5
  bar charts, 18
  pie, 5
Graph paper:
  log-log, 339
  probability, 177
  semilog, 318
Group mean, 404
Grouped data, 38
Grouping error, 39

H statistic, 448
Harmonic mean, 65
  relation to arithmetic and geometric means, 66
  weighted, 86
Histograms, 39
  computing medians for, 64
  percentage or relative frequency, 39
  probability, 153
Homogeneity of variance test, 285
Hyperbola, 318
Hyperplane, 385
Hypotheses, 245
  alternative, 245
  null, 245

Identity, 5
Independent events, 140
Independent variable, 4
Individuals chart, 489
Inductive statistics, 1
Inefficient estimates and estimators, 228
Inequalities, 5
Inequality symbols, 5
Inspection unit, 489
Interaction, 411
Interaction plot, 429
Intercepts, 319
Interest, compound, 85
Interpolation, 7
Interquartile range, 96
  semi-, 96
Intersection of sets, 146
Interval estimates, 228

Kruskal–Wallis H test, 448
Kurtosis, 125
  moment coefficient of, 125
  of binomial distribution, 173
  of normal distribution, 125
  of Poisson distribution, 176
  percentile coefficient of, 125

Latin squares, 413
  orthogonal, 413
Least squares:
  curve, 319
  line, 319
  parabola, 320
  plane, 321
Leptokurtic, 125
Level of significance, 246
Logarithms, 6
  base of, 6
  common, 6
  computations using, 7
  natural, 6
Logistic curve, 318

Lower capability index, 485
Lower control limit, 480
Lower specification limit, 484

Main effects plot, 429
Mann–Whitney U test, 454
Marginal frequencies, 294
Mean deviation, 95
  for grouped data, 96
  of normal distribution, 174
Measurements, 2
Measures of central tendency, 61
Median, 64
  computed from histogram or percentage ogive, 64
  effect of extreme values on, 78
  for grouped data, 64
  relation to arithmetic mean and mode, 64
Median chart, 489
Mesokurtic, 125
Modal class, 43
Mode, 64
  for grouped data, 64
  formula for, 64
  relation to arithmetic mean and median, 65
Model or theoretical distribution, 177
Moment coefficient of kurtosis, 125
Moment coefficient of skewness, 125
Moments, 123
  Charlier's check for computing, 124
  coding method for computing, 124
  definition of, 123
  dimensionless, 124
  for grouped data, 123
  relations between, 124
  Sheppard's corrections for, 124
Multimodal frequency curve, 41
Multinomial distribution, 177
Multinomial expansion, 177
Multiple correlation, 382
Multiple determination, coefficient of, 384
Mutually exclusive events, 141

Natural logarithms, base of, 6
Nonconformance rates, 484
Nonconforming item, 486
Nonlinear:
  correlation and regression, 346
  equations reducible to linear form, 320
  multiple regression, 382
  relationship between variables, 316
Nonparametric tests, 446
  for correlation, 450
  Kruskal–Wallis H test, 448
  Mann–Whitney U test, 447
  runs test, 449
  sign test, 440
Nonrandom, 449

Nonsense correlation, 350
Normal approximation to the binomial distribution, 174
Normal curve, 173
  areas under, 173
  ordinates of, 187
  standard form of, 173
Normal distribution, 173
  proportions of, 174
  relation to binomial distribution, 174
  relation to Poisson distribution, 176
  standard form of, 174
Normal equations:
  for least-squares line, 320
  for least-squares parabola, 320
  for least-squares plane, 321
Normality test, 196
NP-chart, 487
*n*th degree curve, 317
Null hypothesis, 245
Null set, 146

Observed frequencies, 294
OC curves (see Operating characteristic curve)
Ogives, 40
  deciles, percentiles, and quartiles obtained from, 87
  "less than", 52
  median obtained from, 78
  "or more", 52
  percentage, 53
  smoothed, 53
One-factor experiments, 403
One-sided or one-tailed tests, 247
One-way classification, 403
"Or more" cumulative distribution, 40
Ordinates, 4
Origin, 5
Orthogonal Latin square, 413

Parabola, 320
Parameters, estimation of, 227
Pareto chart, 504
Partial correlation, 382
Partial regression coefficients, 382
Parts per million (ppm), 484
Pascal's triangle, 180
P-chart, 487
Pearson's coefficients of skewness, 125
Percentage:
  cumulative distributions, 39
  cumulative frequency, 39
  distribution, 39
  histogram, 38
  ogives, 39
Percentile coefficient of kurtosis, 125
Percentile range, 96
Percentiles, 66
Permutations, 145
Pie graph or chart, 5
Plane, 4
Platykurtic, 125
Point estimates, 228
Poisson distribution, 175
  fitting of data by, 198
  properties of, 175
  relation to binomial and normal distribution, 176
Polynomials, 318
Population, 1
Population parameters, 228
Positive correlation, 346
Probability, 139
  axiomatic, 140
  classic definition of, 139
  combinatorial analysis and, 143
  conditional, 140
  distributions, 142
  empirical, 140
  fundamental rules of, 146
  relation to point set theory, 146
  relative frequency definition of, 140
Probability function, 142
Probable error, 230
Process capability index, 485
Process spread, 484
Product-moment formula for correlation coefficient, 350
Proportions, 205
  confidence interval for, 229
  sampling distribution of, 205
  tests of hypothesis for, 245
p-value, 248

Quadrants, 4
Quadratic:
  curve, 317
  equation, 35
  function, 17
Quantiles, 66
Quartic curve, 317
Quartile coefficient of relative dispersion, 116
Quartile coefficient of skewness, 125
Quartiles, 66
Quintiles, 87

Random:
  errors, 403
  numbers, 203
  sample, 203
  variable, 142
Randomization, complete, 413
Randomized blocks, 413
Range, 95
  10–90 percentile, 96

Range (*Contd.*):
- interquartile, 96
- semi-interquartile, 96

Rank correlation, coefficient of, 450
Raw data, 37
Regression, 321
- curve, 321
- line, 321
- multiple, 345
- plane, 321
- sampling theory of, 352
- simple, 343
- surface, 322

Relative dispersion or variance, 95
Relative frequency, 39
- curves, 39
- definition of probability, 139
- distribution, 39
- table, 39

Residual, 319
Residual variation, 409
Reverse J-shaped distribution, 41
Root mean square, 66
Rounding errors, 2
Rounding of data, 2
Row means, 403
Runs, 449

Sample, 1
Sample space, 146
Sample statistics, 203
Sampling:
- with replacement, 204
- without replacement, 204

Sampling distributions, 204
- experimental, 208
- of differences and sums, 205
- of means, 204
- of proportions, 205
- of various statistics, 204

Sampling numbers, 213
Sampling theory, 203
- large samples, 207
- of correlation, 351
- of regression, 352
- use of in estimation, 227
- use of in tests of hypotheses and significance, 245

Scientific notation, 2
Semi-interquartile range, 96
Semilog graph paper, 318
Sheppard's correction for moments, 124
Sheppard's correction for variance, 100
Sign test, 446
Significant digits or figures, 3
Simple correlation, 382
Simultaneous equations, 5
Single-valued functions, 4
Skewed frequency curves, 41
Skewness, 41
- 10–90 percentile coefficient of, 125
- binomial distribution, 172
- negative, 41
- normal distribution, 173
- Pearson's coefficients of, 125
- Poisson distribution, 175
- quartile coefficient of, 125

Slope of a line, 318
Small sampling theory, 275
Solution of equations, 5
Spearman's formula for rank correlation, 450
Special causes, 480
Specification limits, 484
Spurious correlation, 349
Standard deviation, 96
- coding method for, 98
- confidence interval for, 230
- from grouped data, 98
- minimal property of, 98
- of probability distribution, 155
- of sampling distributions, 204
- properties of, 98
- relation of population and sample, 97
- relation to mean deviation, 100
- short methods for computing, 98

Standard error of estimate, 348
- modified, 348

Standard errors of sampling distributions, 206
Standard scores, 101
Standardized variable, 101
Statistical decisions, 245
Statistics, 1
- deductive or descriptive, 1
- definition of, 1
- inductive, 1
- sample, 1

Stirling's approximation to $n!$, 146
Straight line, 317
- equation of, 317
- least-squares, 318
- regression, 321
- slope of, 318

Subgroups, 481
Subscript notation, 61
Success, 139
Summation notation, 61
Symmetric or bell-shaped curve, 41

t distribution, 275
t score or statistic, 275
Table, 4
Table entry, 42
Tally sheet, 39
Ten to ninety percentile range, 96